Perspectives in Partial Differential Equations, Harmonic Analysis and Applications

Vladimir Gilelevich Maz'ya

Proceedings of Symposia in
PURE MATHEMATICS

Volume 79

Perspectives in Partial Differential Equations, Harmonic Analysis and Applications

A Volume in Honor of
Vladimir G. Maz'ya's 70th Birthday

Dorina Mitrea
Marius Mitrea
Editors

American Mathematical Society
Providence, Rhode Island

2000 *Mathematics Subject Classification.* Primary 01A50, 26D10, 31B15, 34L40, 35J25, 35Q53, 42B25, 46–06, 46E35, 74J15.

Photo on page ii courtesy of Tatyana Shaposhnikova

Library of Congress Cataloging-in-Publication Data

Perspectives in partial differential equations, harmonic analysis, and applications : a volume in honor of Vladimir G. Maz'ya's 70th birthday / Dorina Mitrea, Marius Mitrea, editors.

p. cm. — (Proceedings of symposia in pure mathematics : v. 79)

Includes bibliographical references.

ISBN 978-0-8218-4424-3 (alk. paper)

1. Maz'ya, V. G. 2. Differential equations, Partial. 3. Harmonic analysis. I. Mitrea, Dorina, 1965– II. Mitrea, Marius.

QA377.P378 2008

515′.353—dc22 2008030028

∞ The paper used in this book is acid-free and falls within the guidelines
established to ensure permanence and durability.
Visit the AMS home page at `http://www.ams.org/`

10 9 8 7 6 5 4 3 2 1 13 12 11 10 09 08

Contents

On the scientific work of V.G. Maz'ya: a personalized account

Vladimir Gilelevich[1] Maz'ya, one of the most distinguished analysts of our time, has recently celebrated his 70th birthday. This personal landmark is also a great opportunity to reflect upon the depth and scope of his vast, multi-faceted scientific work, as well as on its impact on contemporary mathematics.

It is no easy task to re-introduce to the general public a persona of the caliber of Vladimir Maz'ya. Nonetheless, the narrative of his life is such an inspirational epic of triumph against adversity and seemingly insurmountable odds, of sheer perseverance and dazzling success, that such an endeavor is worth undertaking even while fully aware that the present abridged account will have severe inherent limitations.

Siméon Poisson once famously said that "*life is good for only two things: discovering mathematics and teaching mathematics*". Considering the sheer volume of his scientific work and scholarly activities, one might be tempted to regard Vladimir Maz'ya as the perfect embodiment of this credo. However, with his larger-than-life personality, boundless energy, strong opinions and keen interest in a diverse range of activities, Vladimir Maz'ya transcends such a cliche: he is a remarkable man by any reasonable measure. His life, however, cannot be separated from mathematics, regarded as a general human endeavor: much as his own destiny has been prefigured by his deep affection for mathematics, so has Vladimir Maz'ya helped shape the mathematics of our time. Meanwhile, his views on mathematical ability are rooted in a brand of stoic pragmatism: he regards the latter not unlike the skill and sensitivity expected of a musician, or the stamina and endurance required of an athlete. In [**6**], I. Gohberg remarks: "*Whatever he writes is beautiful, his love for art, music and literature seeming to feed his mathematical aesthetic feeling*".

I. Rough childhood. Vladimir Maz'ya was born on December 31, 1937, in Leningrad (present day St. Petersburg) in the former USSR, roughly two years before World War II broke out in Europe. USSR was subsequently attacked and the capture of Leningrad was one of three strategic goals in Hitler's initial plans for Operation Barbarossa ("Leningrad first, the Donetsk Basin second, Moscow third"), with the goal of "Celebrating New Year's Eve 1942 in the Tsar's Palaces." It is in this context that Vladimir Maz'ya's early life was marred by profound personal tragedy: his father was killed on the World War II front in December 1941, and all four of his grandparents perished during the subsequent siege of Leningrad, which lasted from September 9, 1941 to January 27, 1944. Vladimir was brought up by his mother, alone, who worked as a state accountant. They lived on her

[1] patronymic after his father Hillel

meager salary in a cramped (nine square meter) room of a big communal apartment. These days, it is perhaps difficult to imagine the hardship in which a young Vladimir was finding his feet, and yet he spoke of occasional glimmers in this desolate atmosphere. He once recounted a touching story about the lasting impression a children's botanical book he had received, about the fruits of the world, made on him: how the pictures he gazed upon over and over still vividly live in his memory, and how it took many long years before he had a chance to actually see and taste some of the fruits depicted there. Resolute and driven, Vladimir rose above these challenges. At the same time, his talent and ability were apparent from early on: he earned a gold medal in secondary school and, as a high-schooler, he was a frequent winner of city olympiads in mathematics and physics.

II. The formative years. While 17 years of age, Vladimir Maz'ya entered the Faculty of Mathematics and Mechanics (Mathmech) of Leningrad State University (LSU) as a student. His first publication, "On the criterion of de la Vallée-Poussin", was in ordinary differential equations and appeared in a rota-printed collection of student papers when he was in his third year of undergraduate studies. In the following year, while he was a fourth-year student, his article on the Dirichlet problem for second order elliptic equations was published in *Doklady Akad. Nauk SSSR.* Upon finishing his undergraduate studies at Mathmech-LSU, Vladimir Maz'ya secured a position as a junior research fellow at the Research Institute of Mathematics and Mechanics of Leningrad State University. Two years later he successfully defended his Ph.D. thesis on "Classes of sets and embedding theorems for function spaces". This remarkable piece of work was based on ideas emerging from his talks in Smirnov's seminar. In their reviews, the examiners noted that the level of quality and technical mastery far exceeded the standard requirements of the Higher Certification Commission for Ph.D. theses. Testament to the outstanding nature of his thesis, Vladimir Maz'ya was awarded the Leningrad Mathematical Society's prize for young scientists. Subsequently, Vladimir Maz'ya was a volunteer director of the Mathematical School for High School Students at Mathmech, an institution born out of his own initiative. Interestingly, Vladimir Maz'ya never had a formal scientific adviser, both for his diploma paper (master's thesis), and for his Ph.D. thesis. Indeed, in each instance, he chose the problems considered in his work by himself. However, starting with his undergraduate years, he became acquainted with S.G. Mikhlin, and their relationship turned into a long-lasting friendship that had a great influence on the mathematical development of Vladimir Maz'ya. According to I. Gohberg, [**6**], "*Maz'ya never was a formal student of Mikhlin, but Mikhlin was for him more than a teacher. Maz'ya had found the topics of his dissertations by himself, while Mikhlin taught him mathematical ethics and rules of writing, refereeing and reviewing.*"

III. Becoming established. During 1961-1986, Vladimir Maz'ya held a senior research fellow position at the Research Institute of Mathematics and Mechanics of LSU. Four years into that tenure, he defended his D.Sc. thesis, entitled "Dirichlet and Neumann problems in domains with non-regular boundaries", at Leningrad State University. From 1968 to 1978, he lectured at the Leningrad Shipbuilding Institute, where he became a professor in 1976. In 1986 he departed the university for the Leningrad Division of the Institute of Engineering Studies of the Academy of Sciences of the USSR, where he created and headed the Laboratory of Mathematical Models in Mechanics. At the same time, he also founded the influential Consultation Center in Mathematics for Engineers, serving as its head for several years. In 1990 Vladimir Maz'ya relocated to Sweden and became a professor at Linköping University. At this stage in his career, in recognition of his fundamental contributions to the field of mathematics, Vladimir Maz'ya has become the recipient of a series of distinguished awards in relatively quick succession. In 1990 he received an honorary doctorate from the University of Rostock, Germany. In 1999 he was the recipient of the Humboldt Prize, and in 2000 was elected a corresponding member of the Royal Society of Edinburgh (Scotland's National Academy). Two years later he became a full member of the Royal Swedish Academy of Sciences. In 2003 he received the Verdaguer Prize of the French Academy of Sciences, and in 2004 the Celsius Gold Medal of the Royal Society of Sciences at Uppsala. A number of international conferences in his honor have been organized during this period of time, such as the conference in Kyoto, Japan, in 1993, the conferences at the University of Rostock, Germany, and at École Polytechnique, France, in 1998, and the conferences in Rome, Italy, and Stockholm, Sweden, in the summer of 2008. In 2002 Vladimir Maz'ya was an invited speaker at the International Congress of Mathematicians in Beijing, China. More recently, he has held appointments at the University of Liverpool, England, and at the Ohio State University, USA, while continuing to be a Professor Emeritus at Linköping University, Sweden.

IV. The mathematical work. By any standards, Vladimir Maz'ya has been extraordinarily prolific, as his 50 years of research activities have culminated in about a couple dozen research monographs, and more than 450 articles, containing fundamental results and powerfully novel techniques. Besides being remarkably deep and innovative, his work is also incredibly diverse. Drawing upon several sources, most notably [**1**], [**2**], [**5**] and [**9**], below we briefly survey some of the main topics covered by Vladimir Maz'ya's publications. The references labeled [Ma-X] refer to the list of books published by Vladimir Maz'ya, which is included following the current subsection.

◇ `Boundary integral equations on non-smooth surfaces.` One of the early significant contributions of Vladimir Maz'ya was his 1967 monograph [Ma-25] with Yuri D. Burago, where they developed a theory of boundary integral equations (involving operators such as the harmonic single- and double-layer potentials) in the space C^0, of continuous functions, on irregular surfaces. The book contains two parts: the first of which concerns the higher-dimensional potential theory and the solutions of the boundary problems for regions with irregular boundaries, while the second part deals with spaces of functions whose derivatives are measures. This was happening around the time the Calderón-Zygmund program, one of its goals being a re-thinking of the finer aspects of Partial Differential Equations from the perspective of Harmonic Analysis, was becoming of age. In the early 60's, the solvability properties of elliptic multidimensional singular integral operators were well-understood, due to the fundamental contributions of people such as Tricomi, Mikhlin, Giraud, Calderón and Zygmund, and Gohberg, among others; but very little was known about the degenerate and/or non-elliptic case. Influenced by Mikhlin, Vladimir Maz'ya began in the mid 60's a life-long research program (part of which has been a collaboration effort) aimed at shedding light on this challenging and important problem. These innovative ideas did not get instantaneous recognition as a certain degree of skepticism has long accompanied efforts to understand non-smooth calculus. One well-known quotation attributed to H. Poincaré, which typifies the aforementioned distrust, goes as follows: "*Autrefois quand on invantait une fonction nouvelle, c'était en vue de quelque but practique; aujourd'hui on les invente tout exprès pour mettre en défaut les raisonnements de nos péres et on n'en tirera jamais que celà*". Such a point of view was by no means isolated. Even S.G. Mikhlin, years later, referring to the perspective of studying PDE's under minimal smoothness assumptions on the boundary, opined to the effect that "*no mother would ever let her child play in such ravines*".

The subject of analysis in non-smooth settings permeates through much of the work of Vladimir Maz'ya, who has had a most significant contribution in ensuring the eventual acceptance of this, nowadays fashionable, area of research. In collaboration with his Ph.D. student N.V. Grachev (1991), Vladimir Maz'ya solved the classical problem of inverting the boundary integral operators naturally associated with the Dirichlet problem for the Laplacian, in the space C^0, on a polyhedral surface. Also, Maz'ya and A. A. Solov'ev were the first to consider (in 1990) boundary integral equations on a curve with cusps. Subsequently, they developed a logarithmic potential theory which is applicable to integral equations in elasticity theory in a plane domain with inward or outward peaks on the boundary (2001). More recently, in collaboration with T. Shaposhnikova, Vladimir Maz'ya has studied the classical boundary integral equations of the harmonic potential theory on Lipschitz surfaces, and obtained higher fractional Sobolev regularity results for their solutions under optimal regularity conditions on the boundary. The method employed, going back to work of Maz'ya in the early 80's, consists of establishing well-posedness results for certain auxiliary boundary value and transmission problems for the Laplace equation in weighted Sobolev spaces.

◇ **Counterexamples related to Hilbert's 19th and 20th problems.** In his famous plenary address at the International Congress of Mathematicians in 1900, held at the Sorbonne, Paris, David Hilbert put forth a list of twenty-three open problems in mathematics, many of which turned out to be very influential for 20th century mathematics (strictly speaking, Hilbert presented ten of the problems: 1, 2, 6, 7, 8, 13, 16, 19, 21 and 22, at the conference, and the full list was published later). The 19th problem read: *Are the solutions of regular problems in the calculus of variations always analytic*? Originally, Hilbert was referring to regular variational problems of first order in two-dimensional domains, but the issue of (local) regularity makes sense in higher dimensions and for higher-order problems as well. Hilbert's 19th and 20th problems, the latter asking "*Is it not the case that every regular variational problem has a solution, provided certain assumptions on the boundary conditions are satisfied, and provided also, if need be, that the concept of solution is suitably extended?*" have generated a large amount of attention and, in the second half of the 20th century, proofs were obtained in sufficient generality. It was therefore natural to speculate that the conjectures continue to hold for *higher-order* variational problems. However, in 1968 Vladimir Maz'ya proved that this is not the case. In [**8**], Maz'ya constructed higher-order quasi-linear elliptic equations with analytic coefficients whose solutions are not smooth.

Other counterexamples constructed in [**8**] (and, independently, by De Georgi [**4**]) concern the celebrated De Giorgi-Nash Hölder regularity result for solutions of the second order linear elliptic equations in divergence form with bounded measurable coefficients. Maz'ya showed that this property fails for higher-order equations which may admit variational solutions which are not locally bounded. The counterexamples in [**8**] stimulated the development of the theory of partial regularity of solutions to nonlinear equations, i.e., the study of regularity properties outside of a sufficiently small, exceptional set.

◇ `The oblique derivative problem.` The oblique derivative problem was first formulated by Poincaré in his studies related to the theory of tides, and by the late 60's the two-dimensional setting was well-understood. At that time, much of the work in the multidimensional case has been restricted to the situation when the direction field of the derivatives is transversal to the boundary at each point, a condition which ensures that the ellipticity is nowhere violated. However, when the ellipticity degenerates, this problem turned out to be considerably more difficult and subtle. This case came under scrutiny in 60's when a series of papers were published in which the degenerate oblique derivative problem was considered in the scenario when the vector field is tangent to the boundary along a submanifold of codimension one, to which this vector is not tangent. This line of work received a big impetus when in 1970 Vladimir Maz'ya initiated a deep investigation of the problem in the case in which the boundary contains a nested family of submanifolds $\Gamma_1 \supset \Gamma_2 \supset \cdots \supset \Gamma_s$ with the property that the vector field is tangent to Γ_k at points belonging to Γ_{k+1}, and is transversal to Γ_s. By employing a new technique, Vladimir Maz'ya was able to prove in this setting the unique solvability of the problem in a formulation which includes an additional Dirichlet condition on the entry set of the vector field and allows the possibility of discontinuities of the solution at points of the exit set. Up to now, this is the only known result pertaining to the oblique derivative problem in the generic situation in the sense of V. Arnold, who has considered this problem as an illustration of his calculus of infinite co-dimensions (see [**3**], §29 B). According to a hypothesis of Arnold, all submanifolds $\Gamma_1, \ldots, \Gamma_s$, induce infinite dimensional kernels or co-kernels for the oblique derivative problem. Nonetheless, Maz'ya's striking theorem reveals that Arnold's hypothesis is inadequate, since it turns out that submanifolds of co-dimension greater than one in the boundary are negligible, in the sense that they play the same type of role as removable singularities.

◇ `Boundary-value problems in domains with piecewise-smooth boundaries.` Vladimir Maz'ya has started working in this field at the beginning of the 1960's and from his early publications he was able to establish deep and unexpected results regarding second-order elliptic equations. For example, in studying selfadjointness conditions for the Laplace operator with zero Dirichlet data on contours of class C^1 (but not C^2), he discovered a surprising instability effect for the index under affine coordinate transformations. Following the emergence of Kondrat'ev's well-known 1967 paper on elliptic boundary-value problems in domains with conic singularities, Vladimir Maz'ya began working actively in this field and, in collaboration with B.A. Plamenevskii, and later with V.A. Kozlov and J. Rossmann, has produced a string of papers which contain a fascinating theory of boundary-value problems in domains with piecewise smooth boundary, including regularity estimates, asymptotic representations of solutions, well-posedness theorems, and methods for computing the coefficients in the asymptotics of solutions near boundary singularities. The theory thus developed, together with important applications to problems arising in mechanics, engineering and mathematical physics, is presented in the monographs [Ma-7], [Ma-8], [Ma-15], and [Ma-16].

The aforementioned body of results complements the theory of elliptic boundary value problems in Lipschitz domains, as initiated by A. Calderón, B. Dahlberg, E. Fabes, N. Riviere, M. Jodeit, C. Kenig, D. Jerison, J. Pipher, G. Verchota starting in the late 70's and early 80's. An authoritative account of the state of the art in this field, up to the mid 90's can be found in C. Kenig's book [**7**]. Compared with the latter, the former setting of domains with piecewise smooth boundaries allows for a wide range of non-Lipschitz domains. A simple example is offered by Maz'ya's "two-brick domain":

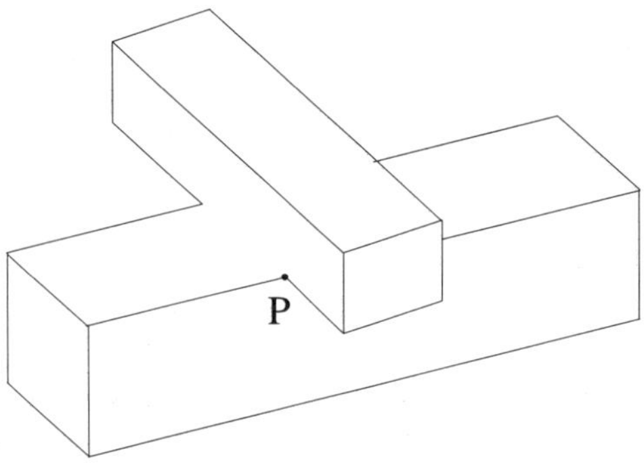

FIGURE 1

Indeed, a moment's reflection shows that near the point P, the boundary of the above domain is not the graph of any function (as it fails the vertical line test) even after applying a rigid motion. Most recently, progress in understanding such configurations from the Harmonic Analysis perspective has been recorded in [**12**], [**13**], [**14**].

◇ `Multipliers between spaces of differentiable functions.` In the late 70's, Vladimir Maz'ya and Tatyana Shaposhnikova initiated a systematic study of multipliers in pairs of various spaces of differentiable functions. This resulted in their joint book [Ma-19], which for the time being, is the only monograph on this topic. The forthcoming book [Ma-1] by the same authors reports on the more recent progress in this area. The obvious motivation for a thorough investigation of properties of multipliers stems from the study of partial differential equations of the type

$$Lu := \sum_{|\alpha|,|\beta|\leq m} \partial^\alpha(a_{\alpha,\beta}(X)\partial^\beta u) = f \quad \text{in} \quad \Omega, \tag{1}$$

in which the data and the solution belong to appropriate Sobolev spaces in the domain $\Omega \subset \mathbb{R}^n$. It is then of interest to understand how multiplication by the coefficients $a_{\alpha,\beta}$ transforms these classes of functions. A similar perspective comes from treating (1) via localization and flattening of the boundary of the domain Ω, for the purpose of transforming the original PDE into a problem in the upper half-space $\mathbb{R}^n_+$. In this scenario, the multiplier properties of the functions $\varphi : \mathbb{R}^{n-1} \to \mathbb{R}$ which locally describe $\partial\Omega$ come into play. For example, in the case of the Poisson problem for the Laplacian with a Dirichlet boundary condition, i.e.,

$$\begin{cases} \Delta u = f & \text{in } \Omega, \\ \operatorname{Tr} u = 0 & \text{on } \partial\Omega, \end{cases} \tag{2}$$

such techniques allow for a sharp description of the analytical properties of Ω required for the implication

$$f \in L^p(\Omega) \implies u \in W^{2,p}(\Omega) \tag{3}$$

to hold (where $1 < p < \infty$ is given). One other route through which multipliers take center-stage in a natural fashion is when one considers PDE's on manifolds, in which case the transformational properties of (1) under changes of variables are of central focus.

For general multipliers, Maz'ya and Shaposhnikova have established a wealth of basic results on the spectrum, traces and extensions, implicit functions, and two-sided estimates for the essential norm. They have also identified various classes of mappings and classes of non-smooth manifolds on which these multiplier spaces are invariantly defined. In addition, a calculus of singular integral operators with symbols in the space of multipliers was developed. These efforts have been amply rewarded by the fact that such a theory permits for deep applications to elliptic boundary value problems in domains with non-smooth boundaries.

◇ `Isoperimetric and integral inequalities, and theory of capacities.` While a fourth-year student at LSU, Vladimir Maz'ya made the remarkable discovery that integral inequalities of Sobolev type are actually equivalent to certain isoperimetric and isocapacitary inequalities for subsets of the domain where a function is defined. Even today, Vladimir likes to recall that special moment of inspiration, and he can artfully and fluidly reproduce the original calculations, to the delight of an interested interlocutor. These results, which eventually became part of his Ph.D. thesis, appeared in press in 1960-61. This original approach enabled him to obtain sharp constants in the aforementioned integral inequalities. In particular, the sharp constant in Gagliardo's inequality

$$\|u\|_{L^{\frac{n}{n-1}}(\mathbb{R}^n)} \le C_n \|\nabla u\|_{L^1(\mathbb{R}^n)}, \qquad u \in C_0^\infty(\mathbb{R}^n), \tag{4}$$

proved to be equal to that in the classical isoperimetric inequality: $C_n = n^{-1} v_n^{-1/n}$, where v_n is the volume of the unit ball in $\mathbb{R}^n$ (this was also found simultaneously and independently by G. Federer and W.H. Fleming). More importantly, as Maz'ya himself emphasized in 1966, his proofs did not make use of any specific properties of the Euclidean space and, hence, could be carried over to the setting of Riemannian manifolds.

An important inequality proved by Maz'ya (1964, 1972), and which later became known as the strong type capacitary inequality, allowed him to obtain capacitary criteria for Sobolev-type estimates. In more recent papers (2005, 2006), he has also obtained some important generalizations of this inequality. He also discovered (2003) that embeddings in fractional Besov spaces, or Riesz potential spaces, are equivalent with the validity of a certain new type of isoperimetric inequalities. The 1964, 1972 papers of Maz'ya, mentioned above, have motivated a thorough study of different aspects of the theory of Sobolev spaces, and have decisively influenced the development of this branch of mathematics. Currently, the methods in those papers are the driving force in the study of Sobolev spaces on metric spaces. The collection of results obtained up to 1985 are diligently presented in [Ma-20], arguably the most popular book authored by Vladimir Maz'ya (scheduled to appear in a new edition shortly).

The systematic use of the notion of capacity of a set eventually became a recurrent theme of a sizable number of Maz'ya's papers. As early as 1963 he introduced the polyharmonic capacity and successfully employed it in order to find optimal conditions for the well-posedness of the Dirichlet problem in the energy space for higher-order elliptic equations. At the beginning of 70's, V. Maz'ya and V.P. Khavin considered non-linear potentials and systematically studied their properties. Presently, the theory of non-linear potentials (naturally viewed as an extension of the classical linear theory) is a main-stream, active and fast-growing area of research, which has helped produce answers to many basic questions in the theory of functions, particularly for those concerning the nature of exceptional sets.

◇ `Theory of the Schrödinger operator.` By making essential use of his previously developed capacitary criteria, Vladimir Maz'ya was able to obtain (in 1962, 1964), sharp conditions ensuring the validity of various spectral properties of the Schrödinger operator. More recently, in their masterful 2002 Acta Mathematica paper, Vladimir G. Maz'ya and Igor E. Verbitsky have identified the correct class of complex-valued potentials for which the Schrödinger operator $-\Delta + V$ maps the energy space into its own dual. Subsequently, V. Maz'ya, V.A. Kondrat'ev and M.A. Shubin (2004) have proved necessary and sufficient conditions for the spectrum of the Schrödinger operator with a magnetic potential to be positive and discrete, thus generalizing the well-known work of A.M. Molchanov on this topic (who has treated the case when the magnetic field is absent). In 2005, V. Maz'ya and M. Shubin succeeded in characterizing the sets which are negligible in Molchanov's criterion, thereby solving an long-standing open problem, originally posed by I.M. Gel'fand in 1953.

◇ `Boundary behavior and maximum principles for elliptic and parabolic systems.` One of the prevalent themes of research throughout Vladimir Maz'ya's career, is the issue of regularity of a boundary point in the sense of Wiener. As early as 1962, he has proved an estimate for the modulus of continuity of a harmonic function, formulated in terms of the Wiener integral which, in turn, has found important applications in the qualitative theory of linear and non-linear elliptic equations. Then in 1970 he formulated a condition for regularity, in the sense of Wiener, of a boundary point for a certain class of quasi-linear second-order elliptic operators, which includes the p-Laplacian. Conspicuously, all these years virtually nothing was known about the Wiener type regularity of a boundary point for higher-order equations. The breakthrough came in 2002 when Vladimir Maz'ya succeeded in generalizing the Wiener test to elliptic equations of arbitrary order. Subsequently, this fundamental result made the subject of Maz'ya's talk at the International Congress of Mathematicians in Beijing.

In collaboration with G.I. Kresin, Vladimir Maz'ya has produced, in a series of papers starting around mid 80's, a necessary and sufficient condition formulated in algebraic terms guaranteeing the validity of the classical maximum modulus principle for second-order elliptic and parabolic systems. Next, in 1992, V. Maz'ya and J. Rossmann proved that the classical Miranda-Agmon maximum principle actually holds for any strongly elliptic operator of arbitrary order in a plane domain with a piece-wise smooth boundary, without peaks. While a similar result holds in the three-dimensional setting, in dimensions four and higher this principle fails for certain domains with conical vertices. For the polyharmonic (and biharmonic) equations in Lipschitz and C^1 domains, this issue has been further investigated in [**10**], [**11**].

◇ `Theory of water waves.` During his tenure at the Leningrad Shipbuilding Institute, Vladimir Maz'ya became interested in the mathematical theory of linear surface waves and, in 1973, wrote two articles in collaboration with B.R. Vainberg, in which the basic boundary value problems of this theory are studied. Four years later, Vladimir Maz'ya was the first to obtain a rather general uniqueness condition for the problem of oscillations of a body fully immersed in a liquid, which was originally stated by F. John as far back as 1950. The papers produced by Vladimir Maz'ya and his collaborators on this topic eventually led to the monograph [Ma-6].

Even from this brief review it is amply clear that Vladimir Maz'ya's work has an astonishing range and depth. However, he has left a lasting mark of originality and technical virtuosity in many more other branches of mathematics, such as *estimates for general differential and pseudodifferential operators in a half-space*, an area in which he has co-authored with I.V. Gel'man the monograph [Ma-21]; *Sobolev spaces and asymptotic theory of elliptic boundary-value problems on singularly perturbed domains*, in which Maz'ya has developed a rather sophisticated theory, first in collaboration with S.V. Poborchi, then jointly with S.A. Nazarov and B. Plamenevskii, as well as V.A. Kozlov and A.B. Movchan, which makes the subject of [Ma-8], [Ma-9], and [Ma-11], respectively, *numerical analysis* (cf. [Ma-2] written with G. Schmidt); *history of mathematics*, an area in which he has co-authored with Tatyana Shaposhnikova a delightful and highly informative book about the life and work of J. Hadamard ([Ma-5],[Ma-12]); *asymptotic theory of solutions to differential equations with operator coefficients* [Ma-10], written jointly with V.A. Kozlov; and *estimates for analytic functions with a bounded real part*, described in the book [Ma-3], based on the joint research with G. Kresin. This list should also include pointwise interpolation inequalities for derivatives, approximation by analytic and harmonic functions, degenerate elliptic pseudodifferential operators, uniqueness theorems for certain boundary value problems with data prescribed on only a portion of the boundary, characteristic Cauchy problems for hyperbolic equations, iterative procedures for solving ill-posed boundary value problems, etc.

Always animated by large, important ideas, magnanimous in sharing his expertise with other, particularly younger, people, one can only wonder what other magnificent contributions Vladimir Maz'ya will make in the future; we wish him many more years ahead, in good health.

V. Books (co-)authored by Vladimir Maz'ya.

[Ma-1] *Theory of Sobolev Multipliers with Applications to Differential and Integral Operators*, with T. Shaposhnikova, Grundlehren der Mathematischen Wissenschaften, vol. 337, Springer, 2008.

[Ma-2] *Approximate Approximations*, with G. Schmidt, American Mathematical Society, 2007.

[Ma-3] *Sharp Real-Part Theorems. A Unified approach*, with G. Kresin, Lecture Notes in Mathematics, No. 1903, Springer, 2007.

[Ma-4] *Imbedding and Extension Theorems for Functions in Non-Lipschitz Domains*, with S. Poborchi, St-Petersburg University Publishers, 2007.

[Ma-5] *Jacques Hadamard, un Mathématicien Universel*, with T. Shaposhnikova, EDP Sciences, Paris, 2005 (revised and extended translation from English).

[Ma-6] *Linear Water Waves. A Mathematical Approach*, with N. Kuznetsov and B. Vainberg, Cambridge University Press, 2002.

[Ma-7] *Spectral Problems Associated with Corner Singularities of Solutions to Elliptic Equations*, with V. Kozlov and J. Rossmann, Mathematical Surveys and Monographs, Vol. 85, American Mathematical Society, 2000.

[Ma-8] *Asymptotic Theory of Elliptic Boundary Value Problems in Singularly Perturbed Domains*, Vol. 2, with S. Nazarov and B. Plamenevskij, Operator Theory. Advances and Applications, Vol. 112, Birkhäuser, 2000.

[Ma-9] *Asymptotic Theory of Elliptic Boundary Value Problems in Singularly Perturbed Domains*, Vol. 1, with S. Nazarov and B. Plamenevskij, Operator Theory. Advances and Applications, Vol. 111, Birkhäuser, 2000.

[Ma-10] *Differential Equations with Operator Coefficients*, with V. Kozlov, Springer Monographs in Mathematics, Springer-Verlag, 1999.

[Ma-11] *Asymptotic Analysis of Fields in Multistructures*, with V. Kozlov and A. Movchan, Oxford Science Publications, 1999

[Ma-12] *Jacques Hadamard, a Universal Mathematician*, with T. Shaposhnikova, American Mathematical Society and London Mathematical Society, 1998.

[Ma-13] *Differentiable Functions on Bad Domains*, with S. Poborchi, World Scientific, 1997.

[Ma-14] *Theory of a Higher-order Sturm-Liouville Equation*, with V. Kozlov, Springer-Verlag, Lecture Notes in Mathematics, 1997.

[Ma-15] *Elliptic Boundary Value Problems in Domains with Point Singularities*, with V. Kozlov and J. Rossmann, American Mathematical Society, 1997.

[Ma-16] *Elliptic Boundary Value Problems*, with N. Morozov, B. Plamenevskii, L. Stupyalis, American Mathematical Society Translations, Ser. 2, Vol. 123, 1984, AMS.

[Ma-17] *Encyclopaedia of Mathematical Sciences, Vol. 27, Analysis IV, Linear and Boundary Integral Equations*, S.M. Nikol'skii (Eds.), Contributors: V.G. Maz'ya, S. Prössdorf, 233 pages, Springer-Verlag, 1991, V. G. Maz'ya: Boundary Integral Equations, pp. 127–222.

[Ma-18] *Encyclopaedia of Mathematical Sciences, Vol. 26, Analysis III, Spaces of Differentiable Functions*, S.M. Nikol'skii (Ed.), Contributors: L.D. Kudryavtsev, V.G. Maz'ya, S.M. Nikol'skii, 218 pages, Springer-Verlag, 1990, V.G. Maz'ya: Classes of Domains, Measures and Capacities in the Theory of Differentiable Functions, pp. 141–211.

[Ma-19] *Theory of Multipliers in Spaces of Differentiable Functions*, with T. Shaposhnikova, Pitman, 1985 (Russian version: Leningrad University Press, 1986).

[Ma-20] *Sobolev Spaces, Springer-Verlag*, 1985 (Russian version: Leningrad University Press, 1985).

[Ma-21] *Abschätzungen fur Differentialoperatoren in Halbraum*, with I. Gelman, Berlin, Akademie Verlag, 1981; Birkhäuser, 1982.

[Ma-22] *Zur Theorie Sobolewsche Räume*, Series: Teubner-Texte zur Mathematik BSB B. G. Teubner Verlagsgesellschaft, Leipzig, 1981.

[Ma-23] *Einbettungssätze für Sobolewsche Räume*, Teil 2, Series: Teubner-Texte zur Mathematik, Band 28, BSB B. G. Teubner Verlagsgesellschaft, Leipzig, 1980.

[Ma-24] *Einbettungssätze für Sobolewsche Räume*, Teil 1, Series: Teubner-Texte zur Mathematik BSB B. G. Teubner Verlagsgesellschaft, Leipzig, 1979.

[Ma-25] *Potential Theory and Function Theory for Irregular Regions*, with Yu. Burago, Seminars in Mathematics, Steklov Institute, Leningrad, Vol. 3, Consultants Bureau, New York, 1969 (Russian version: 1967).

Bibliography

[1] M.S. Agranovich, Yu.D. Burago, V.P. Khavin, V.A. Kondratiev, V.P. Maslov, S.M. Nikol'skii, Yu.G. Reshetnyak, M.A. Shubin, B.R. Vainberg, M.I. Vishik, L.R. Volevich, *Vladimir G.*

Maz'ya, On the occasion of his 65th birthday, Russian Journal of Mathematical Physics **10** (2003), no. 3, 239–244.

[2] M.S. Agranovich, Yu.D. Burago, B.R. Vainberg, M.I. Vishik, S.G. Gindikin, V.A. Kondrat'ev, V.P. Maslov, S.V. Poborchii, Yu.G. Reshetnyak, V.P. Khavin, M.A. Shubin, *Vladimir Gilelevich Maz'ya (on his 70th birthday)*, Russian Math. Surveys **63:1** (2008), 189–196, Uspekhi Mat. Nauk **63:1** (2008), 183–189.

[3] V. I. Arnold, *Geometrical Methods in the Theory of Ordinary Differential Equations*, Grundlehren der mathematischen Wissenschaften, Springer, 1983.

[4] E. De Giorgi, *Un esempio di estremali discontinue per un problema variazionale di tipo ellittico*, Boll. Un. Mat. Ital. **4** (1968), 135–137.

[5] D. Eidus, A. Khvoles, G. Kresin, E. Merzbach, S. Prössdorf, T. Shaposhnikova, P. Sobolevskii, M. Solomiak, *Mathemathical Work of Vladimir Maz'ya (on the occasion of his 60-th birthday)*, Functional Differential Equations, **4** (1997), no. 1-2, 3–11.

[6] I. Gohberg, *Vladimir Maz'ya: friend and mathematician. Recollections*, in J. Rossmann, P. Takác and G. Wildenhain (eds.), The Maz'ya Anniversary Collection, Birkhäuser Verlag, Basel, 1999, pp. 1–5.

[7] C.E. Kenig, *Harmonic Analysis Techniques for Second Order Elliptic Boundary Value Problems*, CBMS Regional Conference Series in Mathematics, Vol. 83, AMS, Providence, RI, 1994.

[8] V. Maz'ya, *Examples of nonregular solutions of quasilinear elliptic equations with analytic coefficients*, Funkcional. Anal. i Prilozen., 2 (1968) no. 3, 53–57; English translation: Functional Anal. Appl. **2** (1968), 230–234.

[9] *The Maz'ya Anniversary Collection*, edited by J. Rossmann, P. Takác, and G. Wildenhain, University of Rostock, Germany, Birkhäuser Verlag, Switzerland, 1999.

[10] J. Pipher and G.C. Verchota, *Maximum principles for the polyharmonic equation on Lipschitz domains*, Potential Anal. **4** (1995), no. 6, 615–636.

[11] J. Pipher and G.C. Verchota, *A maximum principle for biharmonic functions in Lipschitz and C^1 domains*, Comment. Math. Helv., **68** (1993), no. 3, 385–414.

[12] G.C. Verchota, *The use of Rellich identities on certain nongraph boundaries*, pp. 127–138 in "Harmonic Analysis and Boundary Value Problems", Contemp. Math., Vol. 277, Amer. Math. Soc., Providence, RI, 2001.

[13] G.C. Verchota and A.L. Vogel, *A multidirectional Dirichlet problem*, J. Geom. Anal. **13** (2003), no. 3, 495–520.

[14] G.C. Verchota and A.L. Vogel, *The multidirectional Neumann problem in $\mathbb{R}^4$*, Math. Ann. **335** (2006), no. 3, 571–644.

Dorina Mitrea and Marius Mitrea
Columbia, Missouri

Proceedings of Symposia in Pure Mathematics
Volume **79**, 2008

Capacity, Carleson Measures, Boundary Convergence, and Exceptional Sets

N. Arcozzi, R. Rochberg, and E. Sawyer

Dedicated to Prof. V. Maz'ya on the occasion of his 70th birthday. He has opened many doors.

ABSTRACT. We study the relationships between a capacity theoretic description of Carleson measures for the Dirichlet space (as well as for a discrete model space, a function space on a dyadic tree) and a description based on a testing condition. Using those relationships we show how analysis in the model space can be used to establish boundary convergence results for both the model space and also the classical Dirichlet space.

1. Introduction and Summary

There is a fundamental relation between the capacity of a set and energy integrals of probability measures supported on that set. If the capacity is small the energy integral will be large; in particular sets of capacity zero cannot support probability measures of finite energy. Here we develop similar ideas relating capacity to Carleson measures. We show that if a set has small capacity then any probability measure supported on it must have large Carleson embedding constant. In particular sets of capacity zero are exactly the simultaneous null set for all nontrivial Carleson measures.

Functions having limited smoothness often exhibit attractive or convenient behavior at most points, the exceptional set being of capacity zero; that is, the good behavior holds *quasi-everywhere,* henceforth *q.e.*. Using the relationship between capacity and Carleson measures such a conclusion can be reformulated by saying the function exhibits the good behavior $\mu - a.e.$ for every Carleson measure μ. This can be useful because sometimes it is relatively easy, even tautological, to establish that a property holds $\mu - a.e.$. We will use this viewpoint to give a new approach to results related to boundary behavior of holomorphic and harmonic functions.

2000 *Mathematics Subject Classification.* 30A40, 46E35, 46E39.

The first author partially supported by the COFIN project Analisi Armonica, funded by the Italian Minister for Research.

The second author's work was supported by the National Science Foundation under Grant No. 0070642.

The third author's work was supported by the National Science and Engineering Council of Canada.

The dyadic Dirichlet space is a Hilbert space of functions on a dyadic tree. In many ways that space models the classical Dirichlet space, the space of holomorphic functions, f, on the unit disk $\mathbb{D}$ for which $\int_{\mathbb{D}} |f'|^2 < \infty$. In Sections 3 and 4 we present background material on the dyadic Dirichlet space and the associated theory of Carleson measures. In Section 5 we present our new results on Carleson measures for the dyadic Dirichlet space. Those include the use of Carleson measures to measure capacity and a direct proof of the equivalence of the measure theoretic and capacity theoretic criteria for a Carleson embedding.

In Section 6 we show how those results can be used to study boundary behavior and exceptional sets. Roughly, the idea is to work with the tree geometry to construct a majorant of the variation of the function being studied. If this majorant is in a discrete function space, X, then the boundary set on which the majorant is infinite must be a null set for every $X-$Carleson measure. We then appeal to results of Section 4 to recast this as a statement about the capacity of the exceptional set.

The discrete case is a model case in which the proofs are relatively straightforward and the geometric issues we wish to highlight are particularly clear; and in this paper we will focus almost exclusively on that model case. However the ideas we present are very general and we believe they can be extended to the study of smooth functions whenever there is control of local oscillation. The details of extending these techniques to classes of smooth functions, and to function spaces that are not Hilbert spaces, requires some further ideas and technical machinery. In Section 7 we briefly discuss the types of extensions we have in mind and our approach to them; however we postpone a detailed presentation of that work to a later paper [**ARSp**].

2. Embedding Maps

2.1. Capacity Conditions. Here we record some results which go back to classic work by Maz'ya. For proofs and references we refer to [**AH**] and [**AE**].

For a function f defined on $\mathbb{R}^2$ we define its Riesz potential of order one by $I_1 f = (-\Delta)^{-1/2} f$. We define the associated ($p = 2$) capacity of a compact $E \subset \mathbb{R}^2$ by

$$\begin{aligned}\operatorname{Cap}(E) &= \operatorname{Cap}_{1,2}(E) \\ &= \inf\left\{\int_{\mathbb{R}^2} g^2 dx : I_1 g \geq 1 \text{ on } E,\ g \geq 0,\ g \in L^2\left(\mathbb{R}^2\right)\right\} \\ &\approx \inf\left\{\int_{\mathbb{R}^2} |\nabla u|^2 dx : u \geq 1 \text{ on } E,\ u \in C_0^\infty\right\}.\end{aligned}$$

(Below we will tacitly assume all sets discussed are capacitable.) The basic trace theorem in this context is:

THEOREM 2.1. *Suppose ω is a nonnegative Borel measure on $\mathbb{R}^2$. The following are equivalent:*

(1) *There is a constant C so that $\forall u \in C_0^\infty$*

$$\|u\|_{L^2(d\omega)} \leq C \|\nabla u\|_{L^2(dx)}.$$

(2) *There is a constant C so that $\forall f \in L^2(dx)$*

$$\|I_1 f\|_{L^2(d\omega)} \leq C \|f\|_{L^2(dx)}.$$

(3) *There is a constant C so that for all compact $E \subset \mathbb{R}^2$*

$$\omega(E) \leq C \operatorname{Cap}(E).$$

There are similar results for functions defined in regions of $\mathbb{R}^2$ and for spaces of holomorphic functions. In particular we recall the result of Stegenga [**St**] for $\mathcal{D}$, the Dirichlet space of holomorphic functions on the unit disk $\mathbb{D}$;

$$\mathcal{D} = \left\{ f \in \operatorname{Hol}(\mathbb{D}) : \|f\|_{\mathcal{D}}^2 = |f(0)|^2 + \int_{\mathbb{D}} |f'|^2 \, dV < \infty \right\}.$$

Recall that functions in $\mathcal{D}$ have boundary values *a.e.* on $\mathbb{T}$.

For any $z \in \mathbb{D}$ we define the associated interval I_z in $\mathbb{T}$, the boundary of $\mathbb{D}$, to be the interval with center $z/|z|$ and length $2(1-|z|)$. For any open set $\mathcal{O}$ contained in $\mathbb{T}$ we define the *tent over* $\mathcal{O}$, $T(\mathcal{O})$ by

$$T(\mathcal{O}) = \{z \in \mathbb{D} : I_z \subset \mathcal{O}\}.$$

For such an $\mathcal{O}$ we define its $\mathcal{D}$–capacity by

$$\operatorname{Cap}_{\mathcal{D}}(\mathcal{O}) = \inf \left\{ \|f\|_{\mathcal{D}}^2 : \operatorname{Re} f \geq 1 \text{ on } \mathcal{O} \right\}.$$

In particular, for an interval $I \subset T$ we have the estimate

$$\operatorname{Cap}_{\mathcal{D}}(I) \sim \frac{1}{|\log |I||}.$$

THEOREM 2.2. [**St**] *The following are equivalent for a positive measure ω supported on $\overline{\mathbb{D}}$:*

(1) *There is a constant C so that $\forall f \in \mathcal{D}$*

$$\|f\|_{L^2(d\omega)} \leq C \|f\|_{\mathcal{D}}. \tag{2.1}$$

(2) *There is a constant C so that for any open $\mathcal{O} \subset \mathbb{T}$*

$$\omega(T(\mathcal{O})) \leq C \operatorname{Cap}_{\mathcal{D}}(\mathcal{O}). \tag{2.2}$$

2.2. Testing conditions. One approach to results such as the previous two is to find a good collection of testing functions and evaluate the desired inequality, for instance (2.1), on those functions. The inequalities obtained are of course necessary in order to have (2.1) hold for all functions and, in the favorable cases, those conditions are also sufficient.

A famous example where this approach succeeds is the following. Let H^2 be the Hardy space, the subspace of $L^2(\mathbb{T}, d\theta)$ consisting of functions, f, whose negative Fourier coefficients vanish. Such a function, f, extends to a holomorphic function in $\mathbb{D}$ which we also denote by f.

THEOREM 2.3. *(Carleson) The following are equivalent for a positive measure ω supported on $\overline{\mathbb{D}}$:*

(1) *There is a constant C so that $\forall f \in H^2(\mathbb{T})$*

$$\|f\|_{L^2(d\omega)} \leq C \|f\|_{L^2(\mathbb{T}, d\theta)} \tag{2.3}$$

(2) *There is a constant C so that for any open* interval *$I \subset \mathbb{T}$*

$$\omega(T(I)) \leq C |I|. \tag{2.4}$$

H^2 is a Hilbert space with reproducing kernels; the functions $k_z(e^{i\theta}) = (1 - \bar{z}e^{i\theta})^{-1}$ are the kernel functions. Testing (2.3) on k_z gives the estimate (2.4) for the interval I_z.

A digression on terminology: The issues in the previous theorems are certainly similar to each other in spirit. However the questions arose in different contexts and are discussed using different words. The result in Theorem 2.1 is referred to as a *trace theorem.* On the other hand the measures in Theorem 2.3 are referred to as *Carleson measures (for the Hardy space)* and the results in Theorems 2.2 and 2.3 are described as *characterizing the Carleson measures* for the respective spaces.

(Actually there is a lack of consensus on terminology here. Some call a measure a "Carleson measure" if it satisfies a measure theoretic condition in the style of (2.4); the measure of a region of simple shape is controlled by a function of the size of that region. Other people use the name to denote measures for which conclusions such as (2.1) and (2.3) hold; there is a continuous embedding of the space being studied into a Lebesgue space associated with the measure. We use the second style and call the measures of Theorem 2.2 Carleson measures. Also we call the embeddings whose continuity is insured by (2.1) or (2.3) Carleson embeddings and refer to the norms of those embeddings as the embedding constants of the measures or as the Carleson measure norms of the measures.)

Although the testing philosophy is a guide to the correct conjecture in this case, that is not always so. Similar ideas for the space $\mathcal{D}$ would lead to speculation that it was sufficient to have (2.1) for all intervals I, rather than all open sets. However that is not the correct answer and Stegenga provides the appropriate examples. Yet there is a slightly more sophisticated testing scheme which can be used effectively to study Carleson measures on the Dirichlet space. The idea, which goes back to work by Kerman and Sawyer [**KS**], is to test the embedding operator and also test its adjoint.

One main theme of this paper is the relationship between results such as Theorems 2.1 and 2.2 formulated in terms of capacity and the more measure theoretic results suggested by the testing scheme just described.

For more about capacity in general, and in particular the fact that set functionals which appear to be quite different can be used to measure capacity, we refer to [**AH**], [**AE**], [**V**], and [**KV**].

3. The Dyadic Dirichlet Space

We will work with functions on the dyadic tree T which we now define. T is a rooted, directed, loopless graph. It contains a root vertex, o, which is connected by two edges to vertices o_r and o_l (its right and left children). Every other element α is connected to three vertices; one, its parent, α^-, is on the path connecting α to o, the other two are its children, α_r and α_l. For α, $\beta \in T$ we denote by $[\alpha, \beta]$ the set of vertices on the geodesic path from α to β. We let $d(\alpha, \beta)$ denote the length of that path, $d(\alpha, \beta) = |[\alpha, \beta]| - 1$, and we will abbreviate $d(o, \alpha)$ to $d(\alpha)$. For $\alpha, \beta \in T$ we write $\beta \prec \alpha$ if $\beta \in [o, \alpha]$ and we define the successor set of β to be $S(\beta) = \{\alpha : \alpha \succeq \beta\}$

We will work with several operators defined on functions on T. For f a function on T we define functions Df, If, and I^*f by

$$Df(\alpha) = \begin{cases} f(o) & \text{if } \alpha = o \\ f(\alpha) - f(\alpha^+) & \text{otherwise,} \end{cases}$$
$$If(\alpha) = \sum_{\gamma \in [0,\alpha]} f(\gamma),$$
$$I^*f(\alpha) = \sum_{\gamma \in S(\alpha)} f(\gamma).$$

We have $ID = DI = Id$, the identity operator, and I^* is the adjoint of I with respect to the inner product on $l^2(T)$.

We envision T as a subset of $\mathbb{D}$ with o at the origin and with the 2^n vertices in $\{\alpha : d(\alpha) = n\}$ located at equally spaced points on the circle centered at the origin and of radius $1 - 2^{-n}$. With this in mind we define the *dyadic Dirichlet space,* $\mathcal{D}_d$, to be the space of functions F on T for which

$$\|F\|_{\mathcal{D}_d}^2 = |F(o)|^2 + \sum\nolimits_T |DF|^2 < \infty.$$

Informally we think of T as sitting in $\mathbb{D}$ and of holomorphic functions $F \in \mathcal{D}$ as being modeled by their restrictions to T. Oversimplifying a bit we have these restriction are in $\mathcal{D}_d$ and there is an informal correspondence between the operators $\int_0^z$ and $(1 - |z|^2)\frac{d}{dz}$ acting on holomorphic functions and the operators I and D operating on functions on the tree. This point of view is developed in detail in [**Ar**], [**AR**], [**ARS1**], and [**ARS2**].

In this context we will associate to any measure, μ, defined on $\mathbb{D}$ a measure μ_T which is defined on T by setting $\mu_T(\alpha)$ equal to the μ measure of the set of $z \in \mathbb{D}$ which are hyperbolically closer to α (thought of as sitting in D) than to any other point of T (with any convenient choice in cases of ties).

We will also be interested in functions and measures defined on the (abstract) boundary of T, ∂T, which we now introduce. A point of ∂T is an equivalence class of geodesics in T each infinite in one direction. The geodesics Γ and Γ' are equivalent exactly if their symmetric difference, $(\Gamma \cup \Gamma') \setminus (\Gamma \cap \Gamma')$, is finite. A convenient representative for a class is the unique geodesic in the class starting at the root. For each $\alpha \in T$ let $\partial S(\alpha)$ be the set of τ in ∂T with the property that every representative geodesic for τ intersects $S(\alpha)$. The topology of ∂T is defined by declaring the sets $\partial S(\alpha)$ to be a basis for the open neighborhoods. We set $\overline{T} = T \cup \partial T$. Note that any representative geodesic Γ for a point of ∂T, when viewed as a subset of $\mathbb{D}$, is a sequence that converges to a point of $\mathbb{T}$. Hence when we think of T as a subset of $\mathbb{D}$ it is also convenient to think of ∂T as identified with $\mathbb{T}$ with its usual topology. This lets us identify open intervals of $\mathbb{T}$ with open sets ∂T. Using this identification we will regard measures defined on $\mathbb{T}$ as also defined on ∂T and vice versa. (The pairing of ∂T with $\mathbb{T}$ has a countable set of ambiguous points, but these cause no difficulty in the discussion which follows.) Finally, if a function F on T has a limit along a geodesic Γ then we extend it by continuity to the corresponding point of ∂T. In particular if f has finite support then If extends to the entire boundary.

We define capacity for subsets E of $\overline{T}$ by

$$\mathrm{Cap}_T(E) = \inf\{\|\varphi\|_{\ell^2}^2 : I\varphi|_E \geq 1\}. \tag{3.1}$$

Alternatively, there is a straightforward variation which we will use later

$$\mathrm{Cap}_T(E) = \inf\{\|I^*\mu\|_{\ell^2}^2 : \ \mu \text{ a positive measure on } E,\ II^*\mu|_E \geq 1\}. \qquad (3.2)$$

The quantity $\|I^*\mu\|_{\ell^2}^2$ is sometimes called the *energy integral of* μ and denoted $\mathcal{E}(\mu)$.

4. Carleson Measures for $\mathcal{D}_d$ and $\mathcal{D}$

Here is the characterization of Carleson measures for $\mathcal{D}_d$ and $\mathcal{D}$ from [**ARS1**] which complements Stegenga's description of Carleson measures for the Dirichlet space. (Actually the final statement of the theorem, a capacitary condition in the style of Stegenga, is not in [**ARS1**] but can be included, for instance, by virtue of Theorem 5.6 below.)

We will also be interested in the radial variation of functions in $\mathcal{D}$ and we now introduce a function which measures that. For f in $\mathcal{D}$ we define

$$V(f)(re^{i\theta}) = \int_0^r \left|f'(te^{i\theta})\right| dt.$$

THEOREM 4.1. *Let μ be a positive Borel measure on $\overline{\mathbb{D}}$. Then, the following conditions are equivalent:*

(1) *μ is a Carleson measure for $\mathcal{D}$; i.e. (2.1) holds.*
(2) *μ satisfies a Carleson inequality for the radial variation. There is a constant C so that $\forall f \in \mathcal{D}$*

$$\int_{\overline{\mathbb{D}}} |V(f)(z)|^2 d\mu \leq C\|f\|_{\mathcal{D}}^2.$$

(3) *μ_T is a Carleson measure for $\mathcal{D}_d$. There is a constant C so that $\forall f \in \mathcal{D}_d$*

$$\sum_{\alpha\in T} |F(\alpha)|^2 \mu_T(\alpha) \leq C \|F\|_{\mathcal{D}_d}^2 . \qquad (4.1)$$

(4) *μ_T satisfies the tree condition. There is a constant C so that $\forall \alpha \in T$*

$$I^*(I^*\mu_T)^2(\alpha) \leq CI^*\mu_T(\alpha) < \infty. \qquad (4.2)$$

(5) *μ_T satisfies a tree capacity condition. There is a constant C so that for all sets $E = \cup_j \partial S(x_j)$ in ∂T,*

$$\mu_T\left(\cup_j\overline{S(x_j)}\right) \leq C\,\mathrm{Cap}_T(\cup_j\partial S(x_j)). \qquad (4.3)$$

The condition (4.1) states that the "integration map" I is bounded from $l^2(T)$ into $l^2(T,\mu_T)$. We could analyze that condition by testing it on simple functions as follows. Pick $\alpha \in T$ and set $f = d(\alpha)^{-1}\chi_{(o.\alpha]}$. If (4.1) holds then we find

$$\begin{aligned}(I^*\mu_T)(\alpha) = \mu(S(\alpha)) &\leq \|If\|_{l^2(T,\mu)}^2 \\ &\leq C\|If\|_{\mathcal{D}_d}^2 = d(\alpha)^{-1}.\end{aligned}$$

The condition obtained, $d(\alpha)\mu(S(\alpha)) \leq C$, is equivalent to the condition (4.3) restricted to the case where $\{x_j\}$ has only one element. As was true in Stegenga's theorem that condition is necessary for the boundedness of I but not sufficient. However if I is bounded then so is its adjoint, I^*, mapping $l^2(T,\mu_T)$ to $l^2(T)$.

Condition (4.2) tests the boundedness of I^* on the functions $\chi_{S(\alpha)}$; (4.2) is a slightly weakened version of the statement that

$$\left\|I^* \chi_{S(\alpha)}\right\|_{l^2(T)} \leq C \left\|\chi_{S(\alpha)}\right\|_{l^2(T,\mu_T)}.$$

Let μ be a positive, Borel measure on $\overline{T}$. We denote the best constant in the testing condition (4.2) by $\|\mu\|_{CM(T)}$;

$$\|\mu\|_{CM(T)} = \sup_{\alpha \in T} \frac{I^*(I^*\mu)^2(\alpha)}{I^*\mu(\alpha)}. \tag{4.4}$$

By the previous theorem this quantity is comparable to the norm of the Carleson embedding on $\mathcal{D}_d$.

5. Capacity and Carleson Measures on Trees.

5.1. Computing Capacity Using Carleson Measures. The ordinary definition of capacity for a subset E of $\overline{T}$ is that in (3.1). We now give an alternative way of computing capacity for subsets of $\overline{T}$ using Carleson measures. To minimize the notational burden we give the proof below for $E \subset \partial T$ which is the case of primary interest. The extension to the general case is straightforward.

THEOREM 5.1. *Let $E \subset \overline{T}$ be compact. Then,*

$$\mathrm{Cap}_T(E) = \sup_{\mathrm{Supp}\,(\mu) \subset E} \frac{\mu(E)}{\|\mu\|_{CM(T)}}. \tag{5.1}$$

Before proceeding with the proof we note that the quantities which appear in the definition $\|\mu\|_{CM(T)}$ are in fact related to quantities which show up in the study of capacity. From the definitions we have that $I^*\mu\,(\alpha) = \mu(\overline{S(\alpha)})$. If we multiply out the square in $I^*(I^*\mu)^2\,(\alpha)$ we find the comparable quantity

$$I^*(I^*\mu)^2\,(\alpha) \sim \sum_{\beta,\beta' \succ \alpha} d(\alpha, \beta \wedge \beta')\mu(\beta)\mu(\beta').$$

Thus, completely informally, the ratio in (4.2) is comparing the mass of the measure $\mu\chi_{S(\alpha)}$ to a type of energy integral of the same measure.

PROOF. We start with inequality [$\leq$]. For fixed $\alpha \in T$ we denote by $\bar{T}_\alpha = \bar{S}(\alpha)$ the subtree of $\bar{T}$ having root α and we add a subscript α to the corresponding tree objects: Cap_α is the capacity of subsets of $\bar{T}_\alpha$, $\mathcal{E}_\alpha$ is the energy in $\bar{T}_\alpha$, ω_α is the extremal measure in the definition of capacity, and so on. Let $E_\alpha = E \cap \bar{T}_\alpha$. The extremal measure ω_α and the function $\varphi_\alpha = I^*_\alpha \omega$ satisfy

$$Cap_\alpha(E_\alpha) = \omega_\alpha(E_\alpha) = \mathcal{E}_\alpha(\omega_\alpha) = \|\varphi_\alpha\|^2_{L^2(\bar{T}_\alpha)}. \tag{5.2}$$

We **claim** that ω_α is a rescaling of the extremal measure for E in $\bar{T}$, ω, restricted to E_α:

$$\omega_\alpha = \frac{\omega|_{E_\alpha}}{1 - II^*\omega(\alpha^{-1})}. \tag{5.3}$$

Here, α^{-1} is the predecessor of α in T. In fact, ω_α minimizes $\mathcal{E}_\alpha(\mu)$ over all measures μ such that $I_\alpha I^*_\alpha \mu(\xi) \geq 1$ on E_α (with the possible exception of a set having null-capacity). On the other hand, we claim that $\omega|_{E_\alpha}$ minimizes $\mathcal{E}_\alpha(\nu)$ among all measures ν on E_α such that $I_\alpha I^*_\alpha \nu(\xi) \geq 1 - II^*\omega(\alpha^{-1})$ *p.p.* on E_α.

Suppose this is not the case. Then there exists a measure ν on E_α such that $I_\alpha I_\alpha^*\nu(\xi) \geq 1 - II^*\omega(\alpha^{-1})$ for *p.p.* $\xi \in E_\alpha$ and

$$\mathcal{E}_\alpha(\nu) = \sum_{T_\alpha}(I^*\nu)^2 < \sum_{T_\alpha}(I^*\omega)^2 = \mathcal{E}_\alpha(\omega|_{E_\alpha}).$$

Consider the functions $I^*\omega$ in $\bar{T}$ and $I_\alpha^*\nu = I^*\nu$ in $\bar{T}_\alpha$. Define a new function ψ on T:

$$\psi(x) = \begin{cases} I^*\nu(x) \text{ if } x \in T_\alpha, \\ I^*\omega(x) \text{ if } x \in T \setminus T_\alpha. \end{cases}$$

We have

$$I\psi(\xi) \geq 1 \text{ } q.e. \text{ on } E,$$

hence $\|\psi\|_{L^2}^2 \geq Cap(E)$. On the other hand,

$$\|\psi\|_{L^2}^2 = \mathcal{E}_\alpha(\nu) + [\mathcal{E}(\omega) - \mathcal{E}_\alpha(\omega)] < \mathcal{E}(\omega) = Cap(E),$$

and we have reached a contradiction.

The measure

$$\lambda = \frac{\omega|_{E_\alpha}}{1 - II^*\omega(\alpha^{-1})},$$

then, minimizes $\mathcal{E}_\alpha(\mu)$ over the set of the measures μ such that $II^*\mu(\xi) \geq 1$ for *p.p.* ξ in E_α, hence $\lambda = \omega_\alpha$. The claim is proved.

By the homogeneity of the energy,

$$\begin{aligned}\mathcal{E}_\alpha(\omega|_{E_\alpha}) &= (1 - II^*\omega(\alpha^{-1}))^2\mathcal{E}_\alpha(\omega_\alpha) \\ &= (1 - II^*\omega(\alpha^{-1}))^2\omega_\alpha(E_\alpha) \\ &= (1 - II^*\omega(\alpha^{-1}))\omega(E_\alpha).\end{aligned}$$

As a consequence,

$$\frac{\sum_{x\geq\alpha}(I^*\omega)^{p'}\rho^{1-p'}}{I^*\omega} = \frac{\mathcal{E}_\alpha(\omega|_{E_\alpha})}{\omega(E_\alpha)} = (1 - V(\omega)(\alpha^{-1}))^{p'-1} \leq 1, \tag{5.4}$$

with equality if and only if $\alpha = o$ (we set the default value $II^*\omega(o^{-1}) = 0$). Hence, $\|\omega\|_{CM} = 1$ and

$$Cap(E) = \omega(E) = \frac{\omega(E)}{\|\omega\|_{CM}}.$$

We now prove [$\geq$]. By definition of $\|\cdot\|_{CM}$, $\mathcal{E}(\mu) \leq \|\mu\|_{CM}\mu(E)$ for all measures μ. Then,

$$\frac{\mu(E)}{\|\mu\|_{CM}} \leq \frac{\mu(E)}{\left(\frac{\mathcal{E}(\mu)}{\mu(E)}\right)} = \frac{\mu(E)^2}{\mathcal{E}(\mu)} \leq Cap(E),$$

as wished. □

COROLLARY 5.2. *E has capacity zero if and only if it is a null set for all Carleson measures.*

5.2. On a Question of Maz'ya. Conditions 4 and 5 of Theorem 4 are, by that theorem, equivalent to each other. Some time ago Prof. Maz'ya asked if we could give a direct proof of that equivalence, one not relying on the fact that both conditions characterize the same class of Carleson measures. We do that now. In our proof we will use the fact that if a measure satisfies condition 4 then so does any smaller positive measure. That fact follows from knowing that condition 4 characterizes a class of Carleson measures. However, in the spirit of this section, we also give a proof of that fact which is relatively elementary and which does not involve the theory of Carleson measures. We begin with the monotonicity.

For a measure μ on $\overline{T}$, let $\sigma_\mu = (I^*\mu)^2$.

THEOREM 5.3. *Let μ be a measure on $\overline{T}$ and let λ be a measurable function on $\overline{T}$, $0 \leq \lambda \leq 1$. If $I^*\sigma_\mu \leq I^*\mu$ on T, then $I^*\sigma_{\lambda\mu} \leq 2 \cdot I^*(\lambda\mu)$.*

COROLLARY 5.4. *If $\nu \leq \mu$ and μ, ν are measures on $\overline{T}$, then $\|\nu\|_{CM(T)} \leq 2\|\mu\|_{CM(T)}$.*

PROOF. By rescaling, it suffices to verify the conclusion at the root. We use a simple argument based on distribution functions. Let

$$\mathcal{M}_\mu\lambda(\alpha) = \max_{o\leq y\leq\alpha} \frac{I^*(\lambda\mu)(y)}{I^*\mu(y)}$$

be the discrete maximal function we used in [**ARS3**]. If necessary, we can extend the definition to $\alpha \in \partial T$ in the obvious way. Then,

$$\begin{aligned} I^*\sigma_{\lambda\mu}(o) &= \sum_{\alpha\in T} \left[\frac{I^*(\lambda\mu)(\alpha)}{I^*\mu(\alpha)}\right]^2 (I^*\mu(\alpha))^2 \\ &\leq \sum_{\alpha\in T} [\mathcal{M}_\mu\lambda(\alpha)]^2 \sigma_\mu(\alpha) \\ &= 2\int_0^1 t\sigma_\mu(\zeta \in \overline{T} : \mathcal{M}_\mu\lambda(\zeta) > t)dt. \end{aligned}$$

Now, $\{\zeta \in \overline{T} :\ M_\mu\lambda(\zeta) > t\} = \cup S(\alpha_j)$ is the disjoint union in $\overline{T}$ (by the definition of the maximal function, we do not need to consider the closure of $S(\alpha_j)$ in $\overline{T}$). Then,

$$\begin{aligned} t\sigma_\mu(\zeta \in \overline{T} :\ \mathcal{M}_\mu\lambda(\zeta) > t) &= \sum_j t\sigma_\mu(S(\alpha_j)) \\ &\leq \sum_j tI^*\mu(\alpha_j) \\ &\leq \sum_j I^*(\lambda\mu)(\alpha_j) \\ &\leq I^*(\lambda\mu)(o). \end{aligned}$$

Inserting this estimate in the previous one, we have

$$I^*\sigma_{\lambda\mu}(o) \leq 2 \cdot I^*\sigma_\mu(o).$$

□

We now give a direct proof that the tree condition is equivalent to the capacitary condition, Theorem 5.6 below. That the capacitary condition implies the tree condition was noted in [**KS**] and earlier as Theorem 4 in [**Ad**]. We proceed to the opposite implication. We need an estimate for measures supported in $\overline{T}$.

LEMMA 5.5. *We have*

$$\mathrm{Cap}_T(S(E)) \leq 4\,\mathrm{Cap}_T(E),$$

for $E = \cup \partial S(\alpha_j)$, *where* $S(E) = \cup \overline{S(\alpha_j)}$.

Note that $\mathrm{Cap}_T(S(E)) = \mathrm{Cap}_T(\{\alpha_j\})$. Obviously, $\mathrm{Cap}_T(S(E)) \geq \mathrm{Cap}_T(E)$.

PROOF. Let φ be the extremal function for $\mathrm{Cap}_T(E)$:

$$I\varphi(\zeta) = 1 \text{ on } E,\ \mathrm{Cap}_T(E) = \sum \varphi^2.$$

We show that it is near extremal for $\overline{S(E)}$.

The function φ can be recovered from the equilibrium measure, $\varphi = I^*\mu$, and μ is clearly constant on each $\partial S(\alpha_j)$: there is $\Gamma_j > 0$ such that

$$\varphi = \Gamma_j 2^{-d(\beta)}\ \forall \beta \in S(\alpha_j).$$

Now, for all $\zeta \in \partial S(\alpha_j)$,

$$\begin{aligned} 1 - I\varphi(\alpha_j) &= I\varphi(\zeta) - I\varphi(\alpha) \\ &= \sum_{\beta \in [\alpha_j, \zeta]} \varphi(\beta) \\ &= \Gamma_j \sum_{[\alpha_j,\zeta]} 2^{-d(\beta)} \\ &= \Gamma_j 2^{-d(\alpha_j)} \varphi(\alpha_j). \end{aligned}$$

Hence, $1 - \varphi(\alpha_j) = I\varphi(\alpha_j)$. Note that φ, the extremal function, is monotone increasing with respect to the partial ordering in T, thus $\varphi(\alpha_j) \geq 1/d(\alpha_j)$. Hence,

$$I\varphi(\alpha_j) \geq 1 - 1/d(\alpha_j) \geq 1/2.$$

This means that 2φ is an admissible function for E:

$$\mathrm{Cap}_T(S(E)) \leq 4\,\mathrm{Cap}_T(E).$$

□

THEOREM 5.6. *Let* μ *be a positive, Borel measure on* $\overline{T}$. *Then* μ *satisfies*

$$\sup_{\alpha \in T} \frac{I^*[I^*\mu]^2(\alpha)}{I^*\mu(\alpha)} \leq C_1(\mu). \tag{5.5}$$

if and only if μ *satisfies, for all sets* $E = \cup_j \partial S(\alpha_j)$ *in* ∂T,

$$\mu\left(\cup_j \overline{S(\alpha_j)}\right) \leq C_2(\mu)\,\mathrm{Cap}_T(\cup_j \partial S(\alpha_j)). \tag{5.6}$$

Moreover, $C_2(\mu) \sim C_1(\mu)$.

PROOF. Suppose that μ satisfies (5.5) with $C_1(\mu) = 1$. Recall that $\overline{S(E)} = \cup \overline{S(\alpha_j)}$. Then, $\mu_E = \mu|_{\overline{S(E)}} \le \mu$ satisfies $\|\mu_E\|_{CM(T)} \le 2$ by Theorem 5.3. Hence,

$$\begin{aligned}\operatorname{Cap}_T(\overline{S(E)}) &= \sup_{\sup(\nu)\subseteq\overline{S(E)}} \frac{\nu(\overline{S(E)})}{\|\nu\|_{CM(T)}} \\ &\ge \frac{\mu(\overline{S(E)})}{\|\mu_E\|_{CM(T)}} \\ &\ge \frac{\mu(\overline{S(E)})}{2}.\end{aligned}$$

i.e., $\mu(\overline{S(E)}) \le 2\operatorname{Cap}_T(\overline{S(E)}) \le 8\operatorname{Cap}_T(E)$, by Lemma 5.5. As mentioned, the opposite inequality is already known. □

6. Boundary Behavior and Exceptional Sets

In this section we give a number of results about boundary behavior and exceptional sets for the dyadic Dirichlet space. In several cases we show that certain behavior occurs with an exceptional set that is a null set for a class of Carleson measures. Then by Corollary 5.2, or a variation of that corollary, we conclude that the possible exceptional set has capacity zero.

The results we present now are discrete analogs of established results about boundary convergence of smooth functions and about the associated exceptional sets. The literature on those problems is extensive. We offer [**A**], [**AC**], [**Ca**], [**DB**], [**GP**], [**Ki**], [**M**], [**NRS**], [**NS**], [**Tw**], and [**W**] as recent representatives as well as places where the reader can get more information. In particular the versions of our next few results for harmonic functions are in [**Tw**]. A main theme here is to show that when appropriate oscillation estimates are available then there is a unified approach to such results. In particular this approach highlights the basic geometry of the tree model, or, what is roughly the same thing, the geometry of a Whitney covering of the domain.

6.1. Boundary Values. It is not clear at first glance that functions in $\mathcal{D}_d$ must have boundary values on a large subset of ∂T. We now establish that with an argument whose basic form is at the core of the later discussions. For each positive integer n let χ_n be the characteristic function of the set $\{\alpha \in T : d(\alpha) \le n\}$. Given $F \in \mathcal{D}_d$, set, for each n

$$F_n^* = I(|DF| \cdot \chi_n).$$

It is immediate that the sequence of functions $\{F_n^*\}$ is increasing in n, that each function extends by continuity to all of ∂T, and that each function is in $\mathcal{D}_d$ and has norm at most $\|F\|$. By monotonicity the extended limit $\lim F_n^* = F^*$ is defined everywhere. It then follows using Fatou's lemma that $|F^*|^2$ has finite integral with respect to any Carleson measure μ for $\mathcal{D}_d$. Hence F^* is finite $\mu - a.e.$. With F^* in hand as a majorant for the variation of F along each geodesic it is easy to see that F also has boundary limits $\mu - a.e.$ Hence, by Corollary 5.2, F has finite boundary limits $q.e.$

6.2. Beurling's Theorem. The prototypical result in this area is Beurling's theorem that any $f \in \mathcal{D}$ has radial boundary values *q.e.* He did this by showing that the boundary values of the radial variation, $V(f)(e^{it})$ were finite *q.e.* and noting that at such points f must have radial boundary values. In fact the argument in the previous paragraph is essentially a complete proof of a discrete analog of Beurling's theorem. For $F \in \mathcal{D}_d$ we measure its radial variation by $V_T(F)(\alpha) = I(|DF|)(\alpha)$.

THEOREM 6.1 (Discrete Beurling's Theorem). *For $F \in \mathcal{D}_d$,*

$$\mathrm{Cap}_T(\{\Gamma \in \partial T : V_T(F)(\Gamma) = +\infty\}) = 0.$$

Hence

$$\mathrm{Cap}_T(\{\Gamma \in \partial T : \lim_{\alpha \in \Gamma} F(\alpha) \text{ does not exist}\}) = 0.$$

PROOF. Start with $F \in \mathcal{D}_d$. Hence $DF \in l^2(T)$, and hence $|DF| \in l^2(T)$, and thus $V_T(F) \in \mathcal{D}_d$. This insures that the set on which $V_T(F) = \infty$ is a null set for any $\mathcal{D}_d$ Carleson measure and hence, by Corollary 5.2, has vanishing capacity. Finally, one easily checks that if $V_T(F)(\Gamma) < \infty$ then $\lim_{\alpha \in \Gamma} F(\alpha)$ exists. □

6.3. Algebraic Approach Regions. We continue to focus on $\mathcal{D}_d$ and now consider boundary limits through more general approach regions. For any subset $S \subset T$ we define the boundary limit through S of a function F defined on T by

$$\lim_S F(\beta) = \lim_{\substack{\beta \in S \\ d(\beta) \to \infty}} F(\beta).$$

At the level of metaphor, convergence along the geodesics $\Gamma \subset T$ is similar to both radial convergence in the disk and to non-tangential convergence through a narrow wedge with apex at the boundary point corresponding to Γ. The analog of non-tangential convergence with wider wedges is obtained by also including points that are at most a fixed distance, k, from Γ. For $\alpha \in T, n \in \mathbb{N}$ let

$$\{\alpha + n\} = \{(\beta \in T : \beta \succ \alpha,\ d(\alpha, \beta) = n)\}$$

and we define

$$\Gamma_1(k) = \bigcup_{\alpha \in \Gamma} \{\alpha + k\}.$$

THEOREM 6.2. *Fix $k \geq 1$. For $F \in \mathcal{D}_d$, F has $\Gamma_1(k)$ limits q.e.; that is*

$$\mathrm{Cap}_T(\{\Gamma \in \partial T : \lim_{\Gamma_1(k)} F(\alpha) \text{ does not exist}\}) = 0.$$

PROOF. For $F \in \mathcal{D}_d$ we define $DF^{(1)}$ by

$$DF^{(1)}(\alpha) = |DF(\alpha)| + \sum_{\beta \in \{\alpha + k\}} |DF(\beta)|.$$

Because $DF \in l^2(T)$ we also have $DF^{(1)} \in l^2(T)$, the reason is that each value $|DF(\delta)|$ shows up as part of $DF^{(1)}(\delta)$ and as part of at most one other $DF^{(1)}(\delta')$. Hence $G = I(DF^{(1)}) \in \mathcal{D}_d$ and thus G has boundary values *q.e.* Suppose now that Γ is a geodesic representing a point in the boundary at which G has boundary limit $G(\Gamma)$. We claim that it follows that F has a boundary limit through the approach region $\Gamma_1(k)$. We are assuming

$$\lim_{\Gamma} G(\alpha) - G(\Gamma) = 0, \tag{6.1}$$

Now pick α in Γ and let β, β' be two points of $\Gamma_1(k)$ which are further from the root than α. Let α^{-k} be the direct ancestor of α of order k. Now we have that

$$\begin{aligned}|F(\beta)-F(\beta')| &\leq |F(\beta)-F(\alpha)|+|F(\alpha)-F(\beta')| \\ &\leq \sum_{\gamma\in[\alpha,\beta]}|DF(\gamma)|+\sum_{\gamma\in[\alpha,\beta']}|DF(\gamma')| \\ &\leq 4\left(G(\Gamma)-G(\alpha^{-k})\right). \end{aligned} \tag{6.2}$$

To see this last inequality we concentrate on the first sum. The geodesic segment $[\alpha,\beta]$ has two regions, the first consisting of γ's in Γ, the second part consisting of the γ''s not in Γ. In the first case $|DF(\gamma)| \leq DG(\gamma)$. In the second case set $\beta(\gamma') = \gamma'^{-k}$, the ancestor of γ' of order k. We have $\beta(\gamma') \in [\alpha^{-k},\alpha]$ and thus $|DF(\gamma')| \leq DG(\beta(\gamma'))$. Thus the sum of the $|DF|$ along $[\alpha,\beta]$ is dominated by twice the sum of the DG along $[\alpha^{-k},\alpha]$. Finally, because G is increasing we get (6.2). Furthermore (6.1) insures that the right hand side of (6.2) will tend to zero as α tends to the boundary; and hence F has the desired limit through $\Gamma_1(k)$. □

We now consider larger regions. Fix an integer k and set

$$\Gamma_2(k) = \bigcup_{\alpha\in\Gamma}\{\alpha + kd(\alpha)\}.$$

We will consider limits over the sets $\Gamma_2(k)$. To do this we start with F and construct a majorant for its variation. Define

$$DF^{(2)}(\alpha) = |DF(\alpha)| + \sum_{j=0}^{k-1}\max\{|DF(\beta)| : \beta\in\{\alpha+kd(\alpha)+j\}\}$$

If $F\in\mathcal{D}_d$ then $DF\in l^2(T)$. We now claim that $DF^{(2)}\in l^2(T)$. Each $DF^{(2)}(\alpha)$ is a sum of $k+1$ terms taken from the square summable sequence $\{DF(\alpha)\}$ so we only need show that there is an upper bound on how many times an individual element of that latter sequence is used in this construction. However, in fact, no term is used more than twice. The term $|DF(\beta)|$ does appear in $DF^{(2)}(\beta)$. The only other time that term can be used as a summand is in $DF^{(2)}(\alpha)$ for the unique α which satisfies the two conditions $\alpha\in[o,\beta]$ and $(k+1)\,d(\alpha)\leq d(\beta) < (k+2)\,d(\alpha)$. This observation together with an application of the Cauchy-Schwarz inequality to the sum used in defining $DF^{(2)}(\alpha)$ shows that $DF^{(2)}\in l^2(T)$.

A minor modification of the argument we used before now shows that F converges along $\Gamma_2(k)$. The only required change is where before we backed up from α to an α^{-k} now we have to back up from α to $\alpha'\in[o,\alpha]$ with $(k+1)\,d(\alpha') > d(\alpha)$.

THEOREM 6.3. *Fix $k\geq 1$. For $F\in\mathcal{D}_d$, F has $\Gamma_2(k)$ limits q.e.; that is*

$$\mathrm{Cap}_T(\{\Gamma\in\partial T : \lim_{\Gamma_2(k)} F(\alpha) \text{ does not exist}\}) = 0.$$

Again at the level of metaphor we can describe the geometry of the regions which correspond to these types of approach. We give the description in the upper half plane with the positive imaginary axis as the geodesic of interest. The geodesic convergence in the tree corresponds to convergence in the Stoltz region $y > |x|$. The thickened geodesics $\Gamma_1(k)$ correspond to the wider, but still non-tangential, approach regions $y > 2^{-k}|x|$. The regions $\Gamma_2(k)$ correspond to regions which are tangent to the boundary with the tangency being of finite order; roughly, $\Gamma_2(k)$

corresponds to the region $y > |x|^{k+1}$. Results on tangential convergence and the size of the associated exceptional sets go back to Kinney [**Ki**] and more general versions are in [**Tw**].

6.4. Beyond Algebraic Approach Regions. The previous result can be extended to regions which are tangent of infinite order to the boundary but at a cost; the convergence will be quasi-everywhere but now quasi-everywhere with respect to a *different capacity.* The capacities will be those associated with the spaces $\mathcal{D}_{d,\varepsilon}$, the Hilbert space of F functions on T for which

$$\|F\|^2_{\mathcal{D}_{d,\varepsilon}} = |F(o)|^2 + \sum\nolimits_T |DF(\alpha)|^2\, 2^{-\varepsilon d(\alpha)} < \infty. \tag{6.3}$$

The approach regions of interest are these. For $0 < \varepsilon \leq 1$ and Γ a geodesic in T which defines an element of ∂T we set

$$\Gamma_3(\varepsilon) = \bigcup_{\alpha\in\Gamma} \left\{\alpha + \left[2^{\varepsilon d(\alpha)}\right]\right\}.$$

(Hereafter we will regard the nearest integer brackets as implicit and will not write them.)

We begin by a straightforward modification of the argument which gave Theorem 6.3. We start with $F \in \mathcal{D}_d$ and construct a majorant for its variation. Define

$$DF^{(3)}(\alpha) = |DF(\alpha)| + \sum_{j=2^{\varepsilon d(\alpha)}}^{j=2\cdot 2^{\varepsilon d(\alpha)}} \max\left\{|DF(\beta)| : \beta \in \{\alpha + j\}\right\} \tag{6.4}$$

If $F \in \mathcal{D}_d$ then $DF \in l^2(T)$, but now it need not hold that $DF^{(3)} \in l^2(T)$. For each α let $\{\beta(\alpha)_j\}$ be the vertices of T which appear on the right hand side of (6.4); that is, α and the selected elements where $\max|DF(\beta)|$ is attained. Thus

$$DF^{(3)}(\alpha) = \sum_j |DF(\beta(\alpha)_j)|.$$

Hence

$$\begin{aligned}\left|DF^{(3)}(\alpha)\right|^2 &\leq \Big(\sum_j |DF(\beta(\alpha)_j)|^2\Big) \cdot (\text{number of } j\text{'s})\\ &\leq C 2^{\varepsilon d(\alpha)} \Big(\sum_j |DF(\beta(\alpha)_j)|^2\Big).\end{aligned}$$

Hence the sequence of numbers $\left\{2^{-\varepsilon d(\alpha)}\left|DF^{(3)}(\alpha)\right|^2\right\}$ is summable because, again, no vertex shows up as a $\beta(\alpha)_j$ more than a few times. Thus $DF^{(3)} \in l^2(T, 2^{-\varepsilon d(\alpha)})$. We now use the same arguments as before. Set $G = I(DF^{(3)})$; G will have finite radial limits along every geodesic Γ with the possible exception of a set which is a null set for every Carleson measure for the space $\mathcal{D}_{d,\varepsilon}$. Also as in the previous proof, any boundary point Γ at which $I(DF^{(3)})(\Gamma) < \infty$ will be a boundary point where we have good convergence of F; in this case the good convergence meaning convergence over $\Gamma_3(\varepsilon)$.

The description of the Carleson measures for these spaces is given in [**AR**]. Here is the description for $0 < \varepsilon < 1$.

THEOREM 6.4. *Suppose* $0 < \varepsilon \leq 1$. *Let* μ *be a positive Borel measure on* $\overline{T}$. *Then, the following are equivalent:*

(1) *μ is a Carleson measure for $\mathcal{D}_{d,\varepsilon}$; i.e. there is a constant C so that $\forall f \in \mathcal{D}_{d,\varepsilon}$*

$$\sum_{\alpha \in \overline{T}} |F(\alpha)|^2 \mu(\alpha) \leq C \|F\|^2_{\mathcal{D}_{d,\varepsilon}}. \tag{6.5}$$

(2) *μ satisfies the $\varepsilon-$tree condition. There is a constant C so that $\forall \alpha \in T$*

$$I^* \left[(I^*\mu)^2(\cdot) 2^{\varepsilon d(\cdot)} \right] (\alpha) \leq C I^* \mu(\alpha). \tag{6.6}$$

Hence we have

THEOREM 6.5. *Fix ε, $0 < \varepsilon < 1$. For $F \in \mathcal{D}_d$, F has $\Gamma_3(\varepsilon)$ limits for all $\Gamma \in \partial T$ with the possible exception of a set which is a null set for every measure μ which satisfies the condition (*6.6*).*

In this case the approach regions have infinite order tangency, in fact subexponential contact. The Euclidean analogs of these regions shaped like the part of the upper halfplane where $y > \exp(-|x|^{-\varepsilon})$.

Finally if a variant of Theorem 5.1 is available in this context the result can be reformulated as *q.e.* convergence with respect to the appropriate capacity. In fact such a theorem does holds; its statement and proof are similar to the $\varepsilon = 0$ case considered earlier; details will be in [**ARSp**]

6.5. The Result of Nagel, Rudin, and Shapiro. It is a result of Nagel, Rudin, and Shapiro [**NRS**] that, with a possible exceptional set of Lebesgue measure zero, functions in the Dirichlet space approach their radial boundary values through approach regions of full exponential contact; that is, with the shape of the set $y > \exp(-|x|^{-1})$. Further work in that direction is in [**NS**], [**DB**], and [**Tw**]. It would be interesting to know if an analogous result holds for $\mathcal{D}_d$ and we leave that as a question.

The proof just given shows that any $F \in \mathcal{D}_d$ has limits along regions $\Gamma_3(1)$ with an exceptional set that is a null set for all the measures which satisfy (6.6) with $\varepsilon = 1$. However the full boundary is, in fact, such a set. The quickest way to see that is to note $G = Ig$ with $g(a) = d(\alpha)^{-1}$ has boundary values identically $+\infty$. We could consider the subspace of $\mathcal{D}_{d,1}$ consisting of martingales; the Carleson measures for that subspace have null sets which are exactly the sets of Lebesgue, for that see [**AR**]. However it is not clear how to use that result in this context.

One difference between the two cases is that the proofs for harmonic functions make systematic use of the reconstruction of the interior values of functions from the boundary values; in contrast the values of an $F \in \mathcal{D}_d$ on T are not determined by the boundary values.

6.6. Boundary Convergence for a *BMO*−type Space. The Carleson measures for $\mathcal{D}$ are the positive measures on the disk which satisfy the equivalent conditions of Theorem 4. There is an interesting subspace $\mathcal{X}$ of $\mathcal{D}$ consisting of those $f \in \mathcal{D}$ which generate Carleson measures in the following sense:

$$\mathcal{X} = \left\{ f \in \mathcal{D} : |f'(z)|^2 \, dxdy \text{ is a Carleson measure for } \mathcal{D} \right\}.$$

The discrete analog is

$$\mathcal{X}_d = \left\{ F \in \mathcal{D}_d : \mu_F = |DF|^2 \text{ is a Carleson measure for } \mathcal{D}_d \right\}.$$

As discussed briefly in [**AR**], the space $\mathcal{X}$ has a relation to $\mathcal{D}$ similar to the relation the space BMO has to the Hardy space H^2. In fact one of the characterizations of functions in BMO is that $f \in BMO$ if and only if $|f'(z)|^2 (1-|z|^2)dxdy$ is a Carleson measure for the Hardy space. The functions in BMO are both smaller than and smoother than generic functions in H^2. Similarly functions $\mathcal{X}$ are smaller than generic functions in $\mathcal{D}$. Specifically if $f \in \mathcal{D}$ then for some small ε_f it holds that $\exp(\varepsilon_f |f|^2)$ has integrable boundary values; however $f \in \mathcal{X}$ insures that the boundary values of $\exp(\exp(\varepsilon_f |f|))$ are integrable. Similar results also hold for the model spaces on trees, for all this see [**LL**]. Here we obtain a different result, but one in the same spirit, the functions in $\mathcal{X}_d$ have nicer properties than the generic elements of $\mathcal{D}_d$.

For comparison recall that Theorem 6.1 states that

$$\forall F \in \mathcal{D}_d \lim_{\alpha\in\Gamma} F(\alpha) = \sum_{\alpha\in\Gamma} DF(\alpha) \text{ exists for quasi-every } \Gamma. \tag{6.7}$$

Suppose now that $F \in \mathcal{X}_d$ is fixed $\mu(\alpha) = \mu_F(\alpha) = |DF(\alpha)|^2$. Recall that the tree condition for μ is that there is a C so that for all α

$$I^*(I^*\mu)^2(\alpha) \le CI^*\mu(\alpha).$$

THEOREM 6.6. *Suppose μ is a Carleson measure $\mathcal{D}_d$ then*

$$\sum_{\alpha\in\Gamma} \mu(S(\alpha)) < \infty \text{ for quasi-every } \Gamma.$$

Equivalently

$$\int \log^+ \frac{1}{|z-e^{i\theta}|} d\mu(z) < \infty \text{ for quasi-every } \theta.$$

In particular, if $F \in \mathcal{X}_d$ then

$$\sum_{\alpha\in\Gamma} (I^*(DF)^2)(\alpha) < \infty \text{ for quasi-every } \Gamma.$$

PROOF. The argument proving Theorem 6.1 applies to Ih for any square summable function h defined on T.The tree condition evaluated at the origin insures that $h(\alpha) = I^*\mu(\alpha) = \mu(S(\alpha))$ is such a sequence. That gives the first statement. The second follows from the first by estimating how often each value $\mu(\beta)$ occurs in the sum. By writing out all the terms in $(I^*(DF)^2)(\alpha)$ and then discarding those corresponding to vertices not on Γ we obtain a weaker, but more transparent, corollary. □

COROLLARY 6.7. *If $F \in \mathcal{X}_d$ then*

$$\sum_{\alpha\in\Gamma} d(\alpha)|DF|^2(\alpha) < \infty \text{ for quasi-every } \Gamma.$$

These results describe radial convergence and, as with the results for functions in $\mathcal{D}$, they can be extended to larger convergence regions for both

$$\lim_{\beta} II^*(|DF|^2)(\beta) \text{and}$$

$$\lim_{w} \int \log^+ \frac{1}{|1-\bar{w}z|} d\mu(z).$$

7. Possible Extensions

Here we briefly and very informally discuss how some of these ideas will be taken further in [**ARSp**].

7.1. Other Function Spaces on Trees. Various function spaces on T have been studied both on their own and as discrete models for spaces of smooth functions such as Besov spaces. This view is developed among other places in [**Ar**], [**AR**], [**ARS1**], [**ARS2**] where in addition to $l^2(T)$ study is also made of various weighted $l^p(T)$ spaces. The arguments of the previous sections adapt directly to show that such functions converge to boundary values through various approach regions with exceptional sets that are null sets for classes of Carleson measures. One way to get further insight is to develop geometric characterizations of the relevant classes of Carleson measures. For the spaces mentioned that is done in the earlier work by the authors. To go to results involving capacity we need a result such as Corollary 5.2. For the function space described in (6.3) the proof we gave in the case $\alpha = 0$ continues to work with straightforward changes. However for $p \neq 2$, for instance for the dyadic Besov spaces of [**AR**], one needs to work with the nonlinear potential theory appropriate for l^p spaces and the arguments are more complicated. That work will be presented in [**ARSp**]

7.2. Holomorphic, Harmonic, and other Smooth Functions. Results such as those we described for model spaces such as $\mathcal{D}_d$ can be used to obtain results for spaces of smooth functions. Suppose for instance that we want to derive a version of Beurling's theorem [**Beu**].

THEOREM 7.1 (Beurling, 1940). *For all* $f \in \mathcal{D}$

$$\mathrm{Cap}_{\mathcal{D}}(\{e^{it} : \lim_r f(re^{i\theta}) \textit{ fails to exist}) = 0.$$

First select and fix $f \in \mathcal{D}$. Also select R so large that the hyperbolic disks of radius R centered at points of the tree, $\{D(\alpha, R) : \alpha \in T\}$ is a cover for $\mathbb{D}$ and so that for all $\alpha \in T$ we have $\{\alpha^-, \alpha_r, \alpha_l\} \subset D(\alpha, R)$. For each $\alpha \in T$ we measure the local oscillation of f by

$$\mathrm{Osc}(\alpha) = \mathrm{Osc}(\alpha, f) = \sup\{|f(z) - f(z')| : z, z' \in D(\alpha.R)\}.$$

Straightforward considerations of the geometry of the placement of T in $\mathbb{D}$ show that the disks $\{D(\alpha, 5R) : \alpha \in T\}$, have the property that there is an M so that no point is in more than M disks. This insures that $\{\mathrm{Osc}(\alpha)\} \in l^2(T)$ because function theoretic estimates yield

$$\mathrm{Osc}(\alpha)^2 \leq C \int_{D(\alpha,5R)} |f'|^2 dxdy.$$

The finite overlap of the disks and the definition $\mathcal{D}$ insures that the integrals on the right can be summed.

Theorem 6.1 then insures that $\lim_\Gamma I^* \mathrm{Osc}(\alpha)$ is finite except for a set of Γ of $\mathcal{D}_T$ capacity zero. This insures that f has a limit along the path connecting the vertices in Γ and that in turn is enough to insure that f also has a limit along the radius which terminates at the point of the circle corresponding to the boundary element determined by Γ.

This outline gives convergence off of an exceptional set is of $\mathcal{D}_T$ capacity zero rather than $\mathcal{D}$ capacity zero. We hope to return to the general question of the relationship between null sets for discrete capacities and for continuous capacities. In this particular case however, the capacities associated with $\mathcal{D}_T$ and with $\mathcal{D}$, the two collections of null sets are known to agree, as is shown by Benjamini and Peres [**BP**].

A limitation of the proof we outlined is that it established the existence of boundary limits rather than finiteness of the variation functional $V(f)(e^{i\theta})$. However that was just for convenience of presentation. A slightly more elaborate definition of oscillation together with a similar argument, but using the fact that functions in $\mathcal{D}_T$ have nontangential limits, would establish the variation result.

We would like to emphasize that large parts of the previous argument do not involve holomorphy at all. If one has local oscillation estimates and knows that the oscillation numbers live in a space X, for instance X could be a weighted $l^p(T)$, then the argument shows that limits exist along Γ with an exceptional set of Γ that is a null set for all the Carleson measures for X. So, for instance, these arguments can certainly be used with harmonic functions or holomorphic functions of several variables. Also, there are other, rather different types of function spaces such as $\mathcal{A}-$harmonic functions and monotone Sobolev functions where such oscillation estimates are available; see for instance [**KMV**], [**MV**]. These types of variations have not been explored

7.3. Final Questions. One of the themes in the study of boundary value results for, say, harmonic functions is consideration of whether the description of the exceptional sets is sharp. That is also a natural question in this context but we haven't considered it. We conclude by mentioning two areas where we do not know if the approach we have been describing can be used but the possibility is intriguing.

First, the study of radial variation for bounded holomorphic (or harmonic, $p-$harmonic, etc.) functions on the unit disk (or ball, tree, etc.) is a very active research area. The indications so far are that the results there are deeper than and different from the results for, for instance, general Hardy or Besov spaces. The paper [**CFPR**] includes some general discussion of the area and references.

Second, although we considered various types of approach regions for boundary convergence, they were all of the same sort, a geodesic Γ with a symmetrical enveloping shell which, in the Euclidean sense, narrowed as the region approached the boundary. These are all versions of having boundary limits along a collection of paths, the geometry of the envelope controlling the type of paths. However one can also consider convergence to boundary values through a collection of sets which is (in some appropriate sense approximately) translation invariant and contains no paths. This theme has a long history, recently it shows up in the alternative approach to the results of Nagel, Rudin, and Shapiro given by Nagel and Stein [**NS**]. In this more general context it also makes sense to look for descriptions of approach regions that are optimal in various senses. These topics are treated fully by DiBiasi in [**DB**]. Particularly interesting to us is that a substantial part of the work there proceeds through analysis on model spaces defined on dyadic trees.

References

[Ad] D. Adams, *On the existence of capacitary strong type estimates in R^n*, Ark. Mat. **14** (1976), no. 1, 125–140.

[AH] D. Adams, L. Hedberg, *Function spaces and potential theory*, Grundlehren der Mathematischen Wissenschaften, 314, Springer-Verlag, Berlin, 1996.

[AC] P. Ahern, W. Cohn, *Weighted maximal functions and derivatives of invariant Poisson integrals of potentials*, Pacific J. Math. **163** (1994), no. 1, 1–16.

[A] H. Aikawa, *Capacity and Hausdorff content of certain enlarged sets*, Mem. Fac. Sci. Eng. Shimane Univ. Ser. B Math. Sci. **30** (1997), 1–21.

[AE] H. Aikawa, M. Essén, *Potential theory– selected topics*, Lecture Notes in Mathematics, vol. 1633. Springer-Verlag, Berlin, 1996.

[Ar] N. Arcozzi, *Carleson measures for the analytic Besov spaces: the upper triangle case*, J. Inequal. Pure Appl. Math. **6** (2005), no. 1, Article 60,

[AR] N. Arcozzi, R. Rochberg, *Topics in dyadic Dirichlet spaces*, New York J. Math. **10** (2004), 45–67

[ARS1] N. Arcozzi, R. Rochberg, E. Sawyer, *Carleson measures for analytic Besov spaces*, Revista Math. Iberoamericana **18** (2002), 443—510.

[ARS2] N. Arcozzi, R. Rochberg, E. Sawyer, *Carleson measures and interpolating sequences for Besov spaces on complex balls*, Mem. Amer. Math. Soc., Vol. 182, 2006, no. 859, vi+163 pp.

[ARS3] N. Arcozzi, R. Rochberg, E. Sawyer, *The Characterization of Carleson measures for analytic Besov spaces: a simple proof, Complex and Harmonic Analysis*, A. Carbery, P. Duren, D. Khavinson, A. Sistakis, Eds., Destech Publ. 2007, 167-178.

[ARSp] N. Arcozzi, R. Rochberg, E. Sawyer, *Carleson Measures, Capacity, and Exceptional Sets*, in preparation.

[BP] I. Benjamini, Itai; Y. Peres, *Random walks on a tree and capacity in the interval*, Ann. Inst. H. Poincaré Probab. Statist. **28** (1992), no. 4, 557–592.

[Beu] A. Beurling, *Ensembles exceptionnels*, (French) Acta Math. **72** (1940). 1–13.

[CFPR] A. Cantón, J. L. Fernández, D. Pestana, J. M. Rodríguez, *On harmonic functions on trees*, Potential Anal. **15** (2001), no. 3, 199-244.

[Ca] L. Carleson, *Selected problems on exceptional sets*, Van Nostrand Mathematical Studies, No. 13 D. Van Nostrand Co., Inc., Princeton, N.J.-Toronto, Ont.-London 1967.

[DB] F. Di Biase, *Fatou type theorems. Maximal functions and approach regions*, Progress in Mathematics vol. 147 Birkhäuser Boston, Inc., Boston, MA, 1998.

[GP] D. Girela, J. A. Peláez, *Boundary behaviour of analytic functions in spaces of Dirichlet type*, J. Inequal. Appl. 2006, Art. ID 92795, 12 pp.

[KV] N. Kalton, I. Verbitsky. *Nonlinear equations and weighted norm inequalities*, Trans. Amer. Math. Soc. **351** (1999), no. 9, 3441–3497.

[KS] R. Kerman, E.Sawyer, *The trace inequality and eigenvalue estimates for Schrödinger operators*, Ann. Inst. Fourier (Grenoble) **36** (1986), no. 4, 207–228.

[Ki] J. Kinney, *Tangential limits of functions of the class S_α*, Proc. Amer. Math. Soc. **14** (1963), 68–70.

[KMV] P. Koskela, J. Manfredi, E. Villamor, *Regularity theory and traces of $\mathcal{A}$-harmonic functions*, Trans. Amer. Math. Soc. **348** (1996), no. 2, 755–766.

[LL] Y. Lin, Thesis, Washington University, in preparation, 2008.

[MV] J. Manfredi, E. Villamor, *Traces of monotone Sobolev functions*, J. Geom. Anal. **6** (1996), no. 3, 433–444 (1997).

[M] Y. Mizuta, *Existence of tangential limits for -harmonic functions on half spaces*, Potential Anal. **25** (2006), no. 1, 29–36.

[NRS] A. Nagel, W. Rudin, J. Shapiro, *Tangential boundary behavior of functions in Dirichlet-type spaces*, Ann. of Math. (2) **116** (1982), no. 2, 331–360.

[NS] A. Nagel, E. Stein, *On certain maximal functions and approach regions*, Adv. in Math. **54** (1984), no. 1, 83–106.

[St] D. Stegenga, *Multipliers of the Dirichlet space*, Illinois J. Math. **24** (1980), no. 1, 113–139.

[Tw] J. Twomey, *Tangential boundary behaviour of harmonic and holomorphic functions*, J. London Math. Soc. (2) **65** (2002), no. 1, 68–84.

[V] I. Verbitsky, *Nonlinear potentials and trace inequalities*, The Maz'ya anniversary collection, Vol. 2 (Rostock, 1998), 323–343, Oper. Theory Adv. Appl., 110, Birkhäuser, Basel, 1999.

[W] D. Walsh, *Radial variation of functions in Besov spaces*, Publ. Mat. **50** (2006), no. 2, 371–399.

(Arcozzi) Dipartimento do Matematica, Universita di Bologna, 40127 Bologna, ITALY
E-mail address: `arcozzi@dm.unibo.it`

(Rochberg) Department of Mathematics, Washington University, St. Louis, MO 63130, U.S.A.
E-mail address: `rr@math.wustl.edu`

(Sawyer) Department of Mathematics & Statistics, McMaster University, Hamilton, Ontairo, L8S 4K1, CANANDA
E-mail address: `Saw6453CDN@aol.com`

Proceedings of Symposia in Pure Mathematics
Volume **79**, 2008

On the absence of dynamical localization in higher dimensional random Schrödinger operators

Jean Bourgain

Dedicated to V.G. Maz'ya

ABSTRACT. It is shown that dynamical localization fails in random Schrödinger operators with a slowly decaying potential in dimension at least 5, exhibiting a higher dimensional phenomenon. The method is perturbative and uses suitable renormalization of the Bohr series.

§1. Introduction

In what follows, we consider lattice Schrödinger operators on $\mathbb{Z}^d$, of the form

$$H_\omega = \Delta + V_\omega \tag{1.1}$$

where Δ is the lattice Laplacian on $\mathbb{Z}^d$, i.e.

$$\Delta(n,n') = \begin{cases} 1 & \text{if } |n_1 - n_1'| + \cdots + |n_d - n_d'| = 1 \\ 0 & \text{otherwise} \end{cases} \tag{1.2}$$

and V_ω is a random potential, which we take of the form

$$V_\omega(n) = \omega_n v_n \tag{1.3}$$

or

$$V_\omega(n) = \omega_n v_n + w_n \tag{1.3'}$$

with $\{\omega_n | n \in \mathbb{Z}^d\}$ independent Bernoulli or normalized Gaussian random variables and $\{v_n | n \in \mathbb{Z}^d\}, \{w_n | n \in \mathbb{Z}^d\}$ decaying deterministic sequences. More precisely,

$$v_n = \frac{\kappa}{(|n|+1)^\alpha} \quad \text{for some } \alpha \geq 0 \tag{1.4}$$

and w_n will satisfy

$$|w_n| < O(\kappa^2 |n|^{-2\alpha}). \tag{1.5}$$

2000 *Mathematics Subject Classification.* Primary 4606, 5206.

Let us start with the well-known Anderson model (of transport in inhomogeneous media) where $V_\omega(n) = \kappa\omega_n$ $(\alpha = 0, w_n = 0)$ with coupling constant $\kappa > 0$. In the 1-dimensional case $(d = 1)$, the spectral theory of H_ω has been extensively studied and is well-understood.

Rewriting the equation $H\psi = E\psi$ as

$$\psi_{n+1} + \psi_{n-1} + (V_\omega)_n\psi_n = E\psi_n \ (n \in \mathbb{Z}) \tag{1.6}$$

the main tool is the transfer matrix formulation of (1.6)

$$\begin{pmatrix} \psi_{N+1} \\ \psi_N \end{pmatrix} = M_N(E) \begin{pmatrix} \psi_1 \\ \psi_0 \end{pmatrix} \tag{1.7}$$

where

$$M_N(E) = \begin{pmatrix} E - V_n & -1 \\ 1 & 0 \end{pmatrix} \cdots \begin{pmatrix} E - V_1 & -1 \\ 1 & 0 \end{pmatrix} \tag{1.8}$$

and the Lyapounov exponent

$$L(E) = \lim_{N\to\infty} \frac{1}{N} \mathbb{E} \log \|M_N(E)\|. \tag{1.9}$$

For all $\kappa > 0$, H_ω satisfies almost surely Anderson localization (AL) (pure point spectrum with exponentially decaying eigenfunctions) and dynamical localization (DL). This last property relates to the associated Schrödinger group $(e^{itH})_{t\in\mathbb{R}}$ and means that for any given exponent $A > 0$

$$\sup_t \Big(\sum_{n\in\mathbb{Z}} |n|^{2A} |(e^{itH}\xi)_n|^2 \Big)^{1/2} < \infty \tag{1.10}$$

whenever ξ decays rapidly for $|n| \to \infty$.

Recall that (DL) implies pure point spectrum but not conversely.

Despite quite precise and widely believed conjectures, the higher dimensional case $d > 1$ is still poorly understood. The belief is that for $d = 2$, the spectrum remains pure point with localized eigenfunctions (but for small κ a different behaviour of the localization length at the edge and in the bulk of the spectrum) and for $d \geq 3$ and small κ, there is a component of absolutely continuous spectrum. But at this time, we only dispose of the Frohlich-Spencer multiscale analysis, which enables us to produce point spectrum (also (AL) and (DL)) for large κ and at the edge of the spectrum. We do not know of a method to produce continuous spectrum. The only rigorous result distinguishing $d = 1$ and $d > 1$ is due to [ESY] and relates to localization length for small κ. For $d = 1$, the behaviour is $\sim \frac{1}{\kappa^2}$ for $\kappa \to 0$ (i.e. the reciprocal of the Lyapounov exponent) and [ESY] establishes in $d = 3$ localization lengths at least $\frac{1}{\kappa^{2+\delta}}$, $\delta > 0$ some constant, in the bulk of the spectrum.

Let us now turn to (1.1) with decaying random potential. For $d = 1$ and using again transfer matrix methods, it is shown in [DSS] that if $V_\omega(n) = \omega_n|n|^{-\alpha}$, then (see [DSS], Theorem 1.1).

- If $0 < \alpha < \frac{1}{2}$, H_ω satisfies a.s. (AL) and (DL). In fact the eigenfunctions ψ_E have a decay rate

$$|\psi_E(n)| < C(E) \exp\{-c'|n|^{1-2\alpha}\}. \tag{1.11}$$

•• If $\alpha > \frac{1}{2}$, H_ω has pure a.c.-spectrum, a. s.

The situation $\alpha = \frac{1}{2}$ is more complicated and we do not discuss it here (see [DSS] and later papers).

Our purpose is to point out a higher dimensional phenomenon (we need $d \geq 5$ in fact) regarding (DL) with a random decaying potential. We will sketch a proof of the following

THEOREM. *Let $d \geq 5$ and H_ω be given by (1.1) with*

$$V_\omega(n) = \frac{\kappa}{(|n|+1)^\alpha}\omega_n + w_n \ (n \in \mathbb{Z}^d) \tag{1.12}$$

where $\alpha > 0$ is arbitrary, κ is sufficiently small and w_n is some deterministic potential of size

$$|w_n| = O(\kappa^2 |n|^{-2\alpha}). \tag{1.13}$$

Then, with high probability in ω, H_ω fails (DL). More precisely

$$\sup_{\substack{t\in\mathbb{R}\\ n\in\mathbb{Z}^d}} |n|^A |e^{itH}(0,n)| = \infty \ \textit{if}\ A > d. \tag{1.14}$$

For $d = 1$, this behaviour is indeed impossible if $\alpha < \frac{1}{2}$.

Remarks.

(1). In what follows, we take for simplicity $\{\omega_n\}$ to be Bernoulli, but we could take other symmetric i.i.d random variables satisfying suitable moment estimates (we do not intend to specify).

(2). The deterministic potential $\{w_n\}$ is used when renormalizing the Bohr expansion. It does not play an essential role in the statement of the Theorem and likely can be removed (this would require replacing the free Laplacian Δ by $\Delta + w$ with w a deterministic smooth decaying potential with decay (1.13)).

(3). The proof of the Theorem relies heavily on [B] and is in fact a corollary of this result. In [B], we consider $H_\omega = \Delta + V_\omega$ with V_ω as in (1.12) and show that at a specific energy E (the lower edge of the spectrum of Δ), H_ω has an extended state and the Green's function $G(E)$ of H_ω behaves like the free Green's function $G_0(E)$, i.e.

$$|G(E)(n,n')| \sim \frac{1}{|n-n'|^{d-2}}. \tag{1.15}$$

Statement (1.14) is then easily deduced from (1.15).

The analysis in [B] is actually only carried out for $\alpha > \frac{1}{3}$ but may be generalized to arbitrary $\alpha > 0$ (it requires considering more terms in the multi-linear expansion of $G(E)$).

We will recall the main ideas below.

(4). [B] was inspired by an unpublished preprint of T. Spencer and W. Wang [S - W] on Lipschitz trails. Note that for E in the bulk of the spectrum, $|G_0(E)(n,n')| \sim \frac{1}{|n-n'|^{\frac{d-1}{2}}}$ (with an essential phase factor), while at the edge E_-, we have $|G_0(E_-)(n,n')| \sim \frac{1}{|n-n'|^{d-2}}$. The latter is square integrable for $d \geq 5$, which explains our assumption. The renormalization for E in the bulk is quite different and much harder (cf. [ESY]).

(5). It is conceivable that an analogue of [B] and the above Theorem remains valid with $\alpha = 0$ (i.e. no decay). This would require a different strategy however, as an infinite multilinear Bohr extension in κ is unlikely to be tractable (even after renormalization).

(6). Instead of lattice models, one could also consider a model on $\mathbb{R}^n$, of the form $H_\omega = -\Delta + \sum_{n\in\mathbb{Z}^d} V_\omega(n)\varphi(\cdot - n)$ where Δ is the usual Laplacian, V_ω a random potential as above and $\varphi(x)$ a localizing bumpfunction on $\mathbb{R}^d$.

The remainder of the paper is organized as follows. We will first briefly sketch the argument from [B] leading to (1.15). We then deduce (1.14), which is a rather elementary spectral consideration (and may well be known).

§2. The Green's function estimate

(i) Redefine the Laplacian by subtracting $2d$ from the lattice Laplacian, thus

$$-\hat{\Delta}(\xi) = 2d - 2\Big(\sum_{j=1}^{d} \cos 2\pi\xi_j\Big) = c|\xi|^2 + O(|\xi|^4) \tag{2.1}$$

and consider the Green's function $G_0 = (-\Delta + io)^{-1}$ at 0-energy. Hence

$$G_0(n,n') = \int \frac{e^{-2\pi i(n-n').\xi}}{-\hat{\Delta}(\xi)} d\xi \sim \frac{1}{|n-n'|^{d-2}}. \tag{2.2}$$

(ii) Write

$$H = -\Delta + \tilde{V} \tag{2.3}$$

with Δ as above and where

$$\tilde{V} = V_\omega + w \tag{2.4}$$

with

$$V(n) = V_\omega(n) = \frac{\kappa}{(1+|n|)^\alpha}\,\omega_n \quad (\omega_n \text{ Bernoulli}) \tag{2.5}$$

and ω_n deterministic satisfying (1.13).

The role of w will become clear later on.

Denote G the Green's function of H at 0-energy.

The basic idea is to write a Bohr series (with finitely many terms) for G and make probabilistic estimates on the terms.

Thus iterating the resolvent identity (for the time we let $\tilde{V} = V$, dropping the w-potential)

$$G = G_0 - GVG_0 \tag{2.6}$$

we obtain

$$G = G_0 - G_0VG_0 + G_0VG_0VG_0 - G_0VG_0VG_0VG_0 + \cdots \pm GVG_0\cdots VG_0. \tag{2.7}$$

Our aim is to bound $\|G_0VG_0\|, \|G_0VG_0VG_0\|, \ldots$ using the randomness of $V = V_\omega$ and probabilistic considerations. In an ideal situation of square cancellations, the estimate $|G_0(n,n')| \lesssim \frac{1}{|n-n'|^{d-2}}$ (with $\frac{1}{|n|^{d-2}}$ square-summable for $d \geq 5$), together with some decay of V_n, would easily imply bounds of the form

$$|(G_0 V G_0 \cdots V G_0)(n,n')| \ll \frac{\kappa^s}{|n-n'|^{d-2}} \tag{2.8}$$

and also

$$|(V G_0 V \cdots V G_0)(n,n')| \ll \frac{\kappa^s}{|n-n'|^{d-2}\min(|n|^{s\alpha}, |n'|^{s\alpha})} \tag{2.9}$$

with s the number of V-factors in (2.8), (2.9).

Letting $\bar{s} > \frac{2}{\alpha}$ and taking in (2.7) an expansion of length 3, one derives that G is a perturbation of G_0,

$$|G(n,n') - G_0(n,n')| < \frac{O(\kappa)}{|n-n'|^{d-2}}. \tag{2.10}$$

The difficulty with the terms in (2.7) is that the V-factors are not independent however and the ω_n-factors may cancel out. We need therefore to proceed with more care when iterating the resolvent identity and certain renormalization is needed, requiring the additional deterministic potential w in (2.4).

(iii) As the V-factors are not independent, we will need an appropriate probabilistic estimate which we describe next.

Recall that $\{\omega_n | n \in \mathbb{Z}^d\}$ are independent Bernoulli variables (this is not essential for what follows however).

Considering a s-tuple $(n_1, \dots, n_s)$, we say that there is 'cancellation' if

$$\omega_{n_1} \cdots \omega_{n_s} = 1. \tag{2.11}$$

Note that this property is invariant under translation.

Say that $(n_1, \dots, n_s)$ is 'admissible' if for any segment $1 \leq s_1 < s_2 \leq s$, the sub-complex $(n_{s_1}, n_{s_1+1}, \dots, n_{s_2})$ does not cancel.

Use the notation $\sum^*_{n_1,\dots,n_s}$ to indicate summation restricted to admissible s-tuples.

The interest of this notion is clear from the following estimate.

LEMMA 1. *For $s \geq 2$*

$$\Big\| \sum^*_{n_1,\dots,n_s} \omega_{n_1} \cdots \omega_{n_s} a^{(0)}_{n,n_1} a^{(1)}_{n_1 n_2} \cdots a^{(s)}_{n_s,n'} \Big\|_{L^2_\omega}$$
$$\leq C_s \Big[\sum_{n_1,\dots,n_s} |a^{(0)}_{n_s,n_1} \cdots a^{(s)}_{n_s,n'}|^2 \Big]^{1/2}. \tag{2.12}$$

Proof. We may clearly assume $a^{(j)}_{m,n} \geq 0$.

Since in the $\sum^*$ summation no $(n_1, \dots, n_s)$ cancels, there is some index $n_{s'}$ which is not repeated or repeated an odd number of times.

Specifying a subset I of $\{1, \dots, s\}$ of odd size (at most 2^s possibilities), we consider now s-tuples of the form

$$(\nu^{(1)}, m, \nu^{(2)}, m, \nu^{(3)}, m \cdots)$$

where $m \in \mathbb{Z}^d$ appears on the I-places and $\nu^{(1)}, \nu^{(2)}, \dots$ are admissible complexes indexed by sub-intervals of $\{1, \dots, s\}$ determined by I.

Thus, enlarging the $\sum^*$-sum, (which we may by the positivity assumption), it follows that

$$\Big\| \sum_{n_1,\dots,n_s}^{*} \omega_{n_1}\dots\omega_{n_s} a^{(0)}_{n,n_1}\cdots a^{(s)}_{n_s,n'}\Big\|_{L^2_\omega}$$

$$\leq \sum_I \Big\| \sum_{m\in\mathbb{Z}^d} \omega_m \Big[\sum^{*}_{\substack{\nu^{(1)}\\ \| \\ (n_1,\dots,n_{s_1})}} \omega_{n_1}\dots\omega_{n_{s_1}} a^{(0)}_{n,n_1}\cdots a^{(s_1)}_{n_{s_1},m}\Big]$$

$$\Big[\sum^{*}_{\substack{\nu^{(2)}\\ \| \\ (n_{s_1+2},\dots,n_{s_2})}} \omega_{n_{s_1+2}}\cdots\omega_{n_{s_2}} a^{(s_1+1)}_{m,n_{s_1+2}}\cdots a^{(s_2)}_{n_{s_2},m}\Big]\cdots\Big\|_{L^2_\omega} \tag{2.13}$$

where, for fixed m,

$$m \notin \{n_1,\dots,n_{s_1},n_{s_1+2},\dots,n_{s_2},n_{s_2+2},\dots\}.$$

We used here that if $(n_1,\dots,n_s)$ is admissible, then so is $(n_{s_1},n_{s_1+1},\dots,n_{s_2})$ for all $1\leq s_1\leq s_2\leq s$.

Enlargement of the original sum $\sum^*$ enables thus to get the product structure in (2.15). The number of factors is at least 2.

Thus the preceding and a standard decoupling argument implies

$$(2.13) \leq \sum_I \Big[\sum_{m\in\mathbb{Z}^d} \Big\| \Big[\sum_{\nu^{(1)}}^{*}\Big]\Big[\sum_{\nu^{(2)}}^{*}\Big]\cdots\Big\|^2_{L^2_\omega}\Big]^{1/2}. \tag{2.14}$$

Next, by Hölder's inequality and moment-equivalence, we get

$$\sum_I \Big[\sum_{m\in\mathbb{Z}^d} \Big\| \sum_{\nu^{(1)}}^{*}\Big\|^2_{L^2_\omega} \Big\| \sum_{\nu^{(2)}}^{*}\Big\|^2_{L^2_\omega}\cdots\Big]^{1/2}. \tag{2.15}$$

Proceeding by induction on s, we obtain thus

$$\sum_I \Bigg[\sum_{m\in\mathbb{Z}^d} \Big[\sum_{n_1,\dots,n_{s_1}} |a^{(0)}_{n,n_1}\cdots a^{(s_1)}_{n_{s_1},m}|^2\Big]$$

$$\Big[\sum_{n_{s_1+2},\dots,n_{s_2}} |a^{(s_1+1)}_{m,n_{s_1+2}}\cdots a^{(s_2)}_{n_{s_2},m}|^2\Big]\cdots\Bigg]^{1/2} < (2.12).$$

This proves Lemma 1.

Expressions considered in Lemma 1 appear when writing out matrix elements of products

$$(A^{(0)}V_\omega A^{(1)}V_\omega\cdots A^{(s)})(n,n')$$

where the $A^{(s')}$ are matrices and V_ω a random potential

$$V_\omega(n) = \omega_n v_n.$$

We use the notation

$$(A^{(0)}V_\omega A^{(1)}V_\omega\cdots A^{(s)})^{(*)} \tag{2.16}$$

to indicate that, when writing out the matrix product as a sum over multi-indices, we do restrict the sum to the admissible multi indices.

Lemma 1 then implies that

$$\mathbb{E}_\omega[|(A^{(0)}V_\omega A^{(1)}V_\omega \cdots A^{(s)})^{(*)}(n,n')|]$$

(2.17)

$$< C_s\Big[\sum_{n_1,\dots n_s} |v_{n_1}|^2 \cdots |v_{n_s}|^2 |A^{(0)}(n,n_1)|^2 \cdots |A^{(s)}(n_s,n')|^2\Big]^{1/2}.$$

(iv) Returning to (2.3), (2.4), write

$$G = G_0 - G\tilde{V}G_0 = G_0 - GVG_0 - GwG_0$$

and iterating once

$$G = G_0 - G_0VG_0 + GVG_0VG_0 - G_0wG_0 + GwG_0VG_0. \tag{2.18}$$

In order to replace GVG_0VG_0 by $(GVG_0VG_0)^*$, we need to remove σGv^2G_0, where

$$\sigma = G_0(0,0) \tag{2.19}$$

since

$$\int V_\omega G_0 V_\omega d\omega = G_0(0,0)v^2.$$

Hence, form (2.18)

$$G = G_0 - G_0VG_0 + (GVG_0VG_0)^* - G(w - \sigma v^2)G_0 + GwG_0VG_0. \tag{2.20}$$

Continuing the iteration, write

$$\begin{aligned}(GVG_0VG_0)^* &= (G_0VG_0VG_0)^* - GV(G_0VG_0VG_0)^* - Gw(G_0VG_0VG_0)^* \\ &= (G_0VG_0VG_0)^* - (GVG_0VG_0VG_0)^* - \sigma Gv^2G_0VG_0 + \sigma^2Gv^2VG_0 \\ &\quad - Gw(G_0VG_0VG_0)^*\end{aligned}$$

and substituting in (2.20)

(2.21)

$$\begin{aligned}G &= G_0 - G_0VG_0 + (G_0VG_0VG_0)^* - (GVG_0VG_0VG_0)^* + \sigma^2Gv^2VG_0 \\ &\quad - G(w-\sigma v^2)G_0 + G(w-\sigma v^2)G_0VG_0 - Gw(G_0VG_0VG_0)^*.\end{aligned}$$

Next

(2.22)

$$(GVG_0VG_0VG_0)^* = (G_0VG_0VG_0VG_0)^* - GV(G_0VG_0VG_0VG_0)^* - Gw(G_0VG_0VG_0VG_0)^*$$

and

(2.23)

$$\begin{aligned}&GV(G_0VG_0VG_0VG_0)^* = \\ &(GVG_0VG_0VG_0VG_0)^* + \sigma Gv^2(G_0VG_0VG_0)^* - \sigma^2(Gv^2VG_0VG_0)^* + GWG_0\end{aligned}$$

where W is the non-diagonal operator

$$W(n,n') = \begin{cases} v_n^2 v_{n'}^2 G_0(n,n')^3 & \text{if } n \neq n' \\ 0 & \text{otherwise.}\end{cases} \tag{2.24}$$

Substituting (2.22), (2.23) in (2.21) and expanding the $\sigma^2 Gv^2 VG_0$-term gives

$$\begin{aligned} G =& G_0 - G_0VG_0 + (G_0VG_0VG_0)^* - (G_0VG_0VG_0VG_0)^* + \sigma^2 G_0 v^2 VG_0 \\ &+ (GVG_0VG_0VG_0VG_0)^* - \sigma^2(Gv^2VG_0VG_0)^* - \sigma^2(GVG_0v^2VG_0)^* \\ &- G(w-\sigma v^2 - \sigma^3 v^4)G_0 + G(w-\sigma v^2)G_0VG_0 - G(w-\sigma v^2)(G_0VG_0VG_0)^* \\ &+ Gw(G_0VG_0VG_0VG_0)^* - \sigma^2 GwG_0v^2VG_0 \\ &+ GWG_0. \end{aligned} \tag{2.25}$$

We need a more appropriate expression for W. Write

$$W = \frac{1}{2}v^4 M + \frac{1}{2} M v^4 + P \tag{2.26}$$

where M is the convolution operator

$$M(n,n') = \begin{cases} G_0(n,n')^3 & \text{if } n \neq n' \\ 0 & \text{otherwise} \end{cases} \tag{2.27}$$

and the operator P has matrix elements bounded by

$$|P(n,n')| \leq \kappa^4 \big|\, |n|^{-2\alpha} - |n'|^{-2\alpha} \big|^2 \, |n-n'|^{-3(d-2)} < \frac{\kappa^4}{|n-n'|^{d+\alpha}(|n|^{2+\alpha}+|n'|^{2+\alpha})} \tag{2.28}$$

since $d \geq s$ (and assuming $\alpha < \frac{1}{2}$).

The convolution operator M is symmetric and even in each variable with behaviour $M(n) \sim |n|^{-3(d-2)}$. Hence

$$\partial_\xi^\alpha \hat{M} \in L^1_\xi \text{ for } |\alpha| < 3(d-2) \tag{2.29}$$

and

$$\hat{M}(\xi) = \hat{M}(0) + \hat{M}(2)|\xi|^4 + 0(|\xi|^4). \tag{2.30}$$

Start by subtracting $\hat{M}(0)$ from $\hat{M}(\xi)$, introducing an additional potential $\hat{M}(0)v^4$ in (2.26). We factor then

$$M - \hat{M}(0) = M_1\Delta = \Delta M_1 \tag{2.31}$$

where the convolution operator satisfies in particular

$$|M_1(n)| < \frac{C}{|n|^{d+1}}. \tag{2.32}$$

This brings W in the form

$$W = \hat{M}(0)v^4 + \frac{1}{2}v^4 M_1 \Delta + \frac{1}{2}\Delta M_1 v^4 + P. \tag{2.33}$$

Note that $\Delta G_0 = -I$ and $G\Delta = -I + G\tilde{V}$. Substitution in (2.25) gives

$$G = (1 - \frac{1}{2}Mv^4)G_0 - G_0VG_0 + (G_0VG_0VG_0)^* - (G_0VG_0VG_0VG_0)^* + \sigma^2 G_0 v^2 V G_0$$

(2.34)

$$- \sigma^2(GVG_0Vv^2G_0)^* - \sigma^2(Gv^2VG_0VG_0)^* + \frac{1}{2}GVM_1v^4G_0 + (GVG_0VG_0VG_0VG_0)^*$$

$$- G(w - \sigma v^2 - \sigma^3 v^4 - \hat{M}(0)v^4)G_0 + G(w - \sigma v^2)(G_0VG_0)$$

(2.35)

$$- G(w - \sigma v^2)(G_0VG_0VG_0)^* + Gw(G_0VG_0VG_0VG_0)^* + \frac{1}{2}GwM_1v^4G_0 - \sigma^2 GwG_0v^2VG_0$$

(2.36)

$$- \frac{1}{2}Gv^4M_1 + GPG_0.$$

In continuing the process, we only expand G further in the (2.34) terms. In general, when replacing $GV(\cdots)^*$ by $\big(GV(\cdots)\big)^*$, there will be additional terms of the form $GW(\cdots)^*$ where the operator W appears from ω_n- cancellations. These operators W need to be processed further. Observe first that, dismissing the v-factors, we certainly have an estimate

$$|W(n, n')| < \frac{1}{|n - n'|^{3(d-2)}} \tag{2.37}$$

which is obtained by removing the repeated variables iteratively, using the estimate

$$\sum_n |x-n|^{-d+2}|n-y|^{-d+2}|z-n|^{-d+2}|n-w|^{-d+2} < C|x-y|^{-d+2}|z-w|^{-d+2}. \tag{2.38}$$

Our aim is to bring W in the form (2.33).

The first step is to move the v-factors outside the product in order to remain with a convolution operator.

Decompose v_n as follows

$$\frac{1}{|n|^\alpha} = \frac{1}{|n'|^\alpha} - \alpha\frac{n'.(n-n')}{|n'|^{2+\alpha}} + 0\big(|n-n'|^2(|n|^{-2-\alpha} + |n'|^{-2-\alpha})\big)$$

$$= \frac{1}{|n'|^\alpha} - \alpha\frac{n'(n-n')}{|n'|^{2+\alpha}} + 0\Big(\frac{|n-n'|^{2+\alpha}}{|n'|^{2+\alpha}}\Big). \tag{2.39}$$

This decomposition is applied to each v-factor (n' being the left matrix element). All contributions with at least one $\frac{|n-n'|^{2+\alpha}}{|n'|^{2+\alpha}}$ or at least two $\frac{n'.(n-n')}{|n'|^{2+\alpha}}$ are absorbed by P satisfying

$$|P(n', n'')| < \kappa |n' - n''|^{-3d+8+\alpha}(|n'|^{-2-\alpha} + |n''|^{-2-\alpha}). \tag{2.40}$$

Remain the contributions

$$v^s_{n'} M(n' - n'') \tag{2.41}$$

and

$$v^s_{n'} \frac{n'}{|n'|^2} \cdot K(n' - n'') \tag{2.42}$$

where M satisfies (2.29), (2.30) and K is odd ($\mathbb{R}^d$-valued), satisfying

$$|K(n)| < |n|^{-3d+7}. \tag{2.43}$$

Symmetrizing the operator W, the preceding gives

$$W = \frac{1}{2} v^s M + \frac{1}{2} M v^s + P' \tag{2.44}$$

with P' still satisfying (2.40).

Treating M as before, we obtain with the same notation as earlier

$$GW(G_0 \cdots)^* = \hat{M}(0) G v^s (G_0 V \dots)^* \tag{2.45}$$

$$- \frac{1}{2} G v^s M_1 (V \dots)^* \tag{2.46}$$

$$- \frac{1}{2} M_1 v^s (G_0 V \dots)^* \tag{2.47}$$

$$+ \frac{1}{2} G V M_1 v^s (G_0 V \dots)^* \tag{2.48}$$

$$+ \frac{1}{2} G w M_1 v^s (G_0 V \dots)^* \tag{2.49}$$

$$+ GP(\dots)^*. \tag{2.50}$$

Term (2.45) is added to (2.35), (2.46) to (2.34) or (2.36); (2.47), (2.48) to (2.34); (2.49) to (2.35); (2.50) to (2.36).

Note that by (2.17)

$$\mathbb{E}_\omega [|(\underbrace{G_0 V G_0 \dots}_{j\ V-\text{factors}})^*(n, n')|] \ll \kappa^j |n - n'|^{-d+2} (|n| \wedge |n'|)^{-\min(j\alpha, d-2)} \tag{2.51}$$

with large probability.

We carry out the preceding iteration s times, with $s = s(\alpha)$ large enough.

Denoting the sum of the terms in (2.34) that do not involve G, by $G_0 + A, A = A_\omega$, it follows from (2.51) that

$$\mathbb{E}_\omega [|A(n, n')|] < \frac{\kappa}{|n - n'|^{d-2}}. \tag{2.52}$$

Write the remainder terms in (2.34) as GB, $B = B_\omega$ where

$$\mathbb{E}_\omega [|B(n, n')|] < \frac{\kappa}{|n - n'|^{d-2}} (|n|^{-3} + |n'|^{-3}) \tag{2.53}$$

by our choice of s and since $d \geq 5$.

The potential w is chosen as to make (2.35) vanish up to sufficiently high order to capture the remainder by a term of the form GB with B satisfying (2.53).

Finally, the term (2.36) are of the form GB_1, GB_2 with

$$|B_1(n, n')| < \frac{\kappa}{|n - n'|^{d+1}} \tag{2.54}$$

and, cf. (2.28), (2.40)

$$|B_2(n, n')| < \frac{\kappa}{|n - n'|^{d-2}} (|n'|^{-2-\alpha} + |n'|^{-2-\alpha}). \tag{2.55}$$

Summarizing, we obtain

$$G = G_0 + A + G(B + B_1) \tag{2.56}$$

hence

$$G = (G_0 + A)(I - B')^{-1} \tag{2.57}$$

with

$$|B'(n,n')| < \kappa\big(|n-n'|^{-d-1} + |n-n'|^{-d+2}(|n|^{-2-\alpha} + |n'|^{-2-\alpha})\big). \tag{2.58}$$

The inverse $(I - B')^{-1}$ may be estimated by a Neumann series and (2.58). One verifies that

$$(I - B')^{-1} = I + B'' \tag{2.59}$$

with B'' again satisfying the estimate (2.58).

It follows that

$$|G(n,n') - G_0(n,n')| < \frac{\kappa}{|n-n'|^{d-2}} + \frac{\kappa}{|n-n'|^{d-4}|n'|^{2+\alpha}} \tag{2.60}$$

and by symmetry considerations

$$|G(n,n') - G_0(n,n')| < \frac{0(\kappa)}{|n-n'|^{d-2}} \tag{2.61}$$

which is (2.10).

From the preceding we may also derive a bound

$$|G(0+i\varepsilon)(n,n')| \lesssim |n-n'|^{-d+2}. \tag{2.62}$$

§3. Diffusion for the Schrödinger group

It remains to derive (1.14) from the Green's function behaviour

$$|G(0)(n,n')| \sim \frac{1}{|n-n'|^{d-2}} \tag{3.1}$$

at $E = 0$.

Write from the spectral theorem

$$e^{itH}(0,x) = \int e^{it\lambda}\mu_{0,x}(d\lambda) \tag{3.2}$$

where

$$\frac{d\mu_{0,x}}{d\lambda} = \operatorname{Im} G(\lambda + io)(0,x) = \operatorname{Im}(H - \lambda - io)^{-1}(0,x). \tag{3.3}$$

Denote P_ε the Poisson kernel on $\mathbb{R}$ (thus $\hat{P}_\varepsilon(t) = e^{-\varepsilon|t|}$) and Q_ε the conjugate Poisson kernel. From (3.2)

$$|e^{itH}(0,x)| = e^{\varepsilon|t|}|(\mu_{0,x} * P_\varepsilon)^\wedge(t)|. \tag{3.4}$$

Now assume dynamical localization in the form

$$\sup_t |e^{itH}(0,x)| \ll |x|^{-A} \text{ for } |x| \to \infty \tag{3.5}$$

for some exponent $A > 0$. We will show that $A \leq d$.

It follows from (3.4), (3.5) that

$$|(\mu_{0,x} * P_\varepsilon)^\wedge(t)| < Ce^{-\varepsilon|t|}|x|^{-A} \tag{3.6}$$

for some constant C.

Next, we have from the resolvent identity

$$G(0+i\varepsilon) = G(0) - i\varepsilon G(0)G(0+i\varepsilon)$$

implying by (3.1), (2.62)

$$|G(0+i\varepsilon)(0,x) - G(0)(0,x)| \lesssim \varepsilon \sum_y \frac{1}{|x-y|^{d-2}} \frac{1}{|y|^{d-2}} \lesssim \frac{\varepsilon}{|x|^{d-4}}. \tag{3.7}$$

Again using (3.1), this gives

$$|G(0+i\varepsilon)(0,x)| \sim \frac{1}{|x|^{d-2}} \text{ for } |x| \ll \frac{1}{\sqrt{\varepsilon}}. \tag{3.8}$$

Estimate

$$\begin{aligned} |G(0+i\varepsilon)(0,x)| &\le \|\mu_{0,x} * P_\varepsilon\|_\infty + \|\mu_{0,x} * Q_\varepsilon\|_\infty \\ &\le (\|P_{\frac{\varepsilon}{2}}\|_2 + \|Q_{\frac{\varepsilon}{2}}\|_2)\|\mu_{0,x} * P_{\frac{\varepsilon}{2}}\|_2 \\ &< C\varepsilon^{-1}|x|^{-A} \end{aligned} \tag{3.9}$$

invoking (3.6).

Combining (3.8), (3.9), it follows indeed that $A \le d$.

References

[B]. J. Bourgain, *Random lattice Schrödinger operators with decaying potential: Some higher dimensional phenomena*, Springer LNM.

[DSS]. F. Delyon, B. Simon, B. Souillard, *From power pure point to continuous spectrum in disordered systems*, Ann. Inst. H. Poincare, Vol. 42, n3 (1985), 283–309.

[ESY]. L. Erdös, M. Salmhofer, H.T. Yau, *Quantum diffusion for the Anderson model in the scaling limit*, Ann. Henri Poincaré 8 (2007), no. 4, 553–555.

[S-W]. T. Spencer, W-M. Wang, *Lipschitz tails*, (unpublished ms).

Institute for Advanced Study, Princeton, NJ 08540
E-mail address: bourgain@ias.edu

Proceedings of Symposia in Pure Mathematics
Volume **79**, 2008

Circulation integrals and critical Sobolev spaces: problems of optimal constants

Haïm Brezis and Jean Van Schaftingen

Dedicated to Vladimir Maz'ya on the occasion of his 70th birthday, with high esteem and friendship

ABSTRACT. We study various questions related to the best constants in the following inequalities established in [**1, 2, 3**];

$$\left|\int_\Gamma \vec{\varphi}\cdot\vec{t}\right| \le C_n \|\nabla\varphi\|_{\mathrm{L}^n} |\Gamma| ,$$

and

$$\left|\int_{\mathbf{R}^n} \vec{\varphi}\cdot\vec{\mu}\right| \le C_n \|\nabla\varphi\|_{\mathrm{L}^n} \|\vec{\mu}\| ,$$

where Γ is a closed curve in $\mathbf{R}^n$, $\vec{\varphi}\in C_c^\infty(\mathbf{R}^n;\mathbf{R}^n)$ and $\vec{\mu}$ is a bounded measure on $\mathbf{R}^n$ with values into $\mathbf{R}^n$ such that $\operatorname{div}\vec{\mu}=0$. In 2d the answers are rather complete and closely related to the best constants for Sobolev and isoperimetric inequalities.

1. Estimates for curves and divergence-free vector fields

In general, functions in the Sobolev space $\mathrm{W}^{1,n}(\mathbf{R}^n)$ do not need be bounded continuous functions. However, it was shown recently that in circulation integrals, such functions behave as if they were bounded continuous functions:

THEOREM 1 (Bourgain, Brezis and Mironescu [**4**]). *There exists $C>0$ depending only on n such that if Γ is a closed rectifiable oriented curve whose tangent vector is $\vec{t}$, for every $\vec{\varphi}\in C_c^\infty(\mathbf{R}^n;\mathbf{R}^n)$,*

$$\left|\int_\Gamma \vec{\varphi}\cdot\vec{t}\right| \le C\|\nabla\vec{\varphi}\|_{\mathrm{L}^n(\mathbf{R}^n)}|\Gamma| , \tag{1.1}$$

1991 *Mathematics Subject Classification.* 26D10, 35J25, 46E35, 49Q15, 51M16, 58C35.

Key words and phrases. Circulation integrals, optimal constants, Sobolev inequalities, isoperimetric inequalities, divergence-free vector fields.

The second author (JVS) was partially supported by the Fonds de la Recherche Scientifique–FNRS (Belgium) and by the Fonds Spécial de Recherche (Université catholique de Louvain). He acknowledges the hospitality of the Laboratoire J.-L. Lions, Université de Paris VI and of the Rutgers University Mathematics Department, at which part of this research was carried.

Here and in the sequel, the subscript c in $C_c^\infty(\mathbf{R}^n; \mathbf{R}^n)$ means compact support. The original proof of Theorem 1 relied on a Littlewood–Paley decomposition; an elementary proof was given by the second author [**18**].

In two-dimensional space, this estimate follows easily from the classical isoperimetric theorem. This is not anymore the case in higher dimensions, but the inequality still retains some isoperimetric flavour.

The aim of this paper is to present various problems concerning optimal constants in Theorem 1. A first family of problems consists in fixing the curve Γ and finding the optimal constant A_Γ and the optimal vector field $\vec{\varphi}$. Such a vector field always exists. In two dimensions, the value of the associated optimal constant can even be given in terms of the area enclosed by Γ.

The next problem consists in optimizing A_Γ over all possible curves i.e., studying $A = \sup_\Gamma A_\Gamma$. In $\mathbf{R}^2$, one merely needs to maximize the area under a constraint on the perimeter; the classical isoperimetric theorem then yields a circle as the optimal curve. Similar results in higher dimensions would be desirable. However, none of the two-dimensional arguments work. The question whether the optimal constant is attained is even open.

As explained by Bourgain and Brezis [**2**], the optimal constant A in Theorem 1 is the same as the optimal constant in

THEOREM 2 (Bourgain and Brezis [**2, 3**]). *There exists $C > 0$ such that if $\vec{\mu}$ is a divergence-free bounded measure, then for every $\vec{\varphi} \in C_c^\infty(\mathbf{R}^n; \mathbf{R}^n)$,*

$$\left| \int_{\mathbf{R}^n} \vec{\varphi} \cdot \vec{\mu} \right| \le C \|\nabla \vec{\varphi}\|_{\mathrm{L}^n(\mathbf{R}^n)} \|\vec{\mu}\| \, . \tag{1.2}$$

Theorem 2 also has a direct proof by Littlewood–Paley decompostion [**3**], and an elementary proof [**17**]. The latter inequality provides thus a relaxed problem for the optimal constants, which could be more tractable.

This paper is organized as follows. In section 2, we explain how for a given curve, one can find an optimal vector field, how this leads to the optimal problem, how the problem of the optimal constant for divergence-free measures is the same as the problem for closed curves. We end this sections by some problems related to the confinement of the inequalities to domains. In section 3, we show that for a given divergence-free vector measure, one can find an optimal vector field. We then raise various questions concerning the optimal constant in the inequality.

2. The two-dimensional case

2.1. Curves in 2d. Throughout this section, we assume that Γ is a simple, closed, smooth, oriented curve with tangent vector $\vec{t}$. The optimal constant for the inequality of Theorem 1 is given by

$$A_\Gamma = \sup\Bigl\{ \int_\Gamma \vec{\varphi} \cdot \vec{t} \; : \; \vec{\varphi} \in C_c^\infty(\mathbf{R}^2; \mathbf{R}^2) \text{ and } \int_{\mathbf{R}^2} |\nabla \vec{\varphi}|^2 \le 1 \Bigr\} ,$$

where $|\cdot|$ denotes the Euclidean norm (of a vector or a matrix). By Theorem 1, this can be rewritten as

$$A_\Gamma = \sup\Bigl\{ \int_\Gamma \vec{\varphi} \cdot \vec{t} \; : \; \vec{\varphi} \in \dot{\mathrm{H}}^1(\mathbf{R}^2; \mathbf{R}^2) \text{ and } \int_{\mathbf{R}^2} |\nabla \vec{\varphi}|^2 \le 1 \Bigr\} ,$$

where

$$\dot{\mathrm{H}}^1(\mathbf{R}^2; \mathbf{R}^2) = \Bigl\{ \vec{\varphi} \in \mathrm{W}^{1,1}_{\mathrm{loc}}(\mathbf{R}^2; \mathbf{R}^2) \; : \; \nabla \vec{\varphi} \in \mathrm{L}^2(\mathbf{R}^2; \mathbf{R}^{2\times 2}) \Bigr\}$$

is the completion of $C_c^\infty(\mathbf{R}^2;\mathbf{R}^2)$ equipped with the norm $\|\nabla\vec{\varphi}\|_{\mathrm{L}^2}$. Note that there are two possible definitions for $\int_\Gamma \vec{\varphi}\cdot\vec{t}$ when $\vec{\varphi}\in\dot{\mathrm{H}}^1(\mathbf{R}^2;\mathbf{R}^2)$. The first one is in the sense of traces; secondly, starting from (1.1), one can define $\int_\Gamma \vec{\varphi}\cdot\vec{t}$ abstractly via density. Both definitions coincide.

Since the space $\dot{\mathrm{H}}^1(\mathbf{R}^2;\mathbf{R}^2)$ modulo constants is a Hilbert space, one has

PROPOSITION 2.1. *There exists a unique* $\vec{\varphi}\in\dot{\mathrm{H}}^1(\mathbf{R}^2;\mathbf{R}^2)$ *(modulo constants) such that*

$$\int_\Gamma \vec{\varphi}\cdot\vec{t} = A_\Gamma\ , \tag{2.1}$$

$$\int_{\mathbf{R}^2} |\nabla\vec{\varphi}|^2 = 1\ . \tag{2.2}$$

The vector field $\vec{\varphi}$ *satisfies the equation,*

$$-\Delta\vec{\varphi} = \frac{\vec{t}}{A_\Gamma}\mathcal{H}^1\llcorner\Gamma \qquad \text{in } \mathbf{R}^2\ , \tag{2.3}$$

$$\operatorname{div}\vec{\varphi} = 0 \qquad \text{in } \mathbf{R}^2\ , \tag{2.4}$$

$$\vec{\varphi}\in(\mathrm{L}^\infty\cap C)(\mathbf{R}^2;\mathbf{R}^2)\ , \tag{2.5}$$

where $\mathcal{H}^1\llcorner\Gamma$ *denotes the* 1*–dimensional Hausdorff measure restricted to* Γ.

Note that the additional property (2.5) allows to give a classical meaning to $\int_\Gamma \vec{\varphi}\cdot\vec{t}$.

PROOF OF PROPOSITION 2.1. Since $\dot{\mathrm{H}}^1(\mathbf{R}^2;\mathbf{R}^2)$ is a Hilbert space, there exists a unique $\vec{\varphi}$ such that

$$\int_\Gamma \vec{\varphi}\cdot\vec{t} = A_\Gamma \qquad \text{and} \qquad \int_{\mathbf{R}^2}|\nabla\vec{\varphi}|^2 = 1\ .$$

The vector field $\vec{\varphi}$ satisfies the Euler equation

$$-\lambda\Delta\vec{\varphi} = \vec{t}\mathcal{H}^1\llcorner\Gamma$$

for some Lagrange multiplier $\lambda\in\mathbf{R}$. Obviously $\lambda = A_\Gamma$. In particular, $\operatorname{div}\vec{\varphi}$ satisfies

$$-\Delta\operatorname{div}\vec{\varphi} = 0\ .$$

Since $\operatorname{div}\vec{\varphi}\in\mathrm{L}^2(\mathbf{R}^2)$, one has necessarily $\operatorname{div}\vec{\varphi} = 0$. Finally, since $-\Delta\vec{\varphi}$ is a divergence-free bounded measure, the boundedness and the continuity of $\vec{\varphi}$ follow from the results of Bourgain and Brezis (see [**2**, Remark 5] and [**3**, Theorem 3]). □

The constant A_Γ can be determined explicitely:

PROPOSITION 2.2. *If* $\Gamma = \partial\Sigma$, *then*

$$A_\Gamma = |\Sigma|^{\frac{1}{2}}\ .$$

PROOF. To fix the ideas, we shall assume that Γ is positively oriented. Applying successively Stokes' theorem and the Cauchy–Schwarz inequality, we obtain

$$\begin{aligned}\int_\Gamma \vec{\varphi}\cdot\vec{t} &= \int_\Sigma \nabla\wedge\vec{\varphi} \le |\Sigma|^{1/2}\Big(\int_{\mathbf{R}^2}|\nabla\wedge\vec{\varphi}|^2\Big)^{\frac{1}{2}}\\ &\le |\Sigma|^{1/2}\Big(\int_{\mathbf{R}^2}|\nabla\wedge\vec{\varphi}|^2 + |\operatorname{div}\vec{\varphi}|^2\Big)^{\frac{1}{2}} = |\Sigma|^{1/2}\Big(\int_{\mathbf{R}^2}|\nabla\vec{\varphi}|^2\Big)^{\frac{1}{2}}\ ,\end{aligned} \tag{2.6}$$

where $\nabla \wedge \vec{\varphi} = \partial_1 \varphi_2 - \partial_2 \varphi_1$. This proves that $A_\Gamma \leq |\Sigma|^{\frac{1}{2}}$.

In order to obtain the equality, choose $\vec{\varphi} \in \dot{\mathrm{H}}^1(\mathbf{R}^2; \mathbf{R}^2)$ to be the solution of

$$\begin{cases} \nabla \wedge \vec{\varphi} = 1 & \text{in } \Sigma\,, \\ \nabla \wedge \vec{\varphi} = 0 & \text{in } \mathbf{R}^2 \setminus \Sigma\,, \\ \operatorname{div} \vec{\varphi} = 0 & \text{in } \mathbf{R}^2\,. \end{cases}$$

By construction, $\vec{\varphi}$ realizes the equality in (2.6); hence, $A_\Gamma = |\Sigma|^{\frac{1}{2}}$. □

The main result is

PROPOSITION 2.3. *One has*

$$A_\Gamma \leq \frac{|\Gamma|}{2\sqrt{\pi}}\,. \tag{2.7}$$

Moreover, equality holds in (2.7) *if, and only if,* Γ *is a circle.*

PROOF. Writing $\Gamma = \partial\Sigma$, we combine Proposition 2.2 together with the classical isoperimetric inequality, in order to obtain

$$A_\Gamma = |\Sigma|^{\frac{1}{2}} \leq \frac{|\Gamma|}{2\sqrt{\pi}}.$$

Moreover, one has equality if and only if $\Gamma = \partial\Sigma$ is a circle. □

2.2. A relaxed problem. Let $\vec{\mu}$ be a bounded $\mathbf{R}^2$–valued vector measure such that $\operatorname{div} \vec{\mu} = 0$ in the sense of distributions, and set

$$A_{\vec{\mu}} = \sup\Bigl\{ \int_{\mathbf{R}^2} \vec{\varphi} \cdot \vec{\mu} \,:\, \vec{\varphi} \in C_c^\infty(\mathbf{R}^2; \mathbf{R}^2) \text{ and } \int_{\mathbf{R}^2} |\nabla \vec{\varphi}|^2 \leq 1 \Bigr\}\,.$$

A particular case is when Γ is as in Section 2.1 and the measure is $\vec{\mu} = \vec{t}\mathcal{H}^1 \llcorner \Gamma$; one has then $A_{\vec{\mu}} = A_\Gamma$.

By Theorem 2, the integral $\int_{\mathbf{R}^2} \vec{\varphi} \cdot \vec{\mu}$ is well-defined when $\vec{\varphi} \in \dot{\mathrm{H}}^1(\mathbf{R}^2; \mathbf{R}^2)$ as an extension of a linear functional[1]. One can thus relax the supremum to

$$A_{\vec{\mu}} = \sup\Bigl\{ \int_{\mathbf{R}^2} \vec{\varphi} \cdot \vec{\mu} \,:\, \vec{\varphi} \in \dot{\mathrm{H}}^1(\mathbf{R}^2; \mathbf{R}^2) \text{ and } \int_{\mathbf{R}^2} |\nabla \vec{\varphi}|^2 \leq 1 \Bigr\}\,.$$

As previously, one can show that if $\vec{\mu} \neq 0$, $A_{\vec{\mu}}$ is attained by a vector field $\vec{\varphi} \in (\dot{\mathrm{H}}^1 \cap \mathrm{L}^\infty)(\mathbf{R}^2; \mathbf{R}^2)$ that satisfies

$$-\Delta \vec{\varphi} = \frac{1}{A_{\vec{\mu}}} \vec{\mu}\,, \tag{2.8}$$

so that in particular $\operatorname{div} \vec{\varphi} = 0$. By the result of Bourgain and Brezis mentioned above, $\vec{\varphi}$ is a bounded continuous vector field.

One can also compute explicitely the value $A_{\vec{\mu}}$. Recall that by the Sobolev–Nirenberg embedding, if $\vec{\mu}$ is divergence-free, it can be written as $\vec{\mu} = \nabla^\perp \zeta$, for some unique $\zeta \in \mathrm{L}^2(\mathbf{R}^2)$, where $\nabla^\perp \zeta = (\partial_2 \zeta, -\partial_1 \zeta)$.

PROPOSITION 2.4. *Let* $\vec{\mu}$ *and* ζ *be as above, then*

$$A_{\vec{\mu}} = \Bigl(\int_{\mathbf{R}^2} |\zeta|^2 \Bigr)^{\frac{1}{2}}\,.$$

---

[1]In fact, one can even show using Theorem 2 that null sets for the H^1 capacity in $\mathbf{R}^2$ are null sets for the variation $|\vec{\mu}|$ of $\vec{\mu}$, so that every H^1– quasicontinuous function is $\vec{\mu}$–measurable. In particular, the integral makes sense for every $\vec{\varphi} \in (\dot{\mathrm{H}}^1 \cap \mathrm{L}^\infty)(\mathbf{R}^2; \mathbf{R}^2)$.

PROOF. One has, by the Cauchy-Schwarz inequality, for every $\vec{\varphi} \in C_c^\infty(\mathbf{R}^2; \mathbf{R}^2)$,

$$\begin{aligned}\int_{\mathbf{R}^2} \vec{\varphi} \cdot \vec{\mu} &= \int_{\mathbf{R}^2} \nabla^\perp \zeta \cdot \vec{\varphi} = \int_{\mathbf{R}^2} \zeta (\nabla \wedge \vec{\varphi}) \\ &\le \Big(\int_{\mathbf{R}^2} |\zeta|^2 \Big)^{\frac{1}{2}} \Big(\int_{\mathbf{R}^2} |\nabla \wedge \vec{\varphi}|^2 \Big)^{\frac{1}{2}} \\ &\le \Big(\int_{\mathbf{R}^2} |\zeta|^2 \Big)^{\frac{1}{2}} \Big(\int_{\mathbf{R}^2} |\nabla \wedge \vec{\varphi}|^2 + |\operatorname{div} \vec{\varphi}|^2 \Big)^{\frac{1}{2}} \\ &= \Big(\int_{\mathbf{R}^2} |\zeta|^2 \Big)^{\frac{1}{2}} \Big(\int_{\mathbf{R}^2} |\nabla \vec{\varphi}|^2 \Big)^{\frac{1}{2}} ,\end{aligned}$$

whence one obtains an upper bound

$$A_{\vec{\mu}} \le \Big(\int_{\mathbf{R}^2} |\zeta|^2 \Big)^{\frac{1}{2}} .$$

For the equality, observe that the solution $\vec{\varphi} \in \dot{\mathrm{H}}^1(\mathbf{R}^2; \mathbf{R}^2)$ to

$$\begin{cases} \nabla \wedge \vec{\varphi} = \zeta , \\ \operatorname{div} \vec{\varphi} = 0 , \end{cases}$$

achieves equality in the above inequality. □

PROPOSITION 2.5. *One has*

$$A_{\vec{\mu}} \le \frac{\|\vec{\mu}\|}{2\sqrt{\pi}} , \tag{2.9}$$

where

$$\|\vec{\mu}\| = \sup \Big\{ \int_{\mathbf{R}^2} \vec{\varphi} \cdot \vec{\mu} \; : \; \forall x \in \mathbf{R}^2, \; |\vec{\varphi}(x)| \le 1 \Big\} .$$

Moreover, equality holds in (2.9) *if, and only if,*

$$\vec{\mu} = \lambda \nabla^\perp \chi_{B(x,r)}$$

for some $\lambda \in \mathbf{R}$, $x \in \mathbf{R}^2$ *and* $r > 0$.

PROOF. By the optimal Sobolev–Nirenberg inequality, one has

$$\|\zeta\|_{\mathrm{L}^2(\mathbf{R}^2)} \le \frac{1}{2\sqrt{\pi}} \|\nabla \zeta\|_{\mathrm{L}^1(\mathbf{R}^2;\mathbf{R}^2)} = \frac{1}{2\sqrt{\pi}} \|\vec{\mu}\| ,$$

with equality if and only if ζ is a multiple of a characteristic function of a ball; this result is originaly due to H. Federer and W. H. Fleming [**8**], and independently to V. G. Maz'ya [**13**], see e.g. [**14**, Lemma 3.2.3/1] and also [**6, 9**]. □

REMARK 2.6. In view of the result of Smirnov [**16**] representing any divergence-free measure $\vec{\mu}$ with $\|\vec{\mu}\| = 1$ as a "convex combination" of measures of the form

$$\frac{\vec{t} \mathcal{H}^1 \llcorner \Gamma}{|\Gamma|} ,$$

it is clear that (2.7) implies (2.9). The main interest of Proposition 2.5 is that the relaxation to Lipschitz curves (having possibly self-intersections), or to divergence-free measures, does not introduce new maximizers.

There is a slight improvement of Proposition 2.5

COROLLARY 2.7. *For every* $\vec{\varphi} \in \dot{\mathrm{H}}^1(\mathbf{R}^2;\mathbf{R}^2)$,

$$\int_{\mathbf{R}^2} \vec{\varphi}\cdot\vec{\mu} \le \frac{\|\vec{\mu}\|}{2\sqrt{\pi}}\Big(\int_{\mathbf{R}^2} |\nabla\wedge\vec{\varphi}|^2\Big)^{\frac{1}{2}} .$$

Moreover there is equality in the nontrivial case if, and only if,

$$\vec{\mu} = \lambda\nabla^{\perp}\chi_{B(x,r)} \tag{2.10}$$

and

$$\nabla\wedge\vec{\varphi} = \nu\lambda\chi_{B(x,r)} , \tag{2.11}$$

for some $\lambda \in \mathbf{R}$ *and* $\nu > 0$, $x \in \mathbf{R}^2$ *and* $r > 0$.

PROOF. Given $\vec{\varphi} \in \dot{\mathrm{H}}^1(\mathbf{R}^2;\mathbf{R}^2)$, let $\vec{\psi}$ solve

$$\begin{cases} \nabla\wedge\vec{\psi} = \nabla\wedge\vec{\varphi} , \\ \operatorname{div}\vec{\psi} = 0 . \end{cases}$$

One then has

$$\begin{aligned}\int_{\mathbf{R}^2} \vec{\varphi}\cdot\vec{\mu} = \int_{\mathbf{R}^2} \vec{\psi}\cdot\vec{\mu} &\le \frac{\|\vec{\mu}\|}{2\sqrt{\pi}}\Big(\int_{\mathbf{R}^2} |\nabla\vec{\psi}|^2\Big)^{\frac{1}{2}} \\ &= \frac{\|\vec{\mu}\|}{2\sqrt{\pi}}\Big(\int_{\mathbf{R}^2} |\nabla\wedge\vec{\psi}|^2\Big)^{\frac{1}{2}} = \frac{\|\vec{\mu}\|}{2\sqrt{\pi}}\Big(\int_{\mathbf{R}^2} |\nabla\wedge\vec{\varphi}|^2\Big)^{\frac{1}{2}} .\end{aligned}$$

If equality holds, one has by Proposition 2.5,

$$\vec{\mu} = \lambda\nabla^{\perp}\chi_{B(x,r)} ,$$

so that, by (2.8),

$$-\Delta\vec{\psi} = \nu\vec{\mu} = \nu\lambda\nabla^{\perp}\chi_{B(x,r)} .$$

On the other hand,

$$\nabla^{\perp}(\nabla\wedge\vec{\varphi}) = \nabla^{\perp}(\nabla\wedge\vec{\psi}) = -\Delta\vec{\psi}$$

since $\operatorname{div}\vec{\psi} = 0$. Therefore,

$$\nabla^{\perp}(\nabla\wedge\vec{\varphi}) = \nu\lambda\nabla^{\perp}\chi_{B(x,r)}$$

and consequently $\nabla\wedge\vec{\varphi} = \nu\lambda\chi_{B(x,r)}$. □

2.3. Confinement to domains. Let $\Omega \subset \mathbf{R}^2$ be a smooth, bounded, simply connected domain with normal vector field $\vec{n}$. We will work with the class of measures on $\bar{\Omega}$ satisfying $\operatorname{div}\vec{\mu} = 0$ in Ω and $\vec{\mu}\cdot\vec{n} = 0$ on $\partial\Omega$. This class is defined as

$$\mathcal{M}_0^{\#}(\bar{\Omega},\mathbf{R}^2) = \Big\{\vec{\mu} \in C(\bar{\Omega};\mathbf{R}^2)^* \ : \ \forall\zeta \in C^1(\bar{\Omega}),\ \int_{\bar{\Omega}} \nabla\zeta\cdot\vec{\mu} = 0\Big\} .$$

Note that $\int_{\bar{\Omega}} \vec{\mu} = 0$ for every $\vec{\mu} \in \mathcal{M}_0^{\#}(\bar{\Omega},\mathbf{R}^2)$ (just take $\zeta(x) = x_i$, $i = 1,2$ in the definition).

PROPOSITION 2.8. *For every* $\vec{\varphi} \in \mathrm{H}^1(\Omega)\cap C(\bar{\Omega})$ *and* $\vec{\mu} \in \mathcal{M}_0^{\#}(\bar{\Omega},\mathbf{R}^2)$

$$\int_{\Omega} \vec{\varphi}\cdot\vec{\mu} \le \frac{\|\vec{\mu}\|}{2\sqrt{\pi}}\|\nabla\wedge\vec{\varphi}\|_{\mathrm{L}^2(\Omega)} . \tag{2.12}$$

Equality holds in the nontrivial cases if $\vec{\mu} = \lambda\vec{t}\mathcal{H}^1 \llcorner \partial B(x,r)$ *with* $\partial B(x,r) \subset \bar{\Omega}$ *and* $\vec{\varphi}$ *satisfies* (2.11) *for* $\nu > 0$.

PROOF. Let $\vec{\psi} \in \dot{H}^1(\mathbf{R}^2;\mathbf{R}^2)$ satisfy

$$\begin{cases} \nabla \wedge \vec{\psi} = \nabla \wedge \vec{\varphi} & \text{in } \Omega\,, \\ \nabla \wedge \vec{\psi} = 0 & \text{in } \mathbf{R}^2 \setminus \Omega\,, \\ \operatorname{div} \vec{\psi} = 0 & \text{in } \mathbf{R}^2\,, \end{cases}$$

and define the measure $\vec{\eta} \in \mathcal{M}(\mathbf{R}^2)$ by

$$\int_{\mathbf{R}^2} \vec{\vartheta} \cdot \vec{\eta} = \int_{\bar{\Omega}} \vec{\vartheta} \cdot \vec{\mu}\,,$$

for every $\vec{\vartheta} \in C_c(\mathbf{R}^2;\mathbf{R}^2)$. One has, by Corollary 2.7,

$$\int_{\bar{\Omega}} \vec{\varphi} \cdot \vec{\mu} = \int_{\mathbf{R}^2} \vec{\psi} \cdot \vec{\eta} \le \frac{\|\vec{\eta}\|}{2\sqrt{\pi}} \Big(\int_{\mathbf{R}^2} |\nabla \wedge \vec{\psi}|^2 \Big)^{\frac{1}{2}} = \frac{\|\vec{\mu}\|}{2\sqrt{\pi}} \Big(\int_{\Omega} |\nabla \wedge \vec{\varphi}|^2 \Big)^{\frac{1}{2}}\,.$$

The equality cases follow again from the conclusion of Corollary 2.7. □

REMARK 2.9. If Ω is not simply connected, the inequality

$$\int_{\Omega} \vec{\varphi} \cdot \vec{\mu} \le C \|\vec{\mu}\|\, \|\nabla \wedge \vec{\varphi}\|_{\mathrm{L}^2(\Omega)}$$

cannot be true. Indeed, assume that 0 belongs to a bounded connected component of $\mathbf{R}^2 \setminus \Omega$. Take $\Gamma \subset \Omega$ to be any closed curve of index 1 with respect to 0, and set $\vec{\mu} = \vec{t}\,\mathcal{H}^1 \llcorner \Gamma$ and $\vec{\varphi}(x) = \frac{x^\perp}{|x|^2}$. One then has

$$\int_{\Omega} \vec{\varphi} \cdot \vec{\mu} = 2\pi\,,$$

while

$$\nabla \wedge \vec{\varphi} = 0$$

in Ω. □

From (2.12), we also have

$$\int_{\Omega} \vec{\varphi} \cdot \vec{\mu} \le \frac{\|\vec{\mu}\|}{\sqrt{2\pi}} \|\nabla \vec{\varphi}\|_{\mathrm{L}^2(\Omega)} \tag{2.13}$$

since

$$\|\nabla \wedge \vec{\varphi}\|_{\mathrm{L}^2(\Omega)} \le \sqrt{2} \|\nabla \vec{\varphi}\|_{\mathrm{L}^2(\Omega)}.$$

For every $\vec{\mu} \in \mathcal{M}_0^{\#}(\bar{\Omega},\mathbf{R}^2)$, let

$$A_{\vec{\mu},\Omega} = \sup\Big\{ \int_{\bar{\Omega}} \vec{\varphi} \cdot \vec{\mu} \ :\ \vec{\varphi} \in \mathrm{H}^1(\mathbf{R}^2;\mathbf{R}^2) \text{ and } \int_{\Omega} |\nabla \vec{\varphi}|^2 \le 1 \Big\}\,; \tag{2.14}$$

of course the supremum in (2.14) is achieved by some unique $\vec{\varphi} \in \mathrm{H}^1(\Omega;\mathbf{R}^2)$ modulo constants; moreover, $\vec{\varphi} \in (\mathrm{L}^\infty \cap C)(\bar{\Omega};\mathbf{R}^2)$ by Theorem 2 in **[5]** (when $\vec{\mu}$ is an L^1 function — the case of measures is similar).

Thus we have for every $\vec{\mu} \in \mathcal{M}_0^{\#}(\bar{\Omega},\mathbf{R}^2)$

$$A_{\vec{\mu},\Omega} \le \frac{\|\vec{\mu}\|}{\sqrt{2\pi}}\,.$$

Set

$$A_{\Omega} = \sup\big\{ A_{\vec{\mu},\Omega} \ :\ \vec{\mu} \in \mathcal{M}_0^{\#}(\bar{\Omega},\mathbf{R}^2) \text{ and } \|\vec{\mu}\| \le 1 \big\}\,. \tag{2.15}$$

PROPOSITION 2.10. *One has*

$$\frac{1}{2\sqrt{\pi}} < A_\Omega \le \frac{1}{\sqrt{2\pi}} .$$

Moreover, if Ω *is a disc, then* $A_\Omega = \frac{1}{\sqrt{2\pi}}$ *and supremum in* (2.15) *is achieved.*

PROOF. Let $\vec{\mu} = \lambda \vec{t}\mathcal{H}^1 \llcorner \partial B(x_0, r)$ with $\partial B(x_0, r) \subset \Omega$ and

$$\vec{\varphi}(x) = \begin{cases} (x - x_0)^\perp & \text{if } |x - x_0| \le r , \\ (x - x_0)^\perp \frac{r^2}{|x - x_0|^2} & \text{if } |x - x_0| > r . \end{cases}$$

One then has

$$\int_\Omega \vec{\varphi} \cdot \vec{\mu} = \int_{\mathbf{R}^2} \vec{\varphi} \cdot \vec{\mu} = \frac{\|\vec{\mu}\|}{2\sqrt{\pi}} \Big(\int_{\mathbf{R}^2} |\nabla \vec{\varphi}|^2 \Big)^{\frac{1}{2}} > \frac{\|\vec{\mu}\|}{2\sqrt{\pi}} \Big(\int_\Omega |\nabla \vec{\varphi}|^2 \Big)^{\frac{1}{2}} ,$$

so that $A_\Omega > \frac{1}{2\sqrt{\pi}}$. The inequality $A_\Omega \le \frac{1}{\sqrt{2\pi}}$ follows immediately from (2.13). Finally, assume Ω is a disc; without loss of generality, $\Omega = B(0, 1)$. One sets then $\Gamma = \partial B(0, 1)$, $\vec{\mu} = \vec{t}\mathcal{H}^1 \llcorner \Gamma$ and $\vec{\varphi}(x) = x^\perp$; immediate computations give

$$\int_\Omega \vec{\varphi} \cdot \vec{\mu} = 2\pi ,$$

$$\|\vec{\mu}\| = 2\pi ,$$

$$\int_\Omega |\nabla \vec{\varphi}|^2 = 2\pi . \qquad \square$$

We have no clue about the dependence of A_Ω on Ω and whether the supremum in (2.15) is achieved. The only information we have is

PROPOSITION 2.11. *Assume that* $A_\Omega = \frac{1}{\sqrt{2\pi}}$ *and* A_Ω *is achieved. Then* Ω *is a disc.*

There are two extreme scenarios:

SCENARIO 1. $A_\Omega = \frac{1}{\sqrt{2\pi}}$ only when Ω is a disc.

SCENARIO 2. $A_\Omega = \frac{1}{\sqrt{2\pi}}$ for every domain $\Omega \subset \mathbf{R}^2$.

PROBLEM 1. Decide between Scenario 1, Scenario 2 and intermediate scenarios.

PROBLEM 2. Is it true that for every domain Ω, A_Ω is achieved?

By Proposition 2.11, a positive answer to Problem 2 would lead to Scenario 1. This scenario would be reminiscent of the situation of the balls who have the worst best Sobolev inequalities [**12**].

There is a variant of Proposition 2.8 where the boundary condition $\vec{\mu} \cdot \vec{n} = 0$ is replaced by the condition that $\vec{\varphi}$ should vanish on $\partial\Omega$. Set

$$\mathcal{M}^{\#}(\Omega, \mathbf{R}^2) = \Big\{ \vec{\mu} \in C(\Omega; \mathbf{R}^2)^* \ : \ \forall \zeta \in C^1_c(\Omega),\ \int_\Omega \nabla\zeta \cdot \vec{\mu} = 0 \Big\}.$$

PROPOSITION 2.12. *For every* $\vec{\mu} \in \mathcal{M}^{\#}(\Omega; \mathbf{R}^2)$ *and for every* $\vec{\varphi} \in \mathrm{H}^1_0(\Omega; \mathbf{R}^2) \cap C(\bar{\Omega}; \mathbf{R}^2)$, *one has*

$$\int_\Omega \vec{\varphi} \cdot \vec{\mu} \le S_\Omega \|\vec{\mu}\| \Big(\int_\Omega |\nabla \wedge \vec{\varphi}|^2 \Big)^{\frac{1}{2}} , \tag{2.16}$$

for every $\vec{\varphi} \in \mathrm{H}_0^1(\Omega)$ *where*

$$S_\Omega = \sup\Big\{ \|u\|_{\mathrm{L}^2(\Omega)} \ : \ u \in \mathrm{BV}(\Omega),\ \|\nabla u\| \le 1 \text{ and } \int_\Omega u = 0 \Big\}\,, \tag{2.17}$$

and $\|\nabla u\|$ *denotes the total mass of the measure* ∇u. *Moreover the constant* S_Ω *in* (2.16) *cannot be improved.*

PROOF. Inequality (2.16) is established as above, see also Theorem 2.1 in [**5**]. For the last statement, assume that for every $\vec{\mu} \in \mathcal{M}^{\#}(\Omega; \mathbf{R}^2)$ and for every $\vec{\varphi} \in \mathrm{H}_0^1(\Omega; \mathbf{R}^2) \cap C(\bar{\Omega}; \mathbf{R}^2)$, one has

$$\int_\Omega \vec{\varphi} \cdot \vec{\mu} \le A\|\vec{\mu}\| \Big(\int_\Omega |\nabla \wedge \vec{\varphi}|^2 \Big)^{\frac{1}{2}}\,,$$

for some constant A. We claim that for every $u \in \mathrm{BV}(\Omega)$ with $\int_\Omega u = 0$, we have

$$\|u\|_{\mathrm{L}^2} \le A\|\nabla u\|\,.$$

Indeed, set $\vec{\mu} = \nabla^\perp u$, and choose any function $\vec{\varphi} \in \mathrm{H}_0^1(\Omega) \cap C(\bar{\Omega})$ such that $\nabla \wedge \vec{\varphi} = u$ in Ω [**1**, Theorem 3]. □

PROBLEM 3. Is the supremum in (2.17) achieved by some $u \in \mathrm{BV}(\Omega)$? Or equivalently, does equality hold in (2.16) in the nontrivial cases?

The problem has been treated on the sphere [**19**] and on the unit ball [**11**]. For a general domain $\Omega \subset \mathbf{R}^n$, with $n \ge 3$ and when $\mathrm{BV}(\Omega)$ and $\mathrm{L}^2(\Omega)$ are replaced by the spaces $\mathrm{H}^1(\Omega)$ and $\mathrm{L}^{\frac{2n}{n-2}}(\Omega)$, an affirmative answer has been given [**10**, Proposition 1.2].

REMARK 2.13. As is well known, there is no universal bound on S_Ω, even when replacing the constraint $\|\nabla u\| \le 1$ by the constraint $\|\nabla u\|_{\mathrm{L}^2} \le 1$. This is related to the eigenvalue problem for the Laplacian with Neumann boundary condition. In the similar inequality

$$\inf_{c \in \mathbf{R}} \|u - c\|_{\mathrm{L}^{\frac{n}{n-1}}} \le S'_\Omega \|\nabla u\|_{\mathrm{L}^1}\,,$$

the best constant S'_Ω is proportional to a relative isoperimetric constant of Ω [**14**, Theorem 3.2.3 and § 6.1.7].

A consequence of Proposition 2.12 is the inequality

$$\int_\Omega \vec{\varphi} \cdot \vec{\mu} \le S_\Omega \|\vec{\mu}\| \Big(\int_\Omega |\nabla \vec{\varphi}|^2 \Big)^{\frac{1}{2}}\,,$$

for every $\vec{\mu} \in \mathcal{M}^{\#}(\Omega; \mathbf{R}^2)$ and for every $\vec{\varphi} \in \mathrm{H}_0^1(\Omega; \mathbf{R}^2) \cap C(\bar{\Omega}; \mathbf{R}^2)$, since

$$\int_\Omega |\nabla \wedge \vec{\varphi}|^2 \le \int_\Omega |\nabla \vec{\varphi}|^2\,.$$

By analogy with the above, for $\vec{\mu} \in \mathcal{M}^{\#}(\Omega, \mathbf{R}^2)$, set

$$A'_{\vec{\mu},\Omega} = \sup\Big\{ \int_{\bar{\Omega}} \vec{\varphi} \cdot \vec{\mu} \ : \ \vec{\varphi} \in \mathrm{H}_0^1(\mathbf{R}^2; \mathbf{R}^2) \text{ and } \int_\Omega |\nabla \vec{\varphi}|^2 \le 1 \Big\}\,; \tag{2.18}$$

and

$$A'_\Omega = \sup\{ A'_{\vec{\mu},\Omega} \ : \ \vec{\mu} \in \mathcal{M}^{\#}(\Omega, \mathbf{R}^2) \text{ and } \|\vec{\mu}\| \le 1 \}\,, \tag{2.19}$$

so that

$$A'_\Omega \le S_\Omega\,.$$

The supremum in (2.18) is uniquely achieved since $\mathcal{M}^{\#}(\Omega, \mathbf{R}^2) \subset \mathrm{H}^{-1}(\Omega; \mathbf{R}^2)$, and the maximizer is bounded and continuous [**5**].

In general, we do not expect having $A'_\Omega = S_\Omega$. Indeed, the maximizing vector fields $\vec{\varphi}$ in Proposition 2.12 need not be divergence-free.

One has

PROPOSITION 2.14. *There exists $\alpha > 0$ such that for every domain $\Omega \subset \mathbf{R}^2$,*

$$A'_\Omega \geq \alpha \, .$$

PROOF. Simply take some compactly supported divergence-free measure $\vec{\mu} \in \mathcal{M}(\Omega; \mathbf{R}^2)$, and some compactly supported vector field $\vec{\psi} \in \mathcal{C}_c^\infty(\mathbf{R}^2; \mathbf{R}^2)$ such that $\int_{\mathbf{R}^2} \vec{\psi} \cdot \vec{\mu} \neq 0$. By translation and dilation, one has that

$$A'_\Omega \geq \frac{\int_{\mathbf{R}^2} \vec{\psi} \cdot \vec{\mu}}{\|\vec{\mu}\| \|\nabla \vec{\psi}\|_{\mathrm{L}^2}} \, . \qquad \square$$

This raises the question

PROBLEM 4. Compute $\inf_\Omega A'_\Omega$ and $\inf_\Omega S_\Omega$. Are they achieved?

In [**19**, Question 4.1], the question was asked whether $\inf_\Omega S_\Omega = S_{B(0,1)}$.

Remember that A_Ω does not have an upper bound independent of the geometry. If we allow Ω to be multiply connected, A'_Ω has no upper bound. On the other hand, we do not know whether A'_Ω has an upper bound independently of the geometry for simply connected domains.

PROBLEM 5. Does one have

$$\sup\{A'_\Omega \, : \, \Omega \subset \mathbf{R}^2 \text{ is a simply connected domain}\} < \infty?$$

3. Higher dimensions

3.1. Inequalities for curves. Throughout this section $\Gamma \subset \mathbf{R}^n$ is a simple, closed, rectifiable curve. The optimal constant in Theorem 2 is

$$A_\Gamma = \sup\Big\{ \int_\Gamma \vec{\varphi} \cdot \vec{t} \, : \, \vec{\varphi} \in C_c^\infty(\mathbf{R}^n; \mathbf{R}^n) \text{ and } \int_{\mathbf{R}^n} |\nabla \vec{\varphi}|^n \leq 1 \Big\} \, .$$

As in 2d, we obtain

PROPOSITION 3.1. *There exists a unique $\vec{\varphi} \in \dot{\mathrm{W}}^{1,n}(\mathbf{R}^n; \mathbf{R}^n)$ modulo constants such that*

$$\int_\Gamma \vec{\varphi} \cdot \vec{t} = A_\Gamma \, ,$$

$$\|\nabla \vec{\varphi}\|_{\mathrm{L}^n} = 1 \, .$$

The vector field $\vec{\varphi}$ satisfies

$$-\sum_{i=1}^n \partial_i(|\nabla \vec{\varphi}|^{n-2} \partial_i \vec{\varphi}) = \frac{1}{A_\Gamma} \vec{t} \mathcal{H}^1 \llcorner \Gamma \, .$$

In contrast with Proposition 2.1, we do not in general expect to have $\operatorname{div} \vec{\varphi} = 0$ (find a counterexample) and we do not know whether $\vec{\varphi}$ is a bounded continuous vector field. This is an interesting open problem:

PROBLEM 6. Does one have $\vec{\varphi} \in (\mathrm{L}^\infty \cap C)(\mathbf{R}^n;\mathbf{R}^n)$. More generally, let $\vec{\mu} \in \mathcal{M}(\mathbf{R}^n;\mathbf{R}^n)$ be a bounded measure such that $\operatorname{div}\vec{\mu} = 0$. From [**2, 3**], we know that $\vec{\mu} \in \left(\dot{\mathrm{W}}^{1,n}(\mathbf{R}^n;\mathbf{R}^n)\right)^*$ and hence there exists a unique $\vec{\varphi} \in \dot{\mathrm{W}}^{1,n}(\mathbf{R}^n;\mathbf{R}^n)$ modulo constants that solves

$$-\sum_{i=1}^{n} \partial_i(|\nabla\vec{\varphi}|^{n-2}\partial_i\vec{\varphi}) = \vec{\mu}\,.$$

Does one have $\vec{\varphi} \in (\mathrm{L}^\infty \cap C)(\mathbf{R}^n;\mathbf{R}^n)$?

In two dimensions, Proposition 2.2 gives the exact value of A_Γ in terms of $|\Sigma|$, the area of the surface spanned by Γ. This is not anymore the case in higher dimensions.

Let us examine what happens when $\Gamma \subset \mathbf{R}^n$ is planar, i.e., if $\Gamma \subset \Pi$ where $\Pi \subset \mathbf{R}^n$ is a (two-dimensional) plane. Recall that the trace on Π of a function $u \in \dot{\mathrm{W}}^{1,n}(\mathbf{R}^n)$ belongs to $\dot{\mathrm{W}}^{\frac{2}{n},n}(\Pi)$ with

$$\|u|_\Pi\|_{\dot{\mathrm{W}}^{\frac{2}{n},n}(\Pi)} \le C\|u\|_{\dot{\mathrm{W}}^{1,n}(\mathbf{R}^n)}.$$

Conversely, given any $g \in \dot{\mathrm{W}}^{\frac{2}{n},n}(\Pi)$, there is an extension u to $\mathbf{R}^n$ such that

$$\|u\|_{\dot{\mathrm{W}}^{1,n}(\mathbf{R}^n)} \le C\|g\|_{\dot{\mathrm{W}}^{\frac{2}{n},n}(\Pi)}$$

(one proceeds for example by successive harmonic extensions). As a consequence, we have, when $\Gamma \subset \Pi \subset \mathbf{R}^n$,

$$A_\Gamma \simeq \sup\Big\{\int_\Gamma \vec{\varphi}\cdot\vec{t}\ :\ \vec{\varphi} \in C_c^\infty(\Pi;\mathbf{R}^2) \text{ and } \|\vec{\varphi}\|_{\dot{\mathrm{W}}^{\frac{2}{n},n}(\Pi)} \le 1\Big\}\,, \tag{3.1}$$

Since $\dot{\mathrm{W}}^{1,2}(\mathbf{R}^2;\mathbf{R}^2) \subset \dot{\mathrm{W}}^{\frac{2}{n},n}(\mathbf{R}^2;\mathbf{R}^2)$ [**15**, Theorem 2.2.3], we see, by Proposition 2.2, that

$$|\Sigma|^{\frac{1}{2}} \le CA_\Gamma\,, \tag{3.2}$$

for every planar curve $\Gamma \subset \mathbf{R}^n$, where $\Sigma \subset \Pi$ is the surface spanned by Γ. This leads to the problem

PROBLEM 7. Let $\Gamma \subset \mathbf{R}^n$ be a simple, closed, rectifiable curve, and let $|\Sigma|$ be the area of an area-minizing surface spanned by Γ. Does (3.2) hold?

On the other hand, one cannot find an upper bound on A_Γ in terms of $|\Sigma|$:

PROPOSITION 3.2. *There exists a sequence of planar surfaces Σ_k and planar curves $\Gamma_k = \partial\Sigma_k$ such that, as $k \to \infty$,*

$$A_{\Gamma_k} \to \infty$$

while

$$|\Sigma_k| \le C\,.$$

In view of (3.1), the conclusion follows from

LEMMA 3.3. *If $p > 2$ and $s = 2/p$, then there exists a sequence of planar surfaces Σ_k and planar curves $\Gamma_k = \partial\Sigma_k$, and a sequence of vector fields $\vec{\varphi}_k \in C_c^\infty(\mathbf{R}^2;\mathbf{R}^2)$ such that, as $k \to \infty$,*

$$\int_{\Gamma_k} \vec{\varphi}_k\cdot\vec{t} \to \infty$$

while

$$|\Sigma_k| \le C$$

and

$$\|\vec{\varphi}_k\|_{W^{s,p}} \le C\,.$$

PROOF. Consider $\psi \in C_c^\infty(\mathbf{R}^2)$ and $\Gamma = \partial\Sigma \subset \mathbf{R}^2$ such that

$$\int_\Gamma \psi\, t_2 \neq 0\,,$$

where $t_2 = \vec{t}\cdot\vec{e}_2$. Set

$$T_k(x_1,x_2) = (\frac{1}{k}x_1, kx_2)\,.$$

and define

$$\begin{aligned}\Gamma_k &= T_k(\Gamma)\,,\\ \Sigma_k &= T_k(\Sigma)\,,\\ \vec{\psi}_k &= k^{-s}\vec{e}_2(\psi\circ T_k^{-1})\,.\end{aligned}$$

One has

$$\int_{\Gamma_k} \vec{\psi}_k\cdot\vec{t}_k = k^{1-s}\int_\Gamma \psi\, t_2 \to \infty\,.$$

On the other hand, one has

$$|\Sigma_k| = |\Sigma|\,,$$

and

$$\|\vec{\psi}_\varepsilon\|_{\dot{W}^{s,p}} \le C\|\psi\|_{\dot{W}^{s,p}}\,.$$

□

Let $\Gamma \subset \mathbf{R}^n$ be a simple closed rectifiable curve. Recall that (see [**2, 18**])

$$\int_\Gamma \vec{\varphi}\cdot\vec{t} \le C_{s,p}|\Gamma|\,\|\vec{\varphi}\|_{\dot{W}^{s,p}}$$

for every $0 < s < n$ and $sp = n$. Recall also that (see [**7**, 4.2.10])

$$|\Gamma| \ge C|\Sigma|^{\frac{1}{2}}\,,$$

where $|\Sigma|$ is the area of an area-minimizing surface spanned by Γ. A natural question is whether for every $\vec{\varphi} \in \dot{W}^{s,p}(\mathbf{R}^n;\mathbf{R}^n)$,

$$\int_\Gamma \vec{\varphi}\cdot\vec{t} \le C_{s,p}|\Sigma|^{\frac{1}{2}}\|\vec{\varphi}\|_{\dot{W}^{s,p}}\,, \tag{3.3}$$

with $0 < s < n$ and $sp = n$. From Lemma 3.3, we deduce that (3.3) fails when $p > 2$. We ask the question whether (3.3) holds when $p = 2$ and $s = \frac{n}{2}$:

PROBLEM 8. Does (3.3) holds when $p = 2$ and $s = \frac{n}{2}$?

The final question concerning the inequality for curves is finding the optimal constant among all curves:

PROBLEM 9. When $n \ge 3$, is

$$A = \sup\bigl\{A_\Gamma\,:\,\Gamma \subset \mathbf{R}^n \text{ is a closed rectifiable curve}\bigr\}$$

attained? By which curve?

The answer to Problem 9 is open even when Γ is a planar curve of $\mathbf{R}^n$, $n \geq 3$. There are numerous variants of Problem 9. In particular, one can define

$$\tilde{A}_\Gamma = \sup\Big\{ \int_\Gamma \vec{\varphi} \cdot \vec{t} \,:\, \vec{\varphi} \in C_c^\infty(\mathbf{R}^n; \mathbf{R}^n) \text{ and } \int_{\mathbf{R}^n} |\nabla \wedge \vec{\varphi}|^n \leq 1 \Big\},$$

or one could work with different norms on $\dot{\mathrm{W}}^{1,n}(\mathbf{R}^n; \mathbf{R}^n)$, or even on $\dot{\mathrm{W}}^{s,p}(\mathbf{R}^n; \mathbf{R}^n)$, with $0 < s < n$ and $sp = n$.

3.2. Inequalities for measures. As in two dimensions, we can also consider the relaxed problem with measures. When $\vec{\mu}$ is a finite vector measure such that $\operatorname{div} \vec{\mu} = 0$, define

$$A_{\vec{\mu}} = \sup\Big\{ \int_{\mathbf{R}^2} \vec{\varphi} \cdot \vec{\mu} \,:\, \vec{\varphi} \in C_c^\infty(\mathbf{R}^n; \mathbf{R}^n) \text{ and } \|\nabla \vec{\varphi}\|_{\mathrm{L}^n} \leq 1 \Big\}.$$

As explained in [**2**] and in Remark 2.6, the optimal constants in Theorems 1 and 2 are the same, i.e.

PROPOSITION 3.4. *One has*

$$A = \sup\Big\{ A_{\vec{\mu}} \,:\, \vec{\mu} \textit{ is a measure}, \operatorname{div} \vec{\mu} = 0 \textit{ and } \|\vec{\mu}\| \leq 1 \Big\}.$$

In view of Proposition 3.4, Problem 9 can be relaxed to

PROBLEM 10. Is the supremum in Proposition 3.4 attained? By what measure?

The advantage of the formulation of Problem 10 versus Problem 9, is that while Γ was taken among closed curves, $\vec{\mu}$ is taken in the vector space of divergence-free measures. One could then hope that some kind of concentration-compactness could provide the existence of an optimizer. The divergence-free condition however is quite rigid for this kind of approach.

In two dimensions, the maximizing measures are integrals along circles. In higher dimensions, we ask

PROBLEM 11. Let $\vec{\mu}$ be a maximizing measure in Proposition 3.4 (assuming that the supremum is achived). Is $\vec{\mu}$ an integral along a curve?

A partial answer is given by

PROPOSITION 3.5. *If $\vec{\mu}$ achieves the supremum of Proposition 3.4, then $\vec{\mu}$ is an extremal point of the unit ball in $\mathcal{M}^\#(\mathbf{R}^n; \mathbf{R}^n)$.*

Proposition 3.5 means that maximizing measures are atomic, i.e., they do not have any nontrivial decomposition preserving the mass into divergence-free measures. One might be tempted to claim that atomic divergence-free measures are circulation integrals. However, as explained by Smirnov [**16**], there are divergence-free atomic measure that are not circulation integrals: Consider for $k \geq 2$ a k–dimensional torus $\mathbf{T}^k$ and a constant vector field $\vec{v}$ on $\mathbf{T}^k$ such that the equation $\dot{x} = \vec{v}$ does not have periodic solutions. If $\Phi : \mathbf{T}^k \to \mathbf{R}^n$ maps $\mathbf{T}^k$ on Θ and $\vec{v}$ on $\vec{\vartheta}$, one has that $\vec{\mu}$ defined by

$$\int_{\mathbf{R}^n} \vec{\varphi} \cdot \vec{\mu} = \int_\Theta \vec{\varphi} \cdot \vec{\vartheta}$$

is atomic but is clearly not a circulation integral.

Proof of Proposition 3.5. Since one has clearly $\|\vec{\mu}\| = 1$, assume by contradiction that $\vec{\mu} = \lambda\vec{\mu}_1 + (1-\lambda)\vec{\mu}_2$, where $\lambda \in (0,1)$, $\vec{\mu}_1, \vec{\mu}_2 \in \mathcal{M}^\#(\mathbf{R}^n; \mathbf{R}^n)$, $\|\vec{\mu}_i\| = 1$ and $\vec{\mu}_i \neq \vec{\mu}$. Let $\vec{\varphi} \in \dot{\mathrm{W}}^{1,n}(\mathbf{R}^n; \mathbf{R}^n)$ such that $\|\nabla\vec{\varphi}\|_{\mathrm{L}^n} = 1$ and

$$\int_{\mathbf{R}^n} \vec{\varphi} \cdot \vec{\mu} = A_{\vec{\mu}} \,.$$

Because $\vec{\varphi}$ is a maximizer for $A_{\vec{\mu}}$,

$$-\sum_{i=1}^n \partial_i(|\nabla\vec{\varphi}|^{n-2}\partial_i\vec{\varphi}) = \frac{\vec{\mu}}{A_{\vec{\mu}}} \neq \frac{\vec{\mu}_i}{A_{\vec{\mu}_i}} \,.$$

Therefore, $\vec{\varphi}$ cannot be a maximizer for $A_{\vec{\mu}_i}$, and

$$\int_{\mathbf{R}^n} \vec{\varphi} \cdot \vec{\mu}_i < A_{\vec{\mu}_1} \,,$$

whence

$$\begin{aligned} A = A_{\vec{\mu}} = \int_{\mathbf{R}^n} \vec{\varphi} \cdot \vec{\mu} = \lambda \int_{\mathbf{R}^n} \vec{\varphi} \cdot \vec{\mu}_1 + (1-\lambda) \int_{\mathbf{R}^n} \vec{\varphi} \cdot \vec{\mu}_2 \\ < \lambda A_{\vec{\mu}_1} + (1-\lambda) A_{\vec{\mu}_1} \leq A \,, \end{aligned}$$

which is a contradiction. □

References

1. J. Bourgain and H. Brezis, *On the equation* div $Y = f$ *and application to control of phases*, J. Amer. Math. Soc. **16** (2003), no. 2, 393–426 (electronic).
2. ______, *New estimates for the Laplacian, the div-curl, and related Hodge systems*, C. R. Math. Acad. Sci. Paris **338** (2004), no. 7, 539–543.
3. ______, *New estimates for elliptic equations and Hodge type systems*, J. Eur. Math. Soc. (JEMS) **9** (2007), no. 2, 277–315.
4. J. Bourgain, H. Brezis, and P. Mironescu, $\mathrm{H}^{1/2}$ *maps with values into the circle: minimal connections, lifting, and the Ginzburg–Landau equation*, Publ. Math. Inst. Hautes Études Sci. (2004), no. 99, 1–115.
5. H. Brezis and J. Van Schaftingen, *Boundary estimates for elliptic systems with* L^1*-data*, Calc. Var. Partial Differential Equations **30** (2007), no. 3, 369–388.
6. Andrea Cianchi, *A quantitative Sobolev inequality in* BV, J. Funct. Anal. **237** (2006), no. 2, 466–481.
7. H. Federer, *Geometric measure theory*, Springer-Verlag, New York, 1969.
8. H. Federer and W.H. Fleming, *Normal and integral currents*, (1960).
9. N. Fusco, F. Maggi, and A. Pratelli, *The sharp quantitative Sobolev inequality for functions of bounded variation*, J. Funct. Anal. **244** (2007), no. 1, 315–341.
10. P. Girão and T. Weth, *The shape of extremal functions for Poincaré-Sobolev-type inequalities in a ball*, J. Funct. Anal. **237** (2006), no. 1, 194–223.
11. M. Leckband, *On the existence of extremals for the Sobolev inequality on the ball* B^n *for functions with mean value zero*, preprint.
12. F. Maggi and Cédric V., *Balls have the worst best Sobolev inequalities*, J. Geom. Anal. **15** (2005), no. 1, 83–121.
13. V.G. Maz'ya, *Classes of domains and imbedding theorems for function spaces.*, Sov. Math., Dokl. **1** (1960), 882–885.
14. ______, *Sobolev spaces*, Springer Series in Soviet Mathematics, Springer-Verlag, Berlin, 1985.
15. T. Runst and W. Sickel, *Sobolev spaces of fractional order, Nemytskij operators, and nonlinear partial differential equations*, de Gruyter Series in Nonlinear Analysis and Applications, vol. 3, Walter de Gruyter & Co., Berlin, 1996.
16. S. K. Smirnov, *Decomposition of solenoidal vector charges into elementary solenoids, and the structure of normal one-dimensional flows*, Algebra i Analiz **5** (1993), no. 4, 206–238.
17. J. Van Schaftingen, *Estimates for* L^1*-vector fields*, C.R.Math. **339** (2004), no. 3, 181–186.

18. ______, *A simple proof of an inequality of Bourgain, Brezis and Mironescu*, C.R.Math. **338** (2004), no. 1, 23–26.
19. M. Zhu, *Sharp Poincaré-Sobolev inequalities and the shortest length of simple closed geodesics on a topological two sphere*, Commun. Contemp. Math. **6** (2004), no. 5, 781–792.

Added in proof. We have been informed by N. Fusco that he has obtained jointly with A. Pratelli partial answers to our Problems 3 and 4. More precisely, $\inf_\Omega S_\Omega = S_{B(0,1)}$ and if $\partial\Omega$ is Lipschitz, the supremum in (2.17) is achieved by the characteristic function of a subdomain (paper in preparation).

DEPARTMENT OF MATHEMATICS – HILL CENTER, RUTGERS, THE STATE UNIVERSITY OF NEW JERSEY, 110 FRELINGHUYSEN RD., PISCATAWAY, NJ 08854-8019
E-mail address: brezis@math.rutgers.edu

DÉPARTMENT DE MATHÉMATIQUE, UNIVERSITÉ CATHOLIQUE DE LOUVAIN, CHEMIN DU CYCLOTRON 2, 1348 LOUVAIN-LA-NEUVE, BELGIUM
E-mail address: Jean.VanSchaftingen@uclouvain.be

Proceedings of Symposia in Pure Mathematics
Volume **79**, 2008

Mutual absolute continuity of harmonic and surface measures for Hörmander type operators

Luca Capogna, Nicola Garofalo, and Duy-Minh Nhieu

Dedicated to Professor Maz'ya, on his 70th birthday

ABSTRACT. In this paper, we consider the Sub-Laplacian $\mathcal{L}$ which consists of a sum of squares of smooth vector fields satisfying the Hörmander finite rank condition. We study the Dirichlet problem associated with this operator on domains that satisfy certain geometric conditions. For such domains, several key results are established. These results consist of 1) A reversed Hölder inequality for the Poisson kernel 2) The $\mathcal{L}$-Harmonic measure and the surface measure (as well as the H-Perimeter measure) are mutually absolutely continuous 3) A representation (hence solvability of the Dirichlet problem) for solutions to the Dirichlet problem.

1. Introduction

In this paper we study the Dirichlet problem for the sub-Laplacian associated with a system $X = \{X_1, ..., X_m\}$ of C^∞ real vector fields in $\mathbb{R}^n$ satisfying Hörmander's finite rank condition

$$\text{rank } Lie[X_1, ..., X_m] \equiv n. \tag{1.1}$$

Throughout this paper $n \geq 3$, and X_j^* denotes the formal adjoint of X_j. The sub-Laplacian associated with X is defined by

$$\mathcal{L}u = \sum_{j=1}^{m} X_j^* X_j u . \tag{1.2}$$

A distributional solution of $\mathcal{L}u = 0$ is called $\mathcal{L}$*-harmonic*. Hörmander's hypoellipticity theorem [**H**] guarantees that every $\mathcal{L}$-harmonic function is C^∞, hence it is a classical solution of $\mathcal{L}u = 0$. We consider a bounded open set $D \subset \mathbb{R}^n$, and study

2000 *Mathematics Subject Classification.* 31C05, 35C15, 65N99.

First author supported in part by NSF Career grant DMS-0134318.
Second author supported in part by NSF Grant DMS-07010001.

the Dirichlet problem

$$\begin{cases} \mathcal{L}u = 0 & \text{in } D \ , \\ u = \phi & \text{on } \partial D \ . \end{cases} \tag{1.3}$$

Using Bony's maximum principle [**B**] one can show that for any $\phi \in C(\partial D)$ there exists a unique Perron-Wiener-Brelot solution H_ϕ^D to (1.3). We focus on the boundary regularity of the solution. In particular, we identify a class of domains, which are referred to as ADP_X domains (*admissible for the Dirichlet problem*), for which we prove the mutual absolute continuity of the $\mathcal{L}$-harmonic measure $d\omega^x$ and of the so-called horizontal perimeter measure $d\sigma_X = P_X(D;\cdot)$ on ∂D. The latter constitutes the appropriate replacement for the standard surface measure on ∂D and plays a central role in sub-Riemannian geometry. Moreover, we show that a reverse Hölder inequality holds for a suitable Poisson kernel which is naturally associated with the system X. As a consequence of such reverse Hölder inequality we then derive the solvability of (1.3) for boundary data $\phi \in L^p(\partial D, d\sigma_X)$, for $1 < p \leq \infty$. If instead the domain D belongs to the smaller class $\sigma - ADP_X$ introduced in Definition 8.10 below, we prove that $\mathcal{L}$-harmonic measure is mutually absolutely continuous with respect to the standard surface measure, and we are able to solve the Dirichlet problem for (1.2) for boundary data $\phi \in L^p(\partial D, d\sigma)$, for $1 < p \leq \infty$.

The connection between harmonic and surface measure is a central question in the study of boundary value problems for second order partial differential equations. As it is well-known a basic result of Brelot allows to solve the Dirichlet problem for the standard Laplacian when the boundary datum is in L^1 with respect to the harmonic measure. However, since the latter is difficult to pin down, it becomes important to know for what domains one can solve the Dirichlet problem when the boundary data are in some L^p space with respect to the ordinary surface measure $d\sigma$. In his ground-breaking 1977 paper [**Da1**] Dahlberg was able to settle the long standing conjecture that in a Lipschitz domain in $\mathbb{R}^n$ harmonic measure for the Laplacian and Hausdorff measure H^{n-1} restricted to the boundary are mutually absolutely continuous. One should also see the sequel paper [**Da2**] where the mutual absolute continuity was obtained as a consequence of the reverse Hölder inequality for the kernel function $k = d\omega/d\sigma$. For C^1 domains Dahlberg's result was also independently proved by Fabes, Jodeit and Rivière [**FJR**] by the method of layer potentials.

The results in this paper should be considered as a subelliptic counterpart of Dahlberg's results in [**Da1**], [**Da2**]. There are however four aspects which substantially differ from the analysis of the ordinary Laplacian, and they are all connected with the presence of the so-called characteristic points on the boundary. In order to describe these aspects we recall that given a C^1 domain $D \subset \mathbb{R}^n$, a point $x_o \in \partial D$ is called *characteristic* for the system $X = \{X_1, ..., X_m\}$ if indicating with $\boldsymbol{N}(x_o)$ a normal vector to ∂D in x_o one has

$$< \boldsymbol{N}(x_o), X_1(x_o) > = \ ... \ = < \boldsymbol{N}(x_o), X_m(x_o) > = \ 0 \ .$$

The characteristic set of D, hereafter denoted by $\Sigma = \Sigma_{D,X}$, is the collection of all characteristic points of ∂D. It is a closed subset of ∂D, and it is compact if D is bounded. We next introduce the most important prototype of a sub-Riemannian space: the Heisenberg group $\mathbb{H}^n$. This is the stratified nilpotent Lie group of step two whose underlying manifold is $\mathbb{C}^n \times \mathbb{R}$ with group law $(z,t)\circ(z't') = (z+z', t+t' - \frac{1}{2}\, Im(z\cdot\overline{z'}))$. If $x = (x_1, ..., x_n), y = (y_1, ..., y_n)$, and we identify $z = x+iy \in \mathbb{C}^n$ with the vector $(x,y) \in \mathbb{R}^{2n}$, then in the real coordinates $(x,y,t) \in \mathbb{R}^{2n+1}$ a basis for the Lie algebra of left-invariant vector fields on $\mathbb{H}^n$ is given by the vector fields

$$(1.4)\quad X_j = \frac{\partial}{\partial x_j} - \frac{y_j}{2}\frac{\partial}{\partial t}, \qquad X_{n+j} = \frac{\partial}{\partial y_j} + \frac{x_j}{2}\frac{\partial}{\partial t}, \quad j = 1, ..., n, \qquad \frac{\partial}{\partial t}\ .$$

In view of the commutation relations

$$[X_j, X_{n+k}] = \delta_{jk}\frac{\partial}{\partial t}\ , \quad j,k = 1, ..., n\ ,$$

the system $X = \{X_1, ..., X_{2n}\}$ generates the Lie algebra of $\mathbb{H}^n$. The real part of the Kohn-Spencer sub-Laplacian on $\mathbb{H}^n$ is given by

$$(1.5)\qquad \mathcal{L}_o = \sum_{j=1}^{2n} X_j^2 = \Delta_z + \frac{|z|^2}{4} D_{tt} + D_t\Big(\sum_{j=1}^{n} x_j D_{y_j} - y_j D_{x_j}\Big)\ .$$

This remarkable operator plays an ubiquitous role in several branches of mathematics and of the applied sciences. We stress that $\mathcal{L}_o$ fails to be elliptic at every point. Concerning the distinctions mentioned above we note:

1) Differently from the classical case, in the subelliptic Dirichlet problem (1.3) the Euclidean smoothness of the ground domain is of no significance from the standpoint of the intrinsic geometry near the characteristic set Σ. In this geometry even a domain with real analytic boundary looks like a cuspidal domain near one of its characteristic points. Since bounded domains typically have non-empty characteristic set it follows that the notion of "Lipschitz domain" is not as important as in the Euclidean setting, and one has to abandon it in favor of a more general one based on purely metrical properties, see [**CG1**]. With these comments in mind, in this paper we will assume that the domain D in (1.3) be NTA_X (*non-tangentially accessible* with respect to the Carnot-Carathéodory distance associated with the system X, see Definition 8.1 below) and C^∞. The former property allows us to use some fundamental results developed in [**CG1**], whereas the smoothness assumption permits to use tools from calculus away from the characteristic set. In this connection we mention that the C^∞ hypothesis guarantees, in view of the results in [**KN1**], that the Green function for (1.3) and singularity at a given point in D is smooth up to the boundary away from Σ, see Theorem 3.12 below.

2) Another striking phenomenon is that in the subelliptic Dirichlet problem nonnegative $\mathcal{L}$-harmonic functions which vanish on a portion of the boundary can do so at very different rates. The dual aspect of this phenomenon is that nonnegative $\mathcal{L}$-harmonic functions which blow up at the boundary (such as for instance the Poisson kernel) have very different rates of blow-up depending on whether the limit point is characteristic or not, see [**GV**]. This is in sharp contrast with the classical setting. It is well-known [**G**] that in a $C^{1,1}$ domain all nonnegative harmonic functions (or solutions to more general elliptic and parabolic equations) vanishing on a portion of the boundary must vanish exactly like the distance to the boundary itself. This fails miserably in the subelliptic setting because of characteristic points on the boundary. For instance, in $\mathbb{H}^n$ the so-called gauge ball

$$B = \{(z,t) \in \mathbb{H}^n \mid |z|^4 + 16t^2 < 1\}$$

is a real analytic domain with two isolated characteristic points $P^{\pm} = (0, \pm\frac{1}{4})$. With $\mathcal{L}_o$ defined by (1.5), the function $u(z,t) = t + \frac{1}{4}$ is a nonnegative $\mathcal{L}_o$-harmonic function in B which along the t-axis vanishes at the (characteristic) boundary point $P^- = (0, -\frac{1}{4})$ as the square of the Carnot-Carathéodory distance to P^-. On the other hand, the function $u(z,t) = x_1 + 1$ is a nonnegative $\mathcal{L}_o$-harmonic function in B which along the x_1-axis vanishes at the (non-characteristic) boundary point $P_1 = (-e_1, 0)$, where $e_1 = (1, 0, ..., 0) \in \mathbb{R}^{2n}$, like the distance to P_1. Thus, there is not one single rate of vanishing for nonnegative $\mathcal{L}_o$-harmonic functions in smooth domains in $\mathbb{H}^n$! Despite this negative phenomenon in [**CG1**] two of us proved that in a NTA_X domain all nonnegative $\mathcal{L}$-harmonic functions vanishing on a portion of the boundary (characteristic or not) must do so at the same rate. This result, known as the *comparison theorem*, plays a fundamental role in the present paper. Returning to the above example of the gauge ball $B \subset \mathbb{H}^n$, the comparison theorem implies in particular that all nonnegative solutions of $\mathcal{L}_o u = 0$ which vanish in a boundary neighborhood of the point $P^- = (0, -\frac{1}{4})$, must vanish non-tangentially like the square of the distance to the boundary (and not linearly like in the classical case)!

3) The third aspect which we want to emphasize is closely connected with the discussion in 1) and leads us to introduce the third main assumption in the present paper paper. In [**J1**], [**J2**] D. Jerison studied the Dirichlet problem (1.3) near characteristic points for $\mathcal{L}_o$. He proved in [**J1**] that for a C^∞ domain $D \subset \mathbb{H}^n$ if the datum ϕ belongs to a Folland-Stein Hölder class Γ^β, then H^D_ϕ is in $\Gamma^\alpha(\overline{D})$, for some α depending on β and on the domain D. It was also shown in [**J1**] that, given any $\alpha \in (0,1)$ there exists $M = M(\alpha) > 0$ for which the real analytic domain

$$\Omega_M = \{(z,t) \in \mathbb{H}^n \mid t > -M|z|^2\} ,$$

admits a $\mathcal{L}_o$-harmonic function u such that $u = 0$ on $\partial\Omega_M$ and which belongs exactly to the Hölder class Γ^α (in the sense that it is not any smoother) in any neighborhood of the (characteristic) boundary point $e = (0,0)$. Once again, this example shows that, despite the (Euclidean) smoothness of the domain and of the boundary datum, near a characteristic point the domain appears quite non-smooth with respect to the intrinsic geometry of the vector fields $X_1, ..., X_{2n}$. In fact, since the paraboloid Ω is a scale invariant region with respect to the non-isotropic group dilations $(z,t) \to (\lambda z, \lambda^2 t)$, the smooth domain Ω_M should be thought of as a non-convex cone from the point of view of the intrinsic geometry of $\mathcal{L}_o$ (for a discussion of Jerison's example see section 4). This suggests that by imposing a condition similar to the classical Poincaré tangent outer sphere [**P**] one should be able to rule out Jerison's negative example and possibly control the intrinsic gradient XG of the Green function near the characteristic set. This intuition was proved successful in the papers [**LU1**], [**CGN1**], which were respectively concerned with the Heisenberg group and with Carnot groups of Heisenberg type. In this paper we generalize this idea and prove the boundedness of the Poisson kernel in a neighborhood of the boundary under the hypothesis that the domain D in (1.3) satisfy what we call a *tangent outer X-ball condition*. It is worth emphasizing that the X-balls in our definition are not metric balls, but instead they are the (smooth) level sets of the fundamental solution of the sub-Laplacian $\mathcal{L}$. The metric balls are not smooth (see [**CG1**]) and therefore it would not be possible to have a notion of tangency based on these sets.

4) In Dahlberg's mentioned theorem on the mutual absolute continuity between harmonic and surface measure in a Lipschitz domain $D \subset \mathbb{R}^n$ there is one important property which, although confined to the background, plays a central role. If we denote by $\sigma = H^{n-1}|_{\partial D}$ the surface measure on the boundary, then there exists constants $\alpha, \beta > 0$ depending on n and on the Lipschitz character of D such that

$$\alpha\, r^{n-1} \ \le\ \sigma(\partial D \cap B(x,r)) \ \le\ \beta\, r^{n-1} ,$$

for any $x \in \partial D$ and any $r > 0$. A property like this is referred to as the 1-Ahlfors regularity of σ, and thanks to it surface measure is the natural measure on ∂D. Things are quite different in the subelliptic Dirichlet problem. Consider in fact the gauge ball B as in 2), with its two (isolated) characteristic points $P^{\pm} = (0, \pm\frac{1}{4})$ of ∂B. Simple calculations show that denoting by $B(P^{\pm}, r)$ a gauge ball centered at one of the points $P^{\pm}$ with radius r, then one has for small $r > 0$

$$\sigma(\partial B \cap B(P^{\pm}, r)) \ \cong\ r^{Q-2} , \tag{1.6}$$

where $Q = 2n + 2$ is the so-called homogeneous dimension of $\mathbb{H}^n$ relative to the non-isotropic dilations $(z,t) \to (\lambda z, \lambda^2 t)$ associated with the grading of the Lie algebra of $\mathbb{H}^n$. The latter equation shows that at the characteristic points $P^{\pm}$ surface measure becomes quite singular and it does not scale correctly with respect to the non-isotropic group dilations. The appropriate "surface measure" in sub-Riemannian geometry is instead the so-called horizontal perimeter $P_X(D;\cdot)$ introduced in [**CDG2**] which on surface metric balls is defined in the following way

$$\sigma_X(\partial D \cap B_d(x,r)) \ \overset{def}{=}\ P_X(D; B_d(x,r)) .$$

To motivate such appropriateness we recall that it was proved in [**DGN1**], [**DGN2**] that for every C^2 bounded domain $D \subset \mathbb{H}^n$ one has for every $x \in \partial D$ and every $0 < r < R_o(D)$

$$\alpha\, r^{Q-1} \ \le\ \sigma_X(\partial D \cap B_d(x,r)) \ \le\ \beta\, r^{Q-1} .$$

Now it was also shown in these papers that the inequality in the right-hand side alone suffices to establish the existence of the traces of Sobolev functions on the boundary. Remarkably, as we prove in Theorem 1.3 below, such a one-sided Ahlfors property also suffices to establish the mutual absolute continuity of $\mathcal{L}$-harmonic and horizontal perimeter measure. Such property will constitute the last basic assumption of our results, to which we finally turn. We need to introduce the relevant class of domains.

DEFINITION 1.1. *Given a system $X = \{X_1, ..., X_m\}$ of smooth vector fields satisfying* (1.1), *we say that a connected bounded open set $D \subset \mathbb{R}^n$ is* admissible for the Dirichlet problem (1.3) *with respect to the system X, or simply ADP_X, if:*

i) D is of class C^∞;

ii) D is non-tangentially accessible (NTA_X) with respect to the Carnot-Caratheodory metric associated to the system $\{X_1, ..., X_m\}$ (see Definition 8.1);

iii) D satisfies a uniform tangent outer X-ball condition (see Definition 6.2);

iv) The horizontal perimeter measure is upper 1-Ahlfors regular. This means that there exist $A, R_o > 0$ depending on X and D such that for every $x \in \partial D$ and $0 < r < R_o$ one has

$$\sigma_X(\partial D \cap B_d(x,r)) \ \le\ A\, \frac{|B_d(x,r)|}{r} .$$

The constants appearing in *iv*) and in Definitions 6.2 and 8.1 will be referred to as the ADP_X-parameters of D. We introduce next a central character in this play, the subelliptic Poisson kernel of D. In fact, we define two such functions, each one playing a different role. Let $G(x,y) = G_D(x,y) = G(y,x)$ indicate the Green function for the sub-Laplacian (1.2) and for an ADP_X domain D [1]. By Hörmander's theorem [**H**] and the results in [**KN1**], see Theorem 3.12 below, for any fixed $x \in D$ the function $y \to G(x,y)$ is C^∞ up to the boundary in a suitably small neighborhood of any non-characteristic point $y_o \in \partial D$. Let $\boldsymbol{\nu}(y)$ indicate the outer unit normal in $y \in \partial D$. At every point $y \in \partial D$ we denote by $\boldsymbol{N}^X(y)$ the vector defined by

$$\boldsymbol{N}^X(y) \;=\; (<\boldsymbol{\nu}(y), X_1(y)>, ..., <\boldsymbol{\nu}(y), X_m(y)>) \ .$$

We also set

$$W(y) \;=\; |\boldsymbol{N}^X(y)| \;=\; \sqrt{<\boldsymbol{\nu}(y), X_1(y)>^2 + ... + <\boldsymbol{\nu}(y), X_m(y)>^2} \ .$$

We note explicitly that it was proved in [**CDG2**] that on ∂D

$$d\sigma_X \;=\; W\, d\sigma \ .$$

Denoting with Σ the characteristic set of D, we remark that the vector $\boldsymbol{N}^X(y) = 0$ if and only if $y \in \Sigma$. For $y \in \partial D \setminus \Sigma$ we define the *horizontal Gauss map* at y by letting

$$\boldsymbol{\nu}^X(y) \;=\; \frac{\boldsymbol{N}^X(y)}{|\boldsymbol{N}^X(y)|} \ .$$

DEFINITION 1.2. *Given a C^∞ bounded open set $D \subset \mathbb{R}^n$, for every $(x,y) \in D \times (\partial D \setminus \Sigma)$ we define* the subelliptic Poisson kernels *as follows*

$$P(x,y) \;=<XG(x,y), \boldsymbol{N}^X(y)> \ , \quad K(x,y) \;=\; \frac{P(x,y)}{W(y)} \;=<XG(x,y), \boldsymbol{\nu}^X(y)> \ .$$

We emphasize here that the reason for which in the definition of $P(x,y)$ and $K(x,y)$ we restrict y to $\partial D \setminus \Sigma$ is that, as we have explained in 3) above (see also section 4), the horizontal gradient $XG(x,y)$ may not be defined at points of Σ. Since as we have observed the function W vanishes on Σ, it should be clear that the function $K(x,y)$ is more singular then $P(x,y)$ at the characteristic points. However, such additional singularity is balanced by the fact that the density W of the measure σ_X with respect to surface measure vanishes at the characteristic points. As a consequence, $K(x,y)$ is the appropriate subelliptic Poisson kernel with respect to the intrinsic measure σ_X, whereas $P(x,y)$ is more naturally attached to the "wrong measure" σ.

Hereafter, for $x \in \partial D$ it will be convenient to indicate with $\Delta(x,r) = \partial D \cap B_d(x,r)$, the boundary metric ball centered at x with radius $r > 0$. The first main result in this paper is contained the following theorem.

THEOREM 1.3. *Let $D \subset \mathbb{R}^n$ be a ADP_X domain. For every $p > 1$ and any fixed $x_1 \in D$ there exist positive constants C, R_1, depending on p, M, R_o, x_1, and on the ADP_X parameters, such that for $x_o \in \partial D$ and $0 < r < R_1$ one has*

$$\left(\frac{1}{\sigma_X(\Delta(x_o,r))}\int_{\Delta(x_o,r)} K(x_1,y)^p d\sigma_X(y)\right)^{\frac{1}{p}} \le C\,\frac{1}{\sigma_X(\Delta(x_o,r))}\int_{\Delta(x_o,r)} K(x_1,y) d\sigma_X(y) \ .$$

[1] In [**B**] it was proved that any bounded open set admits a Green function

Moreover, the measures $d\omega^{x_1}$ *and* $d\sigma_X$ *are mutually absolutely continuous.*

By combining Theorem 1.3 with the results if [**CG1**] we can solve the Dirichlet problem for boundary data in L^p with respect to the perimeter measure $d\sigma_X$. To state the relevant results we need to introduce a definition. Given D as in Theorem 1.3, for any $y \in \partial D$ and $\alpha > 0$ a nontangential region at y is defined by

$$\Gamma_\alpha(y) \ = \ \{x \in D \,|\, d(x,y) \leq (1+\alpha) d(x,\partial D)\} \ .$$

Given a function $u \in C(D)$, the α-nontangential maximal function of u at y is defined by

$$N_\alpha(u)(y) \ = \ \sup_{x\in\Gamma_\alpha(y)} |u(x)| \ .$$

THEOREM 1.4. *Let* $D \subset \mathbb{R}^n$ *be a* ADP_X *domain. For every* $p > 1$ *there exists a constant* $C > 0$ *depending on* D, X *and* p *such that if* $f \in L^p(\partial D, d\sigma_X)$, *then*

$$H_f^D(x) \ = \ \int_{\partial D} f(y)\ K(x,y)\ d\sigma_X(y) \ ,$$

and

$$\|N_\alpha(H_f^D)\|_{L^p(\partial D, d\sigma_X)} \ \leq \ C\ \|f\|_{L^p(\partial D, d\sigma_X)} \ .$$

Furthermore, H_f^D *converges nontangentially* σ_X*-a.e. to* f *on* ∂D.

Theorems 1.3 and 1.4 constitute appropriate sub-elliptic versions of Dahlberg's mentioned results in [**Da1**], [**Da2**]. These theorems generalize those in [**CGN2**] relative to Carnot groups of Heisenberg type. We mention at this point that, as we prove in Theorem 8.3 below, for any $C^{1,1}$ domain $D \subset \mathbb{R}^n$ which is NTA_X the horizontal perimeter measure is lower 1-Ahlfors (this is a basic consequence of the isoperimetric inequality in [**GN1**]). Combining this result with the assumption *iv*) in Definition 1.1, we conclude that for any ADP_X domain the measure σ_X is 1-Ahlfors. In particular, σ_X is also doubling, see Corollary 8.4. This information plays a crucial role in the proof of Theorem 1.4.

On the other hand, even if the ordinary surface measure σ is the "wrong one" in the subelliptic Dirichlet problem, it would still be highly desirable to know if there exist situations in which (1.3) can be solved for boundary data in some L^p with respect to $d\sigma$. To address this question in Definition 8.10 we introduce the class of $\sigma - ADP_X$ domains. The latter differs from that of ADP_X domains for the fact that the assumption *iv*) is replaced by the following *balanced-degeneracy* assumption on σ: there exist $B, R_o > 0$ depending on X and D such that for every $x_o \in \partial D$ and $0 < r < R_o$ one has

$$\left(\max_{y\in\Delta(x_o,r)} W(y) \right) \sigma(\Delta(x_o,r)) \ \leq \ B\ \frac{|B_d(x_o,r)|}{r} \ .$$

As we have previously observed surface measure becomes singular near a characteristic point. On the other hand, the angle function W vanishes, thus balancing the singularities of σ. For $\sigma - ADP_X$-domains we obtain the following two results which respectively establish the mutual absolute continuity of $\mathcal{L}$-harmonic and surface measure $d\sigma$, and the solvability of the Dirichlet problem with data in $L^p(\partial D, d\sigma)$.

THEOREM 1.5. *Let $D \subset \mathbb{R}^n$ be a $\sigma - ADP_X$ domain. Fix $x_1 \in D$. For every $p > 1$ there exist positive constants C, R_1, depending on p, M, R_o, x_1, and on the $\sigma - ADP_X$ parameters, such that for every $y \in \partial D$ and $0 < r < R_1$ one has*

$$\left(\frac{1}{\sigma(\Delta(x_o,r))}\int_{\Delta(x_o,r)} P(x_1,y)^p d\sigma(y)\right)^{\frac{1}{p}} \leq C\,\frac{1}{\sigma(\Delta(x_o,r))}\int_{\Delta(x_o,r)} P(x_1,y) d\sigma(y)\,.$$

Moreover, the measures $d\omega^x$ and $d\sigma$ are mutually absolutely continuous.

We mention explicitly that a basic consequence of Theorem 1.5 is that the standard surface measure on the boundary of a $\sigma - ADP_X$ domain is doubling.

THEOREM 1.6. *Let $D \subset \mathbb{R}^n$ be a $\sigma - ADP_X$ domain. For every $p > 1$ there exists a constant $C > 0$ depending on D, X and p such that if $f \in L^p(\partial D, d\sigma)$, then*

$$H_f^D(x) \ = \ \int_{\partial D} f(y)\ P(x,y)\ d\sigma(y)\ ,$$

and

$$\|N_\alpha(H_f^D)\|_{L^p(\partial D,d\sigma)} \ \leq \ C\ \|f\|_{L^p(\partial D,d\sigma)}.$$

Furthermore, H_f^D converges nontangentially σ-a.e. to f on ∂D.

Concerning Theorems 1.3, 1.4, 1.5 and 1.6 we mention that large classes of domains to which they apply were found in [**CGN2**], but one should also see [**LU1**] for domains satisfying assumption *iii)* in Definition 1.1. The discussion of these examples is taken up in section 9.

In closing we briefly describe the organization of the paper. In section 2 we collect some known results on Carnot-Carathéodory metrics which are needed in the paper. In section 3 we discuss some known results on the subelliptic Dirichlet problem which constitute the potential theoretic backbone of the paper. In section 4 we discuss Jerison's mentioned example.

Section 5 is devoted to proving some new interior a priori estimates of Cauchy-Schauder type. Such estimates are obtained by means of a family of subelliptic mollifiers which were introduced by Danielli and two of us in [**CDG1**], see also [**CDG2**]. The main results are Theorems 5.1, 5.5, and Corollary 5.3. We feel that, besides being instrumental to the present paper, these results will prove quite useful in future research on the subject.

In section 6 we use the interior estimates in Theorem 5.1 to prove that if a domain satisfies a uniform outer tangent X-ball condition, then the horizontal gradient of the Green function G is bounded up to the boundary, hence, in particular, near Σ, see Theorem 6.6. The proof of such result rests in an essential way on the linear growth estimate provided by Theorem 6.3. Another crucial ingredient is Lemma 6.1 which allows a delicate control of some ad-hoc subelliptic barriers. In the final part of the section we show that, by requesting the uniform outer X-ball condition only in a neighborhood of the characteristic set Σ, we are still able to obtain the boundedness of the horizontal gradient of G up to the characteristic set, although we now loose the uniformity in the estimates, see Theorem 6.9, 6.10 and Corollary 6.11.

In section 7 we establish a Poisson type representation formula for domains which satisfy the uniform outer X-ball condition in a neighborhood of the characteristic set. This result generalizes a similar Poisson type formula in the Heisenberg group $\mathbb{H}^n$ obtained by Lanconelli and Uguzzoni in [**LU1**], and extended in [**CGN2**] to Carnot groups of Heisenberg type. If generically the Green function of a smooth domain had bounded horizontal gradient *up to the characteristic set*, then such Poisson formula would follow in an elementary way from integration by parts. As we previously stressed, however, things are not so simple and the boundedness of XG fails in general near the characteristic set. However, when $D \subset \mathbb{R}^n$ satisfies the uniform outer X-ball condition in a neighborhood of the characteristic set, then combining Theorem 6.6 with the estimate

$$K(x,y) \ \leq \ |XG(x,y)| \ , \quad x \in D, y \in \partial D \ ,$$

see (7.7), we prove the boundedness of the Poisson kernel $y \to K(x,y)$ on ∂D. The main result in section 7 is Theorem 7.10. This representation formula with the estimates of the Green function in sections 5 and 6 lead to a priori estimates in L^p for the solution to (1.3) when the datum $\phi \in C(\partial D)$. Solvability of (1.3) with data in Lebesgue classes requires, however, a much deeper analysis.

The first observation is that the outer ball condition alone does not guarantee the development of a rich potential theory. For instance, it may not be possible to find: a) Good nontangential regions of approach to the boundary from within the domain; b) Appropriate interior Harnack chains of nontangential balls. This is where the basic results on NTA_X domains from [**CG1**] enter the picture. In the opening of section 8 we recall the definition of NTA_X-domain along with those results from [**CG1**] which constitute the foundations of the present study. Using these results we establish Theorem 8.9. The remaining part of the section is devoted to proving Theorems 1.3, 1.4, 1.5 and 1.6.

Finally, section 9 is devoted to the discussion of examples of ADP_X and $\sigma - ADP_X$ domains and of some open problems.

2. Preliminaries

In $\mathbb{R}^n$, with $n \geq 3$, we consider a system $X = \{X_1, ..., X_m\}$ of C^∞ vector fields satisfying Hörmander's finite rank condition (1.1). A piecewise C^1 curve $\gamma : [0,T] \to \mathbb{R}^n$ is called *sub-unitary* [**FP**] if whenever $\gamma'(t)$ exists one has for every $\xi \in \mathbb{R}^n$

$$< \gamma'(t), \xi >^2 \ \leq \ \sum_{j=1}^{m} < X_j(\gamma(t)), \xi >^2 \ .$$

We note explicitly that the above inequality forces $\gamma'(t)$ to belong to the span of $\{X_1(\gamma(t)), ..., X_m(\gamma(t))\}$. The sub-unit length of γ is by definition $l_s(\gamma) = T$. Given $x, y \in \mathbb{R}^n$, denote by $\mathcal{S}_\Omega(x,y)$ the collection of all sub-unitary $\gamma : [0,T] \to \Omega$ which join x to y. The accessibility theorem of Chow and Rashevsky, [**Ra**], [**Ch**], states that, given a connected open set $\Omega \subset \mathbb{R}^n$, for every $x, y \in \Omega$ there exists $\gamma \in \mathcal{S}_\Omega(x,y)$. As a consequence, if we pose

$$d_\Omega(x,y) \ = \ \inf \{l_s(\gamma) \mid \gamma \in \mathcal{S}_\Omega(x,y)\},$$

we obtain a distance on Ω, called the *Carnot-Carathéodory distance on* Ω, associated with the system X. When $\Omega = \mathbb{R}^n$, we write $d(x,y)$ instead of $d_{\mathbb{R}^n}(x,y)$. It is clear that $d(x,y) \le d_\Omega(x,y)$, $x,y\in\Omega$, for every connected open set $\Omega\subset\mathbb{R}^n$. In [**NSW**] it was proved that for every connected $\Omega\subset\subset\mathbb{R}^n$ there exist $C,\epsilon>0$ such that

$$C\,|x-y|\ \le d_\Omega(x,y)\ \le C^{-1}\,|x-y|^\epsilon,\qquad x,y\in\Omega. \tag{2.1}$$

This gives $d(x,y)\ \le C^{-1}|x-y|^\epsilon$, $x,y\in\Omega$, and therefore

$$i:(\mathbb{R}^n,|\cdot|)\to(\mathbb{R}^n,d)\qquad \textit{is continuous.}$$

It is easy to see that also the continuity of the opposite inclusion holds [**GN1**], hence the metric and the Euclidean topology are compatible.

For $x\in\mathbb{R}^n$ and $r>0$, we let $B_d(x,r) = \{y\in\mathbb{R}^n \mid d(x,y)<r\}$. The basic properties of these balls were established by Nagel, Stein and Wainger in their seminal paper [**NSW**]. Denote by $Y_1,...,Y_l$ the collection of the X_j's and of those commutators which are needed to generate $\mathbb{R}^n$. A formal "degree" is assigned to each Y_i, namely the corresponding order of the commutator. If $I=(i_1,...,i_n), 1\le i_j\le l$ is a n-tuple of integers, following [**NSW**] we let $d(I)=\sum_{j=1}^n deg(Y_{i_j})$, and $a_I(x)=\det\,(Y_{i_1},...,Y_{i_n})$. The *Nagel-Stein-Wainger polynomial* is defined by

$$\Lambda(x,r)\ =\ \sum_I\ |a_I(x)|\ r^{d(I)},\qquad r>0. \tag{2.2}$$

For a given bounded open set $U\subset\mathbb{R}^n$, we let

$$Q\ =\ \sup\,\{d(I)\mid\ |a_I(x)|\neq 0, x\in U\},\qquad Q(x)\ =\ \inf\,\{d(I)\mid |a_I(x)|\neq 0\}, \tag{2.3}$$

and notice that $n\le Q(x)\le Q$. The numbers Q and $Q(x)$ are respectively called the *local homogeneous dimension* of U and the homogeneous dimension at x with respect to the system X.

THEOREM 2.1 ([**NSW**]). *For every bounded set $U\subset\mathbb{R}^n$, there exist constants $C,R_o>0$ such that, for any $x\in U$, and $0<r\le R_o$,*

$$C\ \Lambda(x,r)\ \le\ |B_d(x,r)|\ \le\ C^{-1}\ \Lambda(x,r). \tag{2.4}$$

As a consequence, one has with $C_1=2^Q$

$$|B_d(x,2r)|\ \le\ C_1\ |B_d(x,r)|\qquad \textit{for every}\quad x\in U\quad \textit{and}\quad 0<r\le R_o. \tag{2.5}$$

The numbers C_1,R_o in (2.5) will be referred to as the *characteristic local parameters* of U. Because of (2.2), if we let

$$E(x,r)\ =\ \frac{\Lambda(x,r)}{r^2}, \tag{2.6}$$

then the function $r\to E(x,r)$ is strictly increasing. We denote by $F(x,\cdot)$ the inverse function of $E(x,\cdot)$, so that $F(x,E(x,r))=r$. Let $\Gamma(x,y)=\Gamma(y,x)$ be the positive fundamental solution of the sub-Laplacian

$$\mathcal{L}\ =\ \sum_{j=1}^m X_j^*X_j,$$

and consider its level sets

$$\Omega(x,r)\ =\ \left\{y\in\mathbb{R}^n\mid\Gamma(x,y)>\frac{1}{r}\right\}.$$

The following definition plays a key role in this paper.

DEFINITION 2.2. *For every $x \in \mathbb{R}^n$, and $r > 0$, the set*

$$B(x,r) = \left\{ y \in \mathbb{R}^n \mid \Gamma(x,y) > \frac{1}{E(x,r)} \right\}$$

*will be called the X-*ball*, centered at x with radius r.*

We note explicitly that

$$B(x,r) = \Omega(x, E(x,r)), \qquad \text{and that} \qquad \Omega(x,r) = B(x, F(x,r)).$$

One of the main geometric properties of the X-balls, is that they are equivalent to the Carnot-Carathéodory balls. To see this, we recall the following important result, established independently in [**NSW**], [**SC**]. Hereafter, the notation $Xu = (X_1 u, ..., X_m u)$ indicates the sub-gradient of a function u, whereas $|Xu| = (\sum_{j=1}^m (X_j u)^2)^{\frac{1}{2}}$ will denote its length.

THEOREM 2.3. *Given a bounded set $U \subset \mathbb{R}^n$, there exists R_o, depending on U and on X, such that for $x \in U$, $0 < d(x,y) \le R_o$, one has for $s \in \mathbb{N} \cup \{0\}$, and for some constant $C = C(U, X, s) > 0$*

$$(2.7) \qquad |X_{j_1} X_{j_2} ... X_{j_s} \Gamma(x,y)| \le C^{-1} \frac{d(x,y)^{2-s}}{|B_d(x, d(x,y))|},$$

$$\Gamma(x,y) \ge C \frac{d(x,y)^2}{|B_d(x, d(x,y))|}.$$

In the first inequality in (2.7), one has $j_i \in \{1, ..., m\}$ for $i = 1, ..., s$, and X_{j_i} is allowed to act on either x or y.

In view of (2.5), (2.7), it is now easy to recognize that, given a bounded set $U \subset \mathbb{R}^n$, there exists $a > 1$, depending on U and X, such that

$$(2.8) \qquad B_d(x, a^{-1} r) \subset B(x,r) \subset B_d(x, ar),$$

for $x \in U, 0 < r \le R_o$. We observe that, as a consequence of (2.4), and of (2.7), one has

$$(2.9) \qquad C\, d(x,y) \le F\left(x, \frac{1}{\Gamma(x,y)}\right) \le C^{-1}\, d(x,y),$$

for all $x \in U, 0 < d(x,y) \le R_o$.

We observe that for a Carnot group $\mathbb{G}$ of step k, if $\mathfrak{g} = V_1 \oplus ... \oplus V_k$ is a stratification of the Lie algebra of $\mathbb{G}$, then one has $\Lambda(x,r) = const\ r^Q$, for every $x \in \mathbb{G}$ and every $r > 0$, with $Q = \sum_{j=1}^k j\ dim V_j$, the homogeneous dimension of the group $\mathbb{G}$. In this case $Q(x) \equiv Q$.

In the sequel the following properties of a Carnot-Carathéodory space will be useful.

PROPOSITION 2.4. *$(\mathbb{R}^n, d)$ is locally compact. Furthermore, for any bounded set $U \subset \mathbb{R}^n$ there exists $R_o = R_o(U) > 0$ such that the closed balls $\bar{B}(x_o, R)$, with $x_o \in U$ and $0 < R < R_o$, are compact.*

REMARK 2.5. *Compactness of balls of large radii may fail in general, see* [**GN1**]. *However, there are important cases in which Proposition 2.4 holds globally, in the sense that one can take U to coincide with the whole ambient space and $R_o = \infty$. One example is that of Carnot groups. Another interesting case is that when the vector fields X_j have coefficients which are globally Lipschitz, see* [**GN1**], [**GN2**]. *Henceforth, for any given bounded set $U \subset \mathbb{R}^n$ we will always assume that the local parameter R_o has been chosen so to accommodate Proposition 2.4.*

3. The Dirichlet problem

In what follows, given a system $X = \{X_1, ..., X_m\}$ of C^∞ vector fields in $\mathbb{R}^n$ satisfying (1.1), and an open set $D \subset \mathbb{R}^n$, for $1 \le p \le \infty$ we denote by $\mathcal{L}^{1,p}(D)$ the Banach space $\{f \in L^p(D) \mid X_j f \in L^p(D), j = 1, ..., m\}$ endowed with its natural norm

$$||f||_{\mathcal{L}^{1,p}(D)} = ||f||_{L^p(D)} + \sum_{j=1}^m ||X_j f||_{L^p(D)} .$$

The local space $\mathcal{L}^{1,p}_{loc}(D)$ has the usual meaning, whereas for $1 \le p < \infty$ the space $\mathcal{L}^{1,p}_0(D)$ is defined as the closure of $C_0^\infty(D)$ in the norm of $\mathcal{L}^{1,p}(D)$. A function $u \in \mathcal{L}^{1,2}_{loc}(D)$ is called *harmonic* in D if for any $\phi \in C_0^\infty(D)$ one has

$$\int_D \sum_{j=1}^m X_j u X_j \phi \, dx = 0 ,$$

i.e., a harmonic function is a weak solution to the equation $\mathcal{L}u = \sum_{j=1}^m X_j^* X_j u = 0$. By Hörmander's hypoellipticity theorem [**H**], if u is harmonic in D, then $u \in C^\infty(D)$. Given a bounded open set $D \subset \mathbb{R}^n$, and a function $\phi \in \mathcal{L}^{1,2}(D)$, the Dirichlet problem consists in finding $u \in \mathcal{L}^{1,2}_{loc}(D)$ such that

$$\begin{cases} \mathcal{L}u = 0 & \text{in } D , \\ u - \phi \in \mathcal{L}^{1,2}_0(D) . \end{cases} \tag{3.1}$$

By adapting classical arguments, see for instance [**GT**], one can show that there exists a unique solution $u \in \mathcal{L}^{1,2}(D)$ to (3.1). If we assume, in addition, that $\phi \in C(D)$, in general we cannot say that the function u takes up the boundary value ϕ with continuity. A Wiener type criterion for sub-Laplacians was proved in [**NS**]. Subsequently, using the Wiener series in [**NS**], Citti obtained in [**Ci**] an estimate of the modulus of continuity at the boundary of the solution of (3.1). In [**D**] an integral Wiener type estimate at the boundary was established for a general class of quasilinear equations having $p-$growth in the sub-gradient. Since such estimate is particularly convenient for the applications, we next state it for the special case $p = 2$ of linear equations.

THEOREM 3.1. *Let $\phi \in \mathcal{L}^{1,2}(D) \cap C(\overline{D})$. Consider the solution u to (3.1). There exist $C = C(X) > 0$, and $R_o = R_o(D, X) > 0$, such that given $x_o \in \partial D$, and $0 < r < R < R_o/3$, one has*

$$osc\ \{u, D \cap B_d(x_o, r)\} \le osc\ \{\phi, \partial D \cap \overline{B}_d(x_o, 2R)\}$$
$$+ \ osc\ (\phi, \partial D) \ \exp\left\{ - C \int_r^R \left[\frac{cap_X\ (D^c \cap \overline{B}_d(x_o, t), B_d(x_o, 2t))}{cap_X\ (\overline{B}_d(x_o, t), B_d(x_o, 2t))} \right] \frac{dt}{t} \right\}.$$

In Theorem 3.1, given a condenser (K,Ω), we have denoted by $cap_X(K,\Omega)$ its Dirichlet capacity with respect to the subelliptic energy $\mathcal{E}_X(u) = \int_\Omega |Xu|^2 dx$ associated with the system $X = \{X_1, ..., X_m\}$. For the relevant properties of such capacity we refer the reader to [**D**], [**CDG4**]. A point $x_o \in \partial D$ is called regular if, for any $\phi \in \mathcal{L}^{1,2}(D) \cap C(\overline{D})$, one has

$$\lim_{x \to x_o} u(x) = \phi(x_o) . \tag{3.2}$$

If every $x_o \in \partial D$ is regular, we say that D is regular. Similarly to the classical case, in the study of the Dirichlet problem an important notion is that of generalized, or Perron-Wiener-Brelot (PWB) solution to (3.1). For operators of Hörmander type the construction of a PWB solution was carried in the pioneering work of Bony [**B**], where the author also proved that sub-Laplacians satisfy an elliptic type strong maximum principle. We state next one of the main results in [**B**] in a form which is suitable for our purposes.

THEOREM 3.2. *Let $D \subset \mathbb{R}^n$ be a connected, bounded open set, and $\phi \in C(\partial D)$. There exists a unique harmonic function H_ϕ^D which solves* (1.3) *in the sense of Perron-Wiener-Brelot. Moreover, H_ϕ^D satisfies*

$$\sup_D |H_\phi^D| \le \sup_{\partial D} |\phi| . \tag{3.3}$$

Theorem 3.2 allows to define the harmonic measure $d\omega^x$ for D evaluated at $x \in D$ as the unique probability measure on ∂D such that for every $\phi \in C(\partial D)$

$$H_\phi^D(x) = \int_{\partial D} \phi(y)\, d\omega^x(y), \qquad x \in D.$$

A uniform Harnack inequality was established, independently, by several authors, see [**X**], [**CGL**], [**L**]: If u is $\mathcal{L}-$harmonic in $D \subset \mathbb{R}^n$ and non-negative then there exists $C, a > 0$ such that for each ball $B(x, ar) \subset D$ one has

$$\sup_{B(x,r)} u \le C \inf_{B(x,r)} u. \tag{3.4}$$

Using such Harnack principle one sees that for any $x, y \in D$, the measures $d\omega^x$ and $d\omega^y$ are mutually absolutely continuous. For the basic properties of the harmonic measure we refer the reader to the paper [**CG1**]. Here, it is important to recall that, thanks to the results in [**B**], [**CG1**], the following result of Brelot type holds.

THEOREM 3.3. *A function ϕ is resolutive if and only if $\phi \in L^1(\partial D, d\omega^x)$, for one (and therefore for all) $x \in D$.*

The following definition is particularly important for its potential-theoretic implications. In the sequel, given a condenser (K,Ω), we denote by $cap(K,\Omega)$ the sub-elliptic capacity of K with respect to Ω, see [**D**].

DEFINITION 3.4. *An open set $D \subset \mathbb{R}^n$ is called* thin *at $x_o \in \partial D$, if*

$$\liminf_{r \to 0} \frac{cap_X(D^c \cap \overline{B}_d(x_o, r), B_d(x_o, 2r))}{cap_X(\overline{B}_d(x_o, r), B_d(x_o, 2r))} > 0. \tag{3.5}$$

THEOREM 3.5. *If a bounded open set $D \subset \mathbb{R}^n$ is thin at $x_o \in \partial D$, then x_o is regular for the Dirichlet problem.*

Proof. If D is thin at $x_o \in \partial D$, then

$$\int_0^R \left[\frac{cap_X(D^c \cap \overline{B}_d(x_o,t), B_d(x_o,2t))}{cap_X(\overline{B}_d(x_o,t), B_d(x_o,2t))}\right] \frac{dt}{t} = \infty .$$

Thanks to Theorem 3.1, the divergence of the above integral implies for $0 < r < R/3$

$$osc\ \{u, D \cap B_d(x_o, r)\} \le osc\ \{\phi, \partial D \cap \overline{B}_d(x_o, 2R)\}\ .$$

Letting $R \to 0$ we infer the regularity of x_o.

□

A useful, and frequently used, sufficient condition for regularity is provided by the following definition.

DEFINITION 3.6. *An open set* $\Omega \subset \mathbb{R}^n$ *is said to have* positive density *at* $x_o \in \partial\Omega$, *if one has*

$$\liminf_{r\to 0} \frac{|\Omega \cap B_d(x_o,r)|}{|B_d(x_o,r)|} > 0.$$

PROPOSITION 3.7. *If* D^c *has positive density at* x_o, *then* D *is thin at* x_o.

Proof. We recall the Poincaré inequality

$$\int_\Omega |\phi|^2\, dx \le C\ (diam(\Omega))^2 \int_\Omega |X\phi|^2\, dx\ ,$$

valid for any bounded open set $\Omega \subset \mathbb{R}^n$, and any $\phi \in C_o^1(\Omega)$, where $diam(\Omega)$ represents the diameter of Ω with respect to the distance $d(x,y)$, and $C = C(\Omega, X) > 0$. From the latter, we obtain

$$\frac{cap_X(D^c \cap \overline{B}_d(x_o,r), B_d(x_o,2r))}{cap_X(\overline{B}_d(x_o,r), B_d(x_o,2r))} \ge \frac{C}{r^2}\ \frac{|D^c \cap \overline{B}_d(x_o,r)|}{cap_X(\overline{B}_d(x_o,r), B_d(x_o,2r))}. \tag{3.6}$$

Now the capacitary estimates in [**D**], [**CDG3**] give

$$C\ r^{Q-2} \le cap_X(\overline{B}_d(x_o,r), B_d(x_o,2r)) \le C^{-1}\ r^{Q-2}\ ,$$

for some constant $C = C(\Omega, X) > 0$. Using these estimates in (3.6) we find

$$\frac{cap_X(D^c \cap \overline{B}_d(x_o,r), B_d(x_o,2r))}{cap_X(\overline{B}_d(x_o,r), B_d(x_o,2r))} \ge C^*\ \frac{|D^c \cap \overline{B}_d(x_o,r)|}{|B_d(x_o,r)|},$$

where $C^* = C^*(\Omega, X) > 0$. The latter inequality proves that if D^c has positive density at x_o, then D is thin at the same point.

□

A basic example of a class of regular domains for the Dirichlet problem is provided by the (Euclidean) $C^{1,1}$ domains in a Carnot group of step $r = 2$. It was proved in [**CG1**] that such domains possess a scale invariant region of non-tangential approach at every boundary point, hence they satisfy the positive density condition in Proposition 3.7. Thus, in particular, every such domain is regular for the Dirichlet problem for any fixed sub-Laplacian on the group. Another important example is provided by the non-tangentially accessible domains (NTA domains, henceforth) studied in [**CG1**]. Such domains constitute a generalization of those introduced by Jerison and Kenig in the Euclidean setting [**JK**], see Section 8.

DEFINITION 3.8. *Let $D \subset \mathbb{R}^n$ be a bounded open set. For $0 < \alpha \le 1$, the class $\Gamma_d^{0,\alpha}(D)$ is defined as the collection of all $f \in C(D) \cap L^\infty(D)$, such that*

$$\sup_{x,y\in D, x\neq y} \frac{|f(x) - f(y)|}{d(x,y)^\alpha} < \infty.$$

We endow $\Gamma_d^{0,\alpha}(D)$ with the norm

$$||f||_{\Gamma_d^{0,\alpha}(D)} = ||f||_{L^\infty(D)} + \sup_{x,y\in D, x\neq y} \frac{|f(x) - f(y)|}{d(x,y)^\alpha}.$$

The meaning of the symbol $\Gamma_{loc}^{0,\alpha}(D)$ is the obvious one, that is, $f \in \Gamma_{loc}^{0,\alpha}(D)$ if, for every $\omega \subset\subset D$, one has $f \in \Gamma_d^{0,\alpha}(\omega)$. If $F \subset \mathbb{R}^n$ denotes a bounded closed set, by $f \in \Gamma_d^{0,\alpha}(F)$ we mean that f coincides on the set F with a function $g \in \Gamma_d^{0,\alpha}(D)$, where D is a bounded open set containing F. The Lipschitz class $\Gamma_d^{0,1}(D)$ has a special interest, due to its connection with the Sobolev space $\mathcal{L}^{1,\infty}(D)$. In fact, we have the following theorem of Rademacher-Stepanov type, established in [**GN1**], which will be needed in the proof of Lemma 6.1.

THEOREM 3.9. ***(i)** Given a bounded open set $U \subset \mathbb{R}^n$, there exist $R_o = R_o(U,X) > 0$, and $C = C(U,X) > 0$, such that if $f \in \mathcal{L}^{1,\infty}(B_d(x_o,3R))$, with $x_o \in U$ and $0 < R < R_o$, then f can be modified on a set of dx-measure zero in $\bar{B}_d = \bar{B}_d(x_o,R)$, so as to satisfy for every $x,y \in \bar{B}_d(x_o,R)$*

$$|f(x) - f(y)| \le C\, d(x,y)\, ||f||_{\mathcal{L}^{1,\infty}(B_d(x_o,3R))}.$$

If, furthermore, $f \in C^\infty(B_d(x_o,3R))$, then in the right-hand side of the previous inequality one can replace the term $||f||_{\mathcal{L}^{1,\infty}(B_d(x_o,3R))}$ with $||Xf||_{L^\infty(B_d(x_o,3R))}$.
***(ii)** Vice-versa, let $D \subset \mathbb{R}^n$ be an open set such that $\sup_{x,y\in D} d(x,y) < \infty$. If $f \in \Gamma_d^{0,1}(D)$, then $f \in \mathcal{L}^{1,\infty}(D)$.*

We note explicitly that part (i) of Theorem 3.9 asserts that every function $f \in \mathcal{L}^{1,\infty}(B_d(x_o,3R))$ has a representative which is Lipschitz continuous in $B_d(x_o,R)$ with respect to the metric d, i.e., continuing to denote with f such representative, one has $f \in \Gamma^{0,1}(B_d(x_o,R))$. Part (ii) was also obtained independently in [**FSS**]. The following result was established in [**D**].

THEOREM 3.10. *Let $D \subset \mathbb{R}^n$ be a bounded open set which is thin at every $x_o \in \partial D$. If $\phi \in \Gamma^{0,\beta}(\overline{D})$, for some $\beta \in (0,1)$, then there exists $\alpha \in (0,1)$, with $\alpha = \alpha(D,X,\beta)$, such that*

$$\sup_{x,y\in \overline{D}, x\neq y} \frac{|H_\phi^D(x) - H_\phi^D(y)|}{d(x,y)^\alpha} < \infty.$$

Given a bounded open set $D \subset \mathbb{R}^n$, consider the positive Green function $G(x,y) = G(y,x)$ for $\mathcal{L}$ and D, constructed in [**B**]. For every fixed $x \in D$, one can represent $G(x,\cdot)$ as follows

$$G(x,\cdot) = \Gamma(x,\cdot) - h_x, \qquad \text{where} \quad h_x = H^D_{\Gamma(x,\cdot)}. \tag{3.7}$$

Since, by Hörmander's hypo-ellipticity theorem, $\Gamma(x,\cdot) \in C^\infty(\mathbb{R}^n \setminus \{x\})$, we conclude that, if D is thin at every $x_o \in \partial D$, then there exists $\alpha \in (0,1)$ such that, for every $\epsilon > 0$, one has

$$G(x,\cdot) \in \Gamma_d^{0,\alpha}(\bar{D} \setminus B(x_o,\epsilon)). \tag{3.8}$$

We close this section with recalling an important consequence of the results of Kohn and Nirenberg [**KN1**] (see Theorem 4), and of Derridj [**De1**], [**De2**], about smoothness in the Dirichlet problem at non-characteristic points. We recall the following definition.

DEFINITION 3.11. *Given a C^1 domain $D \subset \mathbb{R}^n$, a point $x_o \in \partial D$ is called characteristic for the system $X = \{X_1, ..., X_m\}$ if for $j = 1, ..., m$ one has*

$$< X_j(x_o), \boldsymbol{N}(x_o) > \; = \; 0 \; ,$$

where $\boldsymbol{N}(x_o)$ indicates a normal vector to ∂D at x_o. We indicate with $\Sigma = \Sigma_{D,X}$ the collection of all characteristic points. The set Σ is a closed subset of ∂D.

THEOREM 3.12. *Let $D \subset \mathbb{R}^n$ be a C^∞ domain which is regular for (1.3). Consider the harmonic function H^D_ϕ, with $\phi \in C^\infty(\partial D)$. If $x_o \in \partial D$ is a non-characteristic point for $\mathcal{L}$, then there exists an open neighborhood V of x_o such that $H^D_\phi \in C^\infty(\overline{D} \cap V)$.*

REMARK 3.13. *We stress that, as we indicated in the introduction, the conclusion of Theorem 3.12 fails in general at characteristic points. In fact, it fails so completely that even if the domain D and the boundary datum ϕ are real analytic, in general the solution of the Dirichlet problem H^D_ϕ may be not better that Hölder continuous up to the boundary, see Theorem 3.10. An example of such negative phenomenon in the Heisenberg group $\mathbb{H}^n$ was constructed by Jerison in* [**J1**]. *The next section is dedicated to it. For a related example concerning the heat equation see* [**KN2**].

4. The example of D. Jerison

Consider the Heisenberg group (discussed in the introduction) with its left-invariant generators (1.4) of its Lie algebra. Recall that $\mathbb{H}^n$ is equipped with the non-isotropic dilations

$$\delta_\lambda(z,t) \;=\; (\lambda z, \lambda^2 t) \; ,$$

whose infinitesimal generator is given by the vector field

$$\mathcal{Z} \;=\; \sum_{i=1}^n \left(x_i \frac{\partial}{\partial x_i} + y_i \frac{\partial}{\partial y_i} \right) \; + \; 2 \frac{\partial}{\partial t} \; .$$

We say that a function $u : \mathbb{H}^n \to \mathbb{R}$ is *homogeneous of degree* $\alpha \in \mathbb{R}$ if for every $(z,t) \in \mathbb{H}^n$ and every $\lambda > 0$ one has

$$u(\delta_\lambda(z,t)) \;=\; \lambda^\alpha \, u(z,t) \; .$$

One easily checks that if $u \in C^1(\mathbb{H}^n)$ then u is homogeneous of degree α if and only if

$$\mathcal{Z} u \;=\; \alpha \, u \; .$$

We also consider the vector field

$$\Theta \;=\; \sum_{i=1}^n \left(x_i \frac{\partial}{\partial y_i} - y_i \frac{\partial}{\partial x_i} \right) \; , \tag{4.1}$$

which is the infinitesimal generator of the one-parameter group of transformations $R_\theta : \mathbb{H}^n \to \mathbb{H}^n$, $\theta \in \mathbb{R}$, given by

$$R_\theta(z,t) \;=\; (e^{i\theta} z, t), \quad z = x + iy \in \mathbb{C}^n \; .$$

Notice that when $n = 1$, then in the z-plane R_θ is simply a counterclockwise rotation of angle θ, and in such case in the standard polar coordinates (r, θ) in $\mathbb{C}$ we have

$$\Theta = \frac{\partial}{\partial \theta} .$$

In the sequel we will tacitly identify $z = x + iy \simeq (x, y) \in \mathbb{R}^{2n}$, and so $|z| = \sqrt{|x|^2 + |y|^2}$. We note explicitly that in the real coordinates (x, y, t) the real part of the Kohn-Spencer sub-Laplacian (1.5) on $\mathbb{H}^n$ is given by

$$\mathcal{L}_o = \sum_{i=1}^{2n} X_i^2 = \Delta_z + \frac{|z|^2}{4} \frac{\partial^2}{\partial t^2} + \frac{\partial}{\partial t} \Theta .$$

It is easy to see that if u has *cylindrical symmetry*, i.e., if

$$u(z, t) = f(|z|, t) ,$$

then

$$\Theta u \equiv 0 .$$

Consider the gauge in $\mathbb{H}^n$

$$N = N(z, t) = (|z|^4 + 16t^2)^{1/4} .$$

The following formula follows from an explicit calculation

$$\psi \stackrel{def}{=} |\nabla^H N|^2 = \frac{|z|^2}{N^2} , \quad \Delta_H N = \frac{Q-1}{N} , \tag{4.2}$$

where

$$Q = 2n + 2$$

is the so-called *homogeneous dimension* associated with the non-isotropic dilations $\{\delta_\lambda\}_{\lambda > 0}$. As a consequence of (4.2), if $u = f \circ N$ for some function $f : [0, \infty) \to \mathbb{R}$, then one has the beautiful formula

$$\mathcal{L}_o u = \psi \left[f''(N) + \frac{Q-1}{N} f'(N) \right] . \tag{4.3}$$

Since $f(t) = t^{2-Q}$ satisfies the ode in the right-hand side of (4.3) one can show that a fundamental solution of $-\mathcal{L}_o$ with pole at the group identity $e = (0, 0) \in \mathbb{H}^n$ is given by

$$\Gamma(z, t) = \frac{C_Q}{N(z, t)^{Q-2}} , \quad (z, t) \neq e , \tag{4.4}$$

where $C_Q > 0$ needs to be appropriately chosen.

The following example due to D. Jerison [**J1**] shows that, even when the domain and the boundary data are real analytic, in general the solution to the subelliptic Dirichlet problem (1.3) may not be any better than $\Gamma^{0,\alpha}$ near a characteristic boundary point. Consider the domain

$$\Omega_M = \{(z, t) \in \mathbb{H}^n \mid t > M|z|^2\} , \qquad M \in \mathbb{R} .$$

Since Ω_M is scale invariant with respect to $\{\delta_\lambda\}_{\lambda > 0}$ we might think of Ω_M as the analogue of a *convex cone* ($M \geq 0$), or a *concave cone* ($M < 0$). Introduce the variable

$$\tau = \tau(z, t) = \frac{4t}{N^2} , \quad (z, t) \neq e .$$

It is clear that τ is homogeneous of degree zero and therefore

$$\mathcal{Z}\tau \ = \ 0 \ .$$

Moreover, with Θ as in (4.1) , one easily checks that

$$\Theta\tau \ = \ 0 \ .$$

It is important to observe the level sets $\{\tau = \gamma\}$ are constituted by the t-axis when $\gamma = 1$, and by the paraboloids

$$t \ = \ \frac{\gamma}{4\sqrt{1-\gamma^2}}|z|^2 \ ,$$

if $|\gamma| < 1$. Furthermore, the function τ takes the constant value

$$\tau \ = \ \frac{4M}{\sqrt{1+16M^2}} \ ,$$

on $\partial\Omega_M$. We now consider a function of the form

$$v \ = \ v(z,t) \ = \ N^\alpha \ u(\tau) \ , \tag{4.5}$$

where the number $\alpha > 0$ will be appropriately chosen later on. One has the following result whose verification we leave to the reader.

PROPOSITION 4.1. *For any $\alpha > 0$ one has*

$$\begin{aligned}\mathcal{L}_o v \ &= \ 4\psi N^{\alpha-2}\left\{(1-\tau^2)u''(\tau) - \frac{Q}{2}\tau u'(\tau) + \frac{\alpha(\alpha+Q-2)}{4}u(\tau)\right\} \\ &= \ 4\psi N^{\alpha-2}\left\{(1-\tau^2)u''(\tau) - (n+1)\tau u'(\tau) + \frac{\alpha(\alpha+2n)}{4}u(\tau)\right\}.\end{aligned}$$

Using Proposition 4.1 we can now construct a positive harmonic function in Ω_M which vanishes on the boundary (this function is a Green function with pole at an interior point).

PROPOSITION 4.2. *For any $\alpha \in (0,1]$ there exists a number $M = M(\alpha) < 0$ such that the nonconvex cone Ω_M admits a positive solution of $\mathcal{L}_o v = 0$ of the form (4.5) which vanishes on $\partial\Omega_M$.*

Proof. From Proposition 4.1 we see that if v of the form (4.5) has to solve the equation $\mathcal{L}_o v = 0$, then the function u must be a solution of the Jacobi equation

$$(1-\tau^2)u''(\tau) - (n+1)\tau u'(\tau) + \frac{\alpha(\alpha+2n)}{4}u(\tau) \ = \ 0 \ . \tag{4.6}$$

As we have observed the level $\{\tau = 1\}$ is degenerate and corresponds to the t-axis $\{z = 0\}$. One solution of (4.6) which is smooth as $\tau \to 1$ (remember, the t-axis is inside Ω_M and thus we want our function v to be smooth around the t-axis since by hypoellipticity v has to be in $C^\infty(\Omega_M)$) is the hypergeometric function

$$g_\alpha(\tau) \ = \ F\left(-\frac{\alpha}{2}, n+\frac{\alpha}{2}; \frac{n+1}{2}; \frac{1-\tau}{2}\right) \ .$$

When $0 < \alpha < 2$ one can verify that

$$g_\alpha(1) = 1 \ , \quad \text{and that } \ g_\alpha(\tau) \to -\infty \text{ as } \tau \to -1^+ \ .$$

Therefore, g_α has a zero τ_α. One can check (see Erdelyi, Magnus, Oberhettinger and Tricomi, vol.1, p.110 (14)), that as $\alpha \to 0^+$, then $\tau_\alpha \to -1^+$. We infer that for $\alpha > 0$ sufficiently close to 0 there exists $-1 < \tau_\alpha < 0$ such that

$$g_\alpha(\tau_\alpha) \;=\; 0\ .$$

If we choose

$$M \;=\; M(\alpha) \;=\; \frac{\tau_\alpha}{\sqrt{1-\tau_\alpha^2}} \;<\; 0\ ,$$

then it is clear that on $\partial\Omega_M$ we have $\tau \equiv \tau_\alpha$, and therefore the function v of the form (6.10), with $u(\tau) = g_\alpha(\tau)$, has the property of being harmonic and nonnegative in Ω, and furthermore on $\partial\Omega_M$ we have that $v = N^\alpha g_\alpha(\tau_\alpha) \equiv 0$. This completes the proof.

□

Since α belongs the interval $(0,1)$, then it is clear that $v \;=\; N^\alpha(z,t)g_\alpha(\tau)$ belongs at most to the Folland-Stein Hölder class $\Gamma^{0,\alpha}(\overline{\Omega}_M)$, but is not any better than metrically Hölder in any neighborhood of $e \;=\; (0,0)$. What produces this negative phenomenon is the fact that the point $e \in \partial\Omega_M$ is characteristic for Ω_M.

5. Subelliptic interior Schauder estimates

In this section we establish some basic interior Schauder type estimates that, besides from playing an important role in the sequel, also have an obvious independent interest. Such estimates are tailored on the intrinsic geometry of the system $X = \{X_1, ..., X_m\}$, and are obtained by means of a family of sub-elliptic mollifiers which were introduced in [**CDG1**], see also [**CDG2**]. For convenience, we state the relevant results in terms of the X-balls $B(x,r)$ introduced in Definition 2.2, but we stress that, thanks to (2.8), we could have as well employed the metric balls $B_d(x,r)$. Since in this paper our focus is on $\mathcal{L}$-harmonic functions, we do not explicitly treat the non-homogeneous equation $\mathcal{L}u = f$ with a non-zero right-hand side. Estimates for solutions of the latter equation can, however, be obtained by relatively simple modifications of the arguments in the homogeneous case.

The following is the main result in this section.

THEOREM 5.1. *Let $D \subset \mathbb{R}^n$ be a bounded open set and suppose that u is harmonic in D. There exists $R_o > 0$, depending on D and X, such that for every $x \in D$ and $0 < r \le R_o$ for which $\overline{B}(x,r) \subset D$, one has for any $s \in \mathbb{N}$*

$$|X_{j_1}X_{j_2}...X_{j_s}u(x)| \;\le\; \frac{C}{r^s} \max_{\overline{B}(x,r)} |u|,$$

for some constant $C = C(D,X,s) > 0$. In the above estimate, for every $i = 1, ..., s$, the index j_i runs in the set $\{1, ..., m\}$.

REMARK 5.2. *We emphasize that Theorem 5.1 cannot be established similarly to its classical ancestor for harmonic functions, where one uses the mean-value theorem coupled with the trivial observation that any derivative of a harmonic function is harmonic. In the present non-commutative setting, derivatives of harmonic functions are no longer harmonic!*

A useful consequence of Theorem 5.1 is the following.

COROLLARY 5.3. *Let $D \subset \mathbb{R}^n$ be a bounded, open set and suppose that u is a non-negative harmonic function in D. There exists $R_o > 0$, depending on D and X, such that for any $x \in D$ and $0 < r \le R_o$ for which $\overline{B}(x, 2r) \subset D$, one has for any given $s \in \mathbb{N}$*

$$|X_{j_1} X_{j_2} ... X_{j_s} u(x)| \ \le \ \frac{C}{r^s} \ u(x),$$

for some $C = C(D, X, s) > 0$.

Proof. Since $u \ge 0$, we immediately obtain the result from Theorem 5.1 and from the Harnack inequality (3.4). □

To prove Theorem 5.1, we use the family of sub-elliptic mollifiers introduced in [**CDG1**], see also [**CDG2**]. Choose a nonnegative function $f \in C_o^\infty(\mathbb{R})$, with *supp* $f \subset [1, 2]$, and such that $\int_{\mathbb{R}} f(s)ds = 1$, and let $f_R(s) = R^{-1} f(R^{-1}s)$. We define the kernel

$$K_R(x, y) \ = \ f_R\left(\frac{1}{\Gamma(x, y)}\right) \ \frac{|X_y \Gamma(x, y)|^2}{\Gamma(x, y)^2}.$$

Given a function $u \in L^1_{loc}(\mathbb{R}^n)$, following [**CDG1**] we define the *subelliptic mollifier* of u by

$$J_R \ u(x) \ = \ \int_{\mathbb{R}^n} u(y) \ K_R(x, y) \ dy, \qquad R > 0. \tag{5.1}$$

We note that for any fixed $x \in \mathbb{R}^n$,

$$supp \ K_R(x, \cdot) \ \subset \ \Omega(x, 2R) \ \setminus \ \Omega(x, R). \tag{5.2}$$

One of the important features of $J_R \ u$ is expressed by the following theorem.

THEOREM 5.4. *Let $D \subset \mathbb{R}^n$ be open and suppose that u is harmonic in D. There exists $R_o > 0$, depending on D and X, such that for any $x \in D$, and every $0 < R \le R_o$ for which $\overline{\Omega}(x, 2R) \subset D$, one has*

$$u(x) \ = \ J_R \ u(x).$$

Proof. Let u and $\Omega(x, R)$ be as in the statement of the theorem. We obtain for $\psi \in C^\infty(D)$ and $0 < t \le R$, see [**CGL**],

(5.3)

$$\psi(x) \ = \ \int_{\partial\Omega(x,t)} \psi(y) \ \frac{|X_y \Gamma(x, y)|^2}{|D\Gamma(x, y)|} \ dH_{n-1}(y) \ + \ \int_{\Omega(x,t)} \mathcal{L}\psi(y) \ [\Gamma(x, y) - \frac{1}{t}] \ dy.$$

Taking $\psi = u$ in (5.3), we find

$$u(x) \ = \ \int_{\partial\Omega(x,t)} u(y) \ \frac{|X_y \Gamma(x, y)|^2}{|D\Gamma(x, y)|} \ dH_{n-1}(y). \tag{5.4}$$

We are now going to use (5.4) to complete the proof. The idea is to start from the definition of $J_R \ u(x)$, and then use Federer co-area formula [**Fe**]. One finds

$$J_R \ u(x) \ = \ \int_0^\infty \ f_R(t) \ \left[\int_{\partial\Omega(x,t)} \ u(y) \ \frac{|X_y \Gamma(x, y)|^2}{|D\Gamma(x, y)|} \ dH_{n-1}(y)\right] \ dt.$$

The previous equality, (5.4), and the fact that $\int_{\mathbb{R}} f_R(s)ds = 1$, imply the conclusion.

□

The essence of our main a priori estimate is contained in the following theorem.

THEOREM 5.5. *Fix a bounded set $U \subset \mathbb{R}^n$. There exists a constant $R_o > 0$, depending only on U and on the system X, such that for any $u \in L^1_{loc}(\mathbb{R}^n)$, $x \in U, 0 < R \leq R_o$, and $s \in \mathbb{N}$ one has for some $C = C(U, X, s) > 0$,*

$$|X_{j_1} X_{j_2} ... X_{j_s} \; J_R \; u(x)| \;\leq\; \frac{C}{R} \frac{1}{F(x,R)^{2+s}} \int_{\Omega(x,R)} |u(y)| \; dy.$$

Proof. We first consider the case $s = 1$. From (2.7), and from the support property (5.2) of $K_R(x, \cdot)$, we can differentiate under the integral sign in (5.1), to obtain

$$|X \; J_R \; u(x)| \;\leq\; \int_{B(x,2R)} |u(y)| \; |X_x K_R(x,y)| \; dy.$$

By the definition of $K_R(x,y)$ it is easy to recognize that the components of its sub-gradient $X_x K_R(x,y)$ are estimated as follows

$$\begin{aligned} |X_j \; K_R(x,y)| \;&\leq\; C \; R^{-2} \; |X\Gamma(x,y)|^3 \; \Gamma(x,y)^{-4} \\ &+\; C \; R^{-1} \; \Gamma(x,y)^{-2} \sum_{k=1}^{m} |X_j X_k \Gamma(x,y)| \; |X_k \Gamma(x,y)| \\ &+\; C \; R^{-1} \; |X\Gamma(x,y)|^3 \; \Gamma(x,y)^{-3} \\ &=\; I^1_R(x,y) \;+\; I^2_R(x,y) \;+\; I^3_R(x,y). \end{aligned}$$

To control the three terms in the right-hand side of the above inequality, we use the size estimates (2.7), along with the observation that, due to the fact that on the support of $K_R(x,\cdot)$ one has

$$\frac{1}{2R} \;<\; \Gamma(x,y) \;\leq\; \frac{1}{R},$$

then Theorem 2.3, and (2.9), give for all $x \in U, 0 < R \leq R_o$, and $y \in \Omega(x,2R) \setminus \Omega(x,R)$

$$C \;\leq\; \frac{d(x,y)}{F(x,R)} \;\leq\; C^{-1}. \tag{5.5}$$

Using (2.7), (5.5), one obtains that for $i = 1,2,3$

$$\sup_{y \in \Omega(x,2R) \setminus \Omega(x,R)} |I^i_R(x,y)| \leq \frac{C}{RF(x,R)^3}$$

for any $x \in U$, provided that $0 < R \leq R_o$. This completes the proof in the case $s = 1$. The case $s \geq 2$ is handled recursively by similar considerations based on Theorem 2.3, and we omit details. It may be helpful for the interested reader to note that Theorem 2.3 implies

$$|X_{j_1} X_{j_2} ... X_{j_s} \; \Gamma(x,y)| \;\leq\; C \; d(x,y)^{-s} \; \Gamma(x,y),$$

so that by (5.5) one obtains

$$\sup_{y \in \Omega(x,2R) \setminus \Omega(x,R)} |X_{j_1} X_{j_2} ... X_{j_s} \; \Gamma(x,y)| \;\leq\; \frac{C}{RF(x,R)^s}. \tag{5.6}$$

□

We are finally in a position to prove Theorem 5.1.

Proof of Theorem 5.1. We observe explicitly that the assumption states that with $R = E(x,r)/2$, then $\overline{\Omega}(x,2R) = \overline{B}(x,r) \subset D$. By Theorem 5.4, and by (5.6), we find

$$\begin{aligned} |X_{j_1}X_{j_2}...X_{j_s}\, u(x)| &= |X_{j_1}X_{j_2}...X_{j_s}(J_R\, u)(x)| \\ &\le \frac{C}{RF(x,R)^{2+s}} \int_{\Omega(x,R)} |u(y)|\, dy \le C\, \frac{|\Omega(x,R)|}{RF(x,R)^{2+s}} \max_{\Omega(x,R)} |u|. \end{aligned}$$

To complete the proof we only need to observe that $\Omega(x,R) = B(x,r)$, and that, thanks to Theorem 2.3, (2.9), one has

$$\frac{C}{r^s} \le \frac{|B(x,r)|}{RF(x,R)^{2+s}} \le \frac{C^{-1}}{r^s}.$$

□

REMARK 5.6. *We observe explicitly that when $\mathbb{G}$ is a Carnot group with $X_1, ..., X_m$ being a fixed basis of the horizontal layer of its Lie algebra, then the constant C in Theorem 5.1 and Corollary 5.3 can be taken independent of the open set D.*

6. Lipschitz boundary estimates for the Green function

In this section we establish some basic estimates at the boundary for the Green function associated to a sub-Laplacian, when the relevant domain possesses an appropriate analogue of the outer tangent sphere condition introduced by Poincaré in his famous paper [**P**]. Analyzing the domain Ω_M in Remark 3.13 one recognizes that Jerison's negative example fails to possess a tangent outer gauge sphere at its characteristic point. We thus conjectured that by imposing such condition one should be able to establish the boundedness near the boundary of the horizontal gradient of the Green function (see for instance [**G**] for the classical case of elliptic or parabolic operators). This intuition has proved correct. In their paper [**LU1**] Lanconelli and Uguzzoni have proved the boundedness of the Poisson kernel for a domain satisfying the outer sphere condition in the Heisenberg group, whereas in [**CGN2**] a similar result was successfully combined with those in [**CG1**] to obtain a complete solution of the Dirichlet problem for a large class of domains in groups of Heisenberg type.

The objective of this section is to generalize the cited results in [**LU1**] and [**CGN2**] to the Poisson kernel associated with an operator of Hörmander type. Namely, *if $D \subset \mathbb{R}^n$ is a bounded domain satisfying an intrinsic uniform outer sphere condition with respect to a system $X = \{X_1, ..., X_m\}$ satisfying* (1.1), *and having Green function $G(x,y) = G_D(x,y)$, if we fix the singularity at an interior point $x_1 \in D$, then the function $x \to |XG(x_1,x)|$, which is well defined for $x \in D \setminus \{x_1\}$, belongs to L^∞ in a neighborhood of ∂D.* The exact statements are contained in Corollaries 6.7 and 6.11.

We emphasize that, in view of Theorem 3.12, the main novelty of this result lies in that we do allow the boundary point to be characteristic. As it will be clear from the analysis below, the passage from the group setting to the case of general sub-Laplacians involves overcoming various non-trivial obstacles.

Our first task is to obtain a growth estimate at the boundary for harmonic functions which vanish on a distinguished portion of the latter. We show that any such function grows at most linearly with respect to the Carnot-Carathéodory distance associated to the system X. The proof of this result ultimately relies on delicate estimates of a suitable barrier whose construction is inspired to that given by Poincaré [**P**], see also [**G**]. We begin with a lemma which plays a crucial role in the sequel. The function $\Gamma(x,y) = \Gamma(y,x)$ denotes the positive fundamental solution of the sub-Laplacian associated with the system X, see Section 2.

LEMMA 6.1. *For any bounded set $U \subset \mathbb{R}^n$, there exist $R_o, C > 0$, depending on U and X, such that for every $x_o \in U$, and $x, y \in \mathbb{R}^n \setminus B_d(x_o, r)$, one has*

$$|\Gamma(x_o,x) - \Gamma(x_o,y)| \le C \frac{r}{|B_d(x_o,r)|} d(x,y).$$

Proof. We distinguish two cases: (i) $d(x,y) > \theta r$; (ii) $d(x,y) \le \theta r$. Here, $\theta \in (0,1)$ is to be suitably chosen. Case (i) is easy. Using (2.7) we find

$$\begin{aligned}|\Gamma(x_o,x) &- \Gamma(x_o,y)| \le \Gamma(x_o,x) + \Gamma(x_o,y)\\ &\le C\left[\frac{d(x_o,x)^2}{|B_d(x_o,d(x_o,x))|} + \frac{d(x_o,y)^2}{|B_d(x_o,d(x_o,y))|}\right]\\ &\le C\left\{\frac{1}{E(x_o,d(x_o,x))} + \frac{1}{E(x_o,d(x_o,y))}\right\} \le C\,\frac{1}{E(x_o,r)} < \frac{C}{\theta}\,\frac{r}{|B_d(x_o,r)|}\,d(x,y).\end{aligned}$$

We next consider case (ii), and let $\rho = d(x,y) \le \theta r$. Let γ be a sub-unitary curve joining x to y with length $l_s(\gamma) \le \rho + \rho/16$. The existence of such a curve is guaranteed by the definition of $d(x,y)$. Consider the function $g(P) \overset{def}{=} d(x,P) - d(y,P)$. By the continuity of $g : \{\gamma\} \to \mathbb{R}$, and by the intermediate value theorem, we can find $P \in \{\gamma\}$ such that $d(x,P) = d(y,P)$. For such point P, we must have

$$d(x,P) = d(y,P) \le \frac{3}{4}\rho. \tag{6.1}$$

If (6.1) were not true, we would in fact have

$$\frac{3}{4}\rho + \frac{3}{4}\rho < d(x,P) + d(y,P) \le l_s(\gamma) \le \rho + \frac{\rho}{16},$$

which is a contradiction. From (6.1) we conclude that $x, y \in B_d(P, 3\rho/4)$. Moreover,

$$d(P,x_o) \ge d(x,x_o) - d(x,P) \ge r - \frac{3}{4}\rho \ge \left(1 - \frac{3}{4}\theta\right) r.$$

We claim that

$$B_d(P, \frac{9}{4}\rho) \subset \mathbb{R}^n \setminus B_d(x_o, r/2), \tag{6.2}$$

provided that we take $\theta = \frac{1}{6}$. In fact, let $z \in B_d(P, \frac{9}{4}\rho)$, then

$$d(z,x_o) \ge d(P,x_o) - d(z,P) \ge \left(1 - \frac{3}{4}\theta\right) r - \frac{9}{4}\,\theta\, r = (1 - \frac{3}{4}\theta - \frac{9}{4}\theta)\, r = \frac{r}{2}.$$

This proves (6.2). The above considerations allow to apply Theorem 3.9, which, keeping in mind that $\Gamma(x_o,\cdot) \in C^\infty(B_d(P, \frac{9}{4}\rho))$, presently gives

$$|\Gamma(x_o,x) - \Gamma(x_o,y)| \le C\rho \sup_{\xi \in B_d(P,\frac{9}{4}\rho)} |X\Gamma(x_o,\xi)|. \tag{6.3}$$

Using (2.7) we obtain for $\xi \in B_d(P, \frac{9}{4}\rho)$

$$|X\Gamma(x_o,\xi)| \le C \frac{1}{d(x_o,\xi)\, E(x_o,d(x_o,\xi))},$$

where $t \to E(x_o, t)$ is the function introduced in (2.6). Since by (6.2) we have $d(x_o, \xi) \geq r/2$, the latter estimate, combined with the increasingness of $E(x_o, \cdot)$, leads to the conclusion

$$\sup_{\xi \in B_d(P, \frac{9}{4}\rho)} |X\Gamma(x_o, \xi)| \leq C \frac{1}{rE(x_o, r)}.$$

Inserting this inequality in (6.3), and observing that $\frac{1}{rE(x_o,r)} \leq C \frac{r}{|B_d(x_o,r)|}$, we find

$$|\Gamma(x_o, x) - \Gamma(x_o, y)| \leq C \frac{r}{|B_d(x_o, r)|} d(x, y).$$

This completes the proof of the lemma.

□

The following definition plays a crucial role in the subsequent development.

DEFINITION 6.2. *A domain $D \subset \mathbb{R}^n$ is said to possess an* outer X-ball tangent *at $x_o \in \partial D$ if for some $r > 0$ there exists a X-ball $B(x_1, r)$ such that:*

$$x_o \in \partial B(x_1, r), \qquad B(x_1, r) \cap D = \varnothing. \tag{6.4}$$

We say that D possesses the uniform outer X-ball *if one can find $R_o > 0$ such that for every $x_o \in \partial D$, and any $0 < r < R_o$, there exists a X-ball $B(x_1, r)$ for which (6.4) holds.*

Some comments are in order. First, it should be clear from (2.8) that the existence of an outer X-ball tangent at $x_o \in \partial D$ implies that D is thin at x_o (the reverse implication is not necessarily true). Therefore, thanks to Theorem 3.5, x_o is regular for the Dirichlet problem. Secondly, when $X = \{\frac{\partial}{\partial x_1}, \ldots, \frac{\partial}{\partial x_n}\}$, then the distance $d(x, y)$ is just the ordinary Euclidean distance $|x - y|$. In such case, Definition 6.2 coincides with the notion introduced by Poincaré in his classical paper [**P**]. In this setting a X-ball is just a standard Euclidean ball, then every $C^{1,1}$ domain and every convex domain possess the uniform outer X-ball condition. When we abandon the Euclidean setting, the construction of examples is technically much more involved and we discuss them in the last section of this paper.

We are now ready to state the first key boundary estimate.

THEOREM 6.3. *Let $D \subset \mathbb{R}^n$ be a connected open set, and suppose that for some $r > 0$, D has an outer X-ball $B(x_1, r)$ tangent at $x_o \in \partial D$. There exists $C > 0$, depending only on D and on X, such that if $\phi \in C(\partial D)$, $\phi \equiv 0$ in $B(x_1, 2r) \cap \partial D$, then we have for every $x \in D$*

$$|H_\phi^D(x)| \leq C \frac{d(x, x_o)}{r} \max_{\partial D} |\phi|.$$

Proof. Without loss of generality we assume $\max_{\partial D} |\phi| = 1$. Following the idea in [**P**] we introduce the function

$$f(x) = \frac{E(x_1, r)^{-1} - \Gamma(x_1, x)}{E(x_1, r)^{-1} - E(x_1, 2r)^{-1}}, \qquad x \in D, \tag{6.5}$$

where $x \to \Gamma(x_1, x)$ denotes the positive fundamental solution of $\mathcal{L}$, with singularity at x_1, and $t \to E(x_1, t)$ is defined as in (2.6). Clearly, f is $\mathcal{L}$-harmonic in $\mathbb{R}^n \setminus \{x_1\}$. Since $\Gamma(x_1, \cdot) \leq E(x_1, r)^{-1}$ outside $B(x_1, r)$, we see that $f \geq 0$ in $\mathbb{R}^n \setminus B(x_1, r)$, hence in particular in D. Moreover, $f \equiv 1$ on $\partial B(x_1, 2r) \cap D$, whereas $f \geq 1$ in $(\mathbb{R}^n \setminus B(x_1, 2r)) \cap \overline{D}$. By Theorem 3.2 we infer

$$|H_\phi^D(x)| \leq f(x) \qquad \text{for every} \quad x \in D.$$

The proof will be completed if we show that

$$f(x) \leq C \frac{d(x, x_o)}{r}, \qquad \text{for every} \quad x \in D. \tag{6.6}$$

Consider the function $h(t) = E(x_1, t)^{-1}$. We have for $0 < s < t < R_o$,

$$h(s) - h(t) = (t - s) \frac{E'(x_1, \tau)}{E(x_1, \tau)^2},$$

for some $s < \tau < t$. Using the increasingness of the function $r \to rE(x_1, r)$, which follows from that of $E(x_1, \cdot)$, and the crucial estimate

$$C \leq \frac{rE'(x_1, r)}{E(x_1, r)} \leq C^{-1},$$

which is readily obtained from the definition of $\Lambda(x_1, r)$ in (2.2), we find

$$C \frac{t - s}{tE(x_1, t)} \leq h(s) - h(t) \leq C^{-1} \frac{t - s}{sE(x_1, s)}. \tag{6.7}$$

Keeping in mind the definition (6.5) of f, from (6.7), and from the fact that $E(x_1, \cdot)$ is doubling, we obtain

$$f(x) \leq C\, E(x_1, r) \{\Gamma(x_1, x_o) - \Gamma(x_1, x)\},$$

where we have used the hypothesis that $x_o \in \partial B(x_1, r)$. The proof of (6.6) will be achieved if we show that for $x \in \mathbb{R}^n \setminus B(x_1, r)$

$$\Gamma(x_1, x_o) - \Gamma(x_1, x) \leq C\, d(x, x_o) \frac{1}{rE(x_1, r)}.$$

In view of (2.8), the latter inequality follows immediately from Lemma 6.1. This completes the proof.

□

Let $D \subset \mathbb{R}^n$ be a domain. Consider the positive Green function $G(x, y)$ associated to $\mathcal{L}$ and D. From Theorem 3.2 and from the estimates (2.7) one easily sees that there exists a positive constant C_D such that for every $x, y \in D$

$$0 \leq G(x, y) \leq C_D \frac{d(x, y)^2}{|B_d(x, d(x, y))|}, \tag{6.8}$$

for each $x, y \in D$. Our next task is to obtain more refined estimates for G.

THEOREM 6.4. *Suppose that $D \subset \mathbb{R}^n$ satisfy the uniform outer X-ball condition. There exists a constant $C = C(X, D) > 0$ such that*

$$G(x, y) \leq C \frac{d(x, y)}{|B_d(x, d(x, y))|} d(y, \partial D)$$

for each $x, y \in D$, with $x \neq y$.

Proof. Consider $a > 1$ as in (2.8), and let R_0 be the constant in Definition 6.2 of uniform outer X-ball condition. The estimate that we want to prove is immediate if one of the points is away from the boundary. In fact, if either $d(y, \partial D) \geq \frac{d(x,y)}{a^2(3+a)}$, or $d(y, \partial D) \geq R_o$, then the conclusion follows from (6.8). We may thus assume that

$$a\, d(y, \partial D) < \frac{d(x,y)}{a(a+3)}, \quad \text{and} \quad d(y, \partial D) < R_o. \tag{6.9}$$

We now choose

$$r = min\left(\frac{d(x,y)}{2a(a+3)}, \frac{aR_o}{2}\right).$$

One easily verifies from (6.9) that $ad(y, \partial D) < 2r$. Let x_o be the point in ∂D such that $d(y, \partial D) = d(y, x_o)$ and consider the outer X-ball $B(x_1, r/a)$ tangent to the boundary of D in x_o. We claim that

$$y \in D \cap B(x_1, (a+3)r).$$

To see this observe that by (2.8) $x_o \in \overline{B}(x_1, \frac{r}{a}) \subset \overline{B}_d(x_1, r)$, and therefore

$$d(y, x_1) \leq d(y, x_o) + d(x_o, x_1) = d(y, \partial D) + d(x_o, x_1) \leq \frac{a+2}{a} r < \frac{a+3}{a} r.$$

This shows $y \in B_d(x_1, a^{-1}(a+3)r)$. Another application of (2.8) implies the claim. Next, the triangle inequality gives

$$d(x, x_1) \geq d(x,y) - d(x_1, y) \geq d(x,y) - \frac{a+3}{a} r \geq d(x,y)(1 - \frac{1}{2a^2}),$$

and consequently

$$x \in \mathbb{R}^n \setminus B_d(x_1, (1 - \frac{1}{2a^2}) d(x,y)).$$

On the other hand (2.8) implies

$$\mathbb{R}^n \backslash B_d(x_1, (1-\frac{1}{2a^2})d(x,y)) \subset \mathbb{R}^n \backslash B(x_1, \frac{1}{a}(1-\frac{1}{2a^2})d(x,y)) \subset \mathbb{R}^n \backslash B(x_1, (a+3)r),$$

the last inclusion being true since $a > 1$.

We now consider the Perron-Wiener-Brelot solution v to the Dirichlet problem $\mathcal{L}v = 0$ in $B(x_1, (a+3)r) \cap D$, with boundary datum a function $\phi \in C(\partial(B(x_1, (a+3)r) \cap D))$, such that $0 \leq \phi \leq 1$, $\phi = 1$ on $\partial B(x_1, (3+a)r) \cap D$, and $\phi = 0$ on $\partial D \cap B(x_1, (1+a)r)$. We observe in passing that, thanks to the assumptions on D, we can only say that v is continuous up to the boundary in that portion of $\partial(B(x_1, (a+3)r) \cap D)$ that is common to ∂D. However such continuity is not needed to implement Theorem 3.2 and deduce that $0 \leq v \leq 1$. We observe that $D \cap B(x_1, (a+3)r)$ satisfies the outer $\mathcal{L}$-ball condition at the point $x_o \in \partial D$. Applying Theorem 6.3 one infers for every $y \in D \cap B(x_1, (a+3)r)$

$$|v(y)| \leq C\, \frac{d(y, \partial D)}{r}. \tag{6.10}$$

Let C_D be as in (6.8) and define $w(z) = C_D^{-1} E(x, \beta d(x,y)) G(x,z)$, where $\beta = (1 - \frac{1}{2a^2} - \frac{1}{2a})$. Since $x \notin B(x_1, (a+3)r)$, then $\mathcal{L}w = 0$ in $B(x_1, (a+3)r) \cap D$. Observe that if $z \in \partial B(x_1, (a+3)r)$, then

$$d(x,z) \geq d(x, x_1) - d(z, x_1) \geq (1 - \frac{1}{2a^2}) - (a+3)r \geq \beta d(x,y),$$

from our choice of r and β. Consequently, in view of the monotonicity of $r \to E(x,r)$ and (6.8), we have that $w \le C_D^{-1} E(x, d(x,z)) G(x,z) \le 1$ on $\partial(B(x_1,(a+3)r) \cap D)$. By Theorem 3.2 one concludes that $w(y) \le v(y)$ in $D \cap B(x_1,(a+3)r)$. The estimate of v established above, along with (2.1), completes the proof.

□

It was observed in [**LU2**, Theorem 50] that in a Carnot group, by exploiting the symmetry of the Green function $G(y,x) = G(x,y)$, one can actually improve the estimate in Theorem 6.4 as follows

$$G(x,y) \ \le\ C\ d(x,y)^{-Q} d(x,\partial D) d(y,\partial D)\ , \quad x,y \in D\ ,\ x \ne y\ ,$$

where Q represents the homogeneous dimension of the group. An analogous improvement can be obtained in the more general setting of this paper. To see this, note that the symmetry of G and the estimate in Theorem 6.4 give for every $x, y \in D$

$$G(y,x) = \ G(x,y) \le C \frac{d(x,y)}{|B_d(x, d(x,y))|}\ d(y,\partial D)\ , \tag{6.11}$$

where $C > 0$ is the constant in the statement of Theorem 6.4. We now argue exactly as in the case in which (6.9) holds in the proof of Theorem 6.4, except that we now define

$$w(z) = C^{-1} d(x,\partial D)^{-1} \frac{|B_d(x, d(x,y))|}{d(x,y)} G(z,x)\ , \quad z \in \overline{B(x_1,(a+3)r) \cap D}\ .$$

Using (6.11) instead of (6.8) we reach the conclusion that

$$w(z) \le 1\ , \quad \text{for every } z \in \partial(B(x_1,(a+3)r) \cap D)\ .$$

Since $\mathcal{L}w = 0$ in $B(x_1,(a+3)r) \cap D$, by Theorem 3.2 we conclude as before that $w(y) \le v(y)$ in $D \cap B(x_1,(a+3)r)$. Combining this estimate with (6.10) we have proved the following result.

COROLLARY 6.5. *Suppose that $D \subset \mathbb{R}^n$ satisfy the uniform outer X-ball condition. There exists a constant $C = C(X,D) > 0$ such that*

$$G(x,y)\ \le\ C\ \frac{d(x,\partial D) d(y,\partial D)}{|B_d(x, d(x,y))|}\ ,$$

for each $x, y \in D$, with $x \ne y$.

We now turn to estimating the horizontal gradient of the Green function up to the boundary. The next result plays a central role in the rest of the paper.

THEOREM 6.6. *Assume the uniform outer X-ball condition for $D \subset \mathbb{R}^n$. There exists a constant $C = C(X,D) > 0$ such that*

$$|XG(x,y)| \le C\ \frac{d(x,y)}{|B_d(x, d(x,y))|},$$

for each $x, y \in D$, with $x \ne y$.

Proof. Let R_o be as in Definition 6.2. Fix $x, y \in D$ and choose $0 < r < R_o$ such that $x \notin B_d(y, ar) \subset \overline{D}$. Applying Corollary 5.3 and (2.8) to $G(x,\cdot)$ we obtain for every $z \in B(y,r)$

$$|XG(x,z)| \le\ \frac{C}{r}\ G(x,z).$$

If $d(y, \partial D) \leq 2ad(x, y)$, we choose $r = min\left(\frac{d(y,\partial D)}{2a}, \frac{R_o}{2}\right)$ and then the latter inequality implies the conclusion via Theorem 6.4. If $d(y, \partial D) > 2ad(x, y)$, then keeping in mind that $G(x, \cdot) = \Gamma(x, \cdot) - h_x$, we use (2.7) to bound $|X\Gamma|$, and, with $r = min\left(\frac{d(y,\partial D)}{2a}, \frac{R_o}{2}\right)$, we apply Corollary 5.3 and the maximum principle to obtain

$$|Xh_x(y)| \leq \frac{C}{r} h_x(y) = \frac{C}{r} h_y(x) \leq \frac{C}{r} \sup_{w \in \partial D} \Gamma(y, w) = \frac{C}{r} \Gamma(y, z)$$

for some $z \in \partial D$. On the other hand, one has

$$d(x, y) < \frac{d(y, \partial D)}{2a} \leq \frac{d(y, z)}{2a}$$

so that using (2.7) one more time

$$\Gamma(y, z) \leq C \frac{1}{E(y, d(y, z))} \leq C \frac{1}{E(y, 2ad(x, y))} \leq C\,\Gamma(y, x) \leq C \frac{d(x, y)^2}{|B_d(x, d(x, y))|}.$$

Replacing this inequality in the estimate for $|Xh_x(y)|$ we reach the desired conclusion.

□

COROLLARY 6.7. *If $D \subset \mathbb{R}^n$ satisfies the uniform outer X-ball condition, then for any $x_o \in D$ and every open neighborhood U of ∂D, such that $x_o \notin \overline{U}$, one has $G(x_o, \cdot) \in \mathcal{L}^{1,\infty}(U)$. Moreover, its $\mathcal{L}^{1,\infty}(U)$ norm depends on D, X and U but it is independent of x_o.*

Localizing the hypothesis. It is interesting to note that one can still prove that $G(x_o, \cdot) \in \mathcal{L}^{1,\infty}(U)$ under the weaker hypothesis that the uniform outer X-ball condition be satisfied only in a neighborhood of the characteristic set of D. In this case, however, the uniform estimates in x_o will be lost. We devote the last part of this section to the proof of this result. Let $\Sigma = \Sigma_D \subset \partial D$ denote the compact set of all characteristic points.

DEFINITION 6.8. *Let D be a C^1 domain. We say that D possesses the* uniform outer X-ball *in a neighborhood of Σ if for a given choice of an open set V containing Σ, one can find $R_o > 0$ such that for every $Q \in V \cap \partial D$ and $0 < r < R_o$ there exists a X-ball $B(x_1, r)$ for which* (6.4) *holds. More in general, we say that D possesses the* uniform outer X-ball *along the set $V \cap \partial D$ if one can find $R_o > 0$ such that for every $x_o \in V \cap \partial D$ and $0 < r < R_o$ there exists a X-ball $B(x_1, r)$ for which* (6.4) *holds.*

Our first step consists in proving "localized" versions of Theorems 6.4 and 6.6.

THEOREM 6.9. *Let $D \subset \mathbb{R}^n$ be a domain that is regular for the Dirichlet problem. Let $P \in \partial D$ and assume that for some $\epsilon > 0$ the set D possesses the uniform outer X-ball along $B_d(P, 2\epsilon) \cap \partial D$. There exists a constant $C = C(X, D) > 0$ such that*

$$G(x, y) \leq C \frac{d(x, y)}{|B_d(x, d(x, y))|} d(y, \partial D)$$

for each $y \in B_d(P, \epsilon) \cap D$, and $x \in D$, with $x \neq y$.

Proof. The proof follows closely the one of Theorem 6.4 and we will adopt the same notation as in that proof. Let x_o be the point in ∂D closest to y. In order to apply the arguments in the proof of Theorem 6.4 we need to show that the set D has an outer $\mathcal{L}$-ball $B(x_1, r/a)$ at x_o for every $0 < r < R_0$. Given our hypothesis it suffices to show that $x_o \in B_d(P, 2\epsilon) \cap \partial D$. Observe that $d(y, x_o) \leq d(y, P) < \epsilon$, and consequently $d(P, x_o) < 2\epsilon$. Since D has an outer X-ball $B(x_1, r/a)$ at x_o for every $0 < r < R_0$, then so does the subset $B(x_1, (a+3)r) \cap D$. The rest of the proof is a word by word repetition of the one for Theorem 6.4.

□

THEOREM 6.10. *Let $D \subset \mathbb{R}^n$ be a domain that is regular for the Dirichlet problem. Let $P \in \partial D$ and assume that for some $\epsilon > 0$ the set D possesses the uniform outer X-ball along $B_d(P, 2\epsilon) \cap \partial D$. There exists a constant $C = C(X, D) > 0$ such that*

$$|XG(x,y)| \leq C \, \frac{d(x,y)}{|B_d(x, d(x,y))|},$$

for each $y \in B_d(P, \frac{1}{2}\epsilon) \cap D$, and $x \in D$, with $x \neq y$.

Proof. In the proof of Theorem 6.6 there is only one point where the outer X-ball condition is used. Consider $y \in B_d(P, \frac{1}{2}\epsilon) \cap D$ and assume that $d(y, \partial D) \leq d(x,y)$. Choose $2r = \frac{d(y,\partial D)}{a}$ and observe that if $z \in B(y, r)$ then $d(z, y) < ar \leq \epsilon/2$. Consequently $d(z, P) \leq d(z, y) + d(y, P) \leq \epsilon$, and we can apply Theorem 6.9 to the function $G(x, z)$ concluding the proof in the same way as before.

□

COROLLARY 6.11. *Let $D \subset \mathbb{R}^n$ be a C^∞ domain. If D satisfies the uniform outer X-ball condition in a neighborhood V of Σ, then for any $x_o \in D$ and every open neighborhood U of ∂D, such that $x_o \notin \overline{U}$, one has $||G(x_o, \cdot)||_{\mathcal{L}^{1,\infty}(U)} \leq C(x_o, D, V, U, X)$.*

Proof. Observe that D is regular for the Dirichlet problem. The regularity away from the characteristic set follows by Theorem 3.12 and the regularity in a neighborhood of Σ is a consequence of the uniform outer X-ball condition and of the cited results in [**Ci**],[**D**], [**NS**] and [**CDG3**]. Denote by V the neighborhood of Σ where the uniform outer X-ball condition holds. In view of the compactness of Σ, we have that $W = \bigcup_{P \in \Sigma} B(P, 2\epsilon) \subset V$, for some $\epsilon > 0$. We will consider also the set $A = \bigcup_{P \in \Sigma} B(P, \frac{1}{2}\epsilon) \subset W$. In view of Theorem 3.12, we have that $G(x_o, \cdot) \in C^\infty(\bar{D} \setminus \{A \cup \{x_0\}\})$. In particular, $G(x_o, \cdot)$ is smooth in $U \setminus A$. This implies the estimate $||G(x_o, \cdot)||_{\mathcal{L}^{1,\infty}(U \setminus A)} \leq C_0 = C_0(x_o, D, V, X)$. To complete the proof of the corollary we consider $y \in A$ and observe that there must be a $P \in \Sigma$ such that $y \in B(P, \frac{1}{2}\epsilon)$. Denote by Q the homogeneous dimension associated to the system X in a neighborhood of D. In view of Theorem 6.10 we have that $|XG(x_o, y)| \leq C d(y, x_o)^{1-Q} \leq C_1$, with C_1 depending only on X, D and U. At this point we choose $C(x_o, D, V, U, X) = \min\{C_0, C_1\}$, and the proof is concluded. □

7. The subelliptic Poisson kernel and a representation formula for $\mathcal{L}$-harmonic functions

In this section we establish a basic Poisson type representation formula for smooth domains that satisfy the outer X-ball condition in a neighborhood of the characteristic set. This results generalizes an analogous representation formula for the Heisenberg group $\mathbb{H}^n$ obtained by Lanconelli and Uguzzoni in [**LU1**] and extended in [**CGN2**] to groups of Heisenberg type. Consider a domain D which is regular for the Dirichlet problem. For a fixed point $x_o \in D$ we respectively denote by $\Gamma(x) = \Gamma(x, x_o)$ and $G(x) = G(x, x_o)$, the fundamental solution of $\mathcal{L}$, and the Green function for D and $\mathcal{L}$ with pole at x_o. Recall that $G(x) = \Gamma(x) - h(x)$, where h is the unique $\mathcal{L}$-harmonic function with boundary values Γ. We also note that due to the assumption that D be regular, G, h are continuous in any relatively compact subdomain of $\overline{D} \setminus \{x_o\}$. We next consider a C^∞ domain $\Omega \subset \overline{\Omega} \subset \overline{D}$ containing the point x_o. For any $u, v \in C^\infty(\overline{D})$ we obtain from the divergence theorem

$$\int_\Omega [u\ \mathcal{L}v - v\ \mathcal{L}u]\ dx \ = \ \sum_{j=1}^m \int_{\partial\Omega} [v\ X_j u - u\ X_j v] < X_j, \boldsymbol{\nu} > \ d\sigma\ , \tag{7.1}$$

where $\boldsymbol{\nu}$ denotes the outer unit normal and $d\sigma$ the surface measure on $\partial\Omega$. By Hörmander's hypoellipticity theorem [**H**] the function $x \to \Gamma(x_o, x)$ is in $C^\infty(D \setminus \{x_o\})$. By Sard's theorem there exists a sequence $s_k \nearrow \infty$ such that the sets $\{x \in \mathbb{R}^n \mid \Gamma(x_o, x) = s_k\}$ are C^∞ manifolds. Since by (2.7) the fundamental solution has a singularity at x_o, we can assume without restriction that such manifolds are strictly contained in Ω. Set $\epsilon_k = F(x_o, s_k^{-1})$, where $F(x_o, \cdot)$ is the inverse function of $E(x_o, \cdot)$ introduced in section two. The sets $B(\epsilon_k) = B(x_o, \epsilon_k) \subset \overline{B}(x_o, \epsilon_k) \subset \Omega$ are a sequence of smooth X-balls shrinking to the point x_o. We note explicitly that the outer unit normal on $\partial B(\epsilon_k)$ is $\boldsymbol{\nu} = -\frac{D\Gamma(\cdot, x_o)}{|D\Gamma(\cdot, x_o)|}$.

Applying (7.1) with $v(x) = G(x)$, and Ω replaced by $\Omega_{\epsilon_k} = \Omega \setminus \overline{B}(\epsilon_k)$, where one has $\mathcal{L}G = 0$, we find

$$\begin{aligned}\int_{\Omega_{\epsilon_k}} G\ \mathcal{L}u\ dx \ &= \ \sum_{j=1}^m \int_{\partial\Omega} [u\ X_j G - G\ X_j u] < X_j, \boldsymbol{\nu} > d\sigma \\ &+ \sum_{j=1}^m \int_{\partial B(\epsilon_k)} [G\ X_j u - u\ X_j G] < X_j, \boldsymbol{\nu} > d\sigma\ .\end{aligned}$$

Again the divergence theorem gives

$$\int_{B(\epsilon_k)} \mathcal{L}u\ dx \ = \ - \sum_{j=1}^m \int_{\partial B(\epsilon_k)} X_j u < X_j, \boldsymbol{\nu} > \ d\sigma\ . \tag{7.2}$$

Using (7.2), and the fact that $G = \Gamma - h$, we find

(7.3)

$$\sum_{j=1}^{m} \int_{\partial B(\epsilon_k)} [G\ X_j u - u\ X_j G] < X_j, \boldsymbol{\nu} > \ d\sigma$$

$$= \frac{1}{E(x_o, \epsilon_k)} \sum_{j=1}^{m} \int_{\partial B(\epsilon_k)} X_j u < X_j, \boldsymbol{\nu} > d\sigma \ - \ \sum_{j=1}^{m} \int_{\partial B(\epsilon_k)} h\ X_j u < X_j, \boldsymbol{\nu} > \ d\sigma$$

$$- \sum_{j=1}^{m} \int_{\partial B(\epsilon_k)} u\ X_j \Gamma < X_j, \boldsymbol{\nu} > \ d\sigma \ + \ \sum_{j=1}^{m} \int_{\partial B(\epsilon_k)} u\ X_j h < X_j, \boldsymbol{\nu} > \ d\sigma$$

$$= \ - \ \frac{1}{E(x_o, \epsilon_k)} \int_{B(\epsilon_k)} \mathcal{L}u\ dx \ + \ \int_{\partial B(\epsilon_k)} u\ \frac{|X\Gamma|^2}{|D\Gamma|}\ d\sigma$$

$$+ \sum_{j=1}^{m} \int_{\partial B(\epsilon_k)} u\ X_j h < X_j, \boldsymbol{\nu} > \ d\sigma \ - \ \sum_{j=1}^{m} \int_{\partial B(\epsilon_k)} h\ X_j u < X_j, \boldsymbol{\nu} > \ d\sigma\ .$$

Using (5.3) we find

$$\int_{\partial B(\epsilon_k)} u\ \frac{|X\Gamma|^2}{|D\Gamma|}\ d\sigma \ = \ u(x_o) \ - \ \int_{B(\epsilon_k)} \mathcal{L}u\ \left[\Gamma - \frac{1}{E(x_o, \epsilon_k)}\right]\ dx\ .$$

Keeping in mind that $u, h \in C^\infty(\overline{\Omega})$, from the estimates (2.7) and the fact that

$$\frac{|B(\epsilon_k)|}{E(x_o, \epsilon_k)} \ \leq \ C\ \epsilon_k^2,$$

letting $k \to \infty$, so that $\epsilon_k \to 0$, we conclude from (7.2), (7.3),

$$u(x_o) \ = \ \sum_{j=1}^{m} \int_{\partial\Omega} [G\ X_j u - u\ X_j G]\ < X_j, \boldsymbol{\nu} > \ d\sigma \ + \ \int_{\Omega} G\ \mathcal{L}u\ dx\ . \tag{7.4}$$

To summarize what we have found we introduce the following definition.

DEFINITION 7.1. *Given a bounded open set $\Omega \subset \overline{\Omega} \subset \mathbb{R}^n$ of class C^1, at every point $y \in \partial\Omega$ we let*

$$\boldsymbol{N}^X(y) \ = \ (< \boldsymbol{\nu}(y), X_1(y) >, ..., < \boldsymbol{\nu}(y), X_m(y) >)\ ,$$

where $\boldsymbol{\nu}(y)$ is the outer unit normal to Ω in y. We also set

$$W(y) \ = \ |\boldsymbol{N}^X(y)| \ = \ \sqrt{\sum_{j=1}^{m} < \boldsymbol{\nu}(y), X_j(y) >^2}\ .$$

If $y \in \partial\Omega \setminus \Sigma$, we set

$$\boldsymbol{\nu}^X(y) \ = \ \frac{\boldsymbol{N}^X(y)}{|\boldsymbol{N}^X(y)|}\ .$$

One has $|\boldsymbol{\nu}^X(y)| = 1$ for every $y \in \partial D \setminus \Sigma$.

We note explicitly from Definitions 3.11 and 7.1 that one has for the characteristic set Σ of Ω

$$\Sigma \ = \ \{y \in \partial\Omega \mid W(y) = 0\}\ .$$

Using the quantities introduced in this definition we can express (7.4) in the following suggestive way.

PROPOSITION 7.2. *Let $D \subset \mathbb{R}^n$ be a bounded open set with (positive) Green function G of the sub-Laplacian (1.2) and consider a C^2 domain $\Omega \subset \overline{\Omega} \subset D$. For any $u \in C^\infty(D)$ and every $x \in \Omega$ one has*

$$u(x) = \int_{\partial\Omega} G(x,y) < Xu(y), \mathbf{N}^X(y) > d\sigma(y) - \int_{\partial\Omega} u(y) < XG(x,y), \mathbf{N}^X(y) > d\sigma(y) + \int_\Omega G(x,y)\, \mathcal{L}u(y)\, dy \ .$$

If moreover $\mathcal{L}u=0$ in D, then

$$u(x) = \int_{\partial\Omega} G(x,y) < Xu(y), \mathbf{N}^X(y) > d\sigma(y) - \int_{\partial\Omega} u(y) < XG(x,y), \mathbf{N}^X(y) > d\sigma(y) \ .$$

In particular, the latter equality gives for every $x \in \Omega$

$$\int_{\partial\Omega} < XG(x,y), \mathbf{N}^X(y) > d\sigma(y) = -1 \ .$$

REMARK 7.3. *If $u \in C^\infty(\overline{D})$, then we can weaken the hypothesis on Ω and require only $\overline{\Omega} \subset \overline{D}$ rather than $\overline{\Omega} \subset D$.*

We consider next a C^∞ domain $D \subset \mathbb{R}^n$ satisfying the uniform outer X-ball condition in a neighborhood of Σ. Our purpose is to pass from the interior representation formula in Proposition 7.2 to one on the boundary of ∂D. The presence of characteristic points becomes important now. The following result due to Derridj [**De1**, Theorem 1] will be important in the sequel.

THEOREM 7.4. *Let $D \subset \mathbb{R}^n$ be a C^∞ domain. If Σ denotes its characteristic set, then $\sigma(\Sigma) = 0$.*

We now define two functions on $D \times (\partial D \setminus \Sigma)$ which play a central role in the results of this paper. They constitutes subelliptic versions of the Poisson kernel from classical potential theory. The former function $P(x,y)$ is the Poisson kernel for D and the sub-Laplacian (1.2) with respect to surface measure σ. The latter $K(x,y)$ is instead the Poisson kernel with respect to the perimeter measure σ_X. This comment will be clear after we prove Theorem 7.10 below.

DEFINITION 7.5 (Subelliptic Poisson kernels). *With the notation of Definition 7.1, for every $(x,y) \in D \times (\partial D \setminus \Sigma)$ we let*

$$P(x,y) = - < XG(x,y), \mathbf{N}^X(y) > \ . \tag{7.5}$$

We also define

$$K(x,y) = \frac{P(x,y)}{W(y)} = - < XG(x,y), \boldsymbol{\nu}^X(y) > \ . \tag{7.6}$$

We extend the definition of P and K to all $D \times \partial D$ by letting $P(x,y) = K(x,y) = 0$ for any $x \in D$ and $y \in \Sigma$. According to Theorem 7.4 the extended functions coincide σ-a.e. with the kernels in (7.5), (7.6).

It is important to note that if we fix $x \in D$, then in view of Theorem 3.12 the functions $y \to P(x,y)$ and $y \to K(x,y)$ are C^∞ up to $\partial D \setminus \Sigma$. The following estimates, which follow immediately from (7.5) and (7.6), will play an important role in the sequel. For $(x,y) \in D \times (\partial D \setminus \Sigma)$ we have

$$P(x,y) \le W(y)\, |XG(x,y)| \ , \quad K(x,y) \le |XG(x,y)| \ . \tag{7.7}$$

We now introduce a new measure on ∂D by letting

$$d\sigma_X \;=\; W\; d\sigma\;. \tag{7.8}$$

We observe that since we are assuming that $D \in C^\infty$ the density W is smooth and bounded on ∂D and therefore (7.8) implies that $d\sigma_X \ll d\sigma$. In view of this observation Theorem 7.4 implies that also $\sigma_X(\Sigma) = 0$.

REMARK 7.6. *We mention explicitly that the measure $d\sigma_X$ in (7.8) is the X-perimeter measure $P_X(D;\cdot)$ (following De Giorgi) concentrated on ∂D. To explain this point we recall that for any open set $\Omega \subset \mathbb{R}^n$*

$$P_X(D;\Omega) \;=\; Var_X(\chi_D;\Omega)\;, \tag{7.9}$$

where Var_X indicates the sub-Riemannian X-variation introduced in [**CDG2**] *and also developed in* [**GN1**]. *Given a bounded C^2 domain $D \subset \mathbb{R}^n$ one obtains from* [**CDG2**] *that*

$$P_X(D;\Omega) \;=\; \int_{\partial D\cap\Omega} W\; d\sigma\;. \tag{7.10}$$

From (7.10) *one concludes that for every $y \in \partial D$ and every $r > 0$*

$$\sigma_X(\partial D \cap B_d(y,r)) \;=\; P_X(D; B_d(y,r))\;, \tag{7.11}$$

which explains the remark. The measure $\sigma_X = P_X(D;\cdot)$ on ∂D plays a pervasive role in the analysis and geometry of sub-Riemannian spaces, and its intrinsic properties have many deep implications both in subelliptic pde's and in geometric measure theory. For an account of some of these aspects we refer the reader to [**DGN2**].

PROPOSITION 7.7. *Let $D \subset \mathbb{R}^n$ be a bounded C^∞ domain satisfying the uniform outer X-ball condition in a neighborhood of its characteristic set Σ. For every $x \in D$ we have*

$$\int_{\partial D} P(x,y)d\sigma(y) \;=\; 1 \;=\; \int_{\partial D} K(x,y)d\sigma_X(y)\;.$$

Proof. We fix $x \in D$ and recall that Σ is a compact set. In view of Theorem 7.4 we can choose an exhaustion of D with a family of C^∞ connected open sets $\Omega_k \subset \overline{\Omega}_k \subset \overline{D}$, with $\Omega_k \nearrow D$ as $k \to \infty$, such that $\partial\Omega_k = \Gamma_k^1 \cup \Gamma_k^2$, with $\Gamma_k^1 \subset \partial D\backslash\Sigma$, $\Gamma_k^1 \nearrow \partial D$, $\sigma(\Gamma_k^2) \to 0$. By Proposition 7.2 (and the remark following it) we obtain for every $k \in \mathbb{N}$

(7.12)

$$\begin{aligned}
-1 &= \int_{\partial\Omega_k} < XG(x,y), \boldsymbol{N}^X(y) > d\sigma(y)\\
&= \int_{\partial\Gamma_k^1} < XG(x,y), \boldsymbol{N}^X(y) > d\sigma(y) \;+\; \int_{\partial\Gamma_k^2} < XG(x,y), \boldsymbol{N}^X(y) > d\sigma(y)\;.
\end{aligned}$$

We now pass to the limit as $k \to \infty$ in the above integrals. Using Corollary 6.11 and $\sigma(\Gamma_k^2) \to 0$, we infer that

$$\lim_{k\to\infty} \int_{\partial\Gamma_k^2} < XG(x,y), \boldsymbol{N}^X(y) > d\sigma(y) \;=\; 0\;.$$

Theorem 3.12, Corollary 6.11, and the fact that $\Gamma_k^1 \nearrow \partial D$, allow to use dominated convergence and obtain

$$\lim_{k\to\infty} \int_{\partial\Gamma_k^1} < XG(x,y), \boldsymbol{N}^X(y) > d\sigma(y) \ = \ \int_{\partial D} < XG(x,y), \boldsymbol{N}^X(y) > d\sigma(y) \ .$$

In conclusion, we have found

$$-\,1 \ = \ \int_{\partial D} < XG(x,y), \boldsymbol{N}^X(y) > d\sigma(y) \ ,$$

which, in view of (7.5), proves the first identity. To establish the second identity we return to (7.12), which in view of (7.6), (7.8) we can rewrite as follows

$$\begin{aligned} 1 = & -\int_{\partial\Gamma_k^1} < XG(x,y), \boldsymbol{\nu}^X(y) > d\sigma_X(y) \ - \ \int_{\partial\Gamma_k^2} < XG(x,y), \boldsymbol{N}^X(y) > d\sigma(y) \\ = & \int_{\partial\Gamma_k^1} K(x,y)\ d\sigma_X(y) \ - \ \int_{\partial\Gamma_k^2} < XG(x,y), \boldsymbol{N}^X(y) > d\sigma(y) \ . \end{aligned}$$

Since as we have observed $d\sigma_X \ll d\sigma$, in view of the second estimate $K(x,y) \le |XG(x,y)|$ in (7.7), we can again use Theorem 3.12, Corollary 6.11 and dominated convergence (with respect to σ_X) to conclude that

$$\lim_{k\to\infty} \int_{\partial\Gamma_k^1} K(x,y)\ d\sigma_X(y) \ = \ \int_{\partial D} K(x,y)\ d\sigma_X(y) \ .$$

This completes the proof. □

THEOREM 7.8. *Let D satisfy the assumptions in Proposition 7.7. If $\phi \in C^\infty(\partial D)$ assumes a single constant value in a neighborhood of Σ, then $H_\phi^D \in \mathcal{L}^{1,\infty}(D)$. Furthermore, if for $\phi \in C(\partial D)$ we have $H_\phi^D \in \mathcal{L}^{1,\infty}(D)$, then*

$$H_\phi^D(x) \ = \ \int_{\partial D} P(x,y)\ \phi(y)\ d\sigma(y) \ = \ \int_{\partial D} K(x,y)\ \phi(y)\ d\sigma_X(y) \ , \qquad x \in D \ .$$

Proof. We start with the proof of the regularity result. Let ϕ be as in the first part of the statement. We mention explicitly that, by definition, ϕ is C^∞ in a neighborhood of ∂D. Denote by U a neighborhood of Σ in which the function ϕ is constant and along which the domain D satisfies the uniform outer X-ball condition. As in the proof of Corollary 6.11, we can assume that $U = \bigcup_{P\in\Sigma} B_d(P,\epsilon)$, for some $\epsilon = \epsilon(U,X) > 0$. If we denote by R_0 the constant involved in the definition of the uniform outer X-ball (see Definition 6.8), then we can always select a smaller constant so that $\epsilon = 2aR_0$ (here $a > 1$ is the constant from (2.8)). In view of Proposition 7.7 we can assume without loss of generality that ϕ vanishes in a neighborhood of Σ and $\underset{\partial D}{max}|\phi| = 1$. We want to show that the horizontal gradient of H_ϕ^D is in L^∞ in such neighborhood. By Theorem 3.12 the conclusion $H_\phi^D \in \mathcal{L}^{1,\infty}(D)$ will follow. Fix $x_o \in \Sigma$, and $0 < r < R_o$, where R_o is as in Definition 6.2. Theorem 6.3 implies

$$|H_\phi^D(y)| \le \ C\ \frac{d(y,x_o)}{r} \tag{7.13}$$

for every $y \in D$. Let now $x \in B(x_o, r/2) \cap D$ and consider the metric ball $B_d(x, a^{-1}\tau) \subset B(x,\tau)$, see (2.8), where $\tau = \frac{d(x,\partial D)}{4}$. Corollary 5.3 implies

$$|XH^D_\phi(x)| \leq \frac{C}{d(x,\partial D)}\, H^D_\phi(x). \tag{7.14}$$

Pick $P \in \partial D$ such that $d(x,P) = d(x,\partial D)$. Observe that $d(P,\Sigma) \leq d(P,x_o) \leq d(P,x) + d(x,x_o) \leq 2d(x,x_o) \leq aR_0 = \epsilon/2$. In particular we can apply once more Theorem 6.3, and obtain (7.13) with P in place of x_o. Arguing in this way we find

$$H^D_\phi(x) \leq C\, \frac{d(x,P)}{r} = C\, \frac{d(x,\partial D)}{r}.$$

The latter inequality and (7.14) imply

$$|XH^D_\phi(x)| \leq \frac{C}{r}.$$

This proves that $|XH^D_\phi| \in L^\infty(B(x_o, r/2) \cap D)$. To establish the second part of the theorem, we take a function $\phi \in C(\partial D)$ for which $H^D_\phi \in \mathcal{L}^{1,\infty}(D)$. We fix $x \in D$ and consider the sequence of C^∞ domains Ω_k as in the proof of Proposition 7.7. Proposition 7.2 gives

(7.15)

$$H^D_\phi(x) = \int_{\partial\Omega_k} G(x,y) < X(H^D_\phi)(y), \boldsymbol{N}^X(y) > d\sigma(y) - \int_{\partial\Omega_k} H^D_\phi(y) < XG(x,y), \boldsymbol{N}^X(y) > d\sigma(y)\,.$$

At this point the conclusion follows along the lines of the proof of Proposition 7.7.

□

PROPOSITION 7.9. *Let D be a C^∞ domain. i) If D satisfies the uniform outer X-ball condition in a neighborhood of Σ, then $P(x,y) \geq 0$ and $K(x,y) \geq 0$ for each $(x,y) \in D \times \partial D$; ii) If D satisfies the uniform outer X-ball condition, then there exists a constant $C_D > 0$ such that for $(x,y) \in D \times \partial D$*

$$0 \leq P(x,y) \leq C_D\, W(y)\, \frac{d(x,y)}{|B_d(x,d(x,y))|}, \qquad 0 \leq K(x,y) \leq C_D\, \frac{d(x,y)}{|B_d(x,d(x,y))|}\,.$$

In particular, if we fix $x \in D$, then for any open set U containing ∂D, such that $x \notin \overline{U}$, one has $K(x,\cdot) \in L^\infty(\overline{D} \cap U)$.

Proof. We start with the proof of part (i). We argue by contradiction. If for some $x \in D$ and $x_o \in \partial D$ we had $P(x,x_o) = \alpha < 0$, then $x_o \notin \Sigma$. By Theorem 3.12 there exists a sufficiently small $r > 0$ such that $P(x,x') \leq \alpha/2$ for every $x' \in B(x_o, 2r) \cap \partial D$. We can also assume that $d(x_o,\Sigma) > 2r$. We now choose $\phi \in C^\infty(\partial D)$ such that $0 \leq \phi \leq 1$, $\phi \equiv 1$ on $B(x_o,r) \cap \partial D$ and $\phi \equiv 0$ outside $B(x_o, 3r/2) \cap \partial D$. Theorem 3.2 implies $H^D_\phi \geq 0$ in D. By the Harnack inequality we must have $H^D_\phi(x) > 0$. On the other hand, Theorem 7.8 gives

$$H^D_\phi(x) \leq \frac{\alpha}{2} \int_{B(x_o,3r/2)\cap\partial D} \phi(y)\, d\sigma(y) \leq 0\,,$$

which gives a contradiction. The proof of part (ii) is an immediate consequence of (7.7) and of Theorem 6.6. The estimate for $K(x,y)$ follows from (7.6) and from the one for $P(x,y)$.

□

We now fix $x \in D$. For every σ-measurable $E \subset \partial D$ we set

$$\nu^x(E) = \int_E K(x,y)\, d\sigma_X(y) .$$

According to Proposition 7.9, $d\nu^x$ defines a Borel measure on ∂D. Using Theorems 7.4 and 7.8 we can now establish the main result of this section.

THEOREM 7.10. *Let $D \subset \mathbb{R}^n$ be a C^∞ domain possessing the uniform outer X-ball condition in a neighborhood of the characteristic set Σ. For every $x \in D$, we have $\omega^x = \nu^x$, i.e., for every $\phi \in C(\partial D)$ one has*

$$H^D_\phi(x) = \int_{\partial D} \phi(y)\, K(x,y)\, d\sigma_X(y) = \int_{\partial D} \phi(y)\, P(x,y)\, d\sigma(y) , \quad x \in D .$$

In particular, $d\omega^x$ is absolutely continuous with respect to $d\sigma_X$ and $d\sigma$, and for every $(x,y) \in D \times \partial D$ one has

$$\frac{d\omega^x}{d\sigma_X}(y) = K(x,y), \qquad \frac{d\omega^x}{d\sigma}(y) = P(x,y) . \tag{7.16}$$

Proof. We begin with proving (7.16). Let $F \subset \partial D$ be a Borel set. If $F = \partial D$ then the result follows from Proposition 7.7. We now consider the case when the inclusion $F \subset \partial D$ is strict. Choose $\epsilon > 0$. Since both $K(x,y)$ and $W(y)$ are bounded, there exists open sets $E_\epsilon, F_\epsilon \subset \partial D$ such that $F \subset F_\epsilon \subset \overline{F}_\epsilon \subset E_\epsilon$, and $\nu^x(E_\epsilon \setminus F) < \epsilon/2$. Theorem 7.4 guarantees the existence of open sets $\Sigma_\epsilon, U_\epsilon$ such that $\Sigma \subset \Sigma_\epsilon \subset \overline{\Sigma_\epsilon} \subset U_\epsilon$ and $\nu^x(U_\epsilon) < \epsilon/2$. We now choose a function $\phi \in C^\infty_o(\partial D)$ and $0 \le \phi \le 1$ with $\phi \equiv 1$ on U_ϵ and $\nu^x(supp\, \phi) < \frac{3}{4}\epsilon$. We have

$$\begin{aligned} \omega^x(U_\epsilon) &= \int_{U_\epsilon} d\omega^x(y) \le \int_{\partial D} \phi(y)\, d\omega^x(y) = H^D_\phi(x) \\ \text{(by Theorem 7.8)} &= \int_{\partial D} \phi(y)\, K(x,y) d\sigma_X(y) \le \nu^x(supp\, \phi) < \frac{3}{4}\epsilon . \end{aligned} \tag{7.17}$$

Let now $\psi_o, \psi_1 \in C^\infty_o(\partial D)$ such that $0 \le \psi_o, \psi_1 \le 1$ and

$$\psi_o \equiv 1 \text{ in } \partial D \setminus U_\epsilon, \quad \psi_o \equiv 0 \text{ in } \Sigma_\epsilon,$$
$$\psi_1 \equiv 1 \text{ in } F, \quad \psi_1 \equiv 0 \text{ in } \partial D \setminus E_\epsilon .$$

One has

$$\begin{aligned} \omega^x(F) &\le \omega^x(U_\epsilon) + \omega^x(F \setminus U_\epsilon) \quad \text{(by (7.17))} \\ &\le \frac{3}{4}\epsilon + \int_{\partial D} \psi_o(y)\psi_1(y)\, d\omega^x(y) \\ &= \frac{3}{4}\epsilon + H^D_{\psi_o\psi_1}(x) \quad \text{(by Theorem 7.8)} \\ &= \frac{3}{4}\epsilon + \int_{\partial D} \psi_o(y)\psi_1(y) K(x,y)\, d\sigma_X(y) \le \frac{3}{4}\epsilon + \nu^x(E_\epsilon) \\ &= \frac{3}{4}\epsilon + \nu^x(F) + \nu^x(E_\epsilon \setminus F) < \nu^x(F) + \frac{7}{4}\epsilon . \end{aligned}$$

Since $\epsilon > 0$ is arbitrary, we infer that $\omega^x(F) \le \nu^x(F)$. If we repeat the same argument with $E_\epsilon \setminus F$ playing the role of the set F, we can prove $\omega^x(E_\epsilon \setminus F) \le \nu^x(E_\epsilon \setminus F)$. This allows to exchange the role of ω^x and ν^x in the computations above and conclude $\nu^x(F) \le \omega^x(F)$.

To complete the proof of the theorem we now use (7.16). From the definition of harmonic measure we know that for each $\phi \in C(\partial D)$ and $x \in D$ we have

$$H_\phi^D(x) = \int_{\partial D} \phi(y) d\omega^x(y). \tag{7.18}$$

On the other hand (7.16) yields $d\omega^x(y) = K(x,y) d\sigma_X(y)$. If we substitute the latter in (7.18) we reach the conclusion.

□

8. Reverse Hölder inequalities for the Poisson kernel

This section is devoted to proving the main results of this paper, namely Theorems 1.3, 1.4, 1.5 and 1.6. In the course of the proofs we will need some basic results about NTA_X domains from the paper [**CG1**]. We begin by recalling the relevant definitions.

DEFINITION 8.1. *We say that a connected, bounded open set $D \subset \mathbb{R}^n$ is a non-tangentially accessible domain with respect to the system $X = \{X_1, ..., X_m\}$ (NTA_X domain, hereafter) if there exist M, $r_o > 0$ for which:*

(i) *(Interior corkscrew condition) For any $x_o \in \partial D$ and $r \leq r_o$ there exists $A_r(x_o) \in D$ such that $\frac{r}{M} < d(A_r(x_o), x_o) \leq r$ and $d(A_r(x_o), \partial D) > \frac{r}{M}$. (This implies that $B_d(A_r(x_o), \frac{r}{2M})$ is $(3M, X)$-nontangential.)*
(ii) *(Exterior corkscrew condition) $D^c = \mathbb{R}^n \setminus D$ satisfies property (i).*
(iii) *(Harnack chain condition) There exists $C(M) > 0$ such that for any $\epsilon > 0$ and $x, y \in D$ such that $d(x, \partial D) > \epsilon$, $d(y, \partial D) > \epsilon$, and $d(x,y) < C\epsilon$, there exists a Harnack chain joining x to y whose length depends on C but not on ϵ.*

We note the following lemma which will prove useful in the sequel and which follows directly from Definition 8.1.

LEMMA 8.2. *Let $D \subset \mathbb{R}^n$ be NTA_X domain, then there exist constants C, R_1 depending on the NTA_X parameters of D such that for every $y \in \partial D$ and every $0 < r < R_1$ one has*

$$C\ |B_d(y,r)| \ \leq\ \min\{|D \cap B_d(y,r)|, |D^c \cap B_d(y,r)|\} \ \leq\ C^{-1}\ |B_d(y,r)|\ .$$

In particular, every NTA_X domain has positive density at every boundary point and therefore it is regular for the Dirichlet problem (see Definition 3.6, Proposition 3.7, and Theorem 3.5).

In the sequel, for $y \in \partial D$ and $r > 0$ we denote by

$$\Delta(y,r) \ =\ \partial D\ \cap\ B_d(y,r)$$

the surface metric ball centered at y with radius r. We next prove a basic non-degeneracy property of the horizontal perimeter measure $d\sigma_X$ in (7.8).

THEOREM 8.3. *Let $D \subset \mathbb{R}^n$ be a NTA_X domain of class C^2, then there exist $C^*, R_1 > 0$ depending on D, X and on the NTA_X parameters of D such that for every $y \in \partial D$ and every $0 < r < R_1$*

$$\sigma_X(\Delta(y,r)) \ \geq\ C^*\ \frac{|B_d(y,r)|}{r}\ .$$

*In particular, σ_X is lower 1-Ahlfors according to [**DGN2**] and $\sigma_X(\Delta(y,r)) > 0$.*

Proof. According to (I) in Theorem 1.15 in [**GN1**] every metric ball $B_d(y,r)$ is a PS_X (Poincaré-Sobolev) domain with respect to the system X. We can thus apply the isoperimetric inequality Theorem 1.18 in [**GN1**] to infer the existence of $R_1 > 0$ such that for every $y \in \partial D$ and every $0 < r < R_1$

$$\min\{|D \cap B_d(y,r)|, |D^c \cap B_d(y,r)|\}^{\frac{Q-1}{Q}} \le C_{iso} \frac{diam\ B_d(y,r)}{|B_d(y,r)|^{\frac{1}{Q}}} P_X(D; B_d(y,r)) ,$$

where Q is the homogeneous dimension of a fixed bounded set U containing $\overline{D}$. On the other hand, every NTA_X domain is a PS_X domain. We can thus combine the latter inequality with (7.11) and Lemma 8.2 to finally obtain

$$\sigma_X(\Delta(y,r)) \ge C^* \frac{|B_d(y,r)|}{r} .$$

This proves the theorem.

□

COROLLARY 8.4. *Let $D \subset \mathbb{R}^n$ be a NTA_X domain of class C^2 satisfying the upper 1-Ahlfors assumption in iv) of Definition 1.1. Then the measure σ_X is 1-Ahlfors, in the sense that there exist $\tilde{A}, R_1 > 0$ depending on the NTA_X parameters of D and on $A > 0$ in iv), such that for every $y \in \partial D$, and every $0 < r < R_1$, one has*

$$\tilde{A} \frac{|B_d(y,r)|}{r} \le \sigma_X(\Delta(y,r)) \le \tilde{A}^{-1} \frac{|B_d(y,r)|}{r} . \tag{8.1}$$

In particular, the measure σ_X is doubling, i.e., there exists $C > 0$ depending on $\tilde{A}$ and on the constant C_1 in (2.5), such that

$$\sigma_X(\Delta(y,2r)) \le C\, \sigma_X(\Delta(y,r)) . \tag{8.2}$$

for every $y \in \partial D$ and $0 < r < R_1$.

Proof. According to Theorem 8.3 the measure σ_X is lower 1-Ahlfors. Since by *iv*) of Definition 1.1 it is also upper 1-Ahlfors, the conclusion (8.1) follows. From the latter and the doubling condition (2.5) for the metric balls, we reach the desired conclusion (8.2).

□

The following results from [**CG1**] play a fundamental role in this paper.

THEOREM 8.5. *Let $D \subset \mathbb{R}^n$ be a NTA_X domain with relative parameters M, r_o. There exists a constant $C > 0$, depending only on X and on the NTA_X parameters of D, M and r_o, such that for every $x_o \in \partial D$ one has*

$$\omega^{A_r(x_o)}(\Delta(x_o,r)) \ge C .$$

THEOREM 8.6 (Doubling condition for $\mathcal{L}$-harmonic measure). *Consider a NTA_X domain $D \subset \mathbb{R}^n$ with relative parameters M, r_o. Let $x_o \in \partial D$ and $r \le r_o$. There exist $C > 0$, depending on X, M and r_o, such that*

$$\omega^x(\Delta(x_o,2r)) \le C\omega^x(\Delta(x_o,r))$$

for any $x \in D \setminus B_d(x_o,Mr)$.

THEOREM 8.7 (Comparison theorem). *Let $D \subset \mathbb{R}^n$ be a $X - NTA$ domain with relative parameters M, r_o. Let $x_o \in \partial D$ and $0 < r < \frac{r_o}{M}$. If u, v are $\mathcal{L}$-harmonic functions in D, which vanish continuously on $\partial D \setminus \Delta(x_o, 2r)$, then for every $x \in D \setminus B_d(x_o, Mr)$ one has*

$$C\,\frac{u(A_r(x_o))}{v(A_r(x_o))} \;\le\; \frac{u(x)}{v(x)} \;\le\; C^{-1}\,\frac{u(A_r(x_o))}{v(A_r(x_o))}$$

for some constant $C > 0$ depending only on X, M and r_o.

For any $y \in \partial\Omega$ and $\alpha > 0$ a nontangential region at y is defined by

$$\Gamma_\alpha(y) = \{x \in \Omega \,|\, d(x,y) \le (1+\alpha) d(x, \partial\Omega)\} \ .$$

Given a function u the α-nontangential maximal function of u at $y \in \partial D$ is defined by

$$N_\alpha(u)(y) = \sup_{x \in \Gamma_\alpha(y)} |u(x)| \ .$$

THEOREM 8.8. *Let $D \subset \mathbb{R}^n$ be a NTA_X domain. Given a point $x_1 \in D$, let $f \in L^1(\partial D, d\omega^{x_1})$ and define*

$$u(x) = \int_{\partial D} f(y) d\omega^x(y), \qquad x \in D \ .$$

Then, u is $\mathcal{L}$-harmonic in D, and:

(i) *$N_\alpha(u)(y) \le C M_{\omega^{x_1}}(f)(y)$, $y \in \partial D$;*
(ii) *u converges non-tangentially a.e. $(d\omega^{x_1})$ to f.*

Theorem 8.7 has the following important consequence.

THEOREM 8.9. *Let $D \subset \mathbb{R}^n$ be a ADP_X domain, and let $K(\cdot,\cdot)$ be the Poisson Kernel defined in (7.6). There exists $r_1 > 0$, depending on M and r_o, and a constant $C = C(X, M, r_o, R_o) > 0$, such that given $x_o \in \partial D$, for every $x \in D \setminus B_d(x_o, Mr)$ and every $0 < r < r_1$ one can find $E_{x_o,x,r} \subset \Delta(x_o, r)$, with $\sigma_X(E_{x_o,x,r}) = 0$, for which*

$$K(x,y) \;\le\; C\; K(A_r(x_o), y)\; \omega^x(\Delta(x_o, r))$$

for every $y \in \Delta(x_o, r) \setminus E_{x_o,x,r}$.

Proof. Let $x_o \in \partial D$. For each $y \in \Delta(x_o, r)$ and $0 < s < r/2$ set

$$u(x) \;=\; \omega^x(\Delta(y,s)), \qquad v(x) \;=\; \omega^x(\Delta(x_o, r/2)) \ .$$

The functions u and v are $\mathcal{L}$-harmonic in D and vanish continuously on $\partial D \setminus \Delta(x_o, 2r)$. Theorem 8.7 gives

$$\frac{\omega^x(\Delta(y,s))}{\omega^x(\Delta(x_o, r/2))} \;\le\; C\,\frac{\omega^{A_r(x_o)}(\Delta(y,s))}{\omega^{A_r(x_o)}(\Delta(x_o, r/2))} \tag{8.3}$$

for every $x \in D \setminus B(x_o, Mr)$. Applying (8.3) we thus find

$$\frac{\omega^x(\Delta(y,s))}{\omega^x(\Delta(x_o, r/2))} \;\le\; C\,\frac{\omega^{A_r(x_o)}(\Delta(y,s))}{\omega^{A_r(x_o)}(\Delta(x_o, r/2))} \ . \tag{8.4}$$

Upon dividing by $\sigma_X(\Delta(y,s))$ in (8.4) (observe that in view of Theorem 8.3 the σ_X measure of any surface ball $\Delta(y,s)$ is strictly positive), one concludes

$$\frac{\omega^x(\Delta(y,s))}{\sigma_X(\Delta(y,s))} \le C\, \frac{\omega^{A_r(x_o)}(\Delta(y,s))}{\sigma_X(\Delta(y,s))}\, \frac{\omega^x(\Delta(x_o,r/2))}{\omega^{A_r(x_o)}(\Delta(x_o,r/2))}\,. \tag{8.5}$$

Using Theorem 8.5 in the right-hand side of (8.5) we conclude

$$\frac{\omega^x(\Delta(y,s))}{\sigma_X(\Delta(y,s))} \le C\, \frac{\omega^{A_r(x_o)}(\Delta(y,s))}{\sigma_X(\Delta(y,s))}\, \omega^x(\Delta(x_o,r))\,. \tag{8.6}$$

We now observe that (8.2) in Corollary 8.4 allows to obtain a Vitali covering theorem and differentiate the measure ω^x with respect to the horizontal perimeter measure σ_X. This means that for σ_X-a.e. $y \in \Delta(x_o,r)$ the limit $\lim_{s\to 0}\frac{\omega^x(\Delta(y,s))}{\sigma_X(\Delta(y,s))}$ exists and equals $\frac{d\omega^x}{d\sigma_X}(y)$. This being said, passing to the limit as $s \to 0^+$ in (8.6) we obtain for σ_X-a.e. $y \in \Delta(x_o,r)$

$$\frac{d\omega^x}{d\sigma_X}(y) \le C\, \frac{d\omega^{A_r(x_o)}}{d\sigma_X}(y)\, \omega^x(\Delta(x_o,r))\,.$$

Since by (7.16) in Theorem 7.10 we know that $\frac{d\omega^x}{d\sigma_X}(y) = K(x,y)$, $\frac{d\omega^{A_r(x_o)}}{d\sigma_X}(y) = K(A_r(x_o),y)$, we have reached the desired conclusion. We observe in passing that the exceptional set here depends on x and on $A_r(x_o)$, but this fact will be inconsequential to us since we plan to integrate with respect to σ_X the above inequality on the surface ball $\Delta(x_o,r)$.

□

We now turn to the

Proof of Theorem 1.3. We fix $p > 1$, $x_o \in \partial D$ and $x_1 \in D$. Let R_1 be the minimum of the constants appearing in Definitions 6.2, 8.1, and in Theorem 8.9. Moreover, the constant R_1 should be chosen so small that $d(x_1,x_o) > MR_1$. Let $0 < r < R_1$. If $A_r(x_o)$ is a corkscrew for x_o, then by the definition of a corkscrew, the triangle inequality and (2.3) it is easy to see that we have for all $y \in \Delta(x_o,r)$

$$d(A_r(x_o),y) \sim Cr \quad \text{and} \quad |B_d(x_o,r)| \le C|B_d(A_r(x_o),d(A_r(x_o),y))|. \tag{8.7}$$

Now we have

$$
\begin{aligned}
&\left(\frac{1}{\sigma_X(\Delta(x_o,r))}\int_{\Delta(x_o,r)} K(x_1,y)^p\, d\sigma_X(y)\right)^{\frac{1}{p}} \qquad \text{(by (7.16))}\\
&= \left(\frac{1}{\sigma_X(\Delta(x_o,r))}\int_{\Delta(x_o,r)} K(x_1,y)^{p-1}\, d\omega^{x_1}(y)\right)^{\frac{1}{p}} \qquad \text{(by Theorem 8.9)}\\
&\le C\left(\frac{\omega^{x_1}(\Delta(x_o,r))^{p-1}}{\sigma_X(\Delta(x_o,r))}\int_{\Delta(x_o,r)} K(A_r(x_o),y)^{p-1}\, d\omega^{x_1}(y)\right)^{\frac{1}{p}} \qquad \text{(by (7.7))}\\
&\le C\left(\frac{\omega^{x_1}(\Delta(x_o,r))^{p-1}}{\sigma_X(\Delta(x_o,r))}\int_{\Delta(x_o,r)} |XG(A_r(x_o),y)|^{p-1}\, d\omega^{x_1}(y)\right)^{\frac{1}{p}} \qquad \text{(by Theorem 6.6)}\\
&\le C\left(\frac{\omega^{x_1}(\Delta(x_o,r))^{p-1}}{\sigma_X(\Delta(x_o,r))}\int_{\Delta(x_o,r)} \left(\frac{d(A_r(x_o),y)}{|B_d(A_r(x_o),d(A_r(x_o),y))|}\right)^{p-1} d\omega^{x_1}(y)\right)^{\frac{1}{p}} \qquad \text{(by (8.7))}\\
&\le C\left(\frac{\omega^{x_1}(\Delta(x_o,r))^{p-1}}{\sigma_X(\Delta(x_o,r))}\left(\frac{r}{|B_d(x_o,r)|}\right)^{p-1}\int_{\Delta(x_o,r)} d\omega^{x_1}(y)\right)^{\frac{1}{p}} \qquad \text{(by } iv) \text{ in Definition 1.1)}\\
&\le \frac{C}{\sigma_X(\Delta(x_o,r))}\int_{\Delta(x_o,r)} d\omega^{x_1}(y) \quad \text{(by (7.16))}\\
&= \frac{C}{\sigma_X(\Delta(x_o,r))}\int_{\Delta(x_o,r)} K(x_1,y)d\sigma_X(y)\ .
\end{aligned}
$$

This concludes the proof of the reverse Hölder inequality. Regarding absolute continuity, we already know from (7.16) that $d\omega^{x_1}$ is absolutely continuous with respect to $d\sigma_X$. To prove that $d\sigma_X$ is absolutely continuous with respect to $d\omega^{x_1}$ we only need to observe that the reverse Hölder inequality for K established above and the doubling property for $d\sigma_X$ from (8.2) in Corollary 8.4 allow us to invoke Lemma 5 from [**CF**].

□

We next establish a reverse Hölder inequality for the kernel $P(x,y)$ defined in (7.5). The main trust of this result is that, under a certain balanced-degeneracy assumption on the surface measure σ of ∂D, it implies the mutual absolute continuity of $\mathcal{L}$-harmonic measure and surface measure. Given the fact that, as we have explained in the introduction, surface measure is not the natural measure in the subelliptic Dirichlet problem, being able to isolate a condition which guarantees such mutual absolute continuity has some evident important consequences. To state the main result we modify the class of ADP_X domains in Definition 1.1. Specifically, we pose the following

DEFINITION 8.10. *Given a system $X = \{X_1, ..., X_m\}$ of smooth vector fields satisfying* (1.1), *we say that a connected bounded open set $D \subset \mathbb{R}^n$ is σ-*admissible *for the Dirichlet problem* (1.3), *or simply $\sigma - ADP_X$, if:*

i) D is of class C^∞;

ii) D is non-tangentially accessible (NTA_X) with respect to the Carnot-Caratheodory metric associated to the system $\{X_1, ..., X_m\}$ (see Definition 8.1);

iii) D satisfies a uniform tangent outer X-ball condition (see Definition 6.2);

iv) There exist $B, R_o > 0$ depending on X and D such that for every $x_o \in \partial D$ and $0 < r < R_o$ one has

$$\left(\max_{y\in\Delta(x_o,r)} W(y)\right) \sigma(\Delta(x_o,r)) \ \le\ B\ \frac{|B_d(x_o,r)|}{r}\ .$$

We note that Definitions 1.1 and 8.10 differ only in part *iv*). Also, (7.8) gives

$$\sigma_X(\Delta(x_o,r)) \ =\ \int_{\Delta(x_o,r)} W(y)d\sigma(y) \ \le\ \left(\max_{y\in\Delta(x_o,r)} W(y)\right) \sigma(\Delta(x_o,r))\ .$$

This observation shows that

$$\sigma - ADP_X\ \subset ADP_X\ .$$

The reason for which we have referred to the new assumption on σ as a balanced-degeneracy condition is that, as we have seen in the introduction the measure σ badly degenerates on the characteristic set Σ. On the other hand, the angle function W vanishes on Σ, thus balancing such degeneracy.

Proof of Theorem 1.5. The relevant reverse Hölder inequality for $P(x_1,\cdot)$ is proved starting from the second identity $d\omega^{x_1} = P(x_1,\cdot)d\sigma$ in (7.16) and then arguing in a similar fashion as in the proof of Theorem 1.3 but using the non-degeneracy estimate in *iv*) of Definition 8.10 instead of the upper 1-Ahlfors assumption in Definition 1.1. We leave the details to the interested reader.

□

A consequence of Theorem 1.5 and of Theorem 8.6 is the following result. We stress that such result would be trivial if the surface balls would just be the ordinary Euclidean ones, but this is not the case here. Our surface balls $\Delta(y,r)$ are the metric ones. Another comment is that away from the characteristic set the next result would be already contained in those in [**MM**].

THEOREM 8.11. *Let $D\subset\mathbb{R}^n$ be a $\sigma - ADP_X$ domain. There exist C, $R_1 > 0$ depending on the $\sigma - ADP_X$ parameters of D such that for every $y\in\partial D$ and $0<r<R_1$,*

$$\sigma(\Delta(y,2r)) \ \le\ C\ \sigma(\Delta(y,r))\ .$$

Proof. Applying Theorem 1.5 with $p = 2$, we find

$$\begin{aligned}\frac{1}{\sigma(\Delta(x_o,r))}\int_{\Delta(x_o,r)} P(x_1,y)^2\ d\sigma(y) \ &\le\ \left(\frac{C}{\sigma(\Delta(x_o,r))}\int_{\Delta(x_o,r)} P(x_1,y)\ d\sigma(y)\right)^2\\ &=\ C\left(\frac{\omega^{x_1}(\Delta(x_o,r))}{\sigma(\Delta(x_o,r))}\right)^2\ .\end{aligned}$$

This gives

$$\begin{aligned}\sigma(\Delta(x_o,2r)) \ &\le\ C\frac{\omega^{x_1}(\Delta(x_o,2r))^2}{\int_{\Delta(x_o,2r)} P(x_1,y)^2\ d\sigma(y)} \qquad \text{(by Theorem 8.6)}\\ &\le C\frac{\omega^{x_1}(\Delta(x_o,r))^2}{\int_{\Delta(x_o,r)} P(x_1,y)^2\ d\sigma(y)} \ \le\ C\frac{\left(\int_{\Delta(x_o,r)} P(x_1,y)\, d\sigma(y)\right)^2}{\int_{\Delta(x_o,r)} P(x_1,y)^2\ d\sigma(y)}\\ &\le\ C\frac{\left(\int_{\Delta(x_o,r)} P(x_1,y)^2\, d\sigma(y)\right)\left(\int_{\Delta(x_o,r)} d\sigma(y)\right)}{\int_{\Delta(x_o,r)} P(x_1,y)^2\ d\sigma(y)} \ =\ C\ \sigma(\Delta(x_o,r))\ .\end{aligned}$$

□

Our final goal in this section is to study the Dirichlet problem for sub-Laplacians when the boundary data are in L^p with respect to either the measure σ_X or the surface measure σ. We thus turn to the

Proof of Theorem 1.4. The first step in the proof consists of showing that functions $f \in L^p(\partial D, d\sigma_X)$ are resolutive for the Dirichlet problem (1.3). In view of Theorem 3.3 it is enough to show that $f \in L^1(\partial D, d\omega^{x_1})$ for some fixed $x_1 \in D$. This follows from (7.16) and Proposition 7.9, based on the following estimates

$$\begin{aligned}\int_{\partial D} |f(y)|\, d\omega^{x_1}(y) &= \int_{\partial D} |f(y)| K(x_1,y)\, d\sigma_X(y)\\ &\le \left(\int_{\partial D} |f(y)|^p\, d\sigma_X(y)\right)^{\frac{1}{p}} \left(\int_{\partial D} K(x_1,y)^{p'}\, d\sigma_X(y)\right)^{\frac{1}{p'}}\\ &\le C\left(\int_{\partial D} |f(y)|^p\, d\sigma_X(y)\right)^{\frac{1}{p}} .\end{aligned}$$

This shows that $L^p(\partial D, d\sigma_X) \subset L^1(\partial D, d\omega^{x_1})$ and therefore, in view of Theorem 3.3, for each $f \in L^p(\partial D, d\sigma_X)$ the generalized solution solution H_f^D exists and it is represented by

$$H_f^D(x) = \int_{\partial D} f(y)\, d\omega^x(y) .$$

At this point we invoke Theorem 8.8 and obtain for every $y \in \partial D$

$$N_\alpha(H_f^D)(y) \;\le\; C\; M_{\omega^{x_1}}(f)(y) . \tag{8.8}$$

Moreover, H_f^D converges non-tangentially $d\omega^{x_1}$-*a.e.* to f. By virtue of Theorems 1.3 and 1.5, we also have that H_f^D converges $d\sigma_X$-a.e. to f. To conclude the proof, we need to show that there exists a constant C depending on $1<p<\infty$, D and X such that

$$\|N_\alpha(H_f^D)\|_{L^p(\partial D, d\sigma_X)} \le C\|f\|_{L^p(\partial D, d\sigma_X)} ,$$

for every $f \in L^p(\partial D, d\sigma_X)$. In order to accomplish this we start by proving the following intermediate estimate

$$\|M_{\omega^{x_1}}(f)\|_{L^p(\partial D, d\sigma_X)} \le C\|f\|_{L^p(\partial D, d\sigma_X)}, \quad 1<p\le\infty . \tag{8.9}$$

Since $p>1$, choose β so that $0<\beta<p$ and fix $x_1 \in D$ as in Theorem 1.3. From (7.16) and the reverse Hölder inequality in Theorem 1.3 we have

$$\begin{aligned}&\frac{1}{\omega^{x_1}(\Delta(x_o,r))}\int_{\Delta(x_o,r)} f(y)d\omega^{x_1}(y)\\ &\le \frac{1}{\omega^{x_1}(\Delta(x_o,r))}\left(\int_{\Delta(x_o,r)} |f(y)|^\beta\, d\sigma_X(y)\right)^{\frac{1}{\beta}}\left(\int_{\Delta(x_o,r)} K(x_1,y)^{\beta'}\, d\sigma_X(y)\right)^{\frac{1}{\beta'}}\\ &\le C\frac{\sigma_X(\Delta(x_o,r))^{\frac{1}{\beta'}}}{\omega^{x_1}(\Delta(x_o,r))}\left(\frac{1}{\sigma_X(\Delta(x_o,r))}\int_{\Delta(x_o,r)} K(x_1,y)\, d\sigma_X(y)\right)\|f\|_{L^\beta(\Delta(x_o,r),d\sigma_X)}\\ &= C\left(\frac{1}{\sigma_X(\Delta(x_o,r))}\int_{\Delta(x_o,r)} |f(y)|^\beta\, d\sigma_X(y)\right)^{\frac{1}{\beta}} .\end{aligned}$$

If we now fix $y \in \partial D$ and take the supremum on both sides of the latter inequality by integrating on every surface ball $\Delta(x_o, r)$ containing y, we obtain

$$M_{\omega^{x_1}}(f)(y) \leq C\, M_{\sigma_X}(|f|^\beta)(y)^{\frac{1}{\beta}}\ . \tag{8.10}$$

By the doubling condition (8.2) in Corollary 8.4 we know that the space $(\partial D, d(x,y), d\sigma_X)$ is a space of homogeneous type. This allows us to use the results in [**CW**] and invoke the continuity in $L^p(\partial D, d\sigma_X)$ of the Hardy-Littlewood maximal function obtaining

$$\begin{aligned} \|M_{\omega^{x_1}} f\|^p_{L^p(\partial D, d\sigma_X)} &\leq C\, \|M_{\sigma_X}(|f|^\beta)^{\frac{1}{\beta}}\|^p_{L^p(\partial D, d\sigma_X)} \\ &= \int_{\partial D} M_{\sigma_X}(|f|^\beta)^{\frac{p}{\beta}}\, d\sigma_X \leq C \int_{\partial D} |f|^p\, d\sigma_X = C\, \|f\|^p_{L^p(\partial D, d\sigma_X)}\ , \end{aligned}$$

which proves (8.9). The conclusion of the theorem follows at once from (8.8) and (8.9).

□

Finally, we give the

Proof of Theorem 1.6. If the domain D is a $\sigma - ADP_X$-domain, instead of a ADP_X-domain, then using Theorem 1.5 instead of Theorem 1.3 we can establish the solvability of the Dirichlet problem for boundary data in L^p with respect to the standard surface measure. Since the proof of the following result is similar to that of Theorem 1.4 (except that one needs to use the second identity $d\omega^{x_1} = P(x_1, \cdot)d\sigma$ in (7.16) and also Theorem 8.11), we leave the details to the interested reader.

□

9. A survey of examples and some open problems

In the study of boundary value problems for sub-Laplacians one faces two type of difficulties. On one side there is the elusive nature of the underlying sub-Riemannian geometry which makes most of the classical results hard to establish. On the other hand, any new result requires a detailed analysis of geometrically significant examples, without which the result itself would be devoid of meaning. This task is not easy, the difficulties being mostly related to the presence of characteristic points. In this perspective it becomes important to provide examples of ADP_X-domains. In this section we recall some of the pertinent results from recent literature.

Examples of NTA_X domains. In the classical setting Lipschitz and even BMO_1 domains are NTA domains [**JK**]. In a Carnot-Carathéodory space it is considerably harder to produce examples of such domains, due to the presence of characteristic points on the boundary. In [**CG1**] it was proved that in a Carnot group of step two every $C^{1,1}$ domain with cylindrical symmetry at characteristic points is NTA_X. In particular, the pseudo-balls in the natural gauge of such groups are NTA_X. This result was subsequently generalized by Monti and Morbidelli [**MM**].

THEOREM 9.1. *In a Carnot group of step $r = 2$ every bounded (Euclidean) $C^{1,1}$ domain is NTA_X with respect to the Carnot-Carathéodory metric associated to a system X of generators of the Lie algebra.*

Examples of domains satisfying the uniform outer X-ball property. The following result provides a general class of domains satisfying the uniform X-ball condition, see [**LU1**] and [**CGN2**]. We recall the following definition from [**CGN2**]. Given a Carnot group $\mathbb{G}$, with Lie algebra $\mathfrak{g}$, a set $A \subset \mathbb{G}$ is called *convex*, if $\exp^{-1}(A)$ is a convex subset of $\mathfrak{g}$.

THEOREM 9.2. *Let $\mathbb{G}$ be a step two Carnot group of Heisenberg type with a given orthogonal set $X = \{X_1, ..., X_m\}$ of generators of its Lie algebra, and let $D \subset \mathbb{G}$ be a convex set. For every $R > 0$ and $x_o \in \partial D$ there exists a X-ball $B(x_o, R)$ such that (6.4) is satisfied. From this it follows that every bounded convex subset of $\mathbb{G}$ satisfies the uniform outer X-ball condition. In particular, this is true for the gauge balls.*

We mention explicitly that, thanks to the results in [**K**], in every group of Heisenberg type with an orthogonal system X of generators of $\mathfrak{g} = V_1 \oplus V_2$, the fundamental solution of the sub-Laplacian associated with X is given by

$$\Gamma(x, y) = \frac{C(\mathbb{G})}{N(x^{-1}y)^{Q-2}} ,$$

where $Q = dim(V_1) + 2\ dim(V_2)$ is the homogeneous dimension of $\mathbb{G}$, and

$$N(x, y) = (|x|^4 + 16|y|^2)^{1/4} ,$$

is the non-isotropic Kaplan's gauge. Kaplan's formula for the fundamental solution shows, in particular, that in a group of Heisenberg type the X-balls coincide with the gauge pseudo-balls (incidentally, in this setting the gauge defines an actual distance, see [**Cy**]). As a consequence of this fact, the exterior X-balls in Theorem 9.2 can be constructed explicitly by finding the coordinates of their center through the solution of a linear system and a second order equation.

Ahlfors type estimates for the perimeter measure. Recall that if $D \subset \mathbb{R}^n$ is a standard C^1, or even a Lipschitz domain, then there exist positive constants α, β and R_o depending only on n and on the Lipschitz character of D, such that for every $x_o \in \partial D$, and every $0 < r < R_o$ one has

$$\alpha\, r^{n-1} \le \sigma(\partial D \cap B(x_o, r)) = P(D; B(x_o, r)) \le \beta\, r^{n-1} . \tag{9.1}$$

Here, we have denoted by $P(D, B(x_o, r))$ the perimeter of D in $B(x_o, r)$ according to De Giorgi. Estimates such as (9.1) are referred to as the 1-Ahlfors property of surface measure. They play a pervasive role in Euclidean analysis especially in connection with geometric measure theory and its applications to the study of boundary value problems. In what follows we recall some basic regularity results for the X-perimeter measure which generalize (9.1) and play a central role in the applications of our results. We have mentioned in the introduction that from the standpoint of the Carnot-Carathéodory geometry, Euclidean smoothness of a domain is of no significance. Even for C^∞ domains one should not, therefore, expect 1-Ahlfors regularity in general, see [**CG2**] for various examples. For this reason we introduce the notion of type of a boundary point, and recall a result showing that if a domain possesses such property, then the corresponding X-perimeter satisfies Ahlfors regularity properties with respect to the metric balls.

Given a system of C^∞ vector fields $X = \{X_1, ..., X_m\}$ satisfying (1.1), consider a bounded C^1 domain $D \subset \mathbb{R}^n$ with an outer normal $\boldsymbol{N}$. We say that a point $x_o \in \partial D$ is *of type* ≤ 2 if either there exists $j_o \in \{1, ..., m\}$ such that

$< X_{j_o}(x_o), \boldsymbol{N}(x_o) > \neq 0$ (i.e., x_o is non-characteristic, see Definition 3.11), or there exist indices $i_o, j_o \in \{1, ..., m\}$ such that $< [X_{i_o}, X_{j_o}](x_o), \boldsymbol{N}(x_o) > \neq 0$. We say that D is of type ≤ 2 if every point $x_o \in \partial D$ is of type ≤ 2. We stress that when the system has rank $r \leq 2$, then every C^1 domain is automatically of type ≤ 2. An important instance is given by a Carnot group of step $r = 2$. In such a group, every bounded C^1 domain is of type ≤ 2. The following theorem is from [**CG1**].

THEOREM 9.3. *Consider a bounded $C^{1,1}$ domain $D \subset \mathbb{R}^n$. For every point $x_o \in \partial D$ of type ≤ 2 there exist $A = A(D, x_o) > 0$ and $R_o = R_o(D, x_o) > 0$, depending continuously on x_o, such that for any $0 < r < R_o$ one has*

$$\sigma_X(\Delta(x_o, r)) \leq \left(\max_{y \in \Delta(x_o, r)} W(y) \right) \sigma(\Delta(x_o, r)) \leq A \frac{|B_d(x_o, r)|}{r} . \tag{9.2}$$

The same conclusion holds if ∂D is real analytic in a neighborhood of x_o, regardless of the type of x_o.

If D is a bounded C^2 domain, then for every point $x_o \in \partial D$ of type ≤ 2 there exist $A = A(D, x_o) > 0$ and $R_o = R_o(D, x_o) > 0$, depending continuously on x_o, such that for any $0 < r < R_o$, one has

$$\sigma_X(\Delta(x_o, r)) \geq A^{-1} \frac{|B_d(x_o, r)|}{r} . \tag{9.3}$$

We mention that in Carnot groups of step $r = 2$ the upper 1-Ahlfors estimate (9.2) was first proved in [**DGN1**], whereas for vector fields of rank $r = 2$ the lower estimate (9.3) was first established in [**DGN2**]. In the setting of Hörmander vector fields, upper Ahlfors estimates for the surface measure σ away from the characteristic set were first established in [**MM2**]. As a consequence of Theorem 9.3 we obtain the following

COROLLARY 9.4. *Let $X = \{X_1, ..., X_m\}$ be a set of C^∞ vector fields in $\mathbb{R}^n$ satisfying Hörmander's condition with rank two, i.e. such that*

$$span\{X_1, ..., X_m, [X_1, X_2], ..., [X_{m-1}, X_m]\} = \mathbb{R}^n ,$$

at every point. For every bounded $C^{1,1}$ domain $D \subset \mathbb{R}^n$ the horizontal perimeter measure σ_X is a 1-Ahlfors measure. Moreover the stronger estimate (9.2) *holds.*

As a consequence of the results listed above we obtain the following theorem which provides a large class of domains satisfying the ADP_X or even the stronger $\sigma - ADP_X$ property.

THEOREM 9.5. *Let $\mathbb{G}$ be a Carnot group of Heisenberg type and denote by $X = \{X_1, ..., X_m\}$ a set of generators of its Lie algebra. Every C^∞ convex bounded domain $D \subset \mathbb{G}$ is a ADP_X and also a $\sigma - ADP_X$ domain. In particular, the gauge balls in $\mathbb{G}$ are ADP_X and also $\sigma - ADP_X$ domains.*

To conclude our review of Ahlfors type estimates, we bring up an interesting connection between 1-Ahlfors regularity of the X-perimeter σ_X and the Dirichlet problem for the sub-Laplacian, see [**CG2**]:

THEOREM 9.6. *Let D be a bounded domain in a Carnot group $\mathbb{G}$. If the perimeter measure σ_X is 1-Ahlfors regular, then every $x_o \in \partial D$ is regular for the Dirichlet problem.*

This result, in conjunction with a class of examples for non-regular domain constructed in [**HH**] yields the following

COROLLARY 9.7. *If $r \geq 3$ and $m_1 \geq 3$, or $m_1 = 2$ and $r \geq 4$, then there exist Carnot groups $\mathbb{G}$ of step $r \in \mathbb{N}$, with dim $V_1 = m_1$, and bounded, C^∞ domains $D \subset \mathbb{G}$, whose perimeter measure σ_X is not 1-Ahlfors regular.*

Beyond Heisenberg type groups. The above overview shows that, so far, the known examples of ADP_X domains are relative to group of Heisenberg type. What happens beyond such groups? For instance, what can be said even for general Carnot groups of step two? One of the difficulties here is to find examples of domains satisfying the outer tangent X-ball condition. The explicit construction in Theorem 9.2 above rests on the special structure of a group of Heisenberg type, and an extension to more general groups appears difficult due to the fact that, in a general group, the X-balls are not explicitly known and they may be quite different from the gauge balls. In this connection it would be desirable to replace the uniform outer X-ball condition with a *uniform outer gauge pseudo-ball condition* (clearly the two definitions agree for groups of Heisenberg type). It would be quite interesting to know whether for general Carnot groups a uniform outer gauge pseudo-ball condition would suffice to establish the boundedness of the horizontal gradient of the Green function near the characteristic set (this question is open even for Carnot groups of step two which are not of Heisenberg type!). Concerning the question of examples we have the following special results.

DEFINITION 9.8. *Let $\mathbb{G}$ be a Carnot group and denote by $\mathfrak{g}$ its Lie algebra. We say that a family $\mathcal{F}$of smooth open subsets of $\mathfrak{g}$ is a $T-$family if it satisfies*

(i) For any $F \in \mathcal{F}$, the manifold ∂F is diffeomorphic to the unit sphere in the Lie algebra.

(ii) The family $\mathcal{F}$ is left-invariant, i.e. for any $x \in \mathbb{G}$ and $F \in \mathcal{F}$ we have $\log(x \exp(F)) \in \mathcal{F}$.

If $D \subset \mathfrak{g}$ is a smooth subset and $\mathcal{F}$ is a $T-$family, then we say that D is *tangent to* $\mathcal{F}$ if for every $x \in \partial D$ there exists $F \in \mathcal{F}$ such that $x \in \partial F$ and the tangent hyperplanes to ∂F and ∂D at x are identical, i.e. $T_x\partial F = T_x\partial D$.

THEOREM 9.9. *Let $\mathfrak{g}$ be the Lie algebra of a Carnot group of odd dimension. If $D \subset \mathfrak{g}$ is a smooth open set and $\mathcal{F}$ is a $T-$family, then D is tangent to $\mathcal{F}$.*

Proof. In order to avoid using exp and log maps for all $x, y \in \mathfrak{g}$ we will denote by xy the algebra element $\log(\exp x \exp y)$. We will assume that $\mathfrak{g}$ is endowed with a Euclidean metric, so that notions of orthogonality and projections can be used. Fix $x_o \in \partial D$ and choose any element $F \in \mathcal{F}$. We will show that there exists $z \in \mathfrak{g}$ such that the left-translation zF is tangent to ∂D at x_0. Let $n = dim(\mathbb{G}) = dim(\mathfrak{g})$ be odd, and denote by S^{n-1} the unit (Euclidean) sphere of dimension $n-1$. Define the map $N : \partial F \to S^{n-1}$ as follows: For each point $x \in \partial F$ set $\tilde{D} = xx_o^{-1}D$ and observe that this is a smooth open set with $x \in \partial\tilde{D} \cap \partial F$. Set

$N(x) =$ the outer unit normal to the boundary of the translated set $\partial\tilde{D}$ at the point x.

This amounts to left-translating the point x_o to the point x and considering the unit normal to the translated domain at that point. The smoothness of D and of the group structure of $\mathbb{G}$ implies that N is a smooth vector field in ∂F. In order to prove the theorem we need to show that for some point $x \in \partial F$ the vector $N(x)$ is orthogonal to $T_x\partial F$. In fact in that case the set F would be tangent to the translated set $\tilde{D}$ at the point x_o, and its left translation $x_o x^{-1}F$ could be chosen as the element of $\mathcal{F}$ tangent to D at the point x_o. Recall that left translation, being a diffeomorphism, preserves the property of being tangent. The conclusion comes from the fact that there cannot be any smooth tangent non vanishing vector field on ∂F since it is diffeomorphic to S^{n-1}. Consequently the vector fields obtained by projecting $N(x)$ on $T_x\partial F$ must vanish for some point $x \in \partial F$.

□

COROLLARY 9.10. *Let $\mathbb{G}$ be a Carnot group of step two with odd-dimensional Lie algebra $\mathfrak{g}$ and $D \subset \mathfrak{g}$ be a smooth convex subset. If $\mathcal{F}$ is a $T-$family, composed of convex subsets, and invariant by the transformation $x \to x^{-1}$ then for any $x \in \partial D$ there exists $F \in \mathcal{F}$ such that $F \subset D^c$, and $x \in \partial F$.*

Proof. In a Carnot group of step two the left translation map is affine and hence preserves convexity. The same holds for the inverse map. Consequently at any boundary point $x_0 \in \partial D$ there will be a convex manifold $F \in \mathcal{F}$ tangent to D at x_0. Being D convex as well then D and F must either be on the same side or lay at different sides of the common tangent plane $T_{x_0}\partial D$. By translating x_0 to the origin and considering either F or F^{-1} we can pick the manifold lying on the opposite side of D and hence disjoint from it. □

Choosing appropriate $T-$families of convex sets we can now prove our two main results concerning the uniform outer gauge pesudo-ball and $X-$ball conditions.

COROLLARY 9.11. *Let $\mathbb{G}$ be a Carnot group of step two with odd-dimensional Lie algebra $\mathfrak{g}$. Given a convex set $D \subset \mathbb{G}$, for every $x_o \in \partial D$ and every $r > 0$ there exists a gauge pseudo-ball $B(x_1, r)$ which is tangent to ∂D in x_o from the outside, i.e., such that* (6.4) *is satisfied. Furthermore, every bounded convex set in $\mathbb{G}$ satisfies the uniform outer gauge pseudo-ball condition.*

Proof. If D is smooth then the proof follows from the immediate observation that the gauge balls are convex sets in the Lie algebra and are diffeomorphic to S^{n-1} (see for instance [**F2**]). For non-smooth convex domains D, we consider $x_o \in \partial D$ and a new domain $\hat{D}$ obtained as the half space including D and with boundary $T_{x_o}\partial D$. Since $\hat{D}$ is a smooth convex domain then we can apply to it the previous theorem and find an outer tangent gauge ball at the point x_o with radius $r > 0$. Clearly this ball will also be tangent to the original domain D at x_o, and will be contained entirely in the complement of D. □

COROLLARY 9.12. *Let $\mathbb{G}$ be a Carnot group of step two with odd-dimensional Lie algebra $\mathfrak{g}$. If for every $x \in \mathbb{R}^n$ and for r sufficiently small the $X-$balls $B(x, r)$ are convex, and $B(x^{-1}, r) = B(x, r)^{-1}$ then every bounded convex set in $\mathbb{G}$ satisfies the uniform outer X-ball condition.*

Proof. We need only to show that the family of balls $B(x, r)$ form a $T-$family. In [**DG2**] it is shown that $X-$balls are starlike with respect to the family of homogeneous dilations in the Carnot group. In particular, one has the estimate

$$\langle \nabla\Gamma(\cdot, x), Z\rangle > 0$$

on $\partial B(x, r)$ where we have denoted by Z the generator of the homogeneous dilations. This inequality, coupled with Hörmander's hypoellipticity result, implies that $\partial B(x, r)$ is a smooth manifold, while the starlike property immediately implies that $\partial B(x, r)$ is diffeomorphic to the unit ball. □

We recall from the classical paper of Folland [**F2**] that in a Carnot group the fundamental solution of the sub-Laplacian is a function $\Gamma(x,y) = \Gamma(y^{-1}x)$ and $\Gamma(x^{-1}) = \Gamma^t(x)$, where Γ^t is the fundamental solution of the transpose of the sub-Laplacian $\mathcal{L}$. However, a sub-Laplacian on a Carnot group is self-adjoint, hence $\mathcal{L}^* = -\mathcal{L}$ and $\Gamma(x) = \Gamma(x^{-1})$. Let us denote by $||\cdot||$ the group gauge, if we assume that for all $x, y \in \mathbb{G}$ one has $\Gamma(xy^{-1}) = \Gamma(yx^{-1})$ (this happens for instance if $\Gamma(x) = \Gamma(||x||)$), and set $B(x,r) = \{y | \ \Gamma(y^{-1}x) > c\}$ then

$$B(x,r)^{-1} = \{y^{-1} | \ \Gamma(x^{-1}y) > c\} = B(x^{-1}, r).$$

We conclude by explicitly noting that a serious obstruction to extending the previous results to Carnot groups of higher step consists in the fact that, unlike in the step two case, the group left-translation may not preserve the convexity of the sets.

Beyond linear equations. Another interesting direction of investigation for the subelliptic Dirichlet problem is provided by the study of solutions to the non-linear equations which arise in connection with the case $p \neq 2$ of the Folland-Stein Sobolev embedding. In this direction a first step has been recently taken in [**GNg**] where, among other results, Theorem 6.4 has been extended to the Green function of the nonlinear equation

$$\mathcal{L}_p u = \sum_{j=1}^{2n} X_j(|Xu|^{p-2} X_j u) \ = \ 0 \ , \tag{9.4}$$

in the Heisenberg group $\mathbb{H}^n$. Here is the relevant result.

THEOREM 9.13. *Let $D \subset \mathbb{H}^n$ be a bounded domain satisfying the uniform outer X-ball condition. Given $1 < p \leq Q$, let $G_{D,p}$ denote the Green function associated with* (9.4) *and D. Denote by $g = (z,t), g' = (z',t') \in \mathbb{H}^n$.*

(i) *If $1 < p < Q$ there exists a constant $C = C(\mathbb{G}, D, p) > 0$ such that*

$$G_{D,p}(g',g) \ \leq \ C \ \left(\frac{d(g,g')}{|B(g,d(g,g'))|}\right)^{1/(p-1)} d(g',\partial D) \ , \quad g, g' \in D \ , \ g' \neq g \ .$$

(ii) *If $p = Q$, then there exists $C = C(\mathbb{G}, D) > 0$ such that*

$$G_{D,p}(g',g) \ \leq \ C \ \log\left(\frac{diam(D)}{d(g,g')}\right) \frac{d(g',\partial D)}{d(g,g')} \ , \quad g, g' \in D \ , \ g' \neq g \ .$$

One might naturally wonder about results such as Theorem 6.6 in this setting. However, before addressing this question one has to understand the fundamental open question of the interior local bounds of the horizontal gradient of a solution to (9.4). For recent progress in this direction see the paper [**MZZ**].

References

[Bel] A. Bellaïche, *The tangent space in sub-Riemannian geometry. Sub-Riemannian geometry,,* Progr. Math., **144** (1996), Birkhäuser, 1–78.

[B] J. M. Bony, *Principe du maximum, inégalité de Harnack et unicité du problème de Cauchy pour les operateurs elliptique degeneres*, Ann. Inst. Fourier, Grenoble, 1, **119** (1969), 277-304.

[CFMS] L. Caffarelli, E. Fabes, S. Mortola & S. Salsa, *Boundary behavior of nonnegative solutions of elliptic operators in divergence form*, Indiana J. Math, 4, **30** (1981), 621-640.

[CDG1] L. Capogna, D. Danielli & N. Garofalo, *Subelliptic mollifiers and a characterization of Rellich and Poincaré domains*, Rend. Sem. Mat. Univ. Pol. Torino, 4, **54** (1993), 361-386.

[CDG2] ———, *The geometric Sobolev embedding for vector fields and the isoperimetric inequality*, Comm. Anal. Geom. **2** (1994), no. 2, 203-215.

[CDG3] ———, *Subelliptic mollifiers and a basic pointwise estimate of Poincaré type*, Math. Zeitschrift, **226** (1997), 147-154.

[CDG4] ———, *Capacitary estimates and the local behavior of solutions of nonlinear subelliptic equations*, American J. of Math., **118** (1997), 1153-1196.

[CG1] L. Capogna & N. Garofalo, *Boundary behavior of nonegative solutions of subelliptic equations in NTA domains for Carnot-Carathéodory metrics*, Journal of Fourier Anal. and Appl., 4 **4** (1995).

[CG2] ———, *Ahlfors type estimates for perimeter measures in Carnot-Carathodory spaces*, J. Geom. Anal. **16** (2006), no. 3, 455-497.

[CGN1] L. Capogna, N. Garofalo & D. M. Nhieu, *A version of a theorem of Dahlberg for the subelliptic Dirichlet problem*, Math. Res. Letters, **5** (1998), 541-549.

[CGN2] ———, *Properties of harmonic measures in the Dirichlet problem for nilpotent Lie groups of Heisenberg type*, Amer. J. Math. **124**, vol 2, (2002) 273-306.

[Ch] W.L. Chow, *Über System von linearen partiellen Differentialgleichungen erster Ordnug*, Math. Ann., **117** (1939), 98-105.

[Ci] G. Citti, *Wiener estimates at boundary points for Hörmander's operators*, Boll. U.M.I., **2-B** (1988), 667-681.

[CGL] G. Citti, N. Garofalo & E. Lanconelli, *Harnack's inequality for sum of squares of vector fields plus a potential*, Amer. J. Math., 3,**115** (1993), 699-734.

[CF] R. Coifman & C. Fefferman, *weighted norm inequalities for maximal functions and singular integrals*, Studia Math. **51** (1974), 241-250.

[CW] R. Coifman & G. Weiss, *Analyse harmonique non-commutative sur certains espaces homogenes*, Springer-Verlag, (1971).

[CGr] L. Corwin and F. P. Greenleaf *Representations of nilpotent Lie groups and their applications, Part I: basic theory and examples*, Cambridge Studies in Advanced Mathematics 18, Cambridge University Press, Cambridge (1990).

[CDKR] M. Cowling, A. H. Dooley, A. Korányi and F. Ricci *H-type groups and Iwasawa decompositions*, Adv. in Math., **87** (1991), 1-41.

[Cy] J. Cygan, *Subadditivity of homogeneous norms on certain nilpotent Lie groups*, Proc. Amer. Math. Soc., **83** (1981), 69-70.

[Da1] B. E. J. Dahlberg, *Estimates of harmonic measure*, Arch. Rat. Mech. An., **65** (1977), 272-288.

[Da2] ———, *On the Poisson integral for Lipschitz and C^1-domains*, Studia Math., **66** (1979), 13-24.

[D] D. Danielli, *Regularity at the boundary for solutions of nonlinear subelliptic equations*, Indiana J. of Math., 1, **44** (1995), 269-286.

[DG] D. Danielli & N. Garofalo, *Geometric properties of solutions to subelliptic equations in nilpotent Lie groups*, Lect. Notes in Pure and Appl. Math., "Reaction Diffusion Systems", Trieste, October 1995, Ed. G. Caristi Invernizzi, E. Mitidieri, Marcel Dekker, **194** (1998), 89-105.

[DG2] D.Danielli & N. Garofalo, *Green functions in nonlinear potential theory in Carnot groups and the geometry of their level sets*, preprint 2000.

[DGN1] D. Danielli, N. Garofalo & D. M. Nhieu, *Trace inequalities for Carnot-Carathéodory spaces and applications*, Ann. Sc. Norm. Sup. Pisa, Cl. Sci. (4), 2, **27** (1998), 195-252.

[DGN2] ———, *Non-doubling Ahlfors measures, perimeter measures, and the characterization of the trace spaces of Sobolev functions in Carnot-Carathéodory spaces*, Mem. Amer. Math. Soc. **182** (2006), no. 857, x+119 pp.

[DGN3] ———, *Sub-Riemannian calculus on hypersurfaces in Carnot groups*, Adv. Math. **215** (2007), no. 1, 292-378.

[De1] M. Derridj, *Un probléme aux limites pour une classe d'opérateurs du second ordre hypoelliptiques*, Ann. Inst. Fourier, Grenoble, **21**, 4 (1971), 99-148.

[De2] ———, *Sur un théorème de traces*, Ann. Inst. Fourier, Grenoble, **22**, 2 (1972), 73-83.

[Diaz] R. L. Diaz, *Boundary regularity of a canonical solution of the $\overline{\partial}_b$ problem*, Duke Math. J., 1, **64** (1991), 149-193.

[FJR] E. B. Fabes, M. Jodeit, Jr. & N. Rivière, *Potential techniques for boundary value problems on C^1-domains*, Acta Math. **141** (1978), 165-186.

[Fe] H. Federer, *Geometric measure theory*, Springer-Verlag, (1969).

[FP] C. Fefferman & D.H. Phong, *Subelliptic eigenvalue problems*, Proceedings of the Conference in Harmonic Analysis in Honor of A. Zygmund, Wadsworth Math. Ser., Belmont, CA, (1981), 530-606.

[FSC] C. Fefferman & A. Sanchez-Calle, *Fundamental solutions for second order subelliptic operators*, Ann. Math., **124** (1986), 247–272.

[F1] G. Folland, *A fundamental solution for a subelliptic operator*, **79** (1973), 373-376.

[F2] ———, *Subelliptic estimates and function spaces on nilpotent Lie groups*, Ark. Math., **13** (1975), 161-207.

[FS] G.B. Folland & E.M. Stein, *Hardy Spaces on Homogeneous Groups*, Princeton Univ. Press., (1982).

[FSS] B. Franchi, R. Serapioni & F. Serra Cassano, *Approximation and imbedding theorems for weighted Sobolev spaces associated with Lipschitz continuous vector fields*, Boll. Un. Mat. Ital. B (7) **11** (1997), no. 1, 83-117.

[G] N. Garofalo, *Second order parabolic equations in nonvariational form: Boundary Harnack principle and comparison theorems for nonegative solutions*, Ann. Mat. Pura Appl. (4) **138** (1984), 267-296.

[GN1] N. Garofalo & D.M. Nhieu, *Isoperimetric and Sobolev inequalities for Carnot-Carathéodory spaces and the existence of minimal surfaces*, Comm. Pure Appl. Math., **49** (1996), 1081-1144.

[GN2] ———, *Lipschitz continuity, global smooth approximations and extension theorems for Sobolev functions in Carnot-Carathéodory spaces*, J. d'Analyse Math., **74** (1998), 67-97.

[GNg] N. Garofalo & Nguyen C. P., *Boundary estimates for p-harmonic functions in the Heisenberg group*, preprint, 2007.

[GV] N. Garofalo & D. Vassilev, *Regularity near the characteristic set in the non-linear Dirichlet problem and conformal geometry of sub-Laplacians on Carnot groups*, Math. Ann. **318** (2000), 453-516.

[GT] D. Gilbarg & N. S. Trudinger, "Elliptic Partial Differential Equations of Second Order", 2nd edition, rev. third printing, Springer Verlag, Berlin, Heidelberg ,1998.

[HH] W. Hansen & H. Huber, *The Dirichlet problem for sub-Laplacians on nilpotent groups-Geometric criteria for regularity*, Math. Ann., **246** (1984), 537-547.

[H] H. Hörmander, *Hypoelliptic second-order differential equations*, Acta Math., **119** (1967), 147-171.

[HW1] R. R. Hunt & R. L. Wheeden, *On the boundary values of harmonic functions*, Trans. Amer. Math. Soc., **32** (1968), 307-322.

[HW2] ———, *Positive harmonic functions of Lipschitz domains*, Trans. Amer. Math. Soc., **47** (1970), 507-527.

[J1] D. Jerison, *The Dirichlet problem for the Kohn Laplacian on the Heisenberg group, Parts I and II*, J. Funct. Analysis, **43** (1981), 97-142.

[J2] ———, *Boundary regularity in the Dirichlet problem for* $\Box_b$ *on CR manifolds*, Comm. Pure Appl. Math., **36** (1983), 143-181.

[J3] ———, *The Poincaré inequality for vector fields satisfying Hörmander's condition*, Duke Math. J., **53** (1986), 503-523.

[JK] D. Jerison & C. E. Kenig, *Boundary behavior of harmonic functions in non-tangentially accessible domains*, Adv. Math., **46** 1982, 80-147.

[JK1] ———, *An identity with applications to harmonic measure*, Bull. Amer. Math. Soc., 2, **2** (1980), 447-451.

[K] A. Kaplan, *Fundamental solutions for a class of hypoelliptic PDE generated by composition of quadratic forms*, Trans. Amer. Math. Soc., **258** (1980), 147-153.

[Ke] C. E. Kenig, *Harmonic analysis techniques for second order elliptic boundary value problems*, Amer. Math. Soc., CBMS **83**, 1994.

[KN1] J. J. Kohn & L. Nirenberg, *Non-coercive boundary value problems*, Comm. Pure and Appl. Math., 18, **18** (1965), 443-492.

[KN2] ———, *Degenerate elliptic-parabolic equations of second order*, Comm. Pure Appl. Math. **20** 1967, 797-872.

[Ko] A. Korányi, *Kelvin transform and harmonic polynomials on the Heisenberg group*, Adv. Math. **56** (1985), 28-38.

[LU1] E. Lanconelli & F. Uguzzoni, *On the Poisson kernel for the Kohn Laplacian*, Rend. Mat. Appl. (7) **17** (1997), no. 4, 659–677.

[LU2] ———, *Degree theory for VMO maps on metric spaces and applications to Hrmander operators*, Ann. Sc. Norm. Super. Pisa Cl. Sci. (5) **1** (2002), no. 3, 569-601.

[L] G. Lu, *On Harnack's inequality for a class of strongly degenerate Schrödinger operators formed by vector fields*, Diff. Int. Equations, **7** (1994), no. 1, 73-100.

[MZZ] G. Mingione, A. Zatorska-Goldstein & X. Zhong, *Gradient regularity for elliptic equations in the Heisenberg group*, preprint, 2007.

[MM] R. Monti & D. Morbidelli, *Regular domains in homogeneous groups*, Trans. Amer. Math. Soc., **357** (2005), no. 8, 2975-3011.

[MM2] R. Monti & D. Morbidelli, *Trace theorems for vector fields*, Math. Z., **239** (2002), no. 4, 747-776.

[NSW] A. Nagel, E.M. Stein & S. Wainger, *Balls and metrics defined by vector fields I: basic properties*, Acta Math. **155** (1985), 103-147.

[NS] P. Negrini & V. Scornazzani, *Wiener criterion for a class of degenerate elliptic operators*, J. Diff. Eq., **166** (1987), 151-167.

[P] H. Poincaré, *Sur les equations aux dérivées partielles de la physique mathématique*, Amer. J. of Math., **12** (1890), 211-294.

[Ra] P. K. Rashevsky, *Any two points of a totally nonholonomic space may be connected by an admissible line*, Uch. Zap. Ped. Inst. im. Liebknechta, Ser. Phys. Math., (Russian) **2** (1938), 83-94.

[RS] L. P. Rothschild & E. M. Stein, *Hypoelliptic differential operators and nilpotent groups.* Acta Math. **137** (1976), 247–320.

[SC] A. Sanchez-Calle,*Fundamental solutions and geometry of sum of squares of vector fields*, Inv. Math., **78** (1984), 143-160.

[St] E.M. Stein, *Harmonic Analysis: Real Variable Methods, Orthogonality and Oscillatory Integrals*, Princeton Univ. Press., (1993).

[V] V. S. Varadarajan, *Lie Groups, Lie Algebras, and Their Representations*, Grad. Texts in Math., vol.102, Springer-Verlag, (1984).

[X] C-J, Xu, *On Harnack's inequality for second-order degenerate elliptic operators.* Chinese Ann. Math. Ser. A **10** (1989), no. 3, 359-365.

DEPARTMENT OF MATHEMATICS, UNIVERSITY OF ARKANSAS, FAYETTEVILLE, AR 72701
E-mail address, Luca Capogna: `lcapogna@uark.edu`

DEPARTMENT OF MATHEMATICS, PURDUE UNIVERSITY, WEST LAFAYETTE IN 47907-1968
E-mail address, Nicola Garofalo: `garofalo@math.purdue.edu`

DEPARTMENT OF MATHEMATICS, SAN DIEGO CHRISTIAN COLLEGE, 2100 GREENFIELD DR, EL CAJON CA 92019
E-mail address, Duy-Minh Nhieu: `dnhieu@sdcc.edu`

Proceedings of Symposia in Pure Mathematics
Volume **79**, 2008

Soviet-Russian and Swedish mathematical contacts after the war. A personal account.

Lars Gårding

Dedicated to Prof. V. Maz'ya on the occasion of his 70th birthday.

After the second world war ended it still took a long time for normal communications to be restored but when the change came it could be abrupt. The subject of this note is my personal experience of contacts between Soviet and later Russian mathematicians from the beginning of the 1950's.

It all started in 1946-47 when I spent a year at Princeton University USA on a stipend from the Swedish-American Foundation. I soon came into contact with a group of young American mathematicians sharing their time between the university and the Institute of Advanced Study. A running subject of conversation among them was the theory of normed rings, a subject blending algebra and analysis and later becoming a corner stone of harmonic analysis. The start was a paper in the thirties by the great Soviet mathematician Israel Gelfand. The chief ideologue of our group at the Institute was my friend Irving Segal. He told me that he even had wanted to go to the Soviet Union but his request got a negative answer from Syalin's office.

My chief interest at the the time was hyperbolic differential operators and one day Irving told me that he had seen an interesting hyperbolic paper by one Ivan Georgievich -Petrovsky in the main Soviet periodical the Matematitjeski Sbornik. I spent half the night with this paper without understanding its mixture of analysis and algebraic geometry. The subject was lacunas, regions where the fundamental solution of a hyperbolic operator unexpectedly vanishes. Lacunas in a special but interesting case had been the subject of a paper of mine. Later, in the middle sixties, Michael Atiyah and Roul Bott helped me with the topology to make a complete extension of Petrovsky's paper issued with a dedication to him in his native language.

In Princeton I had met a biologist who had started learning Russian seduced by some interesting paper. I decided to do the same. After this encounter, on the boat home to Sweden I found a prospective teacher who, unfortunately, lived in Stockholm and my home town was Lund in the south. Coming home I started studying Russian with a Russian emigree who had been a lawyer at the tsarist high court. My first success was being able to read Petrovsky's Russian summary of

2000 *Mathematics Subject Classification.* Primary 01A60.

his lacuna paper. I was not alone with my Russian teacher. In the late 1940's the political situation and the reputation of the many classical Russian writers made it interesting to study Russian and many did.

The Soviet mathematical world was opened up to me with a big ban in the form of a meeting in Moscow in the summer of 1956 to which the Russian mathematical society had invited a number of foreign mathematicians and their wives. We lived in one of the big hotels and every national group had its own interpreter. Our stay coincided with the twentieth Soviet party congress of which we knew nothing until the papers one day carried Chrustjev's speech about the cult of the personality of the Stalin era. When we asked out interpreters about the sense of this speech they could only give us a literal translation.

Arriving in Moscow, my wife and I were met by M.I. Vishik and a woman student. Vishik and I had common interests, we became friends and he later was a frequent visitor to Lund. My once imagined encounter took place when we were invited to dinner by Petrovsky together with Gelfand and his wife. Gelfand spoke then no English and very little German which hampered out conversation. Future Russian friends were Olga Ladyzhenskaja and Olga Olejnik. Others were S. Sobolev and the Georgian mathematician Vekua. With time my wife and I were to make many mathematical visits to the Soviet Union, both to Siberia and the South. After the fall of the Soviet Union, mine and the Swede's relations with Russian mathematicians underwent a drastic change. Our former hosts now appeared as emigrants looking for better economic conditions in the West. Some of them could find positions at a Swedish university. Long ago there were besides two Institutes of Technology only two of them, one in Uppsala and one in Lund but from 1950 on many new ones were created. Their somewhat hasty appearances made them in dire need of a competent scientific work force. The university of Linköping profited from the arrival of Vladimir Maz'ya a specialist in nonlinear differential equations and the Institute of Technology adopted Ari Laptev who had had some political difficulties in Leningrad. The university in Blekinge that specialized in computing received competence in analysis with the arrival of Nail N. Ibrahimov whom I had met in Akademgorodok, a Siberian academic town. By now Vladimir Maz'ya retains a connection with Linköping and has position at the University of Liverpool and Laptev is employed by London University.

I had sparse but cordial relations with Maz'ya. He was the opponent of a Lund thesis and I visited him and his family in Linkping. With his wife Shaposhnikova he wrote a book on the French mathematician Hadamard and Shaposhnikova translated my dialogue between God and von Neumann into Russian. As he himself told Vladimir was worried about his ability to support his family, wife and mother-in-law, after his Swedish retirement. As things have turned out, his worries were unsubstantiated.

The Russian mathematicians I have mentioned were my friends. They are only a small part of the many reputable Russian mathematicians who left Russia after the fall of the Soviet State to get jobs in the West. At the same time Russian mathematics seem to regain its previous strength as evidenced by Perelman's proof of the Poincare conjecture. The upheavals after the Second World War reflected in the text above are now being replaced by a world of free communication and travel recalling the conditions of the nineteenth century.

Lunds Universitet Matematiska Institutionen Box 118 221 00 LUND
E-mail address: lars.garding@math.lu.se

Proceedings of Symposia in Pure Mathematics
Volume **79**, 2008

Generalized Robin Boundary Conditions, Robin-to-Dirichlet Maps, and Krein-Type Resolvent Formulas for Schrödinger Operators on Bounded Lipschitz Domains

Fritz Gesztesy and Marius Mitrea

Dedicated with great pleasure to Vladimir Maz'ya on the occasion of his 70th birthday.

ABSTRACT. We investigate generalized Robin boundary conditions, Robin-to-Dirichlet maps, and Krein-type resolvent formulas for Schrödinger operators on bounded Lipschitz domains in $\mathbb{R}^n$, $n \geqslant 2$. We also discuss the case of bounded $C^{1,r}$-domains, $(1/2) < r < 1$.

1. Introduction

This paper is a continuation of the earlier papers [**43**] and [**46**], where we studied general, not necessarily self-adjoint, Schrödinger operators on $C^{1,r}$-domains $\Omega \subset \mathbb{R}^n$, $n \in \mathbb{N}$, $n \geqslant 2$, with compact boundaries $\partial\Omega$, $(1/2) < r < 1$ (including unbounded domains, i.e., exterior domains) with Dirichlet and Neumann boundary conditions on $\partial\Omega$. Our results also applied to convex domains Ω and to domains satisfying a uniform exterior ball condition. In addition, a careful discussion of locally singular potentials V with close to optimal local behavior of V was provided in [**43**] and [**46**].

In this paper we push the envelope in a different direction: Rather than discussing potentials with close to optimal local behavior, we will assume that $V \in L^\infty(\Omega; d^nx)$ and hence essentially replace it by zero nearly everywhere in this paper. On the other hand, instead of treating Dirichlet and Neumann boundary conditions at $\partial\Omega$, we now consider generalized Robin and again Dirichlet boundary conditions, but under minimal smoothness conditions on the domain Ω, that is, we now consider Lipschitz domains Ω. Additionally, to reduce some technicalities, we will assume that Ω is bounded throughout this paper. Occasionally we also discuss the case of bounded $C^{1,r}$-domains, $(1/2) < r < 1$. The principal new result in

---

2000 *Mathematics Subject Classification.* Primary: 35J10, 35J25, 35Q40; Secondary: 35P05, 47A10, 47F05.

Key words and phrases. Multi-dimensional Schrödinger operators, bounded Lipschitz domains, Robin-to-Dirichlet and Dirichlet-to-Neumann maps.

Based upon work partially supported by the US National Science Foundation under Grant Nos. DMS-0400639 and FRG-0456306.

this paper is a derivation of Krein-type resolvent formulas for Schrödinger operators on bounded Lipschitz domains Ω in connection with the case of Dirichlet and generalized Robin boundary conditions on $\partial\Omega$.

In Section 2 we provide a detailed discussion of self-adjoint Laplacians with generalized Robin (and Dirichlet) boundary conditions on $\partial\Omega$. In Section 3 we then treat generalized Robin and Dirichlet boundary value problems and introduce associated Robin-to-Dirichlet and Dirichlet-to-Robin maps. Section 4 contains the principal new results of this paper; it is devoted to Krein-type resolvent formulas connecting Dirichlet and generalized Robin Laplacians with the help of the Robin-to-Dirichlet map. Appendix A collects useful material on Sobolev spaces and trace maps for $C^{1,r}$ and Lipschitz domains. Appendix B summarizes pertinent facts on sesquilinear forms and their associated linear operators. Estimates on the fundamental solution of the Helmholtz equation in $\mathbb{R}^n$, $n \geqslant 2$, are recalled in Appendix C. Finally, certain results on Calderón–Zygmund theory on Lipschitz surfaces of fundamental relevance to the material in the main body of this paper are presented in Appendix D.

While we formulate and prove all results in this paper for self-adjoint generalized Robin Laplacians and Dirichlet Laplacians, we emphasize that all results in this paper immediately extend to closed Schrödinger operators $H_{\Theta,\Omega} = -\Delta_{\Theta,\Omega} + V$, $\mathrm{dom}(H_{\Theta,\Omega}) = \mathrm{dom}(-\Delta_{\Theta,\Omega})$ in $L^2(\Omega; d^n x)$ for (not necessarily real-valued) potentials V satisfying $V \in L^\infty(\Omega; d^n x)$, by consistently replacing $-\Delta$ by $-\Delta + V$, etc. More generally, all results extend directly to Kato–Rellich bounded potentials V relative to $-\Delta_{\Theta,\Omega}$ with bound less than one.

Next, we briefly list most of the notational conventions used throughout this paper. Let $\mathcal{H}$ be a separable complex Hilbert space, $(\,\cdot\,,\,\cdot\,)_{\mathcal{H}}$ the scalar product in $\mathcal{H}$ (linear in the second factor), and $I_{\mathcal{H}}$ the identity operator in $\mathcal{H}$. Next, let T be a linear operator mapping (a subspace of) a Banach space into another, with $\mathrm{dom}(T)$ and $\mathrm{ran}(T)$ denoting the domain and range of T. The spectrum (resp., essential spectrum) of a closed linear operator in $\mathcal{H}$ will be denoted by $\sigma(\,\cdot\,)$ (resp., $\sigma_{\rm ess}(\,\cdot\,)$). The Banach spaces of bounded and compact linear operators in $\mathcal{H}$ are denoted by $\mathcal{B}(\mathcal{H})$ and $\mathcal{B}_\infty(\mathcal{H})$, respectively. Similarly, $\mathcal{B}(\mathcal{H}_1, \mathcal{H}_2)$ and $\mathcal{B}_\infty(\mathcal{H}_1, \mathcal{H}_2)$ will be used for bounded and compact operators between two Hilbert spaces $\mathcal{H}_1$ and $\mathcal{H}_2$. Moreover, $\mathcal{X}_1 \hookrightarrow \mathcal{X}_2$ denotes the continuous embedding of the Banach space $\mathcal{X}_1$ into the Banach space $\mathcal{X}_2$. Throughout this manuscript, if X denotes a Banach space, X^* denotes the *adjoint space* of continuous conjugate linear functionals on X, that is, the *conjugate dual space* of X (rather than the usual dual space of continuous linear functionals on X). This avoids the well-known awkward distinction between adjoint operators in Banach and Hilbert spaces (cf., e.g., the pertinent discussion in [**37**, p. 3–4]).

Finally, a notational comment: For obvious reasons in connection with quantum mechanical applications, we will, with a slight abuse of notation, dub $-\Delta$ (rather than Δ) as the "Laplacian" in this paper.

2. Laplace Operators with Generalized Robin Boundary Conditions

In this section we primarily focus on various properties of general Laplacians $-\Delta_{\Theta,\Omega}$ in $L^2(\Omega; d^n x)$ including Dirichlet, $-\Delta_{D,\Omega}$, and Neumann, $-\Delta_{N,\Omega}$, Laplacians, generalized Robin-type Laplacians, and Laplacians corresponding to classical

Robin boundary conditions associated with open sets $\Omega \subset \mathbb{R}^n$, $n \in \mathbb{N}$, $n \geq 2$, introduced in Hypothesis 2.1 below.

We start with introducing our assumptions on the set Ω and the boundary operator Θ which subsequently will be employed in defining the boundary condition on $\partial\Omega$:

HYPOTHESIS 2.1. *Let $n \in \mathbb{N}$, $n \geq 2$, and assume that $\Omega \subset \mathbb{R}^n$ is an open, bounded, nonempty Lipschitz domain.*

We refer to Appendix A for more details on Lipschitz domains.

For simplicity of notation we will denote the identity operators in $L^2(\Omega; d^n x)$ and $L^2(\partial\Omega; d^{n-1}\omega)$ by I_Ω and $I_{\partial\Omega}$, respectively. Also, we refer to Appendix A for our notation in connection with Sobolev spaces.

HYPOTHESIS 2.2. *Assume Hypothesis* 2.1 *and suppose that a_Θ is a closed sesquilinear form in $L^2(\partial\Omega; d^{n-1}\omega)$ with domain $H^{1/2}(\partial\Omega) \times H^{1/2}(\partial\Omega)$, bounded from below by $c_\Theta \in \mathbb{R}$ (hence, in particular, a_Θ is symmetric). Denote by $\Theta \geqslant c_\Theta I_{\partial\Omega}$ the self-adjoint operator in $L^2(\partial\Omega; d^{n-1}\omega)$ uniquely associated with a_Θ (cf.* (B.27)) *and by $\widetilde{\Theta} \in \mathcal{B}\big(H^{1/2}(\partial\Omega), H^{-1/2}(\partial\Omega)\big)$ the extension of Θ as discussed in* (B.26) *and* (B.32).

Thus one has

$$\big\langle f, \widetilde{\Theta}\, g\big\rangle_{1/2} = \overline{\big\langle g, \widetilde{\Theta}\, f\big\rangle_{1/2}}, \quad f, g \in H^{1/2}(\partial\Omega). \tag{2.1}$$

$$\big\langle f, \widetilde{\Theta}\, f\big\rangle_{1/2} \geq c_\Theta \|f\|^2_{L^2(\partial\Omega; d^{n-1}\omega)}, \quad f \in H^{1/2}(\partial\Omega). \tag{2.2}$$

Here the sesquilinear form

$$\langle\,\cdot\,,\,\cdot\,\rangle_s = {}_{H^s(\partial\Omega)}\langle\,\cdot\,,\,\cdot\,\rangle_{H^{-s}(\partial\Omega)} \colon H^s(\partial\Omega) \times H^{-s}(\partial\Omega) \to \mathbb{C}, \quad s \in [0,1], \tag{2.3}$$

(antilinear in the first, linear in the second factor), denotes the duality pairing between $H^s(\partial\Omega)$ and

$$H^{-s}(\partial\Omega) = \big(H^s(\partial\Omega)\big)^*, \quad s \in [0,1], \tag{2.4}$$

such that

$$\langle f, g\rangle_s = \int_{\partial\Omega} d^{n-1}\omega(\xi)\,\overline{f(\xi)} g(\xi), \quad f \in H^s(\partial\Omega),\ g \in L^2(\partial\Omega; d^{n-1}\omega) \hookrightarrow H^{-s}(\partial\Omega),$$
$$s \in [0,1], \tag{2.5}$$

and $d^{n-1}\omega$ denotes the surface measure on $\partial\Omega$.

Hypothesis 2.1 on Ω is used throughout this paper. Similarly, Hypothesis 2.2 is assumed whenever the boundary operator $\widetilde{\Theta}$ is involved. (Later in this section, and the next, we will occasionally strengthen our hypotheses.)

We introduce the boundary trace operator γ_D^0 (the Dirichlet trace) by

$$\gamma_D^0 \colon C(\overline{\Omega}) \to C(\partial\Omega), \quad \gamma_D^0 u = u|_{\partial\Omega}. \tag{2.6}$$

Then there exists a bounded, linear operator γ_D (cf., e.g., [**69**, Theorem 3.38]),

$$\begin{aligned} &\gamma_D \colon H^s(\Omega) \to H^{s-(1/2)}(\partial\Omega) \hookrightarrow L^2(\partial\Omega; d^{n-1}\omega), \quad 1/2 < s < 3/2, \\ &\gamma_D \colon H^{3/2}(\Omega) \to H^{1-\varepsilon}(\partial\Omega) \hookrightarrow L^2(\partial\Omega; d^{n-1}\omega), \quad \varepsilon \in (0,1), \end{aligned} \tag{2.7}$$

whose action is compatible with that of γ_D^0. That is, the two Dirichlet trace operators coincide on the intersection of their domains. Moreover, we recall that

$$\gamma_D\colon H^s(\Omega)\to H^{s-(1/2)}(\partial\Omega) \text{ is onto for } 1/2<s<3/2. \tag{2.8}$$

While, in the class of bounded Lipschitz subdomains in $\mathbb{R}^n$, the end-point cases $s=1/2$ and $s=3/2$ of $\gamma_D\in\mathcal{B}\big(H^s(\Omega),H^{s-(1/2)}(\partial\Omega)\big)$ fail, we nonetheless have

$$\gamma_D\in\mathcal{B}\big(H^{(3/2)+\varepsilon}(\Omega),H^1(\partial\Omega)\big),\quad \varepsilon>0. \tag{2.9}$$

See Lemma A.4 for a proof. Below we augment this with the following result:

LEMMA 2.3. *Assume Hypothesis* 2.1. *Then for each* $s>-3/2$, *the restriction to boundary operator* (2.6) *extends to a linear operator*

$$\gamma_D:\big\{u\in H^{1/2}(\Omega)\,\big|\,\Delta u\in H^s(\Omega)\big\}\to L^2(\partial\Omega;d^{n-1}\omega), \tag{2.10}$$

is compatible with (2.7), *and is bounded when* $\{u\in H^{1/2}(\Omega)\,|\,\Delta u\in H^s(\Omega)\}$ *is equipped with the natural graph norm* $u\mapsto\|u\|_{H^{1/2}(\Omega)}+\|\Delta u\|_{H^s(\Omega)}$. *In addition, this operator has a linear, bounded right-inverse* (*hence, in particular, it is onto*).

Furthermore, for each $s>-3/2$, *the restriction to boundary operator* (2.6) *also extends to a linear operator*

$$\gamma_D:\big\{u\in H^{3/2}(\Omega)\,\big|\,\Delta u\in H^{1+s}(\Omega)\big\}\to H^1(\partial\Omega), \tag{2.11}$$

which is again compatible with (2.7), *and is bounded when* $\{u\in H^{3/2}(\Omega)\,|\,\Delta u\in H^{1+s}(\Omega)\}$ *is equipped with the natural graph norm* $u\mapsto\|u\|_{H^{3/2}(\Omega)}+\|\Delta u\|_{H^{1+s}(\Omega)}$. *Once again, this operator has a linear, bounded right-inverse* (*hence, in particular, it is onto*).

PROOF. For each $s\in\mathbb{R}$ set $H^s_\Delta(\Omega)=\{u\in H^s(\Omega)\,|\,\Delta u=0\text{ in }\Omega\}$ and observe that this is a closed subspace of $H^s(\Omega)$. In particular, $H^s_\Delta(\Omega)$ is a Banach space when equipped with the norm inherited from $H^s(\Omega)$. Next we recall the nontangential maximal operator M defined in (D.9). According to [**39**], or Corollary 5.7 in [**51**], one has

$$H^{1/2}_\Delta(\Omega)=\big\{u\text{ harmonic in }\Omega\,\big|\,M(u)\in L^2(\partial\Omega;d^{n-1}\omega)\big\} \tag{2.12}$$

and $u\mapsto\|M(u)\|_{L^2(\partial\Omega;d^{n-1}\omega)}$ is an equivalent norm on $H^{1/2}_\Delta(\Omega)$. To continue, fix some $\kappa>0$ and set $d(y)=\operatorname{dist}(y,\partial\Omega)$ for $y\in\Omega$. According to [**28**], the nontangential trace operator

$$(\gamma_{\rm n.t.}u)(x)=\lim_{\substack{\Omega\ni y\to x\\ |x-y|<(1+\kappa)\,d(y)}}u(y) \tag{2.13}$$

is then well-defined when considered as a mapping

$$\gamma_{\rm n.t.}:\big\{u\text{ harmonic in }\Omega\,\big|\,M(u)\in L^2(\partial\Omega;d^{n-1}\omega)\big\}\to L^2(\partial\Omega;d^{n-1}\omega). \tag{2.14}$$

Furthermore, the operator (2.13), (2.14) is bounded.

Granted these results, for a fixed $s>-3/2$ we may then attempt to define

$$\widetilde\gamma_D:H^{1/2}_\Delta(\Omega)+H^{s+2}(\Omega)\to L^2(\partial\Omega;d^{n-1}\omega) \tag{2.15}$$

by setting

$$\widetilde\gamma_D(u+v)=\gamma_{\rm n.t.}u+\gamma_Dv,\quad u\in H^{1/2}_\Delta(\Omega),\ v\in H^{s+2}(\Omega). \tag{2.16}$$

A moment's reflection shows that, in order to establish that the mapping (2.15), (2.16) is well-defined, it suffices to prove that

$$\gamma_{\text{n.t.}} u = \gamma_D u \text{ in } L^2(\partial\Omega; d^{n-1}\omega) \text{ whenever } u \in H^{(1/2)+\varepsilon}(\Omega),\ \varepsilon > 0. \tag{2.17}$$

in the case when Ω is a bounded Lipschitz domain which is star-like with respect to the origin in $\mathbb{R}^n$ (cf. (A.6)).

Assuming that this is the case, pick $u \in H^{(1/2)+\varepsilon}(\Omega)$ for some $\varepsilon > 0$, and for each $t \in (0,1)$ set $u_t(x) = u(tx)$, $x \in \Omega$. We claim that

$$u_t \to u \quad \text{in} \quad H^{(1/2)+\varepsilon}(\Omega) \quad \text{as} \quad t \to 1. \tag{2.18}$$

To justify this, it suffices to prove that this is the case when $u \in C^\infty(\overline{\Omega})$ as the result in its full generality then follows from a standard density argument. However, for every $u \in C^\infty(\overline{\Omega})$ one trivially has $u_t \to u$ as $t \to 1$ in $H^1(\Omega)$, hence in $H^{(1/2)+\varepsilon}(\Omega)$. Having disposed of (2.18), we may then conclude that $\gamma_D u_t \to \gamma_D u$ in $L^2(\partial\Omega; d^{n-1}\omega)$ as $t \to 1$. Since for each $t \in (0,1)$ we have $u_t \in C(\overline{\Omega})$, it follows that $\gamma_D u_t = \gamma_D^0 u_t = u_t|_{\partial\Omega}$. Thus, altogether,

$$u_t|_{\partial\Omega} \to \gamma_D u \quad \text{in} \quad L^2(\partial\Omega; d^{n-1}\omega) \quad \text{as} \quad t \to 1. \tag{2.19}$$

On the other hand, for almost every $x \in \partial\Omega$, and every $t \in (0,1)$, we have that $y = tx$ belongs to Ω, converges to x as $t \to 1$, and $|x-y| \leq (1+\kappa)\,\text{dist}\,(y,\partial\Omega)$ for some sufficiently large $\kappa = \kappa(\Omega) > 0$ (independent of x and t). This implies that

$$u_t(x) \to (\gamma_{\text{n.t.}} u)(x) \text{ pointwise, for a.e. } x \in \partial\Omega, \text{ as } t \to 1. \tag{2.20}$$

Combining (2.19), (2.20) we therefore conclude that the functions $\gamma_{\text{n.t.}} u, \gamma_D u \in L^2(\partial\Omega; d^{n-1}\omega)$ coincide pointwise a.e. on $\partial\Omega$. This proves (2.17) and finishes the justification of the fact that the mapping (2.15), (2.16) is well-defined.

Granted (2.15), (2.16) is well-defined, it is implicit in its own definition that the mapping (2.15), (2.16) is also bounded when we equip $H^{1/2}_\Delta(\Omega) + H^{s+2}(\Omega)$ with the canonical norm

$$w \mapsto \inf_{\substack{w=u+v \\ u\in H^{1/2}_\Delta(\Omega),\, v\in H^{s+2}(\Omega)}} \|u\|_{H^{1/2}_\Delta(\Omega)} + \|v\|_{H^{s+2}(\Omega)}. \tag{2.21}$$

The same type of argument as above (i.e., restricting attention to pieces of Ω which are star-like Lipschitz domains, and using dilations with respect to the respective center of star-likeness) shows the following: If $w \in C(\overline{\Omega})$ can be decomposed as $u+v$ with $u \in H^{1/2}_\Delta(\Omega)$ and $v \in H^{s+2}(\Omega)$ for some $s > -3/2$, then $w|_{\partial\Omega} = \gamma_{\text{n.t.}} u + \gamma_D v$. In other words, the action of the trace operator $\widetilde{\gamma}_D$ in (2.15), (2.16) is compatible with that of (2.6). This completes the study of the nature and properties of $\widetilde{\gamma}_D$ in (2.15), (2.16).

Consider next the claim made about (2.10). As regards its boundedness and the fact that this acts in a compatible fashion with (2.7), it suffices to prove that

$$\{u \in H^{1/2}(\Omega) \,|\, \Delta u \in H^s(\Omega)\} \hookrightarrow H^{1/2}_\Delta(\Omega) + H^{s+2}(\Omega), \quad s > -3/2, \tag{2.22}$$

continuously. To see this, pick $u \in H^{1/2}(\Omega)$ such that $\Delta u \in H^s(\Omega)$ and extend (cf. [**87**]) Δu to a compactly supported distribution $w \in H^s(\mathbb{R}^n)$. Next, set

$$v(x) = \int_{\mathbb{R}^n} d^n y\, E_n(x-y) w(y), \quad x \in \Omega, \tag{2.23}$$

where

$$E_n(x) = \begin{cases} \frac{1}{2\pi}\ln(|x|), & n = 2, \\ \frac{1}{(2-n)\omega_{n-1}}|x|^{2-n}, & n \geq 3, \end{cases} \tag{2.24}$$

is the standard fundamental solution for the Laplacian in $\mathbb{R}^n$ (cf. (C.1) for $z = 0$). Here $\omega_{n-1} = 2\pi^{n/2}/\Gamma(n/2)$ ($\Gamma(\,\cdot\,)$ the Gamma function, cf. [**1**, Sect. 6.1]) represents the area of the unit sphere S^{n-1} in $\mathbb{R}^n$. Then $v \in H^{s+2}(\Omega)$ and $\Delta v = \Delta u$ in Ω. As a consequence, the function $w = u - v$ is harmonic and belongs to $H^{1/2}(\Omega)$, that is, $u = w + v$ with $w \in H^{1/2}_\Delta(\Omega)$, $v \in H^{s+2}(\Omega)$. Furthermore, the estimate

$$\|w\|_{H^{1/2}(\Omega)} + \|v\|_{H^{s+2}(\Omega)} \leq C\big(\|u\|_{H^{1/2}(\Omega)} + \|\Delta u\|_{H^s(\Omega)}\big) \tag{2.25}$$

for some $C = C(\Omega, s) > 0$ is implicit in the above construction. Thus, the inclusion (2.22) is well-defined and continuous, so that the claims about the boundedness of (2.10), as well as the fact that this acts in a compatible fashion with (2.7), follow from this and the fact that $\widetilde{\gamma}_D$ in (2.15), (2.16) is well-defined and bounded. As far as the existence of a linear, bounded, right-inverse is concerned, it suffices to point out (2.12) and recall that the mapping (2.14) is onto (cf. [**28**]).

We now digress momentarily for the purpose of developing an integration by parts formula which will play a significant role shortly. First, if Ω is a bounded star-like Lipschitz domain in $\mathbb{R}^n$ and G is a vector field with components in $H^{1/2}_\Delta(\Omega) + H^{s+2}(\Omega)$, $s > -3/2$, such that $\operatorname{div}(G) \in L^1(\Omega)$, then

$$\int_\Omega dx^n \operatorname{div}(G) = \int_{\partial\Omega} d^{n-1}\omega\, \nu \cdot \widetilde{\gamma}_D G. \tag{2.26}$$

Indeed, if as before $G_t(x) = G(tx)$, $x \in \Omega$, $t \in (0,1)$, then

$$\operatorname{div}(G_t) = t(\operatorname{div}(G))_t \text{ in the sense of distributions in } \Omega. \tag{2.27}$$

Writing (2.26) for G_t in place of G, with $0 < t < 1$, and then passing to the limit $t \to 1$ yields the desired result. As a corollary of (2.26) and (2.22), we also have that (2.26) holds if Ω is a bounded star-like Lipschitz domain in $\mathbb{R}^n$ and G is a vector field with components in $\{u \in H^{1/2}(\Omega) \,|\, \Delta u \in H^s(\Omega)\}$, $s > -3/2$, such that $\operatorname{div}(G) \in L^1(\Omega)$. Since the latter space is a module over $C_0^\infty(\mathbb{R}^n)$ and any Lipschitz domain is locally star-like, a simple argument based on a smooth partition of unity shows that the star-likeness condition on Ω can be eliminated. More precisely,

$$\left.\begin{array}{l} \text{Hypothesis 2.1,} \\ G \in \big\{u \in H^{1/2}(\Omega) \,|\, \Delta u \in H^s(\Omega)\big\}^n, \ s > -3/2, \\ \operatorname{div}(G) \in L^1(\Omega; d^n x) \end{array}\right\} \Longrightarrow \text{(2.26) holds.} \tag{2.28}$$

Moving on, consider the operator (2.11). To get started, we fix $s > -3/2$ and assume that the function $u \in H^{3/2}(\Omega)$ is such that $\Delta u \in H^{1+s}(\Omega)$. Then, by the second line in (2.7),

$$\gamma_D u \in H^{1-\varepsilon}(\partial\Omega) \text{ for every } \varepsilon > 0. \tag{2.29}$$

To continue, we recall the discussion (results and notation) in the paragraph containing (A.11)–(A.16) in Appendix A. For every $j, k \in \{1, ..., n\}$, we now claim that

$$\frac{\partial(\gamma_D u)}{\partial \tau_{j,k}} = \nu_j \gamma_D(\partial_k u) - \nu_k \gamma_D(\partial_j u). \tag{2.30}$$

Since the functions $\partial_j u, \partial_k u$ belong to the space $\{w \in H^{1/2}(\Omega) \,|\, \Delta w \in H^s(\Omega)\}$, we may then conclude from (2.30) and (2.10) that

$$\frac{\partial(\gamma_D u)}{\partial \tau_{j,k}} \in L^2(\partial\Omega; d^{n-1}\omega), \tag{2.31}$$

and, in addition,

$$\left\| \frac{\partial(\gamma_D u)}{\partial \tau_{j,k}} \right\|_{L^2(\partial\Omega; d^{n-1}\omega)} \leq C\big(\|u\|_{H^{3/2}(\Omega)} + \|\Delta u\|_{H^{1+s}(\Omega)}\big), \tag{2.32}$$

for every $j,k \in \{1,...,n\}$. In concert with (2.32) and (2.29), the characterization in (A.16) then entails that $\gamma_D u \in H^1(\partial\Omega)$ and $\|\gamma_D u\|_{H^1(\partial\Omega)} \leq C\big(\|u\|_{H^{3/2}(\Omega)} + \|\Delta u\|_{H^{1+s}(\Omega)}\big)$. In summary, the proof of the claims made about (2.11) is finished, modulo establishing (2.30).

To deal with (2.30), let $\psi \in C_0^\infty(\mathbb{R}^n)$ and fix $j,k \in \{1,...,n\}$. Consider next the vector fields

$$\begin{aligned} F_{j,k} &= \big(0,...,0, u\partial_k\psi, 0...,0, -u\partial_j\psi, 0...,0\big), \\ G_{j,k} &= \big(0,...,0, \psi\partial_k u, 0...,0, -\psi\partial_j u, 0...,0\big), \end{aligned} \tag{2.33}$$

with the nonzero components on the j-th and k-th slots. Then $F_{j,k}$, $G_{j,k}$ have components in the space $\{u \in H^{1/2}(\Omega) \,|\, \Delta u \in H^s(\Omega)\}$ with $s > -3/2$ and satisfy

$$\operatorname{div}(F_{j,k}) = -\operatorname{div}(G_{j,k}) = (\partial_j u \partial_k \psi - \partial_k u \partial_j \psi) \in L^2(\Omega; d^n x), \tag{2.34}$$

in the sense of distributions. Also,

$$\begin{aligned} \nu \cdot \widetilde{\gamma}_D(F_{j,k}) &= (\gamma_D u)\big(\nu_k \partial_j \psi - \nu_j \partial_k \psi\big), \\ \nu \cdot \widetilde{\gamma}_D(G_{j,k}) &= \psi\big(\nu_k \gamma_D(\partial_j u) - \nu_j \gamma_D(\partial_k u)\big). \end{aligned} \tag{2.35}$$

Hence, using (2.28), we obtain

$$\begin{aligned} \int_{\partial\Omega} d^{n-1}\omega\, (\gamma_D u)\big(\nu_k\partial_j\psi - \nu_j\partial_k\psi\big) &= \int_{\partial\Omega} d^{n-1}\omega\, \nu \cdot \widetilde{\gamma}_D(F_{j,k}) \\ &= \int_\Omega d^n x \operatorname{div}(F_{j,k}) = -\int_\Omega d^n x \operatorname{div}(G_{j,k}) \\ &= -\int_{\partial\Omega} d^{n-1}\omega\, \psi\big(\nu_k\gamma_D(\partial_j u) - \nu_j\gamma_D(\partial_k u)\big). \end{aligned} \tag{2.36}$$

This justifies (2.30) and shows that the operator (2.11) is well-defined and bounded. Clearly, this acts in a compatible fashion with (2.7) and (2.10). To finish the proof of Lemma 2.3, there remains to show that this operator also has a bounded, linear, right-inverse. This, however, is a consequence of the well-posedness of the boundary value problem

$$u \in H^{3/2}(\Omega), \quad \Delta u = 0 \text{ in } \Omega, \quad \gamma_D(u) = f \in H^1(\partial\Omega), \tag{2.37}$$

a result which appears in [**101**]. □

Next, we introduce the operator γ_N (the strong Neumann trace) by

$$\gamma_N = \nu \cdot \gamma_D \nabla \colon H^{s+1}(\Omega) \to L^2(\partial\Omega; d^{n-1}\omega), \quad 1/2 < s < 3/2, \tag{2.38}$$

where ν denotes the outward pointing normal unit vector to $\partial\Omega$. It follows from (2.7) that γ_N is also a bounded operator. We seek to extend the action of the

Neumann trace operator (2.38) to other (related) settings. To set the stage, assume Hypothesis 2.1 and recall that the inclusion

$$\iota : H^s(\Omega) \hookrightarrow \big(H^1(\Omega)\big)^*, \quad s > -1/2, \tag{2.39}$$

is well-defined and bounded. We then introduce the weak Neumann trace operator

$$\widetilde{\gamma}_N \colon \big\{u \in H^1(\Omega) \,\big|\, \Delta u \in H^s(\Omega)\big\} \to H^{-1/2}(\partial\Omega), \quad s > -1/2, \tag{2.40}$$

as follows: Given $u \in H^1(\Omega)$ with $\Delta u \in H^s(\Omega)$ for some $s > -1/2$, we set (with ι as in (2.39))

$$\langle \phi, \widetilde{\gamma}_N u\rangle_{1/2} = \int_\Omega d^n x\, \overline{\nabla\Phi(x)} \cdot \nabla u(x) + {}_{H^1(\Omega)}\langle \Phi, \iota(\Delta u)\rangle_{(H^1(\Omega))^*}, \tag{2.41}$$

for all $\phi \in H^{1/2}(\partial\Omega)$ and $\Phi \in H^1(\Omega)$ such that $\gamma_D\Phi = \phi$. We note that this definition is independent of the particular extension Φ of ϕ, and that $\widetilde{\gamma}_N$ is a bounded extension of the Neumann trace operator γ_N defined in (2.38).

The end-point case $s = 1/2$ of (2.38) is discussed separately below.

LEMMA 2.4. *Assume Hypothesis* 2.1. *Then the Neumann trace operator* (2.38) *also extends to*

$$\widetilde{\gamma}_N : \big\{u \in H^{3/2}(\Omega) \,\big|\, \Delta u \in L^2(\Omega; d^n x)\big\} \to L^2(\partial\Omega; d^{n-1}\omega) \tag{2.42}$$

in a bounded fashion when the space $\{u \in H^{3/2}(\Omega) \,|\, \Delta u \in L^2(\Omega; d^n x)\}$ *is equipped with the natural graph norm* $u \mapsto \|u\|_{H^{3/2}(\Omega)} + \|\Delta u\|_{L^2(\Omega;d^n x)}$. *This extension is compatible with* (2.40) *and has a linear, bounded, right-inverse* (*hence, as a consequence, it is onto*).

Moreover, the Neumann trace operator (2.38) *further extends to*

$$\widetilde{\gamma}_N : \big\{u \in H^{1/2}(\Omega) \,\big|\, \Delta u \in L^2(\Omega; d^n x)\big\} \to H^{-1}(\partial\Omega) \tag{2.43}$$

in a bounded fashion when the space $\{u \in H^{1/2}(\Omega) \,|\, \Delta u \in L^2(\Omega; d^n x)\}$ *is equipped with the natural graph norm* $u \mapsto \|u\|_{H^{1/2}(\Omega)} + \|\Delta u\|_{L^2(\Omega;d^n x)}$. *Once again, this extension is compatible with* (2.40) *and has a linear, bounded, right-inverse* (*thus, in particular, it is onto*).

PROOF. Fix $\psi \in C^\infty(\overline{\Omega})$. Applying (2.28) to the vector field $G = \overline{\psi}\nabla u$ yields

$$\int_{\partial\Omega} d^{n-1}\omega\, \overline{\psi}\, \nu \cdot \gamma_D(\nabla u) = \int_\Omega dx^n\, \overline{\nabla\psi} \cdot \nabla u + \int_\Omega dx^n\, \overline{\psi}\, \Delta u. \tag{2.44}$$

Consider now $\phi \in H^{1/2}(\partial\Omega)$ and $\Phi \in H^1(\Omega)$ such that $\gamma_D\Phi = \phi$. Since $C^\infty(\overline{\Omega}) \hookrightarrow H^1(\Omega)$ is dense, it is possible to select a sequence $\psi_j \in C^\infty(\overline{\Omega})$, $j \in \mathbb{N}$, such that $\psi_j \to \Phi$ in $H^1(\Omega)$ as $j \to \infty$. This entails $\nabla\psi_j \to \nabla\Phi$ in $L^2(\Omega; d^n x)$ and $\psi_j|_{\partial\Omega} \to \phi$ in $H^{1/2}(\partial\Omega)$ as $j \to \infty$. Writing (2.44) for ψ_j in place of ψ and passing to the limit $j \to \infty$ then yields

$$\int_{\partial\Omega} d^{n-1}\omega\, \overline{\phi}\, \nu \cdot \gamma_D(\nabla u) = \int_\Omega dx^n\, \overline{\nabla\Phi} \cdot \nabla u + \int_\Omega dx^n\, \overline{\Phi}\, \Delta u. \tag{2.45}$$

This shows that the Neumann trace of u in the sense of (2.40), (2.41) is actually $\nu \cdot \gamma_D(\nabla u)$. In addition,

$$\begin{aligned}
\|\widetilde{\gamma}_N u\|_{L^2(\partial\Omega;d^{n-1}\omega)} &= \|\nu \cdot \gamma_D(\nabla u)\|_{L^2(\partial\Omega;d^{n-1}\omega)} \\
&\le \|\gamma_D(\nabla u)\|_{L^2(\partial\Omega;d^{n-1}\omega)^n} \\
&\le C\big(\|\nabla u\|_{H^{1/2}(\Omega)^n} + \|\Delta(\nabla u)\|_{H^{-1}(\Omega)^n}\big)
\end{aligned}$$

$$= C\big(\|\nabla u\|_{H^{1/2}(\Omega)^n} + \|\nabla(\Delta u)\|_{H^{-1}(\Omega)^n}\big)$$
$$\leq C\big(\|u\|_{H^{3/2}(\Omega)} + \|\Delta u\|_{L^2(\Omega;d^n x)}\big), \tag{2.46}$$

where we have used the boundedness of the Dirichlet trace operator in (2.10) with $s = -1$. This shows that, in the context of (2.42), the Neumann trace operator

$$\widetilde{\gamma}_N u = \nu \cdot \gamma_D(\nabla u) \tag{2.47}$$

has is well-defined, linear, bounded and is compatible with (2.40). The fact that this has a linear, bounded, right-inverse is a consequence of the well-posedness result in Theorem 3.2, proved later.

As far as (2.43) is concerned, let us temporarily introduce

$$\widetilde{\gamma}_n : \{u \in H^{1/2}(\Omega) \,|\, \Delta u \in L^2(\Omega; d^n x)\} \to H^{-1}(\partial\Omega) = \big(H^1(\partial\Omega)\big)^*, \tag{2.48}$$

by setting

$$\langle \phi, \widetilde{\gamma}_n u\rangle_1 = \langle \widetilde{\gamma}_N(\Phi), \gamma_D u\rangle_0 + \langle \Phi, \Delta u\rangle_{L^2(\Omega;d^n x)} - \langle \Delta\Phi, u\rangle_{L^2(\Omega;d^n x)}, \tag{2.49}$$

for all $\phi \in H^1(\partial\Omega)$, where $\Phi \in H^{3/2}(\Omega)$ is such that $\gamma_D \Phi = \phi$ and $\Delta\Phi \in L^2(\Omega; d^n x)$. That such a Φ can be found (with the additional properties that the dependence $\phi \mapsto \Phi$ linear, and that Φ satisfies a natural estimate) is a consequence of the fact that the mapping (2.11) has a linear, bounded, right-inverse. Let us also note that the first pairing in the right hand-side of (2.49) is meaningful, thanks to the first part of Lemma 2.3 and what we have established in connection with (2.42).

We now wish to show that the definition (2.49) is independent of the particular choice of Φ. For this purpose, we recall the following useful approximation result:

$$C^\infty(\overline{\Omega}) \hookrightarrow \{u \in H^s(\Omega) \,|\, \Delta u \in L^2(\Omega; d^n x)\} \text{ densely, whenever } s < 2, \tag{2.50}$$

where the latter space is equipped with the natural graph norm $u \mapsto \|u\|_{H^s(\Omega)} + \|\Delta u\|_{L^2(\Omega;d^n x)}$. When $s = 1$ this appears as Lemma 1.5.3.9 on p. 60 of [**48**], and the extension to $s < 2$ has been worked out, along similar lines, in [**26**]. Returning to the task ast hand, by linearity and density is suffices to show that

$$\langle \widetilde{\gamma}_N(\Phi), \gamma_D u\rangle_0 + \langle \Phi, \Delta u\rangle_{L^2(\Omega;d^n x)} - \langle \Delta\Phi, u\rangle_{L^2(\Omega;d^n x)} = 0 \tag{2.51}$$

whenever $\Phi \in H^{3/2}(\Omega)$ is such that $\gamma_D \Phi = 0$, $\Delta\Phi \in L^2(\Omega; d^n x)$, and $u \in C^\infty(\overline{\Omega})$. Note, however, that by (2.41) with the roles of Φ and u reversed we have

$$\langle \widetilde{\gamma}_N(\Phi), \gamma_D u\rangle_0 = \int_\Omega d^n x\, \overline{\nabla\Phi(x)} \cdot \nabla u(x) + \langle \Delta\Phi, u\rangle_{L^2(\Omega;d^n x)}, \tag{2.52}$$

so matters are reduce to showing that

$$\int_\Omega d^n x\, \overline{\nabla\Phi(x)} \cdot \nabla u(x) = -\langle \Phi, \Delta u\rangle_{L^2(\Omega;d^n x)}. \tag{2.53}$$

Nonetheless, this is a consequence of Green's formula (2.28) written for the vector field $G = \overline{\Phi}\nabla u$ (which has the property that $\gamma_D G = 0$). In summary, the operator (2.48), (2.49) is well-defined, linear and bounded.

Next, we will show that this operator is compatible with (2.40), (2.41). After re-denoting $\widetilde{\gamma}_n$ by $\widetilde{\gamma}_N$, then this becomes the extension of the weak Neumann trace operator, considered in (2.43). To this end, assume that $u \in H^1(\Omega)$ has $\Delta u \in L^2(\Omega; d^n x)$. Our goal is to show that

$$\langle \phi, \widetilde{\gamma}_N u\rangle_{1/2} = \langle \phi, \widetilde{\gamma}_n u\rangle_1 \tag{2.54}$$

for every $\phi \in H^1(\partial\Omega)$ or, equivalently,

$$\int_\Omega d^n x\, \overline{\nabla\Phi(x)} \cdot \nabla u(x) = \langle \widetilde{\gamma}_N(\Phi), \gamma_D u\rangle_0 - \langle \Delta\Phi, u\rangle_{L^2(\Omega;d^n x)}, \tag{2.55}$$

for $\Phi \in H^{3/2}(\Omega)$ such that $\Delta\Phi \in L^2(\Omega; d^n x)$. However,

$$\langle \widetilde{\gamma}_N(\Phi), \gamma_D u\rangle_0 = \langle \widetilde{\gamma}_N(\Phi), \gamma_D u\rangle_{1/2} = \int_\Omega d^n x\, \overline{\nabla\Phi(x)} \cdot \nabla u(x) + \langle \Delta\Phi, u\rangle_{L^2(\Omega;d^n x)}, \tag{2.56}$$

where the first equality is a consequence of what we have proved about the operator (2.42), and the second follows from (2.41) with the roles of u and Φ reversed. This justifies (2.55) and finishes the proof of the lemma. □

For future purposes, we shall need yet another extension of the concept of Neumann trace. This requires some preparations (throughout, Hypothesis 2.1 is enforced). First, we recall that, as is well-known (see, e.g., [**51**]), one has the natural identification

$$\big(H^1(\Omega)\big)^* \equiv \big\{u \in H^{-1}(\mathbb{R}^n) \,\big|\, \mathrm{supp}\,(u) \subseteq \overline{\Omega}\big\}. \tag{2.57}$$

Note that the latter is a closed subspace of $H^{-1}(\mathbb{R}^n)$. In particular, if $R_\Omega u = u|_\Omega$ denotes the operator of restriction to Ω (considered in the sense of distributions), then

$$R_\Omega : \big(H^1(\Omega)\big)^* \to H^{-1}(\Omega) \tag{2.58}$$

is well-defined, linear and bounded. Furthermore, the composition of R_Ω in (2.58) with ι from (2.39) is the natural inclusion of $H^s(\Omega)$ into $H^{-1}(\Omega)$. Next, given $z \in \mathbb{C}$, set

$$W_z(\Omega) = \big\{(u,f) \in H^1(\Omega) \times \big(H^1(\Omega)\big)^* \,\big|\, (-\Delta - z)u = f|_\Omega \text{ in } \mathcal{D}'(\Omega)\big\}, \tag{2.59}$$

equipped with the norm inherited from $H^1(\Omega) \times \big(H^1(\Omega)\big)^*$. We then denote by

$$\widetilde{\gamma}_{\mathcal{N}} \colon W_z(\Omega) \to H^{-1/2}(\partial\Omega) \tag{2.60}$$

the ultra weak Neumann trace operator defined by

$$\begin{aligned} \langle \phi, \widetilde{\gamma}_{\mathcal{N}}(u,f)\rangle_{1/2} &= \int_\Omega d^n x\, \overline{\nabla\Phi(x)} \cdot \nabla u(x) \\ &\quad - z\int_\Omega d^n x\, \overline{\Phi(x)} u(x) - {}_{H^1(\Omega)}\langle \Phi, f\rangle_{(H^1(\Omega))^*}, \quad (u,f) \in W_z(\Omega), \end{aligned} \tag{2.61}$$

for all $\phi \in H^{1/2}(\partial\Omega)$ and $\Phi \in H^1(\Omega)$ such that $\gamma_D \Phi = \phi$. Once again, this definition is independent of the particular extension Φ of ϕ. Also, as was the case of the Dirichlet trace, the ultra weak Neumann trace operator (2.60), (2.61) is onto (this is a corollary of Theorem 4.5). For additional details we refer to equations (A.28)–(A.30).

The relationship between the ultra weak Neumann trace operator (2.60), (2.61) and the weak Neumann trace operator (2.40), (2.41) can be described as follows: Given $s > -1/2$ and $z \in \mathbb{C}$, denote by

$$j_z : \big\{u \in H^1(\Omega) \,\big|\, \Delta u \in H^s(\Omega)\big\} \to W_z(\Omega) \tag{2.62}$$

the injection

$$j_z(u) = (u, \iota(-\Delta u - zu)), \quad u \in H^1(\Omega),\ \Delta u \in H^s(\Omega), \tag{2.63}$$

where ι is as in (2.39). Then

$$\widetilde{\gamma}_{\mathcal{N}} \circ j_z = \widetilde{\gamma}_N. \tag{2.64}$$

Thus, from this perspective, $\widetilde{\gamma}_{\mathcal{N}}$ can also be regarded as a bounded extension of the Neumann trace operator γ_N defined in (2.38).

Moving on, we now wish to discuss generalized Robin Laplacians in Lipschitz subdomains of $\mathbb{R}^n$. Before initiating this discussion in earnest, however, we formulate and prove the following useful result:

LEMMA 2.5. *Assume Hypothesis 2.1. Then for every $\varepsilon > 0$ there exists a $\beta(\varepsilon) > 0$ ($\beta(\varepsilon) \underset{\varepsilon\downarrow 0}{=} O(1/\varepsilon)$) such that*

$$\|\gamma_D u\|^2_{L^2(\partial\Omega;d^{n-1}\omega)} \leqslant \varepsilon\|\nabla u\|^2_{L^2(\Omega;d^nx)^n} + \beta(\varepsilon)\|u\|^2_{L^2(\Omega;d^nx)} \text{ for all } u \in H^1(\Omega). \tag{2.65}$$

PROOF. Since Ω is a bounded Lipschitz domain, there exists an $h \in C_0^\infty(\mathbb{R}^n)^n$ with real-valued components and $\kappa > 0$ such that (cf., [**48**, Lemma 1.5.1.9, p. 40])

$$(\nu \cdot h)_{\mathbb{C}^n} \geqslant \kappa \text{ a.e. on } \partial\Omega. \tag{2.66}$$

Thus,

$$\begin{aligned}
\|\gamma_D u\|^2_{L^2(\partial\Omega;d^{n-1}\omega)} &\leqslant \frac{1}{\kappa}\int_{\partial\Omega} d^{n-1}\omega\,(\nu\cdot h)_{\mathbb{C}^n}|\gamma_D u|^2 \\
&= \frac{1}{\kappa}\int_\Omega d^nx\,\mathrm{div}(|u|^2h), \\
&= \frac{1}{\kappa}\bigg(\int_\Omega d^nx\,\big(\nabla|u|^2,h\big)_{\mathbb{C}^n} + |u|^2\mathrm{div}(h)\bigg), \quad u \in H^1(\Omega),
\end{aligned} \tag{2.67}$$

using the divergence theorem in the second step. Since for arbitrary $\varepsilon > 0$,

$$|2u\nabla u| \leqslant \varepsilon|\nabla u|^2 + (1/\varepsilon)|u|^2, \quad u \in H^1(\Omega), \tag{2.68}$$

and $h \in C_0^\infty(\mathbb{R}^n)^n$, one arrives at (2.65). ☐

Next we describe a family of self-adjoint Laplace operators $-\Delta_{\Theta,\Omega}$ in $L^2(\Omega;d^nx)$ indexed by the boundary operator Θ. We will refer to $-\Delta_{\Theta,\Omega}$ as the generalized Robin Laplacian.

THEOREM 2.6. *Assume Hypothesis 2.2. Then the generalized Robin Laplacian, $-\Delta_{\Theta,\Omega}$, defined by*

$$\begin{aligned}
&-\Delta_{\Theta,\Omega} = -\Delta, \\
&\mathrm{dom}(-\Delta_{\Theta,\Omega}) = \big\{u \in H^1(\Omega)\,\big|\, \Delta u \in L^2(\Omega;d^nx);\ \big(\widetilde{\gamma}_N + \widetilde{\Theta}\gamma_D\big)u = 0 \text{ in } H^{-1/2}(\partial\Omega)\big\},
\end{aligned} \tag{2.69}$$

is self-adjoint and bounded from below in $L^2(\Omega;d^nx)$. Moreover,

$$\mathrm{dom}\big(|-\Delta_{\Theta,\Omega}|^{1/2}\big) = H^1(\Omega). \tag{2.70}$$

PROOF. We introduce the sesquilinear form $a_{-\Delta_{\Theta,\Omega}}(\,\cdot\,,\,\cdot\,)$ with domain $H^1(\Omega)\times H^1(\Omega)$ by

$$a_{-\Delta_{\Theta,\Omega}}(u,v) = a_{-\Delta_{0,\Omega}}(u,v) + \big\langle\gamma_D u, \widetilde{\Theta}\gamma_D v\big\rangle_{1/2}, \quad u,v \in H^1(\Omega), \tag{2.71}$$

where $a_{-\Delta_{0,\Omega}}(\,\cdot\,,\,\cdot\,)$ on $H^1(\Omega)\times H^1(\Omega)$ denotes the Neumann Laplacian form

$$a_{-\Delta_{0,\Omega}}(u,v) = \int_\Omega d^nx\,\overline{(\nabla u)(x)}\cdot(\nabla v)(x), \quad u,v \in H^1(\Omega). \tag{2.72}$$

One verifies that $a_{-\Delta_{\Theta,\Omega}}(\,\cdot\,,\,\cdot\,)$ is well-defined on $H^1(\Omega)\times H^1(\Omega)$ since

$$\gamma_D \in \mathcal{B}\big(H^1(\Omega), H^{1/2}(\partial\Omega)\big), \tag{2.73}$$

$$\widetilde{\Theta} \in \mathcal{B}\big(H^{1/2}(\partial\Omega), H^{-1/2}(\partial\Omega)\big), \tag{2.74}$$

$$(\Theta + (1-c_\Theta)I_{\partial\Omega})^{1/2} \in \mathcal{B}\big(H^{1/2}(\partial\Omega), L^2(\partial\Omega; d^{n-1}\omega)\big) \tag{2.75}$$

(cf. (B.43)). This also implies that

$$(\Theta + (1-c_\Theta)I_{\partial\Omega})^{1/2}\gamma_D \in \mathcal{B}\big(H^1(\Omega), L^2(\partial\Omega; d^{n-1}\omega)\big). \tag{2.76}$$

Employing (2.1) and (2.2), $a_{-\Delta_{\Theta,\Omega}}$ is symmetric and bounded from below by c_Θ. Next, we intend to show that $a_{-\Delta_{\Theta,\Omega}}$ is a closed form in $L^2(\Omega; d^nx)\times L^2(\Omega; d^nx)$. For this purpose we rewrite $\big\langle \gamma_D u, \widetilde{\Theta}\gamma_D v\big\rangle_{1/2}$ as

$$\begin{aligned} &\big\langle \gamma_D u, \widetilde{\Theta}\gamma_D v\big\rangle_{1/2} \\ &\quad = \big((\Theta + (1-c_\Theta)I_{\partial\Omega})^{1/2}\gamma_D u, (\Theta + (1-c_\Theta)I_{\partial\Omega})^{1/2}\gamma_D v\big)_{L^2(\partial\Omega; d^{n-1}\omega)} \\ &\qquad - (1-c_\Theta)(\gamma_D u, \gamma_D v)_{L^2(\partial\Omega; d^{n-1}\omega)}, \quad u, v \in H^1(\Omega), \end{aligned} \tag{2.77}$$

(cf. (B.31), (B.32)), and notice that the last form on the right-hand side of (2.77) is nonclosable in $L^2(\Omega; d^nx)$ since γ_D is nonclosable as an operator defined on a dense subspace from $L^2(\Omega; d^nx)$ into $L^2(\partial\Omega; d^{n-1}\omega)$ (cf. the discussion in connection with (B.44)).

To deal with this noncloseability issue, we now split off the last form on the right-hand side of (2.77) and hence introduce

$$\begin{aligned} b_{-\Delta_{\Theta,\Omega}}(u,v) &= (\nabla u, \nabla v)_{L^2(\Omega; d^nx)^n} \\ &\quad + \big((\Theta + (1-c_\Theta)I_{\partial\Omega})^{1/2}\gamma_D u, (\Theta + (1-c_\Theta)I_{\partial\Omega})^{1/2}\gamma_D v\big)_{L^2(\partial\Omega; d^{n-1}\omega)} \\ &\quad + d_b(u,v)_{L^2(\Omega; d^nx)}, \quad u, v \in H^1(\Omega), \end{aligned} \tag{2.78}$$

for $d_b > 0$. Then due to the nonnegativity of the second form on the right-hand side in (2.78), $b_{-\Delta_{\Theta,\Omega}}$ is $H^1(\Omega)$-coercive, that is, for some $c_1 > 0$,

$$b_{-\Delta_{\Theta,\Omega}}(u,u) \geqslant c_1\|u\|^2_{H^1(\Omega)}, \tag{2.79}$$

where $\|u\|^2_{H^1(\Omega)} = \|\nabla u\|^2_{L^2(\Omega; d^nx)^n} + \|u\|^2_{L^2(\Omega; d^nx)}$. Next, we note that by (2.76),

$$\begin{aligned} &\Big|\big((\Theta + (1-c_\Theta)I_{\partial\Omega})^{1/2}\gamma_D u, (\Theta + (1-c_\Theta)I_{\partial\Omega})^{1/2}\gamma_D v\big)_{L^2(\partial\Omega; d^{n-1}\omega)}\Big| \\ &\quad \leqslant \big\|(\Theta + (1-c_\Theta)I_{\partial\Omega})^{1/2}\gamma_D\big\|^2_{\mathcal{B}(H^1(\Omega), L^2(\partial\Omega; d^{n-1}\omega))}\|u\|_{H^1(\Omega)}\|v\|_{H^1(\Omega)}, \\ &\hspace{22em} u, v \in H^1(\Omega). \end{aligned} \tag{2.80}$$

Since trivially, $\|\nabla u\|^2_{L^2(\Omega; d^nx)} + d_b\|u\|^2_{L^2(\Omega; d^nx)} \leqslant c\|u\|^2_{H^1(\Omega)}$ for some $c > 0$, one infers that $b_{-\Delta_{\Theta,\Omega}}$ is also $H^1(\Omega)$-bounded, that is, for some $c_2 > 0$,

$$b_{-\Delta_{\Theta,\Omega}}(u,u) \leqslant c_2\|u\|^2_{H^1(\Omega)}. \tag{2.81}$$

Thus, the symmetric form $b_{-\Delta_{\Theta,\Omega}}$ is $H^1(\Omega)$-bounded and $H^1(\Omega)$-coercive and hence densely defined and closed in $L^2(\Omega; d^nx)\times L^2(\Omega; d^nx)$ by the discussion following (B.46).

To deal with the nonclosable form $(\gamma_D u, \gamma_D v)_{L^2(\partial\Omega; d^{n-1}\omega)}$, $u, v \in H^1(\Omega)$, it suffices to note that by Lemma 2.5 this form is infinitesimally bounded with respect to the Neumann Laplacian form $a_{-\Delta_{0,\Omega}}$ on $H^1(\Omega)\times H^1(\Omega)$, and since the form

$\big((\Theta+1-c_\Theta)^{1/2}\gamma_D u, (\Theta+1-c_\Theta)^{1/2}\gamma_D v\big)_{L^2(\partial\Omega;d^{n-1}\omega)}$, $u,v \in H^1(\Omega)$, is nonnegative, the form $(\gamma_D u, \gamma_D v)_{L^2(\partial\Omega;d^{n-1}\omega)}$ is also infinitesimally bounded with respect to the form $b_{-\Delta_{\Theta,\Omega}}$. By the discussion in connection with (B.48), (B.49), the form $a_{-\Delta_{\Theta,\Omega}}$ (possibly shifted by an irrelevant real constant) defined on $H^1(\Omega)\times H^1(\Omega)$, is thus densely defined in $L^2(\Omega;d^nx)\times L^2(\Omega;d^nx)$, bounded from below, and closed.

According to (B.34) we thus introduce the operator $-\Delta_{\Theta,\Omega}$ in $L^2(\Omega;d^nx)$ by

$$\begin{aligned}
\mathrm{dom}(-\Delta_{\Theta,\Omega}) = \bigg\{v\in H^1(\Omega)\,\bigg|\, & \text{ there exists an } w_v \in L^2(\Omega;d^nx) \text{ such that}\\
\int_\Omega d^nx\,\overline{\nabla w}\,\nabla v + \big\langle \gamma_D w, \widetilde\Theta\gamma_D v\big\rangle_{1/2} &= \int_\Omega d^nx\,\overline{w}w_v \text{ for all } w\in H^1(\Omega)\bigg\},\\
-\Delta_{\Theta,\Omega}u = w_u, \quad u\in \mathrm{dom}(-\Delta_{\Theta,\Omega}). &
\end{aligned} \tag{2.82}$$

By the formalism displayed in (B.20)–(B.43) (cf., in particular (B.27)), $-\Delta_{\Theta,\Omega}$ is self-adjoint in $L^2(\Omega;d^nx)$ and (2.70) holds. We recall that

$$H_0^1(\Omega) = \{u\in H^1(\Omega)\,|\,\gamma_D u = 0 \text{ on } \partial\Omega\}. \tag{2.83}$$

Taking $v\in C_0^\infty(\Omega)\hookrightarrow H_0^1(\Omega)\hookrightarrow H^1(\Omega)$, one concludes

$$\int_\Omega d^nx\,\overline{v}w_u = -\int_\Omega d^nx\,\overline{v}\,\Delta u \quad \text{for all } v\in C_0^\infty(\Omega), \text{ and hence } w_u = -\Delta u \text{ in } \mathcal{D}'(\Omega), \tag{2.84}$$

with $\mathcal{D}'(\Omega) = C_0^\infty(\Omega)'$ the space of distributions in Ω.

Next, we suppose that $u\in \mathrm{dom}(-\Delta_{\Theta,\Omega})$ and $v\in H^1(\Omega)$. We recall that $\gamma_D\colon H^1(\Omega)\to H^{1/2}(\partial\Omega)$ and compute

$$\begin{aligned}
\int_\Omega d^nx\,\overline{\nabla v}\,\nabla u &= -\int_\Omega d^nx\,\overline{v}\,\Delta u + \langle\gamma_D v, \widetilde\gamma_N u\rangle_{1/2}\\
&= \int_\Omega d^nx\,\overline{v}w_u + \big\langle\gamma_D v, (\widetilde\gamma_N + \widetilde\Theta\gamma_D)u\big\rangle_{1/2} - \big\langle\gamma_D v, \widetilde\Theta\gamma_D u\big\rangle_{1/2}\\
&= \int_\Omega d^nx\,\overline{\nabla v}\,\nabla u + \big\langle\gamma_D v, (\widetilde\gamma_N + \widetilde\Theta\gamma_D)u\big\rangle_{1/2},
\end{aligned} \tag{2.85}$$

where we used the second line in (2.82). Hence,

$$\big\langle\gamma_D v, (\widetilde\gamma_N + \widetilde\Theta\gamma_D)u\big\rangle_{1/2} = 0. \tag{2.86}$$

Since $v\in H^1(\Omega)$ is arbitrary, and the map $\gamma_D\colon H^1(\Omega)\to H^{1/2}(\partial\Omega)$ is actually onto, one concludes that

$$(\widetilde\gamma_N + \widetilde\Theta\gamma_D)u = 0 \text{ in } H^{-1/2}(\partial\Omega). \tag{2.87}$$

Thus,

$$\mathrm{dom}(-\Delta_{\Theta,\Omega}) \subseteq \big\{v\in H^1(\Omega)\,\big|\,\Delta v\in L^2(\Omega;d^nx);\ (\widetilde\gamma_N + \widetilde\Theta\gamma_D)v = 0 \text{ in } H^{-1/2}(\partial\Omega)\big\}. \tag{2.88}$$

Finally, assume that $u\in\big\{v\in H^1(\Omega)\,\big|\,\Delta v\in L^2(\Omega;d^nx);\ (\widetilde\gamma_N + \widetilde\Theta\gamma_D)v = 0\big\}$, $w\in H^1(\Omega)$, and let $w_u = -\Delta u\in L^2(\Omega;d^nx)$. Then,

$$\begin{aligned}
\int_\Omega d^nx\,\overline{w}w_u &= -\int_\Omega d^nx\,\overline{w}\,\mathrm{div}(\nabla u)\\
&= \int_\Omega d^nx\,\overline{\nabla w}\,\nabla u - \langle\gamma_D w, \widetilde\gamma_N u\rangle_{1/2}
\end{aligned}$$

$$= \int_\Omega d^n x\, \overline{\nabla w}\, \nabla u + \langle \gamma_D w, \widetilde{\Theta} \gamma_D u \rangle_{1/2}. \tag{2.89}$$

Thus, applying (2.82), one concludes that $u \in \mathrm{dom}(-\Delta_{\Theta,\Omega})$ and hence

$$\mathrm{dom}(-\Delta_{\Theta,\Omega}) \supseteq \big\{ v \in H^1(\Omega) \,\big|\, \Delta v \in L^2(\Omega; d^n x);\ (\widetilde{\gamma}_N + \widetilde{\Theta} \gamma_D) v = 0 \text{ in } H^{-1/2}(\partial\Omega) \big\}, \tag{2.90}$$

finishing the proof of Theorem 2.6. $\square$

COROLLARY 2.7. *Assume Hypothesis 2.2. Then the generalized Robin Laplacian, $-\Delta_{\Theta,\Omega}$, has purely discrete spectrum bounded from below, in particular,*

$$\sigma_{\mathrm{ess}}(-\Delta_{\Theta,\Omega}) = \emptyset. \tag{2.91}$$

PROOF. Since $\mathrm{dom}\big(|-\Delta_{\Theta,\Omega}|^{1/2}\big) = H^1(\Omega)$, by (2.70), and $H^1(\Omega)$ embeds compactly into $L^2(\Omega; d^n x)$ (cf., e.g., [**37**, Theorem V.4.17]), one infers that $(-\Delta_{\Theta,\Omega} + I_\Omega)^{-1/2} \in \mathcal{B}_\infty(L^2(\Omega; d^n x))$. Consequently, one obtains

$$(-\Delta_{\Theta,\Omega} + I_\Omega)^{-1} \in \mathcal{B}_\infty(L^2(\Omega; d^n x)), \tag{2.92}$$

which is equivalent to (2.91). $\square$

The important special case where Θ corresponds to the operator of multiplication by a real-valued, essentially bounded function θ leads to Robin boundary conditions we discuss next:

COROLLARY 2.8. *In addition to Hypothesis 2.1, assume that Θ is the operator of multiplication in $L^2(\partial\Omega; d^{n-1}\omega)$ by the real-valued function θ satisfying $\theta \in L^\infty(\partial\Omega; d^{n-1}\omega)$. Then Θ satisfies the conditions in Hypothesis 2.2 resulting in the self-adjoint and bounded from below Laplacian $-\Delta_{\theta,\Omega}$ in $L^2(\Omega; d^n x)$ with Robin boundary conditions on $\partial\Omega$ in (2.69) given by*

$$(\widetilde{\gamma}_N + \theta \gamma_D) u = 0 \ \text{ in } \ H^{-1/2}(\partial\Omega). \tag{2.93}$$

PROOF. By Lemma 2.5, the sesquilinear form

$$\langle \gamma_D u, \theta \gamma_D v \rangle_{1/2}, \quad u, v \in H^1(\Omega), \tag{2.94}$$

is infinitesimally form bounded with respect to the Neumann Laplacian form $a_{-\Delta_{0,\Omega}}$. By (B.48) and (B.49) this in turn proves that the form $a_{-\Delta_{\Theta,\Omega}}$ in (2.71) is closed and one can now follow the proof of Theorem 2.6 from (2.82) on, step by step. $\square$

REMARK 2.9. (i) In the case of a smooth boundary $\partial\Omega$, the boundary conditions in (2.93) are also called "classical" boundary conditions (cf., e.g., [**91**]); in the more general case of bounded Lipschitz domains we also refer to [**6**] and [**102**, Ch. 4] in this context. Next, we point out that, in [**62**], the authors have dealt with the case of Laplace operators in bounded Lipschitz domains, equipped with local boundary conditions of Robin-type, with boundary data in $L^p(\partial\Omega; d^{n-1}\omega)$, and produced nontangential maximal function estimates. For the case $p = 2$, when our setting agrees with that of [**62**], some of our results in this section and the following are a refinement of those in [**62**]. Maximal L^p-regularity and analytic contraction semigroups of Dirichlet and Neumann Laplacians on bounded Lipschitz domains were studied in [**106**]. Holomorphic C_0-semigroups of the Laplacian with Robin boundary conditions on bounded Lipschitz domains have been discussed in [**103**]. Moreover, Robin boundary conditions for elliptic boundary value problems on arbitrary open domains were first studied by Maz'ya [**67**], [**68**, Sect. 4.11.6], and subsequently in [**29**] (see also [**30**] which treats the case of the Laplacian). In

addition, Robin-type boundary conditions involving measures on the boundary for very general domains Ω were intensively discussed in terms of quadratic forms and capacity methods in the literature, and we refer, for instance, to [**6**], [**7**], [**17**], [**102**], and the references therein.
(ii) In the special case $\theta = 0$ (resp., $\widetilde{\Theta} = 0$), that is, in the case of the Neumann Laplacian, we will also use the notation

$$-\Delta_{N,\Omega} = -\Delta_{0,\Omega}. \tag{2.95}$$

The case of the Dirichlet Laplacian $-\Delta_{D,\Omega}$ associated with Ω formally corresponds to $\Theta = \infty$ and so we isolate it in the next result:

THEOREM 2.10. *Assume Hypothesis* 2.1. *Then the Dirichlet Laplacian,* $-\Delta_{D,\Omega}$, *defined by*

$$\begin{aligned} &-\Delta_{D,\Omega} = -\Delta, \\ &\operatorname{dom}(-\Delta_{D,\Omega}) = \big\{u \in H^1(\Omega) \,\big|\, \Delta u \in L^2(\Omega; d^n x);\ \gamma_D u = 0 \text{ in } H^{1/2}(\partial\Omega)\big\} \\ &\qquad\qquad\quad = \big\{u \in H_0^1(\Omega) \,\big|\, \Delta u \in L^2(\Omega; d^n x)\big\}, \end{aligned} \tag{2.96}$$

is self-adjoint and strictly positive in $L^2(\Omega; d^n x)$. *Moreover,*

$$\operatorname{dom}\big((-\Delta_{D,\Omega})^{1/2}\big) = H_0^1(\Omega). \tag{2.97}$$

PROOF. We introduce the sesquilinear form $a_{D,\Omega}(\,\cdot\,,\,\cdot\,)$ on the domain $H_0^1(\Omega) \times H_0^1(\Omega)$ by

$$a_{D,\Omega}(u,v) = \int_\Omega d^n x\, \overline{(\nabla u)(x)}\, (\nabla v)(x), \quad u, v \in H_0^1(\Omega). \tag{2.98}$$

Clearly, $a_{D,\Omega}$ is symmetric, nonnegative, and well-defined on $H_0^1(\Omega) \times H_0^1(\Omega)$. Since Ω is bounded, that is, $|\Omega| < \infty$, $H_0^1(\Omega)$-coercivity of $a_{D,\Omega}$ then immediately follows from Poincaré's inequality for $H_0^1(\Omega)$-functions (cf., e.g., [**105**, Theorem I.7.6]).

Next we introduce the operator $-\Delta_{D,\Omega}$ in $L^2(\Omega; d^n x)$ by

$$\begin{aligned} \operatorname{dom}(-\Delta_{D,\Omega}) = \bigg\{ v \in H_0^1(\Omega) \,\bigg|\, &\text{ there exists an } w_v \in L^2(\Omega; d^n x) \text{ such that} \\ &\int_\Omega d^n x\, \overline{\nabla w}\, \nabla v = \int_\Omega d^n x\, \overline{w} w_v \text{ for all } w \in H_0^1(\Omega) \bigg\}, \\ -\Delta_{D,\Omega} u = w_u, \quad & u \in \operatorname{dom}(-\Delta_{D,\Omega}). \end{aligned} \tag{2.99}$$

By the formalism displayed in (B.1)–(B.19), $-\Delta_{D,\Omega}$ is self-adjoint in $L^2(\Omega; d^n x)$ and (2.97) holds. Taking $v \in C_0^\infty(\Omega) \hookrightarrow H_0^1(\Omega)$, one concludes

$$\int_\Omega d^n x\, \overline{v} w_u = -\int_\Omega d^n x\, \overline{v}\, \Delta u \text{ in } \mathcal{D}'(\Omega) \text{ and hence } w_u = -\Delta u \text{ in } \mathcal{D}'(\Omega). \tag{2.100}$$

Since $v \in H_0^1(\Omega)$ if and only if $v \in H^1(\Omega)$ and $\gamma_D v = 0$ in $H^{1/2}(\partial\Omega)$ (cf., e.g., [**48**, Corollary 1.5.1.6]), and $v \in \operatorname{dom}(-\Delta_{D,\Omega})$ implies $\Delta v \in L^2(\Omega; d^n x)$, one computes for $u \in \operatorname{dom}(-\Delta_{D,\Omega})$ and $v \in H_0^1(\Omega)$ that

$$\int_\Omega d^n x\, \overline{\nabla v}\, \nabla u = -\int_\Omega d^n x\, \overline{v} \Delta u = \int_\Omega d^n x\, \overline{v} w_u. \tag{2.101}$$

Thus, $w_u = -\Delta u \in L^2(\Omega; d^n x)$ and hence

$$\operatorname{dom}(-\Delta_{D,\Omega}) \subseteq \big\{v \in H_0^1(\Omega) \,\big|\, \Delta v \in L^2(\Omega; d^n x)\big\}. \tag{2.102}$$

Finally, assume that $u \in \{v \in H_0^1(\Omega) \,|\, \Delta v \in L^2(\Omega; d^n x)\}$, $w \in H_0^1(\Omega)$, and let $w_u = -\Delta u \in L^2(\Omega; d^n x)$. Then,

$$\int_\Omega d^n x \, \overline{w} w_u = -\int_\Omega d^n x \, \overline{w} \operatorname{div}(\nabla u) = \int_\Omega d^n x \, \overline{\nabla w} \, \nabla u, \tag{2.103}$$

since $\gamma_D w = 0$ in $L^2(\partial\Omega; d^{n-1}\omega)$. Thus, applying (2.99), one concludes that $u \in \operatorname{dom}(-\Delta_{D,\Omega})$ and hence

$$\operatorname{dom}(-\Delta_{D,\Omega}) \supseteq \{v \in H_0^1(\Omega) \,|\, \Delta v \in L^2(\Omega; d^n x)\}, \tag{2.104}$$

finishing the proof of Theorem 2.10. □

Since Ω is open and bounded, it is well-known that $-\Delta_{D,\Omega}$ has purely discrete spectrum contained in $(0,\infty)$, in particular,

$$\sigma_{\text{ess}}(-\Delta_{D,\Omega}) = \emptyset \tag{2.105}$$

(this follows from (2.97) since $H_0^1(\Omega)$ embeds compactly into $L^2(\Omega; d^n x)$; the latter fact holds for arbitrary open, bounded sets $\Omega \subset \mathbb{R}^n$, cf., e.g., [**37**, Theorem V.4.18]).

While the principal objective of this paper was to prove the results in this section and the subsequent for minimally smooth domains Ω, it is of interest to study similar problems when Hypothesis 2.1 is further strengthen to:

Hypothesis 2.11. *Let $n \in \mathbb{N}$, $n \geq 2$, and assume that $\Omega \subset \mathbb{R}^n$ is a bounded domain of class $C^{1,r}$ for some $1/2 < r < 1$.*

We refer to Appendix A for some details on $C^{1,r}$-domains.

Correspondingly, the natural strengthening of Hypothesis 2.2 reads:

Hypothesis 2.12. *In addition to Hypothesis 2.2 and 2.11 assume that*

$$\widetilde{\Theta} \in \mathcal{B}_\infty\big(H^{3/2}(\partial\Omega), H^{1/2}(\partial\Omega)\big). \tag{2.106}$$

We note that a sufficient condition for (2.106) to hold is

$$\widetilde{\Theta} \in \mathcal{B}\big(H^{3/2-\varepsilon}(\partial\Omega), H^{1/2}(\partial\Omega)\big) \text{ for some } \varepsilon > 0. \tag{2.107}$$

Notational comment. To avoid introducing an additional sub- or superscript into our notation of $-\Delta_{\Theta,\Omega}$ and $-\Delta_{D,\Omega}$, we will use the same symbol for these operators irrespective of whether the pair of Hypothesis 2.1 and 2.2 or the pair of Hypothesis 2.11 and 2.12 is involved. Our results will be carefully stated so that it is always evident which set of hypotheses is used.

Next, we discuss certain regularity results for fractional powers of the resolvents of the Dirichlet and Robin Laplacians, first in Lipschitz then in smoother domains.

Lemma 2.13. *Assume Hypothesis 2.1 in connection with $-\Delta_{D,\Omega}$ and Hypothesis 2.2 in connection with $-\Delta_{\Theta,\Omega}$. Then the following boundedness properties hold for all $q \in [0,1]$ and $z \in \mathbb{C}\backslash[0,\infty)$,*

$$(-\Delta_{D,\Omega} - zI_\Omega)^{-q/2}, \, (-\Delta_{\Theta,\Omega} - zI_\Omega)^{-q/2} \in \mathcal{B}\big(L^2(\Omega; d^n x), H^q(\Omega)\big). \tag{2.108}$$

The fractional powers in (2.108) (and in subsequent analogous cases) are defined via the functional calculus implied by the spectral theorem for self-adjoint operators. As discussed in [**43**, Lemma A.2] in the closely related situation of Lemma 2.14, the key ingredients in proving Lemma 2.13 are the inclusions

$$\operatorname{dom}(-\Delta_{D,\Omega}) \subset H^1(\Omega), \quad \operatorname{dom}(-\Delta_{\Theta,\Omega}) \subset H^1(\Omega) \tag{2.109}$$

and real interpolation methods. The above results should be compared with its analogue for smoother domains. Specifically, we have:

LEMMA 2.14. *Assume Hypothesis* 2.11 *in connection with* $-\Delta_{D,\Omega}$ *and Hypothesis* 2.12 *in connection with* $-\Delta_{\Theta,\Omega}$. *Then the following boundedness properties hold for all* $q \in [0,1]$ *and* $z \in \mathbb{C}\backslash[0,\infty)$,

$$(-\Delta_{D,\Omega} - zI_\Omega)^{-q},\ (-\Delta_{\Theta,\Omega} - zI_\Omega)^{-q} \in \mathcal{B}\big(L^2(\Omega; d^n x), H^{2q}(\Omega)\big). \tag{2.110}$$

As explained in [**43**, Lemma A.2], the key ingredients in proving Lemma 2.14 are the inclusions

$$\mathrm{dom}(-\Delta_{D,\Omega}) \subset H^2(\Omega), \quad \mathrm{dom}(-\Delta_{\Theta,\Omega}) \subset H^2(\Omega) \tag{2.111}$$

and real interpolation methods.

Moving on, we now consider mapping properties of powers of the resolvents of generalized Robin Laplacians multiplied (to the left) by the Dirichlet boundary trace operator:

LEMMA 2.15. *Assume Hypothesis* 2.1 *and let* $\varepsilon > 0$, $z \in \mathbb{C}\backslash[0,\infty)$. *Then,*

$$\gamma_D(-\Delta_{\Theta,\Omega} - zI_\Omega)^{-(1+\varepsilon)/4} \in \mathcal{B}\big(L^2(\Omega; d^n x), L^2(\partial\Omega; d^{n-1}\omega)\big). \tag{2.112}$$

As in [**43**, Lemma 6.9], Lemma 2.15 follows from Lemma 2.13 and from (2.7) and (2.38). Once again, we wish to contrast this with the corresponding result for smoother domains, recorded below.

LEMMA 2.16. *Assume Hypothesis* 2.11 *in connection with* $-\Delta_{D,\Omega}$ *and Hypothesis* 2.12 *in connection with* $-\Delta_{\Theta,\Omega}$, *and let* $\varepsilon > 0$, $z \in \mathbb{C}\backslash[0,\infty)$. *Then,*

$$\begin{aligned} &\gamma_N(-\Delta_{D,\Omega} - zI_\Omega)^{-(3+\varepsilon)/4} \in \mathcal{B}\big(L^2(\Omega; d^n x), L^2(\partial\Omega; d^{n-1}\omega)\big), \\ &\gamma_D(-\Delta_{\Theta,\Omega} - zI_\Omega)^{-(1+\varepsilon)/4} \in \mathcal{B}\big(L^2(\Omega; d^n x), L^2(\partial\Omega; d^{n-1}\omega)\big). \end{aligned} \tag{2.113}$$

As in [**43**, Lemma 6.9], Lemma 2.16 follows from Lemma 2.14 and from (2.7) and (2.38). In contrast to Lemma 2.16 under the stronger Hypothesis 2.12, we cannot obtain an analog of (2.113) for $-\Delta_{D,\Omega}$ under the weaker Hypothesis 2.1.

The analog of Theorem 2.6 for smoother domains reads as follows:

THEOREM 2.17. *Assume Hypothesis* 2.12. *Then the generalized Robin Laplacian,* $-\Delta_{\Theta,\Omega}$, *defined by*

$$-\Delta_{\Theta,\Omega} = -\Delta, \quad \mathrm{dom}(-\Delta_{\Theta,\Omega}) = \big\{u \in H^2(\Omega) \,\big|\, (\gamma_N + \widetilde{\Theta}\gamma_D)u = 0 \text{ in } H^{1/2}(\partial\Omega)\big\}, \tag{2.114}$$

is self-adjoint and bounded from below in $L^2(\Omega; d^n x)$. *Moreover,*

$$\mathrm{dom}\big(|-\Delta_{\Theta,\Omega}|^{1/2}\big) = H^1(\Omega). \tag{2.115}$$

PROOF. We adapt the proof of [**43**, Lemma A.1], dealing with the special case of Neumann boundary conditions (i.e., in the case $\widetilde{\Theta} = 0$), to the present situation. For convenience of the reader we produce a complete proof below.

By Theorem 2.6, the operator $T_{\Theta,\Omega}$ in $L^2(\Omega; d^n x)$, defined by

$$\begin{aligned} &T_{\Theta,\Omega} = -\Delta, \\ &\mathrm{dom}(T_{\Theta,\Omega}) = \big\{u \in H^1(\Omega) \,\big|\, \Delta u \in L^2(\Omega; d^n x);\ (\widetilde{\gamma}_N + \widetilde{\Theta}\gamma_D)u = 0 \text{ in } H^{-1/2}(\partial\Omega)\big\}, \end{aligned} \tag{2.116}$$

is self-adjoint and bounded from below, and

$$\mathrm{dom}\big(|T_{\Theta,\Omega}|^{1/2}\big) = H^1(\Omega) \tag{2.117}$$

holds. Thus, we need to prove that $\operatorname{dom}(T_{\Theta,\Omega}) \subseteq H^2(\Omega)$.

Consider $u \in \operatorname{dom}(T_{\Theta,\Omega})$ and set $f = -\Delta u + u \in L^2(\Omega; d^n x)$. Viewing f as an element in $\big(H^1(\Omega)\big)^*$, the classical Lax–Milgram Lemma implies that u is the unique solution of the boundary-value problem

$$\begin{cases} (-\Delta + I_\Omega)u = f \in L^2(\Omega) \hookrightarrow \big(H^1(\Omega)\big)^*, \\ u \in H^1(\Omega), \\ (\widetilde{\gamma}_N + \widetilde{\Theta}\gamma_D)u = 0. \end{cases} \tag{2.118}$$

One convenient way to actually show that

$$u \in H^2(\Omega), \tag{2.119}$$

is to use layer potentials. Specifically, let $E_n(z;x)$ be the fundamental solution of the Helmholtz differential expression $(-\Delta - z)$ in $\mathbb{R}^n$, $n \in \mathbb{N}$, $n \geq 2$, that is,

$$E_n(z;x) = \begin{cases} (i/4)\big(2\pi|x|/z^{1/2}\big)^{(2-n)/2} H^{(1)}_{(n-2)/2}\big(z^{1/2}|x|\big), & n \geq 2,\ z \in \mathbb{C}\backslash\{0\}, \\ \frac{-1}{2\pi}\ln(|x|), & n = 2,\ z = 0, \\ \frac{1}{(n-2)\omega_{n-1}}|x|^{2-n}, & n \geq 3,\ z = 0, \end{cases} \tag{2.120}$$
$$\operatorname{Im}\big(z^{1/2}\big) \geq 0,\ x \in \mathbb{R}^n\backslash\{0\}.$$

Here $H^{(1)}_\nu(\,\cdot\,)$ denotes the Hankel function of the first kind with index $\nu \geq 0$ (cf. [**1**, Sect. 9.1]).

We also define the associated single layer potential

$$(\mathcal{S}_z g)(x) = \int_{\partial\Omega} d^{n-1}\omega(y)\, E_n(z; x-y) g(y), \quad x \in \Omega,\ z \in \mathbb{C}, \tag{2.121}$$

where g is an arbitrary measurable function on $\partial\Omega$. As is well-known (the interested reader may consult, e.g., [**73**], [**101**] for jump relations in the context of Lipschitz domains), if

$$(K^\#_z g)(x) = \text{p.v.} \int_{\partial\Omega} d^{n-1}\omega(y)\, \partial_{\nu_x} E_n(z; x-y) g(y), \quad x \in \partial\Omega,\ z \in \mathbb{C}, \tag{2.122}$$

stands for the so-called adjoint double layer on $\partial\Omega$, the following jump formula holds

$$\widetilde{\gamma}_N \mathcal{S}_z g = \big(-\tfrac{1}{2} I_{\partial\Omega} + K^\#_z\big) g. \tag{2.123}$$

It should be noted that

$$K^\#_z \in \mathcal{B}\big(L^2(\partial\Omega; d^{n-1}\omega)\big), \quad z \in \mathbb{C}, \tag{2.124}$$

whenever Ω is a bounded Lipschitz domain. See Lemma D.3.

Now, if we denote by w the convolution of $f \in L^2(\Omega; d^n x)$ with $E_n(-1;\,\cdot\,)$ in Ω, then $w \in H^2(\Omega)$ and the solution u of (2.118) is given by

$$u = w + \mathcal{S}_{-1} g \tag{2.125}$$

for a suitably chosen function g on $\partial\Omega$. Concretely, we shall then require that

$$(\gamma_N + \widetilde{\Theta}\gamma_D)\mathcal{S}_{-1} g = -(\gamma_N + \widetilde{\Theta}\gamma_D) w, \tag{2.126}$$

or equivalently,

$$\big(-\tfrac{1}{2} I_{\partial\Omega} + K^\#_{-1}\big) g + \widetilde{\Theta}\gamma_D \mathcal{S}_{-1} g = -(\gamma_N + \widetilde{\Theta}\gamma_D) w \in H^{1/2}(\partial\Omega). \tag{2.127}$$

By hypothesis, $\widetilde{\Theta} \in \mathcal{B}_\infty\big(H^{3/2}(\Omega), H^{1/2}(\partial\Omega)\big)$ and hence

$$\widetilde{\Theta}\gamma_D \mathcal{S}_{-1} \in \mathcal{B}_\infty\big(H^{1/2}(\Omega), H^{1/2}(\partial\Omega)\big) \tag{2.128}$$

as soon as one proves that $\mathcal{S}_{-1}$ satisfies

$$\mathcal{S}_{-1} \in \mathcal{B}\big(H^{1/2}(\partial\Omega), H^{2}(\Omega)\big). \tag{2.129}$$

To prove this, as a preliminary step we note (cf. [**74**]) that

$$\mathcal{S}_{-1} \colon H^{-s}(\partial\Omega) \to H^{-s+3/2}(\Omega) \tag{2.130}$$

is well-defined and bounded for each $s \in [0,1]$, even when Ω is only a bounded Lipschitz domain. For a fixed, arbitrary $j \in \{1,...,n\}$, consider next the operator $\partial_{x_j}\mathcal{S}_{-1}$ whose integral kernel is $\partial_{x_j}E_n(-1;x-y) = -\partial_{y_j}E_n(-1;x-y)$. We write

$$\partial_{y_j} = \sum_{k=1}^{n} \nu_k(y)\nu_k(y)\partial_{y_j} = \sum_{k=1}^{n} \nu_k(y)\frac{\partial}{\partial\tau_{k,j}(y)} + \nu_j(y)\nu(y)\cdot\nabla_y \tag{2.131}$$

where the tangential derivative operators $\partial/\partial\tau_{k,j} = \nu_k\partial_j - \nu_j\partial_k$, $j,k = 1,\dots,n$, satisfy (A.17). Using the boundary integration by parts formula (A.24) it follows that

$$\partial_j\mathcal{S}_{-1}h = \mathcal{D}_{-1}(\nu_j h) + \sum_{k=1}^{n}\mathcal{S}_{-1}\bigg(\frac{\partial(\nu_k h)}{\partial\tau_{k,j}}\bigg), \quad h \in H^{1/2}(\partial\Omega), \tag{2.132}$$

where, for $z \in \mathbb{C}$,

$$\mathcal{D}_z h(x) = \int_{\partial\Omega} d^{n-1}\omega(y)\,\nu(y)\cdot\nabla_y[E_n(z;x-y)]h(y), \quad x \in \Omega, \tag{2.133}$$

is the so-called (acoustic) double layer potential operator. Its mappings properties on the scale of Sobolev spaces have been analyzed in [**74**] and we note here that

$$\mathcal{D}_{-1} \colon H^{s}(\partial\Omega) \to H^{s+1/2}(\Omega), \quad 0 \le s \le 1, \tag{2.134}$$

requires only that Ω is Lipschitz. Assuming that multiplication by (the components of) ν preserves the space $H^{1/2}(\partial\Omega)$ (which is the case if, e.g., Ω is of class $C^{1,r}$ for some $(1/2) < r < 1$; cf. Lemma A.5), the desired conclusion about the operator (2.129) follows from (2.130), (2.132) and (2.134).

Going further, from Theorem D.8 we know that

$$K^{\#}_{-1} \in \mathcal{B}_\infty\big(H^{1/2}(\partial\Omega)\big), \tag{2.135}$$

so $-\frac{1}{2}I_{\partial\Omega} + K^{\#}_{-1} + \widetilde{\Theta}\gamma_D\mathcal{S}_{-1}$ is a Fredholm operator in $H^{1/2}(\partial\Omega)$ with index zero. This finishes the proof of (2.119). Hence, the fact that $\mathrm{dom}(T_{\Theta,\Omega}) \subseteq H^2(\Omega)$ has been established. □

Again we isolate the Neumann Laplacian $-\Delta_{N,\Omega}$, that is, the special case $\widetilde{\Theta} = 0$ in (2.114), under Hypothesis 2.11,

$$-\Delta_{N,\Omega} = -\Delta, \quad \mathrm{dom}(-\Delta_{N,\Omega}) = \big\{u \in H^2(\Omega) \,\big|\, \widetilde{\gamma}_N u = 0 \text{ in } H^{1/2}(\partial\Omega)\big\}. \tag{2.136}$$

Similarly, one can now treat the case of the Dirichlet Laplacian. This has originally been done under more general conditions on Ω (assuming the boundary of Ω to be compact rather than Ω bounded) in [**43**, Lemmas A.1]. For completeness we repeat the short argument below:

THEOREM 2.18. *Assume Hypothesis 2.11. Then the Dirichlet Laplacian, $-\Delta_{D,\Omega}$, defined by*

$$-\Delta_{D,\Omega} = -\Delta, \quad \mathrm{dom}(-\Delta_{D,\Omega}) = \big\{u \in H^2(\Omega) \,\big|\, \gamma_D u = 0 \text{ in } H^{3/2}(\partial\Omega)\big\}, \tag{2.137}$$

is self-adjoint and strictly positive in $L^2(\Omega; d^n x)$. Moreover,

$$\mathrm{dom}\big((-\Delta_{D,\Omega})^{1/2}\big) = H_0^1(\Omega). \tag{2.138}$$

PROOF. For convenience of the reader we reproduce the short proof of [**43**, Lemma A.1] in the special case of Dirichlet boundary conditions, given the proof of Theorem 2.17.

By Theorem 2.10, the operator $T_{D,\Omega}$ in $L^2(\Omega; d^n x)$, defined by

$$\begin{aligned} &T_{D,\Omega} = -\Delta, \\ &\mathrm{dom}(T_{D,\Omega}) = \big\{u \in H_0^1(\Omega) \,\big|\, \Delta u \in L^2(\Omega; d^n x);\ \gamma_D u = 0 \text{ in } L^2(\partial\Omega; d^{n-1}\omega)\big\}, \end{aligned} \tag{2.139}$$

is self-adjoint and strictly positive, and

$$\mathrm{dom}\big((T_{D,\Omega})^{1/2}\big) = H_0^1(\Omega) \tag{2.140}$$

holds. Thus, we need to prove that $\mathrm{dom}(T_{D,\Omega}) \subseteq H^2(\Omega)$. To achieve this, we follow the proof of Theorem 2.17, starting with the same representation (2.125). This time, the requirement on g is that $\gamma_D \mathcal{S}_{-1} g = h = \gamma_D w \in H^{3/2}(\partial\Omega)$. Thus, it suffices to know that

$$\gamma_D \mathcal{S}_{-1} \colon H^{1/2}(\partial\Omega) \to H^{3/2}(\partial\Omega) \tag{2.141}$$

is an isomorphism. When $\partial\Omega$ is of class C^∞, it has been proved in [**97**, Proposition 7.9] that $\gamma_D \mathcal{S}_{-1} \colon H^s(\partial\Omega) \to H^{s+1}(\partial\Omega)$ is an isomorphism for each $s \in \mathbb{R}$ and, if Ω is of class $C^{1,r}$ with $(1/2) < r < 1$, the validity range of this result is limited to $-1 - r < s < r$, which covers (2.141). The latter fact follows from an inspection of Taylor's original proof of [**97**, Proposition 7.9]. Here we just note that the only significant difference is that if $\partial\Omega$ is of class $C^{1,r}$ (instead of class C^∞), then S is a pseudodifferential operator whose symbol exhibits a limited amount of regularity in the space-variable. Such classes of operators have been studied in, e.g., [**73**], [**96**, Chs. 1, 2]. $\square$

REMARK 2.19. We emphasize that all results in this section extend to closed Schrödinger operators

$$H_{\Theta,\Omega} = -\Delta_{\Theta,\Omega} + V, \quad \mathrm{dom}\big(H_{\Theta,\Omega}\big) = \mathrm{dom}\big(-\Delta_{\Theta,\Omega}\big) \tag{2.142}$$

for (not necessarily real-valued) potentials V satisfying $V \in L^\infty(\Omega; d^n x)$, consistently replacing $-\Delta$ by $-\Delta + V$, etc. More generally, all results extend to Kato–Rellich bounded potentials V relative to $-\Delta_{\Theta,\Omega}$ with bound less than one. Extensions to potentials permitting stronger local singularities, and an extensions to (not necessarily bounded) Lipschitz domains with compact boundary, will be pursued elsewhere.

3. Generalized Robin and Dirichlet Boundary Value Problems and Robin-to-Dirichlet and Dirichlet-to-Robin Maps

This section is devoted to generalized Robin and Dirichlet boundary value problems associated with the Helmholtz differential expression $-\Delta - z$ in connection with the open set Ω. In addition, we provide a detailed discussion of Robin-to-Dirichlet maps, $M^{(0)}_{\Theta,D,\Omega}$, in $L^2(\partial\Omega; d^{n-1}\omega)$.

In this section we strengthen Hypothesis 2.2 by adding assumption (3.1) below:

HYPOTHESIS 3.1. *In addition to Hypothesis* 2.2 *suppose that*

$$\widetilde{\Theta} \in \mathcal{B}_\infty\big(H^1(\partial\Omega), L^2(\partial\Omega; d^{n-1}\omega)\big). \tag{3.1}$$

We note that (3.1) is satisfied whenever there exists some $\varepsilon > 0$ such that

$$\widetilde{\Theta} \in \mathcal{B}\big(H^{1-\varepsilon}(\partial\Omega), L^2(\partial\Omega; d^{n-1}\omega)\big). \tag{3.2}$$

We recall the definition of the weak Neumann trace operator $\widetilde{\gamma}_N$ in (2.40), (2.41) and start with the Helmholtz Robin boundary value problems:

THEOREM 3.2. *Assume Hypothesis* 3.1 *and suppose that* $z \in \mathbb{C}\backslash\sigma(-\Delta_{\Theta,\Omega})$. *Then for every* $g \in L^2(\partial\Omega; d^{n-1}\omega)$, *the following generalized Robin boundary value problem,*

$$\begin{cases} (-\Delta - z)u = 0 \text{ in } \Omega, \quad u \in H^{3/2}(\Omega), \\ \big(\widetilde{\gamma}_N + \widetilde{\Theta}\gamma_D\big)u = g \text{ on } \partial\Omega, \end{cases} \tag{3.3}$$

has a unique solution $u = u_\Theta$. *This solution* u_Θ *satisfies*

$$\begin{aligned} &\gamma_D u_\Theta \in H^1(\partial\Omega), \quad \widetilde{\gamma}_N u_\Theta \in L^2(\partial\Omega; d^{n-1}\omega), \\ &\|\gamma_D u_\Theta\|_{H^1(\partial\Omega)} + \|\widetilde{\gamma}_N u_\Theta\|_{L^2(\partial\Omega;d^{n-1}\omega)}, \leq C\|g\|_{L^2(\partial\Omega;d^{n-1}\omega)} \end{aligned} \tag{3.4}$$

and

$$\|u_\Theta\|_{H^{3/2}(\Omega)} \leq C\|g\|_{L^2(\partial\Omega;d^{n-1}\omega)}, \tag{3.5}$$

for some constant constant $C = C(\Theta, \Omega, z) > 0$. *Finally,*

$$\big[\gamma_D(-\Delta_{\Theta,\Omega} - \overline{z}I_\Omega)^{-1}\big]^* \in \mathcal{B}\big(L^2(\partial\Omega; d^{n-1}\omega), H^{3/2}(\Omega)\big), \tag{3.6}$$

and the solution u_Θ *is given by the formula*

$$u_\Theta = \big(\gamma_D(-\Delta_{\Theta,\Omega} - \overline{z}I_\Omega)^{-1}\big)^* g. \tag{3.7}$$

PROOF. It is clear from Lemma 2.3 and Lemma 2.4 that the boundary value problem (3.3) has a meaningful formulation and that any solution satisfies the first line in (3.4). Uniqueness for (3.3) is an immediate consequence of the fact that $z \in \mathbb{C}\backslash\sigma(-\Delta_{\Theta,\Omega})$. As for existence, as in the proof of Theorem 2.17, we look for a candidate expressed as

$$u(x) = (\mathcal{S}_z h)(x), \quad x \in \Omega \tag{3.8}$$

for some $h \in L^2(\partial\Omega; d^{n-1}\omega)$. This ensures that $u \in H^{3/2}(\Omega)$ and $(-\Delta - z)u = 0$ in Ω. Above, the single layer potential $\mathcal{S}_z$ has been defined in (2.121). The boundary condition $\big(\widetilde{\gamma}_N + \widetilde{\Theta}\gamma_D\big)u = g$ on $\partial\Omega$ is then equivalent to

$$\big(\widetilde{\gamma}_N + \widetilde{\Theta}\gamma_D\big)(\mathcal{S}_z h) = g, \tag{3.9}$$

respectively, to

$$\big(-\tfrac{1}{2}I_{\partial\Omega} + K_z^\#\big)h + \widetilde{\Theta}\gamma_D\mathcal{S}_z h = g. \tag{3.10}$$

Here $K_z^\#$ has been defined in (2.122).

To obtain unique solvability of equation (3.10) for $h \in L^2(\partial\Omega; d^{n-1}\omega)$, given $g \in L^2(\partial\Omega; d^{n-1}\omega)$, at least when $z \in \mathbb{C}\backslash D$, where $D \subset \mathbb{C}$ is a discrete set, we proceed in a series of steps. The first step is to observe that the operator in question is Fredholm with index zero for every $z \in \mathbb{C}$. To see this, we decompose

$$\big(-\tfrac{1}{2}I_{\partial\Omega} + K_z^\#\big) = \big(-\tfrac{1}{2}I_{\partial\Omega} + K_0^\#\big) + \big(K_z^\# - K_0^\#\big), \tag{3.11}$$

and recall that $\big(K_z^{\#} - K_0^{\#}\big) \in \mathcal{B}_\infty(L^2(\partial\Omega; d^{n-1}\omega))$ (cf. Lemma D.3) and that $-\frac{1}{2}I_{\partial\Omega} + K_0^{\#}$ is a Fredholm operator in $L^2(\partial\Omega; d^{n-1}\omega)$ with Fredholm index equal to zero as proven by Verchota [**101**]. In addition, we note that

$$\widetilde{\Theta}\gamma_D\mathcal{S}_z \in \mathcal{B}_\infty(L^2(\partial\Omega; d^{n-1}\omega)), \tag{3.12}$$

which follows from Hypothesis 3.1 and the fact that the following operators are bounded

$$\begin{aligned} &\mathcal{S}_z : L^2(\partial\Omega; d^{n-1}\omega) \to \{u \in H^{3/2}(\Omega) \,|\, \Delta u \in L^2(\Omega; d^n x)\}, \\ &\gamma_D : \{u \in H^{3/2}(\Omega) \,|\, \Delta u \in L^2(\Omega; d^n x)\} \to H^1(\partial\Omega), \end{aligned} \tag{3.13}$$

(where the space $\{u \in H^{3/2}(\Omega) \,|\, \Delta u \in L^2(\Omega; d^n x)\}$ is equipped with the natural graph norm $u \mapsto \|u\|_{H^{3/2}(\Omega)} + \|\Delta u\|_{L^2(\Omega; d^n x)}$). See Lemma 2.3 and Theorem D.7. Thus, $\big(-\frac{1}{2}I_{\partial\Omega} + K_z^{\#}\big) + \widetilde{\Theta}\gamma_D\mathcal{S}_z$ is a Fredholm operator in $L^2(\partial\Omega; d^{n-1}\omega)$ with Fredholm index equal to zero, for every $z \in \mathbb{C}$. In particular, it is invertible if and only if it is injective.

In the second step, we study the injectivity of $\big(-\frac{1}{2}I_{\partial\Omega} + K_z^{\#}\big) + \widetilde{\Theta}\gamma_D\mathcal{S}_z$ on $L^2(\partial\Omega; d^{n-1}\omega)$. For this purpose we now suppose that

$$\big(-\tfrac{1}{2}I_{\partial\Omega} + K_z^{\#}\big)k + \widetilde{\Theta}\gamma_D\mathcal{S}_z k = 0 \text{ for some } k \in L^2(\partial\Omega; d^{n-1}\omega). \tag{3.14}$$

Introducing $w = \mathcal{S}_z k$ in Ω one then infers that w satisfies

$$\begin{cases} (-\Delta - z)w = 0 \text{ in } \Omega, \quad w \in H^{3/2}(\Omega), \\ (\widetilde{\gamma}_N + \widetilde{\Theta}\gamma_D)w = 0 \text{ on } \partial\Omega. \end{cases} \tag{3.15}$$

Thus one obtains,

$$\begin{aligned} 0 \le \int_\Omega d^n x\, |\nabla w|^2 &= \sum_{j=1}^n \int_\Omega d^n x\, \overline{\partial_j w}\partial_j w \\ &= -\int_\Omega d^n x\, \overline{\Delta w} w + \sum_{j=1}^n \int_{\partial\Omega} d^{n-1}\omega\, \big(\gamma_D \overline{\partial_j w}\big)\, \nu_j \gamma_D w \\ &= \overline{z}\int_\Omega d^n x\, |w|^2 + (\gamma_D w, \widetilde{\gamma}_N w)_{L^2(\partial\Omega; d^{n-1}\omega)} \\ &= \overline{z}\int_\Omega d^n x\, |w|^2 + \langle \gamma_D w, \widetilde{\gamma}_N w\rangle_{1/2} \\ &= \overline{z}\int_\Omega d^n x\, |w|^2 - \big\langle \gamma_D w, \widetilde{\Theta}\gamma_D w\big\rangle_{1/2}. \end{aligned} \tag{3.16}$$

At this point we will first consider the case when $z \in \mathbb{C}\backslash\mathbb{R}$ (so that, in particular, $\mathrm{Im}(z) \neq 0$). In this scenario, recalling (2.1) and taking the imaginary parts of the two most extreme sides of (3.16) imply that $\int_\Omega d^n x\, |w|^2 = 0$ and, hence, $w = 0$ in Ω.

To continue, let $\widetilde{\gamma}_N^{ext}$ and γ_D^{ext} denote, respectively, the Neumann and Dirichlet traces for the exterior domain $\mathbb{R}^n\backslash\overline{\Omega}$. Also, parallel to (2.121), set

$$(\mathcal{S}_{ext,z} g)(x) = \int_{\partial\Omega} d^{n-1}\omega(y)\, E_n(z; x-y) g(y), \quad x \in \mathbb{R}^n\backslash\overline{\Omega},\ z \in \mathbb{C}, \tag{3.17}$$

where g is an arbitrary measurable function on $\partial\Omega$. Then, due to the weak singularity in the integral kernel of $\mathcal{S}_z$,

$$\gamma_D \mathcal{S}_z g = \gamma_D^{ext} \mathcal{S}_{ext,z} g \quad \text{for every} \quad g \in L^2(\partial\Omega; d^{n-1}\omega), \tag{3.18}$$

whereas the counterpart of (2.123) becomes

$$\widetilde{\gamma}_N^{ext} \mathcal{S}_{ext,z} g = \big(\tfrac{1}{2} I_{\partial\Omega} + K_z^{\#}\big) g \quad \text{for every} \quad g \in L^2(\partial\Omega; d^{n-1}\omega). \tag{3.19}$$

Compared with (2.123), the change in sign is due to the fact that the outward unit normal to $\mathbb{R}^n \backslash \overline{\Omega}$ is $-\nu$.

Moving on, if we set $w^{ext}(x) = (\mathcal{S}_{ext,z} k)(x)$ for $x \in \mathbb{R}^n \backslash \overline{\Omega}$, then from what we have proved so far

$$0 = \gamma_D w = \gamma_D \mathcal{S}_z k = \gamma_D^{ext} w^{ext} \quad \text{in} \quad L^2(\partial\Omega; d^{n-1}\omega). \tag{3.20}$$

Fix now a sufficiently large $R > 0$ such that $\overline{\Omega} \subset B(0;R)$ and write the analogue of (3.16) for the restriction of w^{ext} to $B(0;R)\backslash\overline{\Omega}$:

$$\begin{aligned} \int_{B(0;R)\backslash\overline{\Omega}} d^n x \, |\nabla w^{ext}|^2 &= \overline{z} \int_{B(0;R)\backslash\overline{\Omega}} d^n x \, |w^{ext}|^2 - \langle \gamma_D^{ext} w^{ext}, \widetilde{\gamma}_N^{ext} w^{ext} \rangle_{1/2} \\ &\quad - \int_{|\xi|=R} d^{n-1}\omega(\xi) \, w^{ext}(\xi) \frac{\xi}{|\xi|} \cdot \nabla w^{ext}(\xi). \end{aligned} \tag{3.21}$$

In view of (3.20), the above identity reduces to

$$\begin{aligned} &\int_{B(0;R)\backslash\overline{\Omega}} d^n x \, |\nabla w^{ext}|^2 \\ &\quad = \overline{z} \int_{B(0;R)\backslash\overline{\Omega}} d^n x \, |w^{ext}|^2 - \int_{|\xi|=R} d^{n-1}\omega(\xi) \, w^{ext}(\xi) \frac{\xi}{|\xi|} \cdot \nabla w^{ext}(\xi). \end{aligned} \tag{3.22}$$

Recall that we are assuming $z \in \mathbb{C}\backslash\mathbb{R}$. Given that, by (C.17) (and the comment following right after it), the integral kernel of $\mathcal{S}_{ext,z} k$ has exponential decay at infinity, it follows that w^{ext} decays exponentially at infinity. Thus, after passing to limit $R \to \infty$, we arrive at

$$\int_{\mathbb{R}^n\backslash\overline{\Omega}} d^n x \, |\nabla w^{ext}|^2 = \overline{z} \int_{\mathbb{R}^n\backslash\overline{\Omega}} d^n x \, |w^{ext}|^2. \tag{3.23}$$

Consequently, taking the imaginary parts of both sides we arrive at the conclusion that $w^{ext} = 0$ in $\mathbb{R}^n\backslash\overline{\Omega}$. With this in hand, we may then invoke (2.123), (3.19) to deduce that

$$k = \widetilde{\gamma}_N^{ext} \mathcal{S}_z k - \widetilde{\gamma}_N \mathcal{S}_z k = \widetilde{\gamma}_N^{ext} w^{ext} - \widetilde{\gamma}_N w = 0, \tag{3.24}$$

given that w, w^{ext} vanish in Ω and $\mathbb{R}^n\backslash\overline{\Omega}$, respectively. Hence, one concludes that $k = 0$ in $L^2(\partial\Omega; d^{n-1}\omega)$. This proves that the operator $\big(-\frac{1}{2} I_{\partial\Omega} + K_z^{\#}\big) + \widetilde{\Theta}\gamma_D \mathcal{S}_z$ is injective, hence, invertible on $L^2(\partial\Omega; d^{n-1}\omega)$ whenever $z \in \mathbb{C}\backslash\mathbb{R}$.

In the third step, the goal is to extend this result to other values of the parameter z. To this end, fix some $z_0 \in \mathbb{C}\backslash\mathbb{R}$, and for $z \in \mathbb{C}$, consider

$$\mathcal{A}_z = \big[\big(-\tfrac{1}{2} I_{\partial\Omega} + K_{z_0}^{\#}\big) + \widetilde{\Theta}\gamma_D \mathcal{S}_{z_0}\big]^{-1} \big[\big(K_z^{\#} - K_{z_0}^{\#}\big) + \widetilde{\Theta}\gamma_D(\mathcal{S}_z - \mathcal{S}_{z_0})\big]. \tag{3.25}$$

Observe that the operator-valued mapping $z \mapsto \mathcal{A}_z \in \mathcal{B}\big(L^2(\partial\Omega; d^{n-1}\omega)\big)$ is analytic and, thanks to Lemma D.3, $\mathcal{A}_z \in \mathcal{B}_\infty\big(L^2(\partial\Omega; d^{n-1}\omega)\big)$. The analytic Fredholm theorem then yields invertibility of $I + \mathcal{A}_z$ except for z in a discrete set $D \subset \mathbb{C}$. Thus,

$$(-\tfrac{1}{2}I_{\partial\Omega} + K_z^{\#}) + \widetilde{\Theta}\gamma_D\mathcal{S}_z = [(-\tfrac{1}{2}I_{\partial\Omega} + K_{z_0}^{\#}) + \widetilde{\Theta}\gamma_D\mathcal{S}_{z_0}][I + \mathcal{A}_z] \tag{3.26}$$

is invertible for $z \in \mathbb{C}\backslash D$ where D is a discrete set which, by the invertibility result proved in the previous paragraph, is contained in $\mathbb{R}$.

The above argument proves unique solvability of (3.3) for $z \in \mathbb{C}\backslash D$, where D is a discrete subset of $\mathbb{R}$. The representation (3.8) and the fact that

$$\gamma_D\mathcal{S}_z \colon L^2(\partial\Omega; d^{n-1}\omega) \to H^1(\partial\Omega) \text{ boundedly,} \tag{3.27}$$

then yields $\gamma_D u_\Theta \in H^1(\partial\Omega)$. Moreover, by (2.123) and (3.8),

$$\widetilde{\gamma}_N u_\Theta = \widetilde{\gamma}_N \mathcal{S}_z h = (-\tfrac{1}{2}I_{\partial\Omega} + K_z^{\#})h \in L^2(\partial\Omega; d^{n-1}\omega) \tag{3.28}$$

since by (2.124), $K_z^{\#} \in \mathcal{B}\big(L^2(\partial\Omega; d^{n-1}\omega)\big)$. This proves (3.4) when $z \in \mathbb{C}\backslash D$. For $z \in \mathbb{C}\backslash D$, the natural estimate (3.5) is a consequence of the integral representation formula (3.8) and (D.28).

Next, fix a complex number $z \in \mathbb{C}\backslash(D \cup \sigma(-\Delta_{\Theta,\Omega}))$ along with two functions, $v \in L^2(\Omega; d^n x)$ and $g \in L^2(\partial\Omega; d^{n-1}\omega)$. Also, let u_Θ solve (3.3). One computes

$$\begin{aligned}
\big(u_\Theta, v\big)_{L^2(\Omega;d^n x)} &= \big(u_\Theta, (-\Delta - \overline{z})(-\Delta_{\Theta,\Omega} - \overline{z}I_\Omega)^{-1}v\big)_{L^2(\Omega;d^n x)} \\
&= \big((-\Delta - z)u_\Theta, (-\Delta_{\Theta,\Omega} - \overline{z}I_\Omega)^{-1}v\big)_{L^2(\Omega;d^n x)} \\
&\quad + \big(\widetilde{\gamma}_N u_\Theta, \gamma_D(-\Delta_{\Theta,\Omega} - \overline{z}I_\Omega)^{-1}v\big)_{L^2(\partial\Omega;d^{n-1}\omega)} \\
&\quad - \big\langle \gamma_D u_\Theta, \widetilde{\gamma}_N\big(-\Delta_{\Theta,\Omega} - \overline{z}I_\Omega\big)^{-1}v\big\rangle_{1/2} \\
&= \big(\widetilde{\gamma}_N u_\Theta, \gamma_D(-\Delta_{\Theta,\Omega} - \overline{z}I_\Omega)^{-1}v\big)_{L^2(\partial\Omega;d^{n-1}\omega)} \\
&\quad + \big\langle \gamma_D u_\Theta, \widetilde{\Theta}\gamma_D(-\Delta_{\Theta,\Omega} - \overline{z}I_\Omega)^{-1}v\big\rangle_{1/2} \\
&= \big(\widetilde{\gamma}_N u_\Theta, \gamma_D(-\Delta_{\Theta,\Omega} - \overline{z}I_\Omega)^{-1}v\big)_{L^2(\partial\Omega;d^{n-1}\omega)} \\
&\quad + \overline{\big\langle \gamma_D(-\Delta_{\Theta,\Omega} - \overline{z}I_\Omega)^{-1}v, \widetilde{\Theta}\gamma_D u_\Theta\big\rangle_{1/2}} \\
&= \big(\widetilde{\gamma}_N u_\Theta, \gamma_D(-\Delta_{\Theta,\Omega} - \overline{z}I_\Omega)^{-1}v\big)_{L^2(\partial\Omega;d^{n-1}\omega)} \\
&\quad + \big\langle \widetilde{\Theta}\gamma_D u_\Theta, \gamma_D(-\Delta_{\Theta,\Omega} - \overline{z}I_\Omega)^{-1}v\big\rangle_{1/2} \\
&= \big\langle (\widetilde{\gamma}_N + \widetilde{\Theta}\gamma_D)u_\Theta, \gamma_D(-\Delta_{\Theta,\Omega} - \overline{z}I_\Omega)^{-1}v\big\rangle_{1/2} \\
&= \big((\widetilde{\gamma}_N + \widetilde{\Theta}\gamma_D)u_\Theta, \gamma_D(-\Delta_{\Theta,\Omega} - \overline{z}I_\Omega)^{-1}v\big)_{L^2(\partial\Omega;d^{n-1}\omega)} \\
&= \big(g, \gamma_D(-\Delta_{\Theta,\Omega} - \overline{z}I_\Omega)^{-1}v\big)_{L^2(\partial\Omega;d^{n-1}\omega)} \\
&= \big(\big(\gamma_D(-\Delta_{\Theta,\Omega} - \overline{z}I_\Omega)^{-1}\big)^* g, v\big)_{L^2(\Omega;d^n x)}.
\end{aligned} \tag{3.29}$$

Since $v \in L^2(\Omega; d^n x)$ was arbitrary, this yields

$$u_\Theta = \big(\gamma_D(-\Delta_{\Theta,\Omega} - \overline{z}I_\Omega)^{-1}\big)^* g \text{ in } L^2(\Omega; d^n x), \quad \text{for } z \in \mathbb{C}\backslash(D \cup \sigma(-\Delta_{\Theta,\Omega})), \tag{3.30}$$

which proves (3.7) for $z \in \mathbb{C}\backslash(D \cup \sigma(-\Delta_{\Theta,\Omega}))$. From this and (3.5), the membership (3.6) also follows when $z \in \mathbb{C}\backslash(D \cup \sigma(-\Delta_{\Theta,\Omega}))$.

The extension to the more general case when $z \in \mathbb{C}\backslash\sigma(-\Delta_{\Theta,\Omega})$ is then done by resorting to analytic continuation with respect to z. More specifically, fix $z_0 \in$

$\mathbb{C}\backslash\sigma(\Delta_{\Theta,\Omega})$. Then there exists $r > 0$ such that

$$\big(\overline{B(z_0,r)}\backslash\{z_0\}\big)\cap D = \emptyset, \quad \overline{B(z_0,r)}\cap\sigma(-\Delta_{\Theta,\Omega}) = \emptyset, \tag{3.31}$$

since D is discrete and $\sigma(-\Delta_{\Theta,\Omega})$ is closed. We may then write

$$[\gamma_D(-\Delta_{\Theta,\Omega} - \overline{z_0}I_\Omega)^{-1}]^* = \frac{1}{2\pi i}\oint_{C(z_0;r)} dz\,(z-z_0)^{-1}[\gamma_D(-\Delta_{\Theta,\Omega} - \overline{z}I_\Omega)^{-1}]^* \tag{3.32}$$

as operators in $\mathcal{B}(H^{-1/2}(\partial\Omega), L^2(\Omega; d^n x))$, where $C(z_0;r) \subset \mathbb{C}$ denotes the counterclockwise oriented circle with center z_0 and radius r. (This follows from dualizing the fact $\gamma_D(-\Delta_{\Theta,\Omega} - z_0 I_\Omega)^{-1} \in \mathcal{B}(L^2(\Omega; d^n x), H^{1/2}(\partial\Omega))$, which in turn follows from the mapping properties $(-\Delta_{\Theta,\Omega} - z_0 I_\Omega)^{-1} \in \mathcal{B}(L^2(\Omega; d^n x), H^1(\partial\Omega))$ and $\gamma_D \in \mathcal{B}\big(H^1(\partial\Omega), H^{1/2}(\partial\Omega)\big)$.) However, granted (3.31), what we have shown so far yields that $[\gamma_D(-\Delta_{\Theta,\Omega} - \overline{z}I_\Omega)^{-1}]^* \in \mathcal{B}(L^2(\partial\Omega; d^{n-1}\omega), H^{3/2}(\Omega))$ whenever $|z - z_0| = r$, with a bound

$$\|[\gamma_D(-\Delta_{\Theta,\Omega} - \overline{z}I_\Omega)^{-1}]^*\|_{\mathcal{B}(L^2(\partial\Omega;d^{n-1}\omega),H^{3/2}(\Omega))} \leq C = C(\Omega,\Theta,z_0,r) \tag{3.33}$$

independent of the complex parameter $z \in \partial B(z_0,r)$. This estimate and Cauchy's representation formula (3.32) then imply that

$$[\gamma_D(-\Delta_{\Theta,\Omega} - \overline{z_0}I_\Omega)^{-1}]^* \in \mathcal{B}(L^2(\partial\Omega; d^{n-1}\omega), H^{3/2}(\Omega)). \tag{3.34}$$

This further entails that $u = [\gamma_D(-\Delta_{\Theta,\Omega} - \overline{z_0}I_\Omega)^{-1}]^* g$ solves (3.3), written with z_0 in place of z, and satisfies (3.5). Finally, the memberships in (3.4) (along with naturally accompanying estimates) follow from Lemma 2.3 and Lemma 2.4. This shows that (3.6), along with the well-posedness of (3.3) and all the desired properties of the solution, hold whenever $z \in \mathbb{C}\backslash\sigma(\Delta_{\Theta,\Omega})$. □

The special case $\Theta = 0$ of Theorem 3.2, corresponding to the Neumann Laplacian, deserves to be mentioned separately.

COROLLARY 3.3. *Assume Hypothesis 2.1 and suppose that* $z \in \mathbb{C}\backslash\sigma(-\Delta_{N,\Omega})$. *Then for every* $g \in L^2(\partial\Omega; d^{n-1}\omega)$, *the following Neumann boundary value problem,*

$$\begin{cases} (-\Delta - z)u = 0 \text{ in } \Omega, \quad u \in H^{3/2}(\Omega), \\ \widetilde{\gamma}_N u = g \text{ on } \partial\Omega, \end{cases} \tag{3.35}$$

has a unique solution $u = u_N$. *This solution* u_N *satisfies*

$$\gamma_D u_N \in H^1(\partial\Omega) \text{ and } \|\gamma_D u_N\|_{H^1(\partial\Omega)} + \|\widetilde{\gamma}_N u_N\|_{L^2(\partial\Omega;d^{n-1}\omega)} \leq C\|g\|_{L^2(\partial\Omega;d^{n-1}\omega)} \tag{3.36}$$

as well as

$$\|u_N\|_{H^{3/2}(\Omega)} \leq C\|g\|_{L^2(\partial\Omega;d^{n-1}\omega)}, \tag{3.37}$$

for some constant constant $C = C(\Theta,\Omega,z) > 0$. *Finally,*

$$\big[\gamma_D(-\Delta_{N,\Omega} - \overline{z}I_\Omega)^{-1}\big]^* \in \mathcal{B}\big(L^2(\partial\Omega; d^{n-1}\omega), H^{3/2}(\Omega)\big), \tag{3.38}$$

and the solution u_N *is given by the formula*

$$u_\Theta = \big(\gamma_D(-\Delta_{N,\Omega} - \overline{z}I_\Omega)^{-1}\big)^* g. \tag{3.39}$$

Next, we turn to the Dirichlet case originally treated in [**46**, Theorem 3.1] but under stronger regularity conditions on Ω. In order to facilitate the subsequent considerations, we isolate a useful technical result in the lemma below.

LEMMA 3.4. *Assume Hypothesis* 2.1 *and suppose that* $z \in \mathbb{C}\backslash\sigma(-\Delta_{D,\Omega})$. *Then*

$$(-\Delta_{D,\Omega} - zI_\Omega)^{-1} : L^2(\Omega; d^n x) \to \{u \in H^{3/2}(\Omega) \,|\, \Delta u \in L^2(\Omega; d^n x)\} \tag{3.40}$$

is a well-defined bounded operator, where the space $\{u \in H^{3/2}(\Omega) \,|\, \Delta u \in L^2(\Omega; d^n x)\}$ *is equipped with the natural graph norm* $u \mapsto \|u\|_{H^{3/2}(\Omega)} + \|\Delta u\|_{L^2(\Omega;d^n x)}$.

PROOF. Consider $z \in \mathbb{C}\backslash\sigma(-\Delta_{D,\Omega})$, $f \in L^2(\Omega; d^n x)$ and set $w = (-\Delta_{D,\Omega} - zI_\Omega)^{-1} f$. It follows that u is the unique solution of the problem

$$(-\Delta - z)w = f \text{ in } \Omega, \quad w \in H_0^1(\Omega). \tag{3.41}$$

The strategy is to devise an alternative representation for w from which it is clear that w has the claimed regularity in Ω. To this end, let $\widetilde{f}$ denote the extension of f by zero to $\mathbb{R}^n$ and denote by E the operator of convolution by $E_n(z; \cdot)$. Since the latter is smoothing of order 2, it follows that $v = \big(E\widetilde{f}\big)\big|_\Omega \in H^2(\Omega)$ and $(-\Delta - z)v = f$ in Ω. In particular, $g = -\gamma_D v \in H^1(\partial\Omega)$.

We now claim that the problem

$$(-\Delta - z)u = 0 \text{ in } \Omega, \quad u \in H^{3/2}(\Omega), \quad \gamma_D u = g \text{ on } \partial\Omega, \tag{3.42}$$

has a solution (satisfying natural estimates). To see this, we look for a solution in the form (3.8) for some $h \in L^2(\partial\Omega; d^{n-1}\omega)$. This guarantees that $u \in H^{3/2}(\Omega)$ by Theorem D.7, and $(-\Delta - z)u = 0$ in Ω. Ensuring that the boundary condition holds comes down to solving $\gamma_D \mathcal{S}_z h = g$. In this regard, we recall that

$$\gamma_D \mathcal{S}_0 \colon L^2(\partial\Omega; d^{n-1}\omega) \to H^1(\partial\Omega) \text{ is invertible} \tag{3.43}$$

(cf. [**101**]). With this in hand, by relying on Theorem D.7 and arguing as in the proof of Theorem 3.2, one can show that there exists a discrete set $D \subset \mathbb{C}$ such that

$$\gamma_D \mathcal{S}_z \colon L^2(\partial\Omega; d^{n-1}\omega) \to H^1(\partial\Omega) \text{ is invertible for } z \in \mathbb{C}\backslash D. \tag{3.44}$$

Thus, a solution of (3.42) is given by

$$u = \mathcal{S}_z\big((\gamma_D \mathcal{S}_z)^{-1}(\gamma_D v)\big) \text{ if } z \in \mathbb{C}\backslash D. \tag{3.45}$$

Moreover, by Theorem D.7, this satisfies

$$\|u\|_{H^{3/2}(\Omega)} \le C(\Omega, z)\|g\|_{H^1(\partial\Omega)} \le C(\Omega, z)\|f\|_{L^2(\Omega;d^n x)}, \quad z \in \mathbb{C}\backslash D. \tag{3.46}$$

Consequently, if $z \in \mathbb{C}\backslash(D \cup \sigma(-\Delta_{D,\Omega}))$, then $u + v$ solves (3.41). Hence, by uniqueness, $w = u + v$ in this case. This shows that $w = (-\Delta_{D,\Omega} - zI_\Omega)^{-1} f$ belongs to $H^{3/2}(\Omega)$ and satisfies $\Delta w \in L^2(\Omega; d^n x)$ with

$$\|w\|_{H^{3/2}(\Omega)} + \|\Delta w\|_{L^2(\Omega;d^n x)} \le C(\Omega, z)\|f\|_{L^2(\Omega;d^n x)}, \quad z \in \mathbb{C}\backslash(D \cup \sigma(-\Delta_{D,\Omega})). \tag{3.47}$$

In summary, the above argument shows that the operator (3.40) is well-defined and bounded whenever $z \in \mathbb{C}\backslash(D \cup \sigma(-\Delta_{D,\Omega}))$. The extension to $z \in \mathbb{C}\backslash\sigma(-\Delta_{D,\Omega})$ is then achieved via analytic continuation (as in the last part of the proof of Theorem 3.2). □

Having established Lemma 3.4, we can now readily prove the following result.

LEMMA 3.5. *Assume Hypothesis* 2.1 *and suppose that* $z \in \mathbb{C}\backslash\sigma(-\Delta_{D,\Omega})$. *Then*

$$\widetilde{\gamma}_N(-\Delta_{D,\Omega} - zI_\Omega)^{-1} \in \mathcal{B}\big(L^2(\Omega; d^n x), L^2(\partial\Omega; d^{n-1}\omega)\big), \tag{3.48}$$

and

$$\big[\widetilde{\gamma}_N(-\Delta_{D,\Omega} - zI_\Omega)^{-1}\big]^* \in \mathcal{B}\big(L^2(\partial\Omega; d^{n-1}\omega), L^2(\Omega; d^n x)\big). \tag{3.49}$$

PROOF. Obviously, it suffices to only prove (3.48). However, this is an immediate consequence of Lemma 3.4 and Lemma 2.4. □

We note that Lemma 3.5 corrects an inaccuracy in the proof of [**46**, Theorem 3.1] in the following sense: The proof of (3.20) and (3.21) in [**46**] relies on [**46**, Lemma 2.4], which in turn requires the stronger assumptions [**46**, Hypothesis 2.1] on Ω than merely the Lipschitz assumption on Ω. However, the current Lemmas 2.15 and 3.5 (and the subsequent Theorem 3.6) show that (3.20) and (3.21) in [**46**], as well as the results stated in [**46**, Theorem 3.1], are actually all correct.

After this preamble, we are ready to state the result about the well-posedness of the Dirichlet problem, alluded to above.

THEOREM 3.6. *Assume Hypothesis* 2.1 *and suppose that* $z \in \mathbb{C}\backslash\sigma(-\Delta_{D,\Omega})$. *Then for every* $f \in H^1(\partial\Omega)$, *the following Dirichlet boundary value problem,*

$$\begin{cases} (-\Delta - z)u = 0 \text{ in } \Omega, \quad u \in H^{3/2}(\Omega), \\ \gamma_D u = f \text{ on } \partial\Omega, \end{cases} \tag{3.50}$$

has a unique solution $u = u_D$. *This solution* u_D *satisfies*

$$\widetilde{\gamma}_N u_D \in L^2(\partial\Omega; d^{n-1}\omega) \quad \text{and} \quad \|\widetilde{\gamma}_N u_D\|_{L^2(\partial\Omega; d^{n-1}\omega)} \leq C_D \|f\|_{H^1(\partial\Omega)}, \tag{3.51}$$

for some constant $C_D = C_D(\Omega, z) > 0$. *Moreover,*

$$\|u_D\|_{H^{3/2}(\Omega)} \leq C_D \|f\|_{H^1(\partial\Omega)}. \tag{3.52}$$

Finally,

$$\big[\widetilde{\gamma}_N(-\Delta_{D,\Omega} - \overline{z}I_\Omega)^{-1}\big]^* \in \mathcal{B}\big(H^1(\partial\Omega), H^{3/2}(\Omega)\big), \tag{3.53}$$

and the solution u_D *is given by the formula*

$$u_D = -\big[\widetilde{\gamma}_N(-\Delta_{D,\Omega} - \overline{z}I_\Omega)^{-1}\big]^* f. \tag{3.54}$$

PROOF. Uniqueness for (3.50) is a direct consequence of the fact that $z \in \mathbb{C}\backslash\sigma(-\Delta_{D,\Omega})$. Existence, at least when $z \in \mathbb{C}\backslash D$ for a discrete set $D \subset \mathbb{C}$, is implicit in the proof of Lemma 3.4 (cf. (3.42)). Note that a solution thus constructed obeys (3.52) and satisfies (3.51) (cf. Lemmas 2.3 and 2.4).

Next, we turn to the proof of (3.54). Assume that $z \in \mathbb{C}\backslash(D \cup \sigma(-\Delta_{D,\Omega}))$ and denote by u_D the unique solution of (3.50). Also, recall (3.48)–(3.49). Based on these and Green's formula, one computes

$$\begin{aligned} (u_D, v)_{L^2(\Omega; d^n x)} &= (u_D, (-\Delta - \overline{z})(-\Delta_{D,\Omega} - \overline{z}I_\Omega)^{-1}v)_{L^2(\Omega; d^n x)} \\ &= ((-\Delta - z)u_D, (-\Delta_{D,\Omega} - \overline{z}I_\Omega)^{-1}v)_{L^2(\Omega; d^n x)} \\ &\quad + (\widetilde{\gamma}_N u_D, \gamma_D(-\Delta_{D,\Omega} - \overline{z}I_\Omega)^{-1}v)_{L^2(\partial\Omega; d^{n-1}\omega)} \\ &\quad - \langle \gamma_D u_D, \widetilde{\gamma}_N(-\Delta_{D,\Omega} - \overline{z}I_\Omega)^{-1}v\rangle_{1/2} \\ &= -\langle f, \widetilde{\gamma}_N(-\Delta_{D,\Omega} - \overline{z}I_\Omega)^{-1}v\rangle_{1/2} \\ &= -\big((\widetilde{\gamma}_N(-\Delta_{D,\Omega} - \overline{z}I_\Omega)^{-1})^* f, v\big)_{L^2(\Omega; d^n x)} \end{aligned} \tag{3.55}$$

for any $v \in L^2(\Omega; d^n x)$. This proves (3.54) with the operators involved understood in the sense of (3.49). Given (3.52), one obtains (3.53) granted that $z \in \mathbb{C}\backslash(D \cup \sigma(-\Delta_{D,\Omega}))$.

Finally, the extension of the above results to the more general case in which $z \in \mathbb{C}\backslash\sigma(-\Delta_{D,\Omega})$ is done using analytic continuation, as in the last part of the proof of Theorem 3.2. □

Assuming Hypothesis 3.1, we introduce the Dirichlet-to-Robin map $M_{D,\Theta,\Omega}^{(0)}(z)$ associated with $(-\Delta - z)$ on Ω, as follows,

$$M_{D,\Theta,\Omega}^{(0)}(z) \colon \begin{cases} H^1(\partial\Omega) \to L^2(\partial\Omega; d^{n-1}\omega), \\ \qquad\qquad f \mapsto -(\widetilde{\gamma}_N + \widetilde{\Theta}\gamma_D)u_D, \end{cases} \quad z \in \mathbb{C}\backslash\sigma(-\Delta_{D,\Omega}), \tag{3.56}$$

where u_D is the unique solution of

$$(-\Delta - z)u = 0 \text{ in } \Omega, \quad u \in H^{3/2}(\Omega), \quad \gamma_D u = f \text{ on } \partial\Omega. \tag{3.57}$$

Continuing to assume Hypothesis 3.1, we next introduce the Robin-to-Dirichlet map $M_{\Theta,D,\Omega}^{(0)}(z)$ associated with $(-\Delta - z)$ on Ω, as follows,

$$M_{\Theta,D,\Omega}^{(0)}(z) \colon \begin{cases} L^2(\partial\Omega; d^{n-1}\omega) \to H^1(\partial\Omega), \\ \qquad\qquad g \mapsto \gamma_D u_\Theta, \end{cases} \quad z \in \mathbb{C}\backslash\sigma(-\Delta_{\Theta,\Omega}), \tag{3.58}$$

where u_Θ is the unique solution of

$$(-\Delta - z)u = 0 \text{ in } \Omega, \quad u \in H^{3/2}(\Omega), \quad (\widetilde{\gamma}_N + \widetilde{\Theta}\gamma_D)u = g \text{ on } \partial\Omega. \tag{3.59}$$

We note that Robin-to-Dirichlet maps have also been studied in [**10**].

We conclude with the following theorem, one of the main results of this paper:

THEOREM 3.7. *Assume Hypothesis* 3.1. *Then*

$$M_{D,\Theta,\Omega}^{(0)}(z) \in \mathcal{B}\big(H^1(\partial\Omega), L^2(\partial\Omega; d^{n-1}\omega)\big), \quad z \in \mathbb{C}\backslash\sigma(-\Delta_{D,\Omega}), \tag{3.60}$$

and

$$M_{D,\Theta,\Omega}^{(0)}(z) = (\widetilde{\gamma}_N + \widetilde{\Theta}\gamma_D)\big[(\widetilde{\gamma}_N + \widetilde{\Theta}\gamma_D)(-\Delta_{D,\Omega} - \overline{z}I_\Omega)^{-1}\big]^*, \quad z \in \mathbb{C}\backslash\sigma(-\Delta_{D,\Omega}). \tag{3.61}$$

Moreover,

$$M_{\Theta,D,\Omega}^{(0)}(z) \in \mathcal{B}\big(L^2(\partial\Omega; d^{n-1}\omega), H^1(\partial\Omega)\big), \quad z \in \mathbb{C}\backslash\sigma(-\Delta_{\Theta,\Omega}), \tag{3.62}$$

and, in fact,

$$M_{\Theta,D,\Omega}^{(0)}(z) \in \mathcal{B}_\infty\big(L^2(\partial\Omega; d^{n-1}\omega)\big), \quad z \in \mathbb{C}\backslash\sigma(-\Delta_{\Theta,\Omega}). \tag{3.63}$$

In addition,

$$M_{\Theta,D,\Omega}^{(0)}(z) = \gamma_D\big[\gamma_D(-\Delta_{\Theta,\Omega} - \overline{z}I_\Omega)^{-1}\big]^*, \quad z \in \mathbb{C}\backslash\sigma(-\Delta_{\Theta,\Omega}). \tag{3.64}$$

Finally, let $z \in \mathbb{C}\backslash(\sigma(-\Delta_{D,\Omega}) \cup \sigma(-\Delta_{\Theta,\Omega}))$. *Then*

$$M_{\Theta,D,\Omega}^{(0)}(z) = -M_{D,\Theta,\Omega}^{(0)}(z)^{-1}. \tag{3.65}$$

PROOF. The membership in (3.60) is a consequence of Theorem 3.6. In this context we note that by the first line in (2.96), $\gamma_D(-\Delta_{D,\Omega} - \overline{z}I_\Omega)^{-1} = 0$, and hence

$$u_D = -\big[\widetilde{\gamma}_N(-\Delta_{D,\Omega} - \overline{z}I_\Omega)^{-1}\big]^* f = -\big[(\widetilde{\gamma}_N + \widetilde{\Theta}\gamma_D)(-\Delta_{D,\Omega} - \overline{z}I_\Omega)^{-1}\big]^* f \tag{3.66}$$

by (3.54). Moreover, applying $-\big(\widetilde\gamma_N+\widetilde\Theta\gamma_D\big)$ to u_D in (3.54) implies formula (3.61). Likewise, (3.62) follows from Theorem 3.2. In addition, since $H^1(\partial\Omega)$ embeds compactly into $L^2(\partial\Omega;d^{n-1}\omega)$ (cf. (A.10) and [**72**, Proposition 2.4]), $M_{\Theta,D,\Omega}^{(0)}(z)$, $z\in\mathbb{C}\backslash\sigma(-\Delta_{\Theta,\Omega})$, are compact operators in $L^2(\partial\Omega;d^{n-1}\omega)$, justifying (3.63). Applying γ_D to u_Θ in (3.7) implies formula (3.64).

There remains to justify (3.65). To this end, let $g\in L^2(\partial\Omega;d^{n-1}\omega)$ be arbitrary. Then

$$\begin{aligned} M_{D,\Theta,\Omega}^{(0)}(z)M_{\Theta,D,\Omega}^{(0)}(z)g = M_{D,\Theta,\Omega}^{(0)}(z)\gamma_D u_\Theta = -\big(\widetilde\gamma_N+\widetilde\Theta\gamma_D\big)u_D,&\\ f=\gamma_D u_\Theta\in H^1(\partial\Omega).& \end{aligned} \tag{3.67}$$

Here u_Θ is the unique solution of $(-\Delta-z)u=0$ with $u\in H^{3/2}(\Omega)$ and $\big(\widetilde\gamma_N+\widetilde\Theta\gamma_D\big)u=g$, and u_D is the unique solution of $(-\Delta-z)u=0$ with $u\in H^{3/2}(\Omega)$ and $\gamma_D u=f\in H^1(\partial\Omega)$. Since $(u_D-u_\Theta)\in H^{3/2}(\Omega)$ and $\gamma_D u_D=f=\gamma_D u_\Theta$, one concludes

$$\gamma_D(u_D-u_\Theta)=0 \text{ and } (-\Delta-z)(u_D-u_\Theta)=0. \tag{3.68}$$

Uniqueness of the Dirichlet problem proved in Theorem 3.6 then yields $u_D=u_\Theta$ which further entails that $-\big(\widetilde\gamma_N+\widetilde\Theta\gamma_D\big)u_D=-\big(\widetilde\gamma_N+\widetilde\Theta\gamma_D\big)u_\Theta=-g$. Thus,

$$M_{D,\Theta,\Omega}^{(0)}(z)M_{\Theta,D,\Omega}^{(0)}(z)g=-\big(\widetilde\gamma_N+\widetilde\Theta\gamma_D\big)u_D=-g, \tag{3.69}$$

implying $M_{D,\Theta,\Omega}^{(0)}(z)M_{\Theta,D,\Omega}^{(0)}(z)=-I_{\partial\Omega}$. Conversely, let $f\in H^1(\partial\Omega)$. Then

$$M_{\Theta,D,\Omega}^{(0)}(z)M_{D,\Theta,\Omega}^{(0)}(z)f=M_{\Theta,D,\Omega}^{(0)}(z)\big(-\big(\widetilde\gamma_N+\widetilde\Theta\gamma_D\big)u_D\big)=\gamma_D u_\Theta, \tag{3.70}$$

and we set

$$g=-\big(\widetilde\gamma_N+\widetilde\Theta\gamma_D\big)u_D\in L^2(\partial\Omega;d^{n-1}\omega). \tag{3.71}$$

Here $u_D,u_\Theta\in H^{3/2}(\Omega)$ are such that $(-\Delta-z)u_\Theta=(-\Delta-z)u_D=0$ in Ω and $\gamma_D u_D=f$, $\big(\widetilde\gamma_N+\widetilde\Theta\gamma_D\big)u_\Theta=g$. Thus $\big(\widetilde\gamma_N+\widetilde\Theta\gamma_D\big)(u_\Theta+u_D)=0$, $(-\Delta-z)(u_\Theta+u_D)=0$ and $(u_D+u_\Theta)\in H^{3/2}(\Omega)$. Uniqueness of the generalized Robin problem proved in Theorem 3.2 then yields $u_\Theta=-u_D$ and hence $\gamma_D u_\Theta=-\gamma_D u_D=-f$. Thus,

$$M_{\Theta,D,\Omega}^{(0)}(z)M_{D,\Theta,\Omega}^{(0)}(z)f=\gamma_D u_\Theta=-f, \tag{3.72}$$

implying $M_{\Theta,D,\Omega}^{(0)}(z)M_{D,\Theta,\Omega}^{(0)}(z)\subseteq-I_{\partial\Omega}$. The desired conclusion now follows. □

REMARK 3.8. In the above considerations, the special case $\Theta=0$ represents the frequently studied Neumann-to-Dirichlet and Dirichlet-to-Neumann maps $M_{N,D,\Omega}^{(0)}(z)$ and $M_{D,N,\Omega}^{(0)}(z)$, respectively. That is, $M_{N,D,\Omega}^{(0)}(z)=M_{0,D,\Omega}^{(0)}(z)$ and $M_{D,N,\Omega}^{(0)}(z)=M_{D,0,\Omega}^{(0)}(z)$. Thus, as a corollary of Theorem 3.7 we have

$$M_{N,D,\Omega}^{(0)}(z)=-M_{D,N,\Omega}^{(0)}(z)^{-1}, \tag{3.73}$$

whenever Hypothesis 2.1 holds and $z\in\mathbb{C}\backslash(\sigma(-\Delta_{D,\Omega})\cup\sigma(-\Delta_{N,\Omega}))$.

REMARK 3.9. We emphasize again that all results in this section extend to Schrödinger operators $H_{\Theta,\Omega}=-\Delta_{\Theta,\Omega}+V$, $\operatorname{dom}\big(H_{\Theta,\Omega}\big)=\operatorname{dom}\big(-\Delta_{\Theta,\Omega}\big)$ in $L^2(\Omega;d^nx)$ for (not necessarily real-valued) potentials V satisfying $V\in L^\infty(\Omega;d^nx)$, or more generally, for potentials V which are Kato–Rellich bounded with respect to $-\Delta_{\Theta,\Omega}$ with bound less than one. Denoting the corresponding M-operators by

$M_{D,N,\Omega}(z)$ and $M_{\Theta,D,\Omega}(z)$, respectively, we note, in particular, that (3.56)–(3.65) extend replacing $-\Delta$ by $-\Delta+V$ and restricting $z\in\mathbb{C}$ appropriately.

Our discussion of Weyl–Titchmarsh operators follows the earlier papers [**43**] and [**46**]. For related literature on Weyl–Titchmarsh operators, relevant in the context of boundary value spaces (boundary triples, etc.), we refer, for instance, to [**3**], [**5**], [**12**], [**13**], [**18**]–[**22**], [**32**]– [**35**], [**42**], [**44**], [**47**, Ch. 3], [**49**, Ch. 13], [**65**], [**66**], [**71**], [**80**], [**81**], [**84**], [**85**], [**88**], [**89**], [**100**].

4. Some Variants of Krein's Resolvent Formula

In this section we present our principal new results, variants of Krein's formula for the difference of resolvents of generalized Robin Laplacians and Dirichlet Laplacians on bounded Lipschitz domains.

We start by weakening Hypothesis 3.1 by using assumption (4.1) below:

HYPOTHESIS 4.1. *In addition to Hypothesis* 2.2 *suppose that*

$$\widetilde{\Theta}\in\mathcal{B}_\infty\big(H^{1/2}(\partial\Omega),H^{-1/2}(\partial\Omega)\big). \tag{4.1}$$

We note that condition (4.1) is satisfied if there exists some $\varepsilon>0$ such that

$$\widetilde{\Theta}\in\mathcal{B}\big(H^{1/2-\varepsilon}(\partial\Omega),H^{-1/2}(\partial\Omega)\big). \tag{4.2}$$

Before proceeding with the main topic of this section, we will comment to the effect that Hypothesis 3.1 is indeed stronger than Hypothesis 4.1, as the latter follows from the former via duality and interpolation, implying

$$\widetilde{\Theta}\in\mathcal{B}_\infty\big(H^{s}(\partial\Omega),H^{s-1}(\partial\Omega)\big),\quad 0\le s\le 1. \tag{4.3}$$

To see this, one first employs the fact that

$$(H^{s_0}(\partial\Omega),H^{s_1}(\partial\Omega))_{\theta,2}=H^{s}(\partial\Omega) \tag{4.4}$$

for $s=(1-\theta)s_0+\theta s_1$, $0<\theta<1$, $0\le s_0,s_1\le 1$, and $s_0\neq s_1$, where $(\cdot,\cdot)_{\theta,q}$ denotes the *real interpolation* method.

Second, one uses the fact that if $T:X_j\to Y_j$, $j=0,1$, is a linear and bounded operator between two pairs of compatible Banach spaces, which is compact for $j=0$, then $T\in\mathcal{B}_\infty((X_0,X_1)_{\theta,p},(Y_0,Y_1)_{\theta,p})$ for every $\theta\in(0,1)$. This is a result due to Cwikel [**27**]:

THEOREM 4.2 ([**27**]). *Let X_j, Y_j, $j=0,1$, be two compatible Banach space couples and suppose that the linear operator $T:X_j\to Y_j$ is bounded for $j=0$ and compact for $j=1$. Then $T:(X_0,X_1)_{\theta,p}\to(Y_0,Y_1)_{\theta,p}$ is compact for all $\theta\in(0,1)$ and $p\in[1,\infty]$.*

(Interestingly, the corresponding result for the complex method of interpolation remains open.)

In our next two results below (Theorems 4.3–4.5) we discuss the solvability of the Dirichlet and Robin boundary value problems with solution in the energy space $H^1(\Omega)$.

THEOREM 4.3. *Assume Hypothesis* 4.1 *and suppose that $z\in\mathbb{C}\backslash\sigma(-\Delta_{\Theta,\Omega})$. Then for every $g\in H^{-1/2}(\partial\Omega)$, the following generalized Robin boundary value problem,*

$$\begin{cases}(-\Delta-z)u=0 \text{ in } \Omega, \quad u\in H^1(\Omega),\\ (\widetilde{\gamma}_N+\widetilde{\Theta}\gamma_D)u=g \text{ on } \partial\Omega,\end{cases} \tag{4.5}$$

has a unique solution $u = u_\Theta$. *Moreover, there exists a constant* $C = C(\Theta, \Omega, z) > 0$ *such that*

$$\|u_\Theta\|_{H^1(\Omega)} \leq C\|g\|_{H^{-1/2}(\partial\Omega)}. \tag{4.6}$$

In particular,

$$\big[\gamma_D(-\Delta_{\Theta,\Omega} - \overline{z}I_\Omega)^{-1}\big]^* \in \mathcal{B}\big(H^{-1/2}(\partial\Omega), H^1(\Omega)\big), \tag{4.7}$$

and the solution u_Θ *of* (4.5) *is once again given by formula* (3.7).

PROOF. The argument follows a pattern similar to that in the proof of Theorem 2.17. In a first stage, we look for a solution for (4.5) in the form

$$u(x) = (\mathcal{S}_z h)(x), \quad x \in \Omega, \tag{4.8}$$

for some $h \in H^{-1/2}(\partial\Omega)$. Here the single layer potential $\mathcal{S}_z$ has been defined in (2.121), and the fundamental Helmholtz solution E_n is given by (2.120) (cf. also (C.1)). Any such choice of h guarantees that u belongs to $H^1(\Omega)$ and satisfies $(-\Delta - z)u = 0$ in Ω. See (D.29). To ensure that the boundary condition in (4.5) is verified, one then takes

$$h = \big[(-\tfrac{1}{2}I_{\partial\Omega} + K_z^{\#}) + \widetilde{\Theta}\gamma_D\mathcal{S}_z\big]^{-1} g. \tag{4.9}$$

That the operator

$$(-\tfrac{1}{2}I_{\partial\Omega} + K_z^{\#}) + \widetilde{\Theta}\gamma_D\mathcal{S}_z : H^{-1/2}(\partial\Omega) \to H^{-1/2}(\partial\Omega) \tag{4.10}$$

is invertible for all but a discreet set of real values of the parameter z, can be established based on Hypothesis 4.1 by reasoning as before. The key result in this context is that the operator $-\frac{1}{2}I_{\partial\Omega} + K_z^{\#} \in \mathcal{B}(H^{-1/2}(\partial\Omega))$ is Fredholm, with Fredholm index zero for every $z \in \mathbb{C}$. □

The special case $\Theta = 0$, corresponding to the Neumann Laplacian, is singled out below.

COROLLARY 4.4. *Assume Hypothesis* 2.1 *and suppose that* $z \in \mathbb{C}\backslash\sigma(-\Delta_{N,\Omega})$. *Then for every* $g \in H^{-1/2}(\partial\Omega)$, *the following Neumann boundary value problem,*

$$\begin{cases} (-\Delta - z)u = 0 \text{ in } \Omega, \quad u \in H^1(\Omega), \\ \widetilde{\gamma}_N u = g \text{ on } \partial\Omega, \end{cases} \tag{4.11}$$

has a unique solution $u = u_N$. *Moreover, there exists a constant* $C = C(\Omega, z) > 0$ *such that*

$$\|u_N\|_{H^1(\Omega)} \leq C\|g\|_{H^{-1/2}(\partial\Omega)}. \tag{4.12}$$

In particular,

$$\big[\gamma_D(-\Delta_{N,\Omega} - \overline{z}I_\Omega)^{-1}\big]^* \in \mathcal{B}\big(H^{-1/2}(\partial\Omega), H^1(\Omega)\big), \tag{4.13}$$

and the solution u_g *of* (4.5) *is given by the formula*

$$u_N = \big(\gamma_D(-\Delta_{N,\Omega} - \overline{z}I_\Omega)^{-1}\big)^* g. \tag{4.14}$$

Finally, as a byproduct of the well-posedness of (4.11), *the weak Neumann trace* $\widetilde{\gamma}_N$ *in* (2.40), (2.41) *is onto.*

In the following we denote by $\widetilde{I}_\Omega$ the continuous inclusion (embedding) map of $H^1(\Omega)$ into $\big(H^1(\Omega)\big)^*$. By a slight abuse of notation, we also denote the continuous inclusion map of $H_0^1(\Omega)$ into $\big(H_0^1(\Omega)\big)^*$ by the same symbol $\widetilde{I}_\Omega$. We recall the ultra weak Neumann trace operator $\widetilde{\gamma}_{\mathcal{N}}$ in (2.60), (2.61). Finally, assuming Hypothesis 4.1, we denote by

$$-\widetilde{\Delta}_{\Theta,\Omega} \in \mathcal{B}\big(H^1(\Omega), \big(H^1(\Omega)\big)^*\big) \tag{4.15}$$

the extension of $-\Delta_{\Theta,\Omega}$ in accordance with (B.26). In particular,

$$\begin{aligned} {}_{H^1(\Omega)}\langle u, -\widetilde{\Delta}_{\Theta,\Omega} v\rangle_{(H^1(\Omega))^*} = \int_\Omega d^n x\, \overline{\nabla u(x)} \cdot \nabla v(x) + \big\langle \gamma_D u, \widetilde{\Theta}\gamma_D v\big\rangle_{1/2}, \\ u, v \in H^1(\Omega), \end{aligned} \tag{4.16}$$

and $-\Delta_{\Theta,\Omega}$ is the restriction of $-\widetilde{\Delta}_{\Theta,\Omega}$ to $L^2(\Omega; d^n x)$ (cf. (B.27)).

THEOREM 4.5. *Assume Hypothesis* 4.1 *and suppose that* $z \in \mathbb{C}\backslash\sigma(-\Delta_{\Theta,\Omega})$. *Then for every* $w \in (H^1(\Omega))^*$, *the following generalized inhomogeneous Robin problem,*

$$\begin{cases} (-\Delta - z)u = w|_\Omega \text{ in } \mathcal{D}'(\Omega), \quad u \in H^1(\Omega), \\ \widetilde{\gamma}_{\mathcal{N}}(u, w) + \widetilde{\Theta}\gamma_D u = 0 \text{ on } \partial\Omega, \end{cases} \tag{4.17}$$

has a unique solution $u = u_{\Theta,w}$. *Moreover, there exists a constant* $C = C(\Theta, \Omega, z) > 0$ *such that*

$$\|u_{\Theta,w}\|_{H^1(\Omega)} \leq C\|w\|_{(H^1(\partial\Omega))^*}. \tag{4.18}$$

In particular, the operator $(-\Delta_{\Theta,\Omega} - zI_\Omega)^{-1}$, $z \in \mathbb{C}\backslash\sigma(-\Delta_{\Theta,\Omega})$, *originally defined as a bounded operator on* $L^2(\Omega; d^n x)$,

$$(-\Delta_{\Theta,\Omega} - zI_\Omega)^{-1} \in \mathcal{B}\big(L^2(\Omega; d^n x)\big), \tag{4.19}$$

can be extended to a mapping in $\mathcal{B}\big(\big(H^1(\Omega)\big)^*, H^1(\Omega)\big)$, *which in fact coincides with*

$$\big(-\widetilde{\Delta}_{\Theta,\Omega} - z\widetilde{I}_\Omega\big)^{-1} \in \mathcal{B}\big(\big(H^1(\Omega)\big)^*, H^1(\Omega)\big). \tag{4.20}$$

PROOF. We recall (2.57). Hence, if $w \in \big(H^1(\Omega)\big)^*$, taking the convolution of w with $E_n(z;\,\cdot\,)$ in (C.1) and then restricting back to Ω yields a function $u_0 \in H^1(\Omega)$ for which $(-\Delta - z)u_0 = w|_\Omega$ in $\mathcal{D}'(\Omega)$. A solution of (4.17) is then given by $u = u_0 + u_1$, where u_1 satisfies

$$\begin{cases} (-\Delta - z)u_1 = 0 \text{ in } \Omega, \quad u_1 \in H^1(\Omega), \\ \big(\widetilde{\gamma}_N + \widetilde{\Theta}\gamma_D\big)u_1 = -\big(\widetilde{\gamma}_{\mathcal{N}}(u_0, w) + \widetilde{\Theta}\gamma_D u_0\big) \in H^{-1/2}(\partial\Omega) \text{ on } \partial\Omega. \end{cases} \tag{4.21}$$

Indeed, we have

$$\begin{aligned} \widetilde{\gamma}_{\mathcal{N}}(u, w) &= \widetilde{\gamma}_{\mathcal{N}}((u_0, w) + (u_1, 0)) = \widetilde{\gamma}_{\mathcal{N}}(u_0, w) + \widetilde{\gamma}_{\mathcal{N}}(u_1, 0) = \widetilde{\gamma}_{\mathcal{N}}(u_0, w) + \widetilde{\gamma}_N u_1 \\ &= -\widetilde{\Theta}\gamma_D u_1 - \widetilde{\Theta}\gamma_D u_0 = -\widetilde{\Theta}\gamma_D u, \end{aligned} \tag{4.22}$$

by (2.64). That the latter boundary problem is solvable is guaranteed by Theorem 4.3. We note that the solution thus constructed satisfies (4.18). Uniqueness for (4.17) follows from the corresponding uniqueness statement in Theorem 4.3.

Next, we observe that the inverse operator in (4.20) is well-defined. To prove that

$$-\widetilde{\Delta}_{\Theta,\Omega} - z\widetilde{I}_\Omega : H^1(\Omega) \to \big(H^1(\Omega)\big)^*, \quad z \in \mathbb{C}\backslash\sigma(-\Delta_{\Theta,\Omega}), \tag{4.23}$$

is onto, assume that $w \in (H^1(\Omega))^*$ is arbitrary and that u solves (4.17). Then, for every $v \in H^1(\Omega)$ we have

$$\begin{aligned}
&{}_{H^1(\Omega)}\langle v, (-\widetilde{\Delta}_{\Theta,\Omega} - z\widetilde{I}_\Omega)u\rangle_{(H^1(\Omega))^*} \\
&\quad = \int_\Omega d^n x\, \overline{\nabla v(x)} \cdot \nabla u(x) - z \int_\Omega d^n x\, \overline{v(x)} u(x) + \big\langle \gamma_D v, \widetilde{\Theta}\gamma_D u\big\rangle_{1/2} \\
&\quad = \int_\Omega d^n x\, \overline{\nabla v(x)} \cdot \nabla u(x) - z \int_\Omega d^n x\, \overline{v(x)} u(x) - \langle \gamma_D v, \widetilde{\gamma}_N(u,w)\rangle_{1/2} \\
&\quad = {}_{H^1(\Omega)}\langle v, w\rangle_{(H^1(\Omega))^*},
\end{aligned} \tag{4.24}$$

on account of (2.61), (4.16), and (4.17). Since the element $v \in H^1(\Omega)$ was arbitrary, this proves that $(-\widetilde{\Delta}_{\Theta,\Omega} - z\widetilde{I}_\Omega)u = w$, hence the operator (4.23) is onto. In fact, this operator is also one-to-one. Indeed, assume that $u \in H^1(\Omega)$ is such that $(-\widetilde{\Delta}_{\Theta,\Omega} - z\widetilde{I}_\Omega)u = 0$. Then, for every $v \in H^1(\Omega)$, formula (4.16) yields

$$\begin{aligned}
0 &= {}_{H^1(\Omega)}\big\langle v, \big(-\widetilde{\Delta}_{\Theta,\Omega} - z\widetilde{I}_\Omega\big)u\big\rangle_{(H^1(\Omega))^*} \\
&= \int_\Omega d^n x\, \overline{\nabla v(x)} \cdot \nabla u(x) - z \int_\Omega d^n x\, \overline{v(x)} u(x) + \big\langle \gamma_D v, \widetilde{\Theta}\gamma_D u\big\rangle_{1/2}.
\end{aligned} \tag{4.25}$$

Specializing (4.25) to the case when $v \in C_0^\infty(\Omega)$ shows that $(-\Delta - z)u = 0$ in the sense of distributions in Ω. Returning with this into (4.25) we then obtain $\big\langle \gamma_D v, (\widetilde{\gamma}_N + \widetilde{\Theta}\gamma_D)u\big\rangle_{1/2} = 0$ for every $v \in H^1(\Omega)$. Given that the Dirichlet trace γ_D maps $H^1(\Omega)$ onto $H^{1/2}(\partial\Omega)$, this proves that $(\widetilde{\gamma}_N + \widetilde{\Theta}\gamma_D)u = 0$ in $H^{-1/2}(\partial\Omega)$ so that ultimately $u = 0$, since $z \in \mathbb{C}\backslash\sigma(-\Delta_{\Theta,\Omega})$. In summary, the operator (4.23) is an isomorphism.

Finally, there remains to show that the operators (4.19), (4.20) act in a compatible fashion. To see this, fix $z \in \mathbb{C}\backslash\sigma(-\Delta_{\Theta,\Omega})$ and assume that $w \in L^2(\Omega; d^n x) \hookrightarrow (H^1(\Omega))^*$. If we then set $u = (-\widetilde{\Delta}_{\Theta,\Omega} - z\widetilde{I}_\Omega)^{-1} w \in H^1(\Omega)$, it follows from (4.16) that

$$\begin{aligned}
{}_{H^1(\Omega)}\langle v, w\rangle_{(H^1(\Omega))^*} &= {}_{H^1(\Omega)}\big\langle v, \big(-\widetilde{\Delta}_{\Theta,\Omega} - z\widetilde{I}_\Omega\big)u\big\rangle_{(H^1(\Omega))^*} \\
&= \int_\Omega d^n x\, \overline{\nabla v(x)} \cdot \nabla u(x) - z \int_\Omega d^n x\, \overline{v(x)} u(x) + \big\langle \gamma_D v, \widetilde{\Theta}\gamma_D u\big\rangle_{1/2},
\end{aligned} \tag{4.26}$$

for every $v \in H^1(\Omega)$. Specializing this identity to the case when $v \in C_0^\infty(\Omega)$ yields $(-\Delta - z)u = w \in L^2(\Omega; d^n x)$. When used back in (4.26), this observation and (2.41) permit us to conclude that

$$\begin{aligned}
\langle \gamma_D v, (\widetilde{\gamma}_N + \widetilde{\Theta}\gamma_D)u\rangle_{1/2} &= \int_\Omega d^n x\, \overline{\nabla v(x)} \cdot \nabla u(x) - z \int_\Omega d^n x\, \overline{v(x)} u(x) \\
&\quad - {}_{H^1(\Omega)}\langle v, \iota(-\Delta u - zu)\rangle_{(H^1(\Omega))^*} + \big\langle \gamma_D v, \widetilde{\Theta}\gamma_D u\big\rangle_{1/2} \\
&= \int_\Omega d^n x\, \overline{\nabla v(x)} \cdot \nabla u(x) - z \int_\Omega d^n x\, \overline{v(x)} u(x) \\
&\quad - {}_{H^1(\Omega)}\langle v, w\rangle_{(H^1(\Omega))^*} + \big\langle \gamma_D v, \widetilde{\Theta}\gamma_D u\big\rangle_{1/2} \\
&= 0,
\end{aligned} \tag{4.27}$$

for every $v \in H^1(\Omega)$. Upon recalling that the Dirichlet trace γ_D maps $H^1(\Omega)$ onto $H^{1/2}(\partial\Omega)$, this shows that $(\widetilde{\gamma}_N + \widetilde{\Theta}\gamma_D)u = 0$ in $H^{-1/2}(\partial\Omega)$. Thus, $u = (-\Delta_{\Theta,\Omega} - zI_\Omega)^{-1}w$, as desired. □

REMARK 4.6. Similar (yet simpler) considerations also show that the operator $(-\Delta_{D,\Omega} - zI_\Omega)^{-1}$, $z \in \mathbb{C}\backslash\sigma(-\Delta_{D,\Omega})$, originally defined as bounded operator on $L^2(\Omega; d^n x)$,

$$(-\Delta_{D,\Omega} - zI_\Omega)^{-1} \in \mathcal{B}\big(L^2(\Omega; d^n x)\big), \tag{4.28}$$

extends to a mapping

$$\big(-\widetilde{\Delta}_{D,\Omega} - z\widetilde{I}_\Omega\big)^{-1} \in \mathcal{B}\big(H^{-1}(\Omega); H^1_0(\Omega)\big). \tag{4.29}$$

Here $-\widetilde{\Delta}_{D,\Omega} \in \mathcal{B}\big(H^1_0(\Omega), H^{-1}(\Omega)\big)$ is the extension of $-\Delta_{D,\Omega}$ in accordance with (B.26). Indeed, the Lax–Milgram lemma applies and yields that

$$\big(-\widetilde{\Delta}_{D,\Omega} - z\widetilde{I}_\Omega\big)\colon H^1_0(\Omega) \to \big(H^1_0(\Omega)\big)^* = H^{-1}(\Omega) \tag{4.30}$$

is, in fact, an isomorphism whenever $z \in \mathbb{C}\backslash\sigma(-\Delta_{D,\Omega})$.

COROLLARY 4.7. *Assume Hypothesis* 4.1 *and suppose that* $z \in \mathbb{C}\backslash\sigma(-\Delta_{\Theta,\Omega})$. *Then the operator* $M^{(0)}_{\Theta,D,\Omega}(z) \in \mathcal{B}\big(L^2(\partial\Omega; d^{n-1}\omega)\big)$ *in* (3.58), (3.59) *extends (in a compatible manner) to*

$$\widetilde{M}^{(0)}_{\Theta,D,\Omega}(z) \in \mathcal{B}\big(H^{-1/2}(\partial\Omega), H^{1/2}(\partial\Omega)\big), \quad z \in \mathbb{C}\backslash\sigma(-\Delta_{\Theta,\Omega}). \tag{4.31}$$

In addition, $\widetilde{M}^{(0)}_{\Theta,D,\Omega}(z)$ *permits the representation*

$$\widetilde{M}^{(0)}_{\Theta,D,\Omega}(z) = \gamma_D\big(-\widetilde{\Delta}_{\Theta,\Omega} - z\widetilde{I}_\Omega\big)^{-1}\gamma_D^*, \quad z \in \mathbb{C}\backslash\sigma(-\Delta_{\Theta,\Omega}). \tag{4.32}$$

The same applies to the adjoint $M^{(0)}_{\Theta,D,\Omega}(z)^* \in \mathcal{B}\big(L^2(\partial\Omega; d^{n-1}\omega)\big)$ *of* $M^{(0)}_{\Theta,D,\Omega}(z)$, *resulting in the bounded extension* $\big(\widetilde{M}^{(0)}_{\Theta,D,\Omega}(z)\big)^* \in \mathcal{B}\big(H^{-1/2}(\partial\Omega), H^{1/2}(\partial\Omega)\big)$, $z \in \mathbb{C}\backslash\sigma(-\Delta_{\Theta,\Omega})$.

PROOF. The claim (4.31) is a direct consequence of Theorem 4.3, while the claim (4.32) follows from the fact that

$$\gamma_D^*\colon \big(H^{1/2}(\partial\Omega)\big)^* = H^{-1/2}(\partial\Omega) \to \big(H^1(\Omega)\big)^* \tag{4.33}$$

in a bounded fashion (cf. (A.32), (4.20) and (3.64)). The rest follows from dualizing these claims. □

The following regularity result for the Robin resolvent will also play an important role shortly.

LEMMA 4.8. *Assume Hypothesis* 3.1 *and suppose that* $z \in \mathbb{C}\backslash\sigma(-\Delta_{\Theta,\Omega})$. *Then*

$$(-\Delta_{\Theta,\Omega} - zI_\Omega)^{-1} : L^2(\Omega; d^n x) \to \big\{u \in H^{3/2}(\Omega) \,\big|\, \Delta u \in L^2(\Omega; d^n x)\big\} \tag{4.34}$$

is a well-defined bounded operator, where the space $\{u \in H^{3/2}(\Omega) \,|\, \Delta u \in L^2(\Omega; d^n x)\}$ *is equipped with the natural graph norm* $u \mapsto \|u\|_{H^{3/2}(\Omega)} + \|\Delta u\|_{L^2(\Omega; d^n x)}$.

PROOF. Consider $f \in L^2(\Omega; d^n x)$ and set $u = (-\Delta_{\Theta,\Omega} - zI_\Omega)^{-1}f$. It follows that u is the unique solution of the problem

$$(-\Delta - z)u = f \text{ in } \Omega, \quad u \in H^1(\Omega), \quad (\widetilde{\gamma}_N + \widetilde{\Theta}\gamma_D)u = 0 \text{ on } \partial\Omega. \tag{4.35}$$

The strategy is to devise an alternative representation for u from which it is clear that u has the claimed regularity in Ω. To this end, let $\widetilde{f}$ denote the extension of f by zero to $\mathbb{R}^n$ and denote by E the operator of convolution by $E_n(z;\,\cdot)$. Since the latter is smoothing of order 2, it follows that $w = (E\widetilde{f})|_\Omega \in H^2(\Omega)$ and $(-\Delta - z)w = f$ in Ω. Also, let v be the unique solution of the problem

$$(-\Delta - z)v = 0 \text{ in } \Omega, \quad v \in H^{3/2}(\Omega), \quad (\widetilde{\gamma}_N + \widetilde{\Theta}\gamma_D)v = -(\gamma_N + \widetilde{\Theta}\gamma_D)w \text{ on } \partial\Omega. \tag{4.36}$$

That (4.36) is solvable is a consequence of Theorem 4.3. Then $v+w$ also solves (4.35) so that, by uniqueness, $u = v + w$. This shows that u has the desired regularity properties and, hence, the operator (4.34) is well-defined and bounded. □

Under the Hypothesis 4.1, (4.20) and (2.7) yield

$$\gamma_D\big(-\widetilde{\Delta}_{\Theta,\Omega} - zI_\Omega\big)^{-1} \in \mathcal{B}\big((H^1(\Omega))^*, H^{1/2}(\partial\Omega)\big). \tag{4.37}$$

Hence, by duality,

$$\big[\gamma_D\big(-\widetilde{\Delta}_{\Theta,\Omega} - zI_\Omega\big)^{-1}\big]^* \in \mathcal{B}\big(H^{-1/2}(\partial\Omega), H^1(\Omega)\big). \tag{4.38}$$

We wish to complement this with the following result.

COROLLARY 4.9. *Assume Hypothesis* 3.1 *and suppose that* $z \in \mathbb{C}\backslash\sigma(-\Delta_{\Theta,\Omega})$. *Then*

$$\gamma_D(-\Delta_{\Theta,\Omega} - zI_\Omega)^{-1} \in \mathcal{B}\big(L^2(\Omega; d^nx), H^1(\partial\Omega)\big). \tag{4.39}$$

In particular,

$$\begin{aligned}\big[\gamma_D(-\Delta_{\Theta,\Omega} - zI_\Omega)^{-1}\big]^* &\in \mathcal{B}\big(H^{-1}(\partial\Omega), L^2(\Omega; d^nx)\big) \\ &\hookrightarrow \mathcal{B}\big(L^2(\partial\Omega; d^{n-1}\Omega), L^2(\Omega; d^nx)\big).\end{aligned} \tag{4.40}$$

In addition, the operator (4.40) *is compatible with* (4.38) *in the sense that*

$$\begin{aligned}\big[\gamma_D(-\Delta_{\Theta,\Omega} - zI_\Omega)^{-1}\big]^* f = \big[\gamma_D\big(-\widetilde{\Delta}_{\Theta,\Omega} - zI_\Omega\big)^{-1}\big]^* f \text{ in } L^2(\Omega; d^nx),& \\ f \in H^{-1/2}(\partial\Omega).&\end{aligned} \tag{4.41}$$

As a consequence,

$$\begin{aligned}\big[\gamma_D(-\Delta_{\Theta,\Omega} - zI_\Omega)^{-1}\big]^* f = \big[\gamma_D\big(-\widetilde{\Delta}_{\Theta,\Omega} - zI_\Omega\big)^{-1}\big]^* f \text{ in } L^2(\Omega; d^nx),& \\ f \in L^2(\partial\Omega; d^{n-1}\omega).&\end{aligned} \tag{4.42}$$

PROOF. The first part of the statement is an immediate consequence of Lemma 4.8 and Lemma 2.3. As for (4.41), pick $f \in H^{-1/2}(\partial\Omega) \hookrightarrow H^{-1}(\partial\Omega)$ and $u \in L^2(\Omega; d^nx) \hookrightarrow (H^1(\Omega))^*$ arbitrary. We may then write

$$\begin{aligned}&\langle[\gamma_D(-\Delta_{\Theta,\Omega} - zI_\Omega)^{-1}]^* f\,, u\rangle_{L^2(\Omega;d^nx)} \\ &\quad = \langle f\,, \gamma_D(-\Delta_{\Theta,\Omega} - \overline{z}I_\Omega)^{-1}u\rangle_1 \\ &\quad = \langle f\,, \gamma_D(-\Delta_{\Theta,\Omega} - \overline{z}I_\Omega)^{-1}u\rangle_{1/2} \\ &\quad = \big\langle f\,, \gamma_D\big(-\widetilde{\Delta}_{\Theta,\Omega} - \overline{z}I_\Omega\big)^{-1}u\big\rangle_{1/2} \\ &\quad = {}_{H^1(\Omega)}\big\langle\big[\gamma_D\big(-\widetilde{\Delta}_{\Theta,\Omega} - zI_\Omega\big)^{-1}\big]^* f\,, u\big\rangle_{(H^1(\Omega))^*} \\ &\quad = \big\langle\big[\gamma_D\big(-\widetilde{\Delta}_{\Theta,\Omega} - zI_\Omega\big)^{-1}\big]^* f\,, u\big\rangle_{L^2(\Omega;d^nx)},\end{aligned} \tag{4.43}$$

since (4.19) and (4.20) are compatible. This gives (4.41). Since $L^2(\partial\Omega; d^{n-1}\omega) \hookrightarrow H^{-1/2}(\partial\Omega)$, (4.42) also follows. □

We will need a similar compatibility result for the composition between the Neumann trace and resolvents of the Dirichlet Laplacian. To state it, recall the restriction operator R_Ω from (2.58). Also, denote by $I_{\mathbb{R}^n}$ the identity operator (for spaces of functions defined in $\mathbb{R}^n$). Finally, recall the space (2.59) and the ultra weak Neumann trace operator in (2.60), (2.61).

LEMMA 4.10. *Assume Hypothesis* 2.1. *Then*

$$\big(\big(-\widetilde{\Delta}_{D,\Omega} - z\widetilde{I}_\Omega\big)^{-1} \circ R_\Omega, I_{\mathbb{R}^n}\big) : (H^1(\Omega))^* \to W_z(\Omega), \quad z \in \mathbb{C}\backslash\sigma(-\Delta_{D,\Omega}), \tag{4.44}$$

is a well-defined, linear and bounded operator. Consequently,

$$\begin{aligned}\widetilde{\gamma}_{\mathcal{N}}\big(\big(-\widetilde{\Delta}_{D,\Omega} - z\widetilde{I}_\Omega\big)^{-1} \circ R_\Omega, I_{\mathbb{R}^n}\big) \in \mathcal{B}\big((H^1(\Omega))^*, H^{-1/2}(\partial\Omega)\big),& \\ z \in \mathbb{C}\backslash\sigma(-\Delta_{D,\Omega}),&\end{aligned} \tag{4.45}$$

and, hence,

$$\begin{aligned}\big[\widetilde{\gamma}_{\mathcal{N}}\big(\big(-\widetilde{\Delta}_{D,\Omega} - z\widetilde{I}_\Omega\big)^{-1} \circ R_\Omega, I_{\mathbb{R}^n}\big)\big]^* \in \mathcal{B}\big(H^{1/2}(\partial\Omega), H^1(\Omega)\big),& \\ z \in \mathbb{C}\backslash\sigma(-\Delta_{D,\Omega}).&\end{aligned} \tag{4.46}$$

Furthermore, the operators (4.45), (4.46) *are compatible with* (3.48) *and* (3.49), *respectively, in the sense that for each* $z \in \mathbb{C}\backslash\sigma(-\Delta_{D,\Omega})$,

$$\begin{aligned}\widetilde{\gamma}_N(-\Delta_{D,\Omega} - zI_\Omega)^{-1} f = \widetilde{\gamma}_{\mathcal{N}}\big(\big(-\widetilde{\Delta}_{D,\Omega} - z\widetilde{I}_\Omega\big)^{-1} \circ R_\Omega, I_{\mathbb{R}^n}\big) f \text{ in } H^{-1/2}(\partial\Omega),& \\ f \in L^2(\Omega; d^n x),&\end{aligned} \tag{4.47}$$

and

$$\begin{aligned}\big[\widetilde{\gamma}_N(-\Delta_{D,\Omega} - zI_\Omega)^{-1}\big]^* f = \big[\widetilde{\gamma}_{\mathcal{N}}\big(\big(-\widetilde{\Delta}_{D,\Omega} - z\widetilde{I}_\Omega\big)^{-1} \circ R_\Omega, I_{\mathbb{R}^n}\big)\big]^* f \text{ in } L^2(\Omega; d^n x),& \\ \text{for every element } f \in H^{1/2}(\partial\Omega).&\end{aligned} \tag{4.48}$$

PROOF. Let $z \in \mathbb{C}\backslash\sigma(-\Delta_{D,\Omega})$. If $f \in (H^1(\Omega))^*$ and $u = \big(-\widetilde{\Delta}_{D,\Omega} - z I_\Omega\big)^{-1}(f|_\Omega)$, then $u \in H^1_0(\Omega)$ satisfies $(-\Delta - z)u = f|_\Omega$ in $\mathcal{D}'(\Omega)$. Hence, $(u, f) \in W_z(\Omega)$ which shows that the operator (4.44) is well-defined and bounded. Then (4.45) is a consequence of this and (2.60), whereas (4.46) follows from (4.45) and duality.

Going further, (4.47) is implied by Lemma 3.4, the compatibility statement in Lemma 2.4, and (2.62)–(2.64). There remains to justify (4.48). To this end, if $f \in H^{1/2}(\partial\Omega) \hookrightarrow L^2(\partial\Omega; d^{n-1}\omega)$ and $u \in L^2(\Omega; d^n x) \hookrightarrow (H^1(\Omega))^*$ are arbitrary, we may write

$$\begin{aligned}&\langle [\widetilde{\gamma}_N(-\Delta_{D,\Omega} - zI_\Omega)^{-1}]^* f\,,\, u\rangle_{L^2(\Omega;d^n x)} \\ &\quad = \langle f\,,\, \widetilde{\gamma}_N(-\Delta_{D,\Omega} - \overline{z}I_\Omega)^{-1} u\rangle_0 \\ &\quad = \langle f\,,\, \widetilde{\gamma}_N\big(-\Delta_{D,\Omega} - \overline{z}I_\Omega\big)^{-1} u\rangle_{1/2} u\rangle_{1/2} \\ &\quad = \langle f\,,\, \widetilde{\gamma}_{\mathcal{N}}\big(\big(-\widetilde{\Delta}_{D,\Omega} - \overline{z}\widetilde{I}_\Omega\big)^{-1} \circ R_\Omega, I_{\mathbb{R}^n}\big) u\rangle_{1/2} \\ &\quad = {}_{H^1(\Omega)}\big\langle \big[\widetilde{\gamma}_{\mathcal{N}}\big(\big(-\widetilde{\Delta}_{D,\Omega} - z\widetilde{I}_\Omega\big)^{-1} \circ R_\Omega, I_{\mathbb{R}^n}\big)\big]^* f\,,\, u\big\rangle_{(H^1(\Omega))^*} \\ &\quad = \big\langle \big[\widetilde{\gamma}_{\mathcal{N}}\big(\big(-\widetilde{\Delta}_{D,\Omega} - z\widetilde{I}_\Omega\big)^{-1} \circ R_\Omega, I_{\mathbb{R}^n}\big)\big]^* f\,,\, u\big\rangle_{L^2(\Omega;d^n x)},\end{aligned} \tag{4.49}$$

where the third equality is based on (4.47). This justifies (4.48) and finishes the proof of the lemma. □

LEMMA 4.11. *Assume Hypothesis 4.1 and suppose that* $z \in \mathbb{C}\backslash(\sigma(-\Delta_{\Theta,\Omega}) \cup \sigma(-\Delta_{D,\Omega}))$. *Then the following resolvent relation holds on* $\big(H^1(\Omega)\big)^*$,

$$\begin{aligned}\big(-\widetilde{\Delta}_{\Theta,\Omega} - z\widetilde{I}_\Omega\big)^{-1} &= \big(-\widetilde{\Delta}_{D,\Omega} - z\widetilde{I}_\Omega\big)^{-1} \circ R_\Omega \\ &\quad + \big(-\widetilde{\Delta}_{\Theta,\Omega} - z\widetilde{I}_\Omega\big)^{-1}\gamma_D^*\widetilde{\gamma}_N\big(\big(-\widetilde{\Delta}_{D,\Omega} - z\widetilde{I}_\Omega\big)^{-1} \circ R_\Omega, I_{\mathbb{R}^n}\big).\end{aligned} \tag{4.50}$$

PROOF. To set the stage, we recall (2.40)–(2.41) and (4.33). Together with (4.20) and (4.29), these ensure that the composition of operators appearing on the right-hand side of (4.50) is well-defined. Next, let $\phi_1, \psi_1 \in L^2(\Omega; d^n x)$ be arbitrary and define

$$\begin{aligned}\phi &= (-\Delta_{\Theta,\Omega} - \overline{z}I_\Omega)^{-1}\phi_1 \in \mathrm{dom}(\Delta_{\Theta,\Omega}) \subset \big(H^1(\Omega) \cap \mathrm{dom}(\widetilde{\gamma}_N)\big),\\ \psi &= (-\Delta_{D,\Omega} - zI_\Omega)^{-1}\psi_1 \in \mathrm{dom}(\Delta_{D,\Omega}) \subset \big(H_0^1(\Omega) \cap \mathrm{dom}(\widetilde{\gamma}_N)\big).\end{aligned} \tag{4.51}$$

As a consequence of our earlier results, both sides of (4.50) are bounded operators from $(H^1(\Omega))^*$ into $H^1(\Omega)$. Since $L^2(\Omega; d^n x) \hookrightarrow (H^1(\Omega))^*$ densely, it therefore suffices to show that the following identity holds:

$$\begin{aligned}&(\phi_1, (-\Delta_{\Theta,\Omega} - zI_\Omega)^{-1}\psi_1)_{L^2(\Omega;d^n x)} - (\phi_1, (-\Delta_{D,\Omega} - zI_\Omega)^{-1}\psi_1)_{L^2(\Omega;d^n x)}\\ &\quad = (\phi_1, (-\Delta_{\Theta,\Omega} - zI_\Omega)^{-1}\gamma_D^*\widetilde{\gamma}_N(-\Delta_{D,\Omega} - zI_\Omega)^{-1}\psi_1)_{L^2(\Omega;d^n x)}.\end{aligned} \tag{4.52}$$

We note that according to (4.51) one has,

$$(\phi_1, (-\Delta_{D,\Omega} - zI_\Omega)^{-1}\psi_1)_{L^2(\Omega;d^n x)} = ((-\Delta_{\Theta,\Omega} - \overline{z}I_\Omega)\phi, \psi)_{L^2(\Omega;d^n x)}, \tag{4.53}$$

$$\begin{aligned}(\phi_1, (-\Delta_{\Theta,\Omega} - zI_\Omega)^{-1}\psi_1)_{L^2(\Omega;d^n x)} &= \big(((-\Delta_{\Theta,\Omega} - zI_\Omega)^{-1})^*\phi_1, \psi_1\big)_{L^2(\Omega;d^n x)}\\ &= ((-\Delta_{\Theta,\Omega} - \overline{z}I_\Omega)^{-1}\phi_1, \psi_1)_{L^2(\Omega;d^n x)}\\ &= (\phi, (-\Delta_{D,\Omega} - zI_\Omega)\psi)_{L^2(\Omega;d^n x)},\end{aligned} \tag{4.54}$$

and, further,

$$\begin{aligned}&(\phi_1, (-\Delta_{\Theta,\Omega} - zI_\Omega)^{-1}\gamma_D^*\widetilde{\gamma}_N(-\Delta_{D,\Omega} - zI_\Omega)^{-1}\psi_1)_{L^2(\Omega;d^n x)}\\ &\quad = {}_{H^1(\Omega)}\langle(-\Delta_{\Theta,\Omega} - \overline{z}I_\Omega)^{-1}\phi_1, \gamma_D^*\widetilde{\gamma}_N(-\Delta_{D,\Omega} - zI_\Omega)^{-1}\psi_1\rangle_{(H^1(\Omega))^*}\\ &\quad = \langle\gamma_D(-\Delta_{\Theta,\Omega} - \overline{z}I_\Omega)^{-1}\phi_1, \widetilde{\gamma}_N(-\Delta_{D,\Omega} - zI_\Omega)^{-1}\psi_1\rangle_{1/2} = \langle\gamma_D\phi, \widetilde{\gamma}_N\psi\rangle_{1/2}.\end{aligned} \tag{4.55}$$

Thus, matters have been reduced to proving that

$$((-\Delta_{\Theta,\Omega} - \overline{z}I_\Omega)\phi, \psi)_{L^2(\Omega;d^n x)} - (\phi, (-\Delta_{D,\Omega} - zI_\Omega)\psi)_{L^2(\Omega;d^n x)} = \langle\gamma_D\phi, \widetilde{\gamma}_N\psi\rangle_{1/2}. \tag{4.56}$$

Using (A.31) for the left-hand side of (4.56) one obtains

$$\begin{aligned}&((-\Delta_{\Theta,\Omega} - \overline{z}I_\Omega)\phi, \psi)_{L^2(\Omega;d^n x)} - (\phi, (-\Delta_{D,\Omega} - zI_\Omega)\psi)_{L^2(\Omega;d^n x)}\\ &\quad = -(\Delta\phi, \psi)_{L^2(\Omega;d^n x)} + (\phi, \Delta\psi)_{L^2(\Omega;d^n x)}\\ &\quad = (\nabla\phi, \nabla\psi)_{L^2(\Omega;d^n x)^n} - \langle\widetilde{\gamma}_N\phi, \gamma_D\psi\rangle_{1/2} - (\nabla\phi, \nabla\psi)_{L^2(\Omega;d^n x)^n} + \langle\gamma_D\phi, \widetilde{\gamma}_N\psi\rangle_{1/2}\\ &\quad = -\langle\widetilde{\gamma}_N\phi, \gamma_D\psi\rangle_{1/2} + \langle\gamma_D\phi, \widetilde{\gamma}_N\psi\rangle_{1/2}.\end{aligned} \tag{4.57}$$

Observing that $\gamma_D\psi = 0$ since $\psi \in \mathrm{dom}(\Delta_{D,\Omega})$, one concludes (4.56). □

The stage is now set for proving the L^2-version of Lemma 4.11.

LEMMA 4.12. *Assume Hypothesis* 3.1 *and suppose that* $z \in \mathbb{C}\backslash(\sigma(-\Delta_{\Theta,\Omega}) \cup \sigma(-\Delta_{D,\Omega}))$. *Then the following resolvent relation holds on* $L^2(\Omega; d^n x)$,

$$\begin{aligned}
&(-\Delta_{\Theta,\Omega} - zI_\Omega)^{-1} \\
&\quad = (-\Delta_{D,\Omega} - zI_\Omega)^{-1} + \big[\gamma_D(-\Delta_{\Theta,\Omega} - \overline{z}I_\Omega)^{-1}\big]^* \big[\widetilde{\gamma}_N(-\Delta_{D,\Omega} - zI_\Omega)^{-1}\big] \quad (4.58)\\
&\quad = (-\Delta_{D,\Omega} - zI_\Omega)^{-1} + \big[\widetilde{\gamma}_N(-\Delta_{D,\Omega} - \overline{z}I_\Omega)^{-1}\big]^* \big[\gamma_D(-\Delta_{\Theta,\Omega} - zI_\Omega)^{-1}\big]. \quad (4.59)
\end{aligned}$$

PROOF. Consider the first equality in (4.58). To begin with, we note that the following operators are well-defined, linear and bounded:

$$(-\Delta_{D,\Omega} - zI_\Omega)^{-1} \in \mathcal{B}\big(L^2(\Omega; d^n x)\big), \quad (-\Delta_{\Theta,\Omega} - zI_\Omega)^{-1} \in \mathcal{B}\big(L^2(\Omega; d^n x)\big), \quad (4.60)$$

$$\widetilde{\gamma}_N(-\Delta_{D,\Omega} - zI_\Omega)^{-1} \in \mathcal{B}\big(L^2(\Omega; d^n x), L^2(\partial\Omega; d^{n-1}\omega)\big), \quad (4.61)$$

$$\big[\gamma_D(-\Delta_{\Theta,\Omega} - \overline{z}I_\Omega)^{-1}\big]^* \in \mathcal{B}\big(L^2(\partial\Omega; d^{n-1}\omega), L^2(\Omega; d^n x)\big)\big). \quad (4.62)$$

Indeed, (4.60) follows from the fact that $z \in \mathbb{C}\backslash\big(\sigma(-\Delta_{D,\Omega}) \cup \sigma(-\Delta_{\Theta,\Omega})\big)$, (4.61) is covered by (3.48), and (4.62) is taken care of by (4.40). Together, these memberships show that both sides of (4.58) are bounded operators on $L^2(\Omega; d^n x)$. Having established this, the first equality in (4.58) follows from Lemma 4.11, granted the compatibility results from Corollary 4.9 and Lemma 4.10. Then the second equality in (4.58) is a consequence of what we have proved so far and of duality. □

We note that the special case $\Theta = 0$ in Lemma 4.12 was discussed by Nakamura [**77**] (in connection with cubic boxes Ω) and subsequently in [**43**, Lemma A.3] (in the case of a Lipschitz domain with a compact boundary).

LEMMA 4.13. *Assume Hypothesis* 4.1 *and suppose that* $z \in \mathbb{C}\backslash\sigma(-\Delta_{\Theta,\Omega})$. *Then*

$$\big[\widetilde{M}_{\Theta,D,\Omega}^{(0)}(z)\big]^* = \widetilde{M}_{\Theta,D,\Omega}^{(0)}(\overline{z}) \quad (4.63)$$

as operators in $\mathcal{B}\big(H^{-1/2}(\partial\Omega); H^{1/2}(\partial\Omega)\big)$. *In particular, assuming Hypothesis* 3.1, *then*

$$\big[M_{\Theta,D,\Omega}^{(0)}(z)\big]^* = M_{\Theta,D,\Omega}^{(0)}(\overline{z}). \quad (4.64)$$

PROOF. Let $f, g \in H^{-1/2}(\partial\Omega)$. Then using the definition of $\widetilde{M}_{\Theta,D,\Omega}^{(0)}(z)$ one infers

$$\big\langle \widetilde{M}_{\Theta,D,\Omega}^{(0)}(z)f, g\big\rangle_{1/2} = \big\langle \gamma_D u, (\widetilde{\gamma}_N + \widetilde{\Theta}\gamma_D)v\big\rangle_{1/2} \quad (4.65)$$

where u, v solve the Robin boundary value problems

$$\begin{cases} (-\Delta - z)u = 0 \text{ in } \Omega, \quad u \in H^1(\Omega), \\ (\widetilde{\gamma}_N + \widetilde{\Theta}\gamma_D)u = f \text{ on } \partial\Omega, \end{cases} \quad (4.66)$$

and

$$\begin{cases} (-\Delta - \overline{z})v = 0 \text{ in } \Omega, \quad v \in H^1(\Omega), \\ (\widetilde{\gamma}_N + \widetilde{\Theta}\gamma_D)v = g \text{ on } \partial\Omega, \end{cases} \quad (4.67)$$

respectively. That this is possible is ensured by Theorem 4.3. Using (A.31) we may then write

$$\begin{aligned}
&\big\langle \gamma_D u, (\widetilde{\gamma}_N + \widetilde{\Theta}\gamma_D)v\big\rangle_{1/2} \\
&\quad = \langle \gamma_D u, \widetilde{\gamma}_N v\rangle_{1/2} + \big\langle \gamma_D u, \widetilde{\Theta}\gamma_D v\big\rangle_{1/2} \\
&\quad = (\nabla u, \nabla v)_{L^2(\Omega; d^n x)^n} + {}_{H^1(\Omega)}\langle u, \Delta v\rangle_{(H^1(\Omega))^*} + \big\langle \gamma_D u, \widetilde{\Theta}\gamma_D v\big\rangle_{1/2}
\end{aligned}$$

$$
\begin{aligned}
&= (\nabla u, \nabla v)_{L^2(\Omega;d^nx)^n} - \overline{z}\,(u,v)_{L^2(\Omega;d^nx)^n} + \big\langle \gamma_D u, \widetilde{\Theta}\gamma_D v\big\rangle_{1/2} \\
&= \overline{(\nabla v, \nabla u)_{L^2(\Omega;d^nx)^n} - z\,(v,u)_{L^2(\Omega;d^nx)^n}} + \big\langle \gamma_D u, \widetilde{\Theta}\gamma_D v\big\rangle_{1/2} \\
&= \overline{(\nabla v, \nabla u)_{L^2(\Omega;d^nx)^n} + {}_{H^1(\Omega)}\langle v, \Delta u\rangle_{(H^1(\Omega))^*}} + \big\langle \gamma_D u, \widetilde{\Theta}\gamma_D v\big\rangle_{1/2} \\
&= \overline{\langle \gamma_D v, \widetilde{\gamma}_N u\rangle_{1/2}} + \big\langle \gamma_D u, \widetilde{\Theta}\gamma_D v\big\rangle_{1/2} \\
&= \overline{\langle \gamma_D v, \widetilde{\gamma}_N u\rangle_{1/2}} + \big\langle \widetilde{\Theta}\gamma_D u, \gamma_D v\big\rangle_{1/2} \\
&= \overline{\langle \gamma_D v, \widetilde{\gamma}_N u\rangle_{1/2} + \big\langle \gamma_D v, \widetilde{\Theta}\gamma_D u\big\rangle_{1/2}} \\
&= \overline{\big\langle \gamma_D v, (\widetilde{\gamma}_N + \widetilde{\Theta}\gamma_D)u\big\rangle_{1/2}} \\
&= \overline{\langle \widetilde{M}^{(0)}_{\Theta,D,\Omega}(\overline{z})g, f\rangle_{1/2}}. \\
&= \langle f, \widetilde{M}^{(0)}_{\Theta,D,\Omega}(\overline{z})g\rangle_{1/2}.
\end{aligned} \tag{4.68}
$$

Now (4.63) follows from (4.65) and (4.68). Finally, (4.64) follows from (4.63) by restriction of the latter to $L^2(\partial\Omega; d^{n-1}\omega)$. ☐

Next we briefly turn to the Herglotz property of the Robin-to-Dirichlet map. We recall that an operator-valued function $M(z) \in \mathcal{B}(\mathcal{H})$, $z \in \mathbb{C}_+$ (where $\mathbb{C}_+ = \{z \in \mathbb{C} \,|\, \mathrm{Im}(z) > 0\}$), for some separable complex Hilbert space $\mathcal{H}$, is called an *operator-valued Herglotz function* if $M(\,\cdot\,)$ is analytic on $\mathbb{C}_+$ and

$$
\mathrm{Im}(M(z)) \geqslant 0, \quad z \in \mathbb{C}_+. \tag{4.69}
$$

Here, as usual, $\mathrm{Im}(M) = (M - M^*)/(2i)$.

LEMMA 4.14. *Assume Hypothesis 4.1 and suppose that $z \in \mathbb{C}_+$. Then for every $g \in H^{-1/2}(\partial\Omega)$, $g \neq 0$,*

$$
\frac{1}{2i}\big\langle g, [\widetilde{M}_{\Theta,D}(z) - \widetilde{M}_{\Theta,D}(z)^*]g\big\rangle_{1/2} = \mathrm{Im}(z)\|u_\Theta\|^2_{L^2(\Omega;d^nx)} > 0, \tag{4.70}
$$

where u_Θ satisfies

$$
\begin{cases}
(-\Delta - z)u = 0 \text{ in } \Omega, \quad u \in H^1(\Omega), \\
(\widetilde{\gamma}_N + \widetilde{\Theta}\gamma_D)u = g \text{ on } \partial\Omega.
\end{cases} \tag{4.71}
$$

In particular, assuming Hypothesis 3.1, then

$$
\mathrm{Im}\big(M^{(0)}_{\Theta,D,\Omega}(z)\big) \geqslant 0, \quad z \in \mathbb{C}_+, \tag{4.72}
$$

and hence $M^{(0)}_{\Theta,D,\Omega}(\,\cdot\,)$ is an operator-valued Herglotz function on $L^2(\partial\Omega; d^{n-1}\omega)$.

PROOF. Let u_Θ be given by the solution of (4.71). Then $M^{(0)}_{\Theta,D,\Omega}g = \gamma_D u_\Theta$ by (3.58), and using self-adjointness of $\widetilde{\Theta}$ (in the sense of (B.7)) and the Green's formula (A.31), one computes,

$$
\begin{aligned}
&\frac{1}{2i}\big\langle g, [\widetilde{M}_{\Theta,D}(z) - \widetilde{M}_{\Theta,D}(z)^*]g\big\rangle_{1/2} \\
&\quad = \frac{1}{2i}\big[\big\langle g, \widetilde{M}_{\Theta,D}(z)g\big\rangle_{1/2} - \big\langle \widetilde{M}_{\Theta,D}(z)g, g\big\rangle_{1/2}\big] \\
&\quad = \frac{1}{2i}\big[\big\langle (\widetilde{\gamma}_N + \widetilde{\Theta}\gamma_D)u_\Theta, \gamma_D u_\Theta\big\rangle_{1/2} - \big\langle \gamma_D u_\Theta, (\widetilde{\gamma}_N + \widetilde{\Theta}\gamma_D)u_\Theta\big\rangle_{1/2}\big]
\end{aligned}
$$

$$
\begin{aligned}
&= \frac{1}{2i}\big[\langle\widetilde{\gamma}_N u_\Theta, \gamma_D u_\Theta\rangle_{1/2} - \langle\gamma_D u_\Theta, \widetilde{\gamma}_N u_\Theta\rangle_{1/2}\big] \\
&\quad + \frac{1}{2i}\big[\big\langle\widetilde{\Theta}\gamma_D u_\Theta, \gamma_D u_\Theta\big\rangle_{1/2} - \big\langle\gamma_D u_\Theta, \widetilde{\Theta}\gamma_D u_\Theta\big\rangle_{1/2}\big] \\
&= \operatorname{Im}(\langle\widetilde{\gamma}_N u_\Theta, \gamma_D u_\Theta\rangle_{1/2}) \\
&= \operatorname{Im}[(\nabla u_\Theta, \nabla u_\Theta)_{L^2(\Omega;d^nx)} + {}_{H^1(\Omega)}\langle\Delta u_\Theta, u_\Theta\rangle_{(H^1(\Omega))^*}] \\
&= \operatorname{Im}(-\overline{z}\, {}_{H^1(\Omega)}\langle u_\Theta, u_\Theta\rangle_{(H^1(\Omega))^*}) \\
&= \operatorname{Im}(z)\, {}_{H^1(\Omega)}\langle u_\Theta, u_\Theta\rangle_{(H^1(\Omega))^*} \\
&= \operatorname{Im}(z)\|u_\Theta\|^2_{L^2(\Omega;d^nx)} > 0
\end{aligned} \tag{4.73}
$$

since $g \neq 0$ implies $u_\Theta \neq 0$. This proves (4.70). Restriction of (4.70) to $g \in L^2(\partial\Omega; d^{n-1}\omega)$ then yields (4.72). □

Returning to the principal goal of this section, we now prove the following variant of a Krein-type resolvent formula relating $\widetilde{\Delta}_{\Theta,\Omega}$ and $\widetilde{\Delta}_{D,\Omega}$:

THEOREM 4.15. *Assume Hypothesis* 4.1 *and suppose that* $z \in \mathbb{C}\backslash(\sigma(-\Delta_{\Theta,\Omega}) \cup \sigma(-\Delta_{D,\Omega}))$. *Then the following Krein formula holds on* $\big(H^1(\Omega)\big)^*$,

$$
\begin{aligned}
\big(-\widetilde{\Delta}_{\Theta,\Omega} - z\widetilde{I}_\Omega\big)^{-1} &= \big(-\widetilde{\Delta}_{D,\Omega} - z\widetilde{I}_\Omega\big)^{-1} \circ R_\Omega \\
&\quad + \big[\widetilde{\gamma}_N\big(\big(-\widetilde{\Delta}_{D,\Omega} - \overline{z}\widetilde{I}_\Omega\big)^{-1} \circ R_\Omega, I_{\mathbb{R}^n}\big)\big]^* \widetilde{M}^{(0)}_{\Theta,D,\Omega}(z)\big[\widetilde{\gamma}_N \\
&\quad \times \big(\big(-\widetilde{\Delta}_{D,\Omega} - z\widetilde{I}_\Omega\big)^{-1} \circ R_\Omega, I_{\mathbb{R}^n}\big)\big].
\end{aligned} \tag{4.74}
$$

PROOF. Applying γ_D from the left to both sides of (4.50) yields

$$
\gamma_D\big(-\widetilde{\Delta}_{\Theta,\Omega} - z\widetilde{I}_\Omega\big)^{-1} = \gamma_D\big(-\widetilde{\Delta}_{\Theta,\Omega} - z\widetilde{I}_\Omega\big)^{-1}\gamma_D^*\widetilde{\gamma}_N\big(\big(-\widetilde{\Delta}_{D,\Omega} - z\widetilde{I}_\Omega\big)^{-1} \circ R_\Omega, I_{\mathbb{R}^n}\big) \tag{4.75}
$$

since $\gamma_D\big(-\widetilde{\Delta}_{D,\Omega} - z\widetilde{I}_\Omega\big)^{-1} = 0$. Thus, by (4.32),

$$
\gamma_D\big(-\widetilde{\Delta}_{\Theta,\Omega} - z\widetilde{I}_\Omega\big)^{-1} = \widetilde{M}^{(0)}_{\Theta,D,\Omega}(z)\widetilde{\gamma}_N\big(\big(-\widetilde{\Delta}_{D,\Omega} - z\widetilde{I}_\Omega\big)^{-1} \circ R_\Omega, I_{\mathbb{R}^n}\big), \tag{4.76}
$$

as operators in $\mathcal{B}\big(\big(H^1(\Omega)\big)^*, H^{1/2}(\partial\Omega)\big)$. Taking adjoints in (4.76) (written with $\overline{z}$ in place of z) then leads to

$$
\begin{aligned}
\big(-\widetilde{\Delta}_{\Theta,\Omega} - z\widetilde{I}_\Omega\big)^{-1}\gamma_D^* &= \big[\gamma_D\big(-\widetilde{\Delta}_{\Theta,\Omega} - \overline{z}\widetilde{I}_\Omega\big)^{-1}\big]^* \\
&= \big[\widetilde{\gamma}_N\big(-\widetilde{\Delta}_{D,\Omega} - \overline{z}\widetilde{I}_\Omega\big)^{-1}\big]^*\big[\widetilde{M}^{(0)}_{\Theta,D,\Omega}(\overline{z})\big]^* \\
&= \big[\widetilde{\gamma}_N\big(\big(-\widetilde{\Delta}_{D,\Omega} - \overline{z}\widetilde{I}_\Omega\big)^{-1} \circ R_\Omega, I_{\mathbb{R}^n}\big)\big]^*\widetilde{M}^{(0)}_{\Theta,D,\Omega}(z),
\end{aligned} \tag{4.77}
$$

by Lemma 4.13. Replacing this back into (4.50) then readily yields (4.74). □

The $L^2(\Omega; d^nx)$-analog of Theorem 4.15 then reads as follows:

THEOREM 4.16. *Assume Hypothesis* 3.1 *and suppose that* $z \in \mathbb{C}\backslash(\sigma(-\Delta_{\Theta,\Omega}) \cup \sigma(-\Delta_{D,\Omega}))$. *Then the following Krein formula holds on* $L^2(\Omega; d^nx)$:

$$
\begin{aligned}
(-\Delta_{\Theta,\Omega} - zI_\Omega)^{-1} &= (-\Delta_{D,\Omega} - zI_\Omega)^{-1} \\
&\quad + \big[\widetilde{\gamma}_N(-\Delta_{D,\Omega} - \overline{z}I_\Omega)^{-1}\big]^* M^{(0)}_{\Theta,D,\Omega}(z)\big[\widetilde{\gamma}_N(-\Delta_{D,\Omega} - zI_\Omega)^{-1}\big].
\end{aligned} \tag{4.78}
$$

PROOF. This follows from Theorem 4.15 and the compatibility results established in Lemma 4.10. □

An attractive feature of the Krein-type formula (4.78) lies in the fact that $M_{\Theta,D,\Omega}^{(0)}(z)$ encodes spectral information about $\Delta_{\Theta,\Omega}$. This will be pursued in future work.

Assuming Hypothesis 2.1, the special case $\Theta = 0$ then connects the Neumann and Dirichlet resolvents,

$$\begin{aligned}\big(-\widetilde{\Delta}_{N,\Omega} - z\widetilde{I}_\Omega\big)^{-1} &= \big(-\widetilde{\Delta}_{D,\Omega} - z\widetilde{I}_\Omega\big)^{-1} \circ R_\Omega \\ &+ \big[\widetilde{\gamma}_N\big(\big(-\widetilde{\Delta}_{D,\Omega} - \overline{z}\widetilde{I}_\Omega\big)^{-1} \circ R_\Omega, I_{\mathbb{R}^n}\big)\big]^* \widetilde{M}_{N,D,\Omega}^{(0)}(z) \\ &\times \big[\widetilde{\gamma}_N\big(\big(-\widetilde{\Delta}_{D,\Omega} - z\widetilde{I}_\Omega\big)^{-1} \circ R_\Omega, I_{\mathbb{R}^n}\big)\big], \quad z \in \mathbb{C}\backslash(\sigma(-\Delta_{N,\Omega}) \cup \sigma(-\Delta_{D,\Omega})),\end{aligned} \tag{4.79}$$

on $\big(H^1(\Omega)\big)^*$, and similarly,

$$\begin{aligned}(-\Delta_{N,\Omega} - zI_\Omega)^{-1} &= (-\Delta_{D,\Omega} - zI_\Omega)^{-1} \\ &\quad + \big[\widetilde{\gamma}_N(-\Delta_{D,\Omega} - \overline{z}I_\Omega)^{-1}\big]^* M_{N,D,\Omega}^{(0)}(z)\big[\widetilde{\gamma}_N(-\Delta_{D,\Omega} - zI_\Omega)^{-1}\big], \\ &\quad z \in \mathbb{C}\backslash(\sigma(-\Delta_{N,\Omega}) \cup \sigma(-\Delta_{D,\Omega})),\end{aligned} \tag{4.80}$$

on $L^2(\Omega; d^n x)$. Here $\widetilde{M}_{N,D,\Omega}^{(0)}(z)$ and $M_{N,D,\Omega}^{(0)}(z)$ denote the corresponding Neumann-to-Dirichlet operators.

REMARK 4.17. In the case when Hypothesis 2.11 is enforced, it can be shown that

$$M_{N,D,\Omega}^{(0)}(z) \in \mathcal{B}\big(H^{1/2}(\partial\Omega), H^{3/2}(\partial\Omega)\big), \quad z \in \mathbb{C}\backslash(\sigma(-\Delta_{N,\Omega}) \cup \sigma(-\Delta_{D,\Omega})), \tag{4.81}$$

and

$$(-\Delta_{D,\Omega} - zI_\Omega)^{-1} \in \mathcal{B}\big(L^2(\Omega; d^n x), H^2(\Omega)\big), \quad z \in \mathbb{C}\backslash\sigma(-\Delta_{D,\Omega}). \tag{4.82}$$

Note that, by duality, the latter membership also entails

$$(-\Delta_{D,\Omega} - zI_\Omega)^{-1} \in \mathcal{B}\big((H^2(\Omega))^*, L^2(\Omega; d^n x)\big), \quad z \in \mathbb{C}\backslash\sigma(-\Delta_{D,\Omega}), \tag{4.83}$$

and, given (2.38),

$$\gamma_N(-\Delta_{D,\Omega} - zI_\Omega)^{-1} \in \mathcal{B}\big(L^2(\Omega; d^n x), H^{1/2}(\partial\Omega)\big), \quad z \in \mathbb{C}\backslash\sigma(-\Delta_{D,\Omega}). \tag{4.84}$$

Since, in the current scenario, we also have

$$\gamma_N^* \in \mathcal{B}\big(H^{-1/2}(\partial\Omega), (H^2(\Omega))^*\big), \tag{4.85}$$

it follows that (4.80) takes the form

$$\begin{aligned}(-\Delta_{N,\Omega} - zI_\Omega)^{-1} &= (-\Delta_{D,\Omega} - zI_\Omega)^{-1} \\ &\quad + (-\Delta_{D,\Omega} - zI_\Omega)^{-1}\gamma_N^* M_{N,D,\Omega}^{(0)}(z)\gamma_N(-\Delta_{D,\Omega} - zI_\Omega)^{-1}, \\ &\quad z \in \mathbb{C}\backslash(\sigma(-\Delta_{N,\Omega}) \cup \sigma(-\Delta_{D,\Omega})),\end{aligned} \tag{4.86}$$

on $L^2(\Omega; d^n x)$, where the composition of the various operators involved is well-defined by the above discussion. Formula (4.86) should be viewed as a variant of (4.78) in which the Neumann trace operator can be decoupled from the two resolvents of $-\Delta_{D,\Omega}$ in the second term on the right-hand side of (4.78).

Due to the fundamental importance of Krein-type resolvent formulas (and more generally, Robin-to-Dirichlet maps) in connection with the spectral and inverse spectral theory of ordinary and partial differential operators, abstract versions, connected to boundary value spaces (boundary triples) and self-adjoint extensions of closed symmetric operators with equal (possibly infinite) deficiency spaces, have received enormous attention in the literature. In particular, we note that Robin-to-Dirichlet maps in the context of ordinary differential operators reduce to the celebrated (possibly, matrix-valued) Weyl–Titchmarsh function, the basic object of spectral analysis in this context. Since it is impossible to cover the literature in this paper, we refer, for instance, to [**2**, Sect. 84], [**4**], [**8**], [**9**], [**12**], [**14**], [**15**], [**20**], [**22**], [**23**], [**41**], [**44**], [**49**, Ch. 13], [**52**], [**54**]–[**61**], [**64**], [**65**], [**71**], [**78**]–[**85**], [**90**], [**93**]–[**95**], and the references cited therein. We add, however, that the case of infinite deficiency indices in the context of partial differential operators (in our concrete case, related to the deficiency indices of the operator closure of $-\Delta \restriction_{C_0^\infty(\Omega)}$ in $L^2(\Omega; d^n x)$), is much less studied and the results obtained in this section, especially, under the assumption of Lipschitz (i.e., minimally smooth) domains, to the best of our knowledge, are new.

Finally, we emphasize once more that Remark 3.9 also applies to the content of this section (assuming that V is real-valued in connection with Lemmas 4.13 and 4.14).

Appendix A. Properties of Sobolev Spaces and Boundary Traces for $C^{1,r}$ and Lipschitz Domains

The purpose of this appendix is to recall some basic facts in connection with Sobolev spaces corresponding to Lipschitz domains $\Omega \subset \mathbb{R}^n$, $n \in \mathbb{N}$, $n \geq 2$, and on domains satisfying Hypothesis 2.11.

In this manuscript we use the following notation for the standard Sobolev Hilbert spaces ($s \in \mathbb{R}$),

$$H^s(\mathbb{R}^n) = \bigg\{U \in \mathcal{S}(\mathbb{R}^n)' \,\bigg|\, \|U\|^2_{H^s(\mathbb{R}^n)} = \int_{\mathbb{R}^n} d^n\xi \, \big|\widehat{U}(\xi)\big|^2 \big(1 + |\xi|^{2s}\big) < \infty\bigg\}, \tag{A.1}$$

$$H^s(\Omega) = \{u \in \mathcal{D}'(\Omega) \,|\, u = U|_\Omega \text{ for some } U \in H^s(\mathbb{R}^n)\}, \tag{A.2}$$

$$H^s_0(\Omega) = \{u \in H^s(\mathbb{R}^n) \,|\, \operatorname{supp}(u) \subseteq \overline{\Omega}\}. \tag{A.3}$$

Here $\mathcal{D}'(\Omega)$ denotes the usual set of distributions on $\Omega \subseteq \mathbb{R}^n$, Ω open and nonempty, $\mathcal{S}(\mathbb{R}^n)'$ is the space of tempered distributions on $\mathbb{R}^n$, and $\widehat{U}$ denotes the Fourier transform of $U \in \mathcal{S}(\mathbb{R}^n)'$. It is then immediate that

$$H^{s_1}(\Omega) \hookrightarrow H^{s_0}(\Omega) \text{ for } -\infty < s_0 \leq s_1 < +\infty, \tag{A.4}$$

continuously and densely.

Next, we recall the definition of a $C^{1,r}$-domain $\Omega \subseteq \mathbb{R}^n$, Ω open and nonempty, for convenience of the reader: Let $\mathcal{N}$ be a space of real-valued functions in $\mathbb{R}^{n-1}$. One calls a bounded domain $\Omega \subset \mathbb{R}^n$ of class $\mathcal{N}$ if there exists a finite open covering $\{\mathcal{O}_j\}_{1 \leq j \leq N}$ of the boundary $\partial\Omega$ of Ω with the property that, for every $j \in \{1, ..., N\}$, $\mathcal{O}_j \cap \Omega$ coincides with the portion of $\mathcal{O}_j$ lying in the over-graph of a function $\varphi_j \in \mathcal{N}$ (considered in a new system of coordinates obtained from the original one via a rigid motion). Two special cases are going to play a particularly important role in the sequel. First, if $\mathcal{N}$ is $\operatorname{Lip}(\mathbb{R}^{n-1})$, the space of real-valued functions satisfying a (global) Lipschitz condition in $\mathbb{R}^{n-1}$, we shall refer to Ω as being a Lipschitz domain; cf. [**92**, p. 189], where such domains are called "minimally smooth". Second, corresponding to the case when $\mathcal{N}$ is the subspace of $\mathrm{Lip}\,(\mathbb{R}^{n-1})$ consisting of functions whose first-order derivatives satisfy a (global) Hölder condition of order $r \in (0,1)$, we shall say that Ω is of class $C^{1,r}$. The classical theorem of Rademacher of almost everywhere differentiability of Lipschitz functions ensures that, for any Lipschitz domain Ω, the surface measure $d^{n-1}\omega$ is well-defined on $\partial\Omega$ and that there exists an outward pointing normal vector ν at almost every point of $\partial\Omega$.

Call a bounded, open set $\Omega \subset \mathbb{R}^n$ a star-like Lipschitz domain with respect to a point x^* (called center of star-likeness) if Ω is Lipschitz domain and

$$x^* + t(x - x^*) \in \Omega \quad \text{for every } x \in \Omega \text{ and } t \in [0.1]. \tag{A.5}$$

The above geometrical characterization of Lipschitz domains can be used to show that, given a bounded Lipschitz domain $\Omega \subset \mathbb{R}^n$ then there exists a finite family of open sets Ω_j, $1 \le j \le N$, such that

$$\Omega = \bigcup_{j=1}^{N} \Omega_j, \qquad \Omega_j \text{ star-like Lipschitz domain, } 1 \le j \le N. \tag{A.6}$$

For a Lipschitz domain $\Omega \subset \mathbb{R}^n$ it is known that

$$\big(H^s(\Omega)\big)^* = H^{-s}(\Omega), \quad -1/2 < s < 1/2. \tag{A.7}$$

See [**99**] for this and other related properties. We also refer to our convention of using the *adjoint* (rather than the dual) space X^* of a Banach space X as described near the end of the introduction.

Next, assume that $\Omega \subset \mathbb{R}^n$ is the domain lying above the graph of a function $\varphi\colon \mathbb{R}^{n-1} \to \mathbb{R}$ of class $C^{1,r}$. Then for $0 \le s < 1 + r$, the Sobolev space $H^s(\partial\Omega)$ consists of functions $f \in L^2(\partial\Omega; d^{n-1}\omega)$ such that $f(x', \varphi(x'))$, as a function of $x' \in \mathbb{R}^{n-1}$, belongs to $H^s(\mathbb{R}^{n-1})$. This definition is easily adapted to the case when Ω is a domain of class $C^{1,r}$ whose boundary is compact, by using a smooth partition of unity. Finally, for $-1 - r < s < 0$, we set $H^s(\partial\Omega) = \big(H^{-s}(\partial\Omega)\big)^*$. The same construction concerning $H^s(\partial\Omega)$ applies in the case when $\Omega \subset \mathbb{R}^n$ is a Lipschitz domain (i.e., $\varphi\colon \mathbb{R}^{n-1} \to \mathbb{R}$ is only Lipschitz) provided $0 \leqslant s \leqslant 1$. In this scenario we set

$$H^s(\partial\Omega) = \big(H^{-s}(\partial\Omega)\big)^*, \quad -1 \leqslant s \leqslant 0. \tag{A.8}$$

It is useful to observe that this entails

$$\|f\|_{H^{-s}(\partial\Omega)} \approx \|\sqrt{1 + |\nabla\varphi(\,\cdot\,)|^2} f(\,\cdot\,, \varphi(\,\cdot\,))\|_{H^{-s}(\mathbb{R}^{n-1})}, \quad 0 \le s \le 1. \tag{A.9}$$

To define $H^s(\partial\Omega)$, $0 \le s \leqslant 1$, when Ω is a Lipschitz domain with compact boundary, we use a smooth partition of unity to reduce matters to the graph case. More precisely, if $0 \le s \le 1$ then $f \in H^s(\partial\Omega)$ if and only if the assignment $\mathbb{R}^{n-1} \ni x' \mapsto (\psi f)(x', \varphi(x'))$ is in $H^s(\mathbb{R}^{n-1})$ whenever $\psi \in C_0^\infty(\mathbb{R}^n)$ and $\varphi\colon \mathbb{R}^{n-1} \to \mathbb{R}$ is a Lipschitz function with the property that if Σ is an appropriate rotation and translation of $\{(x', \varphi(x')) \in \mathbb{R}^n \,|\, x' \in \mathbb{R}^{n-1}\}$, then $(\mathrm{supp}\,(\psi) \cap \partial\Omega) \subset \Sigma$ (this appears to be folklore, but a proof will appear in [**72**, Proposition 2.4]). Then Sobolev spaces with a negative amount of smoothness are defined as in (A.8) above.

From the above characterization of $H^s(\partial\Omega)$ it follows that any property of Sobolev spaces (of order $s \in [-1,1]$) defined in Euclidean domains, which are invariant under multiplication by smooth, compactly supported functions as well as composition by bi-Lipschitz diffeomorphisms, readily extends to the setting of

$H^s(\partial\Omega)$ (via localization and pull-back). As a concrete example, for each Lipschitz domain Ω with compact boundary, one has

$$H^s(\partial\Omega) \hookrightarrow L^2(\partial\Omega; d^{n-1}\omega) \text{ compactly if } 0 < s \leq 1. \tag{A.10}$$

For additional background information in this context we refer, for instance, to [**10**], [**11**], [**37**, Chs. V, VI], [**48**, Ch. 1], [**69**, Ch. 3], [**105**, Sect. I.4.2].

For a Lipschitz domain $\Omega \subset \mathbb{R}^n$ with compact boundary, an equivalent definition of the Sobolev space $H^1(\partial\Omega)$ is the collection of functions in $L^2(\partial\Omega; d^{n-1}\omega)$ with the property that the (pointwise, Euclidean) norm of their tangential gradient belongs to $L^2(\partial\Omega; d^{n-1}\omega)$. To make this precise, consider the first-order tangential derivative operators $\partial/\partial\tau_{j,k}$, $1 \leq j,k \leq n$, acting on a function ψ of class C^1 in a neighborhood of $\partial\Omega$ by

$$\partial\psi/\partial\tau_{j,k} = \nu_j(\partial_k\psi)\big|_{\partial\Omega} - \nu_k(\partial_j\psi)\big|_{\partial\Omega}. \tag{A.11}$$

For every $f \in L^1(\partial\Omega)$ define the functional $\partial f/\partial\tau_{j,k}$ by setting

$$\partial f/\partial\tau_{j,k} : C^1(\mathbb{R}^n) \ni \psi \mapsto \int_{\partial\Omega} d^{n-1}\omega\, f\,(\partial\psi/\partial\tau_{k,j}) \tag{A.12}$$

When $f \in L^1(\partial\Omega; d^{n-1}\omega)$ has $\partial f/\partial\tau_{j,k} \in L^1(\partial\Omega; d^{n-1}\omega)$, the following integration by parts formula holds:

$$\int_{\partial\Omega} d^{n-1}\omega\, f\,(\partial\psi/\partial\tau_{k,j}) = \int_{\partial\Omega} d^{n-1}\omega\,(\partial f/\partial\tau_{j,k})\,\psi, \quad \psi \in C^1(\mathbb{R}^n). \tag{A.13}$$

We then have the Sobolev-type description of $H^1(\partial\Omega)$:

$$H^1(\partial\Omega) = \big\{f \in L^2(\partial\Omega; d^{n-1}\omega) \,\big|\, \partial f/\partial\tau_{j,k} \in L^2(\partial\Omega; d^{n-1}\omega),\ j,k = 1,\dots,n\big\} \tag{A.14}$$

with

$$\|f\|_{H^1(\partial\Omega)} \approx \|f\|_{L^2(\partial\Omega; d^{n-1}\omega)} + \sum_{j,k=1}^n \|\partial f/\partial\tau_{j,k}\|_{L^2(\partial\Omega; d^{n-1}\omega)}, \tag{A.15}$$

($\approx$ denoting equivalent norms), or equivalently,

$$H^1(\partial\Omega) = \bigg\{f \in L^2(\partial\Omega; d^{n-1}\omega) \,\bigg|\, \text{there exists a constant } c > 0 \text{ such that for every}$$
$$v \in C_0^\infty(\mathbb{R}^n),\ \bigg|\int_{\partial\Omega} d^{n-1}\omega f\, \partial v/\partial\tau_{j,k}\bigg| \leq c\,\|v\|_{L^2(\partial\Omega; d^{n-1}\omega)},\ j,k = 1,\dots,n\bigg\}. \tag{A.16}$$

Let us also point out here that if $\Omega \subset \mathbb{R}^n$ is a bounded Lipschitz domain then for any $j,k \in \{1,...,n\}$ the operator

$$\partial/\partial\tau_{j,k} : H^s(\partial\Omega) \to H^{s-1}(\partial\Omega), \quad 0 \leq s \leq 1, \tag{A.17}$$

is well-defined, linear and bounded. This is proved by interpolating the case $s = 1$ and its dual version. In fact, the following more general result (extending (A.14)) is true.

Lemma A.1. *Assume that $\Omega \subset \mathbb{R}^n$ is a bounded Lipschitz domain. Then for every $s \in [0,1]$,*

$$H^s(\partial\Omega) = \{f \in L^2(\partial\Omega; d^{n-1}\omega) \,|\, \partial f/\partial\tau_{j,k} \in H^{s-1}(\partial\Omega),\ 1 \leq j,k \leq n\} \tag{A.18}$$

and

$$\|f\|_{H^s(\partial\Omega)} \approx \|f\|_{L^2(\partial\Omega;d^{n-1}\omega)} + \sum_{j,k=1}^{n} \|\partial f/\partial\tau_{j,k}\|_{H^{s-1}(\partial\Omega)}. \tag{A.19}$$

PROOF. The left-to-right inclusion in (A.18) along with the right-pointing inequality in (A.19) are consequences of the boundedness of (A.17). As for the opposite directions, we note that using a smooth partition of unity and making a rigid transformation of the space, matters can be localized near a boundary point where $\partial\Omega$ coincides with the graph of a Lipschitz function $\varphi : \mathbb{R}^{n-1} \to \mathbb{R}$. Then for each sufficiently nice function $f : \partial\Omega \to \mathbb{R}$ the Chain Rule yields

$$\Big(\frac{\partial f}{\partial\tau_{j,n}}\Big)(x,\varphi(x)) = \frac{1}{\sqrt{1+|\nabla\varphi(x)|^2}}\frac{\partial}{\partial x_j}\big[f(x,\varphi(x))\big], \quad 1 \leq j \leq n-1. \tag{A.20}$$

On account of this and (A.9), we then deduce (upon noticing that $\partial/\partial\tau_{n,n} = 0$) that

$$\sum_{j=1}^{n} \|\partial f/\partial\tau_{j,n}\|_{H^{s-1}(\partial\Omega)} \approx \sum_{j=1}^{n-1} \|\partial_j[f(\,\cdot\,,\varphi(\,\cdot\,))]\|_{H^{s-1}(\mathbb{R}^{n-1})}. \tag{A.21}$$

Furthermore, we also have

$$\|f\|_{L^2(\partial\Omega;d^{n-1}\omega)} \approx \|f(\,\cdot\,,\varphi(\,\cdot\,))\|_{L^2(\mathbb{R}^{n-1};d^{n-1}x)}. \tag{A.22}$$

Next, we recall the general Euclidean lifting result

$$H^s(\mathbb{R}^{n-1}) = \{f \in H^{s-1}(\mathbb{R}^{n-1}) \,|\, \partial_j f \in H^{s-1}(\mathbb{R}^{n-1}),\, 1 \leq j \leq n-1\}, \quad s \in \mathbb{R}, \tag{A.23}$$

which can be found in [**86**, Section 2.1.4]. Now, the right-to-left inclusion in (A.18), as well as the left-pointing inequality in (A.19), follow based on (A.21), (A.22) and the estimate which naturally accompanies (A.23). □

LEMMA A.2. *Assume Hypothesis 2.1. Then for every $s \in [0,1]$ and $j,k \in \{1,...,n\}$*

$$\langle \partial f/\partial\tau_{j,k}\,,\, g\rangle_{1-s} = \langle f\,,\, \partial g/\partial\tau_{k,j}\rangle_{1-s} \tag{A.24}$$

for every $f \in H^s(\partial\Omega)$ and $g \in H^{1-s}(\partial\Omega)$.

PROOF. Since for every $s \in [0,1]$

$$C^\infty(\mathbb{R}^n)\big|_{\partial\Omega} \hookrightarrow H^s(\partial\Omega) \text{ densely}, \tag{A.25}$$

it suffices to prove (A.24) in the case when $f = u|_{\partial\Omega}$ and $g = v|_{\partial\Omega}$ for $u, v \in C^\infty(\mathbb{R}^n)$. In this scenario, we need to establish that

$$\int_{\partial\Omega} d^{n-1}\omega\, (\partial u/\partial\tau_{j,k})v = \int_{\partial\Omega} d^{n-1}\omega\, u(\partial v/\partial\tau_{k,j}), \quad 1 \leq j,k \leq n. \tag{A.26}$$

To this end, we rely on Green's formula (valid for Lipschitz domains) to write

$$\begin{aligned}
\int_{\partial\Omega} d^{n-1}\omega\, (\partial u/\partial\tau_{j,k})v &= \int_{\partial\Omega} d^{n-1}\omega\, (\nu_j\partial_k u - \nu_k\partial_j u)v \\
&= \int_{\Omega} d^n x\, [\partial_j(v\partial_k u) - \partial_k(v\partial_j u)] \\
&= \int_{\Omega} d^n x\, [(\partial_j v)(\partial_k u) - (\partial_k v)(\partial_j u)].
\end{aligned} \tag{A.27}$$

One observes that the right-most integrand above is an antisymmetric expression in the indices j, k. Consequently, so is the left-most integral in (A.27). This, however, is equivalent to (A.26). $\square$

Moving on, we next consider the following bounded linear map

$$\begin{cases}\{(w,f)\in L^2(\Omega;d^nx)^n\times\big(H^1(\Omega)\big)^*\,|\,\operatorname{div}(w)=f|_\Omega\}\to H^{-1/2}(\partial\Omega)=\big(H^{1/2}(\partial\Omega)\big)^* \\ \qquad\qquad\qquad\qquad\qquad\qquad w\mapsto \nu\cdot(w,f)\end{cases} \tag{A.28}$$

by setting

$$_{H^{1/2}(\partial\Omega)}\langle\phi,\nu\cdot(w,f)\rangle_{(H^{1/2}(\partial\Omega))^*}=\int_\Omega d^nx\,\overline{\nabla\Phi(x)}\cdot w(x)+{}_{H^1(\Omega)}\langle\Phi,f\rangle_{(H^1(\Omega))^*} \tag{A.29}$$

whenever $\phi\in H^{1/2}(\partial\Omega)$ and $\Phi\in H^1(\Omega)$ is such that $\gamma_D\Phi=\phi$. Here the pairing ${}_{H^1(\Omega)}\langle\Phi,f\rangle_{(H^1(\Omega))^*}$ in (A.29) is the natural one between functionals in $\big(H^1(\Omega)\big)^*$ and elements in $H^1(\Omega)$ (which, in turn, is compatible with the (bilinear) distributional pairing). It should be remarked that the above definition is independent of the particular extension $\Phi\in H^1(\Omega)$ of ϕ.

Going further, one can introduce the ultra weak Neumann trace operator $\widetilde\gamma_{\mathcal N}$ as follows:

$$\widetilde\gamma_{\mathcal N}\colon\begin{cases}\{(u,f)\in H^1(\Omega)\times\big(H^1(\Omega)\big)^*\,|\,\Delta u=f|_\Omega\}\to H^{-1/2}(\partial\Omega)\\ \qquad\qquad\qquad\qquad u\mapsto\widetilde\gamma_{\mathcal N}(u,f)=\nu\cdot(\nabla u,f),\end{cases} \tag{A.30}$$

with the dot product understood in the sense of (A.28). We emphasize that the ultra weak Neumann trace operator $\widetilde\gamma_{\mathcal N}$ in (A.30) is a re-normalization of the operator γ_N introduced in (2.38) relative to the extension of $\Delta u\in H^{-1}(\Omega)$ to an element f of the space $\big(H^1(\Omega)\big)^*=\{g\in H^{-1}(\mathbb{R}^n)\,|\,\operatorname{supp}(g)\subseteq\overline{\Omega}\}$. For the relationship between the weak and ultra weak Neumann trace operators, see (2.62)–(2.64). In addition, one can show that the ultra weak Neumann trace operator (A.30) is onto (indeed, this is a corollary of Theorem 4.5). We note that (A.29) and (A.30) yield the following Green's formula

$$\langle\gamma_D\Phi,\widetilde\gamma_{\mathcal N}(u,f)\rangle_{1/2}=(\nabla\Phi,\nabla u)_{L^2(\Omega;d^nx)^n}+{}_{H^1(\Omega)}\langle\Phi,f\rangle_{(H^1(\Omega))^*}, \tag{A.31}$$

valid for any $u\in H^1(\Omega)$, $f\in\big(H^1(\Omega)\big)^*$ with $\Delta u=f|_\Omega$, and any $\Phi\in H^1(\Omega)$. The pairing on the left-hand side of (A.31) is between functionals in $\big(H^{1/2}(\partial\Omega)\big)^*$ and elements in $H^{1/2}(\partial\Omega)$, whereas the last pairing on the right-hand side is between functionals in $\big(H^1(\Omega)\big)^*$ and elements in $H^1(\Omega)$. For further use, we also note that the adjoint of (2.7) maps boundedly as follows

$$\gamma_D^*\colon\big(H^{s-1/2}(\partial\Omega)\big)^*\to\big(H^s(\Omega)\big)^*,\quad 1/2<s<3/2. \tag{A.32}$$

REMARK A.3. While it is tempting to view γ_D as an unbounded but densely defined operator on $L^2(\Omega;d^nx)$ whose domain contains the space $C_0^\infty(\Omega)$, one should note that in this case its adjoint γ_D^* is not densely defined: Indeed (cf. [**43**, Remark A.4]), $\operatorname{dom}(\gamma_D^*)=\{0\}$ and hence γ_D is not a closable linear operator in $L^2(\Omega;d^nx)$.

Next we recall the following result from [**46**] (and reproduce its proof for subsequent use in the proofs of Lemmas A.5 and D.3).

LEMMA A.4 (cf. [**46**], Lemma A.6). *Suppose $\Omega \subset \mathbb{R}^n$, $n \geq 2$, is an open Lipschitz domain with a compact, nonempty boundary $\partial\Omega$. Then the Dirichlet trace operator γ_D (originally considered as in (2.7)) satisfies (2.9).*

PROOF. Let $u \in H^{(3/2)+\varepsilon}(\Omega)$, $v \in C_0^\infty(\mathbb{R}^n)$, and $u_\ell \in C^\infty(\overline{\Omega}) \hookrightarrow H^{(3/2)+\varepsilon}(\Omega)$, $\ell \in \mathbb{N}$, be a sequence of functions approximating u in $H^{(3/2)+\varepsilon}(\Omega)$. It follows from (2.7) and (A.4) that $\gamma_D u, \gamma_D(\nabla u) \in L^2(\partial\Omega; d^{n-1}\omega)$. Utilizing (A.13), one computes for all $j, k = 1, \dots, n$,

$$\begin{aligned} \left| \int_{\partial\Omega} d^{n-1}\omega \, \gamma_D u \frac{\partial v}{\partial \tau_{j,k}} \right| &= \left| \lim_{\ell\to\infty} \int_{\partial\Omega} d^{n-1}\omega \, u_\ell \frac{\partial v}{\partial \tau_{j,k}} \right| = \left| \lim_{\ell\to\infty} \int_{\partial\Omega} d^{n-1}\omega \, v \frac{\partial u_\ell}{\partial \tau_{j,k}} \right| \\ &\leq c \left| \lim_{\ell\to\infty} \int_{\partial\Omega} d^{n-1}\omega \, v \, \gamma_D(\nabla u_\ell) \right| \leq c \, \|\gamma_D(\nabla u)\|_{L^2(\partial\Omega; d^{n-1}\omega)} \, \|v\|_{L^2(\partial\Omega; d^{n-1}\omega)} . \end{aligned} \tag{A.33}$$

Thus, it follows from (A.16) and (A.33) that $\gamma_D u \in H^1(\partial\Omega)$. □

Next, we prove the following fact:

LEMMA A.5. *Suppose $\Omega \subset \mathbb{R}^n$, $n \geq 2$, is a bounded Lipschitz domain. Then for each $r \in (1/2, 1)$, the space $C^r(\partial\Omega)$ is a module over $H^{1/2}(\partial\Omega)$. More precisely, if M_f denotes the operator of multiplication by f, then there exists $C = C(\Omega, r) > 0$ such that*

$$M_f \in \mathcal{B}\big(H^{1/2}(\partial\Omega)\big) \text{ and } \|M_f\|_{\mathcal{B}(H^{1/2}(\partial\Omega))} \leq C\|f\|_{C^r(\partial\Omega)} \text{ for every } f \in C^r(\partial\Omega). \tag{A.34}$$

As a consequence, if Ω is actually a bounded $C^{1,r}$-domain with $r \in (1/2, 1)$, then the Neumann and Dirichlet trace operators γ_N, γ_D satisfy

$$\gamma_N \in \mathcal{B}\big(H^2(\Omega), H^{1/2}(\partial\Omega)\big) \tag{A.35}$$

and

$$\gamma_D \in \mathcal{B}\big(H^2(\Omega), H^{3/2}(\partial\Omega)\big). \tag{A.36}$$

PROOF. The first part of the lemma is a direct consequence of general results about pointwise multiplication of functions in Triebel–Lizorkin spaces (a scale which contains both Hölder and Sobolev spaces); see [**86**, Theorem 2 on p. 177]. Then (A.35) follows from this, (2.7), the fact that $\gamma_N = \nu \cdot \gamma_D$, and $\nu \in C^r(\partial\Omega)$. Next, one observes that for each $u \in H^2(\Omega)$ one has $\gamma_D u \in H^1(\partial\Omega)$ by Lemma A.4. In addition,

$$\frac{\partial}{\partial \tau_{j,k}}(\gamma_D u) = \big(\nu_j \gamma_D(\partial_k u) - \nu_k \gamma_D(\partial_j u)\big) \in H^{1/2}(\partial\Omega), \tag{A.37}$$

with a naturally accompanying estimate, by (2.7) and the fact that, as observed in the first part of the current proof, multiplication by ν_j (for $1 \leq j \leq n$) preserves $H^{1/2}(\partial\Omega)$. Consequently, (A.36) follows from this and (A.38), (A.39) below. □

Our next result should be compared with (A.14) and Lemma A.1.

LEMMA A.6. *If $\Omega \subset \mathbb{R}^n$ is a bounded $C^{1,r}$-domain with $r \in (1/2, 1)$ then*

$$H^{3/2}(\partial\Omega) = \{f \in H^1(\partial\Omega) \,|\, \partial f/\partial\tau_{j,k} \in H^{1/2}(\partial\Omega) \, 1 \leq j, k \leq n\} \tag{A.38}$$

and

$$\|f\|_{H^{3/2}(\partial\Omega)} \approx \|f\|_{H^1(\partial\Omega)} + \sum_{j,k=1}^n \|\partial f/\partial\tau_{j,k}\|_{H^{1/2}(\partial\Omega)}. \tag{A.39}$$

PROOF. To justify (A.38) and (A.39) we use a smooth cut-off function to localize the problem near a boundary point where $\partial\Omega$ coincides with the graph of a $C^{1,r}$ function $\varphi : \mathbb{R}^{n-1} \to \mathbb{R}$. In this setting, the desired conclusions follow from (A.20), (A.34), and (A.23) used with $s = 3/2$. $\square$

Appendix B. Sesquilinear Forms and Associated Operators

In this appendix we describe a few basic facts on sesquilinear forms and linear operators associated with them.

Let $\mathcal{H}$ be a complex separable Hilbert space with scalar product $(\,\cdot\,,\,\cdot\,)_{\mathcal{H}}$ (antilinear in the first and linear in the second argument), $\mathcal{V}$ a reflexive Banach space continuously and densely embedded into $\mathcal{H}$. Then also $\mathcal{H}$ embeds continuously and densely into $\mathcal{V}^*$.

$$\mathcal{V} \hookrightarrow \mathcal{H} \hookrightarrow \mathcal{V}^*. \tag{B.1}$$

Here the continuous embedding $\mathcal{H} \hookrightarrow \mathcal{V}^*$ is accomplished via the identification

$$\mathcal{H} \ni u \mapsto (\,\cdot\,, u)_{\mathcal{H}} \in \mathcal{V}^*, \tag{B.2}$$

and we recall the convention in this manuscript (cf. the discussion at the end of the introduction) that if X denotes a Banach space, X^* denotes the *adjoint space* of continuous conjugate linear functionals on X, also known as the *conjugate dual* of X.

In particular, if the sesquilinear form

$${}_{\mathcal{V}}\langle\,\cdot\,,\,\cdot\,\rangle_{\mathcal{V}^*} : \mathcal{V} \times \mathcal{V}^* \to \mathbb{C} \tag{B.3}$$

denotes the duality pairing between $\mathcal{V}$ and $\mathcal{V}^*$, then

$${}_{\mathcal{V}}\langle u, v\rangle_{\mathcal{V}^*} = (u, v)_{\mathcal{H}}, \quad u \in \mathcal{V},\ v \in \mathcal{H} \hookrightarrow \mathcal{V}^*, \tag{B.4}$$

that is, the $\mathcal{V}, \mathcal{V}^*$ pairing ${}_{\mathcal{V}}\langle\,\cdot\,,\,\cdot\,\rangle_{\mathcal{V}^*}$ is compatible with the scalar product $(\,\cdot\,,\,\cdot\,)_{\mathcal{H}}$ in $\mathcal{H}$.

Let $T \in \mathcal{B}(\mathcal{V}, \mathcal{V}^*)$. Since $\mathcal{V}$ is reflexive, $(\mathcal{V}^*)^* = \mathcal{V}$, one has

$$T \colon \mathcal{V} \to \mathcal{V}^*, \quad T^* \colon \mathcal{V} \to \mathcal{V}^* \tag{B.5}$$

and

$${}_{\mathcal{V}}\langle u, Tv\rangle_{\mathcal{V}^*} = {}_{\mathcal{V}^*}\langle T^*u, v\rangle_{(\mathcal{V}^*)^*} = {}_{\mathcal{V}^*}\langle T^*u, v\rangle_{\mathcal{V}} = \overline{{}_{\mathcal{V}}\langle v, T^*u\rangle_{\mathcal{V}^*}}. \tag{B.6}$$

Self-adjointness of T is then defined by $T = T^*$, that is,

$${}_{\mathcal{V}}\langle u, Tv\rangle_{\mathcal{V}^*} = {}_{\mathcal{V}^*}\langle Tu, v\rangle_{\mathcal{V}} = \overline{{}_{\mathcal{V}}\langle v, Tu\rangle_{\mathcal{V}^*}}, \quad u, v \in \mathcal{V}, \tag{B.7}$$

nonnegativity of T is defined by

$${}_{\mathcal{V}}\langle u, Tu\rangle_{\mathcal{V}^*} \geq 0, \quad u \in \mathcal{V}, \tag{B.8}$$

and *boundedness from below of T by $c_T \in \mathbb{R}$* is defined by

$${}_{\mathcal{V}}\langle u, Tu\rangle_{\mathcal{V}^*} \geq c_T \|u\|_{\mathcal{H}}^2, \quad u \in \mathcal{V}. \tag{B.9}$$

(By (B.4), this is equivalent to ${}_{\mathcal{V}}\langle u, Tu\rangle_{\mathcal{V}^*} \geq c_T\, {}_{\mathcal{V}}\langle u, u\rangle_{\mathcal{V}^*}$, $u \in \mathcal{V}$.)

Next, let the sesquilinear form $a(\,\cdot\,,\,\cdot\,) \colon \mathcal{V} \times \mathcal{V} \to \mathbb{C}$ (antilinear in the first and linear in the second argument) be *$\mathcal{V}$-bounded*, that is, there exists a $c_a > 0$ such that

$$|a(u, v)| \leqslant c_a \|u\|_{\mathcal{V}} \|v\|_{\mathcal{V}}, \quad u, v \in \mathcal{V}. \tag{B.10}$$

Then $\widetilde{A}$ defined by

$$\widetilde{A}\colon \begin{cases} \mathcal{V} \to \mathcal{V}^*, \\ v \mapsto \widetilde{A}v = a(\,\cdot\,, v), \end{cases} \tag{B.11}$$

satisfies

$$\widetilde{A} \in \mathcal{B}(\mathcal{V}, \mathcal{V}^*) \text{ and } {}_{\mathcal{V}}\langle u, \widetilde{A}v\rangle_{\mathcal{V}^*} = a(u,v), \quad u,v \in \mathcal{V}. \tag{B.12}$$

Assuming further that $a(\,\cdot\,,\,\cdot\,)$ is *symmetric*, that is,

$$a(u,v) = \overline{a(v,u)}, \quad u,v \in \mathcal{V}, \tag{B.13}$$

and that a is $\mathcal{V}$*-coercive*, that is, there exists a constant $C_0 > 0$ such that

$$a(u,u) \geq C_0 \|u\|_{\mathcal{V}}^2, \quad u \in \mathcal{V}, \tag{B.14}$$

respectively, then,

$$\widetilde{A}\colon \mathcal{V} \to \mathcal{V}^* \text{ is bounded, self-adjoint, and boundedly invertible.} \tag{B.15}$$

Moreover, denoting by A the part of $\widetilde{A}$ in $\mathcal{H}$ defined by

$$\operatorname{dom}(A) = \big\{u \in \mathcal{V} \,\big|\, \widetilde{A}u \in \mathcal{H}\big\} \subseteq \mathcal{H}, \quad A = \widetilde{A}\big|_{\operatorname{dom}(A)}\colon \operatorname{dom}(A) \to \mathcal{H}, \tag{B.16}$$

then A is a (possibly unbounded) self-adjoint operator in $\mathcal{H}$ satisfying

$$A \geq C_0 I_{\mathcal{H}}, \tag{B.17}$$

$$\operatorname{dom}\big(A^{1/2}\big) = \mathcal{V}. \tag{B.18}$$

In particular,

$$A^{-1} \in \mathcal{B}(\mathcal{H}). \tag{B.19}$$

The facts (B.1)–(B.19) are a consequence of the Lax–Milgram theorem and the second representation theorem for symmetric sesquilinear forms. Details can be found, for instance, in [**31**, §VI.3, §VII.1], [**37**, Ch. IV], and [**63**].

Next, consider a symmetric form $b(\,\cdot\,,\,\cdot\,)\colon \mathcal{V} \times \mathcal{V} \to \mathbb{C}$ and assume that b is *bounded from below by* $c_b \in \mathbb{R}$, that is,

$$b(u,u) \geq c_b \|u\|_{\mathcal{H}}^2, \quad u \in \mathcal{V}. \tag{B.20}$$

Introducing the scalar product $(\,\cdot\,,\,\cdot\,)_{\mathcal{V}(b)}\colon \mathcal{V}\times\mathcal{V} \to \mathbb{C}$ (with associated norm denoted by $\|\,\cdot\,\|_{\mathcal{V}(b)}$),

$$(u,v)_{\mathcal{V}(b)} = b(u,v) + (1-c_b)(u,v)_{\mathcal{H}}, \quad u,v \in \mathcal{V}, \tag{B.21}$$

turns $\mathcal{V}$ into a pre-Hilbert space $(\mathcal{V}; (\,\cdot\,,\,\cdot\,)_{\mathcal{V}(b)})$, which we denote by $\mathcal{V}(b)$. The form b is called *closed* if $\mathcal{V}(b)$ is actually complete, and hence a Hilbert space. The form b is called *closable* if it has a closed extension. If b is closed, then

$$|b(u,v) + (1-c_b)(u,v)_{\mathcal{H}}| \leqslant \|u\|_{\mathcal{V}(b)} \|v\|_{\mathcal{V}(b)}, \quad u,v \in \mathcal{V}, \tag{B.22}$$

and

$$|b(u,u) + (1-c_b)\|u\|_{\mathcal{H}}^2| = \|u\|_{\mathcal{V}(b)}^2, \quad u \in \mathcal{V}, \tag{B.23}$$

show that the form $b(\,\cdot\,,\,\cdot\,) + (1-c_b)(\,\cdot\,,\,\cdot\,)_{\mathcal{H}}$ is a symmetric, $\mathcal{V}$-bounded, and $\mathcal{V}$-coercive sesquilinear form. Hence, by (B.11) and (B.12), there exists a linear map

$$\widetilde{B}_{c_b}\colon \begin{cases} \mathcal{V}(b) \to \mathcal{V}(b)^*, \\ \quad v \mapsto \widetilde{B}_{c_b} v = b(\,\cdot\,, v) + (1-c_b)(\,\cdot\,, v)_{\mathcal{H}}, \end{cases} \tag{B.24}$$

with

$$\widetilde{B}_{c_b} \in \mathcal{B}(\mathcal{V}(b), \mathcal{V}(b)^*) \text{ and } {}_{\mathcal{V}(b)}\langle u, \widetilde{B}_{c_b} v\rangle_{\mathcal{V}(b)^*} = b(u,v) + (1-c_b)(u,v)_{\mathcal{H}}, \quad u,v \in \mathcal{V}. \tag{B.25}$$

Introducing the linear map

$$\widetilde{B} = \widetilde{B}_{c_b} + (c_b - 1)\widetilde{I} \colon \mathcal{V}(b) \to \mathcal{V}(b)^*, \tag{B.26}$$

where $\widetilde{I} \colon \mathcal{V}(b) \hookrightarrow \mathcal{V}(b)^*$ denotes the continuous inclusion (embedding) map of $\mathcal{V}(b)$ into $\mathcal{V}(b)^*$, one obtains a self-adjoint operator B in $\mathcal{H}$ by restricting $\widetilde{B}$ to $\mathcal{H}$,

$$\operatorname{dom}(B) = \big\{u \in \mathcal{V} \,\big|\, \widetilde{B}u \in \mathcal{H}\big\} \subseteq \mathcal{H}, \quad B = \widetilde{B}\big|_{\operatorname{dom}(B)} \colon \operatorname{dom}(B) \to \mathcal{H}, \tag{B.27}$$

satisfying the following properties:

$$B \geq c_b I_{\mathcal{H}}, \tag{B.28}$$

$$\operatorname{dom}\big(|B|^{1/2}\big) = \operatorname{dom}\big((B - c_b I_{\mathcal{H}})^{1/2}\big) = \mathcal{V}, \tag{B.29}$$

$$b(u,v) = \big(|B|^{1/2}u, U_B|B|^{1/2}v\big)_{\mathcal{H}} \tag{B.30}$$

$$= \big((B - c_b I_{\mathcal{H}})^{1/2}u, (B - c_b I_{\mathcal{H}})^{1/2}v\big)_{\mathcal{H}} + c_b(u,v)_{\mathcal{H}} \tag{B.31}$$

$$= {}_{\mathcal{V}(b)}\langle u, \widetilde{B}v\rangle_{\mathcal{V}(b)^*}, \quad u,v \in \mathcal{V}, \tag{B.32}$$

$$b(u,v) = (u, Bv)_{\mathcal{H}}, \quad u \in \mathcal{V},\ v \in \operatorname{dom}(B), \tag{B.33}$$

$$\operatorname{dom}(B) = \{v \in \mathcal{V} \,|\, \text{there exists an } f_v \in \mathcal{H} \text{ such that } b(w,v) = (w, f_v)_{\mathcal{H}} \text{ for all } w \in \mathcal{V}\}, \tag{B.34}$$

$$Bu = f_u, \quad u \in \operatorname{dom}(B),$$

$$\operatorname{dom}(B) \text{ is dense in } \mathcal{H} \text{ and in } \mathcal{V}(b). \tag{B.35}$$

Properties (B.34) and (B.35) uniquely determine B. Here U_B in (B.31) is the partial isometry in the polar decomposition of B, that is,

$$B = U_B|B|, \quad |B| = (B^*B)^{1/2}. \tag{B.36}$$

The operator B is called the *operator associated with the form b*.

The norm in the Hilbert space $\mathcal{V}(b)^*$ is given by

$$\|\ell\|_{\mathcal{V}(b)^*} = \sup\{|{}_{\mathcal{V}(b)}\langle u, \ell\rangle_{\mathcal{V}(b)^*}| \,|\, \|u\|_{\mathcal{V}(b)} \leqslant 1\}, \quad \ell \in \mathcal{V}(b)^*, \tag{B.37}$$

with associated scalar product,

$$(\ell_1, \ell_2)_{\mathcal{V}(b)^*} = {}_{\mathcal{V}(b)}\big\langle \big(\widetilde{B} + (1-c_b)\widetilde{I}\big)^{-1}\ell_1, \ell_2\big\rangle_{\mathcal{V}(b)^*}, \quad \ell_1, \ell_2 \in \mathcal{V}(b)^*. \tag{B.38}$$

Since

$$\big\|\big(\widetilde{B} + (1-c_b)\widetilde{I}\big)v\big\|_{\mathcal{V}(b)^*} = \|v\|_{\mathcal{V}(b)}, \quad v \in \mathcal{V}, \tag{B.39}$$

the Riesz representation theorem yields

$$\big(\widetilde{B} + (1-c_b)\widetilde{I}\big) \in \mathcal{B}(\mathcal{V}(b), \mathcal{V}(b)^*) \text{ and } \big(\widetilde{B} + (1-c_b)\widetilde{I}\big) \colon \mathcal{V}(b) \to \mathcal{V}(b)^* \text{ is unitary.} \tag{B.40}$$

In addition,

$$\begin{aligned} {}_{\mathcal{V}(b)}\big\langle u, \big(\widetilde{B} + (1-c_b)\widetilde{I}\big)v\big\rangle_{\mathcal{V}(b)^*} &= \Big(\big(B + (1-c_b)I_{\mathcal{H}}\big)^{1/2}u, \big(B + (1-c_b)I_{\mathcal{H}}\big)^{1/2}v\Big)_{\mathcal{H}} \\ &= (u,v)_{\mathcal{V}(b)}, \quad u,v \in \mathcal{V}(b). \end{aligned} \tag{B.41}$$

In particular,

$$\big\|(B+(1-c_b)I_{\mathcal{H}})^{1/2}u\big\|_{\mathcal{H}} = \|u\|_{\mathcal{V}(b)}, \quad u \in \mathcal{V}(b), \tag{B.42}$$

and hence

$$(B+(1-c_b)I_{\mathcal{H}})^{1/2} \in \mathcal{B}(\mathcal{V}(b),\mathcal{H}) \text{ and } (B+(1-c_b)I_{\mathcal{H}})^{1/2}\colon \mathcal{V}(b) \to \mathcal{H} \text{ is unitary.} \tag{B.43}$$

The facts (B.20)–(B.43) comprise the second representation theorem of sesquilinear forms (cf. [**37**, Sect. IV.2], [**40**, Sects. 1.2–1.5], and [**53**, Sect. VI.2.6]).

A special but important case of nonnegative closed forms is obtained as follows: Let $\mathcal{H}_j$, $j=1,2$, be complex separable Hilbert spaces, and $T\colon \operatorname{dom}(T) \to \mathcal{H}_2$, $\operatorname{dom}(T) \subseteq \mathcal{H}_1$, a densely defined operator. Consider the nonnegative form $a_T\colon \operatorname{dom}(T) \times \operatorname{dom}(T) \to \mathbb{C}$ defined by

$$a_T(u,v) = (Tu,Tv)_{\mathcal{H}_2}, \quad u,v \in \operatorname{dom}(T). \tag{B.44}$$

Then the form a_T is closed (resp., closable) if and only if T is. If T is closed, the unique nonnegative self-adjoint operator associated with a_T in $\mathcal{H}_1$, whose existence is guaranteed by the second representation theorem for forms, then equals T^*T. In particular, one obtains

$$a_T(u,v) = (|T|u,|T|v)_{\mathcal{H}_1}, \quad u,v \in \operatorname{dom}(T) = \operatorname{dom}(|T|). \tag{B.45}$$

In addition, since

$$\begin{aligned} b(u,v) + (1-c_b)(u,v)_{\mathcal{H}} &= \big((B+(1-c_b)I_{\mathcal{H}})^{1/2}u, (B+(1-c_b)I_{\mathcal{H}})^{1/2}v\big)_{\mathcal{H}}, \\ &\qquad u,v \in \operatorname{dom}(b) = \operatorname{dom}\big(|B|^{1/2}\big) = \mathcal{V}, \end{aligned} \tag{B.46}$$

and $(B+(1-c_b)I_{\mathcal{H}})^{1/2}$ is self-adjoint (and hence closed) in $\mathcal{H}$, a symmetric, $\mathcal{V}$-bounded, and $\mathcal{V}$-coercive form is densely defined in $\mathcal{H} \times \mathcal{H}$ and closed (a fact we used in the proof of Theorem 2.6). We refer to [**53**, Sect. VI.2.4] and [**104**, Sect. 5.5] for details.

Next we recall that if a_j are sesquilinear forms defined on $\operatorname{dom}(a_j) \times \operatorname{dom}(a_j)$, $j=1,2$, bounded from below and closed, then also

$$(a_1+a_2)\colon \begin{cases} (\operatorname{dom}(a_1) \cap \operatorname{dom}(a_2)) \times (\operatorname{dom}(a_1) \cap \operatorname{dom}(a_2)) \to \mathbb{C}, \\ (u,v) \mapsto (a_1+a_2)(u,v) = a_1(u,v) + a_2(u,v) \end{cases} \tag{B.47}$$

is bounded from below and closed (cf., e.g., [**53**, Sect. VI.1.6]).

Finally, we also recall the following perturbation theoretic fact: Suppose a is a sesquilinear form defined on $\mathcal{V} \times \mathcal{V}$, bounded from below and closed, and let b be a symmetric sesquilinear form bounded with respect to a with bound less than one, that is, $\operatorname{dom}(b) \supseteq \mathcal{V} \times \mathcal{V}$, and that there exist $0 \leqslant \alpha < 1$ and $\beta \geqslant 0$ such that

$$|b(u,u)| \leqslant \alpha |a(u,u)| + \beta \|u\|_{\mathcal{H}}^2, \quad u \in \mathcal{V}. \tag{B.48}$$

Then

$$(a+b)\colon \begin{cases} \mathcal{V} \times \mathcal{V} \to \mathbb{C}, \\ (u,v) \mapsto (a+b)(u,v) = a(u,v) + b(u,v) \end{cases} \tag{B.49}$$

defines a sesquilinear form that is bounded from below and closed (cf., e.g., [**53**, Sect. VI.1.6]). In the special case where α can be chosen arbitrarily small, the form b is called infinitesimally form bounded with respect to a.

Appendix C. Estimates for the Fundamental Solution of the Helmholtz Equation

The principal aim of this appendix is to recall and prove some estimates for the fundamental solution (i.e., the Green's function) of the Helmholtz equation and its x-derivatives up to the second order.

Let $E_n(z;x)$ be the fundamental solution of the Helmholtz equation $(-\Delta - z)\psi(z;\,\cdot\,) = 0$ in $\mathbb{R}^n$, $n \in \mathbb{N}$, $n \geqslant 2$, already introduced in (2.120), and reproduced for convenience below:

$$E_n(z;x) = \begin{cases} \frac{i}{4}\left(\frac{2\pi|x|}{z^{1/2}}\right)^{(2-n)/2} H^{(1)}_{(n-2)/2}(z^{1/2}|x|), & n \geqslant 2,\ z \in \mathbb{C}\backslash\{0\}, \\ \frac{-1}{2\pi}\ln(|x|), & n = 2,\ z = 0, \\ \frac{1}{(n-2)\omega_{n-1}}|x|^{2-n}, & n \geqslant 3,\ z = 0, \end{cases} \tag{C.1}$$
$$\operatorname{Im}(z^{1/2}) \geq 0,\ x \in \mathbb{R}^n\backslash\{0\},$$

where $H^{(1)}_\nu(\,\cdot\,)$ denotes the Hankel function of the first kind with index $\nu \geq 0$ (cf. [**1**, Sect. 9.1]) and $\omega_{n-1} = 2\pi^{n/2}/\Gamma(n/2)$ ($\Gamma(\,\cdot\,)$ the Gamma function, cf. [**1**, Sect. 6.1]) represents the area of the unit sphere S^{n-1} in $\mathbb{R}^n$.

As $z \to 0$, $E_n(z,x)$, $x \in \mathbb{R}^n\backslash\{0\}$ is continuous for $n \geq 3$,

$$E_n(z,x) \underset{z\to 0}{=} E_n(0,x) = \frac{1}{(n-2)\omega_{n-1}}|x|^{2-n}, \quad x \in \mathbb{R}^n\backslash\{0\},\ n \geqslant 3, \tag{C.2}$$

but discontinuous for $n = 2$ as

$$E_2(z,x) \underset{z\to 0}{=} \frac{-1}{2\pi}\ln(z^{1/2}|x|/2)\left[1 + O(z|x|^2)\right] + \frac{1}{2\pi}\psi(1) + O(z|x|^2), \tag{C.3}$$
$$x \in \mathbb{R}^2\backslash\{0\},\ n = 2.$$

Here $\psi(w) = \Gamma'(w)/\Gamma(w)$ denotes the digamma function (cf. [**1**, Sect. 6.3]). Thus, we simply *define* $E_2(0;x) = \frac{-1}{2\pi}\ln(|x|)$, $x \in \mathbb{R}^2\backslash\{0\}$ as in (C.1).

To estimate E_n we recall that (cf. [**1**, Sect. 9.1])

$$H^{(1)}_{(n-2)/2}(\,\cdot\,) = J_{(n-2)/2}(\,\cdot\,) + iY_{(n-2)/2}(\,\cdot\,), \tag{C.4}$$

with J_ν and Y_ν the regular and irregular Bessel functions, respectively.

We start considering small values of $|x|$ and for this purpose recall the following absolutely convergent expansions (cf. [**1**, Sect. 9.1]):

$$J_\nu(\zeta) = \left(\frac{\zeta}{2}\right)^\nu \sum_{k=0}^\infty \frac{(-1)^k\zeta^{2k}}{4^k k!\Gamma(\nu+k+1)}, \quad \zeta \in \mathbb{C}\backslash(-\infty,0],\ \nu \in \mathbb{R}\backslash(-\mathbb{N}), \tag{C.5}$$

$$J_{-m}(\zeta) = (-1)^m J_m(\zeta), \quad \zeta \in \mathbb{C},\ m \in \mathbb{N}_0, \tag{C.6}$$

$$Y_\nu(\zeta) = \frac{J_\nu(\zeta)\cos(\nu\pi) - J_{-\nu}(\zeta)}{\sin(\nu\pi)}, \quad \zeta \in \mathbb{C}\backslash(-\infty,0],\ \nu \in (0,\infty)\backslash\mathbb{N}, \tag{C.7}$$

$$\begin{aligned} Y_m(\zeta) = &-\frac{\zeta^{-m}}{2^m\pi}\sum_{k=0}^{m-1}\frac{(m-k-1)!}{k!}\frac{\zeta^{2k}}{4^k} + \frac{2}{\pi}J_m(\zeta)\ln(\zeta/2) \\ &-\frac{\zeta^m}{2^m\pi}\sum_{k=0}^\infty[\psi(k+1)+\psi(m+k=1)]\frac{(-1)^k\zeta^{2k}}{4^k k!(m+k)!}, \end{aligned} \tag{C.8}$$
$$\zeta \in \mathbb{C}\backslash(-\infty,0],\ m \in \mathbb{N}_0.$$

We note that all functions in (C.5), (C.7), and (C.8) are analytic in $\mathbb{C}\backslash(-\infty,0]$ and that $J_m(\cdot)$ is entire for $m \in \mathbb{Z}$. In addition, all functions in (C.5)–(C.8) have continuous nontangential limits as $\zeta \to \eta < 0$, with generally different values on either side of the cut $(-\infty,0]$ due to the presence of the functions ζ^ν and $\ln(\zeta)$. (We chose $\nu \in \mathbb{R}$ and subsequently usually $\nu \geqslant 0$ for simplicity only; complex values of ν are discussed in [**1**, Ch. 9].)

Due to the presence of the logarithmic term for even dimensions we next distinguish even and odd space dimensions n:

(*i*) $n = 2m+2$, $m \in \mathbb{N}_0$, and $z \in \mathbb{C}\backslash\{0\}$ fixed:

$$\begin{aligned}
E_{2m+2}(z;x) &= \frac{i}{4}\left(\frac{2\pi|x|}{z^{1/2}}\right)^{-m} H_m^{(1)}\big(z^{1/2}|x|\big) \\
&= \frac{i}{4}\left(\frac{2\pi|x|}{z^{1/2}}\right)^{-m} \big[J_m\big(z^{1/2}|x|\big) + iY_m\big(z^{1/2}|x|\big)\big] \\
&= \frac{i}{4}\left(\frac{2\pi|x|}{z^{1/2}}\right)^{-m} \bigg\{O\big(|x|^m\big) + \frac{2i}{\pi}\ln\bigg(\frac{z^{1/2}|x|}{2}\bigg)O\big(|x|^m\big) \\
&\quad - \frac{i}{\pi}\bigg(\frac{z^{1/2}|x|}{2}\bigg)^{-m}(1-\delta_{m,0})\bigg[(m-1)! + (1-\delta_{m,1})(m-2)!\bigg(\frac{z|x|^2}{4}\bigg) \\
&\quad + O\big(|x|^4\big)\bigg]\bigg\}.
\end{aligned} \tag{C.9}$$

(*ii*) $n = 2m+1$, $m \in \mathbb{N}$, and $z \in \mathbb{C}\backslash\{0\}$ fixed:

$$\begin{aligned}
E_{2m+1}(z;x) &= \frac{i}{4}\left(\frac{2\pi|x|}{z^{1/2}}\right)^{(1/2)-m} H_{m-(1/2)}^{(1)}\big(z^{1/2}|x|\big) \\
&= \frac{i}{4} z^{m/2}\pi^{-m}2^{1-m}|x|^{1-m}h_{m-1}^{(1)}\big(z^{1/2}|x|\big) \\
&= \frac{i}{2z^{1/2}}(2\pi i|x|)^{-m}e^{iz^{1/2}|x|}\sum_{k=0}^{m-1}\frac{(m+k-1)!}{k!(m-k-1)!}\big(-2iz^{1/2}|x|\big)^{-k} \\
&\underset{|x|\to 0}{=} \begin{cases} (4\pi|x|)^{-1}\big[1 + iz^{1/2}|x| + O\big(|x|^2\big)\big], & m=1, \\ [(2m-1)\omega_{2m}]^{-1}|x|^{1-2m}\big[1+O\big(|x|^2\big)\big], & m \geq 2, \end{cases}
\end{aligned} \tag{C.10}$$

with $h_\ell^{(1)}(\cdot)$ defined in [**1**, Sect. 10.1].

Given these expansions we can now summarize the behavior of $E_n(z;x)$ and its derivatives up to the second order as $|x| \to 0$:

Lemma C.1. *Fix $z \in \mathbb{C}\backslash\{0\}$. Then the fundamental solution $E_n(z;\cdot)$ of the Helmholtz equation $(-\Delta - z)\psi(z;\cdot) = 0$ and its derivatives up to the second order satisfy the following estimates for $0 < |x| < R$, with $R > 0$ fixed:*

$$|E_n(z;x) - E_n(0;x)| \leq \begin{cases} C, & n = 2,3, \\ C[|\ln(|x|)| + 1], & n = 4, \\ C\big[|x|^{4-n} + 1\big], & n \geqslant 5, \end{cases} \tag{C.11}$$

$$|\partial_j E_n(z;x) - \partial_j E_n(0;x)| \leq \begin{cases} C, & n = 2,3, \\ C\big[|x|^{3-n} + 1\big], & n \geqslant 4, \end{cases} \tag{C.12}$$

$$|\partial_j\partial_k E_n(z;x) - \partial_j\partial_k E_n(0;x)| \leq \begin{cases} C[|\ln(|x|)|+1], & n=2, \\ C\big[|x|^{2-n}+1\big], & n \geqslant 3. \end{cases} \tag{C.13}$$

Here $C = C(R,n,z)$ *represent various different constants in* (C.11)–(C.13) *and* $\partial_j = \partial/\partial x_j$, $1 \leqslant j \leqslant n$.

PROOF. The estimates in (C.11) follow from combining (C.1), (C.9), and (C.10). The estimates in (C.12) follow from the fact that

$$\partial_j E_n(z;x) = -2\pi x_j E_{n+2}(z;x), \quad z \in \mathbb{C}\backslash\{0\},\ x \in \mathbb{R}^n\backslash\{0\},\ 1 \leqslant j \leqslant n,\ n \geqslant 2, \tag{C.14}$$

which permits one to reduce them essentially to (C.11) with n replaced by $n+2$. The recursion relation (C.14) is a consequence of the well-known identity (cf. [**1**, Sect. 9.1])

$$\frac{d}{d\zeta}\big(\zeta^{-\nu}\mathcal{C}_\nu(\zeta)\big) = -\zeta^{-\nu}\mathcal{C}_{\nu+1}(\zeta), \quad \zeta \in \mathbb{C}\backslash\{0\},\ \nu \in \mathbb{R}, \tag{C.15}$$

where $\mathcal{C}_\nu(\,\cdot\,)$ denotes any linear combination of Bessel functions of order ν with ζ and ν independent coefficients. Iterating (C.14) yields

$$\begin{aligned} \partial_j\partial_k E_n(z;x) = 4\pi^2 x_j x_k E_{n+4}(z;x) - 2\pi\delta_{j,k}E_{n+2}(z;x),& \\ z \in \mathbb{C}\backslash\{0\},\ x \in \mathbb{R}^n\backslash\{0\},\ 1 \leqslant j,k \leqslant n,\ n \geqslant 2.& \end{aligned} \tag{C.16}$$

Combining (C.11) and (C.16) then yields (C.13). □

Finally, we mention for completeness that for large values of $|x|$, (C.1) implies the following simple asymptotic behavior (cf. [**1**, Sect. 9.1]):

$$\begin{aligned} E_n(z;x) \underset{|x|\to\infty}{=} \frac{i}{z^{1/2}}\left(\frac{2\pi|x|}{z^{1/2}}\right)^{(1-n)/2} e^{i\left(z^{1/2}|x|-\pi((n-1)/4)\right)}\big[1+O\big(|x|^{-1}\big)\big],& \\ z \in \mathbb{C}\backslash\{0\},\ \operatorname{Im}\big(z^{1/2}\big) \geqslant 0.& \end{aligned} \tag{C.17}$$

In particular, as long as $z \in \mathbb{C}\backslash[0,\infty)$ (and hence $\operatorname{Im}\big(z^{1/2}\big) > 0$), $E_n(z;x)$ decays exponentially with respect to x as $|x| \to \infty$.

Appendix D. Calderón–Zygmund Theory on Lipschitz Surfaces

This appendix records various useful consequences of the Calderón–Zygmund theory on Lipschitz surfaces.

Our first result, Lemma D.1 below, is modeled upon a more general result in [**50**]. For the sake of completeness we include the full argument.

LEMMA D.1. *Let* $\Omega \subset \mathbb{R}^n$ *be a Lipschitz domain with compact boundary and let* $k(\,\cdot\,,\,\cdot\,)$ *be a real-valued, measurable function on* $\partial\Omega \times \partial\Omega$ *satisfying*

$$|k(x,y)| \leq \frac{\psi(|x-y|)}{|x-y|^{n-1}}, \quad x,y \in \partial\Omega, \tag{D.1}$$

where ψ *is monotone increasing and satisfies*

$$\int_0^1 dt\,\frac{\psi(t)}{t} < \infty. \tag{D.2}$$

Consider

$$(Kf)(x) = \int_{\partial\Omega} d^{n-1}\omega(y)\,k(x,y)f(y), \quad x \in \partial\Omega. \tag{D.3}$$

Then

$$K \in \mathcal{B}_\infty\big(L^p(\partial\Omega; d^{n-1}\omega)\big) \tag{D.4}$$

for each $p \in (1, \infty)$.

PROOF. For a fixed, arbitrary $\varepsilon > 0$, decompose $k(x,y) = k_\varepsilon(x,y) + k_b(x,y)$, where

$$k_\varepsilon(x,y) = \begin{cases} k(x,y), & |x-y| \leq \varepsilon, \\ 0, & |x-y| > \varepsilon. \end{cases} \tag{D.5}$$

Then $K = K_\varepsilon + K_b$, where K_ε, K_b are integral operators on $\partial\Omega$ with integral kernels $k_\varepsilon(x,y)$ and $k_b(x,y)$, respectively. Setting

$$S_j(x) = \big\{y \in \partial\Omega \,|\, 2^{-j-1} \leq |x-y| < 2^{-j}\big\}, \quad x \in \partial\Omega, \ j \in \mathbb{N}, \tag{D.6}$$

for each $x \in \partial\Omega$ we may then compute (with the logarithm taken in base 2)

$$\begin{aligned} \int_{\partial\Omega} d^{n-1}\omega(y)\, |k_\varepsilon(x,y)| &\leq C \sum_{j \geq \log 1/\varepsilon} \int_{S_j(x)} d^{n-1}\omega(y)\, \frac{\psi(|x-y|)}{|x-y|^{n-1}} \\ &\leq C \sum_{j \geq \log 1/\varepsilon} \psi(e^{-j}) \leq C \int_0^\varepsilon dt\, \frac{\psi(t)}{t}. \end{aligned} \tag{D.7}$$

Of course, there is a similar estimate for $\int_{\partial\Omega} d^{n-1}\omega(x)\, |k_\varepsilon(x,y)|$, uniformly for $y \in \partial\Omega$. Schur's lemma then yields

$$\|K_\varepsilon\|_{\mathcal{B}(L^p(\partial\Omega; d^{n-1}\omega))} \leq C \int_0^\varepsilon dt\, \frac{\psi(t)}{t} \to 0 \text{ as } \varepsilon \to 0. \tag{D.8}$$

Thus, it suffices to show that K_b is compact on each $L^p(\partial\Omega; d^{n-1}\omega)$ space, for $p \in (1, \infty)$, under the hypothesis that $k_b(x,y)$ is bounded. First note that K_b is compact on $L^2(\partial\Omega, d^{n-1}\omega)$, since it is Hilbert-Schmidt, due to the fact that $\omega(\partial\Omega) < \infty$. The compactness of K_b on $L^p(\partial\Omega, d^{n-1}\omega)$ for each $p \in (1, \infty)$ then follows from an interpolation theorem of Krasnoselski (see, e.g., [**16**, Theorem 2.9, p. 203]). □

We now record a basic result from the theory of singular integral operators of Calderón–Zygmund-type on Lipschitz domains. To state it, we recall that $\hat{f}$ denotes the Fourier transform of appropriate functions $f\colon \mathbb{R}^n \to \mathbb{C}$. Moreover, given a Lipschitz domain $\Omega \subset \mathbb{R}^n$, set $\Omega_+ = \Omega$, $\Omega_- = \mathbb{R}^n \backslash \overline{\Omega}$, we define the nontangential approach regions $\Gamma^\pm_\kappa(x)$, $x \in \partial\Omega$, by $\Gamma^\pm_\kappa(x) = \{y \in \Omega_\pm \,|\, |x-y| < (1+\kappa)\operatorname{dist}(y, \partial\Omega)\}$, where $\kappa > 0$ is a fixed parameter. Next, at every boundary point the nontangential maximal function of a mapping u (defined in either Ω_+ or Ω_-) is given by

$$(Mu)(x) = \sup\{|u(y)| \,|\, y \in \Gamma^\pm_\kappa(x)\} \tag{D.9}$$

(with the choice of sign depending on whether u is defined in Ω_+, or Ω_-) and, for u defined in $\Omega_\pm$, we set

$$(\gamma_{\text{n.t.}} u)(x) = \lim_{\substack{y \to x \\ y \in \Gamma^\pm_\kappa(x)}} u(y) \ \text{ for a.e. } x \in \partial\Omega. \tag{D.10}$$

For future reference, let us record here a useful estimate proved in [**36**], valid for any Lipschitz domain $\Omega \subset \mathbb{R}^n$ which is either bounded or has an unbounded boundary. In this setting, for any $p \in (0, \infty)$ and any function u defined in Ω,

$$\|u\|_{L^{np/(n-1)}(\Omega; d^n x)} \leq C(\Omega, n, p) \|Mu\|_{L^p(\partial\Omega; d^{n-1}\omega)}. \tag{D.11}$$

THEOREM D.2. *There exists a positive integer $N = N(n)$ with the following significance. Let $\Omega \subset \mathbb{R}^n$ be a Lipschitz domain with compact boundary, and assume that*

$$\begin{aligned} &k \in C^N(\mathbb{R}^n \backslash \{0\}) \ \text{ with } \ k(-x) = -k(x) \\ &\text{and } \ k(\lambda x) = \lambda^{-(n-1)} k(x), \ \lambda > 0, \ x \in \mathbb{R}^n \backslash \{0\}. \end{aligned} \tag{D.12}$$

Define the singular integral operator

$$(\mathcal{T} f)(x) = \int_{\partial\Omega} d^{n-1}\omega(y)\, k(x-y) f(y), \quad x \in \mathbb{R}^n \backslash \partial\Omega. \tag{D.13}$$

Then for each $p \in (1, \infty)$ there exists a finite constant $C = C(p, n, \partial\Omega) > 0$ such that

$$\|M(\mathcal{T} f)\|_{L^p(\partial\Omega; d^{n-1}\omega)} \leq C \|k|_{S^{n-1}}\|_{C^N} \|f\|_{L^p(\partial\Omega; d^{n-1}\omega)}. \tag{D.14}$$

Furthermore, for each $p \in (1, \infty)$, $f \in L^p(\partial\Omega; d^{n-1}\omega)$, the limit

$$(Tf)(x) = \text{p.v.} \int_{\partial\Omega} d^{n-1}\omega(y)\, k(x-y) f(y) = \lim_{\varepsilon \to 0^+} \int_{\substack{y \in \partial\Omega \\ |x-y| > \varepsilon}} d^{n-1}\omega(y)\, k(x-y) f(y) \tag{D.15}$$

exists for a.e. $x \in \partial\Omega$, and the jump-formula

$$\gamma_{\text{n.t.}}(\mathcal{T} f)(x) = \lim_{\substack{z \to x \\ z \in \Gamma_\kappa^{\pm}(x)}} (\mathcal{T} f)(z) = \pm \tfrac{1}{2i} \widehat{k}(\nu(x)) f(x) + (Tf)(x) \tag{D.16}$$

is valid at a.e. $x \in \partial\Omega$, where ν denotes the unit normal pointing outwardly relative to Ω (recall that 'hat' denotes the Fourier transform in $\mathbb{R}^n$).

Finally,

$$\|\mathcal{T} f\|_{H^{1/2}(\Omega)} \leq C \|f\|_{L^2(\partial\Omega; d^{n-1}\omega)}. \tag{D.17}$$

See the discussion in [**24**], [**25**], [**73**].

LEMMA D.3. *Whenever Ω is a Lipschitz domain with compact boundary in $\mathbb{R}^n$,*

$$K_z^{\#} \in \mathcal{B}\big(L^2(\partial\Omega; d^{n-1}\omega)\big), \quad z \in \mathbb{C}, \tag{D.18}$$

and

$$\big(K_{z_1}^{\#} - K_{z_2}^{\#}\big) \in \mathcal{B}_\infty\big(L^2(\partial\Omega; d^{n-1}\omega)\big), \quad z_1, z_2 \in \mathbb{C}, \tag{D.19}$$

$$\gamma_D \mathcal{S}_z \in \mathcal{B}\big(L^2(\partial\Omega; d^{n-1}\omega), H^1(\partial\Omega)\big), \quad z \in \mathbb{C}. \tag{D.20}$$

PROOF. We recall the fundamental solution $E_n(z; \cdot)$ for the Helmholtz equation $(-\Delta - z)\psi(z; \cdot) = 0$ in $\mathbb{R}^n$ introduced in (2.120). Then the integral kernel of the operator $K_z^{\#} - K_0^{\#}$ is given by

$$k(x,y) = \nu(x) \cdot \big(\nabla E_n(z; x-y) - \nabla E_n(0; x-y)\big), \quad x, y \in \partial\Omega. \tag{D.21}$$

By (C.12) we therefore have $|k(x,y)| \leq C|x-y|^{2-n}$, hence (D.1) holds with $\psi(t) = t$. Note that (D.2) is satisfied for this choice of ψ, so (D.19) is a consequence of Lemma D.1. In addition, (D.18) follows from (D.19) and Theorem D.2, according to which $K_0^{\#} \in \mathcal{B}\big(L^2(\partial\Omega; d^{n-1}\omega)\big)$. Finally, the reasoning for (D.20) is similar (here (A.15) is useful). $\square$

LEMMA D.4. *If Ω is a $C^{1,r}$, $r > 1/2$, domain in $\mathbb{R}^n$ with compact boundary, then*

$$\big(K_{z_1}^{\#} - K_{z_2}^{\#}\big) \in \mathcal{B}_\infty\big(H^{1/2}(\partial\Omega)\big), \quad z_1, z_2 \in \mathbb{C}. \tag{D.22}$$

PROOF. The integral kernel of the operator $K_z^\# - K_0^\#$ is given by (D.21). By Lemma A.5, the operator of multiplication by components of $\nu \in [C^r(\partial\Omega)]^n$ belongs to $\mathcal{B}(H^{1/2}(\partial\Omega))$. Hence, it suffices to show that the boundary integral operators whose integral kernels are of the form

$$\partial_j E_n(z; x-y) - \partial_j E_n(0; x-y)), \quad x, y \in \partial\Omega, \ j \in \{1, ..., n\}, \tag{D.23}$$

belong to $\mathcal{B}(L^2(\partial\Omega; d^{n-1}\omega), H^1(\partial\Omega))$. This, however, is a consequence of (A.18), (A.19) (with $s = 1$), (C.13), and Lemma D.1 (with $\psi(t) = t$). □

LEMMA D.5. *Let $0 < \alpha < (n-1)$ and $1 < p < q < \infty$ be related by*

$$\frac{1}{q} = \frac{1}{p} - \Big(\alpha + \frac{1}{p}\Big)\frac{1}{n}. \tag{D.24}$$

Then the the operator J_α defined by

$$J_\alpha f(x) = \int_{\mathbb{R}^{n-1}} d^{n-1}y \, \frac{1}{|x-y|^{n-1-\alpha}} f(y), \quad x \in \overline{\mathbb{R}^n_+}, \ f \in L^p(\mathbb{R}^{n-1}; d^{n-1}x), \tag{D.25}$$

is bounded from $L^p(\mathbb{R}^{n-1}; d^{n-1}x)$ to $L^q(\mathbb{R}^n_+; d^n x)$, that is, for some constant $C_{\alpha,p,q} > 0$,

$$\|J_\alpha f\|_{L^q(\mathbb{R}^n_+; d^n x)} \leq C_{\alpha,p,q} \|f\|_{L^p(\mathbb{R}^{n-1}; d^{n-1}x)}, \quad f \in L^p(\mathbb{R}^{n-1}; d^{n-1}x). \tag{D.26}$$

PROOF. A direct proof appears in [**75**]. An alternative argument is to observe that $M(J_\alpha f)(x) \leq C J_\alpha(|f|)(x)$, uniformly for $x \in \partial\mathbb{R}^n_+$, and then to invoke the general estimate (D.11) in concert with the classical Hardy-Littlewood-Sobolev fractional integration theorem (cf., e.g., [**92**], Theorem 1 on p. 119). □

Next, we record a lifting result for Sobolev spaces in Lipschitz domains in [**51**].

THEOREM D.6. *Let $\Omega \subset \mathbb{R}^n$ be a Lipschitz domain with compact boundary. Then, for every $\alpha > 0$, the following equivalence of norms holds:*

$$\|u\|_{H^{\alpha+1}(\Omega)} \approx \|u\|_{L^2(\Omega)} + \|\nabla u\|_{H^\alpha(\Omega)}. \tag{D.27}$$

THEOREM D.7. *Let $\Omega \subset \mathbb{R}^n$ be a bounded Lipschitz domain. Then for every $z \in \mathbb{C}$,*

$$\mathcal{S}_z \in \mathcal{B}(L^2(\partial\Omega; d^{n-1}\omega), H^{3/2}(\Omega)), \tag{D.28}$$

and

$$\mathcal{S}_z \in \mathcal{B}(H^{-1}(\partial\Omega), H^{1/2}(\Omega)). \tag{D.29}$$

In particular,

$$\mathcal{S}_z \in \mathcal{B}(H^{s-1}(\partial\Omega), H^{s+(1/2)}(\Omega)), \quad 0 \leqslant s \leqslant 1. \tag{D.30}$$

PROOF. Given $f \in L^2(\partial\Omega; d^{n-1}\omega)$, write $\mathcal{S}_z f = \mathcal{S}_0 f + (\mathcal{S}_z - \mathcal{S}_0)f$. From (D.17) and Lemma D.6 we know that $\|\mathcal{S}_0 f\|_{H^{3/2}(\Omega)} \leq C\|f\|_{L^2(\partial\Omega; d^{n-1}\omega)}$, for some constant $C > 0$ independent of f. Using (C.13) and Lemma D.5 (with $\alpha = 1$) one concludes that

$$\nabla^2(\mathcal{S}_z - \mathcal{S}_0) \in \mathcal{B}(L^2(\partial\Omega; d^{n-1}\omega), L^2(\Omega; d^n x)), \tag{D.31}$$

and (D.28) follows from this. The proof of (D.29), is analogous and has as starting point the fact that $\mathcal{S}_0 f \in \mathcal{B}(H^{-1}(\partial\Omega), H^{1/2}(\Omega))$, itself a consequence of (D.17) and the following description of $H^{-1}(\partial\Omega)$:

$$H^{-1}(\partial\Omega) = \Big\{ g + \sum_{1 \leq j,k \leq n} (\partial f_{j,k}/\partial\tau_{j,k}) \, \Big| \, g, f_{j,k} \in L^2(\partial\Omega; d^{n-1}\omega) \Big\}. \tag{D.32}$$

Then (D.31) ensures that

$$(\mathcal{S}_z - \mathcal{S}_0) \in \mathcal{B}\big(H^{-1}(\partial\Omega), H^1(\Omega)\big), \tag{D.33}$$

and (D.29) follows. □

We recall the adjoint double layer on $\partial\Omega$ introduced in (2.122) and denote by

$$(K_z g)(x) = \text{p.v.} \int_{\partial\Omega} d^{n-1}\omega(y)\, \partial_{\nu_y} E_n(z; y-x) g(y), \quad x \in \partial\Omega, \tag{D.34}$$

its adjoint. It is well-known (cf., e.g., [**101**]) that

$$\text{Hypothesis 2.1} \Longrightarrow K \in \mathcal{B}\big(L^2(\partial\Omega; d^{n-1}\omega)\big) \cap \mathcal{B}\big(H^1(\partial\Omega)\big) \tag{D.35}$$

and (cf. [**38**] and (D.19)) that

$$\Omega \text{ a bounded } C^1\text{-domain } \Longrightarrow K_z \in \mathcal{B}_\infty\big(L^2(\partial\Omega; d^{n-1}\omega)\big), \quad z \in \mathbb{C}. \tag{D.36}$$

It follows from (D.35), (D.36), (4.4), and Theorem 4.2 that

$$\Omega \text{ a bounded } C^1\text{-domain } \Longrightarrow K_z \in \mathcal{B}_\infty\big(H^s(\partial\Omega)\big), \quad s \in (0,1),\ z \in \mathbb{C}. \tag{D.37}$$

We wish to complement this with the following compactness result.

THEOREM D.8. *If $\Omega \subset \mathbb{R}^n$ is a bounded $C^{1,r}$-domain with $r \in (1/2, 1)$ then*

$$K_z^\# \in \mathcal{B}_\infty\big(H^{1/2}(\partial\Omega)\big), \quad z \in \mathbb{C}. \tag{D.38}$$

The proof of this result (presented at the end of this section) requires a number of tools from the theory of singular integral operators which we now review, or develop.

THEOREM D.9. *Let $A : \mathbb{R}^n \to \mathbb{R}^m$ be a Lipschitz function, and assume that $F : \mathbb{R}^m \to \mathbb{R}$, $F \in C^1(\mathbb{R}^m)$, F is an odd function. For $x, y \in \mathbb{R}^n$ with $x \neq y$ set $K(x,y) = \frac{1}{|x-y|^n} F\Big(\frac{A(x)-A(y)}{|x-y|}\Big)$, and for $\varepsilon > 0$, f a Lipschitz function with compact support in $\mathbb{R}^n$, define the truncated operator*

$$(T_\varepsilon f)(x) = \int_{\substack{y\in\mathbb{R}^n \\ |x-y|>\varepsilon}} d^n y\, K(x,y) f(y), \quad x \in \mathbb{R}^n. \tag{D.39}$$

Then, for each $1 < p < \infty$, the following assertions hold:

(i) *The maximal operator $(T_* f)(x) = \sup\{|(T_\varepsilon f)(x)| \,|\, \varepsilon > 0\}$ is bounded on $L^p(\mathbb{R}^n; d^n x)$.*

(ii) *If $1 < p < \infty$ and $f \in L^p(\mathbb{R}^n; d^n x)$ then the limit $\lim_{\varepsilon\to 0}(T_\varepsilon f)(x)$ exists for almost every $x \in \mathbb{R}^n$ and the operator*

$$(Tf)(x) = \lim_{\varepsilon \to 0} (T_\varepsilon f)(x) \tag{D.40}$$

is bounded on $L^p(\mathbb{R}^n; d^n x)$.

A proof of this result can be found in [**70**].

THEOREM D.10. *Let $A : \mathbb{R}^n \to \mathbb{R}^m$, $B = (B_1, ..., B_\ell) : \mathbb{R}^n \to \mathbb{R}^\ell$ be two Lipschitz functions and let $F : \mathbb{R}^m \times \mathbb{R}^\ell \to \mathbb{R}$ be a C^N (with $N = N(n,m,\ell)$ a sufficiently large integer) odd function which satisfies the decay conditions*

$$|F(a,b)| \le C(1+|b|)^{-n}, \tag{D.41}$$

$$|\nabla_I F(a,b)| \le C, \tag{D.42}$$

$$|\nabla_{II} F(a,b)| \leq C(1+|b|)^{-1}, \tag{D.43}$$

uniformly for a in compact subsets of $\mathbb{R}^n$ *and arbitrary* $b \in \mathbb{R}^\ell$. *(Above,* ∇_I *and* ∇_{II} *denote the gradients with respect to the first and second sets of variables.) For* $x, y \in \mathbb{R}^n$ *with* $x \neq y$ *and* $t > 0$ *we set*

$$K^t(x,y) = \frac{1}{|x-y|^n} F\left(\frac{A(x)-A(y)}{|x-y|}, \frac{B_1(x)-B_1(y)+t}{|x-y|}, \dots, \frac{B_\ell(x)-B_\ell(y)+t}{|x-y|}\right). \tag{D.44}$$

In addition, for each $t > 0$ *we introduce*

$$(T^t f)(x) = \int_{\mathbb{R}^n} d^n y \, K^t(x,y) f(y), \quad x \in \mathbb{R}^n, \tag{D.45}$$

and, for some fixed, positive κ,

$$(T_{**} f)(x) = \sup\{|(T^t f)(z)| \,|\, z \in \mathbb{R}^n,\ t > 0,\ |x-z| < \kappa t\}, \quad x \in \mathbb{R}^n. \tag{D.46}$$

Then, for each $1 < p < \infty$*, the following assertions are valid:*

(1) *The nontangential maximal operator* T_{**} *is bounded on* $L^p(\mathbb{R}^n; d^n x)$.
(2) *For each* $f \in L^p(\mathbb{R}^n; d^n x)$*, the limit*

$$(\mathcal{T} f)(x) = \lim_{\substack{|x-z|<\kappa t \\ z \to x,\, t \to 0}} (T^t f)(z) \tag{D.47}$$

exists at almost every $x \in \mathbb{R}^n$ *and the operator* $\mathcal{T}$ *is bounded on* $L^p(\mathbb{R}^n; d^n x)$.

PROOF. Fix $p \in (1,\infty)$. For $x, y \in \mathbb{R}^n$ with $x \neq y$ consider the kernel

$$K(x,y) = \frac{1}{|x-y|^n} F\left(\frac{A(x)-A(y)}{|x-y|}, \frac{B(x)-B(y)}{|x-y|}\right), \tag{D.48}$$

and let T, T_* be the operators canonically associated with this integral kernel as in Theorem D.9. The crux of the matter is establishing the a.e. pointwise estimate

$$T_{**} f \leq C T_* f + C \mathcal{M} f \ \text{ in } \mathbb{R}^n, \tag{D.49}$$

uniformly for $f \in L^p(\mathbb{R}^n; d^n x)$, where $\mathcal{M}$ is the Hardy-Littlewood maximal operator in $\mathbb{R}^n$. Then the first claim in the statement of the theorem follows from Theorem D.9 and the well-known fact that $\mathcal{M}$ is bounded on $L^p(\mathbb{R}^n; d^n x)$.

To this end, fix $x, z \in \mathbb{R}^n$, $t > 0$ such that $|x-z| < \kappa t$, and let $\alpha > 0$ be a large constant, to be specified later. Then

$$\left| \int_{\mathbb{R}^n} d^n y \, K^t(z,y) f(y) - \int_{|x-y|>\alpha t} d^n y \, K(x,y) f(y) \right| \tag{D.50}$$

$$\leq \int_{|x-y|<\alpha t} d^n y \, |K^t(z,y)||f(y)| + \int_{|x-y|>\alpha t} d^n y \, |K^t(z,y) - K(x,y)||f(y)|$$

$$= I + II. \tag{D.51}$$

Clearly, it suffices to show that $|I|$, $|II| \leq C\mathcal{M} f$. To see this, first observe that

$$|K^t(z,y)| \leq C t^{-n} \ \text{ uniformly for any } \ z, y \in \mathbb{R}^n,\ z \neq y \tag{D.52}$$

(in fact, this also justifies that T^t is well-defined). Indeed, using the fact that for each $j \in \{1, ..., \ell\}$ one has $Ct \leq |B_j(z) - B_j(y) + t| + |z-y|$ (easily seen by analyzing the cases $|z-y| \geq \frac{t}{2\|\nabla B_j\|_{L^\infty}}$ and $|z-y| \leq \frac{t}{2\|\nabla B_j\|_{L^\infty}}$), we may infer that

$$\left(1+\sum_{j=1}^{\ell}\frac{|B_j(z)-B_j(y)+t|}{|z-y|}\right)^{-n} \leq C\left(\frac{t}{|z-y|}\right)^{-n}. \tag{D.53}$$

With this at hand, the estimate (D.52) is a direct consequence of (D.41). Returning to I, from (D.52), we deduce that $|I| \leq C\mathcal{M}f(x)$.

Thus, we are left with analyzing II in (D.51). To begin with, we shall prove that

$$|K^t(z,y)-K(x,y)| \leq Ct|x-y|^{-n-1} \text{ for } |x-y| > \alpha t. \tag{D.54}$$

Let $G_y(x,t) = K^t(x,y)$. Then

$$|K^t(z,y)-K(x,y)| = |G_y(z,t)-G_y(x,0)| \tag{D.55}$$

can be estimated using the Mean Value Theorem by

$$Ct(|\nabla_I G_y(w,s)| + |\nabla_{II} G_y(w,s)|), \tag{D.56}$$

where $w = (1-\theta)z + \theta x$, $s = (1-\theta)t$ for some $0 < \theta < 1$. Next,

$$\begin{aligned}
&|\nabla_I G_y(w,s)| \\
&\leq \frac{C}{|w-y|^{n+1}}\left|F\left(\frac{A(w)-A(y)}{|w-y|}, \frac{B_1(w)-B_1(y)+s}{|w-y|}, \dots, \frac{B_\ell(w)-B_\ell(y)+s}{|w-y|}\right)\right| \\
&+ \frac{C}{|w-y|^{n+1}}\left|\nabla_I F\left(\frac{A(w)-A(y)}{|w-y|}, \frac{B_1(w)-B_1(y)+s}{|w-y|}, \dots, \frac{B_\ell(w)-B_\ell(y)+s}{|w-y|}\right)\right| \\
&+ \frac{C}{|w-y|^{n}}\left|\nabla_{II} F\left(\frac{A(w)-A(y)}{|w-y|}, \frac{B_1(w)-B_1(y)+s}{|w-y|}, \dots, \frac{B_\ell(w)-B_\ell(y)+s}{|w-y|}\right)\right| \\
&\times \left(\frac{C}{|w-y|} + \sum_{i=1}^{\ell}\frac{|B_i(w)-B_i(y)+s|}{|w-y|^2}\right).
\end{aligned} \tag{D.57}$$

Keeping in mind the restrictions on the size of the derivatives of the function F stated in (D.41)–(D.43), we conclude that the above expression is bounded by $C|w-y|^{-(n+1)}$.

Similarly, it can be shown that

$$|\nabla_{II} G_y(w,s)| \leq \frac{C}{|w-y|^{n+1}}. \tag{D.58}$$

To continue, one observes that if we choose $\alpha > \kappa$ then, in the current context,

$$|w-x| \leq |z-x| \leq \kappa t = \left(\frac{\kappa}{\alpha}\right)\alpha t < \left(\frac{\kappa}{\alpha}\right)|x-y|, \tag{D.59}$$

and $|w-x| + |w-y| \geq |x-y|$. Hence,

$$|w-y| \geq \left(1-\frac{\kappa}{\alpha}\right)|x-y|, \tag{D.60}$$

and, therefore,

$$|K^t(z,y)-K(x,y)| \leq Ct|x-y|^{-n-1}. \tag{D.61}$$

Next, we split the domain of integration of II (appearing in (D.51)) into dyadic annuli of the form $2^j\alpha t \leq |x-y| \leq 2^{j+1}\alpha t$, $j = 0,1,2,\dots$. Then

$$\int_{|x-y|>\alpha t} d^n y\, |K^t(z,y)-K(x,y)||f(y)|$$

$$
\begin{aligned}
&\le \sum_{j=0}^{\infty} \int_{2^j \alpha t<|x-y|<2^{j+1}\alpha t} d^n y \, \frac{t}{|x-y|^{n+1}} |f(y)| \\
&\le C \sum_{j=0}^{\infty} 2^{-j} (\mathcal{M} f)(x) = C(\mathcal{M} f)(x). \qquad \text{(D.62)}
\end{aligned}
$$

This yields the desired inequality, that is, $|II| \le C\mathcal{M} f(x)$.

The proof of last second claim in the statement of the theorem utilizes a well-known principle (cf., e.g., [**38**]) to the effect that pointwise convergence for a dense class along with the boundedness of the maximal operator associated with the type of convergence in question always entails a.e. convergence for the entire space $L^p(\mathbb{R}^n; d^n x)$. Thus, it suffices to identify a dense subspace $\mathcal{V}$ of $L^p(\mathbb{R}^n; d^n x)$ such that for any $f \in \mathcal{V}$ the limit in question exists for almost every $x \in \mathbb{R}^n$. Then the boundedness of the maximal operator associated with the type of convergence under discussion ensures that this limit exists for any $f \in L^p(\mathbb{R}^n; d^n x)$ at almost every $x \in \mathbb{R}^n$.

In our situation, we may take $\mathcal{V} = C_0^1(\mathbb{R}^n)$ and observe that

$$
\lim_{\substack{|x-z|<\kappa t \\ z \to x,\, t \to 0}} (T^t f)(z) = \lim_{\varepsilon \to 0} \lim_{\substack{|x-z|<\kappa t \\ z \to x,\, t \to 0}} \Big[I + II + III \Big], \qquad \text{(D.63)}
$$

where

$$
\begin{aligned}
I &= \int_{|x-y|>1} d^n y \, K^t(z,y) f(y), \\
II &= \int_{1>|x-y|>\varepsilon} d^n y \, K^t(z,y)[f(y) - f(x)], \qquad \text{(D.64)} \\
III &= f(x) \int_{1>|x-y|>\varepsilon} d^n y \, K^t(z,y).
\end{aligned}
$$

Consequently,

$$
\lim_{\varepsilon \to 0} \lim_{\substack{|x-z|<\kappa t \\ z \to x,\, t \to 0}} I = \int_{|x-y|>1} d^n y \, K(x,y) f(y), \qquad \text{(D.65)}
$$

$$
\lim_{\varepsilon \to 0} \lim_{\substack{|x-z|<\kappa t \\ z \to x,\, t \to 0}} II = \int_{1>|x-y|} d^n y \, K(x,y)[f(y) - f(x)], \qquad \text{(D.66)}
$$

whereas

$$
\lim_{\varepsilon \to 0} \lim_{\substack{|x-z|<\kappa t \\ z \to x,\, t \to 0}} III = \lim_{\varepsilon \to 0} \int_{1>|x-y|>\varepsilon} d^n y \, K(x,y). \qquad \text{(D.67)}
$$

Now, this last limit is known to exists at a.e. $x \in \mathbb{R}^n$ (see, e.g., [**76**]). Once the pointwise definition of the operator $\mathcal{T}$ has been shown to be meaningful, the boundedness of this operator on $L^p(\mathbb{R}^n; d^n x)$, $1 < p < \infty$, is implied by that of T_{**}. □

THEOREM D.11. *There exists a positive integer $N = N(n)$ with the following significance: Let $\Omega \subset \mathbb{R}^n$ be a Lipschitz domain with compact boundary, and assume that*

$$
\begin{aligned}
&k \in C^N(\mathbb{R}^n \backslash \{0\}) \ \text{ with } \ k(-x) = -k(x) \\
&\text{and } \ k(\lambda x) = \lambda^{-(n-1)} k(x), \ \lambda > 0, \ x \in \mathbb{R}^n \backslash \{0\}.
\end{aligned} \qquad \text{(D.68)}
$$

Fix $\eta \in C^N(\mathbb{R}^n)$ and define the singular integral operator

$$(Tf)(x) = \text{p.v.} \int_{\partial\Omega} d^{n-1}\omega(y)\, (\eta(x) - \eta(y))k(x-y)f(y), \quad x \in \partial\Omega. \tag{D.69}$$

Then

$$T \in \mathcal{B}\big(L^2(\partial\Omega; d^{n-1}\omega), H^1(\partial\Omega)\big). \tag{D.70}$$

PROOF. Fix an arbitrary $f \in L^2(\partial\Omega; d^{n-1}\omega)$ and consider

$$u(x) = \int_{\partial\Omega} d^{n-1}\omega(y)\, (\eta(x) - \eta(y))k(x-y)f(y), \quad x \in \Omega. \tag{D.71}$$

Since $Tf = u|_{\partial\Omega}$, it suffices to show that

$$\|M(\nabla u)\|_{L^2(\partial\Omega; d^{n-1}\omega)} \leq C\|f\|_{L^2(\partial\Omega; d^{n-1}\omega)}, \tag{D.72}$$

for some finite constant $C = C(\Omega) > 0$ (where the nontangential maximal operator M is as in (D.9)). With this goal in mind, for a given $j \in \{1, ..., n\}$, we decompose

$$(\partial_j u)(x) = u_1(x) + u_2(x), \quad x \in \Omega, \tag{D.73}$$

where

$$u_1(x) = (\partial_j \eta)(x) \int_{\partial\Omega} d^{n-1}\omega(y)\, k(x-y)f(y) \tag{D.74}$$

and

$$u_2(x) = \int_{\partial\Omega} d^{n-1}\omega(y)\, (\eta(x) - \eta(y))(\partial_j k)(x-y)f(y). \tag{D.75}$$

Theorem D.2 immediately gives that

$$\|Mu_1\|_{L^2(\partial\Omega; d^{n-1}\omega)} \leq C\|f\|_{L^2(\partial\Omega; d^{n-1}\omega)}, \tag{D.76}$$

so it remains to prove a similar estimate with u_2 in place of u_1. To this end, we note that the problem localizes, so we may assume that η is compactly supported and Ω is the domain above the graph of a Lipschitz function $\varphi : \mathbb{R}^{n-1} \to \mathbb{R}$. In this scenario, by passing to Euclidean coordinates and denoting $g(y') = f(y', \varphi(y'))$, $y' \in \mathbb{R}^{n-1}$, it suffices to show the following. For $x' \in \mathbb{R}^{n-1}$, $t > 0$, set

$$\begin{aligned} v(x', t) = \int_{\mathbb{R}^{n-1}} & d^{n-1}y'\, (\eta(x', \varphi(x') + t) - \eta(y', \varphi(y'))) \\ & \times (\partial_j k)(x' - y', \varphi(x') - \varphi(y') + t)g(y') \end{aligned} \tag{D.77}$$

and, with $\kappa > 0$ fixed, consider

$$v_{**}(x') = \sup\{|v(z', t)| \,|\, |x' - z'| < \kappa t\}, \quad x' \in \mathbb{R}^{n-1}. \tag{D.78}$$

Then

$$\|v_{**}\|_{L^2(\mathbb{R}^{n-1}; d^{n-1}x')} \leq C\|g\|_{L^2(\mathbb{R}^{n-1}; d^{n-1}x')}. \tag{D.79}$$

To establish (D.79), fix a smooth, even function ψ defined in $\mathbb{R}^n$, with the property that $\psi \equiv 0$ near the origin and $\psi(x) = 1$ for $|x| \geq \frac{1}{2}$. We then further decompose

$$v(x', t) = v^1(x', t) + v^2(x', t) \tag{D.80}$$

where

$$
\begin{aligned}
&v^1(x',t) \\
&= \int_{\mathbb{R}^{n-1}} d^{n-1}y' \, (\eta(x',\varphi(x')) - \eta(y',\varphi(y')))(\partial_j k)(x'-y',\varphi(x')-\varphi(y')+t)g(y') \\
&= \int_{\mathbb{R}^{n-1}} d^{n-1}y' \, \frac{1}{|x'-y'|^{n-1}} \frac{\eta(x',\varphi(x')) - \eta(y',\varphi(y'))}{|x'-y'|} \\
&\qquad \times (\partial_j k)\Big(\frac{x'-y'}{|x'-y'|}, \frac{\varphi(x')-\varphi(y')+t}{|x'-y'|}\Big) g(y') \\
&= \int_{\mathbb{R}^{n-1}} d^{n-1}y' \, \frac{1}{|x'-y'|^{n-1}} \frac{\eta(x',\varphi(x')) - \eta(y',\varphi(y'))}{|x'-y'|} \\
&\qquad \times (\psi\,\partial_j k)\Big(\frac{x'-y'}{|x'-y'|}, \frac{\varphi(x')-\varphi(y')+t}{|x'-y'|}\Big) g(y')
\end{aligned} \tag{D.81}
$$

and

$$
v^2(x',t) = t^{-1}(\eta(x',\varphi(x')+t) - (\eta(x',\varphi(x'))v^3(x',t), \tag{D.82}
$$

where

$$
\begin{aligned}
&v^3(x',t) \\
&= \int_{\mathbb{R}^{n-1}} d^{n-1}y' \, t(\partial_j k)(x'-y',\varphi(x')-\varphi(y')+t)g(y') \\
&= \int_{\mathbb{R}^{n-1}} d^{n-1}y' \, \frac{1}{|x'-y'|^{n-1}} \frac{t}{|x'-y'|} (\partial_j k)\Big(\frac{x'-y'}{|x'-y'|}, \frac{\varphi(x')-\varphi(y')+t}{|x'-y'|}\Big) g(y') \\
&= \int_{\mathbb{R}^{n-1}} d^{n-1}y' \, \frac{1}{|x'-y'|^{n-1}} \frac{t}{|x'-y'|} (\psi\,\partial_j k)\Big(\frac{x'-y'}{|x'-y'|}, \frac{\varphi(x')-\varphi(y')+t}{|x'-y'|}\Big) g(y').
\end{aligned} \tag{D.83}
$$

(In each case, the role of the function ψ is to truncate the singularity of $\partial_j k$ at the origin.) Consequently, if

$$
v^j_{**}(x') = \sup\{|v^j(z',t)| \,|\, |x'-z'| < \kappa t\}, \quad x' \in \mathbb{R}^{n-1}, \ j = 1,3, \tag{D.84}
$$

we have

$$
\|v_{**}\|_{L^2(\mathbb{R}^{n-1};d^{n-1}x')} \leq \|v^1_{**}\|_{L^2(\mathbb{R}^{n-1};d^{n-1}x')} + C\|v^3_{**}\|_{L^2(\mathbb{R}^{n-1};d^{n-1}x')}. \tag{D.85}
$$

As a consequence, it is enough to prove that

$$
\|v^j_{**}\|_{L^2(\mathbb{R}^{n-1};d^{n-1}x')} \leq C\|g\|_{L^2(\mathbb{R}^{n-1};d^{n-1}x')}, \tag{D.86}
$$

for $j = 1,3$. We shall do so by relying on Theorem D.10 (considered with n replaced by $n-1$). When $j = 1$, we apply this theorem with

$$
\begin{aligned}
&m = n, \quad \ell = 1, \quad a = (a_1,a_2) \in \mathbb{R}\times\mathbb{R}^{n-1}, \quad b \in \mathbb{R}, \\
&F(a,b) = a_1(\psi\,\partial_j k)(a_2,b), \quad B = \varphi, \quad A = (\eta(\cdot,\varphi(\cdot)),\cdot).
\end{aligned} \tag{D.87}
$$

When $j = 3$, Theorem D.10 is used with (again $n-1$ in place of n) and

$$
\begin{aligned}
&m = n, \quad \ell = 2, \quad b = (b_1,b_2) \in \mathbb{R}\times\mathbb{R}, \quad a \in \mathbb{R}^{n-1}, \\
&F(a,b) = \zeta(b_1-b_2)b_2(\psi\,\partial_j k)(a,b_1), \quad B_1 = \varphi, \quad B_2 = 0, \quad A = I_{\mathbb{R}^{n-1}},
\end{aligned} \tag{D.88}
$$

where $\zeta \in C_0^\infty(\mathbb{R})$ is an even function with the property that $\zeta \equiv 1$ on $[-M,M]$ (where M is the Lipschitz constant of φ). In each case, the hypotheses on F made in the statement of Theorem D.10 are verified, and part (1) in Theorem D.10 yields the corresponding version of (D.86). This finishes the proof of Theorem D.11. □

After these preparations, we are finally ready to present the following proof:

Proof of Theorem D.8. We work under the assumption that Ω is a $C^{1,r}$-domain, for some $r > 1/2$. In particular, $\nu \in [C^r(\partial\Omega)]^n$. Thanks to Lemma D.4 it suffices to show that (D.38) holds for $z = 0$. To this end, we write

$$K_0^{\#} = K_0 + (K_0^{\#} - K_0) \tag{D.89}$$

and observe that the integral kernel of the operator $R = K_0^{\#} - K_0$ is given by

$$(\nu(x) - \nu(y)) \cdot \nabla E_n(0; x - y), \quad x, y \in \partial\Omega. \tag{D.90}$$

Let $\eta_\alpha \in [C^\infty(\mathbb{R}^n)]^n$, $\alpha \in \mathbb{N}$, be a sequence of vector-valued functions with the property that

$$\eta_\alpha|_{\partial\Omega} \to \nu \text{ in } [C^r(\partial\Omega)]^n \text{ as } \alpha \to \infty, \tag{D.91}$$

and denote by R_α the integral operator with kernel

$$(\eta_\alpha(x) - \eta_\alpha(y)) \cdot \nabla E_n(0; x - y), \quad x, y \in \partial\Omega. \tag{D.92}$$

From (2.129) we know that

$$\gamma_D \nabla \mathcal{S}_0 \in \mathcal{B}\big(H^{1/2}(\partial\Omega), H^{1/2}(\partial\Omega)^n\big), \tag{D.93}$$

which implies that for each $j \in \{1, ..., n\}$, the principal-value boundary integral operator with kernel $\partial_j E_n(0; x - y)$ maps $H^{1/2}(\partial\Omega)$ boundedly into itself. From this, (D.90), (D.91), and Lemma A.5 we may then conclude that

$$R_\alpha \to R \text{ in } \mathcal{B}\big(H^{1/2}(\partial\Omega)\big) \text{ as } \alpha \to \infty. \tag{D.94}$$

Also, from Theorem D.11 we have

$$R_\alpha \in \mathcal{B}\big(L^2(\partial\Omega; d^{n-1}\omega), H^1(\partial\Omega)\big) \hookrightarrow \mathcal{B}_\infty\big(H^{1/2}(\partial\Omega)\big) \text{ for every } \alpha \in \mathbb{N}. \tag{D.95}$$

From (D.94) and (D.95) we may then conclude that

$$R \in \mathcal{B}_\infty\big(H^{1/2}(\partial\Omega)\big), \tag{D.96}$$

hence, ultimately,

$$K_0^{\#} \in \mathcal{B}_\infty\big(H^{1/2}(\partial\Omega)\big), \tag{D.97}$$

by (D.89), (D.96) and (D.37). □

Acknowledgments. We wish to thank Gerd Grubb for questioning an inaccurate claim in an earlier version of the paper and Maxim Zinchenko for helpful discussions on this topic.

References

[1] M. Abramowitz and I. A. Stegun, *Handbook of Mathematical Functions*, Dover, New York, 1972.

[2] N. I. Akhiezer and I. M. Glazman, *Theory of Linear Operators in Hilbert Space*, Dover, New York, 1993.

[3] S. Albeverio, J. F. Brasche, M. M. Malamud, and H. Neidhardt, *Inverse spectral theory for symmetric operators with several gaps: scalar-type Weyl functions*, J. Funct. Anal. **228**, 144–188 (2005).

[4] S. Albeverio, M. Dudkin, A. Konstantinov, and V. Koshmanenko, *On the point spectrum of $\mathcal{H}_{-2}$-singular perturbations*, Math. Nachr. **280**, 20–27 (2007).

[5] W. O. Amrein and D. B. Pearson, *M operators: a generalization of Weyl–Titchmarsh theory*, J. Comp. Appl. Math. **171**, 1–26 (2004).

[6] W. Arendt and M. Warma, *Dirichlet and Neumann boundary conditions: What is in between?*, J. Evolution Eq. **3**, 119–135 (2003).

[7] W. Arendt and M. Warma, *The Laplacian with Robin boundary conditions on arbitrary domains*, Potential Anal. **19**, 341–363 (2003).

[8] Yu. M. Arlinskii and E. R. Tsekanovskii, *Some remarks on singular perturbations of self-adjoint operators*, Meth. Funct. Anal. Top. **9**, No. 4, 287–308 (2003).

[9] Yu. Arlinskii and E. Tsekanovskii, *The von Neumann problem for nonnegative symmetric operators*, Int. Eq. Operator Theory, **51**, 319–356 (2005).

[10] G. Auchmuty, *Steklov eigenproblems and the representation of solutions of elliptic boundary value problems*, Num. Funct. Anal. Optimization **25**, 321–348 (2004).

[11] G. Auchmuty, *Spectral characterization of the trace spaces $H^s(\partial\Omega)$*, SIAM J. Math. Anal. **38**, 894–905 (2006).

[12] J. Behrndt and M. Langer, *Boundary value problems for partial differential operators on bounded domains*, J. Funct. Anal. **243**, 536–565 (2007).

[13] J. Behrndt, M. M. Malamud, and H. Neidhardt, *Scattering matrices and Weyl functions*, preprint, 2006.

[14] S. Belyi, G. Menon, and E. Tsekanovskii, *On Krein's formula in the case of non-densely defined symmetric operators*, J. Math. Anal. Appl. **264**, 598–616 (2001).

[15] S. Belyi and E. Tsekanovskii, *On Krein's formula in indefinite metric spaces*, Lin. Algebra Appl. **389**, 305–322 (2004).

[16] C. Bennett and R. Sharpley, *Interpolation of Operators*, Pure and Applied Mathematics, Vol. 129, Academic Press, Inc., Boston, MA, 1988.

[17] M. Biegert and M. Warma, *Removable singularities for a Sobolev space*, J. Math. Anal. Appl. **313**, 49–63 (2006).

[18] J. F. Brasche, M. M. Malamud, and H. Neidhardt, *Weyl functions and singular continuous spectra of self-adjoint extensions*, in *Stochastic Processes, Physics and Geometry: New Interplays. II. A Volume in Honor of Sergio Albeverio*, F. Gesztesy, H. Holden, J. Jost, S. Paycha, M. Röckner, and S. Scarlatti (eds.), Canadian Mathematical Society Conference Proceedings, Vol. 29, Amer. Math. Soc., Providence, RI, 2000, pp. 75–84.

[19] J. F. Brasche, M. M. Malamud, and H. Neidhardt, *Weyl function and spectral properties of self-adjoint extensions*, Integral Eqs. Operator Theory **43**, 264–289 (2002).

[20] B. M. Brown, G. Grubb, and I. G. Wood, *M-functions for closed extensions of adjoint pairs of operators with applications to elliptic boundary problems*, preprint, 2008, arXiv:0803.3630.

[21] B. M. Brown and M. Marletta, *Spectral inclusion and spectral exactness for PDE's on exterior domains*, IMA J. Numer. Anal. **24**, 21–43 (2004).

[22] B. M. Brown, M. Marletta, S. Naboko, and I. Wood, *Boundary triplets and M-functions for non-selfadjoint operators, with applications to elliptic PDEs and block operator matrices*, J. London Math. Soc., to appear.

[23] J. Brüning, V. Geyler, and K. Pankrashkin, *Spectra of self-adjoint extensions and applications to solvable Schrödinger operators*, preprint, 2007.

[24] R.R. Coifman, G. David and Y. Meyer, *La solution des conjecture de Calderón*, Adv. in Math. **48**, 144–148 (1983).

[25] R.R. Coifman, A. McIntosh and Y. Meyer, *L'intégrale de Cauchy définit un opérateur borné sur L^2 pour les courbes lipschitziennes*, Ann. of Math. (2) **116**, 361–387 (1982).

[26] M. Costabel and M. Dauge, *Un résultat de densité pour les équations de Maxwell régularisées dans un domaine lipschitzien*, C. R. Acad. Sci. Paris Sér. I Math., **327**, 849–854 (1998).
[27] M. Cwikel, *Real and complex interpolation and extrapolation of compact operators*, Duke Math. J. **65**, 333–343 (1992).
[28] B.E.J. Dahlberg, *Estimates of harmonic measure*, Arch. Rational Mech. Anal. **65**, 275–288 (1977).
[29] E. N. Dancer and D. Daners, *Domain perturbation for elliptic equations subject to Robin boundary conditions*, J. Diff. Eq. **138**, 86–132 (1997).
[30] D. Daners, *Robin boundary value problems on arbitrary domains*, Trans. Amer. Math. Soc. **352**, 4207–4236 (2000).
[31] R. Dautray and J.-L. Lions, *Mathematical Analysis and Numerical Methods for Science and Technology, Volume 2, Functional and Variational Methods*, Springer, Berlin, 2000.
[32] V. A. Derkach, S. Hassi, M. M. Malamud, and H. S. V. de Snoo, *Generalized resolvents of symmetric operators and admissibility*, Meth. Funct. Anal. Top. **6**, No. 3, 24–55 (2000).
[33] V. Derkach, S. Hassi, M. Malamud, and H. de Snoo, *Boundary relations and their Weyl families*, Trans. Amer. Math. Soc. **358**, 5351–5400 (2006).
[34] V. A. Derkach and M. M. Malamud, *Generalized resolvents and the boundary value problems for Hermitian operators with gaps*, J. Funct. Anal. **95**, 1–95 (1991).
[35] V. A. Derkach and M. M. Malamud, *The extension theory of Hermitian operators and the moment problem*, J. Math. Sci. **73**, 141–242 (1995).
[36] M. Dindoš and M. Mitrea, *Semilinear Poisson problems in Sobolev–Besov spaces on Lipschitz domains*, Publ. Mat. **46**, 353–403 (2002).
[37] D. E. Edmunds and W. D. Evans, *Spectral Theory and Differential Operators*, Clarendon Press, Oxford, 1989.
[38] E. Fabes, M. Jodeit and N. Rivière, *Potential techniques for boundary value problems on C^1-domains*, Acta Math., **141** (1978), 165–186.
[39] E. Fabes, *Layer potential methods for boundary value problems on Lipschitz domains*, Potential theory—surveys and problems (Prague, 1987), pp. 55–80, Lecture Notes in Math., Vol. **1344**, Springer, Berlin, 1988.
[40] W. G. Faris, *Self-Adjoint Operators*, Lecture Notes in Mathematics, Vol. **433**, Springer, Berlin, 1975.
[41] F. Gesztesy, K. A. Makarov, and E. Tsekanovskii, *An addendum to Krein's formula*, J. Math. Anal. Appl. **222**, 594–606 (1998).
[42] F. Gesztesy, N. J. Kalton, K. A. Makarov, and E. Tsekanovskii, *Some Applications of Operator-Valued Herglotz Functions*, in *Operator Theory, System Theory and Related Topics. The Moshe Livšic Anniversary Volume*, D. Alpay and V. Vinnikov (eds.), Operator Theory: Advances and Applications, Vol. 123, Birkhäuser, Basel, 2001, pp. 271–321.
[43] F. Gesztesy, Y. Latushkin, M. Mitrea, and M. Zinchenko, *Nonselfadjoint operators, infinite determinants, and some applications*, Russ. J. Math. Phys. **12**, 443–471 (2005).
[44] F. Gesztesy and M. M. Malamud, *Boundary value problems for elliptic operators from the extension theory point of view and Weyl–Titchmarsh functions*, in preparation.
[45] F. Gesztesy and M. Mitrea, *Robin-to-Robin maps and Krein-type resolvent formulas for Schrödinger operators on bounded Lipschitz domains*, in *Modern Analysis and Applications. The Mark Krein Centenary Conference*, V. Adamyan, Yu. Berezansky, I. Gohberg, M. Gorbachuk, V. Gorbachuk, A. Kochubei, H. Langer, and G. Popov (eds.), Operator Theory: Advances and Applications, Birkhäuser, to appear.
[46] F. Gesztesy, M. Mitrea, and M. Zinchenko, *Variations on a Theme of Jost and Pais*, J. Funct. Anal. **253**, 399–448 (2007).
[47] V. I. Gorbachuk and M. L. Gorbachuk, *Boundary Value Problems for Operator Differential Equations*, Kluwer, Dordrecht, 1991.
[48] P. Grisvard, *Elliptic Problems in Nonsmooth Domains*, Pitman, Boston, 1985.
[49] G. Grubb, *Distributions and Operators*, Graduate Texts in Mathematics, Springer, in preparation.
[50] S. Hofmann, M. Mitrea and M. Taylor, *Singular integrals and elliptic boundary problems on regular Semmes-Kenig-Toro domains*, preprint, 2007.
[51] D. Jerison and C. Kenig, *The inhomogeneous Dirichlet problem in Lipschitz domains*, J. Funct. Anal. **130**, 161–219 (1995).

[52] W. Karwowski and S. Kondej, *The Laplace operator, null set perturbations and boundary conditions*, in *Operator Methods in Ordinary and Partial Differential Equations*, S. Albeverio, N. Elander, W. N. Everitt, and P. Kurasov (eds.), Operator Theory: Advances and Applications, Vol. 132, Birkhäuser, Basel, 2002, pp. 233–244.

[53] T. Kato, *Perturbation Theory for Linear Operators*, corr. printing of the 2nd ed., Springer, Berlin, 1980.

[54] V. Koshmanenko, *Singular operators as a parameter of self-adjoint extensions*, in *Operator Theory and Related Topics*, V. M. Adamyan, I. Gohberg, M. Gorbachuk, V. Gorbachuk, M. A. Kaashoek, H. Langer, and G. Popov (eds.), Operator Theory: Advances and Applications, Vol. 118, Birkhäuser, Basel, 2000, pp. 205–223.

[55] M. G. Krein, *On Hermitian operators with deficiency indices one*, Dokl. Akad. Nauk SSSR **43**, 339–342 (1944). (Russian).

[56] M. G. Krein, *Resolvents of a hermitian operator with defect index (m, m)*, Dokl. Akad. Nauk SSSR **52**, 657–660 (1946). (Russian).

[57] M. G. Krein and I. E. Ovcharenko, *Q-functions and sc-resolvents of nondensely defined hermitian contractions*, Sib. Math. J. **18**, 728–746 (1977).

[58] M. G. Krein and I. E. Ovčarenko, *Inverse problems for Q-functions and resolvent matrices of positive hermitian operators*, Sov. Math. Dokl. **19**, 1131–1134 (1978).

[59] M. G. Krein, S. N. Saakjan, *Some new results in the theory of resolvents of hermitian operators*, Sov. Math. Dokl. **7**, 1086–1089 (1966).

[60] P. Kurasov and S. T. Kuroda, *Krein's resolvent formula and perturbation theory*, J. Operator Theory **51**, 321–334 (2004).

[61] H. Langer, B. Textorius, *On generalized resolvents and Q-functions of symmetric linear relations (subspaces) in Hilbert space*, Pacific J. Math. **72**, 135–165 (1977).

[62] L. Lanzani and Z. Shen, *On the Robin boundary condition for Laplace's equation in Lipschitz domain*, Comm. Partial Diff. Eqs. **29**, 91–109 (2004).

[63] J. L. Lions, *Espaces d'interpolation et domaines de puissances fractionnaires d'opérateurs*, J. Math. Soc. Japan **14**, 233–241 (1962).

[64] M. M.Malamud, *On a formula of the generalized resolvents of a nondensely defined hermitian operator*, Ukrain. Math. J. **44**, 1522–1547 (1992).

[65] M. M. Malamud and V. I. Mogilevskii, *Krein type formula for canonical resolvents of dual pairs of linear relations*, Methods Funct. Anal. Topology, **8**, No. 4, 72–100 (2002).

[66] M. Marletta, *Eigenvalue problems on exterior domains and Dirichlet to Neumann maps*, J. Comp. Appl. Math. **171**, 367–391 (2004).

[67] V. G. Maz'ja, *Einbettungssätze für Sobolewsche Räume*, Teubner Texte zur Mathematik, Teil 1, 1979; Teil 2, 1980, Teubner, Leipzig.

[68] V. G. Maz'ja, *Sobolev Spaces*, Springer, Berlin, 1985.

[69] W. McLean, *Strongly Elliptic Systems and Boundary Integral Equations*, Cambridge University Press, Cambridge, 2000.

[70] Y. Meyer, *Ondelettes et opérateurs*, Hermann, Paris, 1990.

[71] A. B. Mikhailova, B. S. Pavlov, and L. V. Prokhorov, *Intermediate Hamiltonian via Glazman's splitting and analytic perturbation for meromorphic matrix-functions*, Math. Nachr. **280**, 1376–1416 (2007).

[72] I. Mitrea and M. Mitrea, *Multiple Layer Potentials for Higher Order Elliptic Boundary Value Problems*, preprint, 2008.

[73] D. Mitrea, M. Mitrea, and M. Taylor, *Layer Potentials, the Hodge Laplacian, and Global Boundary Problems in Nonsmooth Riemannian Manifolds*, Mem. Amer. Math. Soc. **150**, No. 713, (2001).

[74] M. Mitrea and M. Taylor, *Potential theory on Lipschitz domains in Riemannian manifolds: Sobolev-Besov space results and the Poisson problem*, J. Funct. Anal. **176**, 1–79 (2000).

[75] M. Mitrea and M. Taylor, *Sobolev and Besov space estimates for solutions to second order PDE on Lipschitz domains in manifolds with Dini or Hölder continuous metric tensors*, Comm. Part. Diff. Eq. **30**, 1–37 (2005).

[76] M. Mitrea and M. Wright, *Layer potentials and boundary value problems for the Stokes system in arbitrary Lipschitz domains*, preprint, 2008.

[77] S. Nakamura, *A remark on the Dirichlet–Neumann decoupling and the integrated density of states*, J. Funct. Anal. **179**, 136–152 (2001).

[78] G. Nenciu, *Applications of the Krein resolvent formula to the theory of self-adjoint extensions of positive symmetric operators*, J. Operator Theory **10**, 209–218 (1983).

[79] K. Pankrashkin, *Resolvents of self-adjoint extensions with mixed boundary conditions*, Rep. Math. Phys. **58**, 207–221 (2006).

[80] B. Pavlov, *The theory of extensions and explicitly-soluble models*, Russ. Math. Surv. **42:6**, 127–168 (1987).

[81] B. Pavlov, *S-matrix and Dirichlet-to-Neumann operators*, Ch. 6.1.6 in *Scattering: Scattering and Inverse Scattering in Pure and Applied Science*, Vol. 2, R. Pike and P. Sabatier (eds.), Academic Press, San Diego, 2002, pp. 1678–1688.

[82] A. Posilicano, *A Krein-like formula for singular perturbations of self-adjoint operators and applications*, J. Funct. Anal. **183**, 109–147 (2001).

[83] A. Posilicano, *Self-adjoint extensions by additive perturbations*, Ann. Scuola Norm. Sup. Pisa Cl. Sci (5) Vol. II, 1–20 (2003).

[84] A. Posilicano, *Boundary triples and Weyl functions for singular perturbations of self-adjoint operators*, Meth. Funct. Anal. Topology **10**, No. 2, 57–63 (2004).

[85] A. Posilicano, *Self-adjoint extensions of restrictions*, Operators and Matrices, to appear; arXiv:math-ph/0703078.

[86] T. Runst and W. Sickel, *Sobolev Spaces of Fractional Order, Nemytskij Operators, and Nonlinear partial Differential Operators*, de Gruyter, Berlin, New York, 1996.

[87] V. S. Rychkov, *On restrictions and extensions of the Besov and Triebel–Lizorkin spaces with respect to Lipschitz domains*, J. London Math. Soc. (2) **60**, 237–257 (1999).

[88] V. Ryzhov, *A general boundary value problem and its Weyl function*, Opuscula Math. **27**, 305–331(2007).

[89] V. Ryzhov, *Weyl–Titchmarsh function of an abstract boundary value problem, operator colligations, and linear systems with boundary control*, Complex Anal. Operator Theory, to appear.

[90] Sh. N. Saakjan, *On the theory of the resolvents of a symmetric operator with infinite deficiency indices*, Dokl. Akad. Nauk Arm. SSR **44**, 193–198 (1965). (Russian.)

[91] B. Simon, *Classical boundary conditions as a tool in quantum physics*, Adv. Math. **30**, 268–281 (1978).

[92] E. M. Stein, *Singular Integrals and Differentiability Properties of Functions*, Princeton Mathematical Series, No. 30, Princeton University Press, Princeton, NJ, 1970.

[93] A. V. Straus, *Generalized resolvents of symmetric operators*, Dokl. Akad. Nauk SSSR **71**, 241–244 (1950). (Russian.)

[94] A. V. Straus, *On the generalized resolvents of a symmetric operator*, Izv. Akad. Nauk SSSR Ser. Math. **18**, 51–86 (1954). (Russian.)

[95] A. V. Straus, *Extensions and generalized resolvents of a non-densely defined symmetric operator*, Math. USSR Izv. **4**, 179–208 (1970).

[96] M. E. Taylor, *Pseudodifferential Operators and Nonlinear PDE*, Progress in Mathematics Vol. 100, Birkhäuser, Boston, MA, 1991.

[97] M. E. Taylor, *Partial Differential Equations, Vol. II*, Springer, New York, 1996.

[98] M. E. Taylor, *Tools for Partial Differential Equations*, American Mathematical Society, Providence, RI, 2000.

[99] H. Triebel, *Function spaces in Lipschitz domains and on Lipschitz manifolds. Characteristic functions as pointwise multipliers*, Rev. Mat. Complut. **15**, 475–524 (2002).

[100] E. R. Tsekanovskii and Yu. L. Shmul'yan, *The theory of bi-extensions of operators on rigged Hilbert spaces. Unbounded operator colligations and characteristic functions*, Russ. Math. Surv. **32:5**, 73–131 (1977).

[101] G. Verchota, *Layer potentials and regularity for the Dirichlet problem for Laplace's equation in Lipschitz domains*, J. Funct. Anal. **59**, 572–611 (1984).

[102] M. Warma, *The Laplacian with general Robin boundary conditions*, Ph.D. Thesis, University of Ulm, 2002.

[103] M. Warma, *The Robin and Wentzell-Robin Laplacians on Lipschitz domains*, Semigroup Forum **73**, 10–30 (2006).

[104] J. Weidmann, *Linear Operators in Hilbert Spaces*, Springer, New York, 1980.

[105] J. Wloka, *Partial Differential Equations*, Cambridge University Press, Cambridge, 1987.

[106] I. Wood, *Maximal L^p-regularity for the Laplacian on Lipschitz domains*, Math. Z. **255**, 855–875 (2007).

Department of Mathematics, University of Missouri, Columbia, MO 65211, USA
E-mail address: fritz@math.missouri.edu
URL: http://www.math.missouri.edu/personnel/faculty/gesztesyf.html

Department of Mathematics, University of Missouri, Columbia, MO 65211, USA
E-mail address: marius@math.missouri.edu
URL: http://www.math.missouri.edu/personnel/faculty/mitream.html

Proceedings of Symposia in Pure Mathematics
Volume **79**, 2008

A local Tb Theorem for square functions

Steve Hofmann

Dedicated to Prof. V. Maz'ya on the occasion of his 70th birthday.

ABSTRACT. We prove a "local" Tb Theorem for square functions, in which we assume only L^q control of the pseudo-accretive system, with $q > 1$. We then give an application to variable coefficient layer potentials for divergence form elliptic operators with bounded measurable non-symmetric coefficients.

1. Introduction, statement of results, history

The Tb Theorems of McIntosh and Meyer [**McM**], and of David, Journé and Semmes [**DJS**], are boundedness criteria for singular integrals, by which the L^2 boundedness of a singular integral operator T may be deduced from sufficiently good behavior of T on some suitable non-degenerate test function b. A "local Tb theorem" is a variant of the standard Tb theorem, in which control of the action of the operator T on a single, globally defined accretive test function b, is replaced by local control, on each dyadic cube Q, of the action of T on a test function b_Q, which satisfies some uniform, scale invariant L^p bound along with the non-degeneracy condition

$$\frac{1}{C_0} \leq |Q|^{-1} \left| \int_Q b_Q \right|, \tag{1.1}$$

for some uniform constant C_0. A collection of such local test functions, ranging over all dyadic cubes Q (or over all cubes or balls) is called a "pseudo-accretive system". The first local Tb theorem, in which the local test functions are assumed to belong uniformly to L^∞, is due to M. Christ [**Ch**], and was motivated in part by applications to the theory of analytic capacity; an extension of Christ's result to the non-doubling setting is due to Nazarov, Treil and Volberg [**NTV**]. A more recent version, in which Christ's L^∞ control of the test functions is relaxed to L^q control, appears in [**AHMTT**], and this sharpened version (see also [**AY**], and the unpublished manuscript [**H**]) has found application to the theory of layer potentials associated to divergence form, variable coefficient elliptic PDE (see [**AAAHK**]).

It is also of interest to consider local Tb theorems for square functions (as opposed to singular integrals). These have found application to the solution of the

2000 *Mathematics Subject Classification.* Primary 42B25; Secondary 35J25 .
S. Hofmann was supported by the National Science Foundation.

Kato problem [**HMc**], [**HLMc**], [**AHLMcT**] (see also [**AT**] and [**S**] for related results), and to variable coefficient layer potentials [**AAAHK**].

In this note, we consider the square function estimate

$$\iint_{\mathbb{R}^{n+1}_+} |\theta_t f(x)|^2 \frac{dx\,dt}{t} \leq C\|f\|^2_{L^2(\mathbb{R}^n)}, \tag{1.2}$$

where

$$\theta_t f(x) := \int_{\mathbb{R}^n} \psi_t(x,y) f(y) dy$$

and $\{\psi_t(x,y)\}_{t\in(0,\infty)}$, satisfies, for some exponent $\alpha > 0$,

$$|\psi_t(x,y)| \leq C \frac{t^\alpha}{(t+|x-y|)^{n+\alpha}} \tag{1.3}$$

and

$$\begin{aligned} &(a) \qquad |\psi_t(x,y+h) - \psi_t(x,y)| \leq C \frac{|h|^\alpha}{(t+|x-y|)^{n+\alpha}} \\ &(b) \qquad |\psi_t(x+h,y) - \psi_t(x,y)| \leq C \frac{|h|^\alpha}{(t+|x-y|)^{n+\alpha}} \end{aligned} \tag{1.4}$$

whenever $|h| \leq t/2$.

Our main result in this paper is the following:

THEOREM 1.1. *Let* $\theta_t f(x) := \int \psi_t(x,y) f(y) dy$*, where* $\psi_t(x,y)$ *satisfies* (1.3) *and* (1.4). *Suppose also that there exists a constant* $C_0 < \infty$*, an exponent* $q > 1$ *and a system* $\{b_Q\}$ *of functions indexed by dyadic cubes* $Q \subseteq \mathbb{R}^n$*, such that for each dyadic cube* Q

(i) $\int_{\mathbb{R}^n} |b_Q|^q \leq C_0 |Q|$

(ii) $\frac{1}{C_0}|Q| \leq \left|\int_Q b_Q\right|$

(iii) $\int_Q \left(\int_0^{\ell(Q)} |\theta_t b_Q(x)|^2 \frac{dt}{t}\right)^{q/2} dx \leq C_0|Q|$.

Then we have the square function bound (1.2).

Here, and in the sequel, we use the notation $\ell(Q)$ to denote the side length of a cube Q.

The case $q = 2$ of this theorem was already known, and requires only the first inequality in (1.4) (smoothness in the y variable). See [**H2**] and [**A**] for explicit formulations in that case, although in fact the result and its proof were already implicit in [**HMc**], [**HLMc**] and [**AHLMcT**]. As mentioned above, analogous results for singular integrals (as opposed to square functions) were obtained for $q = \infty$ in [**Ch**] and [**NTV**], for $q > 1$ in the "perfect dyadic" case treated in [**AHMTT**], and, based on [**AHMTT**], for $q = 2$ (or $q = 2 + \epsilon$) in the case of standard singular integrals in [**AY**] and [**H**]. It remains an open problem to treat the case $q < 2$ for singular integrals that are not of perfect dyadic type, but rather satisfy standard Calderón-Zygmund conditions.

The paper is organized as follows: in the next section we prove Theorem 1.1, and in Section 3 we present an application to the theory of layer potentials for variable coefficient divergence form operators with bounded measurable non-symmetric coefficients.

2. Proof of Theorem 1.1

We begin by recalling the following well known fact, due explicitly to Christ and Journé [**CJ**], but also implicit in the work of Coifman and Meyer [**CM**].

PROPOSITION 2.1. [**CJ**] *Let* $\theta_t f(x) \equiv \int_{\mathbb{R}^n} \psi_t(x,y) f(y)dy$, *where* $\psi_t(x,y)$ *satisfies* (1.3) *and* (1.4) *(a). Suppose that we have the Carleson measure estimate*

$$\sup_Q \frac{1}{|Q|} \int_0^{\ell(Q)} \int_Q |\theta_t 1(x)|^2 \frac{dxdt}{t} \leq C. \tag{2.1}$$

Then we have the square function bound (1.2).

Remark. The converse direction (i.e. that (1.2) implies (2.1)) is essentially due to Fefferman and Stein [**FS**].

Thus, to prove Theorem 1.1, it is enough to establish (2.1). In fact, by covering an arbitrary cube by finitely many dyadic cubes of comparable side length, it is enough to establish a version of (2.1) in which the supremum runs over dyadic cubes only. To this end, we shall use the following lemma of "John-Nirenberg" type.

LEMMA 2.2. *Suppose that there exist* $\eta \in (0,1)$ *and* $C_1 < \infty$, *such that for every dyadic cube* $Q \in \mathbb{R}^n$, *there is a family* $\{Q_j\}$ *of non-overlapping dyadic sub-cubes of* Q, *with*

$$\sum |Q_j| \leq (1-\eta)|Q| \tag{2.2}$$

and

$$\int_Q \left(\int_{\tau_Q(x)}^{\ell(Q)} |\theta_t 1(x)|^2 \frac{dt}{t} \right)^{q/2} dx \leq C_1 |Q|, \tag{2.3}$$

where $\tau_Q(x) := \sum 1_{Q_j}(x)\, \ell(Q_j)$. *Then* (2.1) *holds.*

We shall defer momentarily the proof of Lemma 2.2, and proceed to the proof of Theorem 1.1. We may suppose without loss of generality that $1 < q < 2$, as the case $q > 2$ may be reduced to the known case $q = 2$ by Hölder's inequality. We claim that, in the spirit of [**S**] and [**AT**] (but using also Lemma 2.2), it is enough to prove that for each dyadic cube Q, there is a family $\{Q_j\}$ of non-overlapping dyadic sub-cubes of Q satisfying (2.2) for which

$$\int_Q \left(\int_{\tau_Q(x)}^{\ell(Q)} |\theta_t 1(x)|^2 \frac{dt}{t} \right)^{q/2} dx \leq C \int_Q \left(\int_0^{\ell(Q)} |\theta_t 1(x)\, A_t b_Q(x)|^2 \frac{dt}{t} \right)^{q/2} dx, \tag{2.4}$$

where A_t denotes the usual dyadic averaging operator, i.e.,

$$A_t f(x) := |Q(x,t)|^{-1} \int_{Q(x,t)} f,$$

and $Q(x,t)$ denotes the minimal dyadic cube containing x with side length at least t. Indeed, given (2.4), we may follow [**CM**] and write

$$\theta_t 1 A_t = (\theta_t 1)(A_t - P_t) + (\theta_t 1 P_t - \theta_t) + \theta_t := R_t^{(1)} + R_t^{(2)} + \theta_t,$$

where P_t is a nice approximate identity, of convolution type, with a smooth, compactly supported kernel. By hypothesis (iii) of Theorem 1.1, the contribution of $\theta_t b_Q$, to the right hand side of (2.4), is controlled by $C|Q|$, as desired. Moreover,

$R_t^{(2)}1 = 0$, and its kernel satisfies (1.3) and (1.4). Thus, by standard Littlewood-Paley/vector-valued Calderón-Zygmund theory, we have that

$$\int_Q \left(\int_0^{\ell(Q)} |R^{(2)} b_Q(x)|^2 \frac{dt}{t} \right)^{q/2} dx \leq C_q \|b_Q\|_q^q \leq C|Q|, \tag{2.5}$$

where in the last inequality we have used hypothesis (i) of Theorem 1.1. Furthermore, the same L^q bound holds for $R_t^{(1)}$ (even though (1.4) fails for this term), as may be seen by following the interpolation arguments of [**DRdeF**]. We omit the details. Thus, the right hand side of (2.4) is bounded by $C|Q|$, so that the conclusion of Theorem 1.1 then follows by Lemma 2.2 and Proposition 2.1.

Therefore, it is enough to establish (2.4), for a family of dyadic sub-cubes of Q satisfying (2.2). To this end, we follow the stopping time arguments in [**HMc**], [**HLMc**] and [**AHLMcT**] (but see also [**Ch**], where a similar idea had previously appeared). Our starting point is hypothesis (ii) of Theorem 1.1. Dividing by an appropriate complex constant, we may suppose that

$$\frac{1}{|Q|} \int_Q b_Q = 1. \tag{2.6}$$

We then sub-divide Q dyadically, to select a family of non-overlapping cubes $\{Q_j\}$ which are maximal with respect to the property that

$$\Re e \frac{1}{|Q_j|} \int_{Q_j} b_Q \leq 1/2. \tag{2.7}$$

By the maximality of the cubes in the family $\{Q_j\}$, it follows that

$$\frac{1}{2} \leq \Re e\, A_t b_Q(x), \quad \text{if } t > \tau_Q(x),$$

so that (2.4) holds with $C = 2^q$. It remains only to verify that there exists $\eta > 0$ such that

$$|E| > \eta |Q|, \tag{2.8}$$

where $E \equiv Q \backslash (\cup Q_j)$. By (2.6) we have that

$$|Q| = \int_Q b_Q = \Re e \int_Q b_Q = \Re e \int_E b_Q + \Re e \sum_j \int_{Q_j} b_Q$$
$$\leq |E|^{1/q'} \left(\int_Q |b_Q|^q \right)^{1/q} + \frac{1}{2} \sum |Q_j|,$$

when in the last step we have used (2.7). From hypothesis (i) of Theorem 1.1, we then obtain that

$$|Q| \leq C|E|^{1/q'} |Q|^{1/q} + \frac{1}{2}|Q|,$$

and (2.8) now follows readily. This concludes the proof of Theorem 1.1, modulo Lemma 2.2, whose proof we now give.

Proof of Lemma 2.2. We begin by stating

Lemma 2.3. *Suppose that there exist $N < \infty$ and $\beta \in (0,1)$ such that for every dyadic cube Q,*

$$|\{x \in Q : g_Q(x) > N\}| \leq (1 - \beta)\, |Q|, \tag{2.9}$$

where

$$g_Q(x) := \left(\int_0^{\ell(Q)} |\theta_t 1(x)|^2 \frac{dt}{t} \right)^{1/2}.$$

Then (2.1) *holds.*

We take this lemma for granted momentarily, and prove Lemma 2.2. Fix a dyadic cube Q. For a large, but fixed N to be chosen momentarily, let

$$\Omega_N := \{x \in Q : g_Q(x) > N\}.$$

Under the hypotheses of Lemma 2.2, with $E := Q \setminus (\cup Q_j)$, we have

$$\begin{aligned} |\Omega_N| &\leq \sum |Q_j| + |\{x \in E : g_Q(x) > N\}| \\ &\leq (1-\eta)\,|Q| + |\{x \in Q : \left(\int_{\tau_Q(x)}^{\ell(Q)} |\theta_t 1(x)|^2 \frac{dt}{t} \right)^{1/2} > N\}| \\ &\leq (1-\eta)\,|Q| + \frac{C_1}{N^q}|Q|, \end{aligned}$$

where in the last step we have used Tchebychev's inequality and (2.3). Choosing N so large that $C_1/N^q \leq \eta/2$, we obtain (2.9) with $\beta = \eta/2$. Thus, Lemma 2.3 implies Lemma 2.2.

In turn, to prove Lemma 2.3, we proceed as follows. We momentarily fix $\epsilon \in (0,1)$, and let N, β be as in the hypotheses of Lemma 2.3. For a dyadic cube Q, set

$$g_{Q,\epsilon}(x) := \left(\int_\epsilon^{\min(\ell(Q),1/\epsilon)} |\theta_t 1(x)|^2 \frac{dt}{t} \right)^{1/2},$$

where we take this term to be 0 if $\ell(Q) \leq \epsilon$. Define

$$K(\epsilon) := \sup_Q \frac{1}{|Q|} \int_Q g_{Q,\epsilon}^2,$$

where the supremum runs over all dyadic cubes Q. By the truncation, $K(\epsilon)$ is finite for each fixed ϵ, and our goal is to show that

$$\sup_{0<\epsilon<1} K(\epsilon) < \infty.$$

Now fix a dyadic cube Q, and set

$$\Omega_{N,\epsilon} := \{x \in Q : g_{Q,\epsilon}(x) > N\}.$$

By the truncation, and (1.4) (b), $g_{Q,\epsilon}$ is continuous, so we may make a Whitney decomposition

$$\Omega_{N,\epsilon} = \cup Q_k.$$

Then, if $F_{N,\epsilon} := Q \setminus \Omega_{N,\epsilon}$, we have

$$\begin{aligned} \int_Q g_{Q,\epsilon}^2 &= \int_{F_{N,\epsilon}} g_{Q,\epsilon}^2 + \sum \int_{Q_k} g_{Q,\epsilon}^2 \\ &\leq N^2\,|Q| + \sum \int_{Q_k} g_{Q_k,\epsilon}^2 + \sum \int_{Q_k} \int_{\max(\ell(Q_k),\epsilon)}^{\min(\ell(Q),1/\epsilon)} |\theta_t 1(x)|^2 \frac{dt}{t} dx. \quad (2.10) \end{aligned}$$

But

$$\sum \int_{Q_k} g_{Q_k,\epsilon}^2 \leq K(\epsilon) \sum |Q_k| \leq K(\epsilon)(1-\beta)\,|Q|, \tag{2.11}$$

where in the last step we have used (2.9) and the fact that $g_{Q,\epsilon} \leq g_Q$ for every Q, so that $\Omega_{N,\epsilon} \subseteq \Omega_N$. We now claim that

$$\int_{Q_k} \int_{\max(\ell(Q_k),\epsilon)}^{\min(\ell(Q),1/\epsilon)} |\theta_t 1(x)|^2 \frac{dt}{t} dx \,\leq\, C|Q_k|. \tag{2.12}$$

Assuming momentarily that the claim holds, we deduce from (2.10), (2.11) and (2.12) that

$$\frac{1}{|Q|} \int_Q g_{Q,\epsilon}^2 \leq C + K(\epsilon)\,(1-\beta).$$

Taking the supremum over all dyadic cubes Q, and using that $K(\epsilon)$ is finite for each fixed ϵ, we obtain that

$$K(\epsilon) \leq \frac{C}{\beta},$$

uniformly in ϵ. We may then let $\epsilon \to 0$ to obtain the conclusion of Lemma 2.3.

It therefore remains only to prove (2.12). Since Q_k is a Whitney cube, there is a point $x_k \in F_{N,\epsilon}$ such that

$$\operatorname{dist}(x_k, Q_k) \leq C\ell(Q_k). \tag{2.13}$$

The left hand side of (2.12) is then bounded by an absolute constant times

$$\int_{Q_k} \int_{\ell(Q_k)}^{C_n \ell(Q_k)} |\theta_t 1(x)|^2 \frac{dt}{t} dx + \int_{Q_k} \int_{C_n \ell(Q_k)}^{\ell(Q)} |\theta_t 1(x) - \theta_t 1(x_k)|^2 \frac{dt}{t} dx$$
$$+ \int_{Q_k} \int_{\epsilon}^{\min(\ell(Q),1/\epsilon)} |\theta_t 1(x_k)|^2 \frac{dt}{t} dx \;=: I + II + III,$$

where C_n is a purely dimensional constant that will be chosen momentarily. By (1.3), $\|\theta_t 1\|_\infty \leq C$, so that

$$I \leq C|Q_k|.$$

Since $x_k \in F_{N,\epsilon}$, we have that

$$III \leq N^2\,|Q_k|.$$

Finally, by (1.4) (b) and (2.13), for C_n chosen large enough we have that

$$|\theta_t 1(x) - \theta_t 1(x_k)| \,\leq\, C\left(\frac{\ell(Q_k)}{t}\right)^{\alpha},$$

for all $x \in Q_k$. Hence,

$$II \leq C \int_{Q_k} \int_{C_n \ell(Q_k)}^{\infty} \left(\frac{\ell(Q_k)}{t}\right)^{2\alpha} \frac{dt}{t} \leq C|Q_k|.$$

This concludes the proof. □

3. Application to variable coefficient layer potentials

We consider here the matter of L^2 boundedness of layer potentials associated to divergence form elliptic equations $Lu = 0$, in the domains $\mathbb{R}^{n+1}_{\pm}$, where

$$L = -\operatorname{div} A\nabla \equiv -\sum_{i,j=1}^{n+1} \frac{\partial}{\partial x_i}\left(A_{i,j}\frac{\partial}{\partial x_j}\right)$$

is defined in $\mathbb{R}^{n+1} = \{(x,t) \in \mathbb{R}^n \times \mathbb{R}\}, n \geq 1$ (we use the notational convention that $x_{n+1} = t$), and where $A = A(x)$ is an $(n+1)\times(n+1)$ matrix of real-valued L^∞ coefficients, which we allow to be non-symmetric, defined on $\mathbb{R}^n$ (i.e., independent of the t variable) and satisfying the uniform ellipticity condition

$$\lambda|\xi|^2 \leq \langle A(x)\xi,\xi\rangle = \sum_{i,j=1}^{n+1} A_{ij}(x)\,\xi_j\,\xi_i, \quad \|A\|_{L^\infty(\mathbb{R}^n)} \leq \Lambda, \tag{3.1}$$

for some $\lambda > 0$, $\Lambda < \infty$, and for all $\xi \in \mathbb{R}^{n+1}$, $x \in \mathbb{R}^n$. The divergence form equation is interpreted in the weak sense, i.e., we say that $Lu = 0$ in a domain Ω if $u \in W^{1,2}_{loc}(\Omega)$ and

$$\int A\nabla u \cdot \nabla\Psi = 0$$

for all $\Psi \in C_0^\infty(\Omega)$. Although the case of real symmetric "radially independent" (i.e., in our context, t-independent) coefficients is now rather well understood (see [**JK**], [**KP**], [**K**] and also [**AAAHK**]), in general it remains an open problem to establish solvability results, with L^p data, for boundary value problems associated to *non-symmetric* equations in $\mathbb{R}^{n+1}_{\pm}$. However, in the case $n = 1$, i.e., in the domains $\mathbb{R}^2_{\pm}$, solvability of the L^p Dirichlet problem ("D_p") has been established in [**KKPT**] for p sufficiently large (but finite), while solvability of the L^p Neumann ("N_p") and Regularity ("R_p") problems with p near 1 (in fact, dual to the Dirichlet exponent) was obtained in [**KR**]. We refer to those papers for detailed statements of the boundary value problems (this is not our main emphasis here), but we note that solvability of D_p is equivalent to a scale invariant L^q bound, $p^{-1} + q^{-1} = 1$, for the Poisson kernel. To be precise, fix a boundary cube $Q \subset \mathbb{R}^n$, and let $A_Q^{\pm}$ denote the upper and lower "corkscrew points" associated to Q, i.e., if x_Q denotes the center of Q, then

$$A_Q^{\pm} := (x_Q, \pm\ell(Q)) \in \mathbb{R}^{n+1}_{\pm}.$$

For a point $X \in \mathbb{R}^{n+1}_{\pm}$ (we shall adopt the notational convention that capital letters $X := (x,t)$, $Y := (y,s)$ may be used to denote points in $\mathbb{R}^{n+1}$), let $k^X_{L,\pm}$ denote the Poisson kernel for L with pole at X in $\mathbb{R}^{n+1}_{\pm}$. It turns out that D_p is solvable for L in the domain $\mathbb{R}^{n+1}_{\pm}$ if and only if there is a constant B such that the following scale invariant bound holds for every cube $Q \subset \mathbb{R}^n$:

$$\int_{\mathbb{R}^n} \left(k_{L,\pm}^{A_Q^{\pm}}\right)^q \leq B\,|Q|^{1-q}, \tag{3.2}$$

with $p^{-1} + q^{-1} = 1$ (see, e.g., [**KKPT**] or [**K**]). It is shown in [**KKPT**] that for every L as above in $\mathbb{R}^2_{\pm}$, there is a $q := q(L) > 1$ such that (3.2) holds.

The proof in [**KR**] of the solvability of N_p and R_p, in $\mathbb{R}^2_{\pm}$, with p near 1, uses in a crucial way the L^2 boundedness (but not invertibility) of the layer potentials associated to L. In this paper, we present an alternative (and rather short) proof of this boundedness, based on the local Tb Theorem proved in Section 1. The proof in [**KR**] also uses Tb theory (to be precise, the result of [**DJS**]), but is tied very closely to the 2-dimensionality of the domain. Our proof is in principle not dimension dependent, but rather relies only on the Poisson kernel estimate (3.2). Of course, at present, (3.2) is known to hold for non-symmetric operators only when $n+1=2$. We conjecture that (3.2) remains true for non-symmetric t-independent operators in all dimensions. The idea to use estimates like (3.2) to prove layer potential bounds (in the setting of symmetric coefficients) has appeared previously in [**AAAHK**], but the argument there was limited to the case $q=2$, as it depended on the local Tb theorem for singular integrals [**AHMTT**], as extended to standard Calderón-Zygmund operators in [**H**] or [**AY**]. In the present paper, we will use the square function/non-tangential maximal function estimates of [**DJK**] to reduce matters to square functions treatable via Theorem 1.1.

We now recall the method of layer potentials. For L as above, let $L^* := -\operatorname{div} A^*\nabla$ denote the transpose operator (which is also the adjoint, since we are dealing with real coefficients here), and let $\Gamma(X,Y)$ and $\Gamma^*(X,Y) := \Gamma(Y,X)$ denote the corresponding fundamental solutions in $\mathbb{R}^{n+1}$. Thus,

$$L_X\,\Gamma(X,Y) = \delta_Y, \quad L_Y^*\,\Gamma^*(Y,X) := L_Y^*\,\Gamma(X,Y) = \delta_X,$$

where δ_X denotes the Dirac mass at the point X. By the t-independence of our coefficients, we have that

$$\Gamma(x,t,y,s) = \Gamma(x,t-s,y,0). \tag{3.3}$$

We define the single layer potential operators for L and L^* by

$$\begin{aligned} S_t f(x) &\equiv \int_{\mathbb{R}^n} \Gamma(x,t,y,0)\, f(y)\, dy, \quad t\in\mathbb{R} \\ S_t^* f(x) &\equiv \int_{\mathbb{R}^n} \Gamma^*(x,t,y,0)\, f(y)\, dy, \quad t\in\mathbb{R}, \end{aligned} \tag{3.4}$$

(we apologize for this notation: S_t^* denotes the single layer potential for L^*, and is not, in general, equal to the adjoint of S_t) and our goal is to show that

$$\sup_t \|\nabla_{x,t}\, S_t f\|_{L^2(\mathbb{R}^n)} + \sup_t \|\nabla_{x,t}\, S_t^* f\|_{L^2(\mathbb{R}^n)} \leq C\|f\|_{L^2(\mathbb{R}^n)} \tag{3.5}$$

(the latter estimate implies L^2 bounds for the corresponding double layer potentials via duality). To be precise, we have the following

THEOREM 3.1. *Suppose that L is an operator of the type described above and that there are exponents $q(L), q(L^*) > 1$ and a constant B such that $k_{L,\pm}^{A_Q^\pm}$ and $k_{L^*,\pm}^{A_Q^\pm}$ (the Poisson kernels for L and L^*, respectively) satisfy* (3.2) *for every cube $Q\subset\mathbb{R}^n$. Then the layer potential bound* (3.5) *holds, with a constant depending only on dimension, λ, Λ, B and* $\min(q(L), q(L^*))$.

Remarks. In particular, since (3.2) always holds for such operators when $n=1$ [**KKPT**], we recover the boundedness result of [**KR**]. We also observe that our proof will require that (3.2) hold for both L and L^*, even if we restrict our attention to the bound for S_t.

PROOF. We begin with some preliminary reductions. We treat only S_t in the case $t>0$, as the same argument carries over mutatis mutandi to the case $t<0$

and to S_t^*. By Lemma 5.2 of [**AAAHK**], it suffices to prove

$$\sup_t \|\partial_t\, S_t f\|_{L^2(\mathbb{R}^n)} \leq C\|f\|_{L^2(\mathbb{R}^n)}. \tag{3.6}$$

and

$$\iint_{\mathbb{R}^{n+1}_\pm} |t\,\nabla_{x,t}\partial_t\, S_t f(x)|^2\,\frac{dxdt}{|t|} \leq C\|f\|^2_{L^2(\mathbb{R}^n)}. \tag{3.7}$$

Moreover, the same lemma shows that (3.7) follows from

$$\iint_{\mathbb{R}^{n+1}_\pm} |t\,(\partial_t)^2\, S_t f(x)|^2\,\frac{dxdt}{|t|} \leq C\|f\|^2_{L^2(\mathbb{R}^n)}. \tag{3.8}$$

In addition, from (3.2) for $k_{L,+}^{A_Q^+}$, along with the results of [**DJK**] applied to the solution $u(x,t) := \partial_t S_t f(x)$, we have that (3.7) implies (3.6) (we use here that the arguments of [**DJK**] carry over to the non-symmetric case - see the comments in the introduction to [**KKPT**]). Thus, it is enough to prove (3.8).

To this end, we first note that by [**GW**] (if $n+1 \geq 3$), or (if $n+1 = 2$) as a consequence of the Gaussian bounds and local Hölder continuity of the kernel of the heat semigroup $e^{-\tau L}$ (see, for example [**AMT**]), we have that

$$\psi_t(x,y) := t\,(\partial_t)^2\,\Gamma(x,t,y,0),$$

the kernel of $\theta_t := t\,(\partial_t)^2 S_t$, satisfies (1.3) and (1.4). Thus, it is enough to construct a pseudo-accretive system $\{b_Q\}$ satisfying the hypotheses of Theorem 1.1.

We now set

$$b_Q \equiv |Q|\,k_{L^*,-}^{A_Q^-}.$$

Observe that condition (i) of Theorem 1.1 follows immediately from (3.2). Moreover (ii) is an immediate consequence of the following well known estimate of Caffarelli, Fabes, Mortola and Salsa [**CFMS**], extended to the case of non-symmetric coefficients (as may be done: see the comments in [**KKPT**] concerning the validity of the results of [**CFMS**] in the non-symmetric setting):

$$\int_Q k_{L^*,-}^{A_Q^-}(y)\,dy \geq \frac{1}{C}.$$

It remains to establish condition (iii) of Theorem 1.1. Let $(x,t) \in R_Q^+ \equiv Q \times (0, \ell(Q))$. Then, since for fixed $(x,t) \in \mathbb{R}^{n+1}_+$, we have that $\partial_t^2\Gamma(x,t,\cdot,\cdot)$ is a solution of $L^*u = 0$ in $\mathbb{R}^{n+1}_-$, we obtain

$$\begin{aligned} |\theta_t b_Q(x)| &= |Q|\,t|\int (\partial_t)^2\,\Gamma(x,t,y,0)\,k_{L^*,-}^{A_Q^-}(y)\,dy| \\ &= |Q|\,t\,|(\partial_t)^2\,\Gamma(x,t,A_Q^-)| \\ &= |Q|\,\frac{t}{t+\ell(Q)}\,|\psi_{t+\ell(Q)}(x,x_Q)| \leq C\frac{t}{\ell(Q)}. \end{aligned}$$

where in the last two steps we have used (3.3) and then (1.3). Hypothesis (iii) now follows readily. This concludes the proof of Theorem 3.1.

□

References

[AAAHK] M. Alfonseca, P. Auscher, A. Axelsson, S. Hofmann and S. Kim, *Analyticity of layer potentials and L^2 solvability of boundary value problems for divergence form elliptic equations with complex L^∞ coefficients*, preprint.

[A] P. Auscher, *Lectures on the Kato square root problem*, Surveys in analysis and operator theory (Canberra, 2001), Proc. Centre Math. Appl. Austral. Nat. Univ. **40**, Austral. Nat. Univ., Canberra, 2002, pp. 1–18.

[AHLMcT] P. Auscher, S. Hofmann, M. Lacey, A. McIntosh, and P. Tchamitchian, *The solution of the Kato Square Root Problem for Second Order Elliptic operators on $\mathbb{R}^n$*, Annals of Math., **156** (2002), 633–654.

[AHMTT] P. Auscher, S. Hofmann, C. Muscalu, T. Tao, C. Thiele, *Carleson measures, trees, extrapolation, and $T(b)$ theorems*, Publ. Mat., **46** (2002), no. 2, 257–325.

[AMT] P. Auscher, A. McIntosh and P. Tchamitchian, *Heat kernels of second order complex elliptic operators and applications*, J. Functional Analysis, **152** (1998), 22-73.

[AT] P. Auscher and Ph. Tchamitchian, *Square root problem for divergence operators and related topics*, Astérisque Vol. 249 (1998), Société Mathématique de France.

[AY] P. Auscher and Q. X. Yang, *On local $T(b)$ Theorems*, preprint.

[CFMS] L. Caffarelli, E. Fabes, S. Mortola and S. Salsa, *Boundary behavior of nonnegative solutions of elliptic operators in divergence form*, Indiana Univ. Math. J., **30** (1981), no. 4, 621–640.

[Ch] M. Christ, *A T(b) theorem with remarks on analytic capacity and the Cauchy integral*, Colloquium Mathematicum, LX/LXI (1990) 601-628.

[CJ] M. Christ and J.-L. Journé, *Polynomial growth estimates for multilinear singular integral operators*, Acta Math., **159** (1987), no. 1-2, 51–80.

[CM] R. Coifman and Y. Meyer, *Non-linear harmonic analysis and PDE*, E. M. Stein, editor, Beijing Lectures in Harmonic Analysis, vol. 112, Annals of Math. Studies, Princeton Univ. Press, 1986.

[DJK] B. Dahlberg, D. Jerison and C. Kenig, *Area integral estimates for elliptic differential operators with nonsmooth coefficients*, Ark. Mat., **22** (1984), no. 1, 97–108.

[DJS] G. David, J.-L. Journé, and S. Semmes, *Opérateurs de Calderón -Zygmund, fonctions para-accrétives et interpolation*, Rev. Mat. Iberoamericana, **1** 1–56, 1985.

[DRdeF] J. Duoandikoetxea and J. L. Rubio de Francia, *Maximal and singular integral operators via Fourier transform estimates*, Invent. Math., **84** (1986), 541-561.

[FS] C. Fefferman, and E. M. Stein, *H^p spaces of several variables*, Acta Math., **129** (1972), no. 3-4, 137–193.

[GW] M. Grüter and K. O. Widman, *The Green function for uniformly elliptic equations*, Manuscripta Math., **37** (1982), 303-342.

[H] S. Hofmann, *A proof of the local Tb Theorem for standard Calderón-Zygmund operators*, unpublished manuscript, http://www.math.missouri.edu/~hofmann/

[H2] S. Hofmann, *Local Tb Theorems and applications in PDE*, Proceedings of the ICM Madrid, Vol. **II**, pp. 1375-1392, European Math. Soc., 2006.

[HLMc] S. Hofmann, M. Lacey and A. McIntosh, *The solution of the Kato problem for divergence form elliptic operators with Gaussian heat kernel bounds*, Annals of Math., **156** (2002), 623-631.

[HMc] S. Hofmann and A. McIntosh, *The solution of the Kato problem in two dimensions*, Proceedings of the Conference on Harmonic Analysis and PDE held in El Escorial, Spain in July 2000, Publ. Mat., Vol. extra, 2002, pp. 143-160.

[JK] D. Jerison and C. Kenig, *The Dirichlet problem in nonsmooth domains*, Ann. of Math. (2), **113** (1981), no. 2, 367–382.

[K] C. Kenig, *Harmonic analysis techniques for second order elliptic boundary value problems*, CBMS Regional Conference Series in Mathematics, 83. Published for the Conference Board of the Mathematical Sciences, Washington, DC, American Mathematical Society, Providence, RI, 1994

[KKPT] C. Kenig, H. Koch, H. J. Pipher and T. Toro, *A new approach to absolute continuity of elliptic measure, with applications to non-symmetric equations*, Adv. Math., **153** (2000), no. 2, 231–298.

[KP] C. Kenig and J. Pipher, *The Neumann problem for elliptic equations with nonsmooth coefficients*, Invent. Math., **113** (1993), no. 3, 447–509.

[KR] C. Kenig and D. Rule, *The regularity and Neumann problems for non-symmetric elliptic operators*, preprint.

[McM] A. McIntosh and Y. Meyer, *Algèbres d'opérateurs définis par des intégrales singulières*, C. R. Acad. Sci. Paris, **301** Série 1 395–397, 1985.

[NTV] F. Nazarov, S. Treil and A. Volberg, *Accretive system Tb-theorems on nonhomogeneous spaces*, Duke Math. J., **113** (2002), no. 2, 259–312.

[S] S. Semmes, *Square function estimates and the $T(b)$ Theorem*, Proc. Amer. Math. Soc., **110** (1990), no. 3, 721–726.

Department of Mathematics, University of Missouri, Columbia, Missouri 65211, USA
E-mail address: hofmann@math.missouri.edu

Proceedings of Symposia in Pure Mathematics
Volume **79**, 2008

Partial differential equations, trigonometric series, and the concept of function around 1800: a brief story about Lagrange and Fourier

Jean–Pierre Kahane

Dedicated to Vladimir Maz'ya and Tatyana Shaposhnikova.

Abstract. Functions of real variables, PDE's and trigonometric series have strong relations. A brief history of these relations as they appeared around 1750 and developed around 1800 is given in the first part of the article. The controversy on vibrating strings, involving d'Alembert, Euler, Daniel Bernoulli and Lagrange, is well know, and also the birth of Fourier series with the analytic theory of heat. Among the many references quoted in the article the main source is the thesis of Riemann on trigonometric series.

Riemann showed how difficult is was to accept the idea that an arbitrary function could be expressed by a trigonometric series, and he mentioned the strong opposition of Lagrange to the statements of Fourier. A sentence in Riemann's thesis is the source of the second part of the paper, where the author describes his search of two handwritten pages in the collection of manuscripts of Lagrange, supposed to express this opposition, and tries to explain what he found.

Vladimir Maz'ya and Tatyana Shaposhnikova made a significant contribution to the history of mathematics with their authoritative biography "Jacques Hadamard : A Universal Mathematician" [**11**]. Thus it seems appropriate to include the history of mathematics as one of the themes for the present volume. I will be concerned with the period 1750–1850, and will focus on how the notion of "function" was influenced by the study of partial differential equations (PDEs) and trigonometric series.

There are two parts in this article. The first relies on previous studies, it is just a way to look at a well known story. The second relies on the first, it contains a new material, and it is a tentative answer to a series of puzzling questions about a historical document.

The study of the relations among PDEs, trigonometric series, and the possible notions of function remains active in modern times. A classic example is the theory of distributions developed by Laurent Schwartz in the middle last century. This

2000 *Mathematics Subject Classification.* Primary 01A50; Secondary 01A55.
Key words and phrases. d'Alembert, Euler, Daniel Bernoulli, Lagrange, Fourier, Riemann.

theory extented the notions of functions, derivatives, and Fourier transform [**13**] and it was first applied to the study of PDEs. Todays, classes of functions and their generalizations play an essential role in research on PDEs, while trigonometric series and Fourier transforms enter as important tools. This story is far from over, and I believe that a look at some of the beginning contributes to our appreciation of current research.

1. Part I

I begin with a well known episode, the controversy around vibrating strings after 1747. It deals with PDEs as well as trigonometric series [**16**], and both subjects introduced important ideas and discussions about functions. The principal characters of the story are d'Alembert (1717–1783), Euler (1707–1783), Daniel Bernoulli (1700–1782), and Lagrange (1736–1813). The story was told in many ways, first by the actors themselves, then by many other authors. A volume of the collected works of Euler contains the description of the debate by C. Truesdell [**14**], and this is the most current source of information. The earlier, thorough study by H. Burkhardt on series and PDE (1804 pages) was essentially related to the mathematics in question, and a large part was devoted to the period we consider [**2**]. A short and illuminating article of A.P. Youschkevitch (10 pages) appeared in 1975, with the principal references on the subject of vibrating strings and the use of "discontinuous" functions [**15**]. A French thesis has been defended recently in Lyons by Guillaume Jouve ; it contains much new material, comments, translations, unpublished papers of d'Alembert, together with the relevant part of d'Alembert's Opuscules mathématiques [**8**]. The languages used by these authors are English, German, Russian, and French. I shall just sketch the story, and I recommend Jouve as a source of further information.

The second episode is related to the heat equation. It is a fascinating story, involving mainly Fourier and (again) Lagrange with the participation of many of their contemporaries. I will just sketch the story, but I wish to highlight the appearance of trigonometric series as a tool, as a mathematical object, and as a source of ideas and problems. The main reference here, apart from Fourier's book "Théorie analytique de la chaleur" [**5**], is the historical part of Riemann's dissertation on trigonometric series [**12**]. In fact, this historical part is the best exposition of whole subject, from vibrating strings to Fourier and Dirichlet, that I know. Riemann's dissertation contains many ideas that have been significant for the development of mathematics in general. His appreciation of the people and their work is well informed and accurate. I will devote a section to comments on this dissertation, and this will lead to a comment on Dirichlet's ideas. A statement by Riemann serves as motivations for Part II.

1.1. As I already said, the first part of the story, the controversy about vibrating strings, is well known. It is kind of dramatic play.

The first act begins with d'Alembert in 1747 and Euler in 1748. Both considered a string fixed at two points, say 0 and ℓ on the x–axis, and ordinates $y(t,x)$ above x at time t. Both established the PDE (written here in modern notation)

$$\frac{\partial^2 y}{\partial t^2} = \omega^2 \frac{\partial^2 y}{\partial x^2}\,. \tag{1}$$

Both found a solution

$$F(\omega t + x) - F(\omega t - x)\,, \tag{2}$$

where F is a 2ℓ–periodic function. F is well defined by the initial conditions (position and velocity at time 0) :

$$\begin{aligned} F(x) - F(-x) &= p(x)\,, \\ \omega(F'(x) - F'(-x)) &= v(x)\,. \end{aligned} \tag{3}$$

For d'Alembert the functions of the form (2) were a particular class of functions, and the functions defined in (3) were particular as well. Therefore (2) provided a solution for a special class of initial data. For other initial data, he said that there could be other solutions, and when the data are not regular enough he said that it might become a question of physics, not of mathematics.

For Euler, on the contrary, the solution was general for any kind of initial data. Euler's motivation is clear : for any initial data one can compute F, therefore it has to be the solution of the problem. (At first, Euler assumed that the initial velocity vanishes, and in this case (3) means that F is an odd function such that $2F(x) = p(x)$ when $0 < x < \ell$). No "continuity" is needed for the initial data ; in particular, broken lines could be allowed.

Here is a quotation by d'Alembert, discussing Euler's point of view: *On ne trouve la solution du problème que pour les cas où les différentes figures de la corde vibrante peuvent être renfermée dans une seule et même équation* ([**8**], II, p. 72). (One obtains the solution only in the cases when the different forms of the vibrating string can be expressed by one specific equation.)

Already here there are two conceptions about the functions you can consider in mathematical analysis. For d'Alembert, they should have a well defined expression in terms of known functions. For Euler, they can be defined as well by any graphical representation.

A new scene appeared with Daniel Bernoulli. Since a sound is a superposition of harmonics, the general solution of the problem of vibrating strings should be a series of the form (again, I use modern notations)

$$\sum_k \left(a_k \cos \frac{k\pi\omega t}{\ell} + b_k \sin \frac{k\pi\omega t}{\ell}\right) \sin \frac{k\pi x}{\ell} \tag{4}$$

[**1**]. Now neither d'Alembert nor Euler would agree. For d'Alembert, and for Euler as well, trigonometric series would represent only a very special class of functions.

The last character in this first episode is Lagrange. He was able to treat the problem in a complete form when the string is replaced by equidistant weighted points distributed on a thread. Then, finer and finer discretisations of the initial data result in discrete solutions tending to the solution proposed by Euler. It appeared as a justification for Euler's point of view, although it was rejected by d'Alembert. And that is the end of this part of the story [**9**].

1.2. The second episode involves Joseph Fourier (1768–1830) as a principal character. Fourier sent a memoir to the French Academy of Sciences (then called the first class of the Institut de France) in 1807, on the propagation of heat in solid bodies. The memoir was read by a committee consisting of Lagrange, Laplace, Lacroix and Monge. It was not published. The Academy then proposed the subject for a competition. Fourier extended his study and sent his contribution with the beautiful subtitle “et ignem regunt numeri” (heat also is governed by numbers). Laplace, Lagrange, and Lacroix were again examiners. Fourier won the Prize, but there were severe reservations, and the work again was not published. Only after 1817, when Fourier became a member of the Academy, did a printed version appeared ; an extended version took the form of an important book, la Théorie analytique de la chaleur, the analytic theory of heat, published in 1822 [**5**].

The book consists of an introduction, Discours préliminaire, and nine chapters. The first chapter expounds the physical aspects of heat propagation. The second gives the differential equations, first as examples, then in a general way : inside a homogeneous body the heat propagation is governed by the equation

$$\frac{\partial u}{\partial t} = \frac{K}{CD}\Big(\frac{\partial^2 u}{\partial x^2} + \frac{\partial^2 u}{\partial y^2} + \frac{\partial^2 u}{\partial z^2}\Big) \tag{5}$$

where K, C, D are physical constants depending on the body. If we forget the constant, it is what we now call the heat equation. Moreover there are boundary and initial conditions, as for vibrating strings. The third chapter introduces a method, the use of trigonometric series, for solving a special equilibrium problem. Here is the problem. Take an infinite rectangular body, limited by a horizontal strip and two parallel vertical half–planes, say,

$$-\frac{\pi}{2} \le x \le \frac{\pi}{2}, \qquad 0 \le y < \infty . \tag{6}$$

The strip part is at the temperature of boiling water, the vertical edges at the temperature of melting ice, that is

$$\begin{aligned} &u(x,0) = 1 && \big(-\tfrac{\pi}{2} < x < \tfrac{\pi}{2}\big), \\ &u\big(-\tfrac{\pi}{2}, y\big) = u\big(\tfrac{\pi}{2}, y\big) = 0 && (y > 0), \end{aligned} \tag{7}$$

where we write 1 for the temperature of boiling water.

The temperature inside the body is given by the heat equation in a reduced form :

$$\frac{\partial^2 u}{\partial x^2} + \frac{\partial^2 u}{\partial y^2} = 0 . \tag{8}$$

Nowadays we call this a Dirichlet problem. The treatment by Fourier consists of looking first for solutions of (8) in the form

$$u(x,y) = f(x)\, g(y) ,$$

then, taking into account the boundary conditions on the vertical edges and the fact that temperatures are bounded in the body, considering as general candidates

$$u(x,y) = a\ \exp(-y)\ \cos x + b\ \exp(-3y)\ \cos 3x + c\ \exp(-5y)\ \cos 5x + \cdots .$$

It remains to express that $u(x,0) = 1$ when $-\pi/2 < x < \pi/2$, that is

$$1 = a\ \cos x + b\ \cos 3x + c\ \cos 5x + \cdots \qquad (-\pi/2 < x < \pi/2) .$$

Fourier finds the values of the coefficients and he is pleased with the solution :

$$(9)\quad \frac{\pi}{4}u(x,0) = \cos x - \frac{1}{3}\cos 3x + \frac{1}{5}\cos 5x - \frac{1}{7}\cos 7x + \cdots \quad \Big(-\frac{\pi}{2} < x < \frac{\pi}{2}\Big),$$

$$(10)\quad \begin{aligned} \frac{\pi}{4}u(x,y) = e^{-y}\cos x - \frac{1}{3}e^{-3y}\cos 3x + \frac{1}{5}e^{-5y}\cos 5x - \frac{1}{7}e^{-7y}\cos 7x + \cdots \\ \Big(-\frac{\pi}{2} < x < \frac{\pi}{2}, y > 0\Big). \end{aligned}$$

Fourier knows what convergence means and explains that these series converges, indicating the sum of the series (9) on different intervals (n° 177) with a full proof using the so-called Dirichlet kernel (n° 179) ; "*the limit of the series is positive and negative alternatively. By the way, the convergence is not rapid enough in order to provide an easy approximation, but it suffices for the truth of the equation*" (n° 179). The Fourier's main interest is the second series because it is "extremely convergent" and gives a good estimate of the temperature inside the body by using only a few terms (n° 191).

Then there is a long digression in the book. Before considering the propagation of heat in other domains (chapters 4 to 9), he spends 50 pages playing with particular functions and their expansions into series of cosines or series of sines, thereby giving different extensions of functions defined on an interval. His conclusion is that arbitrary functions can be represented by trigonometric series and that all series converge. (That was a mistake, but a very fruitful mistake.) He observes that his analysis applies to vibrating strings, therefore justifying the approach of Daniel Bernoulli.

As far as the notion of function is concerned, it is clear after Fourier that a function is associated with a domain and that there is no canonical way to extend a function. We shall see in part II how trigonometric series played a role in this clarification.

1.3. The first historical study of these matters is due to Riemann (1826–1866). It is the first chapter of his dissertation on trigonometric series [**12**]. The second chapter introduces the Riemann integral, together with a characterization of the Riemann-integrable functions. The third chapter is a firework of ideas, methods, examples, and general results. It contains a characterization of the functions obtained as sums of everywhere convergent trigonometric series. Starting form the series and not from the function forbids the use of integration for computing the coefficients (or needs another definition of the integral, as Denjoy made much later [**4**]). This is now called the Riemann theory of trigonometric series [**16**]. The Riemann theory was completed by George Cantor (1845–1918) ; he proved that if the sum of the series is zero everywhere, it is the null series. The he extended the theorem and proved that the conclusion still holds when "everywhere" is replaced by "everywhere except on some particular set". This extension is the first paper by Cantor on real numbers and set theory [**3**]. Riemann's third chapter is a jewel mine, but my purpose here is to use and comment the first chapter.

The first chapter is divided into three sections, whose subjects are vibrating strings, Fourier, and Dirichlet (1805–1859). The first section is a very clear exposition of the controversy about vibrating strings : d'Alembert rejecting his own solution when arbitrary initial positions and velocities are given ; Euler claiming that no restriction is needed ; Bernoulli assuming that the motion of vibrating strings is a superposition of harmonic motions ; and Lagrange's approach, from finite to infinite, supporting Euler's claim. D'Alembert did not agree with Euler and Lagrange, and the three of them rejected the claim of Bernoulli.

Riemann says at the beginning of the second section that a new area began with Fourier, namely with the couple of formulas

$$
(11)\qquad
\begin{gathered}
f(x) = \begin{cases} a_1 \ \sin x + a_2 \ \sin 2x + \cdots \\ +\frac{1}{2}b_0 + b_1 \ \cos x + b_2 \ \cos 2x + \cdots , \end{cases} \\
\begin{cases} a_n &= \dfrac{1}{\pi}\displaystyle\int_{-\pi}^{\pi} f(x) \ \sin nx \ dx \, , \\ b_n &= \dfrac{1}{\pi}\displaystyle\int_{-\pi}^{\pi} f(x) \ \cos nx \ dx \, . \end{cases}
\end{gathered}
$$

He then explains that Lagrange strongly opposed Fourier's method. Let me quote Riemann : *Als Fourier in einer seiner ersten Arbeiten über die Wärme, welche er der französische Akademie vorlegte (21 Dec. 1807) zuerst den Satz aussprach, dass eine willkürklich (graphisch) gegebene Function sich durch eine trigonometrische Reihe ausdrücken lasse, war diese Behauptung dem greisen Lagrange so unerwartet, dass er ihr auf das Entschiedenste entgegentrat. Es soll sich hierüber noch ein Schriftstück im Archiv der Pariser Akademie befinden.*

(When Fourier in one of his first works on heat, communicated to the French Academy on Dec 21 1807, stated that an arbitrary function (given in a graphic way) could be expressed by a trigonometric series, this statement was to the old Lagrange so unexpected that he opposed it in the strongest way. There should still be a written document about this in the Archives of the Parisian Academy.)

Let me explain the phrase, *"dem greisen Lagrange."* In December 1807, Lagrange was 71 ; the other members of the committee were much younger : Monge 61, Laplace 58, Lacroix 42, and Fourier was 39. Concerning the "Schriftstück", a footnote explains that the information cames from Dirichlet, who had known Fourier in Paris. I have looked for this document, and the second part of this article describes what I found.

Riemann then discusses matters of priority and concludes :

Durch Fourier was nun zwar die Natur der trigonometrischen Reihen vollkommen richtig erkannt ; sie wurden seitdem in der mathematischen Physik zu Darstellung willkürlicher Funktionen vielfach angewandt, und in jedem einzelne Falle überzeugte man sich leicht, dass die Fourier'sche Reihe wirklich gegen den Werth der Function convergiere ; aber es dauerte lange, ehe dieser wichtige Satz allgemein bewiesen würde.

(Through Fourier indeed was the nature of trigonometric series fully understood ; since then they were applied many times in mathematical physics for representing arbitrary functions, and in each case one was easily convinced that the Fourier series really converges to the function ; but it lasted long before this important theorem was proved in full generality.)

As a last comment on this section, the term of "Fourier series" was not classical when Riemann wrote his thesis. He was the first to emphasize the importance of the notion. Nowadays a Fourier series is a trigonometric series whose coefficients are given by the Fourier formulas. Therefore it depends on the kind of integral we consider, and there are indeed many to consider Fourier–Riemann, Fourier–Stieljes, Fourier–Lebesgue (the most important now), Fourier–Denjoy, Fourier–Wiener, Fourier–Schwartz series, Haar–Fourier series on compact abelian groups, etc.

Before going to Dirichlet in the third section, Riemann mentions several competitors and several mistakes. Cauchy's mistakes were surprising and fruitful : one was about convergent series, another on analytic functions, the very domains where Cauchy made such brilliant contributions. Both were pointed out by Dirichlet and raised important observations by Riemann.

The work of Dirichlet on convergence of the Fourier series was published in 1829. Dirichlet was stimulated by the Cauchy's errors, and his article is a model of rigor in mathematical analysis. Riemann explains the method, and here is his appreciation :

Durch die Arbeit Dirichlet's war einer grossen Menge wichtiger analytischer Untersuchungen eine feste Grundlage gegeben. Es war him gelungen, indem er den Punkt, wo Euler irrte, in volles Licht brachte, eine Frage zu erledigen, die so viele ausgezeichnete Mathematker seit mehr als siebzig Jahren (seit dem Jahre 1753) beschäftig hatte. In der That für alle Fälle der Natur, um welche es sich allein handelte, war sie vollkommen erledigt, denn so gross auch unsere Unwissenheit darüber ist, wie sich die Kräfte and Zustände der Materie nach Ort und Zeit im Unendlichkleinen andern, so können wir doch sicher annehmen, dass die Functionen, auf welche sich die Dirichlet'sche Untersuchung nicht erstreckt, in der Natur nicht vorkommen.

(Dirichlet's work gave a solid ground to a large number of important analytical investigations. It was given to him to fully clarify a point on which Euler was mistaken, and to settle a question investigated by so many eminent mathematicians for more than 60 years (since 1753). Actually the question was settled completely for all cases we can encounter in Nature, since, even ignorant as we are of the evolution of forces and states of matter according to space and time, we can be sure that functions on which Dirichlet's investigation does not apply do not exist in Nature.)

Euler's error was the common error of d'Alembert, Euler and Lagrange, who all considered Bernoulli's approach as hopeless.

There is an inaccuracy in the Riemann's statement that functions occuring in Nature are only of the type studied by Dirichlet (a finite number of maxima and minima on every interval). Brownian motion is a counterexample, which Riemann couldn't guess. But Riemann also says that mathematicians should not restrict the range of functions to those occurring in mathematical physics. For example, functions arising in number theory are also worthy of study. And that is the motivation for the last and most important part of his dissertation, where he investigates the very strange functions that trigonometric series provide.

Riemann did not mention an observation made by Dirichlet at the end of his article. Dirichlet observed that only integrable functions can be considered in his study. For him, an integrable function is necessarily continuous on some interval (it is the notion of an integral described by Cauchy). Then he gives the famous counterexample, a function taking one value on the rationals and another value on the irrationals.

With Fourier, Dirichlet, and Riemann we are not at the end, and the relations between function theory, trigonometric series, and PDE are renewed constantly.

But we have seen the enormous change made between 1750 and 1850 in the conception of what a function is. For d'Alembert, it was an analytic expression, in order for the equation of vibrating strings to make sense ; for Euler, a graphic representation was sufficient, as suggested by the form of a string at initial time ; for Bernoulli, relying to a physical interpretation of sound by means of harmonics, infinite trigonometric series would occur ; for Lagrange, finite to infinite had to start with a discrete analogue of the vibrating strings, but "discontinuous" functions (like broken lines) were not excluded. With Fourier it became clear that the domain has to be specified when a function is introduced, and that there are several natural extensions of a function given on an interval. With Dirichlet a first model is given of a complete proof in the domain of functions of a real variable, and a first example of a function that cannot be defined by a graphical representation. Finally Riemann was able to summarize all the main points and consequences of the discussions about vibrating strings and trigonometric series, and to pave the way to Cantor and others for the foundations of the theory of functions of a real variable.

2. Part II

There are amateurs in mathematics, in the sense that they like to read mathematics and also to play with it ; they enjoy mathematics, and sometimes they make a small contribution. I am an amateur in the history of mathematics, and here is a small contribution.

2.1. Consulting the archives of Académie des sciences and the Library of Institut de France, with the help of the archivists and librarians, I was able to found a document that may be the "Schriftstück" of Lagrange mentioned by Riemann. It is just one page long, and I shall reproduce it below. There is no date on it. Obviously it is negative comment on a formula written by Fourier. It contains elementary computations, leading to an apparent contradiction. Here it is ((a), (b), (c) are in the handwritten paper ; (i), (ii), (iii), (iv), (v) are signs I introduced in order to avoid repeating the formulas later) :

L'équation

$$(a) \qquad \frac{1}{2}x = \sin x - \frac{1}{2}\sin 2x + \frac{1}{3}\sin 3x - \frac{1}{4}\sin 4x + \cdots$$

n'a lieu que depuis $x = 0$ *jusqu'à* $x = \frac{\pi}{2} = 90°$.

Supposons qu'elle ait lieu au delà. Dans (a) *faisons* $x = \pi - y$ y *étant* $< \frac{\pi}{2}$ *on aura*

$$(i) \qquad \frac{\pi - y}{2} = \sin y + \frac{1}{2}\sin 2y + \frac{1}{3}\sin 3y + \frac{1}{4}\sin 4y + \cdots$$

Changeons y *en* x *ce qui est permis on aura donc aussi*

$$(b) \qquad \frac{\pi - x}{2} = \sin x + \frac{1}{2}\sin 2x + \frac{1}{3}\sin 3x + \frac{1}{4}\sin 4x + \cdots$$

Or cette section ne peut s'accorder avec la Ière (a) *car la 1ère différentiée donne*

$$(ii) \qquad \frac{1}{2} = \cos x - \cos 2x + \cos 3x - \cos 4x + \cdots$$

Faisons $x = \pi - y$ *on aura* y *étant* $< 90°$

$$\text{(iii)} \qquad \frac{1}{2} = -\cos y - \cos 2y - \cos 3y - \cos 4y - \cdots$$

Changeant y *en* x *on aura*

$$\text{(iv)} \qquad -\frac{1}{2} = \cos x + \cos 2x + \cos 3x + \cos 4x + \cdots$$

Mul(tiplions) par dx *et intégrons*

$$\text{(v)} \qquad \text{const} - \frac{1}{2}x = \sin x + \frac{1}{2}\sin 2x + \frac{1}{3}\sin 3x + \frac{1}{4}\sin 4x + \cdots$$

Pour déterminer la constante on fait $x = 0$*, on a* $Const = 0$*. Donc*

$$\text{(c)} \qquad -\frac{1}{2}x = \sin x + \frac{1}{2}\sin 2x + \frac{1}{3}\sin 3x + \frac{1}{4}\sin 4x + \cdots$$

En comparant (b) *et* (c) *on voit qu'elles ne peuvent subsister.*

Here is an abbreviated translation.

Equation (a) holds only from $x = 0$ to $x = \frac{\pi}{2}$.
Supposing it extends outside, take $x = \pi - y$. Since $y < \frac{\pi}{2}$ we shall have (i).
Changing y into x we have also (b).
But this cannot agree with (a) for differentiating (a) gives (ii).
Writing $x = \pi - y$ we get (iii), then changing y into x we get (iv), then multiplying by dx and integrating we obtain (v).
In order to define the constant we take $x = 0$, and that gives const $= 0$.
Hence (c).
Comparing (b) and (c) we see that they can't hold together.

Lagrange was mistaken : (b) is not valid for $x = 0$. The root of the mistake lies in the word "au delà" , just after (a) ; by the way "au delà" is not easy to read in Lagrange's handwriting. It means that Lagrange extends (a) from the interval $\left(0, \frac{\pi}{2}\right)$ to what seems natural to him. Actually both sides of (a) have a natural extension, but they are not the same.

2.2. I was not satisfied with this page : it is not as important as I expected, nor does it express what Riemann said about Lagrange's strong opposition towards Fourier. I already knew from the catalogue in the Archives that there were two pages and not one. Would there exist another page of Lagrange more explicit about this opposition ?

Yes and no. The handwritten papers of Lagrange are kept in a series of volumes collecting what Lagrange wrote on different subjects. Volume n° 906 contains mainly contributions to interpolation methods and recurrent series. Each contribution contains a number of sheets of paper, this number is written at the back of the last page and four signatures follow : Lacroix, Legendre (written Le Gendre), Prony (written De Prony) and Poisson, a committee of academicians. Sometimes a title is given at the back of the last page as well. Actually there are two sheets of paper with the title "papiers relatifs au mémoire de Fourier" and "deux feuilles" written at the back of the second. The first is what I just copied. The second has a quite different handwriting, and it is the only piece in the volume that was not written by Lagrange. It is endorsed by the four academicians, and I thought first that they took the decision to react to Lagrange's mistake. I shall reproduce the text below, but I can't reproduce the aspect of this sheet, full of crossings out, easy to read at the beginning, and completely squeezed at the end. It is nothing but a rough draft. This excludes the assumption that it was produced by the academic committee. But why did they endorse this draft ?

Reading the draft I noticed that it was exactly in the spirit of Fourier. Actually there is no doubt that it is his handwriting. Before going further, let us look at what he wrote. On top of the page is the formula

$$\frac{\varpi - x}{2} = \sin x + \frac{1}{2}\sin 2x + \frac{1}{3}\sin 3x + \cdots \cong \frac{1}{2}\int \frac{\sin(m + \frac{1}{2}x)}{\sin(\frac{1}{2}x)} dx\,.$$

Then :

1. *La valeur de la constante dans l'équation*

$$C - \frac{1}{2}x = \sin x + \frac{1}{2}\sin 2x + \frac{1}{3}\sin 3x + \frac{1}{4}\sin 4x + \cdots$$

n'est pas nulle. En effet le calcul fait voir que le reste de la série considérée comme fonction de x et du nombre m de termes ne devient pas nul lorsqu'on fait $x = 0$ et m infini. Mais si l'on donne à x une valeur quelconque plus grande que 0 et plus petite que 2ϖ, le même calcul montre que la valeur du reste devient nulle lorsqu'on suppose m infini. Il suit de là que pour déterminer la constante il faut donner à x une valeur quelconque comprise entre 0 et 2ϖ.

2. *L'équation*

$$\frac{1}{2}x = \sin x - \frac{1}{2}\sin 2x + \frac{1}{3}\sin 3x - \frac{1}{4}\sin 4x + \cdots$$

est vraie pour toutes les valeurs positives moindres que ϖ et pour toutes les valeurs négatives plus grandes que $-\varpi$. C'est-à-dire que si l'on met dans le second membre une valeur de x moindre que ϖ et plus grande que $-\varpi$ ce second membre aura une limite déterminée dont on approchera sans fin à mesure que le nombre de termes augmentera et cette limite sera $\frac{1}{2}x$.

Lorsqu'au lieu de x on met $\varpi - y$ l'équation a lieu entre les limites $\varpi - y = -\varpi$ et $\varpi - y = \varpi$ c'est-à-dire pour toutes les valeurs de y moindres que 2ϖ et plus grandes que 0. C'est pourquoi si dans l'équation

$$\frac{\varpi - x}{2} = \sin x + \frac{1}{2}\sin 2x + \frac{1}{3}\sin 3x + \frac{1}{4}\sin 4x + \cdots \tag{b}$$

on met au lieu de x une valeur plus grande que 0 et moindre que 2ϖ, le second membre aura toujours pour limite la quantité $\frac{\varpi - x}{2}$.

(trois lignes raturées).

Si l'on différentie l'équation (b) *on a*

$$-\frac{1}{2} = \cos x + \cos 2x + \cos 3x + \cos 4x + \cdots$$

comme on trouve dans la note. Si on l'intègre il faut déterminer la constante en donnant à x une valeur quelconque moindre que 2ϖ et plus grande que 0. Ainsi dans l'équation

$$C - \frac{1}{2}x = \sin x + \frac{1}{2}\sin 2x + \frac{1}{3}\sin 3x + \frac{1}{4}\sin 4x + \cdots$$

il faut donner à x une valeur > 0 et $< 2\varpi$. Si par exemple on fait $x = \varpi$ on aura $C = \frac{1}{2}\varpi$ ce qui est la véritable valeur de la constante. Si l'on faisait $x = \frac{1}{2}\varpi$ on aurait

$$C - \frac{1}{4}\varpi = 1 - \frac{1}{3} + \frac{1}{5} - \frac{1}{7} + \cdots$$

où $C = \frac{1}{2}\varpi$ comme précédemment.

L'objection se réduit donc à celle–ci : l'équation

$$\text{(b)} \qquad \frac{\varpi - x}{2} = \sin x + \frac{1}{2}\sin 2x + \frac{1}{3}\sin 3x + \frac{1}{4}\sin 4x + \cdots$$

que fournit le calcul de l'auteur n'est point vraie. En différentiant on a

$$-\frac{1}{2} = \cos x + \cos 2x + \cos 3x + \cos 4x + \cdots$$

multipliant par dx et intégrant et déterminant la constante pour que l'équation (b) *ait lieu pour $x = 0$ on a*

$$\text{(c)} \qquad -\frac{1}{2}x = \sin x + \frac{1}{2}\sin 2x + \frac{1}{3}\sin 3x + \cdots$$

qui ne peut s'accorder avec la précédente.

La réponse consiste à remarquer que l'équation (b) *a lieu pour toutes les valeurs de x qui sont > 0 et $< 2\varpi$* (ici la page est coupée) *et que l'on ne satisfait pas à cette condition des limites en déterminant la constante de manière que l'équation ait lieu lorsque $x = 0$. Ainsi on ferait précisément la même objection si l'on se réduisait à dire : l'équation*

$$\frac{\varpi - x}{2} = \sin x + \frac{1}{2}\sin 2x + \frac{1}{3}\sin 3x + \cdots$$

ne peut pas subsister car elle n'a pas lieu lorsque $x = 0$.

(puis en tout petit, au bas de la page)

En général on ne peut point séparer l'usage d'une équation de ce genre de la considération des limites entre lesquelles les valeurs de la variable doivent être considérées.

Here is an abbreviated translation of the last and most important part :

Here is the objection (of Lagrange) : (b) is not true, because differentiating, then integrating in such a way that the equation holds when $x = 0$ one gets (c), which is not compatible with (b). The answer (of Fourier) is that (b) holds when $0 < x < 2\varpi$ and this condition is not satisfied when $x = 0$

The same objection (of Lagrange) could be made by simply considering (b) and saying that it does not hold when $x = 0$.

It is a general fact that an equation of this type cannot be used without specifying the limits between which the values of the variable have to be considered.

The whole page is rather lenghty, but it is a perfect explanation of the Lagrange's mistake [**10**].

2.3. Now we are faced with a puzzle. Why is this draft of Fourier enclosed in the handwritten papers of Lagrange ? Why was it endorsed by an academic committee ? When and why was it written ? Is there a relation between these two papers and the "entschiedenste" opposition of Lagrange ?

I shall first provide the reader with some documents and then give my interpretation of a possible order of the events.

2.3.1. First, the series under consideration, (a) in the paper of Lagrange, is the matter of n° 182 of the Analytical Heat Theory [**5**]. Fourier checks that the partial sums of even order can be written

$$\frac{x}{2} - \int \frac{\cos(mx + \frac{x}{2})}{2\cos\frac{x}{2}}\, dx$$

and shows that the integral tends to 0 using an integration by parts. It is a perfect proof of the formula. Precisions come later, after considering other examples, in n° 184 : *Il faut observer, à l'égard de toutes ces séries, que les équations qui en sont formées n'ont lieu que lorsque la variable est comprise entre certaines limites.*

(This is almost the same as the end of the draft.)

Moreover, for the series (a), the calculation "donne la valeur $x/2$ toutes les fois que x est plus grand que 0 et moindre que ϖ. Mais elle n'équivaut plus à $x/2$ si l'arc dépasse ϖ."

(Again, this is expressed in the draft.)

2.3.2. According to Fourier, (a) is not new ; it was known before (end of n° 182). Anyway, it was not accepted by Lagrange. There is a long letter of Fourier to Lagrange, published by Hérivel (He), likely 1808 or 1809, the draft of which is kept in the manuscript fund of the French National Library [**6**]. The draft is rough and looks like the paper in the files of Lagrange ; but it is organized in order to be as convincing as possible. About (a) he insists :

Je vous prie de jeter les yeux sur cette dernière note qui établit clairement la convergence de la série et dont la partie essentielle était dans le mémoire (art.). Vous reconnaitrez facilement que cette matière n'est pas du domaine de la foi mais de celui de la géométrie, ce qui est bien différent, et il me semble que si de pareilles démonstrations peuvent être contestées, il faut renoncer à écrire quelque chose d'exact en mathématiques.

(I urge you to look at the last note, which proves the convergence of the series and which was contained essentially in the mémoire (art). You will agree that this is not a matter of faith but of geometry, which is quite different, and that one should give up writing anything exact in mathematics if such a demonstration was not accepted.)

Then he tells how he found the general formula by different methods ; he mentions that he had communicated his results to Biot and Poisson ; he discusses the possible priority of d'Alembert and Euler for the method of integration ; and he insists that nobody before him had tried to develop a constant into a series of cosine, and that one should be cautious about the limits between which the development holds. A note explains that he was not able to consult mathematical books in Grenoble, and that he is willing to quote those who contributed to the subject before him.

There is no mention of Lagrange in the letter, but Hérivel gives very good raisons to believe that it was addressed to him.

At the same time, Fourier wrote to Laplace, again on the subject of trigonometric series. These letters were also published by Hérivel.

2.3.3. Then came the subject proposed for the Prize : *donner la théorie mathématique des lois de propagation de la chaleur, et comparer le résultat de cette théorie à des expériences exactes* (to give the mathematical theory of the propagation of heat, and compare the results to exact experiments).

The memoir of Fourier, *Théorie du mouvement de la chaleur dans les corps solides*, with the subtitle *et ignem regunt numeri (Plato)*, was received on September 28, 1811 and given the number 2, as certified by "le comte Lagrange, le comte Laplace, Malus, Legendre." The Prize was announced on January 6, 1812 :

La classe a décerné le prix, d'une valeur de 3000 F, *au mémoire enregistré sous le n°* 2, *portant cette épitaphe "et ignum regunt numeri (Plato)". Cette pièce renferme les véritables équations différentielles de la transmission de la chaleur, soit à l'intérieur des corps, soit à leur surface ; et la nouveauté du sujet, jointe à son importance, a déterminé la classe à couronner cet Ouvrage, en observant cependant que la manière dont l'Auteur parvient à ses équation n'est pas exemptc de difficultés, et que son analyse, pour les intégrer, laisse encore quelque chose à désirer, soit relativement à la généralité, soit même du coté de la rigueur* [**17**].

Clearly it was not a complete approval of Fourier, and even his analysis (what we now call Fourier analysis) was condemned as non rigorous.

This appreciation had a long run effect in France. When Arago, who succeeded Fourier as secrétaire perpétuel, wrote Fourier's obituary, he was very enthousiastic about many aspects of the life and works of Fourier, but he did not say a word about trigonometric series.

2.3.4. As a final piece of information here are a few facts and dates.

Lagrange died on April 1, 1813. Napoléon, defeated, abdicated on April 4, 1814, was sent to Elba and replaced by Louis XVIII, went back to France on March 1, 1815, took power again on March 20, was defeated at Waterloo on June 18, and abdicated for the second time on June 22. This period, from the beginning of March to June 18, is called *les 100 jours* in France. The years 1812–1815 were a political and military turmoil. But the *Livre des séances* shows that the first class of the Institute (Academy of sciences) met regularly every week. On May 22, 1815, that is, during *les 100 jours*, we can read :

> Le ministre de l'Intérieur annonce qu'il a acquis par ordre de l'Empereur les manuscrits laissés par M. Lagrange, lesquels doivent être déposés à l'Institut qui nommera une commission pour les mettre en ordre et en surveiller l'impression. MM. Legendre, Prony, Lacroix et Poisson sont nommés pour cet objet [**18**].

The minister of the Interior was the mathematician Lazare Carnot. The committee worked until the end of 1815 and succeeded in classifying Lagrange's handwritten papers [**19**].

Fourier was elected as member of the Academy on May 27, 1816, he was turned down by Louis XVIII, elected again and confirmed on May 12, 1817. He was elected as secrétaire perpétuel on November 18, 1822, and he died on May 16, 1830.

2.4. Let me try now to tell the story as I see it. Fourier had fought as much as he could after 1807 in order to let his proofs be known and the objections discussed. The way he obtained the Prize in 1812 was a disappointment for him. He had a high respect for Lagrange, and Lagrange had ignored Fourier's arguments. Lagrange died soon afterwards. He left an enormous amount of unpublished works, and for some reason Fourier had access to the handwritten papers of Lagrange. He saw the page on equation (a), he felt angry, he immediately wrote his comments and joined them to Lagranges's paper. When Lacroix, Legendre, Poisson and Prony found these two sheets of paper together, Prony wrote "deux feuilles" and the others agreed. Fourier never told the story. To Dirichlet and other young people he liked to speak on different matters. These young people knew the works of Fourier and appreciated his treatment of trigonometric series. Fourier had to explain why it was badly received. Then he spoke on vibrating strings, Daniel Bernoulli, Lagrange, and declared that Lagrange strongly opposed his own conclusions and that there was a piece in the Archives proving this opposition. Dirichlet repeated the statement of Fourier to Riemann, and Riemann wrote what we saw.

The second part of this paper, on Lagrange and Fourier, received the constant help of several collaborators of the Archives de l'Académie des sciences and Bibliothèque de l'Institut. I am particularly thankful to Mesdames Florence Greffe, director, and Claudine Pouret at the Archives, and Mireille Pastoureau, director, and Annie Chassagne, conservateur en chef, at the Library.

I also thank Robert Ryan for his careful reading and the many linguistic improvements he made.

The history of mathematics is a pleasant opportunity to go to libraries and read (or try to read) old papers. Shall we leave such opportunities to future mathematicians about what we do now ? This question is my conclusion.

Lagrange Joseph-Louis

Fourier Joseph

Excerpts from the "Schriftstück"

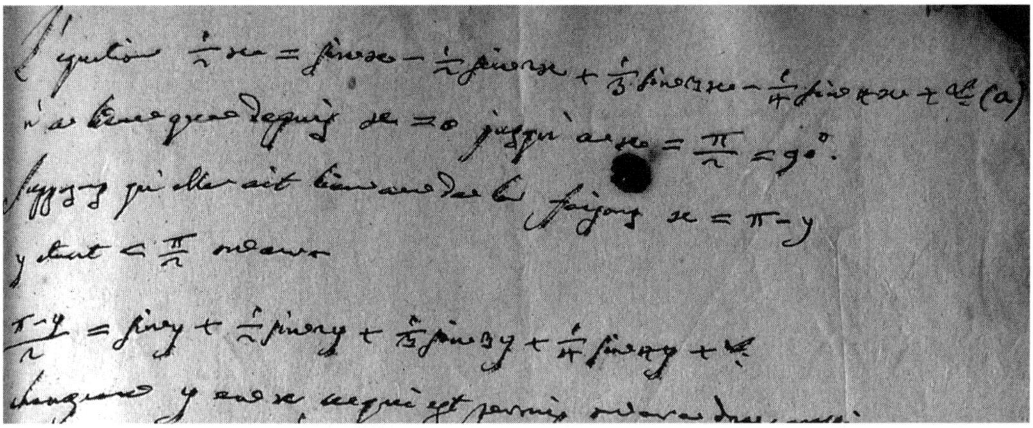
L'équation $\frac{1}{2}x = \sin x - \frac{1}{2}\sin 2x + \frac{1}{3}\sin 3x - \frac{1}{4}\sin 4x + \&c$ (a)
n'a lieu que depuis $x = 0$ jusqu'à $x = \frac{\pi}{2} = 90°$.
Supposons qu'elle ait lieu au delà, faisons $x = \pi - y$
y étant $< \frac{\pi}{2}$ nous aurons
$\frac{\pi - y}{2} = \sin y + \frac{1}{2}\sin 2y + \frac{1}{3}\sin 3y + \frac{1}{4}\sin 4y + \&c$
changeant y en x, ce qui est permis ...

Courtesy of Bibliothèque de l'Institut de France

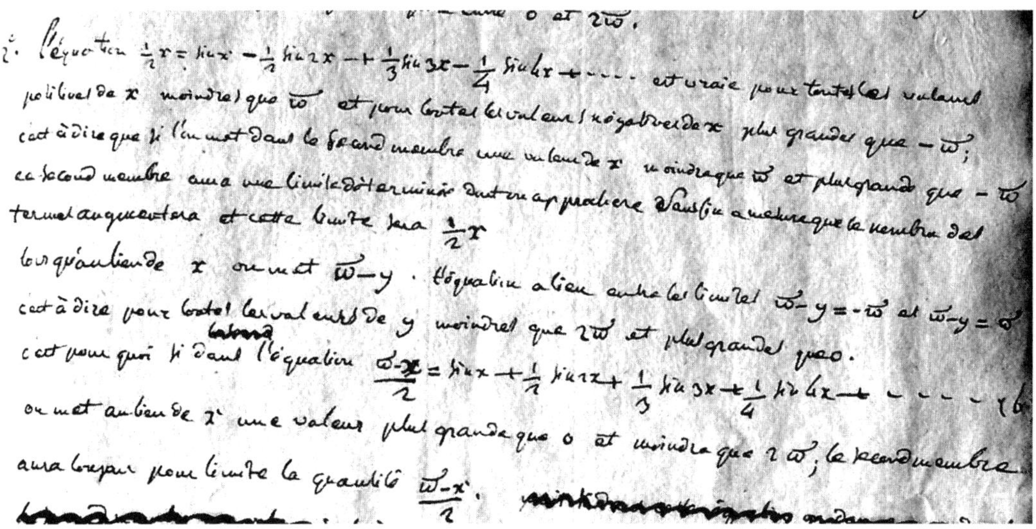
2°. l'équation $\frac{1}{2}x = \sin x - \frac{1}{2}\sin 2x + \frac{1}{3}\sin 3x - \frac{1}{4}\sin 4x + \ldots$ est vraie pour toutes les valeurs positives de x moindres que ϖ et pour toutes les valeurs négatives de x plus grandes que $-\varpi$; c'est à dire que si l'on met dans le second membre une valeur de x moindre que ϖ et plus grande que $-\varpi$ ce second membre aura une limite déterminée dont on approchera d'aussi près que l'on voudra à mesure que le nombre des termes augmentera et cette limite sera $\frac{1}{2}x$

lorsqu'au lieu de x on met $\varpi - y$. l'équation a lieu entre les limites $\varpi - y = -\varpi$ et $\varpi - y = \varpi$ c'est à dire pour toutes les valeurs de y moindres que 2ϖ et plus grandes que 0.
c'est pourquoi si dans l'équation $\frac{\varpi - x}{2} = \sin x + \frac{1}{2}\sin 2x + \frac{1}{3}\sin 3x + \frac{1}{4}\sin 4x + \ldots$ (b)
on met au lieu de x une valeur plus grande que 0 et moindre que 2ϖ, le second membre aura toujours pour limite la quantité $\frac{\varpi - x}{2}$.

Courtesy of Bibliothèque de l'Institut de France

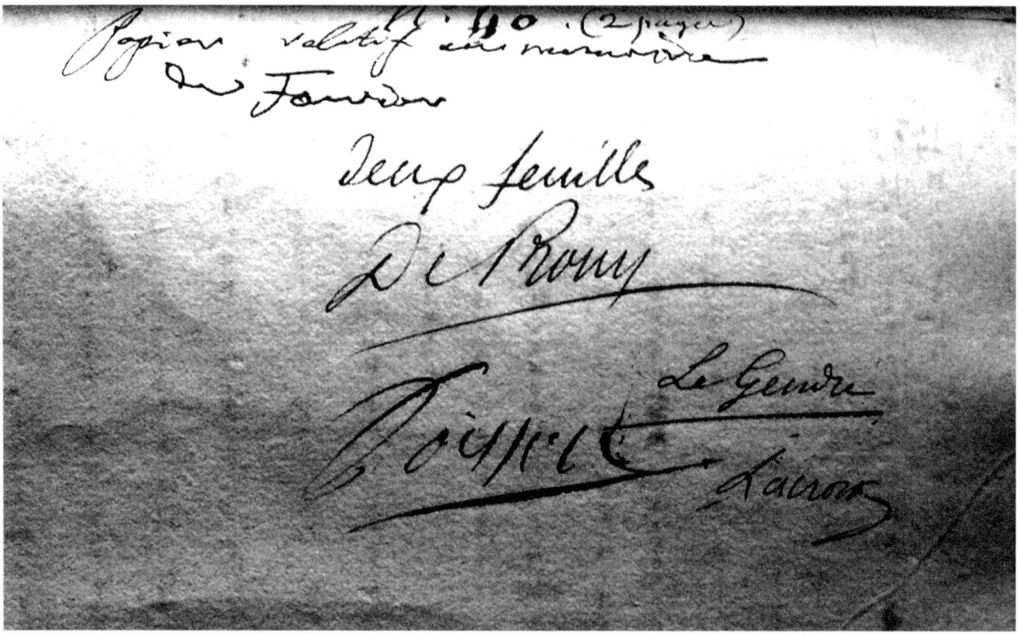

Courtesy of Bibliothèque de l'Institut de France

References

[1] D. Bernoulli, *Sur le mélange de plusieurs espèces de vibrations simples isochrones, qui peuvent exister dans un même système de corps,* Mém. Acad. Berlin 55 (1753), in Die Werke von Daniel Bernoulli, Basel, BIrkhäuser 1982..

[2] H. Burkhardt, *Entwicklungen nach oscillirenden Functionen und integration der Differentialgleichungen der mathematischen Physik,* Erster Hauptteil : die Ausbildung der Methode der Reihenentwicklungen an physikalischen und astronomischen Problemen, Jahresbericht der Deutschen Mathematiker Vereinigung X 2, 1908, 1–1804.

[3] G. Cantor, *Über die Ausdehnung eines Satzes aus der Theorie der trigonometrischen Reihen,* Math. Annalen Leipzig **5**, 123–132, in Georg Cantor gesammelte Abhandlungen, Berlin 1932.

[4] A. Denjoy, *Leçons sur le calcul des coefficients d'une série trigonométrique I, II, III, IV,* Paris, Gauthier–Villars 1941–1949.

[5] J. Fourier, *Théorie analytique de la chaleur,* Paris, Firmin–Didot, 1822 (also in Œuvres, Paris, Gauthier–Villars 1888–1890).

[6] J. Fourier, *Minutes de lettres,* service des manuscrits, Bibliothèque Nationale de France, Fr 22501, fol. 72–74.

[7] J. Herivel, *Joseph Fourier, Lettres inédites,* 1808–1816, Paris 1980.

[8] G. Jouve, *Imprévus et pièges des cordes vibrantes chez d'Alembert (1755–1783), doutes et certitudes sur sur les équations aux dérivées partielles, les séries et les fonctions,* Thèse de doctorat de l'Université Claude Bernard Lyon 1 (10 juillet 2007), I Thèse principale, 166p. II Annexes 398 p.

[9] J. L. Lagrange, *Recherches sur la nature et la propagation du son,* Miscellanea Taurinensia I, 1759, *Nouvelles recherches sur la nature et la propagation du son,* Miscellanea Taurinensia II, 1760, in Œuvres de Lagrange, Tome I, Paris, Gauthier–Villars 1867, 37–150 and 151–332.

[10] J. L. Lagrange, *manuscrits,* Bibliothèque de l'Institut de France, vol. 906, 102–103.

[11] V. Maz'ya and T. Shaposhnikova, *Jacques Hadamard, A Universal Mathematician*, History of Mathematics vol. 14, American Mathematical Society, London Mathematical Society, 1998.

[12] B. Riemann, *Ueber die Darstellbarkeit einer Function durch eine trigonometrische Reihe,* (Habilitation dissertation, Göttingen 1854), in besammelte mathematische Werke, Leipzig 1876.

[13] L. Schwartz, *Généralisation de la notion de fonction, de dérivation, de transformation de Fourier, et applications mathématiques et physiques,* Annales de l'Université de Grenoble, 1946.

[14] C. Truesdell, *The rational mechanics of flexible and elastic bodies 1638–1788,* in L. Euleri Opera Omnia, II 11n sectio altera, Zurich 1960.

[15] A. P. Youschkevitch, *About the history of the debate on vibrating strings (d'Alembert and the use of "discontinuous" functions),* (in Russian) Istoriko–matematitcheskie issledovania **20** 1975, 221-231 (French translation in Jouve II).

[16] A. Zygmund, *Trigonometric series I, II,* Cambridge University Press 1959.

[17] Séance publique de l'Institut Impérial de France du 6 janvier 1812, classe des sciences mathématiques et physiques, proclamation des prix, Archives de l'Académie des sciences.

[18] Livre des séances de la première classe de l'Institut Impérial de France 1812–1815, Archives de l'Académie des sciences.

[19] Procès verbaux des travaux de la commission chargée de mettre en ordre les papiers de Lagrange, 1815–1817, dossier Lagrange (paquet n° 3, pièce n° 40, 8 novembre 1815), Archives de l'Académie des sciences.

Department of Mathematics, University Paris–Sud 91406 Orsay Cedex
E-mail address: `Jean-Pierre.Kahane@math.u-psud.fr`

Proceedings of Symposia in Pure Mathematics
Volume **79**, 2008

Quantitative Unique Continuation, Logarithmic Convexity of Gaussian Means and Hardy's Uncertainty Principle

Carlos E. Kenig

In this paper we describe some recent works on quantitative unique continuation for elliptic, parabolic and dispersive equations. We also discuss recent works on the logarithmic convexity of Gaussian means of solutions to Schrödinger evolutions and the connection with a well-known version of the uncertainty principle, due to Hardy. The elliptic results are joint work with J. Bourgain [BK], while the remainder of the works discussed here are joint works with L. Escauriaza, G. Ponce and L. Vega ([EKPV1], [EKPV2], [EKPV3], [EKPV4], [EKPV5]). The paper is based on lectures presented at WHAPDE 2008, Merida, Mexico. I am grateful to the organizers of WHAPDE 2008 and to the participants in the workshop for the invitation and the very friendly atmosphere of the workshop. For further references and background on the problems discusses here, see [BK], [K1], [K2], [EKPV1], [EKPV2], [EKPV3], [EKPV4], [EKPV5] and the references therein.

1. Some recent quantitative unique continuation theorems

Here I will discuss some quantitative unique continuation theorems for elliptic, parabolic, and dispersive equations. I will start by describing the elliptic situation. This arose as a key step in the work of [BK] which proved Anderson localization at the bottom of the spectrum for the continuous Bernoulli model in higher dimensions, a question originating in Anderson's paper [A]. Briefly, this says the following: consider a random Schrödinger operator on $\mathbb{R}^n$, of the form $H_\epsilon = -\triangle + V_\epsilon$, where

$$V_\epsilon(x) = \sum_{j\in\mathbb{Z}^n} \epsilon_j \phi(x-j), \quad \phi \in C_0^\infty(B(0,1/10)), \quad 0 \le \phi \le 1$$

and $\epsilon_j \in \{0,1\}$ are independent. It is not difficult to see that $\inf \operatorname{spec} H_\epsilon = 0$ a.s. . In this context, Anderson localization means that for energies E near the bottom of the spectrum (i.e. $0 < E < \delta$) H_ϵ has pure point spectrum, with exponentially decaying eigenfunctions, a.s. . When V_ϵ has a continuous site distribution ($\epsilon_j \in [0,1]$) this has been understood for some time ([GMP] $n = 1$, [FS] $n > 1$). For the Anderson-Bernoulli model this was known for $n = 1$ ([CKM]; [SVW]), but not in higher dimensions. We now have:

2000 *Mathematics Subject Classification.* Primary 35Q53; Secondary 35G25, 35D99.
Partially supported by NSF Grant #DMS-0456583.

THEOREM 1.1 ([BK]). *There exists $\delta > 0$ s.t. for $0 < E < \delta$, H_ϵ displays Anderson localization a.s., $n \geq 1$.*

In establishing this result we were lead to the following deterministic quantitative unique continuation theorem: Suppose that u is a solution to $\Delta u + Vu = 0$ in $\mathbb{R}^n$, where $|V| \leq 1$, and $|u| \leq C_0$, $u(0) = 1$. For R large, define

$$M(R) = \inf_{|x_0|=R} \sup_{B(x_0,1)} |u(x)|.$$

Note that by unique continuation, $\sup_{B(x_0,1)} |u(x)| > 0$. How small can $M(R)$ be?

THEOREM 1.2 ([BK]).

$$M(R) \geq C \exp(-CR^{4/3} \log R).$$

REMARK 1.3. In order for our argument to give the desired application to Anderson localization for the Bernoulli model, we would need an estimate of the form $M(R) \geq C \exp(-CR^\beta)$, with $\beta < \frac{1+\sqrt{3}}{2} \approx 1.35$. Note that $4/3 = 1.333\ldots$.

As it turns out, this is a quantitative version of a conjecture of E.M. Landis. He conjectured (late 60's) that if $\Delta u + Vu = 0$ in $\mathbb{R}^n$, where $|V| \leq 1$, $|u| \leq C_0$, and $|u(x)| \leq C \exp(-C|x|^{1+})$, then $u \equiv 0$. This conjecture of Landis was disproved by Meshkov ([M]), who constructed such a V, $u \not\equiv 0$, with $|u(x)| \leq C \exp(-C|x|^{4/3})$. This example also shows the sharpness of our lower bound on $M(R)$. One should note however that in Meshkov's example u, V are complex valued.

Our proof uses a rescaling procedure, combined with well-known Carleman estimates.

Q:. Can $4/3$ be improved to 1 in our lower bound for $M(R)$ for real valued u, V?

Let us now turn our attention to parabolic equations. Thus, consider solutions to

$$\partial_t u - \Delta u + W(x,t) \cdot \nabla u + V(x,t)u = 0$$

in $\mathbb{R}^n \times (0,1]$, with $|W| \leq N$, $|V| \leq M$. Then, as is well-known, the following backward uniqueness result holds: If $|u(x,t)| \leq C_0$ and $u(x,1) \equiv 0$, then $u \equiv 0$ (see [LO]). This result has been extended by Escauriaza-Seregin-Šverák ([ESŠ]) who showed that it is enough to assume that u is a solution on $\mathbb{R}^n_+ \times (0,1]$, where $\mathbb{R}^n_+ = \{x = (x', x_n) : x_n > 0\}$, without any assumption on $u|_{\partial\mathbb{R}^n_+ \times [0,1]}$. This was a crucial ingredient in their proof that weak (Leray–Hopf) solutions of the Navier–Stokes system in $\mathbb{R}^3 \times [0,1)$, which have uniformly bounded L^3_x norm are regular and unique. In 1974, Landis–Oleinik, [LO], in parallel to Landis' conjecture for elliptic equations mentioned earlier, formulated the following conjecture: Let u be as in the backward uniqueness situation mentioned above. Assume that, instead of $u(x,1) \equiv 0$, we assume that $|u(x,1)| \leq C \exp(-C|x|^{2+\epsilon})$, for some $\epsilon > 0$. Is then $u \equiv 0$? Clearly, the exponent 2 is optimal here.

THEOREM 1.4 ([EKPV1]). *The Landis–Oleinik conjecture holds. More precisely, if $||u(\cdot,1)||_{L^2(B(0,1))} \geq \delta$, there exists $R_0 = R_0(\delta, M, N, n) > 0$ s.t. for $|y| \geq R_0$, we have*

$$||u(\cdot,1)||_{L^2(B(0,1))} \geq C \exp(-C|y|^2 \log|y|).$$

Moreover, an analogous result holds for u only defined in $\mathbb{R}^n_+ \times (0,1]$.

The proof of this result uses space-time rescalings and parabolic Carleman estimates, in the spirit of the elliptic case. It holds for both real and complex solutions. We hope that this result will prove useful in control theory.

We now turn our attention to dispersive equations. Ler us consider non-linear Schrödinger equations of the form

$$i\partial_t u + \triangle u + F(u,\overline{u})u = 0, \quad \text{in } \mathbb{R}^n \times [0,1],$$

for suitable non-linearity F, and let us try to understand what (if any) is the analog of the parabolic result we have just explained. The first obstacle is that the Schrödinger equations are time reversible and so "backward" makes no sense here. As is usual for uniqueness questions, we consider linear Schrödinger equations of the form

$$i\partial_t u + \triangle u + Vu = 0, \quad \text{in } \mathbb{R}^n \times [0,1],$$

and deal with suitable $V(x,t)$ so that we can, in the end, set

$$V(x,t) = F(u(x,t), \overline{u}(x,t)).$$

In order to motivate our work, I will first recall the following version of Heisenberg's uncertainty principle, due to Hardy, [SS]: if $f : \mathbb{R} \to \mathbb{C}$, and we have $f(x) = \mathcal{O}(e^{-\pi A x^2})$ and $\hat{f}(\xi) = \mathcal{O}(e^{-\pi B \xi^2})$, $A, B > 0$, if $A \cdot B > 1$, then $f \equiv 0$. For instance, if

$$|f(x)| \le C_\epsilon \exp(-C_\epsilon |x|^{2+\epsilon}), \quad |\hat{f}(\xi)| \le C_\epsilon \exp(-C_\epsilon |\xi|^{2+\epsilon}),$$

then $f \equiv 0$. This can easily be translated into an equivalent formulation for solutions to the free Schrödinger equation. For, if v solves

$$i\partial_t v + \partial_x^2 v = 0 \quad \text{in } \mathbb{R} \times [0,1],$$

with $v(x,0) = v_0(x)$, then

$$v(x,t) = \frac{C}{\sqrt{t}} \int e^{i|x-y|/4t} v_0(y)dy,$$

so that

$$v(x,1) = Ce^{i|x|^2/4} \int e^{-ixy/2} e^{i|y|^2/4} v_0(y)dy.$$

If we then apply the corollary to Hardy's uncertainty principle to $f(y) = e^{iy^2/4}v_0(y)$, we se that if

$$|v(x,0)| \le C_\epsilon \exp(-C_\epsilon |x|^{2+\epsilon}) \quad \text{and} \quad |v(x,1)| \le C_\epsilon \exp(-C_\epsilon |x|^{2+\epsilon}),$$

we must have $v \equiv 0$. Thus, for time-reversible equations, the analog of backward uniqueness should be "uniqueness from behavior at two different times". Thus, we are interested in such results with "data eventually 0" or even with "decaying very fast data". This kind of uniqueness question for "data eventually 0" has been studied for some time. For the 1-d cubic Schrödinger equation

$$i\partial_t u + \partial_x^2 u \mp |u|^2 u = 0 \quad \text{in} \quad \mathbb{R} \times [0,1],$$

B.Y. Zhang ([Z2]) showed that if $u \equiv 0$ on $(-\infty, a] \times \{0,1\}$ or on $[a, +\infty) \times \{0,1\}$, $a \in \mathbb{R}$, then $u \equiv 0$ on $\mathbb{R} \times [0,1]$. His proof used inverse scattering, a non-linear Fourier transform, and analyticity. In 2002, [KPV3] did away with scattering and analyticity, proving corresponding results for solutions to

$$i\partial_t u + \triangle u + V(x,t)u = 0 \quad \text{in} \quad \mathbb{R}^n \times [0,1], \quad n \ge 1.$$

THEOREM 1.5 ([KPV3]). *If* $V \in L^1_t L^\infty_x \cap L^\infty_{loc}$ *and*

$$||V||_{L^1_t L^\infty(|x|>R)} \xrightarrow[R\to\infty]{} 0$$

and there exists a strictly convex cone $\Gamma \subset \mathbb{R}^n$ *and a* $y_0 \in \mathbb{R}^n$ *such that*

$$\operatorname{supp} u(\cdot,0) \subset y_0 + \Gamma, \qquad \operatorname{supp} u(\cdot,1) \subset y_0 + \Gamma,$$

then we must have $u \equiv 0$ *on* $\mathbb{R}^n \times [0,1]$.

Clearly, taking $V(x,t) = |u|^2(x,t)$, we recover Zhang's result mentioned above. This was extended by [IK] who considered more general potentials V and the case when $\Gamma = \mathbb{R}^n_+$. For instance, if $V \in L^{\frac{n+2}{2}}(\mathbb{R}^n \times [0,1])$ or even $V \in L^p_t L^q_x(\mathbb{R}^n \times [0,1])$ with $2/p + n/q \le 2$, $1 < p < \infty$ ($n = 1$, $1 < p < 2$) or $V \in C([0,1]; L^{n/2}(\mathbb{R}^n))$ $n \ge 3$, the result holds with Γ a half-plane. Our extension of Hardy's uncertainty principle, to this context, now is:

THEOREM 1.6 ([EKPV2]). *Let* u *be a solution of*

$$i\partial_t u + \Delta u + Vu = 0, \quad \text{in } \mathbb{R}^n \times [0,1].$$

Assume that $V \in L^\infty(\mathbb{R}^n \times [0,1])$, $\nabla_x V \in L^1_t([0,1]; L^\infty_x(\mathbb{R}^n))$ *and*

$$\lim_{R\uparrow\infty} ||V||_{L^1_t L^\infty(|x|>R)} = 0.$$

If there exists $\alpha > 2$, $a > 0$, *such that* $u(\cdot,0), u(\cdot,1) \in H^1(e^{a|x|^\alpha} dx)$, *then* $u \equiv 0$.

It is conjectured that Theorem 1.6 remains valid assuming only that u, ∇u at times 0, 1 are in $L^2\left((y_0+\Gamma), e^{a|x|^\alpha} dx\right)$, with $y_0 + \Gamma$ as in Theorem 1.5. This extension of Theorem 1.6 would clearly imply Theorem 1.5.

Let me sketch the prof of this result. Our starting point is:

LEMMA 1.7 ([KPV3]). $\exists \epsilon > 0$ *s.t. if* $||V||_{L^1_t L^\infty_x} \le \epsilon$ *and* u *solves*

$$i\partial_t u + \Delta u + Vu = H, \quad \text{in } \mathbb{R}^n \times [0,1],$$

and $u_0(x) = u(x,0)$, $u_1(x) = u(x,1)$ *belong to* $L^2(e^{2\beta x_1} dx) \cap L^2(dx)$ *and*

$$H \in L^1_t(L^2(e^{2\beta x_1} dx) \cap L^2(dx)),$$

then

$$u \in C([0,1]; L^2(e^{2\beta x_1} dx))$$

and

$$\sup_{0\le t\le 1} ||u(\cdot,t)||_{L^2(e^{2\beta x_1} dx)} \le$$

$$\le C\left\{||u_0||_{L^2(e^{2\beta x_1} dx)} + ||u_1||_{L^2(e^{2\beta x_1} dx)} + ||H||_{L^1_t L^2(e^{2\beta x_1} dx)}\right\}$$

with C *independent of* β.

This is a delicate lemma. If we a priori knew that $u \in C([0,1]; L^2(e^{2\beta x_1} dx))$, a variant of the energy method, splitting frequencies into $\xi_1 > 0$, $\xi_1 < 0$, gives the result. But, since we are not free to prescribe both u_0, u_1, we cannot use a priori estimates. This is instead accomplished by "truncating" the weight $2\beta x_1$ and introducing an extra parameter.

Or next step is to deduce, from Lemma 1.7, further weighted estimates:

COROLLARY 1.8. *Assume that we are under the hypothesis of Lemma 1.7 and for some $a > 0$, $\alpha > 1$,*

$$u_0, u_1 \in L^2(e^{a|x|^\alpha}dx), \quad H \in L^1_t L^2_x(e^{a|x|^\alpha}dx).$$

Then $\exists C_\alpha > 0$, $b > 0$ s.t.

$$\sup_{0<t<1} \int_{|x|>C_\alpha} e^{b|x|^\alpha} |u(x,t)|^2 dx < \infty.$$

Idea for the proof of Corollary 1.8: Multiply u by $\eta_R(x) = \eta(x/R)$, $\eta \equiv 0$ for $|x| \le 1$, $\eta \equiv 1$ for $|x| > 2$. We apply Lemma 1.7 to $u_R(x,t) = \eta(x/R)u(x,t)$, with $\beta = \gamma R^{\alpha-1}$, for suitable γ and the corollary follows.

The next step of the proof is to deduce lower bounds for L^2 space-time integrals, in analogy with the elliptic and parabolic arguments. These are "quantitative".

THEOREM 1.9. *Let u solve $i\partial_t u + \triangle u + Vu = 0$, $x \in \mathbb{R}^n, t \in [0,1]$. Assume that*

$$\int_0^1 \int_{\mathbb{R}^n} |u|^2 + |\nabla u|^2 \le A, \quad \text{and that} \quad \int_{\frac{1}{2}-\frac{1}{8}}^{\frac{1}{2}+\frac{1}{8}} \int_{|x|<1} |u|^2 dxdt \ge 1,$$

with $||V||_\infty \le L$. Then there exists $R_0 = R_0(A, L, n) > 0$ and c_n s.t. if $R > R_0$

$$\delta(R) = \left(\int_0^1 \int_{R-1\le|x|\le R} (|u|^2 + |\nabla u|^2) dxdt \right)^{\frac{1}{2}} \ge c_n e^{-c_n R^2}.$$

Clearly, Corollary 1.8 applied to u, ∇u, combined with Theorem 1.9 yield our version of Hardy's uncertainty principle.

In order to prove Theorem 1.9, we use a Carleman estimate which is a variant of one due to V. Isakov [I].

LEMMA 1.10 ([EKPV2]). *Assume that $R > 0$ and $\phi : [0,1] \to \mathbb{R}$ is a smooth real function. Then, there exists $C = C(n, ||\phi'||_\infty, ||\phi''||_\infty) > 0$ s.t.*

$$\frac{\alpha^{3/2}}{R^2} \left\| e^{\alpha\left|\frac{x}{R}+\phi(t)\vec{e}_1\right|^2} g \right\|_{L^2} \le C_n \left\| e^{\alpha\left|\frac{x}{R}+\phi(t)\vec{e}_1\right|^2} (i\partial_t + \triangle) g \right\|_{L^2},$$

for all $\alpha > C_n R^2$, $g \in C_0^\infty(\mathbb{R}^{n+1})$ s.t. $\operatorname{supp} g \subset \{(x,t) : \left|\frac{x}{R} + \phi(t)\vec{e}_1\right| \ge 1\}$.

The proof of Lemma 1.10 follows by conjugating the operator $(i\partial_t + \triangle)$ with the weight $\exp\left(\alpha\left|\frac{x}{R} + \phi(t)\vec{e}_1\right|^2\right)$, and splitting the resulting operator into a Hermitian and an anti-Hermitian part. Then, the commutator between the two parts is positive, for g with the support property above and $\alpha \ge C_n R^2$.

In order to use Lemma 1.10 to prove Theorem 1.9, we choose $\theta_R, \theta \in C^\infty(\mathbb{R}^n)$, $\phi \in C_0^\infty([0,1])$ so that $\theta_R(x) = 1$ if $|x| < R-1$, $\theta_R(x) = 0$, $|x| \ge R$; $\theta(x) \equiv 0$ if $|x| < 1$, $\theta(x) \equiv 1$, when $|x| \ge 2$; $0 \le \phi \le 3$, with $\phi \equiv 3$ on $\left[\frac{1}{2} - \frac{1}{8}, \frac{1}{2} + \frac{1}{8}\right]$ and $\phi \equiv 0$ on $[0, 1/4] \cup [3/4, 1]$. We apply Lemma 1.10 to $g(x,t) = \theta_R(x) \cdot \theta\left(\frac{x}{R} + \phi(t)\vec{e}_1\right) u(x,t)$, $\alpha \approx R^2$, to obtain, after some manipulations, the desired result.

We next turn our attention to corresponding results for the KdV equations. In [Z1] it is proved that if

$$\partial_t u + \partial_x^3 u + u\partial_x u = 0, \quad \text{in } \mathbb{R} \times [0,1],$$

and $u_0(x) = u(x,0)$, $u_1(x) = u(x,1)$ are supported in $(a, +\infty)$ or in $(-\infty, a)$, then $u \equiv 0$. This was later extended by [KPV1], [KPV2], who also showed that if v_1, v_2

are solutions of

$$\partial_t v + \partial_x^3 v + v^k \partial_x v = 0, \quad k \geq 1,$$

and $u_0 = v_1(x,0) - v_2(x,0)$, $u_1 = v_1(x,1) - v_2(x,1)$ are supported in $(a, +\infty)$ or in $(-\infty, a)$, then $v_1 \equiv v_2$.

Further results are due to L. Robbiano ([R]). He considered u a solution to

$$\partial_t u + \partial_x^3 u + a_2(x,t)\partial_x^2 u + a_1(x,t)\partial_x u + a_0(x,t)u = 0 \tag{1.1}$$

with coefficients a_j in suitable function spaces. He showed that, if $u(x,0) = 0$, $x \in (b, \infty)$ some b, and $\exists C_1, C_2 > 0$ s.t.

$$|\partial_x^j u(x,t)| \leq C_1 \exp(-C_2 x^\alpha), \quad (x,t) \in (b, \infty) \times [0,1]$$

for some $\alpha > 9/4$, then $u \equiv 0$.

On the other hand, the Airy function

$$A_i(x) = \int e^{2\pi i x\xi + \xi^3 i} d\xi$$

is the fundamental solution for $\partial_t u + \partial_x^3 u = 0$, and verifies

$$|A_i(x)| \leq C(1 + x_-)^{-1/4} \exp(-C x_+^{3/2}).$$

We now have

THEOREM 1.11 ([EKPV3]). *If u is a solution of (1.1) on $\mathbb{R} \times [0,1]$ such that $u(x,0), u(x,1) \in H^1(e^{a x_+^{3/2}} dx)$ for any $a > 0$, and a_j belong to suitable function spaces, then $u \equiv 0$*

This is clearly optimal for $\partial_t u + \partial_x^3 u = 0$. The same result holds for $e^{a x_-^{3/2}} dx$. The proof of this theorem also has two steps, one consisting of upper bounds, the other of lower bounds. The second step follows closely that used for Schrödinger operators, but the upper bounds can no longer be obtained by any variant of the energy estimates. These are now replaced by suitable "dispersive Carleman estimates". A typical application of Theorem 1.11 is:

THEOREM 1.12 ([EKPV3]). *Let*

$$u_1, u_2 \in C([0,1]; H^3(\mathbb{R})) \cap L^2(|x|^2 dx),$$

solve

$$\partial_t u + \partial_x^3 u + u^k \partial_x u = 0 \quad \text{on } \mathbb{R} \times [0,1].$$

Assume that

$$u_1(\cdot,0) - u_2(\cdot,0), u_1(\cdot,1) - u_2(\cdot,1) \in H^1(e^{a x_+^{3/2}} dx)$$

for any $a > 0$. Then $u_1 \equiv u_2$.

Finally, we end with a result that shows that this result is sharp, even for the non-linear problem.

THEOREM 1.13 ([EKPV3]). *There exists $u \not\equiv 0$, a solution of*

$$\partial_t u + \partial_x^3 u + u^k \partial_x u = 0 \quad \text{in } \mathbb{R} \times [0,1]$$

s.t.

$$|u(x,0)| + |u(x,1)| \leq C \exp(-C x_+^{3/2}).$$

2. Convexity properties of Gaussian means of solutions to Schrödinger equations

As mentioned before, [EKPV2] proved that if $u \in C([0,1]; H^1(\mathbb{R}^n))$ solves

$$\begin{cases} i\partial_t u + \triangle u + V(x,t)u = 0 & \text{in } \mathbb{R}^n \times [0,1] \\ u(0) = u_0 \\ u(1) = u_1 \end{cases}$$

and $u_i \in L^2(e^{a|x|^\theta} dx)$ for some $a > 0$, $\theta > 1$, then $\exists C_\theta > 0$, $b > 0$ s.t.

$$\sup_{0<t<1} \int_{|x|>C_\theta} e^{b|x|^\theta} |u(x,t)|^2 dx < \infty$$

when the (complex) potential verifies $||V||_{L^1_t L^\infty_x} \leq \epsilon$, $\epsilon = \epsilon_n$. We will next re-examine this result and precise it, in the case $\theta = 2$. We will first deal with potentials $V = V(x)$, V real valued; $||V||_\infty \leq M_1$. We will consider $u \in C([0,1]; L^2(\mathbb{R}^n))$ which verifies

$$\partial_t u = i(\triangle u + Vu) \quad \text{in } \mathbb{R}^n \times [0,1].$$

We will assume that there exist positive numbers α and β such that $||e^{|x|^2/\beta^2}u(0)||$, $||e^{|x|^2/\alpha^2}u(1)||$ are finite. Here and in the sequel $||\cdot||$ denotes the L^2 norm in x. Then

$$\left\| e^{|x|^2/(\alpha t+(1-t)\beta)^2} u(t) \right\|^{\alpha t+(1-t)\beta}$$

is "logarithmically convex" in $[0,1]$, i.e.

THEOREM 2.1 ([EKPV5]). *There exists $N = N(\alpha,\beta)$ so that for $0 < s < 1$ we have*

$$\left\| e^{|x|^2/(\alpha s+(1-s)\beta)^2} u(s) \right\| \leq e^{N(M_1+M_1^2)} \times \left\| e^{|x|^2/\beta^2} u(0) \right\|^{\beta(1-s)/(\alpha s+(1-s)\beta)} \left\| e^{|x|^2/\alpha^2} u(1) \right\|^{\alpha s/(\alpha s+(1-s)\beta)}.$$

Moreover ("smoothing effect")

$$\left\| \sqrt{t(1-t)} e^{|x|^2/(\alpha t+(1-t)\beta)^2} \nabla u(t) \right\|_{L^2(\mathbb{R}^n\times[0,1])} \leq N e^{N(M_1+M_1^2)} \left[||e^{|x|^2/\beta^2} u(0)|| + ||e^{|x|^2/\alpha^2} u(1)|| \right].$$

Note that when $\alpha = \beta$, we have $\alpha t + (1-t)\beta = \alpha$ and this gives the precise version (for this case) of the [EKPV2] result. We start with the sketch of the proof in the case $\alpha = \beta$. It turns out that a formal argument giving the proof is not too dificult, but a rigurous justification is tricky. This is an important fact, since, as we will see, the formal arguments actually can lead to false results. To justify the interest of the case $\alpha \neq \beta$, consider the case $V \equiv 0$, i.e. the free particle. Then, if $u_0 = u(0)$,

$$u(x,t) = (e^{-i|\xi|^2 t}\hat{u}_0)\check{} = \int_{\mathbb{R}^n} \frac{e^{i|x-y|^2/4t}}{(4\pi i t)^{n/2}} u_0(y)dy =$$

$$= \frac{e^{i|x|^2/4t}}{(4\pi i t)^{n/2}} \int e^{-2ix\cdot\xi/4t} e^{i|y|^2/4t} u_0(y)dy = \frac{e^{i|x|^2/4t}}{(2it)^{n/2}} (\widehat{e^{i|\cdot|^2/4t}u_0})(x/2t),$$

so that, with $c_t = (2it)^{n/2}$,

$$c_t e^{-i|x|^2/4t} u(x,t) = (e^{i|\cdot|^2/4t} u_0)\hat{}\,(x/2t).$$

In this context the Hardy uncertainty principle says that if

$$u(0) \in L^2(e^{2|x|^2/\beta^2} dx), \quad u(1) \in L^2(e^{2|x|^2/\alpha^2} dx),$$

with $\alpha\beta \leq 4$, then $u_0 \equiv 0$ and 4 is sharp.

KEY CONVEXITY LEMMA 2.2 (abstract). *Let S be a symmetric operator, A an anti-symmetric one (possibly both depending on t), F a positive function, $f(x,t)$ a "reasonable function". Let $H(t) = (f,f)$, $D(t) = (Sf,f)$, $\partial_t S = S_t$ and $N(t) = D(t)/H(t)$ (the "frequency function"). Then*

i) $\partial_t^2 H = 2\partial_t \operatorname{Re}(\partial_t f - Sf - Af, f) +$
$+ 2(S_t f + [S,A]f, f) + ||\partial_t f - Af - Sf||^2 - ||\partial_t f - Af - Sf||^2$
and

ii) $\dot{N}(t) \geq (S_t f + [S,A]f, f)/H - ||\partial_t f - Af - Sf||^2/(2H)$

iii) *Moreover, if*

$$|\partial_t f - Af - Sf| \leq M_1|f| + F \quad \textit{in } \mathbb{R}^n \times [0,1], \quad S_t + [S,A] \geq -M_0$$

and

$$M_2 = sup_{0\leq t\leq 1}||F(t)||/||f(t)|| < \infty,$$

then $H(t)$ is "logarithmically convex" in $[0,1]$ and

$$H(t) \leq e^{N(M_0+M_1+M_1^2+M_2+M_2^2)} H(0)^{1-t} H(1)^t.$$

PROOF.

$$\dot{H}(t) = 2\operatorname{Re}(\partial_t f, f) = 2\operatorname{Re}(\partial_t f - Sf - Af, f) + 2(Sf,f),$$

so

(2.1) $$\dot{H}(t) = 2\operatorname{Re}(\partial_t f - Sf - Af, f) + 2D(t).$$

Also,

$$\dot{H}(t) = \operatorname{Re}(\partial_t f + Sf, f) + \operatorname{Re}(\partial_t f - Sf, f),$$
$$D(t) = \frac{1}{2}\operatorname{Re}(\partial_t f + Sf, f) - \frac{1}{2}\operatorname{Re}(\partial_t f - Sf, f).$$

Multiplying

(2.2) $$\dot{H}(t)D(t) = \frac{1}{2}\operatorname{Re}(\partial_t f + Sf, f)^2 - \frac{1}{2}\operatorname{Re}(\partial_t f - Sf, f)^2.$$

Adding an anti-symmetric part does not change the real parts, so

(2.3) $$\dot{H}(t)D(t) = \frac{1}{2}\operatorname{Re}(\partial_t f + Sf - Af, f)^2 - \frac{1}{2}\operatorname{Re}(\partial_t f - Sf - Af, f)^2.$$

Differentiating $D(t)$, we get

(2.4) $$\begin{aligned}\dot{D}(t) &= (S_t f, f) + (S\partial_t f, f) + (Sf, \partial_t f) = (S_t f, f) + 2\operatorname{Re}(\partial_t f, Sf) = \\ &= (S_t f + [S,A]f, f) + 2\operatorname{Re}(\partial_t f - Af, Sf) = \\ &= (S_t f, [S,A]f, f) + \frac{1}{2}||\partial_t f - Af + Sf||^2 - \frac{1}{2}||\partial_t f - Af - Sf||^2\end{aligned}$$

by polarization. This and (2.1) gives i). Next,

$$\dot{N}(t) = (S_t f + [S, A]f, f)/H +$$
$$+\frac{1}{2}\left[||\partial_t f - Af + Sf||^2||f||^2 - (\operatorname{Re}(\partial_t f - Af + Sf, f))^2\right]/H^2 +$$
$$+\frac{1}{2}\left[(\operatorname{Re}(\partial_t f - Af - Sf, f))^2 - ||\partial_t f - Af - Sf||^2||f||^2\right]/H^2$$

follows from (2.3). Now, the second line is non-negative (Cauchy-Schwartz),

$$(\operatorname{Re}(\partial_t f - Af - Sf, f))^2 \geq 0,$$

so ii) follows.

When we are in the situation of iii),

$$\dot{N}(t) \geq -M_0 + M_1^2 + M_2^2,$$

so that (2.1) now gives $\partial_t[\log H(t)] = \mathcal{O}(1) + 2N(t)$. If $G'(t) = \mathcal{O}(1)$, $G(0) = 0$, we get $\partial_t[\log H(t) - G(t)] = 2N(t)$, so that

$$\partial_t^2[\log H(t) - G(t)] \geq -(M_0 + M_1^2 + M_2^2),$$

so that

$$\partial_t^2\left[\log H(t) - G(t) + (M_0 + M_1^2 + M_2^2)t^2/2\right] \geq 0$$

which gives the desired "log convexity". □

Sketch of Proof ($\alpha = \beta = \gamma$). Let us now indicate how the "formal argument" for the first part of Theorem 2.1 would follow, when $\alpha = \beta = \gamma$. Suppose now (for later use) that

$$\partial_t u = (a + ib)(\triangle u + V(x,t)u + F(x,t)) \quad \text{in } \mathbb{R}^n \times [0,1],$$

$a \geq 0$, $||e^{\gamma|x|^2}u(0)|| < \infty$, $||e^{\gamma|x|^2}u(1)|| < \infty$, $\sup_{[0,1]} ||e^{\gamma|x|^2}F(x,t)||/||u(t)|| = M_2$, V is complex valued, $||V||_\infty \leq M_1$. Let $f = e^{\gamma\phi}u$, where $\phi(x,t)$ is to be chosen. Then, f verifies

$$\partial_t f = Sf + Af + (a + ib)(Vf + e^{\gamma\phi}F),$$

where

$$S = a(\triangle + \gamma^2|\nabla\phi|^2) - ib\gamma(2\nabla\phi \cdot \nabla + \triangle\phi) + \gamma\partial_t\phi,$$
$$A = ib(\triangle + \gamma^2|\nabla\phi|^2) - a\gamma(2\nabla\phi \cdot \nabla + \triangle\phi)$$

are symmetric, anti-symmetric. When $\phi(x,t) = |x|^2$ we obtain

$$S_t + [S, A] = -\gamma(a^2 + b^2)[8\triangle - 32\gamma^2|x|^2],$$

so that

$$S_t + [S, A] \geq 0, \quad |\partial_t f - Sf - Af| \leq \sqrt{a^2 + b^2}(M_1|f| + e^{\gamma|x|^2}|F|)$$

and the Lemma "gives" the (formal) "log convexity" result.

We need to have an argument which gives us the required smoothness and decay to justify the formal argument. Before doing that, we give the "formal" argument for the smoothing estimate: first note that integration by parts shows that

$$\int |\nabla f|^2 + 4\gamma^2|x|^2|f|^2 = \int e^{2\gamma|x|^2}(|\nabla u|^2 - 2n\gamma|u|^2)dx$$

where $f = e^{\gamma|x|^2}u$. Also, since $n = \nabla \cdot x$, integration by parts and Cauchy-Schwartz give

$$\int |\nabla f|^2 + 4\gamma^2|x|^2|f|^2 \geq 2\gamma n \int |f|^2 dx.$$

Adding we obtain

$$2\left(\int |\nabla f|^2 + 4\gamma^2|x|^2|f|^2\right) \geq \int e^{2\gamma|x|^2}|\nabla u|^2 dx. \tag{2.5}$$

Recall

$$\begin{aligned}\partial_t^2 H(t) &= 2\partial_t \operatorname{Re}(\partial_t f - Sf - Af, f) + 2(S_t f + [S,A]f, f) + \\ &\quad + ||\partial_t f + Sf - Af||^2 - ||\partial_t f - Af - Sf||^2 \geq \\ &\geq 2\partial_t \operatorname{Re}(\partial_t f - Sf - Af, f) - ||\partial_t f - Af - Sf||^2 + 2(S_t f + [S,A]f, f).\end{aligned}$$

Multiply by $t(1-t)$ and integrate by parts to obtain

$$\begin{aligned}2\int_0^1 t(1-t)(S_t f + [S,A]f, f)dt + 2\int_0^1 H(t)dt \leq H(1) + H(0) + \\ + 2\int_0^1 (1-2t)\operatorname{Re}(\partial_t f - Sf - Af, f) + \int_0^1 t(1-t)||\partial_t f - Sf - Af||^2 dt.\end{aligned}$$

We now use

$$\begin{aligned}S_t + [S,A] = -\gamma(a^2+b^2)[8\triangle - 32\gamma^2|x|^2], \\ |\partial_t f - Sf - Af| \leq \sqrt{a^2+b^2}\left(M_1|f| + e^{\gamma|x|^2}|F|\right),\end{aligned}$$

to obtain:

$$\begin{aligned}16\gamma(a^2+b^2)\int_0^1 \int t(1-t)\left\{|\nabla f|^2 + 4\gamma^2|x|^2|f|^2\right\} \leq \\ \leq \left[(NM_1^2+1)\sup_{[0,1]}\left\|e^{\gamma|x|^2}u(t)\right\|^2 + \sup_{[0,1]}\left\|e^{\gamma|x|^2}F\right\|^2\right](a^2+b^2).\end{aligned}$$

Finally, $\nabla f = e^{\gamma|x|^2}(\nabla u + 2xu\gamma)$, and (2.5) gives the bound: $(\gamma > 0)$

$$\begin{aligned}||\sqrt{t(1-t)}e^{\gamma|x|^2}\nabla u||_{L^2(\mathbb{R}^n\times[0,1])} + ||\sqrt{t(1-t)}|x|e^{\gamma|x|^2}u||_{L^2(\mathbb{R}^n\times[0,1])} \leq \\ \leq N\left[(1+\sqrt{M_1}+M_1)\sup_{[0,1]}\left\|e^{\gamma|x|^2}|u(t)|\right\| + \sup_{[0,1]}\left\|e^{\gamma|x|^2}F\right\|_{L^2(\mathbb{R}^n\times[0,1])}\right].\end{aligned}$$

How to justify the formal arguments? We first change $i(\triangle+V)$ by $(a+i)(\triangle+V)$, we change $|x|^2$ by $|x|^{2-\epsilon}$, $a > 0$, $\epsilon > 0$ and then pass to the limit. This can be justified when $V = V(x)$, real, bounded. This is how we proceed:

LEMMA 2.3 (Energy method). *Assume that* $u \in L^\infty([0,1];L^2) \cap L^2([0,1];H^1)$ *satisfies*

$$\partial_t u = (a+ib)(\triangle u + V(x,t)u) + F(x,t) \quad \text{in } \mathbb{R}^n \times [0,1],$$

$a > 0$, $b \in \mathbb{R}$. *Then, for* $0 \leq T \leq 1$,

$$\begin{aligned}e^{-M_T}\left\|e^{\gamma a|x|^2/(a+4\gamma(a^2+b^2)T)}u(T)\right\| \leq \\ \leq \left\|e^{\gamma|x|^2}u(0)\right\| + \sqrt{a^2+b^2}\left\|e^{\gamma a|x|^2/(a+4\gamma(a^2+b^2)T)}F\right\|_{L^1([0,T];L^2)},\end{aligned}$$

where $M_T = ||a \operatorname{Re} V - b \operatorname{Im} V||_{L^1([0,T];L^\infty)}$.

PROOF. For ϕ real, to be chosen, $v = e^\phi u$, v verifies

$$\partial_t v = Sv + Av + (a+ib)e^\phi F \quad \text{in } \mathbb{R}^n \times (0,1],$$

where S =sym, A =anti-sym,

$$S = a(\triangle + |\nabla\phi|^2) - ib(2\nabla\phi\cdot\nabla + \triangle\phi) + (\partial_t\phi + a\operatorname{Re} V - b\operatorname{Im} V)$$

and

$$A = ib(\triangle + |\nabla\phi|^2) - a(2\nabla\phi\cdot\nabla + \triangle\phi) + i(b\operatorname{Re} V - a\operatorname{Im} V).$$

$$\partial_t ||v||^2 = 2\operatorname{Re}(Sv,v) + 2\operatorname{Re}((a+ib)e^\phi F, v) \quad \text{(formally).}$$

A (formal) integration by parts gives

$$\operatorname{Re}(Sv,v) = -a\int |\nabla v|^2 + \int (a|\nabla\phi|^2 + \partial_t\phi)|v|^2 + \\ + 2b\operatorname{Im}\int \overline{v}\nabla\phi\cdot\nabla v + \int (a\operatorname{Re} V - b\operatorname{Im} V)|v|^2.$$

Cauchy-Schwartz gives

$$\partial_t ||v(t)||^2 \le 2||a\operatorname{Re} V - b\operatorname{Im} V||_\infty ||v(t)||^2 + 2\sqrt{a^2+b^2}||e^\phi F(t)||\,||v(t)||$$

when

$$\left(a + \frac{b^2}{a}\right)|\nabla\phi|^2 + \partial_t\phi \le 0.$$

When $\phi(x,t) = h(t)\phi(x)$, it suffices that

$$h^2(t)\left(a+\frac{b^2}{a}\right)|\nabla\phi(x)|^2 + h'(t)\phi(x) \le 0.$$

Eventually, we choose $\phi(x) = |x|^2$. We then choose

$$\begin{cases} h'(t) = -4\left(a+\frac{b^2}{a}\right)h^2(t) \\ h(0) = \gamma \end{cases},$$

so that $h(t) = \gamma a/(a + 4\gamma(a^2+b^2)t)$. To formalize the calculations, given $R > 0$, set

$$\phi_R(x) = \begin{cases} |x|^2 & |x| \le R \\ R^2 & |x| \ge R \end{cases},$$

choose a radial mollifier θ_ρ and set

$$\phi_{\rho,R}(x,t) = h(t)\theta_\rho * \phi_R(x), \quad v_{\rho,R} = e^{\phi_{\rho,R}}u.$$

Then, $\theta_\rho * \phi_R \le \theta_\rho * |x|^2 = |x|^2 + C(n)\rho^2$, and our inequality above holds uniformly in ρ and R. We obtain the result for $v_{\rho,R}$, let $\rho \to 0$, then $R \to \infty$, which gives the final estimate. Note that, for $a > 0$, Gaussian decay at $t = 0$ is preserved, with a loss. □

Next, we prove that if $u \in L^\infty([0,1];L^2) \cap L^2([0,1];H^1)$ verifies

$$\partial_t u = (a+ib)(\triangle u + V(x,t)u + F(x,t)),$$

where $||V||_{L^\infty} \leq M_1$, $\sup_{[0,1]} ||e^{\gamma|x|^2}F(t)||/||u(t)|| = M_2 < \infty$, and $||e^{\gamma|x|^2}u(0)||$, $||e^{\gamma|x|^2}u(1)||$ are finite, we have a "log convex" estimate, uniformly in $a > 0$, small. In fact, we now repeat the formal argument, but replace $\phi(x) = |x|^2$ by

$$\phi_\epsilon(x) = \begin{cases} |x|^2 & |x| \leq 1 \\ \frac{2|x|^{2-\epsilon}-\epsilon}{2-\epsilon} & |x| \geq 1 \end{cases}$$

and then by $\phi_{\epsilon,\rho}(x) = \theta_\rho * \phi_\epsilon$, where $\theta_\rho \in C_0^\infty$ is radial. We then have: $\phi_{\epsilon,\rho} \in C^{1,1}$, it is convex and grows at infinity slower that $|x|^{2-\epsilon}$ and $\phi_{\epsilon,\rho} \leq |x|^2 + C(n)\rho^2$. By the "energy estimate", for $a > 0$, $\epsilon > 0$, $\rho > 0$, our argument applies rigurously, since $u(0)e^{\gamma|x|^2} \in L^2 \Rightarrow 0 < t < 1$, $u(t)e^{\gamma|x|^{2-\epsilon}} \in L^2$, and for a t independent ϕ,

$$S_t + [S, A] = -\gamma(a^2+b^2)\left[4\nabla\cdot(D^2\phi\nabla) - 4\gamma^2 D^2\phi\nabla\phi\cdot\nabla\phi + \triangle^2\phi\right].$$

One can see that $||\triangle^2\phi_{\epsilon,\rho}||_\infty \leq C(n,\rho)\epsilon$, which gives the desired log convexity when $\epsilon \to 0$, then $\rho \to 0$, for $a > 0$. Once the log convexity holds, for $a > 0$ again, the "local smoothing" argument applies. The conclusion of these considerations is:

LEMMA 2.4. *Assume that $u \in L^\infty([0,1]; L^2(\mathbb{R}^n)) \cap L^2([0,1]; H^1)$ verifies*

$$\partial_t u = (a+ib)(\triangle u + V(x,t)u + F(x,t)), \quad \text{in } \mathbb{R}^n \times [0,1],$$

$\gamma > 0$ where $a > 0$, $b \in \mathbb{R}$, $||V||_\infty \leq M_1$. Then, $\exists N_\gamma$ s.t.

$$\sup_{[0,1]} ||e^{\gamma|x|^2}u(t)|| \leq$$

$$\leq e^{N_\gamma[(a^2+b^2)[M_1^2+M_2^2]+\sqrt{a^2+b^2}(M_1+M_2)]}||e^{\gamma|x|^2}u(0)||^{1-t}||e^{\gamma|x|^2}u(1)||^t,$$

$$||\sqrt{t(1-t)}e^{\gamma|x|^2}u||_{L^2(\mathbb{R}^n\times[0,1])} \leq N_\gamma(1+M_1+M_2)\left\{\sup_{[0,1]}||e^{\gamma|x|^2}u(t)||\right\},$$

where $M_2 = \sup_{[0,1]} ||e^{\gamma|x|^2}F(t)||/||u(t)|| < \infty$.

Conclusion of the argument when $V(x,t) = V(x)$, real. We now consider the Schrödinger operator $H = \triangle + V$, which is self-adjoint. We consider $u \in C([0,1]; L^2)$ solving

$$\partial_t u = i((\triangle + V)u) \quad \text{in } \mathbb{R}^n \times [0,1]$$

and assume that $||e^{\gamma|x|^2}u(0)|| < \infty$, $||e^{\gamma|x|^2}u(1)|| < \infty$. From spectral theory, $u(t) = e^{iHt}u(0)$. Moreover, for $a > 0$, consider the solution of

$$\partial_t u_a = (a+i)((\triangle+V)u_a) \quad \text{in } \mathbb{R}^n \times [0,1],\ u_a(0) = u(0).$$

We now have

$$u_a(t) = e^{(a+i)tH}u(0) = e^{atH}e^{itH}u(0) = e^{atH}u(t).$$

Clearly

$$||e^{\gamma|x|^2}u_a(0)|| = ||e^{\gamma|x|^2}u(0)||.$$

Also, $u_a(1) = e^{aH}u(1)$. Recall, from the "energy method" that if

$$\begin{cases} \partial_t v = a(\triangle+V)v \\ v(0) = v_0 \end{cases}, \quad V \text{ real},$$

$$\left\|e^{\gamma a|x|^2/(a+4\gamma a^2)}v(1)\right\| \leq \exp(\tilde{M}_1)||e^{\gamma|x|^2}v_0||,$$

where $\tilde{M_1} = ||aV||_{L^1([0,1];L^\infty)}$. Now, if $v_0 = u(1)$, then $v(1) = e^{aH}v_0 = u_a(1)$, so that

$$\left\|e^{\gamma|x|^2/(1+4\gamma a)}u_a(1)\right\| \leq \exp(\tilde{M_1})||e^{\gamma|x|^2}u(1)||.$$

Let $\gamma_a = \gamma/(1+4\gamma a)$ and apply now our log-convexity result for u_a, γ_a. We then obtain

$$||e^{\gamma_a|x|^2}u_a(s)|| \leq e^{NM_1}||e^{\gamma_a|x|^2}u_a(1)||^{1-s}||e^{\gamma_a|x|^2}u_a(0)||^s \leq$$
$$\leq e^{NM_1}\exp(\tilde{M_1})||e^{\gamma|x|^2}u(1)||^{1-s}||e^{\gamma|x|^2}u(0)||^s.$$

We then let $a \to 0$ and obtain the "log convexity" bound. To obtain the "local smoothing" bound, we again use the u_a, let $a \downarrow 0$. This establishes Theorem 2.1 when $\alpha = \beta$.

REMARK 2.5. Solutions so that $e^{\gamma|x|^2}u(0)$, $e^{\gamma|x|^2}u(1) \in L^2$ certainly exist for some γ. In fact, if $h \in L^2(e^{\epsilon|x|^2}dx)$ and $u_0 = e^{\delta(\triangle+V)}h$, our "energy method" gives this for $u(t) = e^{it(\triangle+V)}u_0$, $(V = V(x))$. (We are indebted to R. Killip for this remark.) When $V \equiv 0$, this characterizes such u! (see [EKPV4]).

A misleading convexity argument: Consider now $f = e^{a(t)|x|^2}u$, where u solves the free Schrödinger equation

$$\partial_t u = i\triangle u \quad \text{in } \mathbb{R}\times[-1,1].$$

Then, f verifies

$$\partial_t f = Sf + Af,$$
$$S = -4ia(x\partial_x + \frac{1}{2}) + a'x^2, \quad A = i(\partial_x^2 + 4a^2x^2).$$

In this case we have

$$S_t + [S,A] = 2\frac{a'}{a}S - 8a\partial_x^2 + \left(32a^3 + a'' - 2\frac{(a')^2}{a}\right)x^2.$$

If a is positive, even, and a solution of

$$32a^3 + a'' - 2\frac{(a')^2}{a} = 0 \quad \text{in } [-1,1],$$

then our formal calculations show that

$$\partial_t(a^{-1}\partial_t \log H_a(t)) \geq 0 \quad \text{in } [-1,1].$$

Hence, for $s < t$ we have

$$a(t)\partial_t \log H_a(s) \leq a(s)\partial_t \log H_a(t).$$

Integrating between $[-1,0]$ and $[0,1]$ and using the evenness of a, we conclude

$$H_a(0) \leq H_a(-1)^{1/2}H_a(1)^{1/2}.$$

Now, if a solves

$$\begin{cases} 32a^3 + a'' - 2\frac{(a')^2}{a} = 0 \\ \\ a(0) = 1, a'(1) = 0 \end{cases}$$

a is positive, even, and $\lim_{R\to\infty} Ra(R) = 0$. Also, $a_R(t) = Ra(Rt)$ also solves the equation. If the formal calculation holds for H_{a_R},

$$\left\|e^{Rx^2}u(0)\right\|^2 \leq \left\|e^{Ra(R)x^2}u(-1)\right\|\left\|e^{Ra(R)x^2}u(1)\right\|.$$

In particular, $u \equiv 0$. But $u(x,t) = (t-i)^{-1/2} e^{i|x|^2/4(t-i)}$ is a non-zero free solution, which decays as a quadratic exponential at $t = \pm 1$.

3. The case $\alpha \neq \beta$; the conformal or Appel transformation

LEMMA 3.1. *Assume $u(y,s)$ verifies*

$$\partial_s u = (a+ib)(\Delta u + V(y,s)u + F(y,s)) \quad \text{in } \mathbb{R}^n \times [0,1],$$

$a + ib \neq 0$, $\alpha > 0$, $\beta > 0$, $\gamma \in \mathbb{R}$ and set

$$\tilde{u}(x,t) = \left(\frac{\sqrt{\alpha\beta}}{\alpha(1-t)+\beta t}\right)^{n/2} u\left(\frac{\sqrt{\alpha\beta}\,x}{\alpha(1-t)+\beta t}, \frac{\beta t}{\alpha(1-t)+\beta t}\right) \times \exp\left(\frac{(\alpha-\beta)|x|^2}{4(a+ib)(\alpha(1-t)+\beta t)}\right).$$

Then $\tilde{u}$ verifies

$$\partial_t \tilde{u} = (a+ib)(\Delta \tilde{u} + \tilde{V}(x,t)\tilde{u} + \tilde{F}(x,t)) \quad \text{in } \mathbb{R}^n \times [0,1],$$

$$\tilde{V}(x,t) = \frac{\alpha\beta}{(\alpha(1-t)+\beta t)^2} V\left(\frac{\sqrt{\alpha\beta}\,x}{\alpha(1-t)+\beta t}, \frac{\beta t}{\alpha(1-t)+\beta t}\right),$$

$$\tilde{F}(x,t) = \frac{\sqrt{\alpha\beta}}{(\alpha(1-t)+\beta t)^{\frac{n}{2}+2}} F\left(\frac{\sqrt{\alpha\beta}\,x}{\alpha(1-t)+\beta t}, \frac{\beta t}{\alpha(1-t)+\beta t}\right).$$

Moreover, if $s = \beta t/(\alpha(1-t)+\beta t)$,

$$\left\| e^{\gamma|x|^2} \tilde{u}(t) \right\| = \left\| e^{\left[\frac{\gamma\alpha\beta}{(\alpha s+\beta(1-s))^2} + \frac{(\alpha-\beta)a}{4(a^2+b^2)(\alpha s+\beta(1-s))}\right]|y|^2} u(s) \right\|$$

$$\left\| e^{\gamma|x|^2} \tilde{F}(t) \right\| = \frac{\alpha\beta}{(\alpha(1-t)+\beta t)^2} \left\| e^{\left[\frac{\gamma\alpha\beta}{(\alpha s+\beta(1-s))^2} + \frac{(\alpha-\beta)a}{4(a^2+b^2)(\alpha s+\beta(1-s))}\right]|y|^2} F(s). \right\|$$

The proof is by change of variables.

Conclusion of the proof of Theorem 2.1: We can assume $\alpha \neq \beta$. We can also assume $\alpha < \beta$ (change u for $\overline{u}(1-t)$). (This gives $(\alpha-\beta)a < 0$.) As before,

$$H = (\Delta + V), \quad u_a = e^{(a+i)tH} u(0) = e^{atH} u(t), \quad a > 0.$$

By the "energy estimate" we now have

$$\left\| e^{|x|^2/\alpha^2} u_a(1) \right\| \leq e^{a\|V\|_\infty} \left\| e^{|x|^2/\alpha^2} u(1) \right\|$$

and

$$\left\| e^{|x|^2/\beta^2} u_a(0) \right\| \leq \left\| e^{|x|^2/\beta^2} u(0) \right\|.$$

We now have also

$$\partial_t u_a = (a+i)(\Delta u_a + V u_a),$$

so when we do the Appel transform, we have, with $\gamma_a = 1/\alpha_a\beta_a$,

$$\partial_t \tilde{u}_a = (a+i)((\Delta + \tilde{V}^a)\tilde{u}_a),$$

where

$$\tilde{V}^a(x,t) = \frac{\alpha_a\beta_a}{(\alpha_a(1-t)+\beta_a t)^2} V\left(\frac{\sqrt{\alpha_a\beta_a}\,x}{(\alpha_a(1-t)+\beta_a t}\right).$$

Now, fo r $a > 0$ we have "log convexity" in this last problem. Moreover, by the Appel Lemma and our definitions, we have

$$\left\| e^{\gamma_a|x|^2} \tilde{u}_a(0) \right\| \leq \left\| e^{|x|^2/\beta^2} u(0) \right\|, \quad \left\| e^{\gamma_a|x|^2} \tilde{u}_a(1) \right\| \leq e^{a\|V\|_\infty} \left\| e^{|x|^2/\alpha^2} u(1) \right\|.$$

Thus,

$$\left\| e^{\gamma_a |x|^2} \tilde{u}_a(t) \right\| \le e^{N(1+M_1+M_1^2)} e^{a||V||_\infty} \left\| e^{|x|^2/\beta^2} u(0) \right\|^{1-t} \left\| e^{|x|^2/\alpha^2} u(1) \right\|^t$$

and the corresponding "local smoothing" estimate. But now, letting $a \to 0$ and changing variables our result follows.

Time dependent, complex potentials: We will consider complex potentials $V(x,t)$, $||V||_\infty \le M_0$. We will also assume

$$\lim_{R \to 0} ||V||_{L^1([0,1],L^\infty(|x|>R))} = 0.$$

We first recall a result in [KPV3].

LEMMA 3.2. *There exists $N = N(n)$, $\epsilon_0 = \epsilon_0(n) > 0$ so that, if $\vec{\lambda} \in \mathbb{R}^n$, $V \in L^1([0,1]; L^\infty)$, $||V||_{L^1([0,1];L^\infty)} \le \epsilon_0$, then if $u \in C([0,1]; L^2)$ satisfies*

$$\partial_t u = i(\triangle u + V(x,t)u + F(x,t)) \quad \text{in } \mathbb{R}^n \times [0,1],$$

then

$$\sup_{t \in [0,1]} \left\| e^{\vec{\lambda} x} u(t) \right\| \le N \left[\left\| e^{\vec{\lambda} x} u(0) \right\| + \left\| e^{\vec{\lambda} x} u(1) \right\| + \left\| e^{\vec{\lambda} x} F \right\|_{L_t^1 L_x^2} \right].$$

THEOREM 3.3. *Let $V \in L_t^1 L_x^\infty$, $\lim_{R\to 0} ||V||_{L^1([0,1],L^\infty(|x|>R))} = 0$. Let $u \in C([0,1]; L^2)$ solve*

$$\partial_t u = i(\triangle u + V(x,t)u) \quad \text{in } \mathbb{R}^n \times [0,1].$$

Assume in addition that $V \in L^\infty(\mathbb{R}^{n+1})$, and that

$$\left\| e^{|x|^2/\beta^2} u(0) \right\| < \infty, \quad \left\| e^{|x|^2/\alpha^2} u(1) \right\| < \infty.$$

Then, $\exists N = N(\alpha, \beta)$ s.t.

$$\sup_{[0,1]} \left\| e^{|x|^2/(\alpha t + (1-t)\beta)^2} u(t) \right\| + \left\| \sqrt{t(1-t)} e^{|x|^2/(\alpha t+(1-t)\beta)^2} \nabla u(t) \right\|_{L^2(\mathbb{R}^n \times [0,1])} \le$$

$$\le N e^{N||V||_\infty} \left[\left\| e^{|x|^2/\beta^2} u(0) \right\| + \left\| e^{|x|^2/\alpha^2} u(1) \right\| + \sup_{[0,1]} ||u(t)|| \right].$$

PROOF. We start out by using the Appel transform, $\tilde{u}(x,t)$ and setting $\gamma = 1/\alpha\beta$, $(a+ib) = i$. We now have $\tilde{u} \in C([0,1]; L^2)$,

$$\partial_t \tilde{u} = i(\triangle \tilde{u} + \tilde{V}(x,t)\tilde{u}),$$

and it is easy to check that the potential $\tilde{V}$ verifies

$$||\tilde{V}||_\infty \le \max\left\{ \frac{\alpha}{\beta}, \frac{\beta}{\alpha} \right\} ||V||_\infty$$

and $\lim_{R\to 0} ||\tilde{V}||_{L^1([0,1],L^\infty(\mathbb{R}^n \setminus B_R)} = 0$. Also, we have

$$||\tilde{u}(t)|| = ||u(s)||, \quad ||e^{\gamma |x|^2} \tilde{u}(t)|| = ||e^{|x|^2/(\alpha s + (1-s)\beta)^2} u(s)||, \quad s = \frac{\beta t}{\alpha(1-t) + \beta t}.$$

Choose now $R > 0$ such that $||\tilde{V}||_{L^1([0,1],L^\infty(\mathbb{R}^n \setminus B_R))} \le \epsilon_0$, ϵ_0 as in Lemma 3.2. Then,

$$\partial_t \tilde{u} = i(\triangle \tilde{u} + \tilde{V}_R(x,t)\tilde{u} + \tilde{F}_R(x,t))$$

where $\tilde{V}_R(x,t) = \chi_{\mathbb{R}^n \setminus B_R} \tilde{V}(x,t)$, $\tilde{F}_R = \chi_{B_R} \tilde{V}\tilde{u}$. By the Lemma we have:

$$\sup_{t\in[0,1]} \left\| e^{\vec{\lambda}x} \tilde{u}(t) \right\| \leq N \left[\left\| e^{\vec{\lambda}x} \tilde{u}(0) \right\| + \left\| e^{\vec{\lambda}x} \tilde{u}(1) \right\| + e^{|\vec{\lambda}|R} \sup_{[0,1]} ||\tilde{V}(t)|| \sup_{[0,1]} ||\tilde{u}(t)|| \right].$$

Now, replace $\vec{\lambda}$ by $2\vec{\lambda}\sqrt{\gamma}$, square both sides, multiply be $e^{-|\vec{\lambda}|^2/2}$ and integrate both sides with respect to $\vec{\lambda}$ in $\mathbb{R}^n$. Using this and the identity

$$\int e^{2\sqrt{\gamma}\vec{\lambda}x} e^{-|\vec{\lambda}|^2/2} d\vec{\lambda} = (2\pi)^{n/2} e^{2\gamma|x|^2},$$

we obtain the inequality

$$\sup_{t\in[0,1]} \left\| e^{\gamma|x|^2} \tilde{u}(t) \right\| \leq$$

$$\leq N \left[\left\| e^{|x|^2/\beta^2} u(0) \right\| + \left\| e^{|x|^2/\alpha^2} u(1) \right\| + \sup_{[0,1]} \|V(t)\| \sup_{[0,1]} \|u(t)\| \right]$$

To prove the regularity of u, we proceed as follows: the standard Duhamel formula gives

$$\tilde{u}(t) = e^{it\triangle}\tilde{u}(0) + i\int_0^t e^{i(t-s)\triangle} \left(\tilde{V}(s)\tilde{u}(s) \right) ds.$$

For $0 < a < 1$, set

$$\tilde{F}_a(t) = \frac{i}{a+i} e^{at\triangle} \left(\tilde{V}(t)\tilde{u}(t) \right),$$

$$\tilde{u}_a(t) = e^{(a+i)t\triangle}\tilde{u}(0) + (i+a)\int_0^t e^{(a+i)(t-s)\triangle} \tilde{F}_a(t) ds.$$

Clearly,

$$\tilde{u}_a(t) = e^{at\triangle}\tilde{u}(t).$$

We now have, from the "energy estimates", with $\gamma_a = \frac{\gamma}{(1+4\gamma a)}$,

$$\sup_{[0,1]} \left\| e^{\gamma_a|x|^2} \tilde{u}_a(t) \right\| \leq \sup_{[0,1]} \left\| e^{\gamma|x|^2} \tilde{u}(t) \right\|$$

$$\sup_{[0,1]} \left\| e^{\gamma_a|x|^2} \tilde{F}_a(t) \right\| \leq e^{||\tilde{V}||_\infty} \sup_{[0,1]} \left\| e^{\gamma|x|^2} \tilde{u}(t) \right\|.$$

But then, our formal "smoothing effect" argument applies and gives: (using the first step)(key Lemma)

$$\left\| \sqrt{t(1-t)} \nabla \tilde{u}_a e^{\gamma_a|x|^2} \right\|_{L^2(\mathbb{R}^n\times[0,1])} \leq$$

$$Ne^{N||V||_\infty} \left[\left\| e^{|x|^2/\beta^2} u(0) \right\| + \left\| e^{|x|^2/\alpha^2} u(1) \right\| + \sup_{[0,1]} \|V(t)\| \sup_{[0,1]} \|u(t)\| \right].$$

We now let $a \to 0$. □

4. The Hardy uncertainty principle

Recall that for free evolution, $\partial_t u = i\triangle u$, Hardy's uncertainty principle says that if $u(0) \in L^2(e^{2|x|^2/\beta^2}dx)$, $u(1) \in L^2(e^{2|x|^2/\alpha^2}dx)$, and $\alpha\beta \le 4$, then $u \equiv 0$, and 4 is sharp. We will now show a (weakened) version of this for all our potentials.

THEOREM 4.1. *Let $V = V(x)$, V real, $||V||_\infty < \infty$, or $V = V(x,t)$, V complex, $||V||_\infty < \infty$, $\lim_{R\to 0} ||V||_{L^1([0,1],L^\infty(|x|>R))} = 0$. Assume that $u \in C([0,1];L^2)$ is a solution of*

$$\partial_t u = i(\triangle u + V(x,t)u) \quad \text{in } \mathbb{R}^n \times [0,1],$$

such that $e^{|x|^2/\beta^2}u(0) \in L^2$, $e^{|x|^2/\alpha^2}u(1) \in L^2$, and $\alpha\beta < 2$. Then $u \equiv 0$.

Preliminaries: Let $\gamma = 1/\alpha\beta$. Using the Appel transform and our convexity and "smoothing" estimates we can assume, without loss of generality, that the following holds for $\gamma > 1/2$:

$$\sup_{[0,1]} \left\| e^{\gamma|x|^2} u(t) \right\|_{L^2} + \sup_{[0,1]} \left\| \sqrt{t(1-t)} e^{\gamma|x|^2} \nabla u(t) \right\|_{L^2(\mathbb{R}^n\times[0,1])} < \infty. \tag{4.1}$$

Let me first give a formal argument, in the spirit of our "log convexity" inequalities. If $e_1 = (1,0,\ldots,0)$, $R > 0$, set $f = e^{\mu|x+Re_1t(1-t)|^2}u$, where $0 < \mu < \gamma$, and $H(t) = (f,f)$. At the formal level, it is easy to show (for the free evolution) that

$$\partial_t^2 \log H(t) \ge -R^2/4\mu,$$

so that $H(t)e^{-R^2t(1-t)/8\mu}$ is log convex in $[0,1]$ and so

$$H(1/2) \le H(0)^{1/2}H(1)^{1/2}e^{R^2/32\mu}.$$

Letting $\mu \uparrow \gamma$ we see that

$$\int e^{2\gamma|x+\frac{Re_1}{4}|^2}|u(1/2)|^2 \le \left\| e^{\gamma|x|^2}u(0) \right\| \left\| e^{\gamma|x|^2}u(1) \right\| e^{R^2/32\gamma}.$$

Thus,

$$\int_{B(\epsilon R/4)} |u(1/2)|^2 \le \left\| e^{\gamma|x|^2}u(0) \right\| \left\| e^{\gamma|x|^2}u(1) \right\| e^{[R^2(1-4\gamma^2(1-\epsilon)^2)]/32\gamma},$$

$0 < \epsilon < 1$, which implies $u(1/2) \equiv 0$ as $R \to \infty$, $(\gamma > 1/2)$.

The path from the formal argument to the rigorous one is not easy. We will do it instead with the Carleman inequality:

LEMMA 4.2. *Let $\phi(t)$, $\psi(t)$ be smooth functions on $[0,1]$, $g(x,t) \in C_0^\infty(\mathbb{R}^n \times [0,1])$, $e_1 = (1,0,\ldots,0)$. Then, for $\mu > 0$, we have (for $R \gg 0$),*

$$\iint [\psi''(t) - \frac{R^4}{32\mu}[\phi''(t)]^2]e^{2\psi(t)}e^{2\mu|\frac{x}{R}-\phi(t)e_1|^2}|g|^2 \le$$
$$\le \iint e^{2\psi(t)}e^{2\mu|\frac{x}{R}-\phi(t)e_1|^2}|(i\partial_t + \triangle)g|^2.$$

PROOF. Let $f = e^{\mu|\frac{x}{R}+\phi(t)e_1|^2+\psi(t)}g$. Then

$$e^{\mu|\frac{x}{R}+\phi(t)e_1|^2+\psi(t)}(i\partial_t + \triangle)g = S_\mu f + A_\mu f,$$

where $S_\mu = S_\mu^*$, $A_\mu = -A_\mu^*$ (the adjoints are now with respect to the $L^2(dxdt)$ inner product), and

$$S_\mu = i\partial_t + \triangle + \frac{4\mu^2}{R^2}\left|\frac{x}{R} + \phi e_1\right|^2,$$

$$A_\mu = -\frac{4\mu}{R}\left(\frac{x}{R} + \phi e_1\right)\cdot\nabla - \frac{2\mu n}{R^2} - 2i\mu\phi'\left(\frac{x_1}{R} + \phi e_1\right) - i\psi'.$$

We then have:

$$\iint e^{2\psi(t)} e^{2\mu\left|\frac{x}{R} - \phi(t)e_1\right|^2} |(i\partial_t + \triangle)g|^2 =$$
$$= \langle (S_\mu + A_\mu)f, (S_\mu + A_\mu)f\rangle = \langle S_\mu f, S_\mu f\rangle + \langle A_\mu f, A_\mu f\rangle +$$
$$+ \langle S_\mu f, A_\mu f\rangle + \langle A_\mu f, S_\mu f\rangle \geq \langle [S_\mu, A_\mu]f, f\rangle\,.$$

We now compute $[S_\mu, A_\mu]$ and obtain:

$$[S_\mu, A_\mu] = -\frac{8\mu}{R^2}\triangle + \frac{32\mu^3}{R^4}\left|\frac{x}{R} + \phi e_1\right|^2 +$$
$$+ 2\mu\left(\frac{x_1}{R} + \phi e_1\right)\phi'' + 2\mu(\phi')^2 - \frac{8i\mu\phi'}{R}\partial_{x_1} + \psi''.$$

Thus,

$$\langle [S_\mu, A_\mu]f, f\rangle = \frac{8\mu}{R^2}\int |\nabla_{x'} f|^2 + \frac{32\mu^3}{R^4}\int\left|\frac{x}{R} + \phi e_1\right|^2 |f|^2 +$$
$$+ \frac{8\mu}{R^2}\int |\partial_{x_1} f|^2 + 2\mu\int\left(\frac{x_1}{R} + \phi e_1\right)\phi''|f|^2 +$$
$$+ 2\mu\int(\phi')^2|f|^2 - \frac{8i\mu}{R}\int \phi'\partial_{x_1} f\overline{f} + \int \psi''|f|^2 =$$
$$= \frac{8\mu}{R^2}\int |\nabla_{x'} f|^2 + 8\mu\int\left|\frac{i}{R}\partial_{x_1} f - \frac{\phi'}{2}f\right|^2 +$$
$$+ \frac{32\mu^3}{R^4}\int\left|\frac{x}{R} + \phi e_1\right|^2 |f|^2 + 2\mu\int\left(\frac{x_1}{R} + \phi e_1\right)\phi''|f|^2 + \int \psi''|f|^2 =$$
$$= \frac{8\mu}{R^2}\int |\nabla_{x'} f|^2 + 8\mu\int\left|\frac{i}{R}\partial_{x_1} f - \frac{\phi'}{2}f\right|^2 +$$
$$+ \frac{32\mu^3}{R^4}\int\left|\frac{x}{R} + \left(\phi + \frac{R^4}{32\mu^2}\phi''\right)e_1\right|^2 |f|^2 - \frac{R^4(\phi'')^2}{32\mu}\int |f|^2 + \psi''\int |f|^2,$$

and the Lemma follows. □

Next, choose $\phi(t) = t(1-t)$, $\psi(t) = -(1+\epsilon)\frac{R^4}{16\mu}t(1-t)$. Then

$$\psi''(t) - \frac{R^4}{32\mu}(\phi'')^2(t) = \frac{(1+\epsilon)}{8\mu}R^4 - \frac{R^4}{8\mu} = \frac{\epsilon}{8\mu}R^4$$

and so our inequality reads, for $g \in C_0^\infty(\mathbb{R}^n \times [0,1])$,

$$\frac{\epsilon}{8\mu}R^4 \iint e^{2\psi(t)} e^{2\mu\left|\frac{x}{R} + \phi(t)e_1\right|^2}|g|^2 \quad\leq\quad \iint e^{2\psi(t)} e^{2\mu\left|\frac{x}{R} + \phi(t)e_1\right|^2}|(i\partial_t \;+\; \triangle)g|^2.$$

We next fix $R > 0$, recall that u solves $i\partial_t u + \triangle u = Vu$, and that the estimates (4.1) hold. Choose then $\eta(t)$, $0 \le \eta \le 1$, $\eta \equiv 1$ where $t(1-t) \ge 1/R$, $\eta \equiv 0$ near $t = 1, 0$, so that

$$\operatorname{supp} \eta' \subset \{t(1-t) \le 1/R\}, \quad |\eta'| \le CR.$$

Choose also $M \gg R$, $\theta \in C_0^\infty(\mathbb{R}^n)$, and now set $g(x,t) = \eta(t)\theta(x/M)u(x,t)$, which is compactly supported in $\mathbb{R}^n \times (0,1)$, so that our estimate holds.

$$\begin{aligned}(i\partial_t + \triangle)g &= Vg + i\eta'(t)\theta\left(\tfrac{x}{M}\right)u + \left(\tfrac{1}{M^2}\triangle\theta\left(\tfrac{x}{M}\right)u + \tfrac{2\nabla\theta(x/M)\cdot\nabla u}{M}\right) \\ &= I + II + III.\end{aligned}$$

Finally, let $\mu = (1+\epsilon)^{-3}\gamma R^2$. Our inequality then gives:

$$\frac{\epsilon}{8}\frac{(1+\epsilon)^3}{\gamma}R^2 \iint e^{2\psi(t)}e^{2\mu\left|\frac{x}{R}+\phi(t)e_1\right|^2}|g|^2 \le$$
$$\le \iint e^{2\psi(t)}e^{2\mu\left|\frac{x}{R}+\phi(t)e_1\right|^2}\{I + II + III\}.$$

The contribution of I to the right hand side is bounded by

$$||V||_\infty \iint e^{2\psi(t)}e^{2\mu\left|\frac{x}{R}+\phi(t)e_1\right|^2}|g|^2,$$

so that, if R is very large, we can hide it in the left side, to see that we only have to deal with II and III. Recall that $\psi(t) = (1+\epsilon)\frac{R^4}{16\mu}t(1-t) \le 0$, so $e^{2\psi(t)} \le 1$. On the support of η', we have $t(1-t) \le 1/R$, so that $0 \le \phi(t) \le 1/R$. We now estimate

$$\begin{aligned}2\mu\left|\frac{x}{R} + \phi(t)e_1\right|^2 &= \frac{2\gamma R^2}{(1+\epsilon)^3}\left\{\left|\frac{x}{R}\right|^2 + 2\frac{x_1}{R}\phi(t) + \phi(t)^2\right\} = \frac{2\gamma}{(1+\epsilon)^3}|x|^2 + \\ &+ \frac{2\gamma}{(1+\epsilon)^3}x_1 R\phi(t) + \frac{2\gamma}{(1+\epsilon)^3}R^2\phi(t)^2 \le 2\gamma|x|^2 + C_\epsilon,\end{aligned}$$

on $\operatorname{supp}\eta'$, where $\phi(t) \le 1/R$. Thus, because of (4.1), the contribution of II is bounded by $C_\epsilon R$. The contribution of III is controlled by (recalling that $\eta \equiv 0$ when $t(1-t) \le \frac{1}{2}\frac{1}{R}$)

$$\frac{C}{M^4}\iint_{|x|\le 2M} |u(x,t)|^2 e^{2\psi(t)}e^{2\mu\left|\frac{x}{R}+\phi(t)e_1\right|^2} +$$
$$+ \frac{C}{M^2}\iint_{|x|\le 2M} |\nabla u(x,t)|^2 e^{2\psi(t)}e^{2\mu\left|\frac{x}{R}+\phi(t)e_1\right|^2} t(1-t)R.$$

If we use (4.1), $\psi \le 0$, the bound above for $2\mu\left|x/R + \phi(t)e_1\right|^2$ becomes

$$\le \frac{2\gamma|x|^2}{(1+\epsilon)^3} + \frac{2\gamma|x_1|}{(1+\epsilon)^3}\frac{R}{4} + \frac{2\gamma}{(1+\epsilon)^3}\frac{R^2}{16} \le 2\gamma|x|^2 + C_{\epsilon,R}.$$

Thus, letting $M \to \infty$, we see that, for fixed R, $III \to 0$, so that, since $\eta \equiv 1$ on $t(1-t) \ge 1/R$, we obtain:

$$\frac{\epsilon}{8}\frac{(1+\epsilon)^3}{\gamma}R^2 \iint_{t(1-t)\ge 1/R} e^{2\psi(t)}e^{2\mu\left|\frac{x}{R}+\phi(t)e_1\right|^2}|u|^2 \le C_{\gamma,\epsilon}R.$$

We are now going to restrict to integration over the region $\left|\frac{x}{R}\right| \le \delta$, $|t-1/2| \le \delta$, where δ is small, to be chosen. Then,

$$\left|\frac{x}{R} + \phi(t)e_1\right| \ge \left|\phi\left(\frac{1}{2}\right)\right| - 2\delta = \left(\frac{1}{4} - 2\delta\right),$$

so that

$$\left|\frac{x}{R} + \phi(t)e_1\right|^2 \ge \frac{1}{16} - 2\delta\left(\frac{1}{2} - 2\delta\right).$$

$$\psi(t) = \psi\left(\frac{1}{2}\right) + \left[\psi(t) - \psi\left(\frac{1}{2}\right)\right] \ge \psi\left(\frac{1}{2}\right) - |\psi'(\theta)|\delta \ge$$
$$\ge -(1+\epsilon)\frac{R^4}{16\mu}\frac{1}{4} - \delta(1+\epsilon)\frac{R^4}{16\mu},$$

so that, in our region of integration,

$$\begin{aligned} 2\mu\left|\frac{x}{R} + \phi(t)e_1\right|^2 + 2\psi(t) \ge \\ \ge \frac{2\mu}{16} - \frac{2(1+\epsilon)R^4}{16\mu 4} - 4\delta\mu\left(\frac{1}{2} - 2\delta\right) - \delta(1+\epsilon)\frac{R^4}{16\mu} = \frac{2}{16}\frac{\gamma R^2}{(1+\epsilon)^3} - \\ - \frac{2}{16}\frac{(1+\epsilon)^4}{\gamma 4}R^2 - C\delta R^2 = \frac{2}{16}R^2\left[\frac{\gamma}{(1+\epsilon)^3} - \frac{(1+\epsilon)^4}{4\gamma} - C\delta\right], \end{aligned}$$

since $\mu = \frac{\gamma}{(1+\epsilon)^3}R^2$. But, if $\gamma > 1/2$,

$$\frac{\gamma}{(1+\epsilon)^3} - \frac{(1+\epsilon)^4}{4\gamma} > 0,$$

for some ϵ small, and so, for δ smaller than that we get a lower bound of $C_{\epsilon,\delta}R^2$. We thus have

$$C_{\epsilon,\delta}R\int_{|t-1/2|\le\delta}\int_{\left|\frac{x}{R}\right|\le\delta}|u|^2 e^{C_{\epsilon,\delta}R^2} \le C_{\epsilon,\delta}.$$

But then, since

$$\int_{|t-1/2|\le\delta}\int_{\left|\frac{x}{R}\right|>\delta}|u|^2 = \int_{|t-1/2|\le\delta}\int_{\left|\frac{x}{R}\right|\le\delta} e^{2\gamma|x|^2}e^{-2\gamma|x|^2}|u|^2 \le$$
$$\le e^{-2\gamma\delta^2R^2}\int_{|t-1/2|\le\delta}\int e^{2\gamma|x|^2}|u|^2 \le C_\gamma e^{-2\gamma\delta^2R^2}$$

by (4.1), we see that, for appropriate $C_{\gamma,\epsilon,\delta}$ we have

$$\left(\int_{|t-1/2|\le\delta}\int|u|^2\right)e^{C_{\gamma,\epsilon,\delta}R^2} \le C_{\gamma,\epsilon,\delta}.$$

Letting $R \to \infty$, we see that $u \equiv 0$ on $\{(x,t) : |t-1/2| \le \delta\}$, therefore $u \equiv 0$.

References

[A] P. Anderson, *Absence of diffusion in certain random lattices*, Phys. Review **109** (1958), 1492–1505.

[BK] J. Bourgain and C. Kenig, *On localization in the Anderson–Bernoulli model in higher dimensions*, Invent. Math **161** (2005), 389–426.

[CKM] R. Carmona, A. Klein, and F. Martinelli, *Anderson localization for Bernoulli and other singular potentials*, Commun. Math. Phys. **108** (1987), 41–66.

[EKPV1] L. Escauriaza, C. Kenig, G. Ponce, and L. Vega, *Decay at infinity of caloric functions within characteristic hyperplanes*, Math. Res. Lett. **13** (2006), 441–453.

[EKPV2] ———, *On uniqueness properties of solutions of Schrödinger equations*, Commun. in PDE **31** (2006), 1811–1823.

[EKPV3] ———, *On uniqueness properties of solutions of the k-generalized KdV equations*, Jour. Funct. Anal. **244** (2007), 504–535.

[EKPV4] ———, *Convexity properties of solutions to the free Schrödinger equation with Gaussian decay*, to appear, MRL.

[EKPV5] ———, *Hardy's uncertainty principle, convexity and Schrödinger evolutions*, to appear, JEMS.

[ESŠ] L. Escauriaza, G. Seregin, and V. Šverák, $L^{3,\infty}$ *solutions to the Navier–Stokes equations and backward uniqueness*, Russ. Math. Surv. **58:2** (2003), 211-250.

[FS] J. Frölich and T. Spencer, *Absence of diffusion with Anderson tight binding model for large disorder or low energy*, Commun. Math. Phys. **88** (1983), 151–184.

[GMP] Y. Goldsheid, S. Molchanov, and L. Pastur, *Pure point spectrum of stochastic one dimensional Schrödinger operators*, Funct. Anal. Appl. **11** (1977), 1–10.

[IK] A. Ionescu and C. Kenig, L^p *Carleman inequalities and uniqueness of solutions of non-linear Schrödinger equations*, Acta Math. **193** (2004), 193–239.

[I] V. Isakov, *Carleman type estimates in anisotropic case and applications*, J. Diff. Eqs. **105** (1993), 217–238.

[K1] C. Kenig, *Some recent quantitative unique continuation theorems*, Rend. Accad. Naz. Sci. XL Mem. Mat. Appl. **29** (2005), 231–242.

[K2] ———, *Some recent applications of unique continuation*, Contemp. Math. **439** (2007), 25–56.

[KPV1] C. Kenig, G. Ponce, and L. Vega, *On the support of solutions to generalized KdV equation*, Annales de l'Institut H. Poincaré **19** (2002), 191–208.

[KPV2] ———, *On the unique continuation of solutions to the generalized KdV equation*, Math. Res. Lett **10** (2003), 833–846.

[KPV3] ———, *On unique continuation for nonlinear Schrödinger equations*, Commun. Pure Appl. Math. **60** (2002), 1247–1262.

[LO] E. M. Landis and O. A. Oleinik, *Generalized analyticity and some related properties of solutions of elliptic and parabolic equations*, Russ. Math. Surv. **29** (1974), 195–212.

[M] V. Z. Meshkov, *On the possible rate of decay at infinity of solutions of second order partial differential equations*, Math. USSR Sobornik **72** (1992), 343–360.

[R] L. Robbiano, *Unicité forte à l'infini pour KdV*, ESAIM Control Optim. Calc. Var. **8** (2002), 933–939.

[SVW] C. Shubin, R. Vakilian, and T. Wolff, *Some harmonic analysis questions suggested by Anderson–Bernoulli models*, GAFA **8** (1998), 932–964.

[SS] E. M. Stein and R. Shakarchi, *Complex analysis*, Princeton Lectures in Analysis, II, Princeton University Press, Princeton, NJ, 2003.

[Z1] B. Y. Zhang, *Unique continuation for the Korteweg–de Vries equation*, SIAM J. Math. Anal. **23** (1992), 55–71.

[Z2] ———, *Unique continuation for the nonlinear Schrödinger equation*, Proc. Roy. Soc. Edinburgh Sect. A **127** (1997), 191–205.

Depatment of Mathematics, University of Chicago, Chicago, IL 60637,USA
E-mail address: `cek@math.uchicago.edu`

Proceedings of Symposia in Pure Mathematics
Volume **79**, 2008

Boundary Harnack Inequalities for Operators of p-Laplace Type in Reifenberg Flat Domains

John L. Lewis, Niklas Lundström, and Kaj Nyström

Abstract. In this paper we highlight a set of techniques that recently have been used to establish boundary Harnack inequalities for p-harmonic functions vanishing on a portion of the boundary of a domain which is 'flat' in the sense that its boundary is well-approximated by hyperplanes. Moreover, we use these techniques to establish new results concerning boundary Harnack inequalities and the Martin boundary problem for operators of p-Laplace type with variable coefficients in Reifenberg flat domains.

1. Introduction and statement of main results

In [LN], [LN1], [LN2], see also [LN3] for a survey of these results, a number of results concerning the boundary behaviour of positive p-harmonic functions, $1 < p < \infty$, in a bounded Lipschitz domain $\Omega \subset \mathbf{R}^n$ were proved. In particular, the boundary Harnack inequality, as well as Hölder continuity for ratios of positive p-harmonic functions, $1 < p < \infty$, vanishing on a portion of $\partial\Omega$ were established. Furthermore, the p-Martin boundary problem at $w \in \partial\Omega$ was resolved under the assumption that Ω is either convex, C^1-regular or a Lipschitz domain with small constant. Also, in [LN4] these questions were resolved for p-harmonic functions vanishing on a portion of certain Reifenberg flat and Ahlfors regular NTA-domains.

From a technological perspective the toolbox developed in [LN, LN1-LN4] can be divided into (i) techniques which can be used to establish boundary Harnack inequalities for p-harmonic functions vanishing on a portion of the boundary of a domain which is 'flat' in the sense that its boundary is well-approximated by hyperplanes and (ii) techniques which can be used to establish boundary Harnack inequalities for p-harmonic functions vanishing on a portion of the boundary of a Lipschitz domain or on a portion of the boundary of a domain which can be well approximated by Lipschitz graph domains. Domains in category (i) are called Reifenberg flat domains with small constant or just Reifenberg flat domains. They include domains with small Lipschitz constant, C^1-domains and certain quasi-balls. Domains in category (ii) include Lipschitz domains with large Lipschitz constant and certain Ahlfors regular NTA-domains, which can be well approximated by Lipschitz graph domains in the Hausdorff distance sense. The purpose of this paper is to highlight the techniques labeled as category (i) in the above discussion and to use these techniques to establish boundary Harnack inequalities as well as to

---

2000 *Mathematics Subject Classification.* Primary 35J25, 35J70 .

Key words and phrases. Keywords and phrases: boundary Harnack inequality, p-harmonic function, A-harmonic function, variable coefficients, Reifenberg flat domain, Martin boundary.

Lewis was partially supported by NSF DMS-0139748.

Nyström was partially supported by grant 70768001 from the Swedish Research Council.

resolve the Martin boundary problem for operators of p-Laplace type with variable coefficients in Reifenberg flat domains.

To state our results we need to introduce some notation. Points in Euclidean n-space $\mathbf{R}^n$ are denoted by $x = (x_1, \ldots, x_n)$ or (x', x_n) where $x' = (x_1, \ldots, x_{n-1}) \in \mathbf{R}^{n-1}$. Let $\bar{E}, \partial E$, diam E, be the closure, boundary, diameter, of the set $E \subset \mathbf{R}^n$ and let $d(y, E)$ equal the distance from $y \in \mathbf{R}^n$ to E. $\langle \cdot, \cdot \rangle$ denotes the standard inner product on $\mathbf{R}^n$ and $|x| = \langle x, x \rangle^{1/2}$ is the Euclidean norm of x. Put $B(x, r) = \{y \in \mathbf{R}^n : |x - y| < r\}$ whenever $x \in \mathbf{R}^n$, $r > 0$, and let dx be Lebesgue n-measure on $\mathbf{R}^n$. We let

$$h(E, F) = \max(\sup\{d(y, E) : y \in F\}, \sup\{d(y, F) : y \in E\})$$

be the Hausdorff distance between the sets $E, F \subset \mathbf{R}^n$. If $O \subset \mathbf{R}^n$ is open and $1 \leq q \leq \infty$, then by $W^{1,q}(O)$ we denote the space of equivalence classes of functions f with distributional gradient $\nabla f = (f_{x_1}, \ldots, f_{x_n})$, both of which are q th power integrable on O. Let $\|f\|_{1,q} = \|f\|_q + \| \, |\nabla f| \, \|_q$ be the norm in $W^{1,q}(O)$ where $\| \cdot \|_q$ denotes the usual Lebesgue q-norm in O. Next let $C_0^\infty(O)$ be the set of infinitely differentiable functions with compact support in O and let $W_0^{1,q}(O)$ be the closure of $C_0^\infty(O)$ in the norm of $W^{1,q}(O)$. By $\nabla \cdot$ we denote the divergence operator.

We first introduce the operators of p-Laplace type which we consider in this paper.

Definition 1.1. *Let $p, \beta, \alpha \in (1, \infty)$ and $\gamma \in (0, 1)$. Let $A = (A_1, ..., A_n) : \mathbf{R}^n \times \mathbf{R}^n \to \mathbf{R}^n$, assume that $A = A(x, \eta)$ is continuous in $\mathbf{R}^n \times (\mathbf{R}^n \setminus \{0\})$ and that $A(x, \eta)$, for fixed $x \in \mathbf{R}^n$, is continuously differentiable in η_k, for every $k \in \{1, ..., n\}$, whenever $\eta \in \mathbf{R}^n \setminus \{0\}$. We say that the function A belongs to the class $M_p(\alpha, \beta, \gamma)$ if the following conditions are satisfied whenever x, y, $\xi \in \mathbf{R}^n$ and $\eta \in \mathbf{R}^n \setminus \{0\}$:*

$$(i) \qquad \alpha^{-1} |\eta|^{p-2} |\xi|^2 \leq \sum_{i,j=1}^{n} \frac{\partial A_i}{\partial \eta_j}(x, \eta) \xi_i \xi_j,$$

$$(ii) \qquad \left| \frac{\partial A_i}{\partial \eta_j}(x, \eta) \right| \leq \alpha |\eta|^{p-2}, 1 \leq i, j \leq n,$$

$$(iii) \qquad |A(x, \eta) - A(y, \eta)| \leq \beta |x - y|^\gamma |\eta|^{p-1},$$

$$(iv) \qquad A(x, \eta) = |\eta|^{p-1} A(x, \eta/|\eta|).$$

For short, we write $M_p(\alpha)$ for the class $M_p(\alpha, 0, \gamma)$.

Definition 1.2. *Let $p \in (1, \infty)$ and let $A \in M_p(\alpha, \beta, \gamma)$ for some (α, β, γ). Given a bounded domain G we say that u is A-harmonic in G provided $u \in W^{1,p}(G)$ and*

$$\int \langle A(x, \nabla u(x)), \nabla \theta(x) \rangle \, dx = 0 \tag{1.3}$$

whenever $\theta \in W_0^{1,p}(G)$. If $A(x, \eta) = |\eta|^{p-2}(\eta_1, \ldots, \eta_n)$, then u is said to be p-harmonic in G. As a short notation for (1.3) we write $\nabla \cdot (A(x, \nabla u)) = 0$ in G.

The relevance and importance of the conditions imposed through the assumption $A \in M_p(\alpha, \beta, \gamma)$ will be discussed below. Initially we just note that the class $M_p(\alpha, \beta, \gamma)$ is, see Lemma 2.15, closed under translations, rotations and under dilations $x \to rx$, $r \in (0, 1]$. Moreover, we note that an important class of equations which is covered by Definition 1.1 and 1.2 is the class of equations of the type

$$\nabla \cdot \left[\langle A(x)\nabla u, \nabla u\rangle^{p/2-1} A(x)\nabla u \right] = 0 \text{ in } G \tag{1.4}$$

where $A = A(x) = \{a_{i,j}(x)\}$ is such that the conditions in Definition 1.1 (i) - (iv) are fulfilled.

Next we introduce the geometric notions used in this paper. We define,

Definition 1.5. *A bounded domain Ω is called non-tangentially accessible (NTA) if there exist $M \geq 2$ and $r_0 > 0$ such that the following are fulfilled:*

(i) *corkscrew condition: for any $w \in \partial\Omega, 0 < r < r_0$, there exists $a_r(w) \in \Omega \cap B(w, r/2)$, satisfying $M^{-1}r < d(a_r(w), \partial\Omega)$,*

(ii) $\mathbf{R}^n \setminus \bar{\Omega}$ *satisfies the corkscrew condition,*

(iii) *uniform condition: if $w \in \partial\Omega, 0 < r < r_0$, and $w_1, w_2 \in B(w, r) \cap \Omega$, then there exists a rectifiable curve $\gamma : [0, 1] \to \Omega$ with $\gamma(0) = w_1, \gamma(1) = w_2$, and such that*

(a) $H^1(\gamma) \leq M\,|w_1 - w_2|$,

(b) $\min\{H^1(\gamma([0,t])),\, H^1(\gamma([t,1]))\} \leq M\, d(\gamma(t), \partial\Omega)$.

In Definition 1.5, H^1 denotes length or the one-dimensional Hausdorff measure. We note that (iii) is different but equivalent to the usual Harnack chain condition given in [JK] (see [BL], Lemma 2.5). M will be called the NTA-constant of Ω.

Definition 1.6. *Let $\Omega \subset \mathbf{R}^n$ be a bounded domain, $w \in \partial\Omega$, and $0 < r < r_0$. Then $\partial\Omega$ is said to be uniformly (δ, r_0)-approximable by hyperplanes, provided there exists, whenever $w \in \partial\Omega$ and $0 < r < r_0$, a hyperplane Λ containing w such that*

$$h(\partial\Omega \cap B(w, r), \Lambda \cap B(w, r)) \leq \delta r.$$

We let $\mathcal{F}(\delta, r_0)$ denote the class of all domains Ω which satisfy Definition 1.6. Let $\Omega \in \mathcal{F}(\delta, r_0)$, $w \in \partial\Omega$, $0 < r < r_0$, and let Λ be as in Definition 1.6. We say that $\partial\Omega$ separates $B(w, r)$, if

$$\{x \in \Omega \cap B(w, r) : d(x, \partial\Omega) \geq 2\delta r\} \subset \text{one component of } \mathbf{R}^n \setminus \Lambda. \tag{1.7}$$

Definition 1.8. *Let $\Omega \subset \mathbf{R}^n$ be a bounded domain. Then Ω and $\partial\Omega$ are said to be (δ, r_0)-Reifenberg flat provided $\Omega \in \mathcal{F}(\delta, r_0)$ and (1.7) hold whenever $0 < r < r_0, w \in \partial\Omega$.*

For short we say that Ω and $\partial\Omega$ are δ-Reifenberg flat whenever Ω and $\partial\Omega$ are (δ, r_0)-Reifenberg flat for some $r_0 > 0$. We note that an equivalent definition of a Reifenberg flat domain is given in [KT]. As in [KT] one can show that a δ-Reifenberg flat domain is an NTA-domain with constant $M = M(n)$, provided $0 < \delta < \hat{\delta}$ and $\hat{\delta}$ is small enough.

In this paper we first prove the following theorem.

Theorem 1. *Let $\Omega \subset \mathbf{R}^n$ be a (δ, r_0)-Reifenberg flat domain. Let p, $1 < p < \infty$, be given and assume that $A \in M_p(\alpha, \beta, \gamma)$ for some (α, β, γ). Let $w \in \partial\Omega, 0 < r < r_0$, and suppose that u, v are positive A-harmonic functions in $\Omega \cap B(w, 4r)$, continuous in $\bar{\Omega} \cap B(w, 4r)$, and $u = 0 = v$ on $\partial\Omega \cap B(w, 4r)$. There exists $\tilde{\delta} < \hat{\delta}, \sigma > 0$, and $c_1 \geq 1$, all depending only on $p, n, \alpha, \beta, \gamma$, such that if $0 < \delta < \tilde{\delta}$, then*

$$\left| \log \frac{u(y_1)}{v(y_1)} - \log \frac{u(y_2)}{v(y_2)} \right| \leq c_1 \left(\frac{|y_1 - y_2|}{r} \right)^\sigma$$

whenever $y_1, y_2 \in \Omega \cap B(w, r/c_1)$.

We note that in [LN] we obtained for p-harmonic functions u, v, in a bounded Lipschitz domain Ω,

$$\left| \log \frac{u(y_1)}{v(y_1)} - \log \frac{u(y_2)}{v(y_2)} \right| \leq c$$

whenever $w \in \partial\Omega$, and $y_1, y_2 \in \Omega \cap B(w, r/c)$. Here c depends only on p, n, and the Lipschitz constant for Ω. Moreover, using this result, we showed, in [LN1], that the conclusion of Theorem 1 holds whenever u, v, are p-harmonic, and Ω is Lipschitz. Constants again depend only on p, n, and the Lipschitz constant for Ω.

In this paper we also prove the following theorem.

Theorem 2. *Let $\Omega \subset \mathbf{R}^n$, δ, r_0, p, α, β, γ, and A be as in the statement of Theorem 1. Then there exists $\delta^* = \delta^*(p, n, \alpha, \beta, \gamma) < \hat{\delta}$, such that the following is true. Let $w \in \partial\Omega$ and suppose that $\hat{u}, \hat{v}$ are positive A-harmonic functions in Ω with $\hat{u} = 0 = \hat{v}$ continuously on $\partial\Omega \setminus \{w\}$. If $0 < \delta < \delta^*$, then $\hat{u}(y) = \lambda \hat{v}(y)$ for all $y \in \Omega$ and for some constant λ.*

We remark, using terminology of the Martin boundary problem, that if $\hat{u}$ is as in Theorem 2, then $\hat{u}$ is called a minimal positive A-harmonic function in Ω, relative to $w \in \partial\Omega$. Moreover, the A-Martin boundary of Ω is the set of equivalence classes of positive minimal A-harmonic functions relative to all boundary points of Ω. Two minimal positive A-harmonic functions are in the same equivalence class if they correspond to the same boundary point and one is a constant multiple of the other. Note that the conclusion of Theorem 2 implies that $\hat{u}$ is unique up to constant multiples. Thus, since $w \in \partial\Omega$ is arbitrary, one can say that the A-Martin boundary of Ω is identifiable with $\partial\Omega$.

We remark that in [LN1] the Martin boundary problem for p-harmonic functions was resolved in domains which are either convex, C^1-regular or Lipschitz with sufficiently small constant. Also, in [LN4] the Martin boundary problem was resolved, again for p-harmonic functions, in Reifenberg flat domains and certain Ahlfors regular NTA-domains. Theorem 2 is new in the case of operators of p-Laplace type with variable coefficients.

Recall that Ω is said to be a bounded Lipschitz domain if there exists a finite set of balls $\{B(x_i, r_i)\}$, with $x_i \in \partial\Omega$ and $r_i > 0$, such that $\{B(x_i, r_i)\}$ constitutes a covering of an open neighbourhood of $\partial\Omega$ and such that, for each i,

$$\begin{aligned} \Omega \cap B(x_i, r_i) &= \{x = (x', x_n) \in \mathbf{R}^n : x_n > \phi_i(x')\} \cap B(x_i, r_i), \\ \partial\Omega \cap B(x_i, r_i) &= \{x = (x', x_n) \in \mathbf{R}^n : x_n = \phi_i(x')\} \cap B(x_i, r_i), \end{aligned} \tag{1.9}$$

in an appropriate coordinate system and for a Lipschitz function ϕ_i. The Lipschitz constant of Ω is defined to be $M = \max_i |||\nabla\phi_i|||_\infty$. If Ω is Lipschitz then Ω is NTA with $r_0 = \min r_i/c$, where $c = c(p,n,M) \geq 1$. Moreover, if each $\phi_i : \mathbf{R}^{n-1} \to \mathbf{R}$ can be chosen to be either C^1- or $C^{1,\sigma}$ -regular, then Ω is a bounded C^1- or $C^{1,\sigma}$-domain.

We say that Ω is a quasi-ball provided $\Omega = f(B(0,1))$, where $f = (f_1, f_2, ..., f_n) : \mathbf{R}^n \to \mathbf{R}^n$ is a $K > 1$ quasi-conformal mapping of $\mathbf{R}^n$ onto $\mathbf{R}^n$. That is, $f_i \in W^{1,n}(B(0,\rho)), 0 < \rho < \infty, 1 \leq i \leq n$, and for almost every $x \in \mathbf{R}^n$ with respect to Lebesgue n-measure the following hold,

$$(i) \qquad |Df(x)|^n = \sup_{|h|=1} |Df(x)h|^n \leq K|J_f(x)|,$$

$$(ii) \qquad J_f(x) \geq 0 \text{ or } J_f(x) \leq 0.$$

In this display we have written $Df(x) = (\frac{\partial f_i}{\partial x_j})$ for the Jacobian matrix of f and $J_f(x)$ for the Jacobian determinant of f at x.

Remark 1.10. *Let $\Omega \subset \mathbf{R}^n$ be a bounded Lipschitz domain with constant M. If M is small enough then Ω is (δ, r_0)-Reifenberg flat for some $\delta = \delta(M), r_0 > 0$ with $\delta(M) \to 0$ as $M \to 0$. Hence, Theorems 1-2 apply to any bounded Lipschitz domain with sufficiently small Lipschitz constant. Also, if $\Omega = f(B(0,1))$ where f is a K quasi-conformal mapping of $\mathbf{R}^n$ onto $\mathbf{R}^n$, then one can show that $\partial\Omega$ is δ-Reifenberg flat, with $r_0 = 1$, where $\delta \to 0$ as $K \to 1$ (see [R, Theorems 12.5 -12.7]). Thus Theorems 1, 2, apply when Ω is a quasi-ball and if $K = K(p,n)$ is close enough to 1.*

To state corollaries to Theorems 1-2 we next introduce the notion of Reifenberg flat domains with vanishing constant.

Definition 1.11. *Let $\Omega \subset \mathbf{R}^n$ be a (δ, r_0)-Reifenberg flat domain for some $0 < \delta < \hat{\delta}$, $r_0 > 0$, and let $w \in \partial\Omega$, $0 < r < r_0$. We say that $\partial\Omega \cap B(w,r)$ is Reifenberg flat with vanishing constant, if for each $\epsilon > 0$, there exists $\tilde{r} = \tilde{r}(\epsilon) > 0$ with the following property. If $x \in \partial\Omega \cap B(w,r)$ and $0 < \rho < \tilde{r}$, then there is a plane $P' = P'(x,\rho)$ containing x such that*

$$h(\partial\Omega \cap B(x,\rho), P' \cap B(x,\rho)) \leq \epsilon\rho.$$

The following corollaries are immediate consequences of Theorems 1-2.

Corollary 1. *Let $\Omega \subset \mathbf{R}^n$ be a domain which is Reifenberg flat with vanishing constant. Let p, $1 < p < \infty$, be given and assume that $A \in M_p(\alpha, \beta, \gamma)$ for some (α, β, γ). Let $w \in \partial\Omega$, $0 < r < r_0$. Assume that u, v are positive A-harmonic functions in $\Omega \cap B(w,4r)$, u, v are continuous in $\bar{\Omega} \cap B(w,4r)$ and $u = 0 = v$ on $\partial\Omega \cap B(w,4r)$. There exist $r_1^* = r_1^*(p,n,\alpha,\beta,\gamma) < r$ and $c_2 = c_2(p,n,\alpha,\beta,\gamma) \geq 1$ such that if $w' \in \partial\Omega \cap B(w,r)$ and $0 < r' < r_1^*$, then*

$$\left|\log \frac{u(y_1)}{v(y_1)} - \log \frac{u(y_2)}{v(y_2)}\right| \leq c_2 \left(\frac{|y_1 - y_2|}{r'}\right)^\sigma$$

whenever $y_1, y_2 \in \Omega \cap B(w', r')$.

Corollary 2. *Let $\Omega \subset \mathbf{R}^n$, p, α, β, γ and A be as in the statement of Corollary 1. Let $w \in \partial\Omega$ and suppose that $\hat{u}, \hat{v}$ are positive A-harmonic functions in Ω with*

$\hat{u} = 0 = \hat{v}$ *continuously on* $\partial\Omega \setminus \{w\}$. *Then* $\hat{u}(y) = \lambda\hat{v}(y)$ *for all* $y \in \Omega$ *and for some constant* λ.

Remark 1.12. *We note that if* Ω *is a bounded* C^1*-domain in the sense of (1.9) then* Ω *is also Reifenberg flat with vanishing constant. Hence Corollaries 1-2 apply to any bounded* C^1*-domain.*

Concerning proofs, we here outline the proof of Theorem 1.

Step 0. As a starting point we establish the conclusion of Theorem 1, see Lemma 2.8, when $A \in M_p(\alpha)$, Ω is equal to a truncated cylinder and w is the center on the bottom of $\bar{\Omega}$.

Step A. (Uniform non-degeneracy of $|\nabla u|$ - the 'fundamental inequality'). There exist $\delta_1 = \delta_1(p, n, \alpha, \beta, \gamma)$, $\hat{c}_1 = \hat{c}_1(p, n, \alpha, \beta, \gamma)$ and $\bar{\lambda} = \bar{\lambda}(p, n, \alpha, \beta, \gamma)$, such that if $0 < \delta < \delta_1$, then

$$\bar{\lambda}^{-1}\frac{u(y)}{d(y, \partial\Omega)} \leq |\nabla u(y)| \leq \bar{\lambda}\frac{u(y)}{d(y, \partial\Omega)} \text{ whenever } y \in \Omega \cap B(w, r/\hat{c}_1). \tag{1.13}$$

If (1.13) holds then we say that $|\nabla u|$ satisfies the 'fundamental inequality' in $\Omega \cap B(w, r/\hat{c}_1)$.

Step B. (Extension of $|\nabla u|^{p-2}$ to an A_2-weight). There exist $\delta_2 = \delta_2(p, n, \alpha, \beta, \gamma)$ and $\hat{c}_2 = \hat{c}_2(p, n, \alpha, \beta, \gamma)$ such that if $0 < \delta < \delta_2$, then $|\nabla u|^{p-2}$ extends to an $A_2(B(w, r/(\hat{c}_1\hat{c}_2))$-weight with constant depending only on $p, n, \alpha, \beta, \gamma$.

For the definition of an A_2-weight, see section 4. The 'fundamental inequality' established in Step A is crucial to our arguments and section 3 is devoted to its proof. Armed with the results established in Step A and Step B we introduce certain deformations of A-harmonic functions. In particular, to describe the constructions we let $\Omega \subset \mathbf{R}^n$, δ, r_0, p, α, β, γ, A, w, r, u and v be as in the statement of Theorem 1. Let $\tilde{\delta} = \min\{\delta_1, \delta_2\}$ where δ_1 and δ_2 are given in Step A and Step B respectively. We extend u and v to $B(w, 4r)$ by defining $u \equiv 0 \equiv v$ on $B(w, 4r)\setminus\bar{\Omega}$.

Step C. (Deformation of A-harmonic functions). Let $r^* = r/c$ and assume that

$$\begin{aligned}
&(a) \quad 0 \leq u \leq v/2 \text{ in } \bar{\Omega} \cap \bar{B}(w, 4r^*),\\
&(b) \quad c^{-1} \leq u(a_{r^*}(w)), v(a_{r^*}(w)) \leq c,\\
(1.14)\,&(c) \quad c^{-1}h(a_{r^*}(w)) \leq \max_{\bar{\Omega}\cap\bar{B}(w,4r^*)} h \leq ch(a_{r^*}(w)) \text{ whenever } h = u \text{ or } v.
\end{aligned}$$

Here $c \geq 1$ depends only on $p, n, \alpha, \beta, \gamma$. At the end of section 4 we then show that the assumptions in (1.14) can be easily removed. Hence, to prove Theorem 1 we can without loss of generality assume that (1.14) holds. We let $\tilde{u}(\cdot, \tau)$, $0 \leq \tau \leq 1$, be the A-harmonic function in $\Omega \cap B(w, 4r^*)$ with continuous boundary values,

$$\tilde{u}(y, \tau) = \tau v(y) + (1 - \tau)u(y) \text{ whenever } y \in \partial(\Omega \cap B(w, 4r^*)) \text{ and } \tau \in [0, 1]. \tag{1.15}$$

Using (1.14), (1.15), we see that if $t, \tau \in [0, 1]$, then

$$0 \leq \frac{\tilde{u}(\cdot, t) - \tilde{u}(\cdot, \tau)}{t - \tau} = v - u \leq c(p, n, \alpha, \beta, \gamma) \tag{1.16}$$

on $\partial(\Omega \cap B(w, 4r^*))$. From the maximum principle for A-harmonic functions it then follows that the inequality in (1.16) also holds in $\Omega \cap B(w, 4r^*)$. Therefore, using (1.16) we see that $\tau \to \tilde{u}(y, \tau), \tau \in [0, 1]$, for fixed $y \in \Omega \cap B(w, 4r^*)$, is Lipschitz continuous with Lipschitz norm $\leq c$. Thus $\tilde{u}_\tau(y, \cdot)$ exists, for fixed $y \in \Omega \cap B(w, 4r^*)$, almost everywhere in [0,1]. Let $\{y_\nu\}$ be a dense sequence of $\Omega \cap B(w, 4r^*)$ and let W be the set of all $\tau \in [0, 1]$ for which $u_\tau(y_m, \cdot)$ exists, in the sense of difference quotients, whenever $y_m \in \{y_\nu\}$. We note that $H^1([0, 1] \setminus W) = 0$ where H^1 is one-dimensional Hausdorff measure. Next, applying the 'fundamental inequality', established in Step A, to $\tilde{u}(\cdot, \tau)$, $\tau \in [0, 1]$, we see that there exist constants $\hat{c}$ and $\bar{\lambda}$, which depend on $p, n, \alpha, \beta, \gamma$, but are independent of τ, $\tau \in [0, 1]$, such that if $y \in \Omega \cap B(w, 16r')$, $r' = r^*/\hat{c}$ and $\tau \in [0, 1]$, then

$$\bar{\lambda}^{-1} \frac{\tilde{u}(y, \tau)}{d(y, \partial\Omega)} \leq |\nabla \tilde{u}(y, \tau)| \leq \bar{\lambda} \frac{\tilde{u}(y, \tau)}{d(y, \partial\Omega)}. \tag{1.17}$$

One can then deduce, using the fundamental theorem of calculus and arguing as in [LN4, displays (1.15)-(1.23)], that

$$\log\left(\frac{v(y_m)}{u(y_m)}\right) = \log\left(\frac{\tilde{u}(y_m, 1)}{\tilde{u}(y_m, 0)}\right) = \int_0^1 \frac{f(y_m, \tau)}{\tilde{u}(y_m, \tau)} d\tau \tag{1.18}$$

whenever $y_m \in \{y_\nu\}$, $y_m \in \Omega \cap B(w, r')$, and for a function f which has the following important properties,

$$\begin{array}{ll} (i) & f \geq 0 \text{ is continuous in } \bar{B}(w, r') \text{ with } f \equiv 0 \text{ on } \bar{B}(w, r') \setminus \Omega, \\ (ii) & f(y_m, \tau) = \tilde{u}_\tau(y_m, \tau) \end{array} \tag{1.19}$$

whenever $y_m \in \{y_\nu\}$, $y_m \in \Omega \cap B(w, r'), \tau \in W$. Moreover, f is locally a weak solution in $\Omega \cap B(w, r')$ to the equation

$$\tilde{L}\zeta = \sum_{i,j=1}^{n} \frac{\partial}{\partial y_i} (\tilde{b}_{ij}(y, \tau) \zeta_{y_j}(y)) = 0 \tag{1.20}$$

where

$$\tilde{b}_{ij}(y, \tau) = \frac{\partial A_i}{\partial \eta_j}(y, \nabla \tilde{u}(y, \tau)) \tag{1.21}$$

whenever $y \in \Omega \cap B(w, r')$ and $1 \leq i, j \leq n$. Also, using Definition 1.1 (i) and (ii) we see that

$$\alpha^{-1} \tilde{\lambda}(y, \tau) |\xi|^2 \leq \sum_{i,j} \tilde{b}_{ij}(y, \tau) \xi_i \xi_j \leq \alpha \tilde{\lambda}(y, \tau) |\xi|^2 \tag{1.22}$$

whenever $y \in \Omega \cap B(w, r')$ and where $\tilde{\lambda}(y, \tau) = |\nabla \tilde{u}(y, \tau)|^{p-2}$. Finally, a key observation in this step is that $\zeta = \tilde{u}(\cdot, \tau)$ is also a weak solution to $\tilde{L}$ in $\Omega \cap B(w, r')$. Indeed, using the homogeneity in Definition 1.1 (iv) we see that

$$\begin{aligned} \sum_j \tilde{b}_{ij}(y, \tau) \tilde{u}_{y_j}(y, \tau) &= \sum_j \frac{\partial A_i}{\partial \eta_j}(y, \nabla \tilde{u}(y, \tau)) \tilde{u}_{y_j}(y, \tau) \\ &= (p-1) A_i(y, \nabla \tilde{u}(y, \tau)). \end{aligned} \tag{1.23}$$

We conclude from (1.23) that $\zeta = \tilde{u}(\cdot, \tau)$ is also a weak solution to $\tilde{L}$.

Step D. (Boundary Harnack inequalities for degenerate elliptic equations). Using the deformations introduced in Step C the proof of Theorem 1 therefore boils down to proving boundary Harnack inequalities for the operator $\tilde{L}$. The idea here is to make use of Step B to conclude that $\tilde{\lambda}(\cdot,\tau)$, $\tau\in[0,1]$, can be extended to A_2-weights in $B(w,4r'')$, $r''=r'/(4\hat{c}_2)$. Then the operator $\tilde{L}$ can be considered as a degenerate elliptic operator in the sense of [FKS], [FJK], [FJK1], and we can apply results of these authors. In particular, to do this we first observe that the sequence $\{y_\nu\}$, introduced below (1.16), is a dense sequence in $\Omega\cap B(w,r')$, and $v_1(\cdot)=f(\cdot,\tau)$, $v_2(\cdot)=u(\cdot,\tau)$, are positive solutions to $\tilde{L}$, see (1.20)-(1.22), vanishing continuously on $\Omega\cap B(w,r')$. Second, we observe from Step B that $\tilde{\lambda}(y,\tau)=|\nabla\tilde{u}(y,\tau)|^{p-2}$ can be extended to an $A_2(B(w,4r''))$-weight. Hence, from [FKS], [FJK] and [FJK1], we can conclude that there exist a constant $c=c(p,n,\alpha,\beta,\gamma)$, $1\leq c<\infty$, and $\sigma=\sigma(p,n,\alpha,\beta,\gamma)$, $\sigma\in(0,1)$, such that if $r'''=r''/c$, then

$$\left|\frac{v_1(y_1)}{v_2(y_1)}-\frac{v_1(y_2)}{v_2(y_2)}\right|\leq c\,\frac{v_1(a_{r'''}(w))}{v_2(a_{r'''}(w))}\left(\frac{|y_1-y_2|}{r''}\right)^{\sigma} \tag{1.24}$$

whenever $y_1,y_2\in\Omega\cap B(w,r''')$. Hence, assuming (1.14) we see that Theorem 1 now follows from (1.18), (1.24), as

$$0\leq f(a_{r'''}(w),\tau)\leq c,\ u(a_{r'''}(w),\tau)\geq c^{-1},\ \text{whenever } \tau\in(0,1]. \tag{1.25}$$

(1.25) is a consequence of (1.16) and (1.14) (b).

The proof of Theorem 2 can also be decomposed into steps similar to steps A-D stated above. Still in this case details are more involved and we refer to section 5 for details.

The rest of the paper is organized as follows. In section 2 we state a number of basic estimates for A-harmonic functions in NTA-domains and we obtain the conclusion of Theorem 1 when $A\in M_p(\alpha)$, Ω is equal to a truncated cylinder (see (2.7) and Lemma 2.8), and w is the center of the bottom of $\bar{\Omega}$ (Step 0). In section 3 we establish the 'fundamental inequality' for A-harmonic functions, u, vanishing on a portion of a Reifenberg flat domain (Step A). In section 4 we first state a number of results for degenerate elliptic equations tailored to our situation and we then extend $|\nabla u|^{p-2}$ to an A_2-weight (Step B). In this section we also complete the proof of Theorem 1 by showing that the technical assumption in (1.14) can be removed. In section 5 we prove Theorem 2. Finally in an Appendix to this paper (section 6), we point out an alternative argument to Step C based on an idea in [W].

2. Basic estimates for A-harmonic functions and boundary Harnack inequalities in a prototype case

In this section we first state and prove some basic estimates for non-negative A-harmonic functions in a bounded NTA domain $\Omega\subset\mathbf{R}^n$. We then prove the boundary Harnack inequality for non-negative A-harmonic functions, $A\in M_p(\alpha)$, vanishing on a portion of a hyperplane. Throughout this section we will assume that $A\in M_p(\alpha,\beta,\gamma)$ or $A\in M_p(\alpha)$ for some (α,β,γ) and $1<p<\infty$. Also in this paper, unless otherwise stated, c will denote a positive constant ≥ 1, not necessarily the same at each occurrence, depending only on $p,n,M,\alpha,\beta,\gamma$ where M denotes the NTA-constant for $\Omega\subset\mathbf{R}^n$. In general, $c(a_1,\ldots,a_m)$ denotes a positive constant ≥ 1, which may depend only on $p, n, M, \alpha, \beta, \gamma$ and $a_1, \dots, a_m$, not necessarily the same at each occurrence. If $A \approx B$ then A/B is bounded from above and below by constants which, unless otherwise stated, only depend on $p, n, M, \alpha, \beta, \gamma$. Moreover, we let $\max\limits_{B(z,s)} u$, $\min\limits_{B(z,s)} u$ be the essential supremum and infimum of u on $B(z,s)$ whenever $B(z,s) \subset \mathbf{R}^n$ and whenever u is defined on $B(z,s)$. We put $\Delta(w,r) = \partial\Omega \cap B(w,r)$ whenever $w \in \partial\Omega$, $0 < r$. Finally, $e_i, 1 \leq i \leq n$, denotes the point in $\mathbf{R}^n$ with one in the i th coordinate position and zeroes elsewhere.

Lemma 2.1. *Given $p, 1 < p < \infty$, assume that $A \in M_p(\alpha, \beta, \gamma)$ for some (α, β, γ). Let u be a positive A-harmonic function in $B(w, 2r)$. Then*

$$(i) \qquad r^{p-n} \int\limits_{B(w,r/2)} |\nabla u|^p \, dx \leq c\,(\max_{B(w,r)} u)^p,$$

$$(ii) \qquad \max_{B(w,r)} u \leq c \min_{B(w,r)} u.$$

Furthermore, there exists $\tilde{\sigma} = \tilde{\sigma}(p, n, \alpha, \beta, \gamma) \in (0,1)$ such that if $x, y \in B(w,r)$, then

$$(iii) \qquad |u(x) - u(y)| \leq c\left(\tfrac{|x-y|}{r}\right)^{\tilde{\sigma}} \max_{B(w,2r)} u.$$

Proof: Lemma 2.1 (i), (ii) are standard Caccioppoli and Harnack inequalities while (iii) is a standard Hölder estimate (see [S]). $\Box$

Lemma 2.2. *Let $\Omega \subset \mathbf{R}^n$ be a bounded NTA-domain, suppose that p, $1 < p < \infty$, is given and that $A \in M_p(\alpha, \beta, \gamma)$ for some (α, β, γ). Let $w \in \partial\Omega$, $0 < r < r_0$, and suppose that u is a non-negative continuous A-harmonic function in $\bar{\Omega} \cap B(w, 2r)$ and that $u = 0$ on $\Delta(w, 2r)$. Then*

$$(i) \qquad r^{p-n} \int\limits_{\Omega \cap B(w,r/2)} |\nabla u|^p \, dx \leq c\,(\max_{\Omega \cap B(w,r)} u)^p.$$

Furthermore, there exists $\tilde{\sigma} = \tilde{\sigma}(p, n, M, \alpha, \beta, \gamma) \in (0,1)$ such that if $x, y \in \Omega \cap B(w,r)$, then

$$(ii) \qquad |u(x) - u(y)| \leq c\left(\tfrac{|x-y|}{r}\right)^{\tilde{\sigma}} \max_{\Omega \cap B(w,2r)} u.$$

Proof: Lemma 2.2 (i) is a standard subsolution inequality while (ii) follows from a Wiener criteria first proved in [M] and later generalized in [GZ]. $\Box$

Lemma 2.3. *Let $\Omega \subset \mathbf{R}^n$ be a bounded NTA-domain, suppose that p, $1 < p < \infty$, is given and that $A \in M_p(\alpha, \beta, \gamma)$ for some (α, β, γ). Let $w \in \partial\Omega$, $0 < r < r_0$, and suppose that u is a non-negative continuous A-harmonic function in $\bar{\Omega} \cap B(w, 2r)$ and that $u = 0$ on $\Delta(w, 2r)$. There exists $c = c(p, n, M, \alpha, \beta, \gamma)$, $1 \leq c < \infty$, such that if $\tilde{r} = r/c$, then*

$$\max_{\Omega \cap B(w,\tilde{r})} u \leq c\, u(a_{\tilde{r}}(w)).$$

Proof: A proof of Lemma 2.3 for linear elliptic PDE can be found in [CFMS]. The proof uses only analogues of Lemmas 2.1, 2.2 for linear PDE and Definition 1.5. In particular, the proof also applies in our situation. □

Lemma 2.4. *Let $\Omega \subset \mathbf{R}^n$ be a bounded NTA-domain, suppose that p, $1 < p < \infty$, is given and that $A \in M_p(\alpha, \beta, \gamma)$ for some (α, β, γ). Let $w \in \partial\Omega$, $0 < r < r_0$, and suppose that u is a non-negative continuous A-harmonic function in $\bar{\Omega} \cap \bar{B}(w, 4r)$ and that $u = 0$ on $\Delta(w, 4r)$. Extend u to $B(w, 4r)$ by defining $u \equiv 0$ on $B(w, 4r) \setminus \Omega$. Then u has a representative in $W^{1,p}(B(w, 4r))$ with Hölder continuous partial derivatives of first order in $\Omega \cap B(w, 4r)$. In particular, there exists $\hat{\sigma} \in (0, 1]$, depending only on $p, n, \alpha, \beta, \gamma$ such that if $x, y \in B(\hat{w}, \hat{r}/2)$, $B(\hat{w}, 4\hat{r}) \subset \Omega \cap B(w, 4r)$, then*

$$c^{-1}\,|\nabla u(x) - \nabla u(y)| \leq (|x - y|/\hat{r})^{\hat{\sigma}} \max_{B(\hat{w}, \hat{r})} |\nabla u| \leq c\,\hat{r}^{-1}\,(|x - y|/\hat{r})^{\hat{\sigma}} \max_{B(\hat{w}, 2\hat{r})} u.$$

Proof: Given $\epsilon > 0$ and small, let

$$A(y, \eta, \epsilon) = \int_{\mathbf{R}^n} A(y, \eta - x)\theta_\epsilon(x)dx \text{ whenever } (y, \eta) \in \mathbf{R}^n \times \mathbf{R}^n, \tag{2.5}$$

where $\theta \in C_0^\infty(B(0, 1))$ with $\int_{\mathbf{R}^n} \theta dx = 1$ and $\theta_\epsilon(x) = \epsilon^{-n}\theta(x/\epsilon)$ whenever $x \in \mathbf{R}^n$. From Definition 1.1 and standard properties of approximations to the identity, we deduce for some $c = c(p, n) \geq 1$ that

$$\begin{aligned} &(i) && (c\alpha)^{-1}(\epsilon + |\eta|)^{p-2}|\xi|^2 \leq \sum_{i,j=1}^{n} \frac{\partial A_i}{\partial \eta_j}(y, \eta, \epsilon)\xi_i\xi_j, \\ &(ii) && \left|\frac{\partial A_i}{\partial \eta_j}(y, \eta, \epsilon)\right| \leq c\alpha(\epsilon + |\eta|)^{p-2}, 1 \leq i, j \leq n, \\ (2.6)\quad &(iii) && |A(x, \eta, \epsilon) - A(y, \eta, \epsilon)| \leq c\beta|x - y|^\gamma(\epsilon + |\eta|)^{p-1} \end{aligned}$$

whenever $x, y, \eta \in \mathbf{R}^n$. Moreover, $A(y, \cdot, \epsilon)$ is, for fixed (y, ϵ), infinitely differentiable.

To prove Lemma 2.4 we choose $u(\cdot, \epsilon)$, a weak solution to the PDE with structure as in (2.6), in such a way that $u(\cdot, \epsilon)$ is continuous in $\bar{\Omega} \cap \bar{B}(w, 3r)$ and $u(\cdot, \epsilon) = u$ on $\partial[\Omega \cap B(w, 3r)]$. Existence of $u(\cdot, \epsilon)$ follows from the Wiener criteria in [GZ] mentioned in the proof of Lemma 2.2, the maximum principle for A-harmonic functions, and the fact that the $W^{1,p}$-Dirichlet problem for these functions, in $\Omega \cap B(w, 3r)$, always has a unique solution (see [HKM, Appendix I]). Moreover, from [T], [T1], it follows that $u(\cdot, \epsilon)$ is in $C^{1,\hat{\sigma}}(\Omega \cap B(w, 2r))$ for some $\hat{\sigma} > 0$ with constants independent of ϵ. Letting $\epsilon \to 0$ one can show, using Definition 1.1, that subsequences of $\{u(\cdot, \epsilon)\}, \{\nabla u(\cdot, \epsilon)\}$, converge pointwise to $u, \nabla u$. In view of Lemma 2.1 and the result in [T] it follows that this convergence is uniform on compact subsets of $\Omega \cap B(w, 3r)$. Using this fact we get the last display in Lemma 2.4.

Finally we note that in [T] a stronger assumption, compared to (2.6) *(iii)*, is imposed. However, other authors later obtained the results in [T] under assumption (2.6) (see [Li] for references). □

Next we show that the conclusion of Theorem 1 holds in the case of a truncated cylinder with w the center on the bottom of the cylinder (Step 0). To this end we introduce, for $a, b \in \mathbf{R}^+$ and $w = (w_1, ..., w_n) \in \mathbf{R}^n$, the truncated cylinders,

$$
\begin{aligned}
Q_{a,b}(w) &= \{y = (y', y_n) : |y' - w'| < a, |y_n - w_n| < b\}, \\
Q^+_{a,b}(w) &= \{y = (y', y_n) : |y' - w'| < a, \ 0 < y_n - w_n < b\}, \\
Q^-_{a,b}(w) &= \{y = (y', y_n) : |y' - w'| < a, \ -b < y_n - w_n < 0\}.
\end{aligned} \tag{2.7}
$$

Furthermore, if $a = b$ then we let $Q_a(w) = Q_{a,a}(w)$, $Q^+_a(w) = Q^+_{a,a}(w)$, $Q^-_a(w) = Q^-_{a,a}(w)$.

Lemma 2.8. *Suppose that p, $1 < p < \infty$, is given and that $A \in M_p(\alpha)$ for some α. Assume also, that u, v are non-negative A-harmonic functions in $Q^+_1(0)$, continuous on the closure of $Q^+_1(0)$, and with $u = 0 = v$ on $\partial Q^+_1(0) \cap \{y_n = 0\}$. Then there exist $c = c(p, n, \alpha), 1 \leq c < \infty$, and $\sigma = \sigma(p, n, \alpha) \in (0, 1]$ such that*

$$\left| \log\left(\frac{u(y_1)}{v(y_1)}\right) - \log\left(\frac{u(y_2)}{v(y_2)}\right) \right| \leq c|y_1 - y_2|^\sigma$$

whenever y_1, $y_2 \in Q^+_{1/4}(0)$.

Proof. Let $A = A(\eta)$ be as in Lemma 2.8 and let p be fixed, $1 < p < \infty$. Note that y_n is A-harmonic and that it suffices to prove Lemma 2.8 when $v = y_n$. Define $A(\eta, \epsilon)$ as in (2.5) relative to A and let $u(\cdot, \epsilon)$ be the solution to $\nabla \cdot (A(\nabla u(y, \epsilon), \epsilon)) = 0$ with continuous boundary values equal to u on $\partial Q^+_1(0)$. Let

$$A^*_{ij}(y, \epsilon) = \tfrac{1}{2}(\epsilon + |\nabla u(y, \epsilon)|)^{2-p} \left[\frac{\partial A_i}{\partial \eta_j}(\nabla u(y, \epsilon), \epsilon) + \frac{\partial A_j}{\partial \eta_i}(\nabla u(y, \epsilon), \epsilon)\right]$$

whenever $y \in Q^+_{1/2}(0)$ and $1 \leq i, j \leq n$. From (2.6) (ii) and Schauder type estimates we see that $u(\cdot, \epsilon), y_n$, are classical solutions to the non-divergence form uniformly elliptic equation,

$$L^*\zeta = \sum_{i,j=1}^n A^*_{ij}(y, \epsilon)\zeta_{y_i y_j} = 0, \tag{2.9}$$

for $y \in Q^+_{1/2}(0)$. Note also from (2.6) that the ellipticity constant for $(A^*_{ij}(y, \epsilon))$ and the L^∞-norm for $A^*_{ij}(y, \epsilon), 1 \leq i, j \leq n$, in $Q^+_{1/2}(0)$, depend only on α, p, n. From this note we see that if $z = (z', z_n) \in Q^+_{1/2}(0)$ and $10^{-3} < \rho_1 < \rho_2 < 10^3$, then

$$\psi(y) = \frac{e^{-N|y-z|^2} - e^{-N\rho_2^2}}{e^{-N\rho_1^2} - e^{-N\rho_2^2}} \tag{2.10}$$

is a subsolution to L^* in $Q^+_1(0) \cap [B(z, \rho_2) \setminus B(z, \rho_1)]$, if $N = N(\alpha, p, n)$ is sufficiently large, and $\psi \equiv 1$ on $\partial B(z, \rho_1)$ while $\psi \equiv 0$ on $\partial B(z, \rho_2)$. Using this fact, with $z = (z', 1/16)$, $|z'| < 1/2$, $\rho_1 = 1/64, \rho_2 = 1/16$ and Harnack's inequality for L^* (see [GT, Corollary 9.25]) we get

$$c^{-1} y_n \, u(e_n/4, \epsilon) \leq u(y, \epsilon) \tag{2.11}$$

whenever $y \in Q^+_{1/4}(0)$. Moreover, using $1 - \psi, z = (z', -e_n/64)$, $|z'| < 1/2$, $\rho_1 = 1/64, \rho_2 = 1/16$, in a similar argument it follows that

$$u(y, \epsilon) \leq c y_n \max_{Q^+_{1/4}(0)} u(\cdot, \epsilon) \leq c^2 \, y_n \, u(e_n/4, \epsilon) \tag{2.12}$$

in $Q^+_{1/4}(0)$. In particular, the right-hand inequality in (2.12) follows from the analogue of Lemma 2.3 for L^*.

Fix $x \in \partial Q^+_{1/2}(0) \cap \{y : y_n = 0\}$. From (2.11), (2.12), and linearity of L^* one can deduce (see for example [LN, Lemma 3.27]) that there exists $\theta, 0 < \theta < 1$, such that

$$\text{osc } (\rho/4) \leq \theta \text{ osc } (\rho) \tag{2.13}$$

when $0 < \rho \leq 1/4$, where $\text{osc } (t) = M(t) - m(t)$ and we have put

$$M(t) = \max_{Q^+_t(x)} \tfrac{u(y,\epsilon)}{y_n}, \quad m(t) = \min_{Q^+_t(x)} \tfrac{u(y,\epsilon)}{y_n}.$$

To get (2.13) one can simply apply the same argument as in (2.11), (2.12) to $u - m(\rho)y_n, y_n$ and $M(\rho)y_n - u, y_n$ in $Q^+_\rho(x)$. Iterating (2.13), we obtain for some $\lambda > 0, c > 1$, depending on α, p, n, that

$$\text{osc } (s) \leq c\,(s/t)^\lambda \text{ osc } (t), 0 < s < t \leq 1/4. \tag{2.14}$$

Letting $\epsilon{\to}0$ it follows as in the proof of Lemma 2.4 that $u(\cdot, \epsilon)$ converges uniformly to u on compact subsets of $Q^+_{1/2}(0)$. Thus (2.11), (2.12) and (2.14) also hold for u. Moreover, (2.11), (2.12), (2.14), arbitrariness of x, and interior Harnack - Hölder continuity of u are easily shown to be equivalent to the conclusion of Lemma 2.8 when $v(y) = y_n$. □

We note that boundary Harnack inequalities for non-divergence form linear symmetric operators in Lipschitz domains can be found in either [B] or [FGMS].

We end this section by proving the following lemma.

Lemma 2.15. *Let $G \subset \mathbf{R}^n$ be an open set, suppose that p, $1 < p < \infty$, is given and let $A \in M_p(\alpha, \beta, \gamma)$ for some (α, β, γ). Let $F : \mathbf{R}^n \to \mathbf{R}^n$ be the composition of a translation, a rotation and a dilation $z \to rz$, $r \in (0, 1]$. Suppose that u is A-harmonic in G and define $\hat{u}(z) = u(F(z))$ whenever $F(z) \in G$. Then $\hat{u}$ is $\hat{A}$-harmonic in $F^{-1}(G)$ and $\hat{A} \in M_p(\alpha, \beta, \gamma)$.*

Proof. Suppose that $F(z) = z + w$ for some $w \in \mathbf{R}^n$, i.e., F is a translation. In this case the conclusion follows immediately with $\hat{A}(z, \eta) = A(z + w, \eta)$ and $\hat{A} \in M_p(\alpha, \beta, \gamma)$. Suppose that $F(z) = \Gamma z$, where Γ is an orthogonal matrix with $\det \Gamma = 1$. In this case the conclusion follows with $\hat{A}(z, \eta) = A(\Gamma z, \Gamma \eta)$ and $\hat{A} \in M_p(\alpha, \beta, \gamma)$. Finally, suppose that $F(z) = rz$ for some $r \in (0, 1]$. Then $\hat{u}$ is $\hat{A}$-harmonic in $F^{-1}(G)$ with $\hat{A}(z, \eta) = r^{p-1} A(rz, r^{-1}\eta)$. Moreover, property (i), (ii) and (iv) in Definition 1.1 follow readily. To prove (iii) in Definition 1.1 we see that

$$|\hat{A}(z, \eta) - \hat{A}(y, \eta)| \leq \beta r^\gamma |z - y|^\gamma |\eta|^{p-1} \leq \beta |z - y|^\gamma |\eta|^{p-1}$$

whenever $r \in (0, 1]$. This completes the proof of Lemma 2.15. □

3. Non-degeneracy of $|\nabla u|$

In this section we establish the 'fundamental inequality' referred to as Step A in the introduction. To do this we first prove a few technical results.

Lemma 3.1. *Let $1 < p < \infty$, and assume that $A_1, A_2 \in M_p(\alpha, \beta, \gamma)$ with*

$$|A_1(y, \eta) - A_2(y, \eta)| \leq \epsilon |\eta|^{p-1} \text{ whenever } y \in Q_1^+(0)$$

for some $0 < \epsilon < 1/2$. Let u_2 be a non-negative A_2-harmonic function in $Q_1^+(0)$, continuous on the closure $Q_1^+(0)$, and with $u_2 = 0$ on $\partial Q_1^+(0) \cap \{y_n = 0\}$. Moreover, let u_1 be the A_1-harmonic function in $Q_{1/2}^+(0)$ which is continuous on the closure of $Q_{1/2}^+(0)$ and which coincides with u_2 on $\partial Q_{1/2}^+(0)$. Then there exist, given $\rho \in (0, 1/16)$, c, $\tilde{c}$, θ, and τ, all depending only on $p, n, \alpha, \beta, \gamma$, such that

$$|u_2(y) - u_1(y)| \leq c\epsilon^\theta u_2(e_n/2) \leq \tilde{c}\epsilon^\theta \rho^{-\tau} u_2(y) \text{ whenever } y \in Q_{1/4}^+(0) \setminus Q_{1/4,\rho}^+(0).$$

Proof. To begin the proof of Lemma 3.1 we note that the existence of u_1 in Lemma 3.1 follows from the Wiener criteria in [GZ], see the discussion after Lemma 2.2, the maximum principle for A-harmonic functions, and the fact that the $W^{1,p}$-Dirichlet problem for these functions in $Q_{1/2}^+(0)$ always has a unique solution (see [HKM, Appendix I]). Observe for $x \in \mathbf{R}^n, \lambda \in \mathbf{R}^n, \xi \in \mathbf{R}^n \setminus \{0\}$, and $A \in M_p(\alpha, \beta, \gamma)$, that

$$A_i(x, \lambda) - A_i(x, \xi) = \sum_{j=1}^{n} (\lambda_j - \xi_j) \int_0^1 \frac{\partial A_i}{\partial \eta_j}(x, t\lambda + (1-t)\xi) dt \tag{3.2}$$

for $i \in \{1, .., n\}$. Using (3.2) and Definition 1.1 $(i), (ii)$, we see that

$$c^{-1} (|\lambda| + |\xi|)^{p-2} |\lambda - \xi|^2 \leq \langle A(x, \lambda) - A(x, \xi), \lambda - \xi \rangle \leq c (|\lambda| + |\xi|)^{p-2} |\lambda - \xi|^2. \tag{3.3}$$

Moreover, from (3.3) we deduce that if

$$I = \int_{Q_{1/2}^+(0)} |\nabla u_2 - \nabla u_1|^p dy,$$

then,

$$I \leq cJ, \ J := \int_{Q_{1/2}^+(0)} \langle A_1(y, \nabla u_1(y)) - A_1(y, \nabla u_2(y)), \nabla u_2(y) - \nabla u_1(y) \rangle dy, \tag{3.4}$$

whenever $p \geq 2$. Also, if $1 < p < 2$, we see from (3.3) and Hölder's inequality that

$$I \leq cJ^{p/2} \left(\int_{Q_{1/2}^+(0)} |\nabla u_1|^p + |\nabla u_2|^p dx \right)^{1-p/2} \tag{3.5}$$

where J is as defined in (3.4). As $\nabla \cdot (A_1(y, \nabla u_1(y))) = 0 = \nabla \cdot (A_2(y, \nabla u_2(y)))$ whenever $y \in Q_{1/2}^+(0)$ and as $\theta = u_2 - u_1 \in W_0^{1,p}(Q_{1/2}^+(0))$, we see from the definition of J in (3.4) that

$$J = \int_{Q_{1/2}^+(0)} \langle A_2(y, \nabla u_2(y)) - A_1(y, \nabla u_2(y)), \nabla u_2(y) - \nabla u_1(y) \rangle dy. \tag{3.6}$$

Hence, using (3.4), (3.6), the assumption on the difference $|A_1(y,\eta) - A_2(y,\eta)|$ stated in the lemma and Hölder's inequality we can conclude, for $p \geq 2$, that

$$I \leq c\epsilon \int\limits_{Q^+_{1/2}(0)} (|\nabla u_1|^p + |\nabla u_2|^p)dx. \tag{3.7}$$

Also, for $1 < p < 2$, we can use (3.5) to find that

$$I \leq c\epsilon^{p/2} \int\limits_{Q^+_{1/2}(0)} (|\nabla u_1|^p + |\nabla u_2|^p)dx. \tag{3.8}$$

Now from the observation above (3.6), (3.3) with $\xi = 0$, and Hölder's inequality we see that

$$\begin{aligned}\int\limits_{Q^+_{1/2}(0)} |\nabla u_1|^p dx \;&\leq c \int\limits_{Q^+_{1/2}(0)} \langle A_1(x, \nabla u_1(x)),\, \nabla u_2(x)\rangle dx \\ &\leq (1/2) \int\limits_{Q^+_{1/2}(0)} |\nabla u_1|^p dx + c \int\limits_{Q^+_{1/2}(0)} |\nabla u_2|^p dx.\end{aligned}$$

Thus,

$$\int\limits_{Q^+_{1/2}(0)} |\nabla u_1|^p dx \leq c \int\limits_{Q^+_{1/2}(0)} |\nabla u_2|^p dx. \tag{3.9}$$

Let $a = \min\{1, p/2\}$. Using (3.9) in (3.8), (3.7), and Lemmas 2.1 - 2.3 for u_2 we obtain

$$I \leq c\epsilon^a (u_2(e_n/2))^p. \tag{3.10}$$

Next using the Poincáre inequality for functions in $W^{1,p}_0(Q^+_{1/2}(0))$ we deduce from (3.10) that

$$\int\limits_{Q^+_{1/2}(0)} |u_2 - u_1|^p \, dx \leq c \int\limits_{Q^+_{1/2}(0)} |\nabla u_2 - \nabla u_1|^p \, dx \leq c\epsilon^a (u_2(e_n/2))^p. \tag{3.11}$$

In the following we let $\eta = a/(p+2)$ and we introduce the sets

$$E = \{y \in Q^+_{1/2}(0) : \; |u_2(y) - u_1(y)| \leq \epsilon^\eta u_2(e_n/2)\}, \; F = Q^+_{1/2}(0) \setminus E. \tag{3.12}$$

Moreover, for a measurable function f defined on $Q^+_{1/2}(0)$ we introduce, whenever $y \in Q^+_{1/2}(0)$, the Hardy-Littlewood maximal function

$$M(f)(y) := \sup_{\{r>0,\; Q_r(y) \subset Q^+_{1/2}(0)\}} \frac{1}{|Q_r(y)|} \int\limits_{Q_r(y)} |f(z)|dz. \tag{3.13}$$

Let

$$G = \{y \in Q^+_{1/2}(0) : \; M(\chi_F)(y) \leq \epsilon^\eta\} \tag{3.14}$$

where χ_F is the indicator function for the set F. Then using weak (1,1)-estimates for the Hardy-Littlewood maximal function, (3.11) and (3.12) we see that

$$|Q^+_{1/2}(0) \setminus G| \leq c\epsilon^{-\eta}|F| \leq c\epsilon^{-\eta}\epsilon^{-p\eta}\epsilon^a = c\epsilon^\eta \tag{3.15}$$

by our choice for η. Also, using continuity of $u_2(y) - u_1(y)$ we find for $y \in G$ that

$$(3.16)\qquad |u_2(y) - u_1(y)| = \lim_{r\to 0} \frac{1}{|B(y,r)|} \int_{B(y,r)} |u_2(z) - u_1(z)| dz \le c\epsilon^\eta u_2(e_n/2).$$

If $y \in Q^+_{1/4}(0) \setminus G$, then from (3.15) we see there exists $\hat y \in G$ such that $|y - \hat y| \le c(n)\epsilon^{\eta/n}$. Using Lemmas 2.1, 2.2, we hence get that

$$\begin{aligned} |u_2(y) - u_1(y)| &\le |u_2(\hat y) - u_1(\hat y)| + |u_2(y) - u_2(\hat y)| + |u_1(y) - u_1(\hat y)| \\ (3.17)\qquad &\le c(\epsilon^\eta + \epsilon^{\tilde\sigma\eta/n}) u_2(e_n/2). \end{aligned}$$

This completes the proof of the first inequality stated in Lemma 3.1. Finally, using the Harnack inequality we see that there exists $\tau = \tau(p, n, \alpha, \beta, \gamma) \ge 1$ such that $u_2(e_n/2) \le c\rho^{-\tau} u_2(y)$ whenever $y \in Q^+_{1/4}(0) \setminus Q^+_{1/4,\rho}(0)$. □

We continue by proving the following important technical lemma.

Lemma 3.18. *Let $O \subset \mathbf{R}^n$ be an open set, suppose $1 < p < \infty$, and that $A_1, A_2 \in M_p(\alpha, \beta, \gamma)$. Also, suppose that $\hat u_1, \hat u_2$ are non-negative functions in O, that $\hat u_1$ is A_1-harmonic in O and that $\hat u_2$ is A_2-harmonic in O. Let $\tilde a \ge 1, y \in O$ and assume that*

$$\frac{1}{\tilde a} \frac{\hat u_1(y)}{d(y, \partial O)} \le |\nabla \hat u_1(y)| \le \tilde a \frac{\hat u_1(y)}{d(y, \partial O)}.$$

Let $\tilde\epsilon^{-1} = (c\tilde a)^{(1+\hat\sigma)/\hat\sigma}$, where $\hat\sigma$ is as in Lemma 2.4. If

$$(1 - \tilde\epsilon)\hat L \le \frac{\hat u_2}{\hat u_1} \le (1 + \tilde\epsilon)\hat L \text{ in } B(y, \tfrac{1}{100} d(y, \partial O))$$

for some $\hat L, 0 < \hat L < \infty$, then for $c = c(p, n, \alpha, \beta, \gamma)$ suitably large,

$$\frac{1}{c\tilde a} \frac{\hat u_2(y)}{d(y, \partial O)} \le |\nabla \hat u_2(y)| \le c\tilde a \frac{\hat u_2(y)}{d(y, \partial O)}.$$

Proof. Let $\tilde a \ge 1, y \in O$ be as in the statement of the lemma. Using Lemma 2.4 and the Harnack inequality in Lemma 2.1 (ii) we see that,

$$(3.19)\qquad |\nabla \hat u_2(z_1) - \nabla \hat u_2(z_2)| \le ct^{\hat\sigma} \max_{B(y, td(y,\partial O))} |\nabla \hat u_2(\cdot)| \le c^2 t^{\hat\sigma}\, \hat u_2(y)/d(y, \partial O)$$

whenever $z_1, z_2 \in \bar B(y, td(y, \partial O))$ and $0 < t \le 10^{-3}$. Here c depends only on $p, n, \alpha, \beta, \gamma$. Using (3.19) we see that we only have to prove bounds from below for the gradient of $\hat u_2$ at y. To achieve this we suppose that,

$$(3.20)\qquad |\nabla \hat u_2(y)| \le \zeta\, \hat u_2(y)/d(y, \partial O),$$

for some small $\zeta > 0$ to be chosen. From (3.19) with $z = z_1, y = z_2$ and (3.20) we then deduce that

$$(3.21)\qquad |\nabla \hat u_2(z)| \le [\zeta + c^2 t^{\hat\sigma}]\, \hat u_2(y)/d(y, \partial O)$$

whenever $z \in B(y, td(y, \partial O))$. Integrating, it follows that if $\hat y \in \partial B(y, td(y, \partial O))$, $|y - \hat y| = td(y, \partial O)$, $t = \zeta^{1/\hat\sigma}$, then

$$(3.22)\qquad |\hat u_2(\hat y) - \hat u_2(y)| \le c' \zeta^{1+1/\hat\sigma}\, \hat u_2(y).$$

The constants in (3.21),(3.22) depend only on $p, n, \alpha, \beta, \gamma$.

Next we note that (3.19) also holds with $\hat{u}_2$ replaced by $\hat{u}_1$. Let $\lambda = \frac{\nabla \hat{u}_1(y)}{|\nabla \hat{u}_1(y)|}$. Then from (3.19) for $\hat{u}_1$ and the non-degeneracy assumption on $|\nabla \hat{u}_1|$ in Lemma 3.18, we find that

$$\langle \nabla \hat{u}_1(z), \lambda \rangle \geq (1 - c\,\tilde{a}\zeta)|\nabla \hat{u}_1(y)| \text{ whenever } z \in \bar{B}(y, \zeta^{1/\hat{\sigma}} d(y, \partial O)),$$

for some $c = c(p, n, \alpha, \beta, \gamma)$. If $\zeta \leq (2c\tilde{a})^{-1}$, where c is the constant in the last display, then we get from integration that

$$c^*(\hat{u}_1(\hat{y}) - \hat{u}_1(y)) \geq \tilde{a}^{-1}\zeta^{1/\hat{\sigma}}\hat{u}_1(y) \tag{3.23}$$

with $\hat{y} = y + \zeta^{1/\hat{\sigma}} d(y, \partial O)\lambda$ and where the constant c^* depends only on $p, n, \alpha, \beta, \gamma$. From (3.23), (3.22), we see that if $\tilde{\epsilon}$ is as in Lemma 3.18, then

$$\begin{aligned} (1-\tilde{\epsilon})\hat{L} \;&\leq\; \frac{\hat{u}_2(\hat{y})}{\hat{u}_1(\hat{y})} \leq \left(\frac{1 + c'\zeta^{1+1/\hat{\sigma}}}{1 + \zeta^{1/\hat{\sigma}}/(\tilde{a}c^*)}\right)\frac{\hat{u}_2(y)}{\hat{u}_1(y)} \\ &\leq\; (1+\tilde{\epsilon})\left(\frac{1 + c'\zeta^{1+1/\hat{\sigma}}}{1 + \zeta^{1/\hat{\sigma}}/(\tilde{a}c^*)}\right)\hat{L} < (1-\tilde{\epsilon})\hat{L} \end{aligned} \tag{3.24}$$

provided $1/(\tilde{a}\tilde{c})^{1/\hat{\sigma}} \geq \zeta^{1/\hat{\sigma}} \geq \tilde{a}\tilde{c}\,\tilde{\epsilon}$ for some large $\tilde{c} = \tilde{c}(p, n, \alpha, \beta, \gamma)$. This inequality and (3.23) are satisfied if $\tilde{\epsilon}^{-1} = (\tilde{c}\tilde{a})^{(1+\hat{\sigma})/\hat{\sigma}}$ and $\zeta^{-1} = \tilde{c}\tilde{a}$. Moreover, if the hypotheses of Lemma 3.18 hold for this $\tilde{\epsilon}$, then in order to avoid the contradiction in (3.24) it must be true that (3.20) is false for this choice of ζ. Hence Lemma 3.18 is true. □

Armed with Lemma 3.1 and Lemma 3.18 we prove the 'fundamental inequality' for A-harmonic functions, $A \in M_p(\alpha, \beta, \gamma)$ for some (α, β, γ), vanishing on a portion of $\{y : y_n = 0\}$.

Lemma 3.25. *Let $1 < p < \infty$, and $A \in M_p(\alpha, \beta, \gamma)$ for some (α, β, γ). Suppose that u is a positive A-harmonic function in $Q_1^+(0)$, continuous on the closure of $Q_1^+(0)$, and that $u = 0$ on $\partial Q_1^+(0) \cap \{y_n = 0\}$. Then there exist $\hat{c} = \hat{c}(p, n, \alpha, \beta, \gamma)$ and $\bar{\lambda} = \bar{\lambda}(p, n, \alpha, \beta, \gamma)$, such that*

$$\bar{\lambda}^{-1}\frac{u(y)}{y_n} \leq |\nabla u(y)| \leq \bar{\lambda}\frac{u(y)}{y_n} \text{ whenever } y \in Q^+_{1/\hat{c}}(0).$$

Proof. Let $A \in M_p(\alpha, \beta, \gamma)$, $A = A(y, \eta)$, be given. Put $A_2(y, \eta) = A(y, \eta)$, $A_1(\eta) = A(0, \eta)$. Clearly, $A_1, A_2 \in M_p(\alpha, \beta, \gamma)$. We decompose the proof into the following steps.

Step 1. Lemma 3.25 holds for the operator A_1. To see this we note once again that $\hat{u}_1(y) = y_n$ is A_1-harmonic and $\hat{u}_1 = 0$ on $\partial Q_1^+(0) \cap \{y_n = 0\}$. Let $\hat{u}_2 = u$. Applying Lemma 2.8 to the pair $\hat{u}_1, \hat{u}_2$ we see that

$$\left|\log\left(\frac{\hat{u}_1(y_1)}{\hat{u}_2(y_1)}\right) - \log\left(\frac{\hat{u}_1(y_2)}{\hat{u}_2(y_2)}\right)\right| \leq c|y_1 - y_2|^\sigma \tag{3.26}$$

whenever $y_1,\ y_2 \in Q^+_{1/4}(0)$. Exponentiation of this inequality yields the equivalent inequality

$$\left|\frac{\hat{u}_1(y_1)}{\hat{u}_2(y_1)} - \frac{\hat{u}_1(y_2)}{\hat{u}_2(y_2)}\right| \leq c'\frac{\hat{u}_1(y_2)}{\hat{u}_2(y_2)}|y_1 - y_2|^\sigma \tag{3.27}$$

whenever $y_1,\ y_2 \in Q^+_{1/4}(0)$. Let $O = Q^+_{1/4}(0)$ and note that if $y_2 \in Q^+_{1/8}(0)$ then obviously

$$\frac{1}{\tilde{a}}\frac{\hat{u}_1(y_2)}{d(y_2, \partial O)} \leq |\nabla \hat{u}_1(y_2)| \leq \tilde{a}\frac{\hat{u}_1(y_2)}{d(y_2, \partial O)} \tag{3.28}$$

for some $\tilde{a} = \tilde{a}(n)$. Let r be defined through the relation $c' r^\sigma = \frac{1}{2}\tilde{\epsilon}$ where $\tilde{\epsilon}$ is as in Lemma 3.18. Using (3.27) we then see that

$$(1 - \tilde{\epsilon}/2)\frac{\hat{u}_1(y_2)}{\hat{u}_2(y_2)} \leq \frac{\hat{u}_1(y_1)}{\hat{u}_2(y_1)} \leq (1 + \tilde{\epsilon}/2)\frac{\hat{u}_1(y_2)}{\hat{u}_2(y_2)} \tag{3.29}$$

whenever $y_1 \in B(y_2, r)$. From (3.28), (3.29), and Lemma 3.18 we conclude that Lemma 3.25 holds for the operator A_1.

Step 2. Lemma 3.25 is valid for the operator A_2. We let $\rho \in (0, 1/16)$ and $\bar{\delta} \in (0, 1/8)$ be degrees of freedom to be chosen below. Let $\hat{u}_1$ be the A_1-harmonic function in $Q^+_{\bar{\delta}/2}(0)$ which is continuous on the closure of $Q^+_{\bar{\delta}/2}(0)$ and which satisfies $\hat{u}_1 = u$ on $\partial Q^+_{\bar{\delta}/2}(0)$. Using Step 1 we see there exist $\lambda_1 = \lambda_1(p, n, \alpha)$, $\hat{c}_1 = \hat{c}_1(p, n, \alpha) \geq 1$, such that

$$\lambda_1^{-1}\frac{\hat{u}_1(y)}{y_n} \leq |\nabla \hat{u}_1(y)| \leq \lambda_1\frac{\hat{u}_1(y)}{y_n} \text{ whenever } y \in Q^+_{\bar{\delta}/\hat{c}_1}(0). \tag{3.30}$$

Moreover, using Definition 1.1 *(iii)* we have

$$|A_2(y, \eta) - A_1(y, \eta)| \leq \epsilon|\eta|^{p-2} \text{ with } \epsilon = 2\beta\bar{\delta}^\gamma \text{ whenever } y \in Q^+_{\bar{\delta}}(0). \tag{3.31}$$

Let $\hat{u}_2 = u$. From Lemma 2.15 and Lemma 3.1 we see there exist c', θ, τ, each depending only on $p, n, \alpha, \beta, \gamma$, such that

$$|\hat{u}_2(y) - \hat{u}_1(y)| \leq c'\epsilon^\theta\rho^{-\tau}\,\hat{u}_2(y) \text{ whenever } y \in Q^+_{\bar{\delta}/4}(0) \setminus Q^+_{\bar{\delta}/4, \rho\bar{\delta}}(0). \tag{3.32}$$

Let $\tilde{\epsilon}$ be as in the statement of Lemma 3.18 with $\tilde{a}$ replaced by λ_1 and put $\rho = 1/(32\hat{c}_1)$. Fix $\bar{\delta}$ subject to $c'\epsilon^\theta\rho^{-\tau} = c'\,(2\beta\bar{\delta}^\gamma)^\theta\rho^{-\tau} = \min\{\tilde{\epsilon}/2, 10^{-8}\}$. In particular, we note that $\bar{\delta} = \bar{\delta}(p, n, \alpha, \beta, \gamma)$. Then from (3.32) we see that

$$1 - \tilde{\epsilon} \leq \frac{\hat{u}_2(y)}{\hat{u}_1(y)} \leq 1 + \tilde{\epsilon} \text{ whenever } y \in Q^+_{\bar{\delta}/4}(0) \setminus Q^+_{\bar{\delta}/4, \rho\bar{\delta}}(0). \tag{3.33}$$

Using (3.30), (3.33), and Lemma 3.18 we therefore conclude that

$$\lambda_2^{-1}\frac{\hat{u}_2(y)}{y_n} \leq |\nabla \hat{u}_2(y)| \leq \lambda_2\frac{\hat{u}_2(y)}{y_n} \text{ whenever } y \in Q^+_{\bar{\delta}/\hat{c}_1}(0) \setminus Q^+_{\bar{\delta}/\hat{c}_1, 2\rho\bar{\delta}}(0), \tag{3.34}$$

for some $\lambda_2 = \lambda_2(p, n, \alpha, \beta, \gamma)$. Moreover, if $y \in Q^+_{\bar{\delta}/\hat{c}_1, 2\rho\bar{\delta}}(0)$, then we can also prove that (3.34) is valid at y by iterating the previous argument and by making use of the invariance of the class $M_p(\alpha, \beta, \gamma)$ with respect to translations and dilations, see Lemma 2.15. This completes the proof of Lemma 3.25. □

Finally we use Lemma 3.25 to establish the main result of this section.

Lemma 3.35. *Let $\Omega \subset \mathbf{R}^n$ be a (δ, r_0)-Reifenberg flat domain, $w \in \partial\Omega$, and $0 < r < \min\{r_0, 1\}$. Let $p, 1 < p < \infty$, be given and assume that $A \in M_p(\alpha, \beta, \gamma)$ for some (α, β, γ). Suppose that u is a positive A-harmonic function in $\Omega \cap B(w, 4r)$, that u is continuous in $\bar{\Omega} \cap B(w, 4r)$, and that $u = 0$ on $\Delta(w, 4r)$. There exist $\hat{\delta} =$*

$\hat{\delta}(p,n,\alpha,\beta,\gamma)$, $\hat{c} = \hat{c}(p,n,\alpha,\beta,\gamma)$ *and* $\bar{\lambda} = \bar{\lambda}(p,n,\alpha,\beta,\gamma)$, *such that if* $0 < \delta \leq \hat{\delta}$, *then*

$$\bar{\lambda}^{-1}\frac{u(y)}{d(y,\partial\Omega)} \leq |\nabla u(y)| \leq \bar{\lambda}\frac{u(y)}{d(y,\partial\Omega)} \text{ whenever } y \in \Omega \cap B(w, r/\hat{c}).$$

Proof. Let $A \in M_p(\alpha,\beta,\gamma)$, $A = A(y,\eta)$ be given. Let $w \in \partial\Omega$, $0 < r < r_0$, suppose that u is a positive A-harmonic function in $\Omega \cap B(w,4r)$, that u is continuous in $\bar{\Omega} \cap B(w,4r)$, and that $u = 0$ on $\Delta(w,4r)$. We intend to use Lemma 3.25 and Lemma 3.1 to prove Lemma 3.35. Let $u \equiv 0$ in $B(w,4r) \setminus \Omega$. Then $u \in W^{1,p}(B(w,2r))$ and u is continuous in $B(w,4r)$. Let $c_1 = \hat{c}$ be as in Lemma 3.25 and choose $c' \geq 100c_1$ so that if $\hat{y} \in \Omega \cap B(w, r/c')$, $s = 4c_1 d(\hat{y},\partial\Omega)$, and $z \in \partial\Omega$ with $|\hat{y} - z| = d(\hat{y},\partial\Omega)$, then

$$\max_{B(z,4s)} u \leq c u(\hat{y}) \tag{3.36}$$

for some $c = c(p,n,\alpha,\beta,\gamma)$. Using Definition 1.6 with w, r replaced by $z, 4s$, we see that there exists a hyperplane Λ such that

$$h(\partial\Omega \cap B(z,4s), \Lambda \cap B(z,4s)) \leq 4\delta s. \tag{3.37}$$

For the moment we allow $\hat{\delta}$ in Lemma 3.35 to vary but shall later fix it as a number satisfying several conditions. Using (1.7) we deduce that

$$\{y \in \Omega \cap B(z,4s) : d(y,\partial\Omega) \geq 8\delta s\} \subset \text{ one component of } \mathbf{R}^n \setminus \Lambda.$$

Moreover, using Lemma 2.15 we see that we may without loss of generality assume that $\Lambda = \{(y', y_n) : y' \in \mathbf{R}^{n-1}, y_n = 0\}$ and

$$\{y \in \Omega \cap B(z,4s) : d(y,\partial\Omega) \geq 8\delta s\} \subset \{y \in \mathbf{R}^n : y_n > 0\}. \tag{3.38}$$

From (3.38) we find that if we define

$$\Lambda' = \{(y',0) + 20\delta s e_n,\ y' \in \mathbf{R}^{n-1}\},\ \Omega' = \{y \in \mathbf{R}^n : y_n > 20\delta s\},$$

then

$$\Omega' \cap B(z,2s) \subset \Omega \cap B(z,2s). \tag{3.39}$$

Let v be a A-harmonic function in $\Omega' \cap B(z,2s)$ with continuous boundary values on $\partial(\Omega' \cap B(z,2s))$ and such that $v \leq u$ on $\partial(\Omega' \cap B(z,2s))$. Moreover, we choose v so that

$$\begin{aligned} v(y) &= u(y) \text{ whenever } y \in \partial[\Omega' \cap B(z,2s)] \text{ and } y_n > 40\delta s, \\ v(y) &= 0 \quad \text{ whenever } y \in \partial[\Omega' \cap B(z,2s)] \text{ and } y_n < 30\delta s. \end{aligned}$$

Existence of v follows once again from the Wiener criteria of [GZ], the maximum principle for A-harmonic functions, and the fact that the $W^{1,p}$-Dirichlet problem for these functions in $\Omega' \cap B(z,2s)$ always has a solution. By construction and the maximum principle for A-harmonic functions we have $v \leq u$ in $\Omega' \cap B(z,2s)$. Also, since each point of $\partial[\Omega' \cap B(z,2s)]$ where $u \neq v$ lies within $80\delta s$ of a point where u is zero, it follows from (3.36) and Lemmas 2.2, 2.3 that $u \leq v + c\delta^{\tilde{\sigma}} u(\hat{y})$ on $\partial[\Omega' \cap B(z,2s)]$. In particular, again using the maximum principle for p-harmonic functions we conclude that

$$v \leq u \leq v + c\delta^{\tilde{\sigma}} u(\hat{y}) \text{ in } \Omega' \cap B(z,2s).$$

Thus, using the last inequality and (3.36) we see that

$$1 \leq \frac{u(y)}{v(y)} \leq (1 - c\delta^{\tilde{\sigma}})^{-1} \text{ whenever } y \in \Omega' \cap B(\hat{y}, \tfrac{1}{2}d(\hat{y}, \partial\Omega')) \tag{3.40}$$

provided $\hat{\delta}$ is small enough. Using Lemma 3.25 and the construction we also have

$$\hat{\lambda}^{-1}\frac{v(\hat{y})}{d(\hat{y}, \partial\Omega)} \leq |\nabla v(\hat{y})| \leq \hat{\lambda}\frac{v(\hat{y})}{d(\hat{y}, \partial\Omega)}. \tag{3.41}$$

for some $\hat{\lambda} = \hat{\lambda}(p, n)$. In particular, from (3.40), (3.41) we see for $0 < \delta < \hat{\delta}$, and $\hat{\delta} = \hat{\delta}(p, n, \alpha, \beta, \gamma)$ suitably small, that the hypotheses of Lemma 3.18 are satisfied with $O = \Omega' \cap B(z, 2s)$ and $\tilde{a} = \hat{\lambda}$. We now fix $\hat{\delta}$ and from Lemma 3.18 we conclude that

$$\bar{\lambda}_1^{-1}\frac{u(\hat{y})}{d(\hat{y}, \partial\Omega)} \leq |\nabla u(\hat{y})| \leq \bar{\lambda}_1\frac{u(\hat{y})}{d(\hat{y}, \partial\Omega)}$$

for some $\bar{\lambda}_1 = \bar{\lambda}_1(p, n, \alpha, \beta, \gamma)$. Since $\hat{y} \in \Omega \cap B(w, r/c')$ is arbitrary, the proof of Lemma 3.35 is complete. □

4. Degenerate elliptic equations and extension of $|\nabla u|^{p-2}$ to an A_2-weight

Let $w \in \mathbf{R}^n$, $0 < r$ and let $\lambda(x)$ be a real valued Lebesgue measurable function defined almost everywhere on $B(w, 2r)$. $\lambda(x)$ is said to belong to the class $A_2(B(w, r))$ if there exists a constant Γ such that

$$\tilde{r}^{-2n} \int_{B(\tilde{w},\tilde{r})} \lambda \, dx \cdot \int_{B(\tilde{w},\tilde{r})} \lambda^{-1} dx \leq \Gamma \tag{4.1}$$

whenever $\tilde{w} \in B(w, r)$ and $0 < \tilde{r} \leq r$. If $\lambda(x)$ belongs to the class $A_2(B(w, r))$ then λ is referred to as an $A_2(B(w, r))$-weight. The smallest Γ such that (4.1) holds is referred to as the A_2-constant of λ.

In the following we let $\Omega \subset \mathbf{R}^n$ be a bounded (δ, r_0)-Reifenberg flat domain with NTA-constant M. We let $w \in \partial\Omega$, $0 < r < r_0$, and we consider the operator

$$\hat{L} = \sum_{i,j=1}^{n} \frac{\partial}{\partial x_i}\left(\hat{b}_{ij}(x)\frac{\partial}{\partial x_j}\right) \tag{4.2}$$

in $\Omega \cap B(w, 2r)$. We assume that the coefficients $\{\hat{b}_{ij}(x)\}$ are bounded, Lebesgue measurable functions defined almost everywhere on $B(w, 2r)$. Moreover,

$$c^{-1}\lambda(x)|\xi|^2 \leq \sum_{i,j=1}^{n} \hat{b}_{ij}(x)\xi_i\xi_j \leq c|\xi|^2\lambda(x) \tag{4.3}$$

for almost every $x \in B(w, 2r)$, where $\lambda \in A_2(B(w, r))$. By definition $\hat{L}$ is a degenerate elliptic operator (in divergence form) in $B(w, 2r)$ with ellipticity measured by the function λ. If $O \subset B(w, 2r)$ is open then we let $\tilde{W}^{1,2}(O)$ be the weighted Sobolev space of equivalence classes of functions v with distributional gradient ∇v and norm

$$\|v\tilde{\|}_{1,2}^2 = \int_O v^2 \lambda dx + \int_O |\nabla v|^2 \lambda dx < \infty.$$

Let $\tilde{W}_0^{1,2}(O)$ be the closure of $C_0^\infty(O)$ in the norm of $\tilde{W}^{1,2}(O)$. We say that v is a weak solution to $\hat{L}v = 0$ in O provided $v \in \tilde{W}^{1,2}(O)$ and

$$\int_O \sum_{i,j} \hat{b}_{ij} v_{x_i} \phi_{x_j} dx = 0 \tag{4.4}$$

whenever $\phi \in C_0^\infty(O)$.

The following three lemmas, Lemmas 4.5-4.7, are tailored to our situation and based on the results in [FKS], [FJK] and [FJK1]. We note that these authors assumed $\hat{L}$ was symmetric, i.e., $\hat{b}_{ij} = \hat{b}_{ji}, 1 \leq i,j \leq n$, but this assumption was not needed in the proof of these lemmas. Essentially one can say 'ditto' to the discussion in [KKPT, section 1] for nonsymmetric uniformly elliptic divergence form PDE.

Lemma 4.5. *Let $\Omega \subset \mathbf{R}^n$ be a NTA-domain with constant M, $w \in \partial\Omega$, $0 < r < r_0$, and let λ be an $A_2(B(w,r))$-weight with constant Γ. Suppose that v is a positive weak solution to $\hat{L}v = 0$ in $\Omega \cap B(w,2r)$. Then there exists a constant c, $1 \leq c < \infty$, depending only on n, M and Γ, such that if $\tilde{w} \in \Omega$, $0 < \tilde{r}$, $B(\tilde{w},2\tilde{r}) \subset \Omega \cap B(w,r)$, then*

$$(i) \qquad c^{-1}\tilde{r}^2 \int_{B(\tilde{w},\tilde{r}/2)} |\nabla v|^2 \lambda dx \leq c\Big(\int_{B(\tilde{w},\tilde{r})} \lambda dx \Big) \big(\max_{B(\tilde{w},\tilde{r})} v\big)^2 \leq c \int_{B(\tilde{w},2\tilde{r})} |v|^2 \lambda dx,$$

$$(ii) \qquad \max_{B(\tilde{w},\tilde{r})} v \leq c \min_{B(\tilde{w},\tilde{r})} v.$$

Furthermore, there exists $\tilde{\alpha} = \tilde{\alpha}(n,M,\Gamma) \in (0,1)$ such that if $x, y \in B(\tilde{w},\tilde{r})$ then

$$(iii) \qquad |v(x) - v(y)| \leq c\Big(\frac{|x-y|}{\tilde{r}}\Big)^{\tilde{\alpha}} \max_{B(\tilde{w},2\tilde{r})} v.$$

Lemma 4.6. *Let $\Omega \subset \mathbf{R}^n$ be a NTA-domain with constant M, $w \in \partial\Omega$, $0 < r < r_0$, and let λ be an $A_2(B(w,r))$-weight with constant Γ. Suppose that v is a positive weak solution to $\hat{L}v = 0$ in $\Omega \cap B(w,2r)$ and that $v = 0$ on $\Delta(w,2r)$ in the weighted Sobolev sense. Then there exists $\tilde{c} = \tilde{c}(n,M,\Gamma)$, $1 \leq \tilde{c} < \infty$, such that the following holds with $\tilde{r} = r/\tilde{c}$.*

$$(i) \qquad r^2 \int_{\Omega \cap B(w,r/2)} |\nabla v|^2 \lambda dx \leq \tilde{c} \int_{\Omega \cap B(w,r)} |v|^2 \lambda dx,$$

$$(ii) \qquad \max_{\Omega \cap B(w,\tilde{r})} v \leq \tilde{c} v(a_{\tilde{r}}(w)).$$

Furthermore, there exists $\tilde{\alpha} = \alpha(n,M,\Gamma) \in (0,1)$ such that if $x, y \in \Omega \cap B(w,\tilde{r})$, then

$$(iii) \qquad |v(x) - v(y)| \leq c\Big(\frac{|x-y|}{r}\Big)^{\tilde{\alpha}} \max_{\Omega \cap B(w,2\tilde{r})} v.$$

Lemma 4.7. *Let $\Omega \subset \mathbf{R}^n$ be a NTA-domain with constant M, $w \in \partial\Omega$, $0 < r < r_0$, and let λ be an $A_2(B(w,r))$-weight with constant Γ. Suppose that v_1 and v_2 are two positive weak solutions to $\hat{L}v = 0$ in $\Omega \cap B(w,2r)$ and $v_1 = 0 = v_2$ on $\Delta(w,2r)$ in the weighted Sobolev sense. Then there exist $c = c(n,M,\Gamma)$, $1 \leq c < \infty$, and*

$\sigma = \sigma(n, M, \Gamma) \in (0,1)$ *such that if* $\tilde r = r/c$, $v_1(a_{\tilde r}(w)) = v_2(a_{\tilde r}(w))$, *then* $v_1/v_2 \leq c$ *in* $\Omega \cap B(w, r/c)$ *and if* $y_1, y_2 \in \Omega \cap B(w, r/c)$, *then*

$$\left| \frac{v_1(y_1)}{v_2(y_1)} - \frac{v_1(y_2)}{v_2(y_2)} \right| \leq c \left(\frac{|y_1 - y_2|}{r} \right)^{\sigma}.$$

To continue the proof of Theorem 1, we have the following lemmas.

Lemma 4.8. *Let* $\Omega \subset \mathbf{R}^n$ *be a bounded* (δ, r_0)*-Reifenberg flat domain. Let* $p, 1 < p < \infty$, *be given and assume that* $A \in M_p(\alpha, \beta, \gamma)$ *for some* (α, β, γ). *Let* $w \in \partial\Omega$, $0 < r < r_0$ *and suppose that* u *is a positive* A*-harmonic function in* $\Omega \cap B(w, 4r)$, u *is continuous in* $\bar\Omega \cap \bar B(w, 4r)$, *and* $u = 0$ *on* $\Delta(w, 4r)$. *Then there exist, for* $\epsilon^* > 0$ *given,* $\hat\delta = \hat\delta(p, n, \alpha, \beta, \gamma, \epsilon^*) > 0$ *and* $c = c(p, n, \alpha, \beta, \gamma, \epsilon^*)$, $1 \leq c < \infty$, *such that*

$$c^{-1} \left(\frac{\hat r}{r} \right)^{1+\epsilon^*} \leq \frac{u(a_{\hat r}(w))}{u(a_r(w))} \leq c \left(\frac{\hat r}{r} \right)^{1-\epsilon^*}$$

whenever $0 < \delta \leq \hat\delta$ *and* $0 < \hat r < r/4$.

Lemma 4.9. *Let* $\Omega \subset \mathbf{R}^n$ *be a bounded* (δ, r_0)*-Reifenberg flat domain. Let* $p, 1 < p < \infty$, *be given and assume that* $A \in M_p(\alpha, \beta, \gamma)$ *for some* (α, β, γ). *Let* $w \in \partial\Omega$, $0 < r < \min\{r_0, 1\}$, *and suppose that* u *is a positive* A*-harmonic function in* $\Omega \cap B(w, 2r)$, u *is continuous in* $\bar\Omega \cap B(w, 2r)$, *and* $u = 0$ *on* $\Delta(w, 2r)$. *There exist* $\delta' = \delta'(p, n, \alpha, \beta, \gamma)$, *and* $c = c(p, n, \alpha, \beta, \gamma) \geq 1$ *such that if* $0 < \delta < \delta'$, *and* $\hat r = r/c$, *then* $|\nabla u|^{p-2}$ *extends to an* $A_2(B(w, \hat r))$*-weight with constant depending only on* $p, n, \alpha, \beta, \gamma$.

Proof of Lemma 4.8: Let $A \in M_p(\alpha, \beta, \gamma)$, $A = A(y, \eta)$ be given and set $A_2(y, \eta) = A(y, \eta)$, $A_1(\eta) = A(w, \eta)$. Then $A_1, A_2 \in M_p(\alpha, \beta, \gamma)$. Let u be a A_2-harmonic function as in the statement of the lemma. We extend u to $B(w, 4r) \setminus \Omega$ by putting $u \equiv 0$ in this set and then note that u is continuous in $B(w, 4r)$. We also observe from Definition 1.8 that it suffices to prove Lemma 4.9 for $\delta = \hat\delta$. Also, as discussed after Definition 1.8 we may assume that Ω is a NTA-domain. Moreover, using Lemma 2.15 and Definition 1.1 (iv), we can without loss of generality assume that $r = 4$, $w = 0$ and $u(a_1(0)) = 1$.

In the following we let ξ be a small constant to be chosen below. In particular, ξ will be fixed to depend only on $p, n, \alpha, \beta, \gamma$. For ξ fixed we can, again using Lemma 2.15, without loss of generality also assume that

$$h(P \cap B(0, 4\xi), \partial\Omega \cap B(0, 4\xi)) \leq 4\hat\delta\xi,$$

where $P = \{y \in \mathbf{R}^n : y_n = 0\}$. Furthermore, if $\bar\delta = 4\hat\delta$ is small enough, then we may assume, as in (3.38), that

$$\begin{aligned} B(0,4) \cap \{(y', y_n) : y_n \geq 2\bar\delta\xi\} &\subset \Omega \\ B(0,4) \cap \{(y', y_n) : y_n \leq -2\bar\delta\xi\} &\subset \mathbf{R}^n \setminus \Omega. \end{aligned} \tag{4.10}$$

Moreover, we see that to prove Lemma 4.8 it suffices to show that

$$c^{-1} \hat r^{1+\epsilon^*} \leq u(a_{\hat r}(0)) \leq c \hat r^{1-\epsilon^*} \text{ whenever } 0 < \hat r < \xi. \tag{4.11}$$

In the following we will use the notation introduced in (2.7).

To begin the proof of (4.11) we introduce two auxiliary functions u^+ and u^-. In particular, we define u^+ to be A_2-harmonic in $Q^+_{\xi,(1-8\bar\delta)\xi}(8\bar\delta\xi e_n)$ with continuous boundary values on $\partial Q^+_{\xi,(1-8\bar\delta)\xi}(8\bar\delta\xi e_n)$ defined as follows,

$$u^+(y) = u(y) \qquad \text{if } y \in \partial Q^+_{\xi,(1-8\bar\delta)\xi}(8\bar\delta\xi e_n) \cap \{y : 16\bar\delta\xi \le y_n\},$$
$$u^+(y) = \frac{(y_n - 8\bar\delta\xi)}{8\bar\delta\xi} u(y) \qquad \text{if } y \in \partial Q^+_{\xi,(1-8\bar\delta)\xi}(8\bar\delta\xi e_n) \cap \{y : 8\bar\delta\xi < y_n < 16\bar\delta\xi\},$$
$$u^+(y) = 0 \qquad \text{if } y \in \partial Q^+_{\xi,(1-8\bar\delta)\xi}(8\bar\delta\xi e_n) \cap \{y : y_n = 8\bar\delta\xi\}.$$

Similarly, we define u^- to be the A_2-harmonic function in $Q^+_{\xi,(1+8\bar\delta)\xi}(-8\bar\delta\xi e_n)$ which satisfies $u^- = u$ on $\partial Q^+_{\xi,(1+8\bar\delta)\xi}(-8\bar\delta\xi e_n)$. From the maximum principle for A-harmonic functions and (4.10) we see, by construction, that

$$u^+(y) \le u(y) \le u^-(y) \text{ whenever } y \in Q^+_{\xi,(1-8\bar\delta)\xi}(8\bar\delta\xi e_n). \tag{4.12}$$

Using Definition 1.1 *(iii)* we next note that

$$|A_2(y,\eta) - A_1(y,\eta)| \le \epsilon|\eta|^{p-1} \text{ whenever } y \in Q^+_{\xi,(1+8\bar\delta)\xi}(-8\bar\delta\xi e_n),\ \epsilon = 2\beta\xi^\gamma. \tag{4.13}$$

To proceed we let $\bar u^+$ be the A_1-harmonic function in $Q^+_{\xi/2,(1/2-8\bar\delta)\xi}(8\bar\delta\xi e_n)$ which is continuous on the closure of $Q^+_{\xi/2,(1/2-8\bar\delta)\xi}(8\bar\delta\xi e_n)$ and which coincides with u^+ on $\partial Q^+_{\xi/2,(1/2-8\bar\delta)\xi}(8\bar\delta\xi e_n)$. Similarly, we let $\bar u^-$ be the A_1-harmonic function in $Q^+_{\xi/2,(1/2+8\bar\delta)\xi}(-8\bar\delta\xi e_n)$ which is continuous on the closure of $Q^+_{\xi/2,(1/2+8\bar\delta)\xi}(-8\bar\delta\xi e_n)$ and coincides with u^- on $\partial Q^+_{\xi/2,(1/2+8\bar\delta)\xi}(-8\bar\delta\xi e_n)$. Finally, we define $v^+(y) := y_n - 8\bar\delta\xi$, $v^-(y) := y_n + 8\bar\delta\xi$ whenever $y \in \mathbf{R}^n$. Hence v^+ and v^- are A_1-harmonic functions and grow linearly in the e_n-direction.

We first focus on the right hand inequality in (4.11). Using (4.13), Lemma 2.15, and Lemma 3.1 we see that

$$u^-(y) \le (1 - \tilde c\epsilon^\theta\bar\delta^{-\tau})^{-1}\bar u^-(y) \tag{4.14}$$

for $y \in Q^+_{\xi/4,(1/4+8\bar\delta)\xi}(-8\bar\delta\xi e_n) \cap \{-4\bar\delta\xi < y_n < \xi/2\}$ and for a constant $\tilde c = \tilde c(p,n,\alpha,\beta,\gamma)$. Moreover, using (4.12), the maximum principle and the Harnack inequality for A-harmonic functions, (4.14), as well as Lemma 2.8 applied to the functions $\bar u^-$, v^- we see that there exists a constant $\bar c = \bar c(p,n,\alpha)$, $1 \le \bar c < \infty$, such that

$$u(y) \le u^-(y) \le (1 - \tilde c\epsilon^\theta\bar\delta^{-\tau})^{-1}\bar u^-(y) \le c(1 - \tilde c\epsilon^\theta\bar\delta^{-\tau})^{-1}\,\bar u^-(a_{\xi/8}(0))\frac{v^-(y)}{\xi} \tag{4.15}$$

whenever $y \in \Omega \cap B(0,\xi/\bar c)$. From (4.15) we conclude that

$$u(y) \le c(1 - \tilde c\epsilon^\theta\bar\delta^{-\tau})^{-1}\bar u^-(a_{\xi/8}(0))\ (y_n/\xi) \tag{4.16}$$

whenever $y \in \Omega \cap B(0,\xi/\bar c)$. Let $\bar\delta < 1/(16\bar c)$ and let ξ be defined though the relation

$$1/2 = \tilde c\epsilon^\theta\bar\delta^{-\tau} = \tilde c(2\beta\xi^\gamma)^\theta\,\bar\delta^{-\tau}.$$

Then $\xi = \xi(p,n,\alpha,\beta,\gamma,\bar\delta)$ and from Lemmas 3.1, 2.2, 2.3, as well as the maximum principle for A harmonic functions, we observe that

$$u(a_{\xi/8}(0)) \approx u^-(a_{\xi/8}(0)) \approx \bar u^-(a_{\xi/8}(0)) \approx \max_{Q^+_{\xi/2,(1/2+8\bar\delta)\xi}(-8\bar\delta\xi e_n)} u$$

where proportionality constants depend only on $p, n, \alpha, \beta, \gamma$. Using these displays in (4.16), we get $u(a_{\bar\delta\xi}(0)) \leq \hat c \bar\delta\, u(a_{\xi/8}(0))$. Moreover, suppose by way of induction that we have shown, for some $k \in \{1, 2, ...\}$,

$$u(a_{\bar\delta^k\xi}(0)) \leq (\hat c\bar\delta)^k\, u(a_{\xi/8}(0)) \tag{4.17}$$

where $\hat c$ depends only on $p, n, \alpha, \beta, \gamma$. Then, from Reifenberg flatness we see there exists a plane P' containing 0 such that

$$h(P' \cap B(0, 4\bar\delta^k\xi), \partial\Omega \cap B(0, 4\bar\delta^k\xi)) \leq 4\hat\delta\bar\delta^k\xi.$$

We can now repeat the above argument with P replaced by P' and 4 replaced by $4\bar\delta^k\xi$. Here however we use a cylinder with radius and height $\approx \bar\delta^k\xi$, since we can already apply Lemma 3.1. We get

$$u(a_{\bar\delta^{k+1}\xi}(0)) \leq \hat c\,\bar\delta\, u(a_{\bar\delta^k\xi}(0)) \leq (\hat c\bar\delta)^{k+1} u(a_{\xi/8}(0)).$$

Thus by induction the inequality in (4.17) is true for all positive integers k. Next we fix $\bar\delta$ so that $\bar\delta^{-\epsilon^*} = \hat c$ where $\hat c$ is the constant in the above display. Then $\bar\delta$ and ξ both depend only on $p, n, \alpha, \beta, \gamma$ and ϵ^*. Given $0 < \hat r < \xi$, let k be the smallest integer such that $\bar\delta^k\xi \leq \hat r$. Then from (4.17) and our choice of $\bar\delta$ we see that $u(a_{\hat r}(0)) \leq c\hat r^{1-\epsilon^*}$, for some $c = c(p, n, \alpha, \beta, \gamma, \epsilon^*)$. Here we have also used the fact that $u(a_{\xi/8}(0)) \leq c_* = c_*(p, n, \alpha, \beta, \gamma)$, which follows from Lemmas 2.2, 2.3, and fact that $u(a_1(0)) = 1$. This completes the proof of the right-hand side inequality in (4.11).

Second we focus on the left-hand inequality in (4.11). In this case we first apply Lemma 2.8 to the functions $\bar u^+$, v^+ in $Q^+_{\xi/2,(1/2-8\bar\delta)\xi}(8\bar\delta\xi e_n)$. Indeed, using Lemma 2.8 and the Harnack inequality we see, provided $\bar\delta$ is small enough, that

$$\frac{\bar u^+(a_{32\bar\delta\xi}(0))}{v^+(a_{32\bar\delta\xi}(0))} \approx \frac{\bar u^+(a_{\xi/8}(0))}{v^+(a_{\xi/4}(0))} \approx \frac{\bar u^+(a_{\xi/8}(0))}{\xi}. \tag{4.18}$$

Here $A \approx B$ means that A/B is bounded from above and below by constants which only depend on $p, n, \alpha, \beta, \gamma$. From (4.18) we get

$$\bar u^+(a_{32\bar\delta\xi}(0)) \geq \bar c^{-1}\bar\delta\bar u^+(a_{\xi/8}(0)) \tag{4.19}$$

for some $\bar c = c(p, n, \alpha, \xi)$, $1 \leq \bar c < \infty$. Moreover, using Lemma 3.1 we also see that

$$\bar u^+(y) \leq (1 - \tilde c\,\epsilon^\theta\bar\delta^{-\tau})^{-1} u^+(y) \tag{4.20}$$

for $y \in Q^+_{\xi/2,(1/2-8\bar\delta)\xi}(8\bar\delta\xi\delta e_n) \cap \{16\bar\delta\xi < y_n < \xi/2\}$ and for a constant $\tilde c = \tilde c(p, n, \alpha, \beta, \gamma)$. Using (4.19), (4.20), the fact that the class $M_p(\alpha, \beta, \gamma)$ is closed under translations, rotations, suitable dilations, and multiplication by constants (see Lemma 2.15 and Definition 1.1 *(iv)*) , we can argue as in the proof of the right-hand inequality in (4.11). Thus by induction we obtain

$$u(a_{(32\bar\delta)^k\xi}(0)) \geq (\bar c^{-1}\bar\delta)^k u(a_{\xi/8}(0)) \text{ for } k = 1, 2, ... \tag{4.21}$$

To complete the proof we let $\bar\delta$ be so small that $\bar c^{-1}\bar\delta \geq (32\bar\delta)^{1+\epsilon^*}$ and assume that $\hat r \in [(32\bar\delta)^{k+1}\xi, (32\bar\delta)^k\xi]$. With $\bar\delta(p, n, \alpha, \beta, \gamma, \epsilon^*)$ now fixed, it follows from Harnack's inequality for A-harmonic functions that

$$u(a_{\hat r}(0)) \geq c^{-1}u(a_{(32\bar\delta)^k\xi}(0)) \geq c^{-1}(32\bar\delta)^{k(1+\epsilon^*)}u(a_{\xi/8}(0)) \geq c^{-1}\hat r^{(1+\epsilon^*)}. \tag{4.22}$$

for some $c = c(p, n, \alpha, \beta, \gamma, \epsilon^*)$. In (4.22) we have also used the fact that $u(a_{\xi/8}(0)) \geq 1/c_+(p, n, \alpha, \beta, \gamma, \epsilon^*)$, for some $c_+ \geq 1$, which follows from the definition of $\bar\delta$ in terms of ϵ^*, Harnack's inequality, and the fact that $u(a_1(0)) = 1$. (4.22) completes the proof of (4.11) and hence the proof of Lemma 4.8. □

Proof of Lemma 4.9. Lemma 4.9 follows from Lemma 4.8, in exactly the same way as Lemma 3.30 in [LN4] followed from Lemma 3.15 in [LN4]. For the readers convenience we include the details of the proof. Let $Q_j = Q(x_j, r_j)$, $j = 1, 2, \ldots$ be a Whitney decomposition of $\mathbf{R}^n \setminus \bar{\Omega}$ into open cubes with center at x_j and sidelength r_j. Then $\cup_j \bar{Q}(x_j, r_j) = \mathbf{R}^n \setminus \bar{\Omega}$ and $Q(x_j, r_j) \cap Q(x_i, r_i) = \emptyset$ when $i \neq j$. We furthermore construct the Whitney cubes in such a way that $10^{-4n} d(Q_j, \partial\Omega) \leq r_j \leq 10^{-2n} d(Q_j, \partial\Omega)$. Let $\hat{r} = r/\tilde{c}^2$, where $\tilde{c} = \tilde{c}(p, n, \alpha, \beta, \gamma)$, $1 \leq \tilde{c} < \infty$, is so large that the 'fundamental inequality' in Lemma 3.35 holds in $\Omega \cap B(w, r/\tilde{c})$. From the NTA property of Ω we may also suppose $\tilde{c}$ is so large that if $Q_j \cap B(w, 50\hat{r}) \neq \emptyset$, then there is a $w_j \in \Omega \cap B(w, \tilde{c}\hat{r})$ for which $d(w_j, \partial\Omega) \sim d(w_j, x_j) \sim d(x_j, \partial\Omega)$. Here $A \sim B$ means that A/B is bounded from above and below by constants which only depend on n.

Next we define $\lambda(x) = |\nabla u(x)|^{p-2}$ whenever $x \in \Omega \cap B(w, 50\hat{r})$ and we let Γ be the set of all j such that if $j \in \Gamma$ then $Q_j \cap B(w, 50\hat{r}) \neq \emptyset$. Moreover, if $j \in \Gamma$ then we choose $w_j \in \Omega \cap B(w, \tilde{c}\hat{r})$ as above and define $\lambda(x) = \lambda(w_j)$ when $x \in Q_j$. This defines λ almost everywhere on $B(w, 50\hat{r})$ with respect to Lebesgue n measure, since it follows from (4.27) that for δ small enough, $\partial\Omega \cap B(w, r)$ has Lebesgue n measure zero. From the definition of λ, Lemma 3.35, and the Harnack inequality for A-harmonic functions we see that

$$\lambda(x) = \lambda(w_j) \approx \lambda(z) \text{ whenever } x \in Q_j \text{ and } z \in B(w_j, d(w_j, \partial\Omega)/2). \tag{4.23}$$

Let $\hat{\lambda} = \lambda$ if $p \geq 2$ and $\hat{\lambda} = 1/\lambda$ if $1 < p \leq 2$. If $\tilde{w} \in B(w, r)$ and $d(\tilde{w}, \partial\Omega)/2 < \tilde{r} \leq \hat{r}$, then from Lemmas 2.1 - 2.3, (4.23), and Hölder's inequality it follows that

$$\int_{B(\tilde{w}, \tilde{r})} \hat{\lambda} dx \leq c u(a_{\tilde{r}}(\hat{w}))^{|p-2|} \, \tilde{r}^{n-|p-2|}. \tag{4.24}$$

Here $\hat{w} \in \partial\Omega$ with $|\tilde{w} - \hat{w}| = d(\tilde{w}, \partial\Omega)$. Also, from Lemma 4.8 we get for $\hat{\delta}$ small enough and $y \in \Omega \cap B(\hat{w}, c\tilde{r})$, that

$$c u(y) \geq u(a_{\tilde{r}}(\hat{w})) \left(\frac{d(y, \partial\Omega)}{\tilde{r}} \right)^{1+\epsilon^*}. \tag{4.25}$$

Here $\epsilon^* > 0$ is a small positive number which will be fixed after the display following (4.27). From (4.25) and Lemma 3.35, we see that if $d(\tilde{w}, \partial\Omega)/2 < \tilde{r} \leq \hat{r}$, then

$$\int_{B(\tilde{w}, \tilde{r})} \hat{\lambda}^{-1} dx \leq c \tilde{r}^{(1+\epsilon^*)|p-2|} u(a_{\tilde{r}}(\hat{w}))^{-|p-2|} \int_{\Omega \cap B(\hat{w}, c\tilde{r})} d(y, \partial\Omega)^{-\epsilon^* |p-2|} dy. \tag{4.26}$$

To complete the estimate in (4.26) we need to estimate the integral involving the distance function. To do this we define

$$I(z, s) = \int_{\Omega \cap B(z, s)} d(y, \partial\Omega)^{-\epsilon^* |p-2|} dy$$

whenever $z \in \partial\Omega \cap B(w, r)$, $0 < s < r$. Let

$$E_k = \Omega \cap B(z, s) \cap \{y : d(y, \partial\Omega) \leq \delta^k s\} \text{ for } k = 0, 1, 2, \ldots$$

Then since $\partial\Omega$ is δ-Reifenberg flat we deduce that

$$\int_{E_k} dy \leq c_+^{k+1} \delta^k s^n \tag{4.27}$$

where $c_+ = c_+(p,n)$. Indeed, from δ-Reifenberg flatness it is easily seen that this statement holds for E_0, E_1. Moreover, E_1 can be covered by at most c/δ^{n-1} balls of radius $100\delta s$ with centers in $\partial\Omega \cap B(z,s)$. We can then repeat the argument in each ball to get that (4.27) holds for E_2. Continuing in this way we get (4.27) for all positive integers k. Using (4.27) and writing $I(z,s)$ as a sum over $E_k \setminus E_{k+1}, k = 0, 1, 2, \dots$ we get

$$I(z,s) \leq c s^{n-\epsilon^*|p-2|} \left[1 + \delta^{-\epsilon^*|p-2|} \sum_{k=1}^{\infty} c_+^k \, \delta^{k(1-\epsilon^*|p-2|)} \right] < c_- \, s^{n-\epsilon^*|p-2|},$$

for some $c_- = c_-(p,n) \geq 1$, provided $4\epsilon^*|p-2| \leq 1$ and $\delta' > 0$ is small enough. Using this estimate with $z = \hat{w}, s = c\tilde{r}$, we can continue our calculation in (4.26) and conclude that

$$\int_{B(\tilde{w},\tilde{r})} \hat{\lambda}^{-1} dx \leq c u(a_{\tilde{r}}(\hat{w}))^{-|p-2|} \, \tilde{r}^{n+|p-2|}. \tag{4.28}$$

Combining (4.24), (4.28), we get

$$\int_{B(\tilde{w},\tilde{r})} \hat{\lambda}^{-1} dx \cdot \int_{B(\tilde{w},\tilde{r})} \hat{\lambda} dx \leq c\tilde{r}^{2n}$$

when $d(\tilde{w}, \partial\Omega)/2 \leq \tilde{r} \leq \hat{r}$. This inequality is also valid if $\tilde{r} \leq d(\tilde{w}, \partial\Omega)/2$, as follows easily from Lemma 3.35. We conclude from this inequality and arbitrariness of $\tilde{w}, \tilde{r}$, that Lemma 4.9 is true. □

4.1. Proof of Theorem 1. From the results proved or stated in section 2, 3, 4, we see that Steps 0, A, B, C and D outlined in the introduction are now completed. Hence, to prove Theorem 1 it only remains to remove assumption (1.14). To do this we first note from Definition 1.1 (iv) that Theorem 1 is invariant under multiplication of u, v by constants. Using this note and Lemma 2.3 we see that if $r^* = r/c$, for $c = c(p, n, \alpha, \beta, \gamma)$ large enough, then we may assume that

$$\max_{\Omega \cap B(w,4r^*)} h \approx h(a_{r^*}(w)) = 1 \text{ whenever } h = u \text{ or } v. \tag{4.29}$$

Let $\tilde{u}, \tilde{v}$ be the A-harmonic functions in $\Omega \cap B(w, 4r^*)$ with boundary values $\tilde{u} = \min(u,v)$ and $\tilde{v} = 2\max(u,v)$ respectively on $\partial(\Omega \cap B(w, 4r^*))$. From the maximum principle for A-harmonic functions we then see that $\tilde{u} \leq u, v \leq \tilde{v}/2$ in $\Omega \cap B(w, 4r^*)$. Using this inequality and applying Theorem 1 to $\tilde{u}, \tilde{v}$ with r replaced by r^*, we get

$$\max(u/v, \, v/u) \leq \tilde{v}/\tilde{u} \leq c \text{ in } \Omega \cap B(w, \tilde{r}).$$

Finally we note from boundedness of u/v that (1.14) (a) can be achieved in $\Omega \cap B(w, 4r^*)$ by multiplying v by a suitably large constant which can be chosen to depend depend only on $p, n, \alpha, \beta, \gamma$. Thus Theorem 1 is true. □

5. The Martin boundary problem: preliminary reductions

Let $\Omega \subset \mathbf{R}^n$, δ, r_0, p, α, β, γ, and A be as in the statement of Theorem 2. Moreover, let $w \in \partial\Omega$ and let $0 < r' \ll \tilde{r}_0$, where $\tilde{r}_0 = \min\{r_0, 1\}$. Assume that $\hat{u}$ is an A-harmonic function in $\Omega \setminus B(w, r')$ and that $\hat{u} = 0$ continuously on $\partial\Omega \setminus B(w, r')$. We can apply Lemma 3.35 to conclude that there exist $\delta^*, 0 < \delta^* < 1, \bar{c}, \bar{\lambda} \geq 1$, depending only on $p, n, \alpha, \beta, \gamma$, such that if $0 < \delta \leq \delta^*$, then, for each $\hat{y} \in \partial\Omega \setminus B(w, \bar{c}r')$, the 'fundamental inequality',

$$\bar{\lambda}^{-1} \frac{\hat{u}(y)}{d(y, \partial\Omega)} \leq |\nabla \hat{u}(y)| \leq \bar{\lambda} \frac{\hat{u}(y)}{d(y, \partial\Omega)} \tag{5.1}$$

holds whenever $y \in \partial\Omega \cap B(\hat{y}, |\hat{y} - w|/\bar{c}) \cap B(w, \tilde{r}_0)$. Using this fact we see that if $0 < \delta \leq \delta^*$ then there exists $\tilde{\eta}$, depending only on $p, n, \alpha, \beta, \gamma$, such that if we define a non-tangential approach region at $w \in \partial\Omega$, denoted $\tilde{\Omega}(w, \tilde{\eta})$, by $\tilde{\Omega}(w, \tilde{\eta}) = \{y \in \Omega : \ d(y, \partial\Omega) \geq \tilde{\eta}|y - w|\}$, then

$$\hat{u} \text{ satisfies (5.1) in } [\Omega \setminus \tilde{\Omega}(w, \tilde{\eta})] \cap (B(w, \tilde{r}_0) \setminus B(w, \bar{c}r')). \tag{5.2}$$

We observe that the above argument applies for any small $r' > 0$ if $\hat{u}$ is a minimal positive A-harmonic function with respect to w. We note, in analogy with the proof of Theorem 1, that if we apriori knew that (5.1) held in $\Omega \cap B(w, \tilde{r})$ for some $\tilde{r} > 0$, then we could apply the argument in Steps C, D of the introduction to get an analogue of Theorem 1 in $\Omega \cap B(w, \tilde{r}) \setminus B(w, cr')$. Letting $r' \to 0$ we would then get Theorem 2. Unfortunately though we do not know this apriori and we do not see how to 'deduce' this inequality from simpler functions as in the proof of Lemma 3.35. Still, if (5.1) holds in $\Omega \cap B(w, \tilde{r})$, whenever $A \in M_p(\alpha)$, then we can make use of appropriate versions of Lemmas 3.1 and 3.18, as well as Definition 1.1 *(iii)*, to conclude that (5.1) holds in $\Omega \cap B(w, \tilde{s})$, for some $\tilde{s} < \tilde{r}$, whenever $A \in M_p(\alpha, \beta, \gamma)$. Thus to prove Theorem 2 we first prove Theorem 2 under the assumption that

$$A \in M_p(\alpha). \tag{5.3}$$

In particular, we start by showing that if one such A-harmonic function satisfies the 'fundamental inequality' then all such functions, relative to the given A, have this property. More specifically we prove,

Lemma 5.4. *Let Ω be a bounded (δ, r_0)-Reifenberg flat domain and let $w \in \partial\Omega$. Let $A \in M_p(\alpha)$ for some α and $1 < p < \infty$. Let $\hat{u}, \hat{v} > 0$ be A-harmonic in $\Omega \setminus B(w, r')$, continuous in $\mathbf{R}^n \setminus B(w, r')$, with $\hat{u} \equiv \hat{v} \equiv 0$ on $\mathbf{R}^n \setminus [\Omega \cup B(w, r')]$. Suppose for some $r_1, r' < r_1 < \tilde{r}_0$, and $b \geq 1$, that*

$$b^{-1} \frac{\hat{u}(y)}{d(y, \partial\Omega)} \leq |\nabla \hat{u}(y)| \leq b \frac{\hat{u}(y)}{d(y, \partial\Omega)}$$

whenever $y \in \Omega \cap [B(w, r_1) \setminus B(w, r')]$. There exists $\tilde{\delta}^* > 0, \lambda, c \geq 1$, depending on p, n, α, b, such that if $0 < \delta < \tilde{\delta}^* < \tilde{\delta}$ ($\tilde{\delta}$ as in Theorem 1), then

$$\lambda^{-1} \frac{\hat{v}(y)}{d(y, \partial\Omega)} \leq |\nabla \hat{v}(y)| \leq \lambda \frac{\hat{v}(y)}{d(y, \partial\Omega)}$$

whenever $y \in \Omega \cap [B(w, r_1/c) \setminus B(w, cr')]$. Moreover,

$$\left| \log\left(\frac{\hat{u}(z)}{\hat{v}(z)}\right) - \log\left(\frac{\hat{u}(y)}{\hat{v}(y)}\right) \right| \leq c \left(\frac{r'}{\min(r_1, |z - w|, |y - w|)}\right)^\sigma$$

whenever $z, y \in \Omega \setminus B(w, cr')$.

Proof: We note that to prove the last statement of Lemma 5.4 we can assume that $r'/r_1 << 1$, since otherwise there is nothing to prove. Let $\tilde{r} = \hat{c}r'$. If $\hat{c} = \hat{c}(p, n, \alpha)$ is large enough, we may assume

$$\hat{u} \leq \hat{v}/2 \leq \hat{c}\hat{u} \text{ in } \Omega \setminus B(w, \tilde{r}), \tag{5.5}$$

as we see from Theorem 1, Harnack's inequality, and the maximum principle for A-harmonic functions. As in (1.15) we let $u(\cdot, t), t \in [0, 1]$, be A-harmonic in $\Omega \setminus \bar{B}(w, \tilde{r})$, with continuous boundary values,

$$u(\cdot, t) = (1 - t)\hat{u}(\cdot) + t\hat{v}(\cdot) \text{ on } \partial[\Omega \setminus \bar{B}(w, \tilde{r})]. \tag{5.6}$$

Extend $u(\cdot, t), t \in [0, 1]$, to be continuous on $\mathbf{R}^n \setminus [\Omega \cup \bar{B}(w, \tilde{r})]$ by setting $u(\cdot, t) \equiv 0$ on this set. Next we note from Lemma 3.18 that there exists $\epsilon_0 = \epsilon_0(p, n, \alpha, b)$ such that if s_1 and ρ_1 satisfy $\tilde{r} \leq s_1 < \rho_1/4 \leq r_1/16$, $t \in [0, 1]$, and

$$(1 - \epsilon_0)\tilde{L} \leq u(\cdot, t)/\hat{u}(\cdot) \leq (1 + \epsilon_0)\tilde{L}, \tag{5.7}$$

in $\Omega \cap [B(w, 2\rho_1) \setminus B(w, s_1)]$ for some $\tilde{L}$, then

$$\hat{\lambda}^{-1} \frac{u(y, t)}{d(y, \partial\Omega)} \leq |\nabla u(y, t)| \leq \hat{\lambda} \frac{u(y, t)}{d(y, \partial\Omega)} \tag{5.8}$$

whenever $y \in \Omega \cap [B(w, \rho_1) \setminus B(w, 2s_1)]$ where $\hat{\lambda} = \hat{\lambda}(p, n, \alpha, b)$. Observe from (5.5), (5.6), that if $t_1, t_2 \in [0, 1]$, then

$$\begin{aligned} c^{-1}u(\cdot, t_1) \;&\leq\; U(\cdot, t_1, t_2) = \frac{u(\cdot, t_2) - u(\cdot, t_1)}{t_2 - t_1} \\ &= \;\hat{v}(\cdot) - \hat{u}(\cdot) \leq c\, u(\cdot, t_1) \end{aligned} \tag{5.9}$$

on $\partial[\Omega \setminus B(w, \tilde{r})]$. Moreover, from the maximum principle we see that this inequality also holds in $\Omega \setminus \bar{B}(w, \tilde{r})$. Thus for ϵ_0 as in (5.7), there exists $\epsilon_0', 0 < \epsilon_0' \leq \epsilon_0$, with the same dependence as ϵ_0, such that if $|t_2 - t_1| \leq \epsilon_0'$, then

$$1 - \epsilon_0/2 \leq \frac{u(\cdot, t_2)}{u(\cdot, t_1)} \leq 1 + \epsilon_0/2 \text{ in } \Omega \setminus B(w, \tilde{r}). \tag{5.10}$$

Divide [0,1] into closed intervals, disjoint except for endpoints, of length $\epsilon_0'/2$ except possibly for the interval containing 1 which is of length $\leq \epsilon_0'/2$. Let $\xi_1 = 0 < \xi_2 < \dots < \xi_m = 1$ be the endpoints of these intervals. Thus [0,1] is divided into $\{[\xi_k, \xi_{k+1}]\}_1^m$. Next suppose for some $l, 1 \leq l \leq m - 1$, that (5.8) is valid whenever $t \in [\xi_l, \xi_{l+1}]$ and $y \in \Omega \cap [B(w, \rho_1) \setminus B(w, 2s_1)]$. Under this assumption we claim for some $\hat{c}_1, \hat{c}_2, \sigma$, depending only on p, n, α, b, that

$$\left| \log \frac{u(z, \xi_{l+1})}{u(z, \xi_l)} - \log \frac{u(y, \xi_{l+1})}{u(y, \xi_l)} \right| \leq \hat{c}_1 \left(\frac{s_1}{\min(|z - w|, |y - w|)} \right)^\sigma \tag{5.11}$$

whenever $z, y \in \Omega \cap [B(w, \rho_1/\hat{c}_2) \setminus B(w, \hat{c}_2 s_1)]$. Indeed we can retrace the argument in Step C of the introduction to get, for $z, y \in \Omega \cap [B(w, \rho_1/c) \setminus B(w, cs_1)]$, that there exists f as in (1.19) and $\sigma > 0$ as in (1.24) such that

$$\begin{aligned}\left|\log\frac{u(z,\xi_{l+1})}{u(z,\xi_l)} - \log\frac{u(y,\xi_{l+1})}{u(y,\xi_l)}\right| &\leq \int_{\xi_l}^{\xi_{l+1}} \left|\frac{f(z,t)}{u(z,t)} - \frac{f(y,t)}{u(y,t)}\right| dt \\ &\leq c\left(\frac{s_1}{\min(|z-w|,|y-w|)}\right)^{\sigma}\end{aligned} \tag{5.12}$$

To get the last inequality in (5.12) we have used a slightly more general version of Lemma 4.7.

We now proceed by induction. Observe from (5.10) and $u(\cdot,\xi_1) = \hat u$, that (5.7) (5.8) hold whenever $t \in [\xi_1, \xi_2]$. Thus (5.11) is true for $l = 1$ with $s_1 = \tilde r, \rho_1 = r_1/4$. Let $s_2 = \hat c_2 s_1$, $\rho_2 = \rho_1/\hat c_2$. By induction, suppose for some $2 \leq k < m$,

$$\left|\log\frac{u(z,\xi_k)}{\hat u(z)} - \log\frac{u(y,\xi_k)}{\hat u(y)}\right| \leq (k-1)\hat c_1\left(\frac{s_k}{\min(|z-w|,|y-w|)}\right)^{\sigma} \tag{5.13}$$

whenever $z, y \in \Omega \cap [B(w,\rho_k) \setminus B(w,s_k)]$, where $\sigma, \hat c_1$ are the constants in (5.11). For $\eta > 0$ given and small we choose $s'_k \geq 2s_k$, so that

$$\left|\frac{u(z,\xi_k)}{\hat u(z)} - \frac{u(y,\xi_k)}{\hat u(y)}\right| \leq \eta\frac{u(z,\xi_k)}{\hat u(z)}$$

whenever $z, y \in \Omega \cap [B(w,\rho_k) \setminus B(w,s'_k)]$. Moreover, fix z as in the last display and choose $\eta > 0$ so small that

$$(1-\epsilon_0)\frac{u(z,\xi_k)}{\hat u(z)} \leq \frac{u(y,t)}{\hat u(y)} \leq (1+\epsilon_0)\frac{u(z,\xi_k)}{\hat u(z)}. \tag{5.14}$$

whenever $y \in \Omega \cap [B(w,\rho_k) \setminus B(w,s'_k)]$ and $t \in [\xi_k, \xi_{k+1}]$. To estimate the size of η observe, for $t \in [\xi_k, \xi_{k+1}]$, that

$$\frac{u(y,t)}{\hat u(y)} = \frac{u(y,t)}{u(y,\xi_k)} \cdot \frac{u(y,\xi_k)}{\hat u(y)} \leq (1+\epsilon_0/2)(1+\eta)\frac{u(z,\xi_k)}{\hat u(z)}.$$

Thus if $\eta = \epsilon_0/4$ (ϵ_0 small), then the right hand inequality in (5.14) is valid. A similar argument gives the left hand inequality in (5.14) when $\eta = \epsilon_0/4$. Also since $k \leq 2/\epsilon'_0$, and ϵ'_0, σ depend only on p, n, α, b, we deduce from (5.13) that one can take $s'_k = \hat c_3 s_k$ for $\hat c_3 = \hat c(p,n,\alpha,b)$ large enough. From (5.14) we first find that (5.7) holds with $\tilde L = \frac{u(z,\xi_k)}{\hat u(z)}$ in $\Omega \cap [B(w,\rho_k) \setminus B(w,s'_k)]$ and thereupon that (5.8) also holds. From (5.8) we now get, as in (5.12), that (5.11) is valid for $l = k$ in $\Omega \cap [B(w,\frac{\rho_k}{2\hat c_2}) \setminus B(w, 2\hat c_2 s'_k)]$. Let $s_{k+1} = 2\hat c_3\hat c_2 s_k$ and $\rho_{k+1} = \frac{\rho_k}{2\hat c_2}$. Using (5.11) and the induction hypothesis we have

$$\begin{aligned}\left|\log\frac{u(z,\xi_{k+1})}{\hat u(z)} - \log\frac{u(y,\xi_{k+1})}{\hat u(y)}\right| &\leq \left|\log\frac{u(z,\xi_{k+1})}{u(z,\xi_k)} - \log\frac{u(y,\xi_{k+1})}{u(y,\xi_k)}\right| \\ &\quad + \left|\log\frac{u(z,\xi_k)}{\hat u(z)} - \log\frac{u(y,\xi_k)}{\hat u(y)}\right| \\ &\leq k\hat c_1\left(\frac{s_{k+1}}{\min(|z-w|,|y-w|)}\right)^{\sigma}\end{aligned} \tag{5.15}$$

whenever $z, y \in \Omega \cap [B(w, \rho_{k+1}) \setminus B(w, s_{k+1})]$. From (5.15) and induction we get (5.13) with $k = m$. Since $u(\cdot, \xi_m) = \hat{v}$ and $s_m \leq cr'$, $\rho_m \geq r_1/c$, for some large $c = c(p, n, \alpha)$, we can now argue as in (5.14) to first get (5.7) with $u(\cdot, t)$ replaced by $\hat{v}$ and then (5.8) for $\hat{v}$. We conclude that Lemma 5.4 is valid for $z, y \in \Omega \cap [B(w, r_1/c) \setminus B(w, cr')]$ provided c is large enough. Using the maximum principle for A-harmonic functions it follows that the last display in Lemma 5.4 is also valid for $z, y \in \Omega \setminus B(w, r_1/c)$. □

5.1. Proof of Theorem 2 when $A \in M_p(\alpha)$. Let $\Omega \subset \mathbf{R}^n, w \in \partial\Omega, \delta, p, r_0, \alpha, \beta, \gamma$, be as in Theorem 2. Let $A \in M_p(\alpha)$, and suppose that u, v, are minimal positive A-harmonic functions relative to $w \in \partial\Omega$. If (5.1) holds for u in $\Omega \cap B(w, r_1)$, then we can apply Lemma 5.4 to u, v and let $r' \to 0$. We then get that u/v equals a constant, which is the conclusion of Theorem 2. Thus to complete the proof of Theorem 2 for $A \in M_p(\alpha)$, it suffices to show the existence of a minimal positive A-harmonic function u relative to $w \in \partial\Omega$ and $0 < r_1 < \tilde{r}_0$ for which the 'fundamental inequality' in (5.1) holds in $\Omega \cap B(w, r_1)$. Moreover, it suffices to show that (5.1) holds for some $r_1 = r_1(p, n, \alpha), 0 < r_1 < \tilde{r}_0, \bar{\lambda} = \bar{\lambda}(p, n, \alpha) \geq 1$, in $\tilde{\Omega}(w, \tilde{\eta}) \cap B(w, r_1)$ where $\tilde{\eta} = \tilde{\eta}(p, n, \alpha)$ is as in (5.2). To this end we show there exists $c = c(p, n, \alpha) \geq 1$ such that if $c^2 r' < r < \tilde{r}_0/n$, and $\rho = r/c$, then (5.1) holds for $\hat{u}$ on $\tilde{\Omega}(w, \tilde{\eta}) \cap \partial B(w, \rho)$. Here $\hat{u} > 0$ is A-harmonic in $\Omega \setminus \bar{B}(w, r')$ with continuous boundary values and $\hat{u} \equiv 0$ on $\partial\Omega \setminus \bar{B}(w, r')$. It then follows from arbitrariness of r, r', the above discussion, and Lemma 5.4 that Theorem 2 is valid whenever $A \in M_p(\alpha)$ and u is a minimal positive A-harmonic function relative to $w \in \partial\Omega$. With this game plan in mind, observe from Lemma 2.15 and (1.7), that we may assume $r = 1, w = 0$, and

(5.16) $B(0, 4n) \cap \{y : y_n \geq \mu\} \subset \Omega, \ \ B(0, 4n) \cap \{y : y_n \leq -\mu\} \subset \mathbf{R}^n \setminus \Omega,$

where $\mu = 500n\delta^*, 0 < \mu < 10^{-100}$ and $r' < (\delta^*)^2$. Here δ^* is temporarily allowed to vary but will be fixed after the proof of Lemma 5.19. Extend $\hat{u}$ to be continuous on $\mathbf{R}^n \setminus B(0, r')$, by putting $\hat{u} \equiv 0$ on $\mathbf{R}^n \setminus (\Omega \cup B(0, r'))$. Using the notation in (2.7), let $Q = Q^+_{1,1-\mu}(\mu e_n) \setminus \bar{B}(0, \sqrt{\mu})$ and let v_1 be the A-harmonic function in Q with the following continuous boundary values,

$$
\begin{aligned}
v_1(y) &= \hat{u}(y), \qquad\qquad y \in \partial Q \cap \{y : 2\mu \leq y_n\}, \\
v_1(y) &= \frac{(y_n - \mu)}{\mu} \hat{u}(y), \ y \in \partial Q \cap \{y : \mu \leq y_n < 2\mu\}.
\end{aligned}
$$

Comparing boundary values and using the maximum principle for A-harmonic functions, it follows that

$$
v_1 \leq \hat{u} \text{ in } Q\,. \tag{5.17}
$$

We now set $\mu = \mu(\epsilon) = \exp(-1/\epsilon)$. We shall prove,

Lemma 5.18. *Let $0 < \epsilon \leq \hat{\epsilon}$, $\mu = \mu(\epsilon)$ be as above and let $\tilde{\eta}$ be as in (5.2). If $\hat{\epsilon}$ is small enough, then there exists $\hat{\theta} = \hat{\theta}(p, n, \alpha), 0 < \hat{\theta} \leq 1/2$, such that if $\hat{\rho} = \mu^{1/2 - \hat{\theta}}$, then*

$$
1 \leq \hat{u}(y)/v_1(y) \leq 1 + \epsilon
$$

whenever $y \in \tilde{\Omega}(0, \tilde{\eta}/4) \cap [B(0, \hat{\rho}) \setminus B(0, 2\sqrt{\mu})]$.

Lemma 5.19. *Let $v_1, \epsilon, \hat\epsilon, \hat\theta, \mu$ be as in Lemma 5.18 and let $\tilde\eta$ be as in (5.2). If $\hat\epsilon$ is small enough, there exist $\theta = \theta(p, n, \alpha), 0 < \theta < \hat\theta/4, \tilde\lambda = \tilde\lambda(p, n, \alpha) > 1$, such that if $\rho = \mu^{1/2-2\theta}, a = \mu^{-\theta}$, then*

$$\tilde\lambda^{-1}\frac{v_1(y)}{d(y,\partial\Omega)} \leq |\nabla v_1(y)| \leq \tilde\lambda\frac{v_1(y)}{d(y,\partial\Omega)}$$

whenever $y \in \tilde\Omega(0, \tilde\eta/2) \cap [B(0, a\rho) \setminus B(0, \rho/a)]$.

Assuming Lemmas 5.18, 5.19, are true we complete the proof of Theorem 2 when $A \in M_p(\alpha)$ as follows. From these lemmas and Lemma 3.18 we deduce, for sufficiently small $\hat\epsilon = \hat\epsilon(p, n, \alpha) > 0$, that (5.1) is valid for $\hat u$ and for some $\bar\lambda = \bar\lambda(p, n, \alpha) \geq 1$ in $\Omega(w, \tilde\eta) \cap \partial B(0, \rho)$. With $\hat\epsilon$ now fixed we put $\delta^* = \mu(\hat\epsilon)/(500n)$ and conclude from (5.2), Lemma 2.15, arbitrariness of r, that (5.1) holds in $\Omega \cap [B(w, r_1) \setminus B(w, r')]$ with $r_1 = \tilde r_0/c, r' \leq r_0/c'$, provided c, c' are large enough, depending only on p, n, α. Thus we can apply Lemma 5.4 and proceed as in the discussion after that lemma to get Theorem 2 under the assumption $A \in M_p(\alpha)$.

Proof of Lemma 5.18. To begin the proof of Lemma 5.18 observe from (5.17) that it suffices to prove the righthand inequality in this display. We note that if $y \in \partial Q$ and $\hat u(y) \neq v_1(y)$, then y lies within 4μ of a point in ∂Q. Also $\max_{\partial B(0,t)} \hat u$ is non-increasing as a function of $t \geq r'$, as we see from the maximum principle for A-harmonic functions. Using these facts and Lemmas 2.1- 2.3 we find that

$$\hat u \leq v_1 + c\mu^{\tilde\sigma/2}\,\hat u(\sqrt{\mu}e_n), \tag{5.20}$$

on ∂Q. By the maximum principle this inequality also holds in Q. Here $\tilde\sigma$ is the exponent of Hölder continuity in Lemma 2.2. Using Harnack's inequality, we also find that there exist $\tau = \tau(p, n, \alpha) \geq 1$ and $c = c(p, n, \alpha) > 1$ such that

$$\max\{\psi(z), \psi(y)\} \leq c(d(z, \partial Q)/d(y, \partial Q))^\tau \min\{\psi(z), \psi(y)\} \tag{5.21}$$

whenever $z \in Q$, $y \in Q \cap B(z, 4d(z, \partial Q))$ and $\psi = \hat u$ or v_1. Also from Lemmas 2.1- 2.3 applied to v_1, we get

$$v_1(2\sqrt{\mu}e_n) \geq c^{-1}\,\hat u(\sqrt{\mu}e_n). \tag{5.22}$$

Let $\hat\rho, \hat\theta$ be as in Lemma 5.18. Using (5.20) - (5.22), we see that if $y \in \tilde\Omega(0, \tilde\eta/4) \cap [B(0, \hat\rho) \setminus B(0, 2\sqrt{\mu})]$, then

$$\hat u(y) \leq v_1(y) + c\mu^{\tilde\sigma/2}\,\hat u(\sqrt{\mu}e_n) \leq \left(1 + c^2\mu^{\tilde\sigma/2-\hat\theta\tau}\right) v_1(y) \leq (1+\epsilon)v_1(y) \tag{5.23}$$

provided $\hat\epsilon$ is small enough and $\hat\theta\tau = \tilde\sigma/4$. The proof of Lemma 5.18 is complete. □

Proof of Lemma 5.19. To prove Lemma 5.19 we let $v_1, \epsilon, \hat\epsilon, \hat\theta, \mu$ be as in Lemma 5.18. Using Lemmas 2.2 - 2.3 and Harnack's inequality we see that there exists $\phi = \phi(p, n, \alpha) > 0, 0 < \phi \leq 1/2$, and $c = c(p, n, \alpha) > 1$ with

$$\hat u(y) \leq c(s/t)^\phi \hat u(se_n) \tag{5.24}$$

provided $y \in \mathbf{R}^n \setminus B(0, t), t \geq s \geq 2r'$. Using (5.24) with $t = 1, s = 2\sqrt{\mu}$, and Lemmas 2.1 - 2.3 we see that

$$v_1 \leq c\mu^{\phi/2}\hat u(\sqrt{\mu}e_n) \text{ on } \partial Q \setminus \bar B(0, \sqrt{\mu}), \tag{5.25}$$

where c depends only on p, n, α. Let $\tilde{v}$ be the A-harmonic function in Q with continuous boundary values $\tilde{v} = 0$ on $\partial Q \setminus \bar{B}(0, \sqrt{\mu})$, and $\tilde{v} = v_1$ on $\partial B(0, \sqrt{\mu})$. Then from (5.25) and the maximum principle, we see that

$$\tilde{v} \leq v_1 \leq \tilde{v} + c\mu^{\phi/2}\hat{u}(\sqrt{\mu}e_n) \text{ in } Q. \tag{5.26}$$

Let $\rho = \mu^{1/2-2\theta}, \theta$ small, and $a = \mu^{-\theta}$ be as in Lemma 5.19. Using (5.21) applied to $\psi = \tilde{v}$ we find

$$\tilde{v} \geq c^{-1}(\mu^{1/2}/a\rho)^{\tau}\hat{u}(\sqrt{\mu}e_n) = c^{-1}\mu^{3\theta\tau}\hat{u}(\sqrt{\mu}e_n) \tag{5.27}$$

on $\tilde{\Omega}(0, \tilde{\eta}/8) \cap [B(0, 2a\rho) \setminus B(0, \rho/(2a))]$, where τ is as in (5.21) and the nontangential approach region $\tilde{\Omega}$ was defined above (5.2) relative to $w, \tilde{\eta}$. Also, since $\tilde{\eta}$ depends only on p, n, α, it follows that $c = c(p, n, \alpha)$ in (5.27). If we define θ by $\theta = \min\{\phi/(12\tau), \hat{\theta}/4\}$, then from (5.26), (5.27) we get

$$1 \leq \frac{v_1}{\tilde{v}} \leq 1 + \epsilon \tag{5.28}$$

in $\tilde{\Omega}(0, \tilde{\eta}/8) \cap B(0, 2a\rho) \setminus B(0, \rho/(2a))$, whenever $0 \leq \epsilon \leq \hat{\epsilon}$, provided $\hat{\epsilon}$ is sufficiently small.

Next let v be the A-harmonic function in

$$Q' = Q^+_{1,1-\mu}(\mu e_n) \setminus \bar{B}(2\sqrt{\mu}e_n, \sqrt{\mu})$$

with continuous boundary values $v = 0$ on $\partial Q' \setminus \bar{B}(2\sqrt{\mu}e_n, \sqrt{\mu})$ and $v = 1$ on $\partial B(2\sqrt{\mu}e_n, \sqrt{\mu})$. We claim that

$$v(y) \leq c\langle 2\sqrt{\mu}e_n - y, \nabla v(y)\rangle \tag{5.29}$$

when $y \in Q'$. Assuming claim (5.29) we can complete the proof of Lemma 5.19 in the following manner. First observe that (5.29) implies there exists $c = c(p, n, \eta) \geq 1$, for given $\eta, 0 < \eta \leq 1/2$, with

$$c^{-1}\frac{v(y)}{d(y, \partial Q')} \leq |\nabla v(y)| \leq c\frac{v(y)}{d(y, \partial Q')} \tag{5.30}$$

in $\tilde{Q}(0, \eta) \setminus \bar{B}(0, 10\sqrt{\mu})$, where $\tilde{Q}(0, \eta)$ is the non-tangential approach region defined relative to $0, \eta, Q$, as above (5.2). From the observation in (5.2) with $\hat{u}, \Omega$, replaced by v, Q and (5.30) for suitable $\eta = \eta(p, n, \alpha)$ we deduce that (5.30) in fact holds in $Q \setminus \bar{B}(0, 10\sqrt{\mu})$. We can now use Lemma 5.4 in $Q \setminus \bar{B}(0, 10\sqrt{\mu})$ with $v, \tilde{v}$ playing the role of $\hat{u}, \hat{v}$, respectively. In particular, we get for some large $c = c(p, n, \alpha)$ that

$$c^{-1}\frac{\tilde{v}(y)}{d(y, \partial\Omega)} \leq |\nabla \tilde{v}(y)| \leq c\frac{\tilde{v}(y)}{d(y, \partial\Omega)} \tag{5.31}$$

in $Q \cap B(0, 1/c^*) \setminus B(0, c^*\sqrt{\mu})$ for some $c^* = c^*(p, n, \alpha)$. Finally, note that if $0 \leq \epsilon \leq \hat{\epsilon}$ and if $\hat{\epsilon}$ is sufficiently small, then $1/c^* > 2a\rho > \rho/(2a) > c^*\sqrt{\mu}$. From this fact, (5.31), (5.28), and Lemma 3.18 applied to $\tilde{v}, v_1$, we deduce that Lemma 5.19 is valid subject to claim (5.29).

To prove claim (5.29) we first observe from Lemmas 2.2, 2.3 that $v(z) \leq 1/2$ in $Q' \cap B(0, 10\sqrt{\mu})$ for some z whose distance from ∂Q is at least $c^{-1}\sqrt{\mu}$ where $c = c(p, n, \alpha)$. Using Harnack's inequality it follows for some $c' > 1$ that $v \leq 1 - 1/c'$ on $\partial B(2\sqrt{\mu}e_n, 3\sqrt{\mu}/2)$. If $y \in \bar{B}(2\sqrt{\mu}e_n, 3\sqrt{\mu}/2) \setminus B(2\sqrt{\mu}e_n, \sqrt{\mu})$, set

$$\zeta(y) = \frac{e^{N|y-z|^2/\mu} - e^N}{e^{9N/4} - e^N} \tag{5.32}$$

where $z = 2\sqrt{\mu}e_n$. Then $\zeta \equiv 0$ on $\partial B(2\sqrt{\mu}e_n, \sqrt{\mu})$, and $\zeta \equiv 1$ on $\partial B(2\sqrt{\mu}e_n, 3\sqrt{\mu}/2)$. Also, if $N = N(p,n,\alpha)$ is large enough in (5.32), then from direct calculation and Definition 1.1, we find $\nabla \cdot A(\nabla\zeta) \geq 0$ in $B(2\sqrt{\mu}e_n, 3\sqrt{\mu}/2) \setminus \bar{B}(2\sqrt{\mu}e_n, \sqrt{\mu})$. Moreover, using these facts and the maximum principle we deduce

$$1 - v(y) \geq (c_+\sqrt{\mu})^{-1} d(y, \partial B(2\sqrt{\mu}e_n, \sqrt{\mu})) \tag{5.33}$$

in $B(2\sqrt{\mu}e_n, 3\sqrt{\mu}/2) \setminus B(2\sqrt{\mu}e_n, \sqrt{\mu})$ provided $c_+ = c_+(p,n,\alpha)$ is large enough. Next for fixed $t > 1$ put

$$\begin{aligned} O &= \{y \in Q' : 2\sqrt{\mu}e_n + t(y - 2\sqrt{\mu}e_n) \in Q'\}, \\ F(y) &= F(y,t) = \frac{v(y) - v(2\sqrt{\mu}e_n + t(y - 2\sqrt{\mu}e_n))}{t-1} \quad \text{whenever } y \in O. \end{aligned}$$

From (5.33) for $t > 1$ fixed, t near 1, and basic geometry it follows that

$$F \geq c^{-1} v \text{ on } \partial O. \tag{5.34}$$

We note that *(iv)* of Definition 1.1 and $A \in M_p(\alpha)$ imply that an A-harmonic function remains A-harmonic under scaling, translation, and multiplication by a constant. From this fact we see that F is the difference of two A-harmonic functions in O and one of them is a constant multiple of v. Using this fact, (5.34), and the maximum principle for A-harmonic functions, it follows that $F \geq c^{-1}v$ in O. Letting $t \to 1$, using Lemma 2.4 and the chain rule, we get claim (5.29). The proof of Lemma 5.19 is finished. □

As mentioned earlier, Lemmas 5.18, 5.19 together with Lemma 5.4 imply Theorem 2 when $A \in M_p(\alpha)$.

5.2. Proof of Theorem 2. We are now ready to prove Theorem 2 in the general case.

Lemma 5.35. *Let Ω be a bounded (δ, r_0)-Reifenberg flat domain and let $w \in \partial\Omega$. Let $A \in M_p(\alpha,\beta,\gamma)$ for some (α,β,γ) and $1 < p < \infty$. Let $\hat{u}, \hat{v} > 0$ be A-harmonic in $\Omega \setminus B(w,r')$, continuous in $\mathbf{R}^n \setminus B(w,r')$, with $\hat{u} \equiv \hat{v} \equiv 0$ on $\mathbf{R}^n \setminus [\Omega \cup B(w,r')]$. Then there exists $\delta_*, \sigma > 0, c_+ \geq 1$, depending on p,n,α,β,γ, such that if $0 < \delta < \delta_* < \tilde{\delta}$ ($\tilde{\delta}$ as in Theorem 1) and $r_1 = \tilde{r}_0/c_+$, then*

$$\left|\log\left(\frac{\hat{u}(z)}{\hat{v}(z)}\right) - \log\left(\frac{\hat{u}(y)}{\hat{v}(y)}\right)\right| \leq c_+\left(\frac{r'}{\min(r_1, |z-w|, |y-w|)}\right)^\sigma$$

whenever $z, y \in \Omega \setminus B(w, c_+r')$.

Proof: Once again we assume that $r'/r_1 << 1$, since otherwise there is nothing to prove. As in (5.5) we may assume for some $c = c(p,n,\alpha,\beta,\gamma)$ that

$$\hat{u} \leq \hat{v}/2 \leq c\hat{u} \text{ in } \Omega \setminus B(w, 2r'). \tag{5.36}$$

Let $u(\cdot,t), t \in [0,1]$, be A-harmonic in $\Omega \setminus \bar{B}(w,2r')$, with continuous boundary values,

$$u(\cdot,t) = (1-t)\hat{u}(\cdot) + t\hat{v}(\cdot) \text{ on } \partial[\Omega \setminus \bar{B}(w,2r')]. \tag{5.37}$$

We claim there exists $c, \bar{\lambda} \geq 1$ depending only on $p, n, \alpha, \beta, \gamma$ such that if $t \in [0,1]$, and $y \in \Omega \cap [B(w, \tilde{r}_0/c) \setminus \bar{B}(w, cr')]$, then

$$\bar{\lambda}^{-1} \frac{u(y,t)}{d(y,\partial\Omega)} \leq |\nabla u(y,t)| \leq \bar{\lambda} \frac{u(y,t)}{d(y,\partial\Omega)}. \tag{5.38}$$

Indeed let $A_1(y,\eta) = A(w,\eta)$ whenever $y \in \mathbf{R}^n$ and $\eta \in \mathbf{R}^n \setminus \{0\}$. Let $1 < a < b$ and suppose that ρ is such that $2r' \leq \rho/a < b\rho \leq \tilde{r}_0/2$. Let $v(\cdot,t)$, for $t \in [0,1]$, be A_1-harmonic in $\Omega \setminus \bar{B}(w, \rho/a)$ with continuous boundary values equal to $u(\cdot,t)$. Then, from Lemmas 5.4, 5.18, 5.19 we see that if $a = a(p,n,\alpha)$ is large enough, then

$$|\nabla v(\cdot,t)| \approx v(\cdot,t)/d(\cdot,\partial\Omega) \tag{5.39}$$

on $\Omega \cap \partial B(w,\rho)$. Here $\approx$ means with constants depending only on p, n, α. Let $h(\cdot,t)$ be the A_1-harmonic function in $\Omega_1 = \Omega \cap [B(w, b\rho) \setminus \bar{B}(w, \rho/a)]$ with continuous boundary values equal to $u(\cdot,t)$. We claim that if $b = b(p,n,\alpha,\beta,\gamma)$ is large enough then (5.39) is also valid for h. In fact (5.24) holds with u replaced by $u(\cdot,t)$ for $t \in [0,1]$, where now $\phi = \phi(p,n,\alpha,\beta,\gamma)$ and $s \geq 2r'$. Using (5.24) for $u(\cdot,t)$ we get

$$v(\cdot,t) \leq h(\cdot,t) \leq v(\cdot,t) + c(ab)^{-\phi} u(\rho e_n/a, t) \text{ on } \partial\Omega_1.$$

From the maximum principle this inequality also holds in Ω_1. Moreover, for τ as in (5.21) we deduce,

$$v(\cdot,t) \geq c^{-1} a^{-\tau} u(\rho e_n/a, t)$$

on $\tilde{\Omega}(w, \tilde{\eta}/2) \cap (B(w, 2\rho) \setminus \bar{B}(w, \rho/2))$. Thus, for some $c' = c'(p,n,\alpha,\beta,\gamma) \geq 1$,

$$v(\cdot,t) \leq h(\cdot,t) \leq (1 + c' a^{\tau-\phi} b^{-\phi}) v(\cdot,t) \tag{5.40}$$

on $\tilde{\Omega}(w, \tilde{\eta}/2) \cap (B(w, 2\rho) \setminus \bar{B}(w, \rho/2))$. Choosing $b = b(p,n,\alpha,\beta,\gamma)$ large enough in (5.40), using (5.39), Lemma 3.18, it follows that

$$\lambda_+^{-1} h(y,t)/d(y,\partial\Omega) \leq |\nabla h(y,t)| \leq \lambda_+ h(y,t)/d(y,\partial\Omega) \tag{5.41}$$

whenever $y \in \tilde{\Omega}(w, \tilde{\eta}) \cap \partial B(w,\rho)$ for some $\lambda_+ = \lambda_+(p,n,\alpha,\beta,\gamma) \geq 1$.

From (5.2) we see that (5.41) holds on $\Omega \cap \partial B(w,\rho)$ provided $\lambda_+(p,n,\alpha,\beta,\gamma)$ is large enough. With a, b, now fixed, depending only on $p, n, \alpha, \beta, \gamma$, we can use Lemma 2.15 and argue as in Lemma 3.1 to conclude for given $\epsilon > 0$, the existence of $r_1 = r_1(p,n,\alpha,\beta,\gamma,\epsilon)$ so small that if $b\rho \leq r_1 < \tilde{r}_0$, then

$$1 - \epsilon \leq u(\cdot,t)/h(\cdot,t) \leq 1 + \epsilon$$

on $\tilde{\Omega}(w, \tilde{\eta}/2) \cap (B(w, 2\rho) \setminus \bar{B}(w, \rho/2))$. In view of this inequality, (5.41), and Lemma 3.18, we see that if $\epsilon = \epsilon(p,n,\alpha,\beta,\gamma)$ is small enough, then

$$|\nabla u(\cdot,t)| \approx u(\cdot,t)/d(\cdot,\partial\Omega) \tag{5.42}$$

on $\tilde{\Omega}(w, \tilde{\eta}) \cap \partial B(w,\rho)$, where proportionality constants depend only on $p, n, \alpha, \beta, \gamma$. In view of (5.2), this inequality holds on $\Omega \cap \partial B(w,\rho)$. With r_1, a, b fixed we see from arbitrariness of ρ that (5.38) is true. We can now argue as in Lemma 5.4 or just repeat the argument in (1.18) - (1.25) to conclude Lemma 5.35. □

As pointed out earlier in this section, if u, v are minimal A-harmonic functions relative to $w \in \partial\Omega$, then we can apply Lemma 5.35 and let $r' \to 0$ to get Theorem 2. The proof of Theorem 2 is now complete. □

6. Appendix : an alternative approach to deformations

In this section we show that Step C in Theorem 1 can be replaced by a somewhat different argument based on ideas in [W]. The first author would like to thank Mikhail Feldman for making him aware of the ideas in [W]. In the following all constants will depend only on $p, n, \alpha, \beta, \gamma$ and we suppose that u, v are A-harmonic in $\Omega \cap B(w, 4r)$ and continuous in $B(w, 4r)$ with $u = v = 0$ on $B(w, 4r) \setminus \Omega$. From Lemma 3.35 we see that if δ is small enough, $\hat r = r/c$, and c is large enough, then for some $\mu \geq 1$,

$$\mu^{-1} \frac{h(y)}{d(y, \partial\Omega)} \leq |\nabla h(y)| \leq \mu \frac{h(y)}{d(y, \partial\Omega)} \tag{6.1}$$

whenever $y \in \Omega \cap B(w, 4\hat r)$, $h \in \{u, v\}$. Also from Lemma 4.8 we see that there exists $\mu_* \geq 1$, for $\epsilon^* > 0$ fixed, such that

$$\mu_*^{-1} \left(\frac{s}{\hat r}\right)^{1+\epsilon^*} \leq \frac{h(a_s(w))}{h(a_{\hat r}(w))} \leq \mu_* \left(\frac{s}{\hat r}\right)^{1-\epsilon^*} \tag{6.2}$$

whenever $y \in \Omega \cap B(w, \hat r)$, $h \in \{u, v\}$, where $0 < s \leq 4\hat r$. Observe again, for $x, \lambda \in \mathbf{R}^n, \xi \in \mathbf{R}^n \setminus \{0\}$, that

$$\begin{aligned} A_i(x, \lambda) - A_i(x, \xi) &= \int_0^1 \frac{d}{dt} A_i(x, t\lambda + (1-t)\xi) dt \\ &= \sum_{j=1}^n (\lambda_j - \xi_j) \int_0^1 \frac{\partial A_i}{\partial \eta_j}(x, t\lambda + (1-t)\xi) dt \end{aligned} \tag{6.3}$$

for $i \in \{1, ..., n\}$. In view of (6.3), (6.1), and A-harmonicity of u, v, we deduce that $u - v$ is a weak solution to $\bar L\zeta = 0$ in $\Omega \cap B(w, \hat r)$, where

$$\begin{aligned} \bar L\zeta(x) &= \sum_{i,j=1}^n \frac{\partial}{\partial x_i}(a_{ij}(x)\zeta_{x_j}) \\ \text{and } a_{ij}(x) &= \int_0^1 \frac{\partial A_i}{\partial \eta_j}(t\nabla u(x) + (1-t)\nabla v(x)) dt, \end{aligned} \tag{6.4}$$

for $1 \leq i, j \leq n$. Moreover, from the structure assumptions on A, see Definition 1.1, we find that

$$\begin{aligned} c_+^{-1} (|\nabla u(x)| + |\nabla v(x)|)^{p-2} |\xi|^2 &\leq \sum_{i,j=1}^n a_{ij}(x) \xi_i \xi_j \\ &\leq c_+ (|\nabla u(x)| + |\nabla v(x)|)^{p-2} |\xi|^2 \end{aligned} \tag{6.5}$$

whenever $x \in \Omega \cap B(w, \hat r)$. Next we prove the following lemma.

Lemma 6.6. *There exists $c \geq 1, \delta_0 > 0$, such that if $r^* = \hat r/c$, and $0 < \delta < \delta_0$, then $(|\nabla u| + |\nabla v|)^{p-2}$ extends to an A_2-weight in $B(w, r^*)$ with A_2-constant depending only on $p, n, \alpha, \beta, \gamma$.*

Proof: The proof is essentially the same as the proof of Lemma 4.9. That is, we use a Whitney cube decomposition of $\mathbf{R}^n \setminus \Omega$ to extend $(|\nabla u| + |\nabla v|)^{p-2}$ to a function λ on $B(w, 4r^*)$. Let $\tilde{w} \in B(w, r^*)$ and $0 < \tilde{r} < r^*$. Let $\hat{w} \in \partial\Omega$ with $|\hat{w} - \tilde{w}| = d(\tilde{w}, \partial\Omega)$ and suppose that $|\hat{w} - \tilde{w}|/2 < \tilde{r} < r^*$. We assume, as we may, that

$$\max\{u(a_{\tilde{r}}(\hat{w})), v(a_{\tilde{r}}(\hat{w}))\} = u(a_{\tilde{r}}(\hat{w})). \tag{6.7}$$

Let $\hat{\lambda} = \lambda$ when $p \geq 2$ and $\hat{\lambda} = 1/\lambda$ for $1 < p < 2$. As in (4.25) - (4.28), it follows, for $\epsilon^* > 0$, small enough, that

$$\int\limits_{B(\tilde{w},\tilde{r})} \hat{\lambda} dx \leq c u(a_{\tilde{r}}(\hat{w}))^{|p-2|}\, \tilde{r}^{n-|p-2|} \tag{6.8}$$

and

$$\begin{aligned}
\int\limits_{B(\tilde{w},\tilde{r})} \hat{\lambda}^{-1} dx &\leq c\tilde{r}^{(1+\epsilon^*)(|p-2|)} u(a_{\tilde{r}}(\hat{w}))^{-|p-2|} \int\limits_{\Omega\cap B(\tilde{w},50\tilde{r})} d(y,\partial\Omega)^{-\epsilon^*(|p-2|)} dy \\
&\leq c u(a_{\tilde{r}}(\hat{w}))^{-|p-2|}\, \tilde{r}^{n+|p-2|}. \qquad (6.9)
\end{aligned}$$

These inequalities remain true if $\tilde{r} \leq |\tilde{w} - \hat{w}|/2$, as follows easily from (6.1). Combining (6.8), (6.9), and using arbitrariness of $\tilde{w}, \tilde{r}$, we get Lemma 6.6. □

Using the ideas in [W] we continue by proving the following.

Lemma 6.10. *Given $p, 1 < p < \infty, w \in \partial\Omega$, $0 < r < r_0$, suppose that $\hat{u}$ and $\hat{v}$ are non-negative A-harmonic functions in $\Omega \cap B(w, 2r)$ with $\hat{v} \leq \hat{u}$. Assume also that $\hat{u}, \hat{v}$, are continuous in $B(w, 2r)$ with $\hat{u} \equiv 0 \equiv \hat{v}$ on $B(w, 2r) \setminus \Omega$. Let r^* be as in Lemma 6.6. There exists $c \geq 1$ such that if $\tilde{r} = r^*/c$, then*

$$c^{-1}\frac{\hat{u}(a_{\tilde{r}}(w)) - \hat{v}(a_{\tilde{r}}(w))}{\hat{v}(a_{\tilde{r}}(w))} \leq \frac{\hat{u}(y) - \hat{v}(y)}{\hat{v}(y)} \leq c\frac{\hat{u}(a_{\tilde{r}}(w)) - \hat{v}(a_{\tilde{r}}(w))}{\hat{v}(a_{\tilde{r}}(w))}$$

whenever $y \in \Omega \cap B(w, \tilde{r})$.

Proof: We first prove the lefthand inequality in Lemma 6.10. To do so we show the existence of Λ, $1 \leq \Lambda < \infty$, and $\hat{c} \geq 1$, such that if $r' = r^*/\hat{c}$ and if

$$e(y) = \Lambda\left(\frac{\hat{u}(y) - \hat{v}(y)}{\hat{u}(a_{r^*}(w)) - \hat{v}(a_{r^*}(w))}\right) - \frac{\hat{v}(y)}{\hat{v}(a_{r^*}(w))} \tag{6.11}$$

for $y \in \Omega \cap B(w, r^*)$, then

$$e(y) \geq 0 \text{ whenever } y \in \Omega \cap B(w, 2r'). \tag{6.12}$$

To do this, we initially allow $\Lambda, \hat{c} \geq 1$ to vary in (6.11). $\Lambda, \hat{c}$, are then fixed near the end of the argument. Put

$$\begin{aligned}
u'(y) &= \frac{\Lambda\, \hat{u}(y)}{\hat{u}(a_{r^*}(w)) - \hat{v}(a_{r^*}(w))}, \\
v'(y) &= \frac{\Lambda\, \hat{v}(y)}{\hat{u}(a_{r^*}(w)) - \hat{v}(a_{r^*}(w))} + \frac{\hat{v}(y)}{\hat{v}(a_{r^*}(w))}.
\end{aligned}$$

Observe from (6.11) that $e = u' - v'$. Using Definition 1.1 (iv) we see that u', v' are A-harmonic functions. Let $\bar{L}$ be defined as in (6.4) using u', v', instead of u, v, and let e_1, e_2 be the solutions to $\bar{L}e_i = 0, i = 1, 2$, in $\Omega \cap B(w, r^*)$, with continuous boundary values:

$$e_1(y) = \frac{\hat{u}(y) - \hat{v}(y)}{\hat{u}(a_{r^*}(w)) - \hat{v}(a_{r^*}(w))}, \; e_2(y) = \frac{\hat{v}(y)}{\hat{v}(a_{r^*}(w))}, \tag{6.13}$$

whenever $y \in \partial(\Omega \cap B(w, r^*))$. From Lemma 6.6 we see that Lemma 4.7 can be applied and we get, for some $c_+ \geq 1$ and $r_+ = r^*/c_+$, that

$$c_+^{-1} \frac{e_1(a_{r_+}(w))}{e_2(a_{r_+}(w))} \leq \frac{e_1(y)}{e_2(y)} \leq c_+ \frac{e_1(a_{r_+}(w))}{e_2(a_{r_+}(w))} \tag{6.14}$$

whenever $y \in \Omega \cap B(w, 2r_+)$. We now put

$$\hat{c} = c_+, \; r' = r_+, \; \Lambda = \hat{c}\, \frac{e_2(a_{r'}(w))}{e_1(a_{r'}(w))},$$

and observe from (6.14) that

$$\Lambda e_1(y) - e_2(y) \geq 0 \text{ whenever } y \in \Omega \cap B(w, 2r'). \tag{6.15}$$

Let $\hat{e} = \Lambda\, e_1 - e_2$ and note from linearity of $\bar{L}$ that $\hat{e}, e$, both satisfy the same linear locally uniformly elliptic pde in $\Omega \cap B(w, r^*)$ and also that these functions have the same continuous boundary values on $\partial(\Omega \cap B(w, r^*))$. Hence, using the maximum principle for the operator $\bar{L}$ it follows that $e = \hat{e}$ and then by (6.15) that $e(y) \geq 0$ in $\Omega \cap B(w, 2r')$. To complete the proof of the left-hand inequality in Lemma 6.10 with $\tilde{r} = 2r'$, we observe from Lemmas 4.5, 4.6, that $\Lambda \leq c$. The proof of the right-hand inequality in Lemma 6.11 is similar. We omit the details. □

We note that in [LN5] Lemma 6.10 was proved under the assumptions that $\hat{u}$ and $\hat{v}$ are non-negative p-harmonic functions in $\Omega \cap B(w, 2r)$ and that $\Omega \subset \mathbf{R}^n$ is a Lipschitz domain. In this case the constants in Lemma 6.10 depend only on p, n and the Lipschitz constant of Ω. Moreover, in [LN5] this result is used to prove regularity of a Lipschitz free boundary in a general two-phase free boundary problem for the p-Laplace operator.

Proof of Theorem 1. Let u, v, A, Ω, w, r be as in Theorem 1 and let $\hat{u}, \hat{v}$ be the A-harmonic functions in $\Omega \cap B(w, 2r)$ with

$$\hat{u} = \max\{u, v\} \text{ and } \hat{v} = \min\{u, v\} \text{ on } \partial[\Omega \cap B(w, 2r)].$$

From the maximum principle for A-harmonic functions we have $\hat{u} \geq \hat{v}$ and hence we can apply Lemma 6.10 to conclude that

$$c^{-1} \frac{\hat{u}(a_{\tilde{r}}(w))}{\hat{v}(a_{\tilde{r}}(w))} \leq \frac{\hat{u}(y)}{\hat{v}(y)} \leq c \frac{\hat{u}(a_{\tilde{r}}(w))}{\hat{v}(a_{\tilde{r}}(w))}$$

whenever $y \in \Omega \cap B(w, \tilde{r})$. Moreover, using the definition of $\hat{u}$, $\hat{v}$, and the inequalities in the last display we can conclude that

$$\frac{u(y)}{v(y)} \leq c \frac{u(z)}{v(z)} \text{ whenever } y, z \in \Omega \cap B(w, \tilde{r}). \tag{6.16}$$

Next if $x \in \partial\Omega \cap B(w, \tilde{r}/8)$, then we let

$$M(\rho) = \sup_{B(x,\rho)} \frac{u}{v} \text{ and } m(\rho) = \inf_{B(x,\rho)} \frac{u}{v}$$

when $0 < \rho < \tilde{r}$. If ρ is fixed we can apply Lemma 6.10 with $\hat{u} = u, \hat{v} = m(\rho)v$, and $2r$ replaced by ρ to conclude the existence of c^*, c_*, such that if $\tilde{\rho} = \rho/c^*$, then

$$(6.17) \qquad M(\tilde{\rho}) - m(\rho) \leq c_*(m(\tilde{\rho}) - m(\rho)).$$

Likewise, we can apply Lemma 6.10 with $\hat{u} = M(\rho)\, v$ and $\hat{v} = u$ to conclude

$$(M(\rho)\, v - u)/u \approx \text{ constant on } \Omega \cap B(w, \tilde{\rho}).$$

Using this inequality together with (6.16) it follows that

$$(M(\rho)v - u)/v \approx \text{ constant on } \Omega \cap B(w, \tilde{\rho}).$$

Here we have used heavily the fact that A-harmonic functions remain A-harmonic after multiplication by a constant as follows from Definition 1.1 (iv). Thus if c_* is large enough, then

$$(6.18) \qquad M(\rho) - m(\tilde{\rho}) \leq c_*(M(\rho) - M(\tilde{\rho})).$$

If $\operatorname{osc}(t) = M(t) - m(t)$, then we can add (6.17), (6.18) and we get, after some arithmetic, that

$$(6.19) \qquad \operatorname{osc}(\tilde{\rho}) \leq \frac{c_* - 1}{c_* + 1} \operatorname{osc}(\rho).$$

We can now use (6.19), since c^* is independent of ρ. in an iterative argument. Doing this we can conclude that

$$(6.20) \qquad \operatorname{osc}(s) \leq c(s/t)^\theta \operatorname{osc}(t) \text{ whenever } 0 < s < t \leq r/2$$

for some $\theta > 0, c \geq 1$. (6.20), (6.16), along with arbitrariness of $x \in \partial\Omega \cap B(w, \tilde{r}/8)$ and interior Hölder continuity - Harnack inequalities for u, v, are easily seen to imply Theorem 1. □

References

[B] P. Bauman, *Positive solutions of elliptic equations in non-divergence form and their adjoints,* Ark. Mat. **22** (1984), no.2, 153 - 173.

[BL] B. Bennewitz and J. Lewis, *On the dimension of p-harmonic measure,* Ann. Acad. Sci. Fenn. **30** (2005), 459-505.

[CFMS] L. Caffarelli, E. Fabes, S. Mortola, S. Salsa, *Boundary behavior of nonnegative solutions of elliptic operators in divergence form,* Indiana J. Math. **30** (4) (1981) 621-640.

[FGMS] E. Fabes, N. Garofalo, M. Malave, S. Salsa, *Fatou theorems for some non-linear elliptic equations,* Rev. Mat. Iberoamericana **4** (1988), no. 2, 227 - 251.

[FKS] E. Fabes, C. Kenig, and R. Serapioni, *The local regularity of solutions to degenerate elliptic equations,* Comm. Partial Differential Equations, **7** (1982), no. 1, 77 - 116.

[FJK] E. Fabes, D. Jerison, and C. Kenig, *The Wiener test for degenerate elliptic equations,* Ann. Inst. Fourier (Grenoble) **32** (1982), no. 3, 151-182.

[FJK1] E. Fabes, D. Jerison, and C. Kenig, *Boundary behavior of solutions to degenerate elliptic equations.* Conference on harmonicn analysis in honor of Antonio Zygmund, Vol I, II Chicago, Ill, 1981, 577-589, Wadsworth Math. Ser, Wadsworth Belmont CA, 1983.

[GT] D. Gilbarg and N. S. Trudinger, *Elliptic partial differential equations of second order,* second edition, Springer-Verlag, 1983.

[GZ] R. Gariepy and W. Ziemer, *A regularity condition at the boundary for solutions of quasilinear elliptic equations,* Arch. Rat. Mech. Anal. **67** (1977), no. 1, 25-39.

[JK] D. Jerison and C. Kenig, *Boundary behaviour of harmonic functions in nontangentially accessible domains,* Advances in Math. **46** (1982), 80-147.

[KKPT] C.E Kenig, H. Koch, J. Pipher, T. Toro, *A new approach to absolute continuity of elliptic measure with applications to non-symmetric equations,* Adv. in Math 153 (2000), 231-298.

[KT] C. Kenig and T. Toro, *Harmonic measure on locally flat domains,* Duke Math J. **87** (1997), 501-551.

[LN] J. Lewis and K. Nyström, *Boundary behaviour for p-harmonic functions in Lipschitz and starlike Lipschitz ring domains,* Ann. Sc. École Norm. Sup. (4) **40** (2007), no. 4, 765-813.

[LN1] J. Lewis and K. Nyström, *Boundary behaviour and the Martin boundary problem for p-harmonic functions in Lipschitz domains,* submitted.

[LN2] J. Lewis and K. Nyström, *Regularity and free boundary regularity for the p-Laplacian in Lipschitz and C^1-domains,* Ann. Acad. Sci. Fenn. **33** (2008), 1 - 26.

[LN3] J. Lewis and K. Nyström, *New results for p-harmonic functions,* to appear in Pure and Applied Math Quarterly.

[LN4] J. Lewis and K. Nyström, *Boundary behaviour of p-harmonic functions in domains beyond Lipschitz domains,* Advances in the Calculus of Variations **1** (2008), 1 - 38.

[LN5] J. Lewis and K. Nyström, *Regularity of Lipschitz free boundaries in two-phase problems for the p-Laplace operator,* submitted.

[Li] G. Lieberman, *Boundary regularity for solutions of degenerate elliptic equations,* Nonlinear Anal. **12** (1988), no. 11, 1203-1219.

[M] V.G. Maz'ya, *The continuity at a boundary point of the solutions of quasilinear elliptic equations* (Russian), Vestnik Leningrad. Univ. **25** (1970), no. 13, 42-55.

[R] Y.G Reshetnyak Y.G., *Space mappings with bounded distortion,* Translations of mathematical monographs, 73, American Mathematical Society, 1989.

[S] J. Serrin, *Local behavior of solutions of quasilinear elliptic equations,* Acta Math. **111** (1964), 247-302.

[T] P. Tolksdorf, *Regulariy for a more general class of quasilinear elliptic equations, J. Differential Equations,* **51** (1984), no. 1, 126-150.

[T1] P. Tolksdorf, *Everywhere regularity for some quasilinear systems with a lack of ellipticity,* Ann. Mat. Pura Appl. (4) **134** (1983), 241-266.

[W] P. Wang, *Regularity of free boundaries of two-phase problems for fully non-linear elliptic equations of second order. Part 1: Lipschitz free boundaries are $C^{1,\alpha}$*, Communications on Pure and Applied Mathematics. **53** (2000), 799-810.

Current address: Department of Mathematics, University of Kentucky, Lexington, KY 40506-0027, USA

E-mail address: `john@ms.uky.edu`

Current address: Department of Mathematics, Umeå University, S-90187 Umeå, Sweden

E-mail address: `email: niklas.lundstrom@math.umu.se`

Current address: Department of Mathematics, Umeå University, S-90187 Umeå, Sweden

E-mail address: `kaj.nystrom@math.umu.se`

Proceedings of Symposia in Pure Mathematics
Volume **79**, 2008

Waves on a steady stream with vorticity

M. Lilli and J. F. Toland

ABSTRACT. The existence question for two-dimensional periodic water waves on the surface of a flow with vorticity is one of finding a region upon which the solution of a semi-linear elliptic equation simultaneously satisfies two independent boundary conditions. Here we reduce this problem to a quasi-linear elliptic equation on a fixed domain with one nonlinear boundary condition and study the existence of non-trivial solutions using bifurcation theory. Although our reduction is a very slight variant of the classical one due to Dubreil-Jacotin, it significantly simplifies some of the analysis and extends the scope of the theory. For example, for a large class of laminar flow profiles, we find bifurcating k-modal waves with negative speeds for all $k \in \mathbb{N}$, and with positive speeds for a finite family of k.

1. Introduction

When the wavelength is normalized to be 2π, the existence question for two-dimensional periodic water waves on the surface of a flow with vorticity is one of finding a domain upon which the solution of a semi-linear elliptic equation simultaneously satisfies two independent boundary conditions (1.1d) and (1.1e) below:

$$-\Delta\psi(x,y) = \gamma(\psi(x,y)), \qquad -d < y < \eta(x), \tag{1.1a}$$

$$\psi(\cdot, y) \text{ is } 2\pi\text{-periodic in } x, \tag{1.1b}$$

$$\psi(x,-d) = 0, \tag{1.1c}$$

$$\psi(x,\eta(x)) = c_1, \tag{1.1d}$$

$$\frac{1}{2}|\nabla\psi(x,\eta(x))|^2 + g\eta(x) \equiv c_2, \tag{1.1e}$$

where c_1 (the volume flow rate) and c_2 (the Bernoulli constant, also known as the total head) are constants. Here ψ denotes the stream function and the vorticity in the flow has been assumed to be a given function γ of ψ. The impermeable bottom of the channel is located at $y = -d$, the acceleration due to gravity is g and the

1991 *Mathematics Subject Classification.* Primary 35R35, 74J15; Secondary 74J30, 76D27.

Key words and phrases. Water waves, vorticity, semi-linear elliptic, free-boundary problems.

M. Lilli acknowledges the German Science Foundation which supported his work at the University of Bath.

J. F. Toland holds a Royal Society/Wolfson Merit Award.

curve $\{(x,\eta(x)) : x \in \mathbb{R}\}$, where η is 2π-periodic, is the unknown free boundary upon which the stream function (1.1d) and the pressure (1.1e) must be constants.

In a study of the existence of small amplitude waves, Dubreil-Jacotin [**12**] defined a new function $h(x,p)$ on the fixed domain $R = \mathbb{R} \times [0,1]$ as

$$\psi(x, h(x,p)) = c_1 p, \quad x \in \mathbb{R}, \quad p \in [0,1], \tag{1.2}$$

and observed that the free-boundary problem (1.1) is equivalent to:

$$h_p^2\, h_{xx} + (1+h_x^2)h_{pp} - 2h_p\, h_x\, h_{xp} - c_1^{-1}\gamma(c_1 p)h_p^3 = 0 \text{ on } R, \tag{1.3a}$$

$$h(x,0) = -d, \tag{1.3b}$$

$$1 + h_x^2 + 2c_1^{-2}(gh - c_2)h_p^2 = 0, \quad p = 1, \tag{1.3c}$$

$$h(\cdot, p) \text{ is } 2\pi\text{-periodic in } x. \tag{1.3d}$$

This system has become a cornerstone of the growing literature on large-amplitude water waves with vorticity that began with the work of Constantin & Strauss [**5, 6**], in which the vorticity-stream-function $\gamma \in C^{1,\alpha}$ is prescribed.

In other work on waves with vorticity [**1, 3, 9, 14, 15**], γ is not fixed. Instead, the vorticity of a bifurcating wave is presumed to originate in the parametrized family of laminar streams from which it bifurcates. This is our point of view. To pursue it we introduce variables different from those of [**12**]. As with (1.3), the new system, (2.4) below, involves nonlinear operators that are real-analytic functions of the unknown function $\mathfrak{h}$, and there is an obvious variational structure. Moreover, it has a trivial solution, $\mathfrak{h}(z) = z$, independent of the laminar flow to which it corresponds, and the linearization (2.8) about that solution leads to transparent bifurcation criteria in a large number of situations, see Theorem 5.4. If it is required, this method can readily be adapted to yield an alternative approach to the theory in [**5, 6, 15**].

2. Formulation of the Problem

Consider a laminar running stream for which the vertical distribution of horizontal velocity is given by $U(y)$, $y \in [-d,0]$. Suppose $U \in C^2(-d,0) \cap C^{1,\vartheta}[-d,0]$, $\vartheta \in (0,1)$, and that $U(y) \neq 0$ on $[-d,0]$. The corresponding stream function is then

$$\Psi(y) = \int_{-d}^{y} U(z)\, dz, \tag{2.1}$$

and the dependence of vorticity on the stream function is given by

$$\gamma(\Psi(y)) = -U'(y), \quad y \in [-d,0], \tag{2.2}$$

which is the definition of a C^ϑ-function γ on the interval $\{\Psi(y) : y \in [-d,0]\}$. With

$$c_1 = \Psi(0) \text{ and } c_2 = \frac{1}{2}U(0)^2,$$

Ψ is a solution of (1.1) and the question is whether there are non-laminar solutions of (1.1) for the same vorticity function γ.

REMARK 2.1. Although these hypotheses, and later (5.5) and (5.9), are quite unrestrictive, neither this formulation nor that in [**5**] cover all cases of practical interest. For example, [**3**, page 102] and [**14**] considers waves bifurcating from running streams of the form $U(y) = c + u(y)$, where $u(y) = (d+y)^{1/7}$ and c is a parameter. Here $U \in C^{1/7}$ and the vorticity at the bottom is infinite. This leads to a version of problem (2.4) below with singular coefficients to which we will return. □

This free-boundary problem can be transformed into a problem on a fixed domain by defining $\mathfrak{h} : \mathbb{R} \times [-d, 0] \to \mathbb{R}$ as

$$\psi(x, \mathfrak{h}(x, z)) = \Psi(z), \qquad x \in \mathbb{R}, \quad z \in [-d, 0]. \tag{2.3}$$

Since

$$\Psi'(z) = U(z) \text{ and } \Psi''(z) = U'(z), \quad z \in (-d, 0),$$

the system to be satisfied by $\mathfrak{h}$ is

$$U'(z)\left\{1 - \frac{1+\mathfrak{h}_x^2}{\mathfrak{h}_z^2}\right\} + U(z)\left\{\left(\frac{\mathfrak{h}_x}{\mathfrak{h}_z}\right)_x - \frac{1}{2}\left(\frac{1+\mathfrak{h}_x^2}{\mathfrak{h}_z^2}\right)_z\right\} = 0, \tag{2.4a}$$

$$\mathfrak{h}(x, -d) = -d, \tag{2.4b}$$

$$U(z)^2(1+\mathfrak{h}_x^2) + 2(g\mathfrak{h} - \frac{1}{2}U(0)^2)\mathfrak{h}_z^2 = 0, \quad z = 0, \tag{2.4c}$$

$$\mathfrak{h}(\cdot, z) \text{ is } 2\pi\text{-periodic in } x. \tag{2.4d}$$

The trivial solution of (2.4). Since we seek non-trivial solutions of (1.1), it is important to note that $\mathfrak{h}(z) = z$ is the solution of (2.4) that corresponds to the laminar-flow solution of (1.1), no matter what the given function U may be.

Variational Structure of (2.4). At this point we make the formal observation that the transformed system (2.4) has variational structure. This is no surprise since the original free-boundary problem (1.1) has the variational structure discussed in [**2**] (see also [**10**]) and our change of variables (2.3) leads from there to the functional J below. (An analogous variational formulation [**7**, § 4.1] of the Dubreil-Jacotin equation (1.3) follows similarly.)

For functions $\mathfrak{h}$ which are periodic in x on the semi-infinite strip $S = \mathbb{R} \times [-d, 0]$ in the (x, z)-plane, with $\mathfrak{h}(x, -d) = -d$, let

$$J(\mathfrak{h}) = \frac{1}{2}\int\!\!\int_{S_{2\pi}} U^2\left\{\frac{1+\mathfrak{h}_x^2}{\mathfrak{h}_z} + \mathfrak{h}_z\right\} dxdz - \frac{g}{2}\int_0^{2\pi} \mathfrak{h}(x, 0)^2 dx, \tag{2.5}$$

where $S_{2\pi} = (-\pi, \pi) \times (-d, 0)$. Then critical points of (2.5) satisfy the system (2.4). Moreover, $\mathfrak{h}(z) = z$ is a critical point of J. So let $\mathfrak{h}(x, z) = z + \kappa(x, z)$ in the formula for J. Then the first term has the form

$$\frac{1}{2}\int\!\!\int_{S_{2\pi}} U^2\left\{2 + \frac{\kappa_x^2 + \kappa_z^2}{1+\kappa_z}\right\} dxdz = C + \frac{1}{2}\int\!\!\int_{S_{2\pi}} U^2\left\{\frac{\kappa_x^2 + \kappa_z^2}{1+\kappa_z}\right\} dxdz,$$

where C is independent of κ. Therefore we are interested in critical points of $\mathfrak{J}$, where

$$\mathfrak{J}(\kappa) = \frac{1}{2}\int\!\!\int_{S_{2\pi}} \left(\frac{\kappa_x^2 + \kappa_z^2}{1+\kappa_z}\right) U(z)^2 dxdz \; - \frac{g}{2}\int_{-\pi}^{\pi} \kappa(x, 0)^2\, dx. \tag{2.6}$$

Critical points of $\mathfrak{J}$ satisfy the system

$$\Big(\frac{U^2\kappa_x}{1+\kappa_z}\Big)_x + \Big(\frac{U^2\kappa_z}{1+\kappa_z}\Big)_z - \frac{1}{2}\Big(\frac{U^2(\kappa_x^2 + \kappa_z^2)}{(1+\kappa_z)^2}\Big)_z \;=\; 0, \tag{2.7a}$$

$$\kappa(x, -d) = 0, \tag{2.7b}$$

$$U(0)^2(1 + \kappa_x^2(x, 0)) + (2g\kappa(x, 0) - U(0)^2)(1 + \kappa_z(x, 0))^2 = 0, \tag{2.7c}$$

$$\kappa(\cdot, z) \text{ is } 2\pi\text{-periodic in } x. \tag{2.7d}$$

Linearization of (2.4). The functional $\mathfrak{J}$ has a critical point $\kappa = 0$, irrespective of U and the linearization of the Euler-Lagrange system (2.7), with respect to κ, about this zero solution is

$$(U^2\widehat{\kappa}_x)_x + (U^2\widehat{\kappa}_z)_z = 0, \tag{2.8a}$$

$$\widehat{\kappa}(x,-d) = 0, \tag{2.8b}$$

$$g\widehat{\kappa}(x,0) - U(0)^2\widehat{\kappa}_z(x,0) = 0, \tag{2.8c}$$

$$\widehat{\kappa}(\cdot,z) \text{ is } 2\pi\text{-periodic in } x. \tag{2.8d}$$

We will see that this linear problem is easy to analyze using separation of variables.

3. Parametrized Families of Laminar Streams

Now we consider a parametrized family of laminar running streams,

$$U(y;c),\ y \in [-d,0], \quad c \in I \subset \mathbb{R},$$

where, for $c \in I$ (an open interval), $U(\cdot\,;c) \in C^2(-d,0)$, $U(y;c) \neq 0$ on $[-d,0]$ and $c \mapsto U(\cdot\,;c) \in C^2\big(I; C^{1,\vartheta}[-d,0]\big)$. Here no physical meaning is assigned to the parameter c, the dependence of U on c being quite general. Let the corresponding stream function be denoted by

$$\Psi(y;c) = \int_{-d}^{y} U(z;c)\,dz,$$

and the dependence of vorticity on the stream function by

$$\gamma(\Psi(y;c)) = -U'(y;c), \quad y \in [-d,0].$$

With

$$c_1(c) = \Psi(0;c) \text{ and } c_2(c) = \frac{1}{2}U(0;c)^2,$$

$\Psi(\cdot\,;c)$ is a solution of (1.1) when $U = U(\cdot\,;c)$ and $\gamma = \gamma(\cdot\,;c)$. The corresponding solution of (2.7) is $\kappa = 0$ for all $c \in I$. The question is whether there are other (non-laminar) solutions of (1.1) for the same vorticity function $\gamma(\cdot\,;\,c)$ for certain values of c. This is a global question, but here we regard it as a question of finding bifurcation points on the line of trivial solutions $\{\kappa = 0,\ c \in I\}$ of system (2.7).

4. Bifurcation Theory

We now consider basic bifurcation theory [**8**] for the nonlinear problem

$$\Big(\frac{U(z;c)^2\kappa_x}{1+\kappa_z}\Big)_x + \Big(\frac{U(z;c)^2\kappa_z}{1+\kappa_z}\Big)_z - \frac{1}{2}\Big(\frac{U(z;c)^2(\kappa_x^2+\kappa_z^2)}{(1+\kappa_z)^2}\Big)_z = 0, \tag{4.1a}$$

$$\kappa(x,-d) = 0; \tag{4.1b}$$

$$U(0,c)^2(1+\kappa_x^2(x,0)) + (2g\kappa(x,0) - U(0,c)^2)(1+\kappa_z(x,0))^2 = 0, \tag{4.1c}$$

$$\kappa(\cdot,z) \text{ is } 2\pi\text{-periodic in } x, \tag{4.1d}$$

regarding c as the bifurcation parameter. To simplify matters we will seek solutions κ that are even in x. To this end let

$$X = \big\{\kappa \in C^{2,\vartheta}(\overline{S}) : \kappa \text{ is even and } 2\pi\text{-periodic in } x \text{ and } \kappa(x,-d) = 0\big\},$$

$$Y = \big\{\kappa \in C^{0,\vartheta}(\overline{S}) : \kappa \text{ is even and } 2\pi\text{-periodic in } x\big\},$$

$$Z = \{w \in C^{1,\vartheta}(\mathbb{R}) : w \text{ is even and } 2\pi\text{-periodic}\},$$

which are Banach spaces when endowed with the usual Hölder-space norms. Let B denote the open ball of radius 1 about the origin in $X \times Y$ and define $F : B \times \mathbb{R} \to Y \times Z$ by

$$F(c, \kappa) = \begin{pmatrix} \Big(\frac{U(z;c)^2\kappa_x}{1+\kappa_z}\Big)_x + \Big(\frac{U(z;c)^2\kappa_z}{1+\kappa_z}\Big)_z - \frac{1}{2}\Big(\frac{U(x;c)^2(\kappa_x^2+\kappa_z^2)}{(1+\kappa_z)^2}\Big)_z \\ U(0;c)^2(1+\kappa_x^2(x,0)) + (2g\kappa(x,0) - U(c;0)^2)(1+\kappa_z(x,0))^2 \end{pmatrix}.$$

It is clear that F is twice continuously differentiable from $B \times I$ into $Y \times Z$ and that $F(0,c) = 0 \in Y \times Z$ for all $c \in I$. In order to show that a particular c^* is a bifurcation point for the problem $F(\kappa, c) = 0$, it will suffice to show that, for some $\widehat{\kappa} \in X \setminus \{0\}$,

$$\text{(4.2a)} \qquad \ker d_\kappa F[(0,c^*)] = \operatorname{span}\{\widehat{\kappa}\},$$

$$\text{(4.2b)} \qquad \frac{d}{dc}\,(d_\kappa F[(0,c)]\widehat{\kappa})\Big|_{c=c^*} \notin \operatorname{Range} d_\kappa F[(0,c^*)],$$

$$\text{(4.2c)} \qquad \operatorname{range} d_\kappa F[(0,c^*)] \text{ has codimension 1 in } Y \times Z.$$

REMARK 4.1. It is interesting to note that the parameter c occurs nonlinearly in the linearized problem (2.8). Nevertheless, the system (4.2) coincides with the hypotheses in [**8**] that ensure that c^* is a bifurcation point for (4.1). If U depends real-analytically on c, as it does under the hypotheses of Theorem 5.4, the operator in (4.1) is real-analytic on $B \times \mathbb{R}$. In that case the theory of [**4**] is available to extend the local real-analytic curve that bifurcates from the simple eigenvalue to a *uniquely defined global curve* which has, in a neighbourhood of each of its points, a local real-analytic parametrization. □

Stipulations (4.2a) and (4.2b) mean that the solutions $\widehat{\kappa}$ of the linear problem

$$\text{(4.3a)} \qquad (U(z;c^*)^2\widehat{\kappa}_x)_x + (U(z;c^*)^2\widehat{\kappa}_z)_z = 0,$$

$$\text{(4.3b)} \qquad \widehat{\kappa}(x,-d) = 0,$$

$$\text{(4.3c)} \qquad g\widehat{\kappa}(x,0) - U(0;c^*)^2\widehat{\kappa}_z(x,0) = 0,$$

$$\text{(4.3d)} \qquad \widehat{\kappa}(\cdot, z) \text{ is even and } 2\pi\text{-periodic in } x,$$

form a one-dimensional subspace of X, and

$$\text{(4.4)} \qquad \begin{pmatrix} (2U(z;c^*)\frac{\partial U}{\partial c}(z;c^*)\widehat{\kappa}_x)_x + \big(2U(z;c^*)\frac{\partial U}{\partial c}(z;c^*)\widehat{\kappa}_z\big)_z \\ -2U(0;c^*)\frac{\partial U}{\partial c}(0;c^*)\widehat{\kappa}_z(x,0) \end{pmatrix} \neq \begin{pmatrix} (U(z;c^*)^2\widetilde{\kappa}_x)_x + (U(z;c^*)^2\widetilde{\kappa}_z)_z \\ g\widetilde{\kappa}(x,0) - U(0;c^*)^2\widetilde{\kappa}_z(x,0) \end{pmatrix}$$

for any $\widetilde{\kappa} \in X$. The meaning of (4.2c) is that the set

$$\left\{ \begin{pmatrix} (U(z;c^*)^2\kappa_x)_x + (U(z;c^*)^2\kappa_z)_z \\ g\kappa(x,0) - U(0;c^*)^2\kappa_z(x,0) \end{pmatrix} : \begin{array}{r} \kappa(x,-d) = 0, \\ \kappa(\cdot,z) \text{ even and } 2\pi\text{-periodic,} \end{array} \right\}$$

has codimension 1 in $Y \times Z$. This will follow by standard arguments if we can show that there is a unique solution of (4.3), up to scalar multiplication, because (4.3) is a self-adjoint eigenvalue problem in an L^2 setting. (See the last paragraph of [**13**, § 6.7].) In fact, in that case $(f,h) \in Y \times Z$ is in the range of $d_\kappa F[(0,c^*)]$ if and only if

$$\int\!\!\int_{S_{2\pi}} \widehat{\kappa}(x,z) f(x,z)\,dx dx + \int_{-\pi}^{\pi} \widehat{\kappa}(x,0) h(x)\,dx = 0.$$

Thus c^* will be shown to be a bifurcation point for (4.1) if we can show $\widehat{\kappa}$ is unique up to normalisation and that (4.4) holds. We will study the uniqueness question presently, but first here is an observation that will be useful in checking that (4.4) holds.

Suppose that (4.4) does not hold. More precisely, suppose that $\widehat{\kappa} \neq 0$ satisfies (4.3) and that equality in (4.4) holds for some $\widetilde{\kappa} \in X$. A multiplication of the first component of the equality in (4.4) by $\widehat{\kappa}$ and integration by parts over $S_{2\pi}$, using the periodicity in the x direction, yields

$$\begin{aligned}\int\!\!\int_{S_{2\pi}} & U(z;c^*)\frac{\partial U}{\partial c}(z;c^*)|\nabla \widehat{\kappa}|^2\,dx dz \\ &= \int_{-\pi}^{\pi} U(0;c^*)\frac{\partial U}{\partial c}(0;c^*)\widehat{\kappa}(x,0)\widehat{\kappa}_z(x,0)\,dx \\ &\quad + \frac{1}{2}\int_{-\pi}^{\pi} U(0;c^*)^2\big(\widehat{\kappa}_z(x,0)\widetilde{\kappa}(x,0) - \widehat{\kappa}(x,0)\widetilde{\kappa}_z(x,0)\big)dx = 0,\end{aligned}$$

from equality in the second components of (4.4). Therefore, if (4.4) is false, then

$$\int\!\!\int_{S_{2\pi}} U(z;c^*)\frac{\partial U}{\partial c}(z;c^*)|\nabla \widehat{\kappa}|^2\,dx dz = 0. \tag{4.5}$$

If, for example, if $\big(\partial U/\partial c\big)(z;c^*)$ is not zero on $[-d,0]$, this cannot happen.

Separation of variables. We seek values of c for which there exists a non-trivial solution of (4.3). It is easy to see, by separation of variables and completeness of the eigenvalues in an L^2 setting, that if such a $\widehat{\kappa}$ exists then it must be in the form $\widehat{\kappa}(x,z) = a(x)b(z)$, where a is 2π-periodic. This means that there exists k such that

$$a''(x) + k^2 a(x) = 0 \text{ where } a \text{ is } 2\pi\text{-periodic}, \tag{4.6}$$

$$\big(U(z;c)^2 b'\big)' - k^2 U(z;c)^2 b = 0, \tag{4.7a}$$

$$b(-d) = 0, \quad g b(0) - U(0;c)^2 b_z(0) = 0. \tag{4.7b}$$

The equation for a has constant coefficients and may be solved explicitly if and only if k is an integer. Its only even solution is a multiple of $\cos kx$. In general, we cannot solve the equation for b explicitly and we will study it in greater detail later. However, in one case at least, all its solutions are known in closed form. To see how this is so, suppose that b satisfies (4.7) and let $v(z) = U(z;c)b(z)$. Then v satisfies

$$v'' - \left(k^2 + \frac{U''(z;c)}{U(z;c)}\right) v = 0, \qquad k \in \mathbb{N}, \tag{4.8a}$$

$$\big(g + U(0;c)U'(0;c)\big)\, v(0) = U(0;c)^2\, v'(0). \tag{4.8b}$$

A running stream with constant vorticity, including irrotational flows. An important case of (4.8) is when $U(y;c) := c + \omega_0 y \neq 0$ on $[-d,0]$. Here the vorticity $\gamma \equiv -\omega_0$, a constant, and (4.8) has the form

$$v'' - k^2\, v = 0, \tag{4.9a}$$

$$\big(g + \omega_0 c\big)\, v(0) = c^2\, v'(0). \tag{4.9b}$$

There is a solution for certain values of c:

$$v(z) = \sinh\big(k(d+z)\big) \text{ where } \frac{(g+\omega_0 c)}{k}\tanh(kd) = c^2. \tag{4.9c}$$

The fact that 0 does not lie between c and $c-\omega_0 d$, equivalently that $U(\cdot\,;c)$ does not vanish, is a further restriction, but this problem can be analyzed completely. Let $g>0$ and $k\in\mathbb{N}$ be given. When $\omega_0 = 0$, (4.9c) says that c is uniquely determined up to its sign by $c^2/g = k^{-1}\tanh(dk)$, which is the classical value of c for the kth bifurcation point of irrotational waves from streams of depth d. However, when $\omega_0 \neq 0$, for each $k\in\mathbb{N}$ there exists $c_k^- < 0 < c_k^+ \in \mathbb{R}$ satisfying (4.9c).

If $\omega_0 > 0$, then $0 > c_k^- + \omega_0 y \neq 0$ on $[-d,0]$ for any k. Hence c_k^- is an admissible solution of (4.9c) for all $k\in\mathbb{N}$.

However $c_k^+ + \omega_0 y \neq 0$ on $[-d,0]$ means that $c_k^+ > \omega_0 d > 0$. Therefore c_k^+ is an admissible solution of (4.9c) only for $k\in\mathbb{N}$ with

$$d^2\omega_0^2\left(1 - \frac{\tanh(kd)}{kd}\right) < \frac{g}{k}\tanh(kd). \tag{4.10}$$

The cases $\omega_0 > 0$ and $\omega_0 < 0$ are symmetrical, with the superscripts $+$ and $-$ interchanged. □

5. General Linear Theory of a Running Stream

As an illustration of these general considerations we look at the important case when $U(y;c) = c + u(y)$ for some $u \in C^{1,\vartheta}$. Here we may assume that $u(0) = 0$, without loss of generality, because the value of $u(0)$ can be absorbed in the parameter c. In this set up, u represents the horizontal-velocity profile of a running stream in a fixed frame of reference and, relative to a frame moving with velocity $-c$, the horizontal velocity profile becomes $c+u$. Thus non-zero solutions of (4.1), for some c corresponds to waves travelling with velocity $-c$ on this stream.

Our general theory has reduced the question of bifurcation points to proving that, for a certain value of c, the linear problem

$$\big((u+c)^2 b'\big)' \;=\; \lambda(u+c)^2 b, \tag{5.1a}$$

$$b(-d) = 0, \quad gb(0) = c^2 b_z(0), \tag{5.1b}$$

has a simple eigenvalue $\lambda = k^2$, for some $k\in\mathbb{N}$, where c is such that $u(y)+c\neq 0$ on $[-d,0]$. When $u(y)+c \neq 0$ on $[-d,0]$ the problem for λ is a regular Sturm-Liouvlle problem for which the eigenvalues λ are given by a classical Rayleigh-Ritz minimax principle for the quotient formula

$$Q(v;c) = \frac{\displaystyle\int_{-d}^{0} (u(y)+c)^2 v'(y)^2\,dy - gv(0)^2}{\displaystyle\int_{-d}^{0} (u(y)+c)^2 v(y)^2\,dy}, \tag{5.2}$$

in which g and u are fixed.

LEMMA 5.1. *Suppose that g is fixed and that $u+c\neq 0$ on $[-d,0]$. Then there exists an increasing sequence of eigenvalues $\lambda_j(c)$, $j\in\mathbb{N}$, of (5.1), characterised as follows:*

$$\lambda_j(c)=\inf_{\dim(E)=j}\left\{\sup Q(v,c):v\in E\subset W^{1,2}[-d,0],\ v\neq 0,\ v(-d)=0\right\}.$$

Moreover, $\lambda_1(c)$ is a simple eigenvalue.

PROOF. This minimax characterization of $\lambda_j(c)$ is part of the classical theory, see [**11**, §4.5], for example. Moreover, solutions of the eigenvalue problem (5.1) attain these minimax values. In particular, $\lambda_1(c)$ is attained at a certain function v. Since $Q(|v|;c)\leq Q(v;c)$, we may assume that $\lambda_1(c)$ is also attained at $|v|$. Now suppose that $\lambda_1(c)$ is attained at v_1 and v_2, and therefore that $|v_1|$ and $|v_2|$ are eigenfunctions of (5.1) for the eigenvalue $\lambda_1(c)$. If $|v_1|$ and $|v_2|$ are not linearly dependent, it follows that $\int_{-d}^{0}|v_1||v_2|(u+c)^2dz=0$. Since this is false, $|v_1|$ is a scalar multiple of $|v_2|$. Since both satisfy (5.1), it follows that v_1 is a multiple of v_2, as required for $\lambda_1(c)$ to be a simple eigenvalue. □

Let

$$P_c=\int_{-d}^{0}\frac{dy}{(u(y)+c)^2}>0$$

and consider the eigenvalue problem

(5.3a) $$f''=\mu f,$$

(5.3b) $$f(0)=0,\quad gf(P_c)=f'(P_c),\quad f\not\equiv 0.$$

It is easy to see that there exists a solution with $\mu=\nu^2>0$ if and only if $gP_c>1$, in which case $f(z)=a\sinh\nu z$ for some $a\neq 0$, where ν is uniquely determined by

$$\frac{\tanh\nu P_c}{\nu P_c}=\frac{1}{gP_c}.$$

When $gP_c>1$ all the other eigenvalues μ of (5.3) (there are infinitely many) are negative and determined by

$$\mu=-\nu^2\text{ and }f(z)=\sin\nu z\text{ where }\frac{\tan\nu P_c}{\nu P_c}=\frac{1}{gP_c}.$$

By a similar calculation, every eigenvalue of (5.3) is negative when $gP_c<1$, and when $gP_c=1$ all its eigenvalues are non-positive, exactly one (counting multiplicity) being zero with eigenfunction $f(z)=z$.

As with Q and (5.1), these eigenvalues correspond to minimax values of

$$q(f;c)=\frac{\int_0^{P_c}f'(z)^2\,dz-gf(P_c)^2}{\int_0^{P_c}f(z)^2\,dz}$$

over the class of non-zero functions $f\in W^{1,2}(0,P_c)$ with $f(0)=0$. From the above observations we infer that when $gP_c\leq 1$,

(5.4a) $$\inf\left\{q(f,c):f\in W^{1,2}[0,P_c],\ f\neq 0,\ f(0))=0\right\}\geq 0.$$

However, when $gP_c>1$,

(5.4b) $$\inf\left\{q(f,c):f\in W^{1,2}[0,P_c],\ f\neq 0,\ f(0))=0\right\}<0$$

and, for all $j \geq 2$,

(5.4c) $$\inf_{\dim(F)=j} \{\sup q(f,c) : f \in F \subset W^{1,2}[0,P_c],\ f \neq 0,\ f(0)=0\} > 0.$$

We return now to our study of (5.1). In addition our basic assumption that $u \in C^2(-d,0) \cap C^{1,\vartheta}[-d,0]$ with $u(0)=0$, we now assume that

(5.5) $$u(y) < 0,\ y \in [-d,0).$$

When $f : [0,P_c] \to \mathbb{R}$ is smooth and $f(0)=0$, let

$$v(y) = \int_{-d}^{y} \frac{dt}{(u(t)+c)^2}\ .$$

Then $v(-d)=0$ and, when substituted in (5.2), we infer from (5.4a) that

(5.6) $$\lambda_1(c) > 0 \text{ when } gP_c < 1.$$

Because (4.8) and (5.1) are equivalent, our main results on the eigenvalue problem (5.1) represent a significant simplification and extension of [**15**, Lemma 2.5].

LEMMA 5.2. *Suppose that (5.5) holds and that $c<0$. Then $\lambda_1(c) \to -\infty$ as $c \nearrow 0$ and $\lambda_1(c) > 0$ for all $c<0$ with $|c|$ sufficiently large. Hence, for each $k \in \mathbb{N}$, there exists $c_k^- < 0$ such that $-k^2 = \lambda_1(c_k^-)$.*

PROOF. Note first that

(5.7) $$P_c \to \int_{-d}^{0} \frac{dy}{u(y)^2} = \infty \text{ as } c \nearrow 0,$$

since $u(0)=0$ and $|u'(0)| < \infty$. Let $f : [0,P_c] \to \mathbb{R}$ be such that $f(0)=0$ and substitute

$$v(y) = \int_{-d}^{y} \frac{dt}{(u(t)+c)^2}.$$

Then $v(-d)=0$ and, when substituted in (5.2), we find that

(5.8) $$\begin{aligned}\lambda_1(c) &\leq \frac{P_c - gP_c^2}{\displaystyle\int_{-d}^{0} (u(y)+c)^2 \left(\int_{-d}^{y} \frac{dt}{(u(t)+c)^2}\right)^2 dy} \\ &= \frac{P_c^{-1} - g}{\displaystyle\int_{-d}^{0} (u(y)+c)^2 \left(\frac{1}{P_c}\int_{-d}^{y} \frac{dt}{(u(t)+c)^2}\right)^2 dy}.\end{aligned}$$

Now,

$$1 \geq \frac{1}{P_c}\int_{-d}^{y} \frac{dt}{(u(t)+c)^2} \to 0 \text{ for } y \in [-d,1) \text{ as } c \nearrow 0.$$

by (5.7), and hence $\lambda_1(c) \to -\infty$ as $c \nearrow 0$, by (5.7) and the dominated convergence theorem. It follows from (5.4) that $\lambda_1(c) \leq 0$ if and only if $P_c g \geq 1$, which is true for all $c<0$ with $|c|$ sufficiently small, by (5.7).

Finally note from (5.6) that $\lambda_1(c) > 0$ for all $|c|$ sufficiently large, since $gP_c \to 0$ as $|c| \to \infty$. Since $\lambda_1(c)$ obviously depends continuously on $c<0$, the result follows. □

To consider the behaviour of $\lambda_1(c)$ for positive c suppose that

$$u(-d) < u(y) \text{ for all } y \in (-d, 0] \text{ and let } \underline{u} = -u(-d) > 0. \tag{5.9}$$

Note that

$$P_c \to \infty \text{ as } c \searrow \underline{u}. \tag{5.10}$$

We now restrict attention to $c \in (\underline{u}, \infty)$.

LEMMA 5.3. *Suppose that* (5.9) *holds.*
(a)

$$\liminf_{c \searrow \underline{u}} \lambda_1(c) \leq \frac{-g}{\|u + \underline{u}\|^2_{L^2(-d,0)}}. \tag{5.11}$$

(b) $\lambda_1(c) > 0$ *for all* $c > 0$ *sufficiently large.*
(c) Suppose that $-k^2 > \liminf_{c \searrow \underline{u}} \lambda_1(c)$, $k \in \mathbb{N}$. *There exists* $c_k^+ > \underline{u}$ *such that* $\lambda_1(c_k^+) = -k^2$.

PROOF. Since $P_c \to \infty$ as $c \searrow \underline{u}$,

$$1 \geq \frac{1}{P_c} \int_{-d}^{y} \frac{dt}{(u(t) + c)^2} \to 1 \text{ as } c \searrow \underline{u} \text{ for all } y \in (-d, 0],$$

and (5.11) follows from (5.8). As in the preceding proof, $\lambda_1(c) > 0$ for all c sufficiently large. □

REMARK. An example of Lemma 5.3 (a) arises in the problem of bifurcation of waves on flows of constant vorticity with $d = 1$, in which case $u(y) = \omega_0 y$, $\omega_0 \in \mathbb{R}$. When $\omega_0 > 0$ the hypotheses of Lemma 5.2 are satisfied. Moreover, in Lemma 5.3, $\underline{u} = \omega_0$ and $\|u + \underline{u}\|^2_{L^2(-1,0)} = \omega_0^2/3$. Hence $-k^2 = \lambda_1(c)$ for some $c > \underline{u} > 0$ is an eigenvalue if $\omega_0^2 k^2 < 3g$.

In fact, for this example we have seen from an explicit calculation that $-k^2 = \lambda_1(c)$ for some $c > \underline{u}$ if and only if $\omega_0^2(1 - k^{-1}\tanh k) < gk^{-1}\tanh k$. (In particular, $-1 = \lambda_1(c)$ for some $c > \omega_0$ if and only if $\omega_0^2 < 3.194g$ (which may be compared with $\omega_0^2 < 3g$, the criterion in the Lemma). □

THEOREM 5.4. *The parameter values* $c_k^{\pm}$ *in the preceding two lemmas are bifurcation points for problem* (2.7) *when* $U(z\,; c) = u(z) + c$.

PROOF. We have shown that (4.2a) is satisfied when $c = c_k^{\pm}$, and (4.2c) follows by the self-adjointness of (4.3) in an L^2 setting. Since $\partial U/\partial c(0; c_k^{\pm}) = 1$, it follows from (4.5) that $\widehat{\kappa} = 0$ if (4.2b) is false. Hence hypothesis (4.2) is satisfied, and it follows that the eigenvalues $c_k^{\pm}$ are bifurcation points for the problem of waves on a running stream. □

References

[1] A. J. Abdullah, Wave motion at the surface of a current which has an exponential distribution of vorticity, *Annals New York Acad. Sci.* **51** (1949), 425 – 441.

[2] H. W. Alt and L. A. Caffarelli, Existence and regularity for a minimum problem with free boundary, *J. Reine Angew. Math.*, **325** (1981), 105-144.

[3] T. B. Benjamin, The solitary wave on a stream with an arbitrary distribution of vorticity, *J. Fluid Mech.* **12** (1962), 97 –116.

[4] B. Buffoni and J. F. Toland, *Analytic Theory of Global Bifurcation - An Introduction.* Princeton University Press, Princeton, N. J., 2003.

[5] A. Constantin and W. Strauss, Exact steady periodic water waves with vorticity, *Comm. Pure Appl. Math.*, Vol. **LVII** (2004), 481-527.
[6] A. Constantin and W. Strauss, Exact periodic traveling water waves with vorticity, *C. R. Math. Acad. Sci. Paris* **335** (10) (2002), 797-800.
[7] A. Constantin and W. Strauss, Stability properties of steady water waves with vorticity, *Comm. Pure Appl. Math.*, Vol. **LX** (2007), 911-950.
[8] M. G. Crandall and P. H. Rabinowitz, Bifurcation from a simple eigenvalue, *J. Funct. Anal.* **8** (1971), 321–340.
[9] R. A. Dalrymple and J. C. Cox, Symmetric finite-amplitude rotational water waves, *J. Physical Oceanography*, **6** (1976), 847–852.
[10] I. I. Daniliuk, On integral functionals with a variable domain of integration, *Proc. Steklov Inst. Math.* **118** (1972). In English, Amer. Math. Soc. (1976).
[11] E. B. Davies, *Spectral Theory and Differential Operators.* Cambridge University Press, Cambridge, 1995.
[12] M.-L. Dubreil-Jacotin, Sur la détermination rigoureuse des ondes permanentes périodiques dampleur finie. *J. Math. Pures Appl.* **13** (1934), 217291.
[13] D. Gilbarg and N. S. Trudinger, *Elliptic Partial Differential Equations of Second Order.* 2nd Edition. Springer, New York, 1983.
[14] J. N. Hunt, Gravity waves in flowing water, *Proc. R. Soc. London* A, **231** (1955), 496-504.
[15] V. M. Hur and M. Lin, Unstable surface waves in running water, ArXiv 0708:0541V1 [Math.AP] 3rd Aug 2007, to appear.

UNIVERSITÄT AUGSBURG, INSTITUT FÜR MATHEMATIK, UNIVERSITTSSTRASSE 14, 86159 AUGSBURG, GERMANY

Current address: Department of Mathematical Sciences, University of Bath, Claverton Down, Bath BA2 7AY, UK

E-mail address: lilli@math.uni-augsburg.de

DEPARTMENT OF MATHEMATICAL SCIENCES, UNIVERSITY OF BATH, CLAVERTON DOWN, BATH BA2 7AY, UK

E-mail address: jft@maths.bath.ac.uk

Proceedings of Symposia in Pure Mathematics
Volume **79**, 2008

On analytic capacity of portions of continuum and a question of T. Murai

Fedor Nazarov and Alexander Volberg

This paper is dedicated to Vladimir Maz'ya

ABSTRACT. We give an answer to an old question of T. Murai concerning the characterization of the boundednesss of the Cauchy integral operator on arbitrary sets of finite Hausdorff length. If the set is a continuum, we got a new proof to a theorem of Guy David characterizing the rectifiable curves on the plane for which the Cauchy integral operator is bounded on $L^2(ds)$. In doing that we use also a nonhomogeneous version of a certain Tb theorem first proved by M. Christ in homogeneous spaces. We are going to "compute" in metric terms the analytic capacity of the intersection of an arbitrary continuum and a half-plane (or a disc, or any domain with piecewise smooth boundary).

1. Introduction

Takafumi Murai asked in [**Mu**] the following question:

given a compact set $E \subset \mathbb{C}$ such that its Hausdorff 1-dimensional measure satisfies $0 < H^1(E) < \infty$, is that true that Cauchy integral operator is bounded in $L^2(E, H^1|E)$ if and only if $H^1(E \cap Q) \leq C\,\gamma(E \cap Q)$.

Here γ stands for analytic capacity defined in the next paragraph.

We give a positive answer to this question here.

THEOREM 1.1. Let E be a compact on the complex plane such that $0 < H^1(E) < \infty$. Then the Cauchy integral operator T (and also T^*) is bounded on $L^2(E, H^1)$ if and only if there exists a constant C such that for every square Q on the plane

$$H^1(E \cap Q) \leq C\,\gamma(E \cap Q)\,. \tag{1.1}$$

2000 *Mathematics Subject Classification.* Primary 47B36; Secondary 42C05.

Key words and phrases. Analytic capacity, Hausdorff content, nonhomogeneous harmonic analysis, accretive functions.

The first author was supported in part by NSF Grant 0501067.

The second author was supported in part by NSF Grant 0501067.

Definition. Let K be a compact set in $\mathbb{C}$.

$$\gamma(K) := \sup\{\lim_{z\to\infty} |z\, f(z)| : f \in \mathrm{Hol}(\mathbb{C}\setminus K), |f(z)| \le 1\ \forall z \in \mathbb{C}\setminus K,\ f(\infty) = 0\},$$

$$\gamma_+(K) := \sup\{\lim_{z\to\infty} |z\, f(z)| : f(z) = \int \frac{d\mu(\zeta)}{z-\zeta},\ \mu \in M_+(K), |f(z)| \le 1\ \forall z \in \mathbb{C}\setminus K\}.$$

By definition,

$$\gamma_+(E) \le \gamma(E). \tag{1.2}$$

In [**T4**] Tolsa proved that the opposite inequality also holds with absolute constant. It is a very tough theorem. We discuss its relations with results of this paper in the last section.

A natural question arises: how verifiable is this criterion? Strangely enough it is sometimes verifiable, and this is the second main topic of this article. In Theorem 1.3 we compute (up to the absolute constant) the analytic capacity of certain class of sets. This allows us to observe in Section 3 that a famous theorem of Guy David is a one-line consequence of the above criterion (1.1). David's theorem we are referring to is the one that characterizes all rectifiable curves on the plane on which the Cauchy integral operator is bounded.

Let us recall that there is another criterion of the boundedness of Cauchy integral. It is obtained in [**NTV2**], [**T1**] and we want to formulate it now. To do that we need to recall the reader the notion of Menger's curvature of a measure. Given three points $z_1, z_2, z_3 \in \mathbb{C}$ we call $R(z_1, z_2, z_3)$ the radius of the circle (may be ∞) passing through those points. Then Menger's curvature of a positive measure μ is by definition

$$c_2(\mu) := \left(\int\int\int \frac{d\mu(z_1)d\mu(z_2)d\mu(z_3)}{R^2(z_1,z_2,z_3)}\right)^{\frac{1}{2}}.$$

If $\mu = H^1|E$ for a certain compact $E, 0 < H^1(E) < \infty$, we will use the following notations:

$$c_2(E) := c_2(H^1|E).$$

We are ready to quote the criterion proved by Nazarov-Treil-Volberg in [**NTV2**] and Tolsa in [**T1**].

THEOREM 1.2. 1) Cauchy integral operator is bounded in $L^2(\mu)$ if and only if there exists a finite constant C such that for every square Q

$$\mu(Q) \le C\,\ell(Q), \tag{1.3}$$

where $\ell(Q)$ is the length of the side of Q, and

$$c_2(\mu|Q)^2 \le C\,\mu(Q). \tag{1.4}$$

2) In particular, if $\mu = H^1|E$, for a certain compact $E, 0 < H^1(E) < \infty$, then the boundedness of the Cauchy integral operator in $L^2(E, H^1)$ is equivalent to

$$H^1(E\cap Q) \le C\,\ell(Q), \tag{1.5}$$

and

$$c_2(E\cap Q)^2 \le C\,H^1(E\cap Q). \tag{1.6}$$

Remark. Actually one can sometimes skip assumption (1.3) as Tolsa has shown in Lemma 5.2 of [**T5**]. This is the case for measures μ such that their upper density

$$\limsup_{r\to 0} \mu(B(x,r))/r$$

is uniformly bounded. We are grateful to the referee for this remark.

We will show below that Theorem 1.1 implies easily part 2) of Theorem 1.2. On the other hand, one can deduce Theorem 1.1 from Theorem 1.2, but this requires a much more efforts. This deduction is based on already mentioned very tough result of Tolsa [**T4**]. This deduction is briefly discussed in the last section of this article.

Another interesting feature of our criterion (1.1) is that one can prove its *multi-dimensional analogs*, however, the multi-dimensional analogs of criterion (1.6) from [**NTV2**], [**T1**] are not known now (because they involve the notion of Menger's curvature that did not yet get multi-dimensional understanding).

1.1. Theorem 1.1 implies easily the second part of Theorem 1.2. The difficult part is to prove that (1.5), (1.6) imply the boundedness in $L^2(E, H^1|E)$ of the Cauchy integral operator. We want to use Theorem 1.1. So for our goals it is sufficient to prove the following implication (Q is always a square)

$$(1.7)\quad \forall Q \;\; c_2(E\cap Q)^2 \le C_1\, H^1(E\cap Q) \le C_2\ell(Q) \Rightarrow \;\; H^1(E\cap Q) \le C_2\gamma(E\cap Q)\,.$$

To prove (1.7) we need the trivial inequality $\gamma \ge \gamma_+$ and the following characterization of γ_+ due to Melnikov (see e.g. [**T1**], it can be found in [**Vo**] also):

$$(1.8)\qquad \gamma_+(K) \asymp \sup_{\mu:\, \mu(B(x,r))\le r\, \forall r\, \forall x} \frac{\|\mu\|^2}{c_2(\mu)^2 + \|\mu\|}$$

Obviously the second inequality in the left hand side of (1.7) implies that measure $H^1|E$ satisfies the growth condition:

$$H^1(E\cap B(x,r)) \le C_1\, r\,.$$

Hence, using (1.8) for test measure $\mu := C_1^{-1}\, H^1|E\cap Q$ we obtain

$$\gamma(E\cap Q) \ge \gamma_+(E\cap Q) \ge a\,\frac{H^1(E\cap Q)^2}{c_2(E\cap Q)^2 + H^1(E\cap Q)} \ge a\, H^1(E\cap Q)\,.$$

We got the right hand side of (1.7), which is the reduction we wanted.

1.2. David's characterization of bounded Cauchy integral operator on rectifiable curves: Ahlfors-David curves. If the set E is a rectifiable curve Γ, then a theorem of Guy David describes all such curves on the plane for which the Cauchy integral operator is bounded on $L^2(\Gamma, d\,H^1)$. This is the class of curves Γ (called *Ahlfors-David curves*) satisfying

$$(1.9)\qquad H^1(D(x,R)\cap\Gamma)) \le C\,R$$

for any disc $D(x,R)$. Lipschitz curves and Lavrentiev (chord-arc) curves give us the examples satisfying (1.9). It is slightly strange that for Lipschitz curves (and even for chord-arc curves) there exists a purely analytic proof of the boundedness of the Cauchy operator on $L^2(\Gamma, d\,H^1)$ (see [**CJS**], [**Chr**]), but all the proofs of the theorem of David are the mixtures of analytic and geometric arguments, [**Da1**],

[**DaJ**]. In the present paper we are going to show, in particular, that Guy David's characterization follows from two ingredients:
a) Theorem 1.1,
b) a simple computation in geometric terms of $\gamma(\Gamma \cap Q)$, where Q is a square and Γ is an arbitrary continuum, Theorem 1.3 belows.

In the present paper our main idea is to use a local Tb theorem of M. Christ [**Chr**] (and not the usual Tb theorem or $T1$ theorem). The difference with [**Chr**] is that we have to use a nonhomogeneous version of a local Tb theorem.

This nonhomogeneous version of Christ's local Tb theorem leads us naturally to the "computation" of the analytic capacity of the intersection of our E with a square (or a disc). It is quite well understood now that the analytic capacity of an arbitrary compact cannot be measured in terms of simple geometric characteristics of the compact. The result of Tolsa [**T4**] (see also the exposition of this result in [**Vo**] together with its multi-dimensional analogs) only confirms this because the analytic capacity is proved to be computable in metric terms, but only in quite complicated ones. See also, for example, [**JM**], where it is shown that the Buffon needle probability cannot serve as a metric equivalent of analytic capacity in general. Surely, one can derive that on compact subsets of an Ahlfors-David curve the analytic capacity is equivalent to just H^1 measure. We will discuss this later, but now let us notice that the equivalence constants are not absolute. They depend on the geometry of the ambient curve. Secondly, to establish this equivalence one needs heavy tools: either the theorem of David or geometric arguments of Jones [**J**] and Melnikov [**Me2**].

However, to our surprise, there exists a class of compacts for which one can get the simple geometric measurement equivalent to the analytic capacity up to *absolute constants*. And this class is large enough to enable us to use our version of Tb theorem resulting in a new proof of the theorem of David.

This class consists of intersections of any continuum with any closed half-plane (or any closed disc, or any closed square,...).

Let us introduce some notations. In what follows, a, a', a'', A denote various positive finite absolute constants. Letters Π, Q and D stand for various half-planes, squares and discs respectively. We will use the symbol $h^1(E)$ to denote 1-dimensional Hausdorff content of E, namely,

$$h^1(E) := \inf\{\sum r_j \;:\; E \subset \cup_j D(x_j, r_j)\}\,.$$

It is clear that H^1 is larger than h^1, and they vanish simultaneously.

For any continuum Γ the Hausdorff content is equivalent with the diameter. But the same is true for the analytic capacity $\gamma(\Gamma)$. Thus, for a continuum Γ

$$a\,h^1(\Gamma) \leq \gamma(\Gamma) \leq A\,h^1(\Gamma)\,. \tag{1.10}$$

We are going to prove (1.10) for sets $\Gamma \cap \Pi$, $\Gamma \cap Q$, and $\Gamma \cap D$. We restrict ourselves to the the case of half-planes.

THEOREM 1.3. Let Π be a closed half-plane, and let Γ be a continuum. Then

$$a\,h^1(\Gamma \cap \Pi) \leq \gamma(\Gamma \cap \Pi) \leq A\,h^1(\Gamma \cap \Pi)\,. \tag{1.11}$$

We prove this result below. Our proof was superseded by the proof by John Garnett [**JG**] that is considerably easier than ours. But we still decided to present our proof for the possible future generalizations to higher dimension. The simpler Garnett's proof relies on complex analysis observations.

2. The Cauchy integral operator on sets E, $0 < H^1(E) < \infty$, and nonhomogeneous accretive system Tb theorem

Let us remind that the function $K(x,y)$, $x, y \in \mathbb{R}^n$ is called a Calderón-Zygmund kernel of dimension m if there exist finite and positive constants C, ε such that

$$|K(x,y)| \leq C\text{dist}(x,y)^{-m}\,,$$

$$|K(x,y) - K(x,y')| + |K(y,x) - K(y',x)| \leq \frac{C\text{dist}(y,y')^{\varepsilon}}{\text{dist}(x,y)^{m+\varepsilon}}$$

whenever $\text{dist}(y,y') \leq 1/2\text{dist}(x,y)$.

Constants C, ε are called Calderón-Zygmund constants of the kernel K. A special class of Calderón-Zygmund kernels are antisymmetric Calderón-Zygmund kernels, namely, such that $K(x,y) = -K(y,x)$. For a given antisymmetric Calderón-Zygmund kernel K and a positive measure μ with compact support such that $\mu(D(x,r)) \leq C\,r^m$ one can define an operator T with Calderón-Zygmund kernel K as in [**DaJ**]. One can also define *a maximal singular operator* as follows

$$T^*f(x) := \sup_{\varepsilon>0}\left|\int_{y:|y-x|>\varepsilon} K(x,y)f(y)d\mu(y)\right|$$

THEOREM 2.1. Let μ be a measure with compact support in $\mathbb{R}^n$. Let K be a Calderón-Zygmund kernel of dimension m, and let for any disc $D(x,r)$

$$\mu(D(x,r)) \leq C\,r^m\,.$$

Let $K(x,y) = -K(y,x)$. We denote by T a Calderón-Zygmund operator with kernel K and assume that there exist finite positive constants B, δ such that for every cube Q in $\mathbb{R}^n$ with $\mu(Q) > 0$, we have a function b_Q such that

$$\text{supp}\, b_Q \subset Q\,,\ \|b_Q\|_\infty \leq 1\,,$$

$$|\langle b_Q\rangle_Q| := \frac{1}{\mu(Q)}|\int_Q b_Q\, d\mu| \geq \delta\,,$$

$$\|T^* b_Q\|_\infty \leq B$$

Then T^* is a bounded operator on $L^2(\mu)$, and its norm depends only on B, δ, n and the Calderón-Zygmund constants of the kernel K.

The system of functions $\{b_Q\}$ as above is called *local* accretive system. The statement imitates the statement of the theorem of M. Christ from [**Chr**], but the proof requires considerations from the nonhomogeneous harmonic analysis (see [**NTV3**]).

We are ready to prove Theorem 1.1.

PROOF. First, we are going to reformulate (3.1) in terms of the Cauchy transform with respect to measure H^1 on E. So we call

$$Tf := \int_E \frac{f(\zeta)\, dH^1}{\zeta - z}$$

in the sense of Calderón-Zygmund operators, see [**Da1**] or [**DaJS**]. We also denote

$$T^* f(z) := \sup_{\varepsilon>0} \left| \int_{\zeta\in E\,,|\zeta-z|>\varepsilon} \frac{f(\zeta)\, dH^1}{\zeta - z} \right|$$

This is called maximal Cauchy operator.

LEMMA 2.2. Let E satisfy (3.1). Then there exist positive finite constants K, δ such that for any square Q with $H^1(Q\cap E) > 0$ there exists a function b_Q so that

$$\operatorname{supp} b_Q \subset Q\cap E\,,\ \ \|b_Q\|_\infty \le 1\,,$$
$$|\langle b_Q\rangle_Q| := \frac{1}{H^1(Q\cap E)}\Big|\int_{Q\cap E} b_Q\, dH^1\Big| \ge \delta\,,$$
$$\|T^* b_Q\|_\infty \le K\,.$$

The constants K, δ depend only on the constant in (3.1).

PROOF. Fix a square Q with $H^1(Q\cap E) > 0$. From (3.1) we conclude that there exists a holomorphic function $f_Q \in H^\infty(\mathbb{C}\setminus(E\cap Q))$, such that $\|f_Q\|_\infty \le 1$ and such that $f_Q(\infty) = 0, |f_Q'(\infty))| \ge k\cdot H^1(E\cap Q)$. Clearly (because $E\cap Q$ has a finite H^1 measure), f_Q can be represented as a Cauchy integral of a measure $b_Q\, dH^1$ on $E\cap Q$, where $b_Q \in L^\infty(dH^1)$, $\|b_Q\|_\infty \le 1$. We also know that $f_Q'(\infty) = \int_{E\cap Q} b_Q(\zeta) dH^1(\zeta)$, and thus

$$|\int_{E\cap Q} b_Q\, dH^1| \ge k\cdot H^1(E\cap Q) \tag{2.1}$$

Notice that the first two claims of our theorem are already proved (the second claim follows from (2.1)). To prove the third claim we use the following standard estimates. Fix a point $z\in\mathbb{C}$ and $\varepsilon > 0$. Consider

$$f_{Q,\varepsilon,z}(w) = \int_{\zeta\in E,|\zeta-z|>\varepsilon} \frac{b_Q(\zeta) dH^1(\zeta)}{\zeta - w}$$

This function is holomorphic in $D(z,\varepsilon)$. In particular, putting $\rho = \varepsilon/2$, we get

$$f_{Q,\varepsilon,z}(w) = \frac{1}{\pi\rho^2}\int_{D(z,\rho)} f_{Q,\varepsilon,z}(w) dA(w)$$
$$= \frac{1}{\pi\rho^2}\int_{D(z,\rho)} f_Q(w)\, dA(w) - \frac{1}{\pi\rho^2}\int_{D(z,\rho)}\int_{\zeta\in E\cap D(z,\varepsilon)} \frac{b_Q(\zeta) dH^1(\zeta)}{\zeta - w}\, dA(w)$$
$$= t_1 + t_2\,.$$

The first term is good because it is an average of a bounded function:

$$|t_1| \le \frac{1}{\pi\rho^2}\int_{D(z,\rho)} |f_Q(w)| dA(w) \le \|f_Q\|_\infty \le 1\,.$$

In the second term we put the absolute value inside and interchange the order of integration:

$$\begin{aligned}|t_2| &\le \frac{1}{\pi\rho^2}\int_{E\cap D(z,\varepsilon)} dH^1(\zeta)\int_{D(z,\rho)}\frac{1}{|\zeta-w|}dA(w)\\ &\le \frac{a}{\rho}\int_{E\cap D(z,\varepsilon)} dH^1(\zeta) \le \frac{a}{\rho}H^1(E\cap D(z,\rho))\\ &\le \frac{a}{\rho}H^1(E\cap Q(z,\rho)) \le \frac{a}{c\,\rho}\gamma(E\cap Q(z,\rho))\\ &\le \frac{a}{c\,\rho}\gamma(Q(z,\rho)) \le A\,C\,a\,,\end{aligned}$$

where c denotes the constant from (3.1), $C = c^{-1}$, and $Q(z,\rho)$ denotes the smallest square with the same center as the disc $D(z,\rho)$ and containing this disc. We used here (3.1) and an obvious estimate that the analytic capacity is bounded by the diameter of a set. Lemma 2.2 is proved. □

Now one can apply this Theorem 2.1 to the kernel $K(\zeta,z) = \frac{1}{\zeta-z}$ and measure $\mu = H^1|E$. Of course $m = 1$ now, and the assumption

$$\mu(B(z,r)) = H^1(B(x,r)\cap E) \le C\,r$$

follows from $H^1(B(x,r)\cap E) \le C\gamma(B(x,r)\cap E) \le C\gamma(B(x,r)) \le Cr$ by the obvious estimate that the analytic capacity is bounded by the diameter of a set. In Lemma 2.2 we just proved the existence of a local of functions required in the assumptions of Theorem 2.1.

To finish the proof of our main Theorem 1.1 we just need to apply Theorem 2.1. Theorem 1.1 is completely proved.

□

3. Application: the Cauchy integral operator on Ahlfors-David curves and nonhomogeneous accretive system Tb theorem

THEOREM 3.1. Let C be an Ahlfors-David curve. There exist positive finite constants b, B such that

$$b\,H^1(C\cap Q) \le \gamma(C\cap Q) \le B\,H^1(C\cap Q) \tag{3.1}$$

PROOF. The right inequality is well known. The proof of the left inequality is the corollary of Theorem 1.3. In fact, let $\{D(x_j,r_j)\}$ be a covering of $C\cap Q$ such that $\sum r_j \le 2\,h^1(C\cap Q)$. By the Ahlfors-David property of C, one has

$$H^1(C\cap Q) \le \sum_j H^1(C\cap D(x_j,r_j)) \le B\sum r_j \le 2B\,h^1(C\cap Q)\,.$$

We combine this with (1.11) (for squares Q instead of half-planes Π) and we get (3.1).

□

THEOREM 3.2. If C is an Ahlfors-David curve, then T^* is bounded in $L^2(C,dH^1)$. Conversely, if T^* is a bounded operator in $L^2(C,dH^1)$, where C is a rectifiable curve, then C is an Ahlfors-David curve.

PROOF. In Theorem 3.1 we checked the assumption (3.1). We are now in a position to apply Theorem 1.1 and to get the result. □

Notice that one can replace "if" by "if and only if" in the statement above. But "only if" part is simple and standard.

The original proof of this result [**Da1**] was the mixture of analysis and geometry. Notice that in [**Chr**] another proof of David's theorem has been already given. The proof of [**Chr**] involves a hard geometric construction of "dyadic cubes" on arbitrary space of homogeneous type. We avoid this by working with the usual dyadic squares on the plane, we do not build special "squares" adapted to a problem.

Another proof can be derived from the ideas of Jones [**J**] in conjunction with the Menger's curvature characterization of $L^2(\mu)$ boundedness of the Cauchy operator from [**NTV2**], [**T1**].

It seems to us that our just given proof of David's theorem is somewhat less involved than the other ones we mentioned. It is built on computation of analytic capacity of the portion of a continuum inside an arbitrary square (Theorem 1.3) and on local Tb theorem of Christ [**Chr**] (only the nonhomogeneous version is needed—Theorem 2.1).

We are left to prove Theorem 1.3.

4. Analytic capacity of portions of continuum. The proof of Theorem 1.3

All our considerations will be made for the intersection of a continuum with a half-plane. This is done for the sake of brevity, but one can replace the half-plane by a square or a disc. Of course this will change the absolute constants involved in the estimate, but this will be the only change.

The right inequality in (1.11) is standard. In fact, let $\{D_j\}$ be a finite family of discs of radii r_j covering continuum Γ and such that $\sum r_j \le 2\, h^1(\Gamma)$. Let $K = \cup_j \bar{D}_j$. We use

$$\gamma(K) \le A\, H^1(\partial K) \tag{4.1}$$

and the monotonicity of γ to conclude $\gamma(\Gamma) \le \gamma(K) \le A\, H^1(\partial K) \le A \sum r_j \le 2A\, h^1(\Gamma)$.

In the rest of this section we prove the left inequality of (1.11). Let Γ be a continuum, and $\Pi = \{z : \mathrm{Im} z \ge 0\}$ be the upper half-plane. All our constants will be absolute. So we can assume without the loss of generality that Γ is a real analytic curve. Then the intersection $\Gamma \cap \Pi$ consists of $\Gamma_1, ..., \Gamma_n$ pieces, and $\Gamma_j \cap \Pi \ne \emptyset$, $j = 1, ..., n$.

Let $y_j \in \Gamma_j \cap \mathbb{R}$, $r_j = \mathrm{diam}(\Gamma_j)$, $j = 1, ..., n$. Obviously $h^1(\Gamma \cap \Pi) \le \sum r_j$. Consider $\{D(y_j, r_j)\}$ and choose a subfamily $\{D(y_j, r_j)\}_{j \in \mathcal{F}}$ such that

$$20\, D(y_k, r_k) \cap 20\, D(y_m, r_m) = \emptyset,\ \forall k \ne m,\ k, m \in \mathcal{F}\,, \tag{4.2}$$

$$\Gamma \cap \Pi \subset \cup_{m \in \mathcal{F}} 100\, D(y_m, r_m)\,. \tag{4.3}$$

Clearly,

$$\frac{1}{2} \sum_{m \in \mathcal{F}} r_m \le h^1(\Gamma \cap \Pi) \le 100 \sum_{m \in \mathcal{F}} r_m\,. \tag{4.4}$$

Let $\gamma_i := \gamma(\Gamma_i)$. For every $i \in \mathcal{F}$ let f_i denote a function from $H^\infty(\mathbb{C} \setminus \Gamma_i)$ such that $\|f_i\|_\infty \leq 1$ and such that

$$f_i(\infty) = 0,\ f_i^{'}(\infty) = \gamma_i\,. \tag{4.5}$$

Clearly $1/4\, r_i \leq \gamma_i \leq 2\, r_i$. Let us fix an interval $I_i := [y_i - 1/2\, L_i, y_i + 1/2\, L_i]$ such that $\gamma(I_i) = \gamma_i$. Obviously, $\gamma_i \leq \frac{L_i}{4}$. Thus,

$$r_i \leq L_i \leq 8\, r_i,\ i \in \mathcal{F}\,. \tag{4.6}$$

Let $L = \sum_{i \in \mathcal{F}} L_i$. For every $i \in \mathcal{F}$ let g_i denote a function from $H^\infty(\mathbb{C} \setminus I_i)$ such that $\|g_i\|_\infty \leq 1$ and such that

$$g_i(\infty) = 0,\ g_i^{'}(\infty) = \gamma_i\,. \tag{4.7}$$

Writing

$$g_i(z) = \int_{I_i} \frac{g_i(x)dx}{x - z}\,,$$

we can assume from symmetry that $g_i(x)$ is a real valued function. We also know that $\int g_i(x)dx = \gamma_i$.

Let $\hat{\mathbb{C}}$ denote $\mathbb{C} \cup \infty$ and let $\Omega := \hat{\mathbb{C}} \setminus \cup_{i \in \mathcal{F}} D(y_i, 10\, r_i)$. This is a compact set on the Riemann sphere. We consider $C_{\mathbb{R}}(\Omega)$, the class of real-valued continuous function on Ω. Denote

$$V_1 := \{f \in C_{\mathbb{R}}(\Omega) : \|f\| \leq 1\}\,,$$

$$V_2 := \{\sum_{i \in \mathcal{F}} c_i \mathrm{Im} f_i(z) \,:\, \sum_{i \in \mathcal{F}} c_i \gamma_i \geq \frac{1}{10^6} L,\ c = \{c_i\}_{i \in \mathcal{F}} \in l^\infty_{\mathbb{R}},\ \|c\|_\infty \leq 1\}\,.$$

Suppose that

$$V_1 \cap V_2 \neq \emptyset\,. \tag{4.8}$$

Then we will show that the left inequality in (1.11) holds.

In fact, if we have (4.8) then there exists a collection $c = \{c_i\}_{i \in \mathcal{F}}$ such that the function $f(z) := \sum c_i f_i(z)$ satisfies $|\mathrm{Im} f| \leq 1$ in Ω. But f is a holomorphic function in $\mathbb{C} \setminus \cup_{i \in \mathcal{F}} \Gamma_i$. Fix $i \in \mathcal{F}$. On $\partial D(y_i, 10 r_i)$ we see that $|c_i f_i|$ is bounded by 1. Thus $|\mathrm{Im} \sum_{j \in \mathcal{F}, j \neq i} c_j f_j|$ is bounded by 2 on this circle, and, hence, on the whole disc $D(y_i, 10 r_i)$, where $|c_i f_i|$ is bounded by 1 in its turn. Therefore, $|\mathrm{Im}\, \sum_{j \in \mathcal{F}} c_j f_j| = |\mathrm{Im} f|$ is bounded by 3 in $D(y_i, 10 r_i)$. And this is so for every $i \in \mathcal{F}$. Thus $|\mathrm{Im} f|$ is bounded by 3 everywhere in $\mathbb{C} \setminus \cup_{i \in \mathcal{F}} \Gamma_i$. On the other hand, $f^{'}(\infty) = \sum_{j \in \mathcal{F}} c_j \gamma_j \geq \frac{1}{10^6} L$. We conclude

$$\gamma(\cup_{i \in \mathcal{F}} \Gamma_i) \geq \frac{1}{10^6} L\,. \tag{4.9}$$

In particular, we obtain the left inequality of (1.11):

$$\gamma(\Gamma \cap \Pi) \geq \gamma(\cup_{i \in \mathcal{F}} \Gamma_i) \geq a'\, L \geq a''\, h^1(\Gamma \cap \Pi)\,.$$

We are left to prove that (4.8) holds.

5. Maximal functions, weak type inequalities, and the proof of (4.8)

Suppose (4.8) does not hold. We are going to come to a contradiction. There exists a real measure ν on Ω such that

$$\|\nu\| \le 1\,,$$

but

$$\text{(5.1)} \qquad \operatorname{Im} \int_\Omega (\sum_{i \in \mathcal{F}} c_i f_i(z))\, d\nu(z) > 1$$

for all collections $c = \{c_i\}_{i\in\mathcal{F}} \in l^\infty_{\mathbb{R}}$, $\|c\|_\infty \le 1$, $\sum_{i\in\mathcal{F}} c_i\gamma_i \ge \frac{1}{10^6} L$.

Now we are going to find such a collection c, for which (5.1) fails. This collection will consist only of 1's and 0's. Define for $x \in \mathbb{R}$

$$(M_1\nu)(x) := \sup_{r>0} \frac{|\nu|(D(x,r))}{r}\,.$$

Let us define another maximal function on $I = \cup_{i\in\mathcal{F}} I_i$. Namely, if $x \in I_i$, $i \in \mathcal{F}$

$$\nu^*(x) := \sup_{r>3\,r_i} \frac{|\nu|(D(x,r))}{r}\,.$$

The segment I_i is well inside $D(y_i, 10\,r_i)$ and ν is outside of it, therefore

$$\sup_{x\in I_i} \nu^*(x) \le \inf_{x\in I_i} (M_1\nu)(x)\,.$$

Consider $\mathcal{F}_0 \stackrel{\text{def}}{=} \{i \in \mathcal{F} : \exists x \in I_i, \nu^*(x) \ge \frac{100}{L}\}$.

On $I_0 := \cup_{i\in\mathcal{F}_0} I_i$ we have (this is just (4.2))

$$\text{(5.2)} \qquad M_1\nu \ge \frac{100}{L}$$

On the other hand,

$$|\{x : (M_1\nu)(x) \ge \frac{100}{L}\}| \le \frac{5L}{100}\|\nu\| = \frac{L}{20}\,.$$

This is a usual weak type result.

Consider $\mathcal{F}_1 := \mathcal{F} \setminus \mathcal{F}_0 = \{i \in \mathcal{F} : \forall x \in I_i,\ \nu^*(x) \le \frac{100}{L}\}$. We conclude from (5.2) and the last inequality that ($J_1 \stackrel{\text{def}}{=} \cup_{i\in\mathcal{F}_1} I_i$) :

$$\text{(5.3)} \qquad |J_1| \ge \frac{19}{20} L\,, \text{and}\,, \forall x \in J_1\,, \nu^*(x) \le \frac{100}{L}\,.$$

Without the loss of generality we can think that

$$\text{(5.4)} \qquad \forall i \in \mathcal{F}_1\,, \gamma_i \le \frac{1}{20000} L\,.$$

In fact, if we have the opposite inequality for a certain i_0, then $V_1 \cap V_2 \ne \emptyset$ because we choose just $c_i = 1$ if $i = i_0$, $c_i = 0$ otherwise, and see that $f = \sum c_i f_i = f_{i_0}$ belongs to $V_1 \cap V_2$. But we assumed $V_1 \cap V_2 = \emptyset$. Let us call a subset $\mathcal{F}'$ of $\mathcal{F}_1$ admissible if the following holds

$$\frac{1}{20000} L \le \sum_{i\in\mathcal{F}'} \gamma_i \le \frac{1}{5000} L\,.$$

By (5.3), (5.4) there are plenty of admissible subsets. We call $c = \{c_i\}_{i\in\mathcal{F}}$ admissible if $c_i = 1$ on a certain admissible subset $\mathcal{F}'$ of indices and $c_i = 0$ otherwise. In particular, $c_i = 0$ for all i in $\mathcal{F}_0$. Let $\mathcal{F}'$ be admissible. Consider

$$\operatorname{Im}\int_\Omega (\sum_{i\in\mathcal{F}'} f_i(z))\, d\nu(z) = \operatorname{Im}\sum_{i\in\mathcal{F}'}\int_\Omega (f_i(z) - g_i(z))\, d\nu(z) + \operatorname{Im}\sum_{i\in\mathcal{F}'}\int_\Omega g_i(z)\, d\nu(z) = \sigma_1 + \sigma_2 .$$

Clearly,

$$|f_i(z) - g_i(z)| \le \frac{8\, r_i\, \gamma_i}{|z - y_i|^2} .$$

Thus, using the fact that $i \in \mathcal{F}' \subset \mathcal{F}_1$ and (5.3), we get

$$\int_\Omega |f_i - g_i|\, d|\nu| \le 8\, \nu^*(y_i)\, \gamma_i \le \frac{800}{L}\gamma_i .$$

In particular, combining this with the admissibility of $\mathcal{F}'$, we get

$$|\sigma_1| \le \frac{800}{L}\sum_{i\in\mathcal{F}'}\gamma_i \le \frac{800}{L}\cdot\frac{L}{5000} \le \frac{4}{25} . \tag{5.5}$$

To estimate σ_2 we will write it down as follows:

$$\sigma_2 = -\operatorname{Im}\int \sum_{i\in\mathcal{F}'} g_i(x)\hat{\nu}(x)\, dx ,$$

where $\hat{\nu}(x) := \int_\Omega \frac{d\nu(z)}{z-x}$. The great advantage now is that all $g_i(x)$ are real valued. In particular, as $\|g_i\|_\infty \le 1$, we get

$$|\sigma_2| = |\int \sum_{i\in\mathcal{F}'} g_i(x)\operatorname{Im}\hat{\nu}(x)\, dx| \le \sum_{i\in\mathcal{F}'}\int_{I_i} |\operatorname{Im}\hat{\nu}(x)|\, dx . \tag{5.6}$$

Immediately, $|\operatorname{Im}\hat{\nu}(x)| \le \int_\Omega \frac{|\operatorname{Im} z|\, d|\nu|(z)}{|z-x|^2} = \int_\Omega P(z,x)\, d|\nu|(z)$, where $P(z,x)$ denotes the Poisson kernel.

Consider $\mathcal{F}_2 \subset \mathcal{F}_1$ such that $\mathcal{F}_2 = \{i : \sup_{x\in I_i}\int P(z,x)\, d|\nu|(z) \ge \frac{100}{L}\}$. The segment I_i lies well inside $D(y_i, 10r_i)$, and ν is supported outside of it, therefore,

$$i \in \mathcal{F}_2 \Rightarrow \inf_{x\in I_i}\int P(z,x)\, d|\nu|(z) \ge \frac{10}{L} .$$

And so

$$\sum_{i\in\mathcal{F}_2}\int_{I_i}\int P(z,x)\, d|\nu|(z)dx \ge \frac{10}{L}\sum_{i\in\mathcal{F}_2}|I_i| . \tag{5.7}$$

On the other hand,

$$\sum_{i\in\mathcal{F}_2}\int_{I_i}\int P(z,x)\, d|\nu|(z)dx = \int (\int_{\cup_{i\in\mathcal{F}_2} I_i} P(z,x)\, dx)\, d|\nu|(z) ,$$

and, hence,

(5.8)
$$\sum_{i\in\mathcal{F}_2}\int_{I_i}\int P(z,x)\,d|\nu|(z)dx=\int(\int P(z,x)\chi_{\cup_{i\in\mathcal{F}_2}I_i}\,dx)\,d|\nu|(z)\le\int 1\,d|\nu|(z)\le 1\,.$$

Now (5.7) and (5.8) give us that

$$\sum_{i\in\mathcal{F}_2}|I_i|\le\frac{L}{10}\,. \tag{5.9}$$

Let $\mathcal{F}_3=\mathcal{F}_1\setminus\mathcal{F}_2$. Let $J_3:=\cup_{i\in\mathcal{F}_3}I_i$.
From (5.3) and (5.9) we get

$$|J_3|\ge\frac{17}{20}L\,. \tag{5.10}$$

But also we have for every $x\in J_3$

$$\int_\Omega P(z,x)\,d|\nu|(z)\le\frac{100}{L}\,. \tag{5.11}$$

Let us consider any admissible $\mathcal{F}'$ such that $\mathcal{F}'\subset\mathcal{F}_3'$. From (5.4) and (5.10) it follows that there are plenty of such subsets. Now we can finish the estimate of σ_2:

$$|\sigma_2|=\int_{\cup_{i\in\mathcal{F}'}I_i}dx\int P(z,x)d|\nu|(z)\le\frac{100}{L}\cdot\frac{L}{5000}\le\frac{1}{50}\,. \tag{5.12}$$

Notice that we really used admissibility of $\mathcal{F}'$ here. Combining (5.5) and (5.12), we obtain that for an admissible $c=\{c_i\}$, namely for $c_i=1$, $i\in\mathcal{F}'$, $c_i=0$ otherwise one has

$$|\mathrm{Im}\int\sum c_if_i(z)d\nu(z)|\le\frac{4}{25}+\frac{1}{50}=\frac{9}{50}<\frac{1}{5}$$

But this contradicts (5.1). Theorem 1.3 is completely proved.

6. Discussion.

1) Theorem 1.1 can be derived from Menger's curvature criterion (1.6) if one combines this criterion with a very difficult result of Tolsa ($\gamma\asymp\gamma_+$, see [**T4**]). In fact, one can prove

$$\forall Q\;\; c_2(E\cap Q)^2\le C_1\,H^1(E\cap Q)\Leftarrow\forall Q\;\; H^1(E\cap Q)\le C_2\gamma_+(E\cap Q)\,. \tag{6.1}$$

This is the inversion of (1.7). While (1.7) was based on characterization (1.8), the direct proof of the reverse implication is more subtle. Still suppose that the reverse to (1.7) is already proved.

Still to derive (1.6) from (1.1) (that is to derive Theorem 1.1 from Theorem 1.2) we, however, need to use that $\gamma\le A\gamma_+$ of [**T4**], which is several orders of magnitude harder than anything proved here.

2) We cannot help but quote another very subtle result related to Theorem 1.1.

THEOREM 6.1. Let ν be a complex measure with compact support $K \subset \mathbb{C}$ and let $\nu(K) \neq 0$. Suppose the Cauchy integral of ν is uniformly bounded in $\mathbb{C} \setminus K$:

$$C^{\nu}(z) := \int_K \frac{d\nu(\zeta)}{z-\zeta}, \quad |C^{\nu}(z)| \leq 1 .$$

Then there exists a positive μ on K such that

$$|C^{\mu}(z)| \leq 1 ,$$

and such that

$$\mu(K) \geq a\, |\nu(K)| \left(\frac{|\nu(K)|}{|\nu|(K)|}\right)^{64} ,$$

where $a > 0$ is an absolute constant.

We can apply Theorem 6.1 to the proof of Theorem 1.1. We will meet the situation, where $\|\nu\| = H^1(E \cap Q), |\nu(K)| = \gamma(E \cap Q)$, and the assumption gives us the following inequality that, in its turn, brings the proof of Theorem 1.1 (after the use of the curvature criterion from [**NTV2**], [**T1**]):

$$\gamma_+(E \cap Q) \geq a\gamma(E \cap Q) \,, a > 0 , \tag{6.2}$$

(which is true in general by the abovementioned Tolsa's solution of Vitushkin's problem, [**T4**]).

In (6.2) we obtained Tolsa's conclusion without using difficult result of [**T4**]. We just used Theorem 6.1 proved in [**NTV5**], [**Vo**] and assumption (1.1).

But Theorem 6.1 is itself a pretty difficult one. Actually, it is one of two main ingredients in Tolsa's [**T4**]. So we are back to a very lengthy and difficult proof of our first main result.

3) In this paper we show how to avoid using difficult stuff from [**T4**], [**NTV4**], [**NTV5**], [**Vo**] in the proof of Theorem 1.1. We avoided curvature criterion from [**NTV2**], [**T1**] as well.

4) This, in particular, shows that their exists also a multi-dimensional analog of our Theorem 1.1.

Let us be in $\mathbb{R}^n$ now. Let E be a compact subset of $\mathbb{R}^n$ such that $0 < H^{n-1}(E) < \infty$. Let T denote the vector Riesz transform operator with kernel $R(x-y)$, where

$$R(x) := (\frac{x_1}{|x|^n}, \ldots, \frac{x_n}{|x|^n}) .$$

We recall the reader that there exists in $\mathbb{R}^n$ a full analog of analytic capacity. It is called Lipschitz harmonic capacity, and it was introduced by Mattila and Paramonov. We will call it γ as before, the reader can get acquainted with it by reading [**Vo**].

We have

THEOREM 6.2. Operator T is bounded in $L^2(E, H^{n-1}|E)$ if and only if there exists a finite constant C such that for every cube Q in $\mathbb{R}^n$ we have

$$H^{n-1}(E \cap Q) \leq C\, \gamma(E \cap Q) . \tag{6.3}$$

References

[Chr] M. Christ, *A $T(b)$ theorem with remarks on analytic capacity and the Cauchy integral,* Colloq. Math. **60/61** (1990), 601-628.

[CW] R. R. Coifman, G. Weiss, "Analyse harmonique non-commutative sur certaines espaces homogènes", Lecture Notes in Math. **242**, Springer-Verlag, Berlin, 1971.

[CJS] R. R. Coifman, P. W. Jones, St. Semmes *Two elementary proofs of the L^2 boundedness of Cauchy integrals on Lipschitz curves,* J. of Amer. Math. Soc., **2** 1989, No. 3, 553-564.

[Da1] G. David, *Opérateurs intégraux singuliers sur certaines courbes du plan complexe.* Ann. Sci. Ecole Norm. Sup., **17** (1984), 157-189.

[Da2] G. David, *Completely unrectifiable 1-sets on the plane have vanishing analytic capacity,* Revista Mat. Iberoamericana, v. 14 (1998), 369–479.

[DaJ] G. David, J.-L. Journé, *A boundedness criterion for generalized Calderón-Zygmund operators,* Ann. of Math., **120** (1984), 371-397.

[DaJS] G. David, J.-L. Journé, S. Semmes, *Opérateurs de Calderón-Zygmund, fonctions para-accretive et interpolation,* Rev. Mat. Iberoamer., **1** (1985), 1-56.

[Fa] K. Falkoner "The Geometry of Fractal Sets", Cambridge Univ. Press, 1985.

[Gam] T. Gamelin "Uniform Algebras", Prentice-Hall, Englewood Cliffs, N.J. 1969.

[JG] . J. Garnett Personal communication.

[J] P. W. Jones *Rectifiable sets and the traveling salesman problem,* Invent. Math., **102** (1990), 1-15.

[JM] P. W. Jones and T. Murai, *Positive analytic capacity but zero Buffon needle probability,* Pacific J. Math., **133** (1988), no. 1, 99–114.

[Ma] P. Mattila "Geometry of Sets and Measures in Euclidean Spaces". Cambridge Univ. Press, 1995.

[Mel] M. Melnikov *An estimate of the Cauchy integral over an analytic arc.* Sbornik: Mathematics, **71** (1966), N0 4, 5030514.

[Me2] M. Melnikov *Analytic capacity: discrete approach and curvature of a measure.* Sbornik: Mathematics, **186** (1995), 827-846.

[Mu] T. Murai "A Real Variable Method for the Cauchy Transform, and Analytic Capacity", Lecture Notes in Math., **1307**, Springer Verlag, Berlin-Heidelberg-New York, 1988.

[MMV] P. Mattila, M. Melnikov, J. Verdera, *The Cauchy integral, analytic capacity, and uniform rectifiability,* Ann. of Math., **144** (1996), 127-136.

[NT] F. Nazarov, S. Treil, *The hunt for Bellman function: applications to estimates of singular integral operators and to other classical problems in harmonic analysis,* Algebra i Analysis, **8** (1997), no. 5, 32-162.

[NTV1] F. Nazarov, S. Treil, A. Volberg, *Weak type estimates and Cotlar inequalities for Calderón-Zygmund operators on nonhomogeneous spaces.* **IMRN** International Math. Res. Notices, 1998, no. 9, 463-487.

[NTV2] F. Nazarov, S. Treil, A. Volberg, *Cauchy integral and Calderón-Zygmund operators on nonhomogeneous spaces.* **IMRN** International Math. Res. Notices, 1997, no. 15, 703-726.

[NTV3] F. Nazarov, S. Treil, A. Volberg, *Accretive system Tb theorem of M. Christ for nonhomogeneous spaces,* Duke Math. J., **113** (2002), pp. 259–312

[NTV4] F. Nazarov, S. Treil, A. Volberg, *The Tb-theorem on nonhomogeneous spaces,* Acta Math., **190** (2003), 151–239.

[NTV5] ———, *Nonhomogeneous Tb theorem which proves Vitushkin's conjecture,* Preprint No. 519, CRM, Barcelona, 2002, 1–84.

[St] E. Stein, "Harmonic Analysis: Real-Variable Methods, Orthogonality, and Oscillatory Integrals", with the assistance of Timothy S. Murphy, Princeton Math. Ser. **43**, Monographs in Harmonic analysis, iii, Princeton Univ. Press, Princeton, 1993.

[T1] X. Tolsa, *L^2-boundedness of the Cauchy integral operator for continuous measures,* Duke Math. J., **98** (1999), no. 2, 269–304.

[T2] X. Tolsa, *Cotlar's inequality and the existence of principal values for the Cauchy integral without doubling condition,* J. Reine Angew. Math. 502 (1998), 199–235.

[T3] X. Tolsa, *Curvature of measures, Cauchy singular integral, and analytic capacity,* Thesis, Dept. Math. Univ. Auton. de Barcelona, 1998.

[T4] X. Tolsa, *Painlevé's problem and the semiadditivity of analytic capacity,* Acta Math., **190** (2003), no. 1, 105–149.

[T5] X. Tolsa, *On the analytic capacity* γ_+, Indiana Univ. Math. J., **51** (2), (2002), 317–344.
[Vo] A. Volberg, "Calderón-Zygmund Capacities and Operators on Nonhomogeneous Spaces", CBMS Regional Conference Series in Mathematics, v. 100, 2003, pp. 1-167.

Department of Mathematics, University of Wisconsin, Madison, WI. 53706
E-mail address: nazarov@math.wisc.edu

Department of Mathematics, Michigan State University, East Lansing, Michigan 48823, USA and School of Mathematics, Edinburgh University, EH9 3JZ, UK
E-mail address: volberg@math.msu.edu

Proceedings of Symposia in Pure Mathematics
Volume **79**, 2008

The Christoffel–Darboux Kernel

Barry Simon*

ABSTRACT. A review of the uses of the CD kernel in the spectral theory of orthogonal polynomials, concentrating on recent results.

CONTENTS

2000 *Mathematics Subject Classification.* 34L40, 47-02, 42C05.

Key words and phrases. Orthogonal polynomials, spectral theory.

This work was supported in part by NSF grant DMS-0652919 and U.S.–Israel Binational Science Foundation (BSF) Grant No. 2002068.

1. Introduction

This article reviews a particular tool of the spectral theory of orthogonal polynomials. Let μ be a measure on $\mathbb{C}$ with finite moments, that is,

$$\int |z|^n \, d\mu(z) < \infty \tag{1.1}$$

for all $n = 0, 1, 2, \ldots$ and which is nontrivial in the sense that it is not supported on a finite set of points. Thus, $\{z^n\}_{n=0}^\infty$ are independent in $L^2(\mathbb{C}, d\mu)$, so by Gram–Schmidt, one can define monic orthogonal polynomials, $X_n(z; d\mu)$, and orthonormal polynomials, $x_n = X_n/\|X_n\|_{L^2}$. Thus,

$$\int \bar{z}^j X_n(z; d\mu) \, d\mu(z) = 0 \qquad j = 0, \ldots, n-1 \tag{1.2}$$

$$X_n(z) = z^n + \text{lower order} \tag{1.3}$$

$$\int \overline{x_n(z)} \, x_m(z) \, d\mu = \delta_{nm} \tag{1.4}$$

We will often be interested in the special cases where μ is supported on $\mathbb{R}$ (especially with support compact), in which case we use P_n, p_n rather than X_n, x_n, and where μ is supported on $\partial\mathbb{D}$ ($\mathbb{D} = \{z \mid |z| < 1\}$), in which case we use Φ_n, φ_n. We call these OPRL and OPUC (for "real line" and "unit circle").

OPRL and OPUC are spectral theoretic because there are Jacobi parameters $\{a_n, b_n\}_{n=1}^\infty$ and Verblunsky coefficients $\{\alpha_n\}_{n=0}^\infty$ with recursion relations ($p_{-1} = 0$; $p_0 = \Phi_0 = 1$):

$$z p_n(z) = a_{n+1} p_{n+1}(z) + b_{n+1} p_n(z) + a_n p_{n-1}(z) \tag{1.5}$$

$$\Phi_{n+1}(z) = z\Phi_n(z) - \bar{\alpha}_n \Phi_n^*(z) \tag{1.6}$$

$$\Phi_n^*(z) = z^n \, \overline{\Phi_n(1/\bar{z})} \tag{1.7}$$

We will sometimes need the monic OPRL and normalized OPUC recursion relations:

$$z P_n(z) = P_{n+1}(z) + b_{n+1} P_n(z) + a_n^2 P_{n-1}(z) \tag{1.8}$$

$$z\varphi_n(z) = \rho_n \varphi_{n+1}(z) + \bar{\alpha}_n \varphi_n^*(z) \tag{1.9}$$

$$\rho_n \equiv (1 - |\alpha_n|^2)^{1/2} \tag{1.10}$$

Of course, the use of ρ_n implies $|\alpha_n| < 1$ and all sets of $\{\alpha_n\}_{n=0}^\infty$ obeying this occur. Similarly, $b_n \in \mathbb{R}$, $a_n \in (0, \infty)$ and all such sets occur. In the OPUC case, $\{\alpha_n\}_{n=0}^\infty$ determine $d\mu$, while in the OPRL case, they do if $\sup(|a_n| + |b_n|) < \infty$, and may or may not in the unbounded case. For basics of OPRL, see [**93, 22, 34, 89**]; and for basics of OPUC, see [**93, 37, 34, 80, 81, 79**].

We will use κ_n (or $\kappa_n(d\mu)$) for the leading coefficient of x_n, p_n, or φ_n, so

$$\kappa_n = \|X_n\|_{L^2(d\mu)}^{-1} \tag{1.11}$$

The *Christoffel–Darboux kernel* (named after [**23, 28**]) is defined by

$$K_n(z, \zeta) = \sum_{j=0}^n \overline{x_j(z)} \, x_j(\zeta) \tag{1.12}$$

We sometimes use $K_n(z,\zeta;\mu)$ if we need to make the measure explicit. Note that if $c>0$,

$$K_n(z,\zeta;c\mu)=c^{-1}K_n(z,\zeta;\mu) \tag{1.13}$$

since $x_n(z;cd\mu)=c^{-1/2}x_n(z;d\mu)$.

By the Schwarz inequality, we have

$$|K_n(z,\zeta)|^2\le K_n(z,z)K_n(\zeta,\zeta) \tag{1.14}$$

There are three variations of convention. Some only sum to $n-1$; this is the more common convention but (1.12) is used by Szegő [**93**], Atkinson [**5**], and in [**80, 81**]. As we will note shortly, it would be more natural to put the complex conjugate on $x_n(\zeta)$, not $x_n(z)$—and a very few authors do that. For OPRL with z,ζ real, the complex conjugate is irrelevant—and some authors leave it off even for complex z and ζ.

As a tool in spectral analysis, convergence of OP expansions, and other aspects of analysis, the use of the CD kernel has been especially exploited by Freud and Nevai, and summarized in Nevai's paper on the subject [**68**]. A series of recent papers by Lubinsky (of which [**60, 61**] are most spectacular) has caused heightened interest in the subject and motivated me to write this comprehensive review.

Without realizing they were dealing with OPRL CD kernels, these objects have been used extensively in the spectral theory community, especially the diagonal kernel

$$K_n(x,x)=\sum_{j=0}^{n}|p_j(x)|^2 \tag{1.15}$$

Continuum analogs of the ratios of this object for the first and second kind polynomials appeared in the work of Gilbert–Pearson [**38**] and the discrete analog in Khan–Pearson [**49**] and then in Jitomirskaya–Last [**45**]. Last–Simon [**55**] studied $\frac{1}{n}K_n(x,x)$ as $n\to\infty$. Variation of parameters played a role in all these works and it exploits what is essentially mixed CD kernels (see Section 8).

One of our goals here is to emphasize the operator theoretic point of view, which is often underemphasized in the OP literature. In particular, in describing μ, we think of the operator M_z on $L^2(\mathbb{C},d\mu)$ of multiplication by z:

$$(M_zf)(z)=zf(z) \tag{1.16}$$

If $\operatorname{supp}(d\mu)$ is compact, M_z is a bounded operator defined on all of $L^2(\mathbb{C},d\mu)$. If it is not compact, there are issues of domain, essential selfadjointness, etc. that will not concern us here, except to note that in the OPRL case, they are connected to uniqueness of the solution of the moment problem (see [**77**]). With this in mind, we use $\sigma(d\mu)$ for the spectrum of M_z, that is, the support of $d\mu$, and $\sigma_{\rm ess}(d\mu)$ for the essential spectrum. When dealing with OPRL of compact support (where M_z is bounded selfadjoint) or OPUC (where M_z is unitary), we will sometimes use $\sigma_{\rm ac}(d\mu)$, $\sigma_{\rm sc}(d\mu)$, $\sigma_{\rm pp}(d\mu)$ for the spectral theory components. (We will discuss $\sigma_{\rm ess}(d\mu)$ only in the OPUC/OPRL case where it is unambiguous, but for general operators, there are multiple definitions; see [**31**].)

The basis of operator theoretic approaches to the study of the CD kernel depends on its interpretation as the integral kernel of a projection. In $L^2(\mathbb{C},d\mu)$, the set of polynomials of degree at most n is an $n+1$-dimensional space. We will use

π_n for the operator of orthogonal projection onto this space. Note that

$$(\pi_n f)(\zeta) = \int K_n(z,\zeta) f(z)\, d\mu(z) \tag{1.17}$$

The order of z and ζ is the opposite of the usual for integral kernels and why we mentioned that putting complex conjugation on $x_n(\zeta)$ might be more natural in (1.12).

In particular,

$$\deg(f) \leq n \Rightarrow f(\zeta) = \int K_n(z,\zeta) f(z)\, d\mu(z) \tag{1.18}$$

In particular, since K_n is a polynomial in ζ of degree n, we have

$$K_n(z,w) = \int K_n(z,\zeta) K_n(\zeta,w)\, d\mu(\zeta) \tag{1.19}$$

often called the reproducing property.

One major theme here is the frequent use of operator theory, for example, proving the CD formula as a statement about operator commutators. Another theme, motivated by Lubinsky [**60, 61**], is the study of asymptotics of $\frac{1}{n}K_n(x,y)$ on diagonal ($x = y$) and slightly off diagonal ($(x-y) = O(\frac{1}{n})$).

Sections 2, 3, and 6 discuss very basic formulae, and Sections 4 and 7 simple applications. Sections 5 and 8 discuss extensions of the context of CD kernels. Section 9 starts a long riff on the use of the Christoffel variational principle which runs through Section 23. Section 24 is a final simple application.

Vladimir Maz'ya has been an important figure in the spectral analysis of partial differential operators. While difference equations are somewhat further from his opus, they are related. It is a pleasure to dedicate this article with best wishes on his 70th birthday.

I would like to thank J. Christiansen for producing Figure 1 (in Section 7) in Maple, and C. Berg, F. Gesztesy, L. Golinskii, D. Lubinsky, F. Marcellán, E. Saff, and V. Totik for useful discussions.

2. The ABC Theorem

We begin with a result that is an aside which we include because it deserves to be better known. It was rediscovered and popularized by Berg [**11**], who found it earliest in a 1939 paper of Collar [**24**], who attributes it to his teacher, Aitken—so we dub it the ABC theorem. Given that it is essentially a result about Gram–Schmidt, as we shall see, it is likely it really goes back to the nineteenth century. For applications of this theorem, see [**13, 47**].

K_n is a polynomial of degree n in $\bar{z}$ and ζ, so we can define an $(n+1)\times(n+1)$ square matrix, $k^{(n)}$, with entries $k_{jm}^{(n)}$, $0 \leq j,m \leq n$, by

$$K_n(z,\zeta) = \sum_{j,m=0}^{n} k_{jm}^{(n)} \bar{z}^m \zeta^j \tag{2.1}$$

One also has the moment matrix

$$m_{jk}^{(n)} = \langle z^j, z^k \rangle = \int \bar{z}^j z^k \, d\mu(z) \tag{2.2}$$

$0 \le j, k \le n$. For OPRL, this is a function of $j+k$, so $m^{(n)}$ is a Hankel matrix. For OPUC, this is a function of $j-k$, so $m^{(n)}$ is a Toeplitz matrix.

THEOREM 2.1 (ABC Theorem).

$$(m^{(n)})^{-1} = k^{(n)} \tag{2.3}$$

PROOF. By (1.18) for $\ell = 0, \dots, n$,

$$\int K_n(z,\zeta) z^\ell \, d\mu(z) = \zeta^\ell \tag{2.4}$$

Plugging (2.1) in for K, using (2.2) to do the integrals leads to

$$\sum_{j,q=0}^{n} k_{jq}^{(n)} m_{q\ell}^{(n)} \zeta^j = \zeta^\ell \tag{2.5}$$

which says that

$$\sum_q k_{jq}^{(n)} m_{q\ell}^{(n)} = \delta_{j\ell} \tag{2.6}$$

which is (2.3). □

Here is a second way to see this result in a more general context: Write

$$x_j(z) = \sum_{k=0}^{j} a_{jk} z^k \tag{2.7}$$

so we can define an $(n+1) \times (n+1)$ triangular matrix $a^{(n)}$ by

$$a_{jk}^{(n)} = a_{jk} \tag{2.8}$$

Then (the Cholesky factorization of k)

$$k^{(n)} = a^{(n)} (a^{(n)})^* \tag{2.9}$$

with * Hermitean adjoint. The condition

$$\langle x_j, x_\ell \rangle = \delta_{j\ell} \tag{2.10}$$

says that

$$(a^{(n)})^* m^{(n)} (a^{(n)}) = \mathbf{1} \tag{2.11}$$

the identity matrix. Multiplying by $(a^{(n)})^*$ on the right and $[(a^{(n)})^*]^{-1}$ on the left yields (2.3). This has a clear extension to a general Gram–Schmidt setting.

3. The Christoffel–Darboux Formula

The Christoffel–Darboux formula for OPRL says that

$$K_n(z,\zeta) = a_{n+1} \left(\frac{\overline{p_{n+1}(z)}\, p_n(\zeta) - \overline{p_n(z)}\, p_{n+1}(\zeta)}{\bar{z} - \zeta} \right) \tag{3.1}$$

and for OPUC that

$$K_n(z,\zeta) = \frac{\overline{\varphi_{n+1}^*(z)}\, \varphi_{n+1}^*(\zeta) - \overline{\varphi_{n+1}(z)}\, \varphi_{n+1}(\zeta)}{1 - \bar{z}\zeta} \tag{3.2}$$

The conventional wisdom is that there is no CD formula for general OPs, but we will see, in a sense, that is only half true. The usual proofs are inductive. Our proofs here will be direct operator theoretic calculations.

We focus first on (3.1). From the operator point of view, the key is to note that, by (1.17),

$$\langle g, [M_z, \pi_n] f\rangle = \int \overline{g(\zeta)}\,(\bar\zeta - z) K_n(z,\zeta) f(z)\, d\mu(\zeta) d\mu(z) \tag{3.3}$$

where $[A,B] = AB - BA$. For OPRL, in (3.3), ζ and z are real, so (3.1) for $z, \zeta \in \sigma(d\mu)$ is equivalent to

$$[M_z, \pi_n] = a_{n+1}[\langle p_n, \cdot\,\rangle p_{n+1} - \langle p_{n+1}, \cdot\,\rangle p_n] \tag{3.4}$$

While (3.4) only proves (3.1) for such $\bar z, \zeta$ by the fact that both sides are polynomials in z and ζ, it is actually equivalent. Here is the general result:

THEOREM 3.1 (General Half CD Formula). *Let μ be a measure on $\mathbb{C}$ with finite moments. Then:*

$$(1-\pi_n)[M_z, \pi_n](1-\pi_n) = 0 \tag{3.5}$$

$$\pi_n[M_z, \pi_n]\pi_n = 0 \tag{3.6}$$

$$(1-\pi_n)[M_z, \pi_n]\pi_n = \frac{\|X_{n+1}\|}{\|X_n\|}\,\langle x_n, \cdot\,\rangle x_{n+1} \tag{3.7}$$

REMARK. If μ has compact support, these are formulae involving bounded operators on $L^2(\mathbb{C}, d\mu)$. If not, regard π_n and M_z as maps of polynomials to polynomials.

PROOF. (3.5) follows from expanding $[M_z, \pi_n]$ and using

$$\pi_n(1-\pi_n) = (1-\pi_n)\pi_n = 0 \tag{3.8}$$

If we note that $[M_z, \pi_n] = -[M_z, (1-\pi_n)]$, (3.6) similarly follows from (3.8). By (3.8) again,

$$(1-\pi_n)[M_z, \pi_n]\pi_n = (1-\pi_n)M_z\pi_n \tag{3.9}$$

On $\operatorname{ran}(\pi_{n-1})$, π_n is the identity, and multiplication by z leaves one in π_n, that is,

$$(1-\pi_n)M_z\pi_n \restriction \operatorname{ran}(\pi_{n-1}) = 0 \tag{3.10}$$

On the other hand, for the monic OPs,

$$(1-\pi_n)M_z\pi_n X_n = X_{n+1} \tag{3.11}$$

since $M_z\pi_n X_n = z^{n+1} +$ lower order and $(1-\pi_n)$ takes any such polynomial to X_{n+1}. Since

$$\frac{\|X_{n+1}\|}{\|X_n\|}\,\langle x_n, X_n\rangle x_{n+1} = X_{n+1}$$

we see (3.4) holds on $\operatorname{ran}(1-\pi_n) + \operatorname{ran}(\pi_{n-1}) + [X_n]$, and so on all of L^2. □

From this point of view, we can understand what is missing for a CD formula for general OP. The missing piece is

$$\pi_n[M_z, \pi_n](1-\pi_n) = ((1-\pi_n)M_z^*\pi_n)^* \tag{3.12}$$

The operator on the left of (3.7) is proven to be rank one, but $(1-\pi_n)M_z^*\pi_n$ is, in general, rank n. For $\varphi \in \ker[(1-\pi_n)M_z^*\pi_n] \cap \operatorname{ran}(\pi_n)$ means that φ is a polynomial of degree n and so is $\bar z\varphi$, at least for a.e. z with respect to μ. Two cases where many $\bar z\varphi$ are polynomials of degree n—indeed, so many that $(1-\pi_n)M_z^*\pi_n$ is also rank one—are for OPRL where $\bar z\varphi = z\varphi$ (a.e. $z \in \sigma(d\mu)$) and OPUC where $\bar z\varphi = z^{-1}\varphi$

(a.e. $z \in \sigma(d\mu)$). In the first case, $\bar{z}\varphi \in \operatorname{ran}(\pi_n)$ if $\deg(\varphi) \le n-1$, and in the second case, if $\varphi(0) = 0$.

Thus, only for these two cases do we expect a simple formula for $[M_z, \pi]$.

THEOREM 3.2 (CD Formula for OPRL). *For OPRL, we have*

$$[M_z, \pi_n] = a_{n+1}[\langle p_n, \cdot\,\rangle p_{n+1} - \langle p_{n+1}, \cdot\,\rangle p_n] \tag{3.13}$$

and (3.1) *holds for* $\bar{z} \neq \zeta$.

PROOF. Inductively, one has that $p_n(x) = (a_1 \dots a_n)^{-1} x^n + \dots$, so

$$\|P_n\| = a_1 \dots a_n \mu(\mathbb{R})^{1/2} \tag{3.14}$$

and thus,

$$\frac{\|P_{n+1}\|}{\|P_n\|} = a_{n+1} \tag{3.15}$$

Moreover, since $M_z^* = M_z$ for OPRL and $[A, B]^* = -[A^*, B^*]$, we get from (3.12) that

$$\pi_n[M_z, \pi_n](1 - \pi_n) = -a_{n+1}\langle p_{n+1}, \cdot\,\rangle p_n \tag{3.16}$$

(3.5)–(3.7), (3.14), and (3.16) imply (3.13) which, as noted, implies (3.1). □

For OPUC, the natural object is (note $M_z M_z^* = M_z^* M_z = 1$)

$$B_n = \pi_n - M_z \pi_n M_z^* = -[M_z, \pi_n] M_z^* \tag{3.17}$$

THEOREM 3.3 (CD Formula for OPUC). *For OPUC, we have*

$$\pi_n - M_z \pi_n M_z^* = \langle \varphi_{n+1}^*, \cdot\,\rangle \varphi_{n+1}^* - \langle \varphi_{n+1}, \cdot\,\rangle \varphi_{n+1} \tag{3.18}$$

and (3.2) *holds.*

PROOF. B_n is selfadjoint so $\operatorname{ran}(B_n) = \ker(B_n)^\perp$. Clearly, $\operatorname{ran}(B_n) \subset \operatorname{ran}(\pi_n) + M_z[\operatorname{ran}(\pi_n)] = \operatorname{ran}(\pi_{n+1})$ and $B_n z^\ell = 0$ for $\ell = 1, \dots, n$, so $\operatorname{ran}(B_n) = \{z, z^2, \dots, z^n\}^\perp \cap \operatorname{ran}(\pi_{n+1})$ is spanned by φ_{n+1} and φ_{n+1}^*. Thus, both B_n and the right side of (3.18) are rank two selfadjoint operators with the same range and both have trace 0. Thus, it suffices to find a single vector η in the span of φ_{n+1} and φ_{n+1}^* with $B_n\eta =$ (RHS of (3.18))η, since a rank at most one selfadjoint operator with zero trace is zero!

We will take $\eta = z\varphi_n$, which lies in the span since, by (1.9) and its *,

$$\rho_n \varphi_{n+1} = z\varphi_n - \bar{\alpha}_n \varphi_n^* \qquad \rho_n \varphi_{n+1}^* = \varphi_n^* - \alpha_n z \varphi_n \tag{3.19}$$

By (3.16), (3.17), and

$$\|\Phi_n\| = \rho_0 \dots \rho_{n-1} \mu(\partial\mathbb{D}) \tag{3.20}$$

we have that

$$\begin{aligned} B_n(z\varphi_n) &= [\pi_n, M_z]\varphi_n \\ &= -(1 - \pi_n) M_z \pi_n \varphi_n \\ &= -\rho_n \varphi_{n+1} \end{aligned} \tag{3.21}$$

On the other hand, $\varphi_{n+1}^* \perp \{z, \dots, z^{n+1}\}$, so

$$\langle \varphi_{n+1}^*, z\varphi_n \rangle = 0$$

and, by (3.19),

$$\langle \varphi_{n+1}, z\varphi_n \rangle = \rho_n \langle \varphi_{n+1}, \varphi_{n+1} \rangle + \bar{\alpha}_n \langle \varphi_{n+1}, \varphi_n^* \rangle$$

$$= \rho_n$$

so

$$[\text{LHS of (3.18)}]z\varphi_n = -\rho_n\varphi_{n+1} \tag{3.22}$$

□

Note that (3.2) implies

$$\begin{aligned} K_{n+1}(z,\zeta) &= K_n(z,\zeta) + \overline{\varphi_{n+1}(z)}\,\varphi_{n+1}(\zeta)\left(\frac{1-\bar{z}\zeta}{1-\bar{z}\zeta}\right) \\ &= \frac{\overline{\varphi_{n+1}^*(z)}\,\varphi_{n+1}^*(\zeta) - \bar{z}\zeta\,\overline{\varphi_{n+1}(z)}\,\varphi_{n+1}(\zeta)}{1-\bar{z}\zeta} \end{aligned}$$

so changing index, we get the "other form" of the CD formula for OPUC,

$$K_n(z,\zeta) = \frac{\overline{\varphi_n(z)^*}\,\varphi_n^*(\zeta) - \overline{z\varphi_n(z)}\,\zeta\varphi_n(\zeta)}{1-\bar{z}\varphi} \tag{3.23}$$

We also note that Szegő [**93**] derived the recursion relation from the CD formula, so the lack of a CD formula for general OPs explains the lack of a recursion relation in general.

4. Zeros of OPRL: Basics Via CD

In this section, we will use the CD formula to derive the basic facts about the zeros of OPRL. In the vast literature on OPRL, we suspect this is known but we don't know where. We were motivated to look for this by a paper of Wong [**105**], who derived the basics for zeros of POPUC (paraorthogonal polynomials on the unit circle) using the CD formula (for other approaches to zeros of POPUCs, see [**19, 86**]). We begin with the CD formula on diagonal:

THEOREM 4.1. *For OPRL and x real,*

$$\sum_{j=0}^{n} |p_j(x)|^2 = a_{n+1}[p_{n+1}'(x)p_n(x) - p_n'(x)p_{n+1}(x)] \tag{4.1}$$

PROOF. In (3.1) with $z = x$, $\zeta = y$ both real, subtract $p_{n+1}(y)p_n(y)$ from both products on the left and take the limit as $y \to x$. □

COROLLARY 4.2. *If $p_n(x_0) = 0$ for x_0 real, then*

$$p_{n+1}(x_0)p_n'(x_0) < 0 \tag{4.2}$$

PROOF. The left-hand side of (4.1) is strictly positive since $p_0(x) = 1$. □

THEOREM 4.3. *All the zeros of $p_n(x)$ are real and simple and the zeros of p_{n+1} strictly interlace those of p_n. That is, between any successive zeros of p_{n+1} lies exactly one zero of p_n and it is strictly between, and p_{n+1} has one zero between each successive zero of p_n and it has one zero above the top zero of p_n and one below the bottom zero of p_n.*

PROOF. By (4.2), $p_n(x_0) = 0 \Rightarrow p_n'(x_0) \neq 0$, so zeros are simple, which then implies that the sign of p_n' changes between its successive zeros. By (4.2), the sign of p_{n+1} thus changes between zeros of p_n, so p_{n+1} has an odd number of zeros between zeros of p_n.

p_1 is a real polynomial, so it has one real zero. For x large, $p_n(x) > 0$ since the leading coefficient is positive. Thus, $p_n'(x_0) > 0$ at the top zero. From (4.2), $p_{n+1}(x_0) < 0$ and thus, since $p_{n+1}(x) > 0$ for x large, p_{n+1} has a zero above the top zero of p_n. Similarly, it has a zero below the bottom zero.

We thus see inductively, starting with p_1, that p_n has n real zeros and they interlace those of p_{n-1}. □

We note that Ambroladze [**3**] and then Denisov–Simon [**29**] used properties of the CD kernel to prove results about zeros (see Wong [**105**] for the OPUC analog); the latter paper includes:

THEOREM 4.4. *Suppose* $a_\infty = \sup_n a_n < \infty$ *and* $x_0 \in \mathbb{R}$ *has* $d = \operatorname{dist}(x_0, \sigma(d\mu)) > 0$. *Let* $\delta = d^2/(d + \sqrt{2}\, a_\infty)$. *Then at least one of* p_n *and* p_{n-1} *has no zeros in* $(x_0 - \delta, x_0 + \delta)$.

They also have results about zeros near isolated points of $\sigma(d\mu)$.

5. The CD Kernel and Formula for MOPs

Given an $\ell \times \ell$ matrix-valued measure, there is a rich structure of matrix OPs (MOPRL and MOPUC). A huge literature is surveyed and extended in [**27**]. In particular, the CD kernel and CD formula for MORL are discussed in Sections 2.6 and 2.7, and for MOPUC in Section 3.4.

There are two "inner products," maps from L^2 matrix-valued functions to matrices, $\langle\!\langle \cdot\,, \cdot \rangle\!\rangle_{\mathrm{R}}$ and $\langle\!\langle \cdot\,, \cdot \rangle\!\rangle_{\mathrm{L}}$. The R for right comes from the form of scalar homogeneity, for example,

$$\langle\!\langle f, gA \rangle\!\rangle_{\mathrm{R}} = \langle\!\langle f, g \rangle\!\rangle_{\mathrm{R}} A \tag{5.1}$$

but $\langle\!\langle f, Ag \rangle\!\rangle_{\mathrm{R}}$ is not related to $\langle\!\langle f, g \rangle\!\rangle_{\mathrm{R}}$.

There are two normalized OPs, $p_j^{\mathrm{R}}(x)$ and $p_j^{\mathrm{L}}(x)$, orthonormal in $\langle\!\langle \cdot\,, \cdot \rangle\!\rangle_{\mathrm{R}}$ and $\langle\!\langle \cdot\,, \cdot \rangle\!\rangle_{\mathrm{L}}$, respectively, but a single CD kernel (for z, w real and † is matrix adjoint),

$$K_n(z,w) = \sum_{k=0}^{n} p_k^{\mathrm{R}}(z) p_k^{\mathrm{R}}(w)^\dagger \tag{5.2}$$

$$= \sum_{k=0}^{n} p_k^{\mathrm{L}}(z)^\dagger p_k^{\mathrm{L}}(w) \tag{5.3}$$

One has that

$$\langle\!\langle K_n(\,\cdot\,, z), f(\,\cdot\,) \rangle\!\rangle_{\mathrm{R}} = (\pi_n f)(z) \tag{5.4}$$

where π_n is the projection in the $\operatorname{Tr}(\langle\!\langle \cdot\,, \cdot \rangle\!\rangle_{\mathrm{R}})$ inner product to polynomials of degree n.

In [**27**], the CD formula is proven using Wronskian calculations. We note here that the commutator proof we give in Section 3 extends to this matrix case.

Within the Toeplitz matrix literature community, a result equivalent to the CD formula is called the Gohberg–Semencul formula; see [**10, 35, 39, 40, 48, 100, 101**].

6. Gaussian Quadrature

Orthogonal polynomials allow one to approximate integrals over a measure $d\mu$ on $\mathbb{R}$ by certain discrete measures. The weights in these discrete measures depend on $K_n(x,x)$. Here we present an operator theoretic way of understanding this.

Fix n and, for $b \in \mathbb{R}$, let $J_{n;F}(b)$ be the $n \times n$ matrix

$$J_{n;F}(b) = \begin{pmatrix} b_1 & a_1 & 0 & & \\ a_1 & b_2 & a_2 & & \\ 0 & a_2 & b_3 & & \\ & & & \ddots & \\ & & & & b_n + b \end{pmatrix} \tag{6.1}$$

(i.e., we truncate the infinite Jacobi matrix and change only the corner matrix element b_n to $b_n + b$).

Let $\tilde{x}_j^{(n)}(b)$, $j = 1, \dots, n$, be the eigenvalues of $J_{n;F}(b)$ labelled by $\tilde{x}_1 < \tilde{x}_2 < \dots$. (We shall shortly see these eigenvalues are all simple.) Let $\tilde{\varphi}_j^{(n)}$ be the normalized eigenvectors with components $[\tilde{\varphi}_j^{(n)}(b)]_\ell$, $\ell = 1, \dots, n$, and define

$$\tilde{\lambda}_j^{(n)}(b) = |[\tilde{\varphi}_j^{(n)}(b)]_1|^2 \tag{6.2}$$

so that if e_1 is the vector $(1\,0 \dots 0)^t$, then

$$\sum_{j=1}^n \tilde{\lambda}_j^{(n)}(b)\delta_{\tilde{x}_j^{(n)}(b)} \tag{6.3}$$

is the spectral measure for $J_{n;F}(b)$ and e_1, that is,

$$\langle e_1, J_{n;F}(b)^\ell e_1 \rangle = \sum_{j=1}^n \tilde{\lambda}_j^{(n)}(b)\tilde{x}_j^{(n)}(b)^\ell \tag{6.4}$$

for all ℓ. We are going to begin by proving an intermediate quadrature formula:

THEOREM 6.1. *Let μ be a probability measure. For any b and any $\ell = 0, 1, \dots, 2n-2$,*

$$\int x^\ell \, d\mu = \sum_{j=1}^n \tilde{\lambda}_j^{(n)}(b)\tilde{x}_j^{(n)}(b)^\ell \tag{6.5}$$

If $b = 0$, this holds also for $\ell = 2n-1$.

PROOF. For any measure, $\{a_j, b_j\}_{j=1}^{n-1}$ determine $\{p_j\}_{j=0}^{n-1}$, and moreover,

$$\int x|p_{n-1}(x)|^2 \, d\mu = b_n \tag{6.6}$$

If a measure has finite support with at least n points, one can still define $\{p_j\}_{j=0}^{n-1}$, Jacobi parameters $\{a_j, b_j\}_{j=1}^{n-1}$, and b_n by (6.6).

$d\mu$ and the measure, call it $d\tilde{\mu}_1^{(n)}$, of (6.3) have the same Jacobi parameters $\{a_j, b_j\}_{j=1}^{n-1}$, so the same $\{p_j\}_{j=0}^{n-1}$, and thus by

$$\int x^k p_j(x) \, d\mu = 0 \qquad k = 0, 1, \dots, j-1;\ j = 1, \dots, n-1 \tag{6.7}$$

we inductively get (6.5) for $\ell = 0, 1, 2, \dots, 2n-3$. Moreover,

$$\int p_{n-1}(x)^2 \, d\mu = 1 \tag{6.8}$$

determines inductively (6.5) for $\ell = 2n-2$. Finally, if $b = 0$, (6.6) yields (6.5) for $\ell = 2n-1$. □

As the second step, we want to determine the $\tilde{x}_j^{(n)}(b)$ and $\tilde{\lambda}_j^{(n)}(b)$.

THEOREM 6.2. *Let $K_{n;F} = \pi_{n-1} M_z \pi_{n-1} \restriction \operatorname{ran}(\pi_{n-1})$ for a general finite moment measure, μ, on $\mathbb{C}$. Then*

$$\det\nolimits_{\operatorname{ran}(\pi_{n-1})}(z\mathbf{1} - K_{n;F}) = X_n(z) \tag{6.9}$$

PROOF. Suppose $X_n(z)$ has a zero of order ℓ at z_0. Let $\varphi = X_n(z)/(z-z_0)^\ell$. Then, in $\operatorname{ran}(\pi_n)$,

$$(K_{n;F} - z_0)^j \varphi \neq 0 \qquad j = 0, 1, \dots, \ell - 1 \tag{6.10}$$

$$(K_{n;F} - z_0)^\ell \varphi = 0 \tag{6.11}$$

since $(M_z - z_0)^\ell \varphi = X_n(z)$ and $\pi_{n-1} X_n = 0$. Thus, z_0 is an eigenvalue of $K_{n;F}$ of algebraic multiplicity at least ℓ. Since $X_n(z)$ has n zeros counting multiplicity, this accounts for all the roots, so (6.9) holds because both sides of monic polynomials of degree n with the same roots. □

COROLLARY 6.3. *We have for OPRL*

$$\det(z - J_{n;F}(b)) = P_n(z) - bP_{n-1}(z) \tag{6.12}$$

The eigenvalues $\tilde{x}_j^{(n)}(b)$ are all simple and obey for $0 < b < \infty$ and $j = 1, \dots, n$ (with $\tilde{x}_{n+1}(0) = \infty$),

$$\tilde{x}_j^{(n)}(0) < \tilde{x}_j^{(n)}(b) < \tilde{x}_{j+1}^{(n)}(0) \tag{6.13}$$

and for $-\infty < b < 0$ and $j = 1, \dots, n$ (with $\tilde{x}_{n-1}(0) = -\infty$),

$$\tilde{x}_{j-1}^{(n)}(0) < \tilde{x}_j^{(n)}(b) < \tilde{x}_j^{(n)}(0) \tag{6.14}$$

PROOF. (6.12) for $b = 0$ is just (6.9). Expanding in minors shows the determinant of $(z - J_{n;F}(b))$ is just the value at $b = 0$ minus b times the $(n-1) \times (n-1)$ determinant, proving (6.12) in general.

The inequalities in (6.13)/(6.14) follow either by eigenvalue perturbation theory or by using the arguments in Section 4. □

In fact, our analysis below proves that for $0 < b < \infty$,

$$\tilde{x}_j^{(n)}(0) < \tilde{x}_j^{(n)}(b) < \tilde{x}_j^{(n-1)}(0) \tag{6.15}$$

The recursion formula for monic OPs proves that $p_j(\tilde{x}_j(b))$ is the unnormalized eigenvector for $J_{n;F}(b)$. $K_{n-1}(\tilde{x}_j(b), \tilde{x}_j(b))^{1/2}$ is the normalization constant, so since $p_0 \equiv 1$ (if $\mu(\mathbb{R}) = 1$):

PROPOSITION 6.4. *If $\mu(\mathbb{R}) = 1$, then*

$$\lambda_j^{(n)}(b) = (K_{n-1}(\tilde{x}_j^{(n)}(b), \tilde{x}_j^{(n)}(b)))^{-1} \tag{6.16}$$

Now fix n and $x_0 \in \mathbb{R}$. Define

$$b(x_0) = \frac{P_n(x_0)}{P_{n-1}(x_0)} \tag{6.17}$$

with the convention $b = \infty$ if $P_{n-1}(x_0) = 0$. Define for $b \neq \infty$,

$$x_j^{(n)}(x_0) = \tilde{x}_j^{(n)}(b(x_0)) \qquad j = 1, \dots, n \tag{6.18}$$

and if $b(x_0) = \infty$,

$$x_j^{(n)}(x_0) = \tilde{x}_j^{(n-1)}(0) \qquad j = 1, \dots, n-1 \tag{6.19}$$

and

$$\lambda_j^{(n)}(x_0) = (K_{n-1}(x_j^{(n)}(x_0), x_j^{(n)}(x_0)))^{-1} \tag{6.20}$$

Then Theorem 6.1 becomes

THEOREM 6.5 (Gaussian Quadrature). *Fix n, x_0. Then*

$$\int Q(x)\, d\mu = \sum_{j=1}^{n} \lambda_j^{(n)}(x_0) Q(x_j^{(n)}(x_0)) \tag{6.21}$$

for all polynomials Q of degree up to:
(1) $2n-1$ *if* $P_n(x_0) = 0$
(2) $2n-2$ *if* $P_n(x_0) \neq 0 \neq P_{n-1}(x_0)$
(3) $2n-3$ *if* $P_{n-1}(x_0) = 0$.

REMARKS. 1. The sum goes to $n-1$ if $P_{n-1}(x_0) = 0$.

2. We can define $x_j^{(n)}$ to be the solutions of

$$p_{n-1}(x_0) p_n(x) - p_n(x_0) p_{n-1}(x) = 0 \tag{6.22}$$

which has degree n if $p_{n-1}(x_0) \neq 0$ and $n-1$ if $p_{n-1}(x_0) = 0$.

3. (6.20) makes sense even if $\mu(\mathbb{R}) \neq 1$ and dividing by $\mu(\mathbb{R})$ changes $\int Q(x)\, d\mu$ and $\lambda_j^{(n)}$ by the same amount, so (6.21) holds for all positive μ (with finite moments), not just the normalized ones.

4. The weights, $\lambda_j^{(n)}(x_0)$, in Gaussian quadrature are called Cotes numbers.

7. Markov–Stieltjes Inequalities

The ideas of this section go back to Markov [**63**] and Stieltjes [**92**] based on conjectures of Chebyshev [**21**] (see Freud [**34**]).

LEMMA 7.1. *Fix $x_1 < \cdots < x_n$ in $\mathbb{R}$ distinct and $1 \leq \ell < n$. Then there is a polynomial, Q, of degree $2n-2$ so that*
(i)

$$Q(x_j) = \begin{cases} 1 & j = 1, \dots, \ell \\ 0 & 1 = \ell + 1, \dots, n \end{cases} \tag{7.1}$$

(ii) *For all $x \in \mathbb{R}$,*

$$Q(x) \geq \chi_{(-\infty, x_\ell]}(x) \tag{7.2}$$

REMARK. Figure 1 has a graph of Q and $\chi_{(-\infty, x_\ell]}$ for $n = 5$, $\ell = 3$, $x_j = j - 1$.

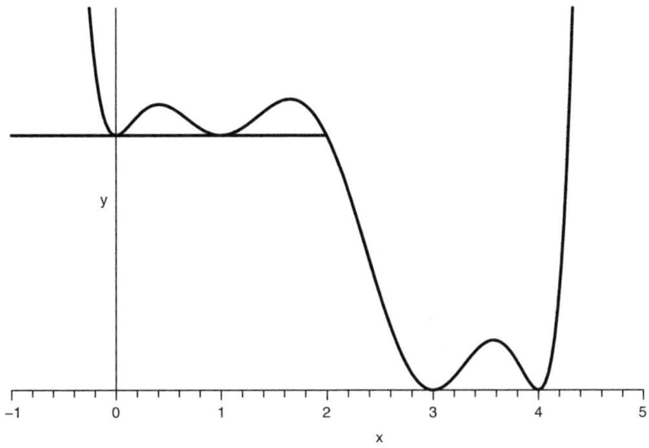

FIGURE 1. An interpolation polynomial

PROOF. By standard interpolation theory, there exists a unique polynomial of degree k with $k+1$ conditions of the form

$$Q(y_j) = Q'(y_j) = \cdots = Q^{(n_j)}(y_j) = 0$$

$\sum_j n_j = k+1$. Let Q be the polynomial of degree $2n-2$ with the n conditions in (7.1) and the $n-1$ conditions

$$Q'(x_j) = 0 \qquad j = 1, \dots, \ell-1, \ell+1, \dots, n \tag{7.3}$$

Clearly, Q' has at most $2n-3$ zeros. $n-1$ are given by (7.3) and, by Snell's theorem, each of the $n-2$ intervals $(x_1, x_2), \dots, (x_{\ell-1}, x_\ell), (x_{\ell+1}, x_{\ell+2}), \dots, (x_{n-1}, x_n)$ must have a zero. Since Q' is nonvanishing on $(x_\ell, x_{\ell+1})$ and $Q(x_\ell) = 1 > Q(x_{\ell+1}) = 0$, $Q'(y) < 0$ on $(x_\ell, x_{\ell+1})$. Tracking where Q' changes sign, one sees that (7.2) holds. $\square$

THEOREM 7.2. *Suppose $d\mu$ is a measure on $\mathbb{R}$ with finite moments. Then*

$$\begin{aligned} \sum_{\{j \mid x_j^{(n)}(x_0) \le x_0\}} \frac{1}{K_{n-1}(x_j^{(n)}(x_0), x_j^{(n)}(x_0))} &\ge \mu((-\infty, x_0]) \\ \ge \mu((-\infty, x_0)) \ge \sum_{\{j \mid x_j^{(n)}(x_0) < x_0\}} &\frac{1}{K_{n-1}(x_j^{(n)}(x_0), x_j^{(n)}(x_0))} \end{aligned} \tag{7.4}$$

REMARKS. 1. The two bounds differ by $K_{n-1}(x_0, x_0)^{-1}$.

2. These imply

$$\mu(\{x_0\}) \le K_{n-1}(x_0, x_0)^{-1} \tag{7.5}$$

In fact, one knows (see (9.21) below)

$$\mu(\{x_0\}) = \lim_{n \to \infty} K_{n-1}(x_0, x_0)^{-1} \tag{7.6}$$

If $\mu(\{x_0\}) = 0$, then the bounds are exact as $n \to \infty$.

PROOF. Suppose $P_{n-1}(x_0) \neq 0$. Let ℓ be such that $x_\ell^{(n)}(x_0) = x_0$. Let Q be the polynomial of Lemma 7.1. By (7.2),

$$\mu((-\infty, x_0]) \leq \int Q(x)\, d\mu$$

and, by (7.1) and Theorem 6.5, the integral is the sum on the left of (7.4).

Clearly, this implies

$$\mu((x_0, \infty)) \geq \sum_{\{j \mid x_j^{(n)}(x_0) > x_0\}} \frac{1}{K_{n-1}(x_j^{(n)}(x_0), x_j^{(n)}(x_0))}$$

which, by $x \to -x$ symmetry, implies the last inequality in (7.4). □

COROLLARY 7.3. *If* $\ell \leq k-1$, *then*

$$\begin{aligned}\sum_{j=\ell+1}^{k-1} \frac{1}{K(x_j^{(n)}(x_0), x_j^{(n)}(x_0))} &\leq \mu([x_\ell^{(n)}(x_0), x_k^{(n)}(x_0)]) \\ &\leq \sum_{j=\ell}^{k} \frac{1}{K(x_j^{(n)}(x_0), x_j^{(n)}(x_0))}\end{aligned} \tag{7.7}$$

PROOF. Note if $x_1 = x_\ell^{(n)}(x_0)$ for some ℓ, then $x_j^{(n)}(x_0) = x_j^{(n)}(x_1)$, so we get (7.7) by subtracting values of (7.4). □

Notice that this corollary gives effective lower bounds only if $k-1 \geq \ell+1$, that is, only on at least three consecutive zeros. The following theorem of Last–Simon [**57**], based on ideas of Golinskii [**41**], can be used on successive zeros (see [**57**] for the proof).

THEOREM 7.4. *If* E, E' *are distinct zeros of* $P_n(x)$, $\widetilde{E} = \frac{1}{2}(E+E')$ *and* $\delta > \frac{1}{2}|E-E'|$, *then*

$$|E - E'| \geq \frac{\delta^2 - (\frac{1}{2}|E-E'|^2)^2}{3n} \left[\frac{K_n(E,E)}{\sup_{|y-\widetilde{E}|\leq\delta} K_n(y,y)}\right]^{1/2} \tag{7.8}$$

8. Mixed CD Kernels

Recall that given a measure μ on $\mathbb{R}$ with finite moments and Jacobi parameters $\{a_n, b_n\}_{n=1}^\infty$, the *second kind polynomials* are defined by the recursion relations (1.5) but with initial conditions

$$q_0(x) = 0 \qquad q_1(x) = a_1^{-1} \tag{8.1}$$

so $q_n(x)$ is a polynomial of degree $n-1$. In fact, if $\tilde{\mu}$ is the measure with Jacobi parameters given by

$$\tilde{a}_n = a_{n+1} \qquad \tilde{b}_n = b_{n+1}$$

then

$$q_n(x; d\mu) = a_1^{-1} p_{n-1}(x; d\tilde{\mu}) \tag{8.2}$$

It is sometimes useful to consider

$$K_n^{(q)}(x,y) = \sum_{j=0}^{n} \overline{q_j(x)}\, q_j(y) \tag{8.3}$$

and the mixed CD kernel

$$K_n^{(pq)}(x,y) = \sum_{j=0}^{n} \overline{q_j(x)}\, p_j(y) \tag{8.4}$$

Since (8.2) implies

$$K_n^{(q)}(x,y;d\mu) = a_1^{-2} K_{n-1}(x,y;d\tilde{\mu}) \tag{8.5}$$

there is a CD formula for $K^{(q)}$ which follows immediately from the one for K. There is also a mixed CD formula for $K_n^{(pq)}$.

OPUC also have second kind polynomials, mixed CD kernels, and mixed CD formulae. These are discussed in Section 3.2 of [**80**].

Mixed CD kernels will enter in Section 21.

9. Variational Principle: Basics

If one thing marks the OP approach to the CD kernel that has been missing from the spectral theorists' approach, it is a remarkable variational principle for the diagonal kernel. We begin with:

LEMMA 9.1. *Fix $(\alpha_1,\dots,\alpha_m) \in \mathbb{C}^m$. Then*

$$\min\biggl(\sum_{j=1}^{m} |z_j|^2 \biggm| \sum_{j=1}^{m} \alpha_j z_j = 1\biggr) = \biggl(\sum_{j=1}^{m} |\alpha_j|^2\biggr)^{-1} \tag{9.1}$$

with the minimizer given uniquely by

$$z_j^{(0)} = \frac{\bar{\alpha}_j}{\sum_{j=1}^{m} |\alpha_j|^2} \tag{9.2}$$

REMARK. One can use Lagrange multipliers to a priori compute $z_j^{(0)}$ and prove this result.

PROOF. If

$$\sum_{j=1}^{m} \alpha_j z_j = 1 \tag{9.3}$$

then

$$\sum_{j=1}^{m} |z_j - z_j^{(0)}|^2 = \sum_{j=1}^{m} |z_j|^2 - \biggl(\sum_{j=1}^{m} |\alpha_j|^2\biggr)^{-1} \tag{9.4}$$

from which the result is obvious. □

If Q has $\deg(Q) \le n$ and $Q_n(z_0) = 1$, then

$$Q_n(z) = \sum_{j=0}^{n} \alpha_j x_j(z) \tag{9.5}$$

with x_j the orthonormal polynomials for a measure $d\mu$, then $\sum \alpha_j x_j(z_0) = 1$ and $\|Q_n\|_{L^2(\mathbb{C},d\mu)}^2 = \sum_{j=0}^{n} |\alpha_j|^2$. Thus the lemma implies:

THEOREM 9.2 (Christoffel Variational Principle). *Let μ be a measure on $\mathbb{C}$ with finite moments. Then for $\bar{z}_0 \in \mathbb{C}$,*

$$\min\biggl(\int |Q_n(z)|^2\, d\mu \biggm| Q_n(z_0) = 1,\ \deg(Q_n) \le n\biggr) = \frac{1}{K_n(z_0,z_0)} \tag{9.6}$$

and the minimizer is given by

$$Q_n(z, z_0) = \frac{K_n(z_0, z)}{K_n(z_0, z_0)} \tag{9.7}$$

One immediate useful consequence is:

THEOREM 9.3. *If $\mu \le \nu$, then*

$$K_n(z, z; d\nu) \le K_n(z, z; d\mu) \tag{9.8}$$

For this reason, it is useful to have comparison models:

EXAMPLE 9.4. Let $d\mu = d\theta/2\pi$ for $z = re^{i\theta}$ and $\zeta = e^{i\varphi}$. We have, since $\varphi_n(z) = z^n$,

$$K_n(z, \zeta) = \frac{1 - r^{n+1}e^{i(n+1)(\varphi-\theta)}}{1 - re^{i(\varphi-\theta)}} \tag{9.9}$$

If $r < 1$, $K_n(z, z_0)$ has a limit as $n \to \infty$, and for $z = e^{i\varphi}$, $z_0 = re^{i\theta}$, $r < 1$,

$$|Q_n(z, z_0)|^2 \,\frac{d\varphi}{2\pi} \to P_r(\theta, \varphi)\,\frac{d\varphi}{2\pi} \tag{9.10}$$

the Poisson kernel,

$$P_r(\theta, \varphi) = \frac{1 - r^2}{1 + r^2 - 2r\cos(\theta - \varphi)} \tag{9.11}$$

For $r = 1$, we have

$$|K_n(e^{i\theta}, e^{i\varphi})|^2 = \frac{\sin^2(\frac{n+1}{2}(\theta - \varphi))}{\sin^2(\theta - \varphi)} \tag{9.12}$$

the Fejér kernel.

For $r > 1$, we use

$$K_n(z, \zeta) = \bar{z}^n \zeta^n K_n\left(\frac{1}{z}, \frac{1}{\zeta}\right) \tag{9.13}$$

which implies, for $z = e^{i\varphi}$, $z_0 = re^{i\theta}$, $r > 1$,

$$|Q_n(z, z_0)|^2 \,\frac{d\varphi}{2\pi} \to P_{r^{-1}}(\theta, \varphi)\,\frac{d\varphi}{2\pi} \tag{9.14}$$

□

EXAMPLE 9.5. Let $d\mu_0$ be the measure

$$d\mu_0(x) = \frac{1}{2\pi}\sqrt{4 - x^2}\,\chi_{[-2,2]}(x)\,dx \tag{9.15}$$

on $[-2, 2]$. Then p_n are the Chebyshev polynomials of the second kind,

$$p_n(2\cos\theta) = \frac{\sin(n+1)\theta}{\sin\theta} \tag{9.16}$$

In particular, if $|x| \le 2 - \delta$,

$$|p_n(x + iy)| \le C_{1,\delta} e^{nC_{2,\delta}|y|} \tag{9.17}$$

and so

$$\frac{1}{n}\,|K_n(x + iy, x + iy)| \le C_{1,\delta}^2 e^{2nC_{2,\delta}|y|} \tag{9.18}$$

□

The following shows the power of the variational principle:

THEOREM 9.6. *Let*

$$d\mu = w(x)\,dx + d\mu_{\rm s} \tag{9.19}$$

Suppose for some x_0, δ, *we have*

$$w(x) \ge c > 0$$

for $x \in [x_0 - \delta, x_0 + \delta]$. *Then for any* $\delta' < \delta$ *and all* $x \in [x_0 - \delta', x_0 + \delta']$, *we have for all a real,*

$$\frac{1}{n} K_n\left(x + \frac{ia}{n}, x + \frac{ia}{n}\right) \le C_1 e^{C_2|a|} \tag{9.20}$$

PROOF. We can find a scaled and translated version of the $d\mu_0$ of (9.15) with $\mu \ge \mu_0$. Now use Theorem 9.3 and (9.18). □

The following has many proofs, but it is nice to have a variational one:

THEOREM 9.7. *Let* μ *be a measure on* $\mathbb{R}$ *of compact support. For all* $x_0 \in \mathbb{R}$,

$$\lim_{n\to\infty} K_n(x_0, x_0) = \mu(\{x_0\})^{-1} \tag{9.21}$$

REMARK. If $\mu(\{x_0\}) = 0$, the limit is infinite.

PROOF. Clearly, if $Q(x_0) = 1$, $\int |Q_n(x)|^2\,d\mu \ge \mu(\{x_0\})$, so

$$K_n(x_0, x_0) \le \mu(\{x_0\})^{-1} \tag{9.22}$$

On the other hand, pick $A \ge \operatorname{diam}(\sigma(d\mu))$ and let

$$Q_{2n}(x) = \left(1 - \frac{(x - x_0)^2}{A^2}\right)^n \tag{9.23}$$

For any a,

$$\sup_{\substack{|x - x_0| \ge a \\ x \in \sigma(d\mu)}} |Q_{2n}(x)| \equiv M_{2n}(a) \to 0 \tag{9.24}$$

so, since $Q_{2n} \le 1$ on $\sigma(d\mu)$,

$$K_n(x_0, x_0) \ge [\mu((x_0 - a, x_0 + a)) + M_{2n}(a)]^{-1} \tag{9.25}$$

so

$$\liminf K_n(x_0, x_0) \ge [\mu((x_0 - a, x_0 + a))] \tag{9.26}$$

for each a. Since $\lim_{a \downarrow 0} \mu((x_0 - a, x_0 + a)) = \mu(\{x_0\})$, (9.22) and (9.26) imply (9.21). □

10. The Nevai Class: An Aside

In his monograph, Nevai [**67**] emphasized the extensive theory that can be developed for OPRL measures whose Jacobi parameters obey

$$a_n \to a \qquad b_n \to b \tag{10.1}$$

for some b real and $a > 0$. He proved such measures have ratio asymptotics, that is, $P_{n+1}(z)/P_n(z)$ has a limit for all $z \in \mathbb{C}\backslash\mathbb{R}$, and Simon [**78**] proved a converse: Ratio asymptotics at one point of $\mathbb{C}_+$ implies there are a, b, with (10.1). The essential spectrum for such a measure is $[b - 2a, b + 2a]$, so the Nevai class is naturally associated with a single interval $\mathfrak{e} \subset \mathbb{R}$.

The question of what is the proper analog of the Nevai class for a set $\mathfrak{e}$ of the form

$$\mathfrak{e} = [\alpha_1, \beta_1] \cup [\alpha_2, \beta_2] \cup \dots [\alpha_{\ell+1}, \beta_{\ell+1}] \tag{10.2}$$

with

$$\alpha_1 < \beta_1 < \cdots < \alpha_{\ell+1} < \beta_{\ell+1} \tag{10.3}$$

has been answered recently and is relevant below.

The key was the realization of López [**8, 9**] that the proper analog of an arc of a circle was $|\alpha_n| \to a$ and $\bar{\alpha}_{n+1}\alpha_n \to a^2$ for some $a > 0$. This is not that α_n approaches a fixed sequence but rather that for each k,

$$\min_{e^{i\theta}\in\partial\mathbb{D}} \sum_{j=n}^{n+k} |\alpha_j - ae^{i\theta}| \to 0 \tag{10.4}$$

as $n \to \infty$. Thus, α_j approaches a set of Verblunsky coefficients rather than a fixed one.

For any finite gap set $\mathfrak{e}$ of the form (10.2)/(10.3), there is a natural torus, $\mathcal{J}_\mathfrak{e}$, of almost periodic Jacobi matrics with $\sigma_{\text{ess}}(J) = \mathfrak{e}$ for all $J \in \mathcal{J}_\mathfrak{e}$. This can be described in terms of minimal Herglotz functions [**90, 89**] or reflectionless two-sided Jacobi matrices [**75**]. All $J \in \mathcal{J}_\mathfrak{e}$ are periodic if and only if each $[\alpha_j, \beta_j]$ has rational harmonic measure. In this case, we say $\mathfrak{e}$ is periodic.

DEFINITION.

$$d_m(\{a_n, b_n\}_{n=1}^\infty, \{\tilde{a}_n, \tilde{b}_n\}_{n=1}^\infty) = \sum_{j=0}^\infty e^{-j}(|a_{m+j} - \tilde{a}_{m+j}| + |b_{m+j} - \tilde{b}_{m+j}|) \tag{10.5}$$

$$d_m(\{a_n, b_n\}, \mathcal{J}_\mathfrak{e}) = \min_{J\in\mathcal{J}_\mathfrak{e}} d_m(\{a_n, b_n\}, J) \tag{10.6}$$

DEFINITION. The *Nevai class* for $\mathfrak{e}$, $N(\mathfrak{e})$, is the set of all Jacobi matrices, J, with

$$d_m(J, \mathcal{J}_\mathfrak{e}) \to 0 \tag{10.7}$$

as $m \to \infty$.

This definition is implicit in Simon [**81**]; the metric d_m is from [**26**]. Notice that in case of a single gap $\mathfrak{e}$ in $\partial\mathbb{D}$, the isospectral torus is the set of $\{\alpha_n\}_{n=0}^\infty$ with $\alpha_n = ae^{i\theta}$ for all n where a is $\mathfrak{e}$ dependent and fixed and θ is arbitrary. The above definition is the López class.

That this is the "right" definition is seen by the following pair of theorems:

THEOREM 10.1 (Last–Simon [**56**]). *If $J \in N(\mathfrak{e})$, then*

$$\sigma_{\text{ess}}(J) = \mathfrak{e} \tag{10.8}$$

THEOREM 10.2 ([**26**] for periodic $\mathfrak{e}$'s; [**75**] in general). *If*

$$\sigma_{\text{ess}}(J) = \sigma_{\text{ac}}(J) = \mathfrak{e}$$

then $J \in N(\mathfrak{e})$.

11. Delta Function Limits of Trial Polynomials

Intuitively, the minimizer, $Q_n(x, x_0)$, in the Christoffel variational principle must be 1 at z_0 and should try to be small on the rest of $\sigma(d\mu)$. As the degree gets larger and larger, one expects it can do this better and better. So one might guess that for every $\delta > 0$,

$$\sup_{\substack{|x-x_0|>\delta \\ x\in\sigma(d\mu)}} |Q_n(x, x_0)| \to 0 \tag{11.1}$$

While this happens in many cases, it is too much to hope for. If $x_1 \in \sigma(d\mu)$ but μ has very small weight near x_1, then it may be a better strategy for Q_n not to be small very near x_1. Indeed, we will see (Example 11.3) that the sup in (11.1) can go to infinity. What is more likely is to expect that $|Q_n(x,x_0)|^2\,d\mu$ will be concentrated near x_0. We normalize this to define

$$d\eta_n^{(x_0)}(x) = \frac{|Q_n(x,x_0)|^2\,d\mu(x)}{\int |Q_n(x,x_0)|^2\,d\mu(x)} \tag{11.2}$$

so, by (9.6)/(9.7), in the OPRL case,

$$d\eta_n^{(x_0)}(x) = \frac{|K_n(x,x_0)|^2}{K_n(x,x_0)}\,d\mu(x) \tag{11.3}$$

We say μ obeys the Nevai δ-convergence criterion if and only if, in the sense of weak (aka vague) convergence of measures,

$$d\eta_n^{(x_0)}(x) \to \delta_{x_0} \tag{11.4}$$

the point mass at x_0. In this section, we will explore when this holds.

Clearly, if $x_0 \notin \sigma(d\mu)$, (11.4) cannot hold. We saw, for OPUC with $d\mu = d\theta/2\pi$ and $z \notin \partial\mathbb{D}$, the limit was a Poisson measure, and similar results should hold for suitable OPRL. But we will see below (Example 11.2) that even on $\sigma(d\mu)$, (11.4) can fail. The major result below is that for Nevai class on $\mathfrak{e}^{\text{int}}$, it does hold. We begin with an equivalent criterion:

DEFINITION. We say Nevai's lemma holds if

$$\lim_{n\to\infty} \frac{|p_n(x_0)|^2}{K_n(x_0,x_0)} = 0 \tag{11.5}$$

THEOREM 11.1. *If $d\mu$ is a measure on $\mathbb{R}$ with bounded support and*

$$\inf_n a_n > 0 \tag{11.6}$$

then for any fixed $x_0 \in \mathbb{R}$,

$$(11.4) \Leftrightarrow (11.5)$$

REMARK. That (11.5) $\Rightarrow$ (11.4) is in Nevai [**67**]. The equivalence is a result of Breuer–Last–Simon [**14**].

PROOF. Since

$$1 - \frac{K_{n-1}(x_0,x_0)}{K_n(x_0,x_0)} = \frac{|p_n(x_0)|^2}{K_n(x_0,x_0)} \tag{11.7}$$

$$(11.5) \Leftrightarrow \frac{K_{n-1}(x_0,x_0)}{K_n(x_0,x_0)} \to 1 \tag{11.8}$$

so

$$(11.5) \Rightarrow \frac{|p_{n+1}(x_0)|^2}{K_n(x_0,x_0)} = \frac{|p_{n+1}(x_0)|^2}{K_{n+1}(x_0,x_0)}\,\frac{K_{n+1}(x_0,x_0)}{K_n(x_0,x_0)} \to 0$$

We thus conclude

$$(11.5) \Leftrightarrow \frac{|p_n(x_0)|^2 + |p_{n+1}(x_0)|^2}{K_n(x_0,x_0)} \to 0 \tag{11.9}$$

By the CD formula and orthonormality of $p_j(x)$,

$$\int |x-x_0|^2 |K_n(x,x_0)|^2\,d\mu = a_{n+1}^2[p_n(x_0)^2 + p_{n+1}(x_0)^2] \tag{11.10}$$

so, by (11.6) and (11.10),

$$\int |x - x_0|^2 \, d\eta_n^{(x_0)}(x) \to 0 \Leftrightarrow (11.5)$$

when a_n is uniformly bounded above and away from zero. But since $d\eta_n$ have support in a fixed interval,

$$(11.4) \Leftrightarrow \int |x - x_0|^2 \, d\eta_n^{(x_0)} \to 0 \qquad \square$$

EXAMPLE 11.2. Suppose at some point x_0, we have

$$\lim_{n\to\infty} (|p_n(x_0)|^2 + |p_{n+1}(x_0)|^2)^{1/n} \to A > 1 \tag{11.11}$$

We claim that

$$\limsup_{n\to\infty} \frac{|p_n(x_0)|^2}{K_n(x_0, x_0)} > 0 \tag{11.12}$$

for if (11.12) fails, then (11.5) holds and, by (11.7), for any ε, we can find N_0 so for $n \geq N_0$,

$$K_{n+1}(x_0, x_0) \leq (1 + \varepsilon) K_n(x_0, x_0) \tag{11.13}$$

so

$$\lim K_n(x_0, x_0)^{1/n} \leq 1$$

So, by (11.5), (11.11) fails. Thus, (11.11) implies that (11.5) fails, and so (11.4) fails. $\square$

REMARK. As the proof shows, rather than a limit in (11.12), we can have a lim inf > 1.

The first example of this type was found by Szwarc [**94**]. He has a $d\mu$ with pure points at $2 - n^{-1}$ but not at 2, and so that the Lyapunov exponent at 2 was positive but 2 was not an eigenvalue, so (11.11) holds. The Anderson model (see [**20**]) provides a more dramatic example. The spectrum is an interval $[a, b]$ and (11.11) holds for a.e. $x \in [a, b]$. The spectral measure in this case is supported at eigenvalues and at eigenvalues (11.8), and so (11.4) holds. Thus (11.4) holds on a dense set in $[a, b]$ but fails for Lebesgue a.e. x_0!

EXAMPLE 11.3. A Jacobi weight has the form

$$d\mu(x) = C_{a,b}(1 - x)^a (1 + x)^b \, dx \tag{11.14}$$

with $a, b > -1$. In general, one can show [**93**]

$$p_n(1) \sim c n^{a+1/2} \tag{11.15}$$

so if $x_0 \in (-1, 1)$ where $|p_n(x_0)|^2 + |p_{n-1}(x_0)|^2$ is bounded above and below, one has

$$\frac{|K_n(x_0, 1)|}{K_n(x_0, x_0)} \sim \frac{n^{a+1/2}}{n} = n^{a-1/2}$$

so if $a > \frac{1}{2}$, $|Q_n(x_0, 1)| \to \infty$. Since $d\mu(x)$ is small for x near 1, one can (and, as we will see, does) have (11.4) even though (11.1) fails. $\square$

With various counterexamples in place (and more later!), we turn to the positive results:

THEOREM 11.4 (Nevai [**67**], Nevai–Totik–Zhang [**69**]). *If $d\mu$ is a measure in the classical Nevai class (i.e., for a single interval, $\mathfrak{e} = [b-2a, b+2a]$), then* (11.5) *and so* (11.4) *holds uniformly on* $\mathfrak{e}$.

THEOREM 11.5 (Zhang [**108**], Breuer–Last–Simon [**14**]). *Let $\mathfrak{e}$ be a periodic finite gap set and let μ lie in the Nevai class for $\mathfrak{e}$. Then* (11.5) *and so* (11.4) *holds uniformly on* $\mathfrak{e}$.

THEOREM 11.6 (Breuer–Last–Simon [**14**]). *Let $\mathfrak{e}$ be a general finite gap set and let μ lie in the Nevai class for $\mathfrak{e}$. Then* (11.5) *and so* (11.4) *holds uniformly on compact subsets of* $\mathfrak{e}^{\text{int}}$.

REMARKS. 1. Nevai [**67**] proved (10.4)/(10.5) for the classical Nevai class for every energy in $\mathfrak{e}$ but only uniformly on compacts of $\mathfrak{e}^{\text{int}}$. Uniformity on all of $\mathfrak{e}$ using a beautiful lemma is from [**69**].

2. Zhang [**108**] proved Theorem 11.5 for any μ whose Jacobi parameters approached a fixed periodic Jacobi matrix. Breuer–Last–Simon [**14**] noted that without change, Zhang's result holds for the Nevai class.

3. It is hoped that the final version of [**14**] will prove the result in Theorem 11.6 on all of $\mathfrak{e}$, maybe even uniformly in $\mathfrak{e}$.

EXAMPLE 11.7 ([**14**]). In the next section, we will discuss regular measures. They have zero Lyapunov exponent on $\sigma_{\text{ess}}(\mu)$, so one might expect Nevai's lemma could hold—and it will in many regular cases. However, [**14**] prove that if $b_n \equiv 0$ and a_n is alternately 1 and $\frac{1}{2}$ on successive very long blocks (1 on blocks of size 3^{n^2} and $\frac{1}{2}$ on blocks of size 2^{n^2}), then $d\mu$ is regular for $\sigma(d\mu) = [-2,2]$. But for a.e. $x \in [-2,2] \setminus [-1,1]$, (10.4) and (10.3) fail. □

CONJECTURE 11.8 ([**14**]). The following is extensively discussed in [**14**]: For general OPRL of compact support and a.e. x with respect to μ, (10.4) and so (10.3) holds.

12. Regularity: An Aside

There is another class besides the Nevai class that enters in variational problems because it allows exponential bounds on trial polynomials. It relies on notions from potential theory; see [**42, 52, 73, 102**] for the general theory and [**91, 85**] for the theory in the context of orthogonal polynomials.

DEFINITION. Let μ be a measure with compact support and let $\mathfrak{e} = \sigma_{\text{ess}}(\mu)$. We say μ is *regular* for $\mathfrak{e}$ if and only if

$$\lim_{n\to\infty} (a_1 \dots a_n)^{1/n} = C(\mathfrak{e}) \tag{12.1}$$

the capacity of $\mathfrak{e}$.

For $\mathfrak{e} = [-1,1]$, $C(\mathfrak{e}) = \frac{1}{2}$ and the class of regular measures was singled out initially by Erdős–Turán [**32**] and extensively studied by Ullman [**103**]. The general theory was developed by Stahl–Totik [**91**].

Recall that any set of positive capacity has an equilibrium measure, $\rho_{\mathfrak{e}}$, and Green's function, $G_{\mathfrak{e}}$, defined by requiring $G_{\mathfrak{e}}$ is harmonic on $\mathbb{C}\setminus\mathfrak{e}$, $G_{\mathfrak{e}}(z) = \log|z| + O(1)$ near infinity, and for quasi-every $x \in \mathfrak{e}$,

$$\lim_{z_n\to x} G_{\mathfrak{e}}(z_n) = 0 \tag{12.2}$$

(quasi-every means except for a set of capacity 0). $\mathfrak{e}$ is called regular for the Dirichlet problem if and only if (12.2) holds for every $x \in \mathfrak{e}$. Finite gap sets are regular for the Dirichlet problem.

One major reason regularity will concern us is:

THEOREM 12.1. *Let $\mathfrak{e} \subset \mathbb{R}$ be compact and regular for the Dirichlet problem. Let μ be a measure regular for $\mathfrak{e}$. Then for any ε, there is $\delta > 0$ and C_ε so that*

$$\sup_{\operatorname{dist}(z,\mathfrak{e})<\delta} |p_n(z, d\mu)| \le C_\varepsilon e^{\varepsilon|n|} \tag{12.3}$$

For proofs, see [**91, 85**]. Since K_n has $n+1$ terms, (12.3) implies

$$\sup_{\substack{\operatorname{dist}(z,\mathfrak{e})<\delta \\ \operatorname{dist}(w,\mathfrak{e})<\delta}} |K_n(z, w)| \le (n+1) C_\varepsilon^2 e^{2\varepsilon|n|} \tag{12.4}$$

and for the minimum (since $K_n(z_0, z_0) \ge 1$),

$$\sup_{\substack{\operatorname{dist}(z,\mathfrak{e})<\delta \\ \operatorname{dist}(z_0,\mathfrak{e})<\delta}} |Q_n(z, z_0)| \le (n+1) C_\varepsilon^2 e^{2\varepsilon|n|} \tag{12.5}$$

The other reason regularity enters has to do with the density of zeros. If $x_j^{(n)}$ are the zeros of $p_n(x, d\mu)$, we define the zero counting measure, $d\nu_n$, to be the probability measure that gives weight to n^{-1} to each $x_j^{(n)}$. For the following, see [**91, 85**]:

THEOREM 12.2. *Let $\mathfrak{e} \subset \mathbb{R}$ be compact and let μ be a regular measure for $\mathfrak{e}$. Then*

$$d\nu_n \to d\rho_{\mathfrak{e}} \tag{12.6}$$

the equilibrium measure for $\mathfrak{e}$.

In (12.6), the convergence is weak.

13. Weak Limits

A major theme in the remainder of this review is pointwise asymptotics of $\frac{1}{n+1} K_n(x, y; d\mu)$ and its diagonal. Therefore, it is interesting that one can say something about $\frac{1}{n+1} K_n(x, x; d\mu)\, d\mu(x)$ without pointwise asymptotics. Notice that

$$d\mu_n(x) \equiv \frac{1}{n+1} K_n(x, x; d\mu)\, d\mu(x) \tag{13.1}$$

is a probability measure. Recall the density of zeros, ν_n, defined after (12.5).

THEOREM 13.1. *Let μ have compact support. Let ν_n be the density of zeros and μ_n given by (13.1). Then for any $\ell = 0, 1, 2, \dots$,*

$$\left| \int x^\ell \, d\nu_{n+1} - \int x^\ell \, d\mu_n \right| \to 0 \tag{13.2}$$

In particular, $d\mu_{n(j)}$ and $d\nu_{n(j)+1}$ have the same weak limits for any subsequence $n(j)$.

PROOF. By Theorem 6.2, the zeros of P_{n+1} are eigenvalues of $\pi_n M_x \pi_n$, so

$$\int x^\ell \, d\nu_{n+1} = \frac{1}{n+1} \operatorname{Tr}((\pi_n M_x \pi_n)^\ell) \tag{13.3}$$

On the other hand, since $\{p_j\}_{j=0}^n$ is a basis for $\operatorname{ran}(\pi_n)$,

$$\begin{aligned} \int x^\ell \, d\mu_n &= \frac{1}{n+1} \sum_{j=0}^n \int x^\ell |p_j(x)|^2 \, d\mu(x) \\ &= \frac{1}{n+1} \operatorname{Tr}(\pi_n M_x^\ell \pi_n) \end{aligned} \tag{13.4}$$

It is easy to see that $(\pi_n M_x \pi_n)^\ell - \pi_n M_x^\ell \pi_n$ is rank at most ℓ, so

$$\text{LHS of (13.2)} \leq \frac{\ell}{n+1} \, \|M_x\|^\ell$$

goes to 0 as $n \to \infty$ for ℓ fixed. □

REMARK. This theorem is due to Simon [**88**] although the basic fact goes back to Avron–Simon [**7**].

See Simon [**88**] for an interesting application to comparison theorems for limits of density of states. We immediately have:

COROLLARY 13.2. *Suppose that*

$$d\mu = w(x) \, dx + d\mu_{\mathrm{s}} \tag{13.5}$$

with $d\mu_{\mathrm{s}}$ Lebesgue singular, and on some open interval $I \subset \mathfrak{e} = \sigma_{\mathrm{ess}}(d\mu)$ we have $d\nu_n \to d\nu_\infty$ and

$$d\nu_\infty \restriction I = \nu_\infty(x) \, dx \tag{13.6}$$

and suppose that uniformly on I,

$$\lim \frac{1}{n} K_n(x,x) = g(x) \tag{13.7}$$

and $w(x) \neq 0$ on I. Then

$$g(x) = \frac{\nu_\infty(x)}{w(x)}$$

PROOF. The theorem implies $d\nu_\infty \restriction I = w(x) g(x)$. □

Thus, in the regular case, we expect that "usually"

$$\frac{1}{n} K_n(x,x) \to \frac{\rho_{\mathfrak{e}}(x)}{w(x)} \tag{13.8}$$

This is what we explore in much of the rest of this paper.

14. Variational Principle: Máté–Nevai Upper Bounds

The Cotes numbers, $\lambda_n(z_0)$, are given by (9.6), so upper bounds on $\lambda_n(z_0)$ mean lower bounds on diagonal CD kernels and there is a confusion of "upper bounds" and "lower bounds." We will present here some very general estimates that come from the use of trial functions in (9.6) so they are called Máté–Nevai upper bounds (after [**65**]), although we will write them as lower bounds on K_n. One advantage is their great generality.

DEFINITION. Let $d\mu$ be a measure on $\mathbb{R}$ of the form

$$d\mu = w(x)\,dx + d\mu_{\rm s} \tag{14.1}$$

where $d\mu_{\rm s}$ is singular with respect to Lebesgue measure. We call x_0 a Lebesgue point of μ if and only if

$$\frac{n}{2}\,\mu_{\rm s}\biggl(\biggl[x_0 - \frac{1}{n}, x_0 + \frac{1}{n}\biggr]\biggr) \to 0 \tag{14.2}$$

$$\frac{n}{2}\int_{x_0-\frac{1}{n}}^{x_0+\frac{1}{n}} |w(x) - w_0(x_0)|\,dx \to 0 \tag{14.3}$$

It is a fundamental fact of harmonic analysis ([**76**]) that for any μ Lebesgue-a.e., x_0 in $\mathbb{R}$ is a Lebesgue point for μ. Here is the most general version of the MN upper bound:

THEOREM 14.1. *Let $\mathfrak{e} \subset \mathbb{R}$ be an arbitrary compact set which is regular for the Dirichlet problem. Let $I \subset \mathfrak{e}$ be a closed interval. Let $d\mu$ be a measure with compact support in $\mathbb{R}$ with $\sigma_{\rm ess}(d\mu) \subset \mathfrak{e}$. Then for any Lebesgue point, x in I,*

$$\liminf_{n\to\infty} \frac{1}{n}\, K_n(x,x) \geq \frac{\rho_{\mathfrak{e}}(x)}{w(x)} \tag{14.4}$$

where $d\rho_{\mathfrak{e}} \restriction I = \rho_{\mathfrak{e}}(x)\,dx$. If w is continuous on I (including at the endpoints as a function in a neighborhood of I) and nonvanishing, then (14.4) *holds uniformly on I. If $x_n \to x \in I$ and $A = \sup_n n|x_n - x| < \infty$ and x is a Lebesgue, then* (14.4) *holds with $K_n(x,x)$ replaced by $K_n(x_n,x_n)$. If w is continuous and nonvanishing on I, then this extended convergence is uniform in $x \in I$ and x_n's with $A \leq A_0 < \infty$.*

REMARKS. 1. If $I \subset \mathfrak{e}$ is a nontrivial interval, the measure $d\rho_{\mathfrak{e}} \restriction I$ is purely absolutely continuous (see, e.g., [**85, 89**]).

2. For OPUC, this is a result of Máté–Nevai [**64**]. The translation to OPRL on $[-1,1]$ is explicit in Máté–Nevai–Totik [**66**]. The extension to general sets via polynomial mapping and approximation (see Section 18) is due to Totik [**96**]. These papers also require a local Szegő condition, but that is only needed for lower bounds on λ_n (see Section 17). They also don't state the $x_n \to x_\infty$ result, which is a refinement introduced by Lubinsky [**60**] who implemented it in certain $[-1,1]$ cases.

3. An alternate approach for Totik's polynomial mapping is to use trial functions based on Jost–Floquet solutions for periodic problems; see Section 19 (and also [**87, 89**]).

One can combine (14.4) with weak convergence and regularity to get

THEOREM 14.2 (Simon [**88**]). *Let $\mathfrak{e} \subset \mathbb{R}$ be an arbitrary compact set, regular for the Dirichlet problem. Let $d\mu$ be a measure with compact support in $\mathbb{R}$ with $\sigma_{\rm ess}(d\mu) = \mathfrak{e}$ and with $d\mu$ regular for $\mathfrak{e}$. Let $I \subset \mathfrak{e}$ be an interval so $w(x) > 0$ a.e. on I. Then*

(i) $$\int_I \biggl|\frac{1}{n}\, K_n(x,x)\, w(x) - \rho_{\mathfrak{e}}(x)\biggr|\, dx \to 0 \tag{14.5}$$

(ii) $$\int_I \frac{1}{n}\, K_n(x,x)\, d\mu_{\rm s}(x) \to 0 \tag{14.6}$$

PROOF. By Theorems 12.2 and 13.1,

$$\frac{1}{n} K_n(x,x)\,d\mu \to d\rho_{\mathfrak{e}} \tag{14.7}$$

Let ν_1 be a limit point of $\frac{1}{n}K_n(x,x)\,d\mu_{\text{s}}$ and

$$d\nu_2 = d\rho_{\mathfrak{e}} - d\nu_1 \tag{14.8}$$

If $f \geq 0$, by Fatou's lemma and (14.4),

$$\int_I f\,d\nu_2 \geq \int_I \rho_{\mathfrak{e}}(x) f(x)\,dx \tag{14.9}$$

that is, $d\nu_2 \restriction I \geq \rho_{\mathfrak{e}}(x)\,dx \restriction I$. By (14.8), $d\nu_2 \restriction I \leq \rho_{\mathfrak{e}}(x)\,dx$. It follows $d\nu_1 \restriction I$ is 0 and $d\nu_2 \restriction I = d\rho_{\mathfrak{e}} \restriction I$.

By compactness, $\frac{1}{n}K_n(x,x)\,d\mu_{\text{s}} \restriction I \to 0$ weakly, implying (14.6). By a simple argument [**88**], weak convergence of $\frac{1}{n}K_n(x,x)w(x)\,dx \to \rho_{\mathfrak{e}}(x)\,dx$ and (14.4) imply (14.5). □

15. Criteria for A.C. Spectrum

Define

$$N = \left\{ x \in \mathbb{R} \;\middle|\; \liminf \frac{1}{n} K_n(x,x) < \infty \right\} \tag{15.1}$$

so that

$$\mathbb{R} \setminus N = \left\{ x \in \mathbb{R} \;\middle|\; \lim \frac{1}{n} K_n(x,x) = \infty \right\} \tag{15.2}$$

Theorem 14.1 implies

THEOREM 15.1. *Let $\mathfrak{e} \subset \mathbb{R}$ be an arbitrary compact set and $d\mu = w(x)\,dx + d\mu_{\text{s}}$ a measure with $\sigma(\mu) = \mathfrak{e}$. Let $\Sigma_{\text{ac}} = \{x \mid w(x) > 0\}$. Then $N \setminus \Sigma_{\text{ac}}$ has Lebesgue measure zero.*

PROOF. If $x_0 \in \mathbb{R} \setminus \Sigma_{\text{ac}}$ and is a Lebesgue point of μ, then $w(x_0) = 0$ and, by Theorem 14.1, $x_0 \in \mathbb{R} \setminus N$. Thus,

$$(\mathbb{R} \setminus \Sigma_{\text{ac}}) \setminus (\mathbb{R} \setminus N) = N \setminus \Sigma_{\text{ac}}$$

has Lebesgue measure zero. □

REMARK. This is a direct but not explicit consequence of the Máté–Nevai ideas [**64**]. Without knowing of this work, Theorem 15.1 was rediscovered with a very different proof by Last–Simon [**55**].

On the other hand, following Last–Simon [**55**], we note that Fatou's lemma and

$$\int \frac{1}{n} K_n(x,x)\,d\mu(x) = 1 \tag{15.3}$$

implies

$$\int \liminf \frac{1}{n} K_n(x,x)\,d\mu(x) \leq 1 \tag{15.4}$$

so

THEOREM 15.2 ([**55**]). *$\Sigma_{\text{ac}} \setminus N$ has Lebesgue measure zero.*

Thus, up to sets of measure zero, $\Sigma_{\rm ac} = N$. What is interesting is that this holds, for example, when $\mathfrak{e}$ is a positive measure Cantor set as occurs for the almost Mathieu operator ($a_n \equiv 1$, $b_n = \lambda\cos(\pi\alpha n + \theta)$, $|\lambda| < 2$, $\lambda \neq 0$, α irrational). This operator has been heavily studied; see Last [**54**].

16. Variational Principle: Nevai Trial Polynomial

A basic idea is that if $d\mu_1$ and $d\mu_2$ look alike near x_0, there is a good chance that $K_n(x_0, x_0; d\mu_1)$ and $K_n(x_0, x_0; d\mu_2)$ are similar for n large. The expectation (13.8) says they better have the same support (and be regular for that support), but this is a reasonable guess.

It is natural to try trial polynomials minimizing $\lambda_n(x_0, d\mu_1)$ in the Christoffel variational principle for $\lambda_n(x_0, d\mu_2)$, but Example 11.3 shows this will not work in general. If $d\mu_1$ has a strong zero near some other x_1, the trial polynomial for $d\mu_1$ may be large near x_1 and be problematical for $d\mu_2$ if it does not have a zero there. Nevai [**67**] had the idea of using a localizing factor to overcome this.

Suppose $\mathfrak{e} \subset \mathbb{R}$, a compact set which, for now, we suppose contains $\sigma(d\mu_1)$ and $\sigma(d\mu_2)$. Pick $A = \text{diam}(\mathfrak{e})$ and consider (with $[\,\cdot\,] \equiv$ integral part)

$$\left(1 - \frac{(x-x_0)^2}{A^2}\right)^{[\varepsilon n]} \equiv N_{2[\varepsilon n]}(x) \tag{16.1}$$

Then for any δ,

$$\sup_{\substack{|x-x_0|>\delta \\ x\in\mathfrak{e}}} N_{2[\varepsilon n]}(x) \leq e^{-c(\delta,\varepsilon)n} \tag{16.2}$$

so if $Q_{n-2[\varepsilon n]}(x)$ is the minimizer for μ_1 and $\mathfrak{e}$ is regular for the Dirichlet problem and μ_1 is regular for $\mathfrak{e}$, then the Nevai trial function

$$N_{2[\varepsilon n]}(x)Q_{n-2[\varepsilon n]}(x)$$

will be exponentially small away from x_0.

For this to work to compare $\lambda_n(x_0, d\mu_1)$ and $\lambda(x_0, d\mu_2)$, we need two additional properties of $\lambda_n(x_0, d\mu_1)$:

(a) $\lambda_n(x_0, d\mu_1) \geq C_\varepsilon e^{-\varepsilon n}$ for each $\varepsilon < 0$. This is needed for the exponential contributions away from x_0 not to matter.

(b)
$$\lim_{\varepsilon\downarrow 0}\limsup_{n\to\infty} \frac{\lambda_n(x_0, d\mu_1)}{\lambda_{n-2[\varepsilon n]}(x, d\mu_1)} = 1$$
so that the change from Q_n to $Q_{n-2[\varepsilon n]}$ does not matter.

Notice that both (a) and (b) hold if

$$\lim_{n\to\infty} n\lambda_n(x_0, d\mu) = c > 0 \tag{16.3}$$

If one only has $\mathfrak{e} = \sigma_{\rm ess}(d\mu_2)$, one can use explicit zeros in the trial polynomials to mask the eigenvalues outside $\mathfrak{e}$.

For details of using Nevai trial functions, see [**87, 89**]. Below we will just refer to using Nevai trial functions.

17. Variational Principle: Máté–Nevai–Totik Lower Bound

In [**66**], Máté–Nevai–Totik proved:

THEOREM 17.1. *Let $d\mu$ be a measure on $\partial\mathbb{D}$*

$$d\mu = \frac{w(\theta)}{2\pi}\,d\theta + d\mu_{\rm s} \tag{17.1}$$

which obeys the Szegő condition

$$\int \log(w(\theta))\,\frac{d\theta}{2\pi} > -\infty \tag{17.2}$$

Then for a.e. $\theta_\infty \in \partial\mathbb{D}$,

$$\liminf n\lambda_n(\theta_\infty) \geq w(\theta_\infty) \tag{17.3}$$

This remains true if $\lambda_n(\theta_\infty)$ is replaced by $\lambda_n(\theta_n)$ with $\theta_n \to \theta_\infty$ obeying $\sup n|\theta_n - \theta_\infty| < \infty$.

REMARKS. 1. The proof in [**66**] is clever but involved ([**89**] has an exposition); it would be good to find a simpler proof.

2. [**66**] only has the result $\theta_n = \theta_\infty$. The general θ_n result is due to Findley [**33**].

3. The θ_∞ for which this is proven have to be Lebesgue points for $d\mu$ as well as Lebesgue points for $\log(w)$ and for its conjugate function.

4. As usual, if I is an interval with w continuous and nonvanishing, and $\mu_{\rm s}(I) = 0$, (17.3) holds uniformly if $\theta_\infty \in I$.

By combining this lower bound with the Máté–Nevai upper bound, we get the result of Máté–Nevai–Totik [**66**]:

THEOREM 17.2. *Under the hypothesis of Theorem 17.1, for a.e. $\theta_\infty \in \partial\mathbb{D}$,*

$$\lim_{n\to\infty} n\lambda_n(\theta_\infty) = w(\theta_\infty) \tag{17.4}$$

This remains true if $\lambda_n(\theta_\infty)$ is replaced by $\lambda_n(\theta_n)$ with $\theta_n \to \theta_\infty$ obeying $\sup n|\theta_n - \theta_\infty| < \infty$. If I is an interval with w continuous on I and $\mu_{\rm s}(I) = 0$, then these results hold uniformly in I.

REMARK. It is possible (see remarks in Section 4.6 of [**68**]) that (17.4) holds if a Szegő condition is replaced by $w(\theta) > 0$ for a.e. θ. Indeed, under that hypothesis, Simon [**88**] proved that

$$\int_0^{2\pi} |w(\theta)(n\lambda_n(\theta))^{-1} - 1|\,\frac{d\theta}{2\pi} \to 0$$

There have been significant extensions of Theorem 17.2 to OPRL on fairly general sets:

1. [**66**] used the idea of Nevai trial functions (Section 16) to prove the Szegő condition could be replaced by regularity plus a local Szegő condition.
2. [**66**] used the Szegő mapping to get a result for $[-1, 1]$.
3. Using polynomial mappings (see Section 18) plus approximation, Totik [**96**] proved a general result (see below); one can replace polynomial mappings by Floquet–Jost solutions (see Section 19) in the case of continuous weights on an interval (see [**87**]).

Here is Totik's general result (extended from $\sigma(d\mu) \subset \mathfrak{e}$ to $\sigma_{\rm ess}(d\mu) \subset \mathfrak{e}$):

THEOREM 17.3 (Totik [**96, 99**]). *Let $\mathfrak{e}$ be a compact subset of $\mathbb{R}$. Let $I \subset \mathfrak{e}$ be an interval. Let $d\mu$ have $\sigma_{\text{ess}}(d\mu) = \mathfrak{e}$ be regular for $\mathfrak{e}$ with*

$$\int_I \log(w)\, dx > -\infty \tag{17.5}$$

Then for a.e. $x_\infty \in I$,

$$\lim_{n\to\infty} \frac{1}{n} K_n(x_\infty, x_\infty) = \frac{\rho_{\mathfrak{e}}(x_\infty)}{w(x_\infty)} \tag{17.6}$$

The same limit holds for $\frac{1}{n}K_n(x_n, x_n)$ if $\sup_n n|x_n - x_\infty| < \infty$. If $\mu_s(I) = \emptyset$ and w is continuous and nonvanishing on I, then those limits are uniform on $x_\infty \in I$ and on all x_n's with $\sup_n n|x_n - x_\infty| \le A$ (uniform for each fixed A).

REMARKS. 1. Totik [**98**] recently proved asymptotic results for suitable CD kernels for OPs which are neither OPUC nor OPRL.

2. The extension to general compact $\mathfrak{e}$ without an assumption of regularity for the Dirichlet problem is in [**99**].

18. Variational Principle: Polynomial Maps

In passing from $[-1, 1]$ to fairly general sets, one uses a three-step process. A finite gap set is an $\mathfrak{e}$ of the form

$$\mathfrak{e} = [\alpha_1, \beta_1] \cup [\alpha_2, \beta_2] \cup \cdots \cup [\alpha_{\ell+1}, \beta_{\ell+1}] \tag{18.1}$$

where

$$\alpha_1 < \beta_1 < \alpha_2 < \beta_2 < \cdots < \alpha_{\ell+1} < \beta_{\ell+1} \tag{18.2}$$

$\mathcal{E}_f$ will denote the family of finite gap sets. We write $\mathfrak{e} = \mathfrak{e}_1 \cup \cdots \cup \mathfrak{e}_{\ell+1}$ in this case with the $\mathfrak{e}_j$ closed disjoint intervals. $\mathcal{E}_p$ will denote the set of what we called periodic finite gap sets in Section 10—ones where each $\mathfrak{e}_j$ has rational harmonic measure. Here are the three steps:

(1) Extend to $\mathfrak{e} \in \mathcal{E}_p$ using the methods discussed briefly below.
(2) Prove that given any $\mathfrak{e} \in \mathcal{E}_f$, there is $\mathfrak{e}^{(n)} \in \mathcal{E}_p$, each with the same number of bands so $\mathfrak{e}_j \subset \mathfrak{e}_j^{(n)} \subset \mathfrak{e}_j^{(n-1)}$ and $\cap_n \mathfrak{e}_j^{(n)} = \mathfrak{e}_j$. This is a result proven independently by Bogatyrëv [**12**], Peherstorfer [**71**], and Totik [**97**]; see [**89**] for a presentation of Totik's method.
(3) Note that for any compact $\mathfrak{e}$, if $\mathfrak{e}^{(m)} = \{x \mid \text{dist}(x, \mathfrak{e}) \le \frac{1}{m}\}$, then $\mathfrak{e}^{(m)}$ is a finite gap set and $\mathfrak{e} = \cap_m \mathfrak{e}^{(m)}$.

Step (1) is the subtle step in extending theorems: Given the Bogatyrëv–Peherstorfer–Totik theorem, the extensions are simple approximation.

The key to $\mathfrak{e} \in \mathcal{E}_p$ is that there is a polynomial $\tilde\Delta\colon \mathbb{C} \to \mathbb{C}$, so $\tilde\Delta^{-1}([-1, 1]) = \mathfrak{e}$ and so that $\mathfrak{e}_j$ is a finite union of intervals $\tilde{\mathfrak{e}}_k$ with disjoint interiors so that $\tilde\Delta$ is a bijection from each $\tilde{\mathfrak{e}}_k$ to $[-1, 1]$. That this could be useful was noted initially by Geronimo–Van Assche [**36**]. Totik showed how to prove Theorem 17.3 for $\mathfrak{e} \in \mathcal{E}_p$ from the results for $[-1, 1]$ using this polynomial mapping.

For spectral theorists, the polynomial $\tilde\Delta = \frac{1}{2}\Delta$ where Δ is the discriminant for the associated periodic problem (see [**43, 53, 104, 95, 89**]). There is a direct construction of $\tilde\Delta$ by Aptekarev [**4**] and Peherstorfer [**70, 71, 72**].

19. Floquet–Jost Solutions for Periodic Jacobi Matrices

As we saw in Section 16, models with appropriate behavior are useful input for comparison theorems. Periodic Jacobi matrices have OPs for which one can study the CD kernel and its asymptotics. The two main results concern diagonal and just off-diagonal behavior:

THEOREM 19.1. *Let μ be the spectral measure associated to a periodic Jacobi matrix with essential spectrum, $\mathfrak{e}$, a finite gap set. Let $d\mu = w(x)\,dx$ on $\mathfrak{e}$ (there can also be up to one eigenvalue in each gap). Then uniformly for x in compact subsets of $\mathfrak{e}^{\text{int}}$,*

$$\frac{1}{n}\,K_n(x,x) \to \frac{\rho_{\mathfrak{e}}(x)}{w(x)} \tag{19.1}$$

and uniformly for such x and a,b in $\mathbb{R}$ with $|a| \le A$, $|b| \le B$,

$$\frac{K_n(x+\frac{a}{n}, x+\frac{b}{n})}{K_n(x,x)} \to \frac{\sin(\pi\rho_{\mathfrak{e}}(x)(b-a))}{\pi\rho_{\mathfrak{e}}(x)(b-a)} \tag{19.2}$$

REMARKS. 1. (19.2) is often called bulk universality. On bounded intervals, it goes back to random matrix theory. The best results using Riemann–Hilbert methods for OPs is due to Kuijlaars–Vanlessen [**51**]. A different behavior is expected at the edge of the spectrum—we will not discuss this in detail, but see Lubinsky [**62**].

2. For $[-1,1]$, Lubinsky [**60**] used Legendre polynomials as his model. The references for the proofs here are Simon [**87, 89**].

The key to the proof of Theorem 19.1 is to use Floquet–Jost solutions, that is, solutions of

$$a_n u_{n+1} + b_n u_n + a_{n-1} u_{n-1} = x u_n \tag{19.3}$$

for $n \in \mathbb{Z}$ where $\{a_n, b_n\}$ are extended periodically to all of $\mathbb{Z}$. These solutions obey

$$u_{n+p} = e^{i\theta(x)} u_n \tag{19.4}$$

For $x \in \mathfrak{e}^{\text{int}}$, u_n and $\bar{u}_n$ are linearly independent, and so one can write $p_{\cdot-1}$ in terms of $u_{\cdot}$ and $\bar{u}_{\cdot}$. Using

$$\rho_{\mathfrak{e}}(x) = \frac{1}{p\pi}\left|\frac{d\theta}{dx}\right| \tag{19.5}$$

one can prove (19.1) and (19.2). The details are in [**87, 89**].

20. Lubinsky's Inequality and Bulk Universality

Lubinsky [**60**] found a powerful tool for going from diagonal control of the CD kernel to slightly off-diagonal control—a simple inequality.

THEOREM 20.1. *Let $\mu \le \mu^*$ and let K_n, K_n^* be their CD kernels. Then for any z, ζ,*

$$|K_n(z,\zeta) - K_n^*(z,\zeta)|^2 \le K_n(z,z)[K_n(\zeta,\zeta) - K_n^*(\zeta,\zeta)] \tag{20.1}$$

REMARK. Recall (Theorem 9.3) that $K_n(\zeta,\zeta) \ge K_n^*(\zeta,\zeta)$.

PROOF. Since $K_n - K_n^*$ is a polynomial $\bar{z}$ of degree n:

$$K_n(z,\zeta) - K_n^*(z,\zeta) = \int K_n(z,w)[K_n(w,\zeta) - K_n^*(w,\zeta)]\,d\mu(w) \tag{20.2}$$

By the reproducing kernel formula (1.19), we get (20.1) from the Schwarz inequality if we show

$$\int |K_n(w,\zeta) - K_n^*(w,\zeta)|^2\, d\mu(w) \le K_n(\zeta,\zeta) - K_n^*(\zeta,\zeta) \tag{20.3}$$

Expanding the square, the K_n^2 term is $K_n(\zeta,\zeta)$ by (1.19) and the $K_nK_n^*$ cross term is $-2K_n^*(\zeta,\zeta)$ by the reproducing property of K_n for $d\mu$ integrals. Thus, (20.3) is equivalent to

$$\int |K_n^*(w,\zeta)|^2\, d\mu(w) \le K_n^*(\zeta,\zeta) \tag{20.4}$$

This in turn follows from $\mu \le \mu^*$ and (1.19) for μ^*! □

This result lets one go from diagonal control on measures to off-diagonal. Given any pair of measures, μ and ν, there is a unique measure $\mu \vee \nu$ which is their least upper bound (see, e.g., Doob [**30**]). It is known (see [**85**]) that if μ, ν are regular for the same set, so is $\mu \vee \nu$. (20.1) immediately implies that (go from μ to μ^* and then μ^* to ν):

COROLLARY 20.2. *Let μ, ν be two measures and $\mu^* = \mu \vee \nu$. Suppose for some $z_n \to z_\infty$, $w_n \to z_\infty$, we have for $\eta = \mu, \nu, \mu^*$ that*

$$\lim_{n\to\infty} \frac{K_n(z_n,z_n;\eta)}{K_n(z_\infty,z_\infty;\eta)} = \lim_{n\to\infty} \frac{K_n(w_n,w_n;\eta)}{K_n(z_\infty,z_\infty;\eta)} = 1$$

and that

$$\lim_{n\to\infty} \frac{K_n(z_\infty,z_\infty;\mu)}{K_n(z_\infty,z_\infty;\mu^*)} = \lim_{n\to\infty} \frac{K_n(z_\infty,z_\infty;\nu)}{K_n(z_\infty,z_\infty;\mu^*)} = 1$$

Then

$$\lim_{n\to\infty} \frac{K_n(z_n,w_n;\mu)}{K_n(z_n,w_n;\nu)} = 1 \tag{20.5}$$

REMARK. It is for use with $x_n = x_\infty + \frac{a}{n}$ or $x_\infty + \frac{a}{\rho_n n}$ that we added $x_n \to x_\infty$ to the various diagonal kernel results. This "wiggle" in x_∞ was introduced by Lubinsky [**60**], so we dub it the "Lubinsky wiggle."

Given Totik's theorem (Theorem 17.3) and bulk universality for suitable models, one thus gets:

THEOREM 20.3. *Under the hypotheses of Theorem 17.3, for a.e. x_∞ in I, we have uniformly for $|a|, |b| < A$,*

$$\lim_{n\to\infty} \frac{K_n(x_\infty + \frac{a}{n}, x_\infty + \frac{b}{n})}{K_n(x_\infty, x_\infty)} = \frac{\sin(\pi\rho_{\mathfrak{e}}(x_\infty)(b-a))}{\pi\rho_{\mathfrak{e}}(x_\infty)(b-a)}$$

REMARKS. 1. For $\mathfrak{e} = [-1,1]$, the result and method are from Lubinsky [**60**].

2. For continuous weights, this is in Simon [**87**] and Totik [**99**], and for general weights, in Totik [**99**].

21. Derivatives of CD Kernels

The ideas in this section come from a paper in preparation with Avila and Last [**6**]. Variation of parameters is a standard technique in ODE theory and used as an especially powerful tool in spectral theory by Gilbert–Pearson [**38**] and in Jacobi matrix spectral theory by Khan–Pearson [**49**]. It was then developed by Jitomirskaya–Last [**44, 45, 46**] and Killip–Kiselev–Last [**50**], from which we take Proposition 21.1.

PROPOSITION 21.1. *For any x, x_0, we have*

$$p_n(x) - p_n(x_0) = (x - x_0) \sum_{m=0}^{n-1} (p_n(x_0)q_m(x_0) - p_m(x_0)q_n(x_0))p_m(x) \tag{21.1}$$

In particular,

$$p_n'(x_0) = \sum_{m=0}^{n-1} (p_n(x_0)q_m(x_0) - p_m(x_0)q_n(x_0))p_m(x_0) \tag{21.2}$$

Here q_n are the second kind polynomials defined in Section 8. For (21.1), see [**44, 45, 46, 50**]. This immediately implies:

COROLLARY 21.2 (Avila–Last–Simon [**6**]).

$$\begin{aligned} &\frac{d}{da} \frac{1}{n} K_n\left(x_0 + \frac{a}{n}, x_0 + \frac{a}{n}\right)\Big|_{a=0} \\ &= \frac{2}{n^2} \sum_{j=0}^{n} \left[p_j(x_0)^2 \left(\sum_{k=0}^{j} p_k(x_0)q_k(x_0) \right) - q_j(x_0)p_j(x_0) \left(\sum_{k=0}^{j} p_k(x_0)^2 \right) \right] \end{aligned} \tag{21.3}$$

This formula gives an indication of why (as we see in the next section is important) $\lim \frac{1}{n} K_n(x_0 + \frac{a}{n}, x_0 + \frac{a}{n})$ has a chance to be independent of a if one notes the following fact:

LEMMA 21.3. *If $\{\alpha_n\}_{n=1}^{\infty}$ and $\{\beta_n\}_{n=1}^{\infty}$ are sequences so that $\lim \frac{1}{N} \sum_{n=1}^{N} \alpha_n = A$ and $\lim \frac{1}{N} \sum_{n=1}^{N} \beta_n = B$ exist and $\sup_N [\frac{1}{N} \sum_{n=1}^{N} |\alpha_n| + |\beta_n|] < \infty$, then*

$$\frac{1}{N^2} \sum_{j=1}^{N} \left[\left(\alpha_j \sum_{k=1}^{j} \beta_k \right) - \left(\beta_j \sum_{k=1}^{j} \alpha_k \right) \right] \to 0 \tag{21.4}$$

This is because

$$\frac{1}{N^2} \sum_{j=1}^{N} \alpha_j \sum_{k=1}^{j} \beta_k \to \frac{1}{2} AB$$

Setting $\alpha_j = p_j(x_0)^2$ and $\beta_j = p_j(x_0)q_j(x_0)$, one can hope to use (21.4) to prove the right side of (21.3) goes to zero.

22. Lubinsky's Second Approach

Lubinsky revolutionized the study of universality in [**60**], introducing the approach we described in Section 20. While Totik [**99**] and Simon [**87**] used those ideas to extend beyond the case of $\mathfrak{e} = [-1, 1]$ treated in [**60**], Lubinsky developed a totally different approach [**61**] to go beyond [**60**]. That approach, as abstracted in Avila–Last–Simon [**6**], is discussed in this section. Here is an abstract theorem:

THEOREM 22.1. *Let $d\mu$ be a measure of compact support on $\mathbb{R}$. Let x_0 be a Lebesgue point for μ and suppose that*

(i) *For any ε, there is a C_ε so that for any R, we have an $N(\varepsilon, R)$ so that for $n \geq N(\varepsilon, R)$,*

$$\frac{1}{n} K_n\left(x_0 + \frac{z}{n}, x_0 + \frac{z}{n}\right) \leq C_\varepsilon e^{\varepsilon |z|^2} \tag{22.1}$$

for all $z \in \mathbb{C}$ with $|z| < R$.

(ii) *Uniformly for real a's in compact subsets of* $\mathbb{R}$,

$$\lim_{n\to\infty} \frac{K_n(x_0+\frac{a}{n},x_0+\frac{a}{n})}{K_n(x_0,x_0)} = 1 \tag{22.2}$$

Let

$$\rho_n = \frac{w(x_0)}{n}\, K_n(x_0,x_0) \tag{22.3}$$

Then uniformly for z,w *in compact subsets of* $\mathbb{C}$,

$$\lim_{n\to\infty} \frac{K_n(x_0+\frac{z}{n\rho_n},x_0+\frac{w}{n\rho_n})}{K_n(x_0,x_0)} = \frac{\sin(\pi(\bar z-w))}{\pi(\bar z-w)} \tag{22.4}$$

REMARKS. 1. If $\rho_n \to \rho_{\mathfrak{e}}(x_0)$, the density of the equilibrium measure, then (22.4) is the same as (19.2). In every case where Theorem 22.1 has been proven to be applicable (see below), $\rho_n \to \rho_{\mathfrak{e}}(x_0)$. But one of the interesting aspects of this is that it might apply in cases where ρ_n does not have a limit. For an example with a.c. spectrum but where the density of zeros has multiple limits, see Example 5.8 of [**85**].

2. Lubinsky [**61**] worked in a situation (namely, x_0 in an interval I with $w(x_0) \geq c > 0$ on I) where (22.1) holds in the stronger form $Ce^{D|z|}$ (no square on $|z|$) and used arguments that rely on this. Avila–Last–Simon [**6**] found the result stated here; the methods seem incapable of working with (22.1) for a fixed ε rather than all ε (see Remark 1 after Theorem 22.2).

Let us sketch the main ideas in the proof of Theorem 22.1:

(1) By (15.1),

$$\liminf \frac{1}{n}\, K_n(x_0,x_0) > 0 \tag{22.5}$$

(2) By the Schwarz inequality (1.14), (22.1), and (22.5), and by the compactness of normal families, we can find subsequences $n(j)$ so

$$\frac{K_{n(j)}(x_0+\frac{z}{n\rho_n},x_0+\frac{w}{n\rho_n})}{K_{n(j)}(x_0,x_0)} \to F(z,w) \tag{22.6}$$

and F is analytic in w and anti-analytic in z.

(3) Note that by (22.2) and the Schwarz inequality (1.14), we have for $a,b\in\mathbb{R}$,

$$F(a,a)=1 \qquad |F(a,b)|\leq 1 \tag{22.7}$$

By compactness, if we show any such limiting F is $\sin(\pi(\bar z-w))/(\bar z-w)$, we have (22.4). By analyticity, it suffices to prove this for $z=a$ real, and we will give details when $z=0$, that is, we consider

$$\frac{K_{n(j)}(x_0,x_0+\frac{z}{n\rho_n})}{K_{n(j)}(x_0,x_0)} \to f(z) \tag{22.8}$$

(22.7) becomes

$$f(0)=1 \qquad |f(x)|\leq 1 \text{ for } x \text{ real}$$

(4) By (1.19),

$$\int |K_n(x_0,x_0+a)|^2 w(a)\,da \leq K_n(x_0,x_0) \tag{22.9}$$

which, by using the fact that x_0 is Lebesgue point, can be used to show

$$\int_{-\infty}^{\infty} |f(x)|^2\,dx \le 1 \tag{22.10}$$

(5) By properties of K_n (see Section 6) and Hurwitz's theorem, f has zeros $\{x_j\}_{j=-\infty, j\neq 0}^{\infty}$ only on $\mathbb{R}$, which we label by

$$\cdots < x_{-1} < 0 < x_1 < x_2 < \cdots \tag{22.11}$$

and define $x_0 = 0$. By Theorem 7.2, using (22.2), we have for any j, k that

$$|x_j - x_k| \ge |j-k| - 1 \tag{22.12}$$

(6) Given these facts, the theorem is reduced to

THEOREM 22.2. *Let f be an entire function obeying*

(1)

$$f(0) = 1 \qquad |f(x)| \le 1 \textit{ for } x \textit{ real} \tag{22.13}$$

(2)

$$\int_{-\infty}^{\infty} |f(x)|^2\,dx \le 1 \tag{22.14}$$

(3) *f is real on $\mathbb{R}$, has only real zeros, and if they are labelled by* (22.11), *then* (22.12) *holds.*

(4) *For any ε, there is a C_ε so*

$$|f(z)| \le C_\varepsilon e^{\varepsilon |z|^2} \tag{22.15}$$

Then

$$f(z) = \frac{\sin \pi z}{\pi z} \tag{22.16}$$

REMARKS. 1. There exist examples ([**6**]) $e^{-az^2+bz}\sin\pi z/\pi z$ that obey (1)–(3) and (22.15) for some but not all ε.

2. We sketch the proof of this in case one has

$$|f(z)| \le C e^{D|z|} \tag{22.17}$$

instead of (22.15); see [**6**] for the general case.

LEMMA 22.3. *If* (1)–(3) *hold and* (22.17) *holds, then for any $\varepsilon > 0$, there is D_ε with*

$$|f(z)| \le D_\varepsilon e^{(\pi+\varepsilon)|\mathrm{Im}\, z|} \tag{22.18}$$

Sketch. By the Hadamard product formula [**2**],

$$f(z) = e^{Dz} \prod_{j\neq 0} \left(1 - \frac{z}{x_j}\right) e^{z x_j}$$

where D is real since f is real on $\mathbb{R}$. Thus, for y real,

$$|f(iy)|^2 = \prod_{j\neq 0} \left(1 + \frac{y^2}{x_j^2}\right)$$

By (22.12), $|x_j| \ge j - 1$, so

$$|f(iy)|^2 \le \left(1 + \frac{y^2}{x_1^2}\right)\left(1 + \frac{y^2}{x_{-1}^2}\right)\left[\prod_{j=1}^{\infty}\left(1 + \frac{y^2}{j^2}\right)\right]^2$$

which, given Euler's formula for $\sin \pi z/z$, implies (22.18) for $z = iy$. By a Phragmén–Lindelöf argument, (22.18) for z real and for z pure imaginary and (22.17) implies (22.18) for all z.

Thus, Theorem 22.2 (under hypothesis (22.17)) is implied by:

LEMMA 22.4. *If f is an entire function that obeys* (22.13), (22.14), *and* (22.18), *then* (22.16) *holds.*

PROOF. Let $\hat{f}$ be the Fourier transform of f, that is,

$$\hat{f}(k) = (2\pi)^{-1/2} \int e^{-ikx} f(x)\, dx \tag{22.19}$$

(in L^2 limit sense). By the Paley–Wiener Theorem [**74**], (22.18) implies $\hat{f}$ is supported on $[-\pi, \pi]$. By (22.14),

$$\|\hat{f}\|_{L^2} = \|(2\pi)^{-1/2} \chi_{[-\pi,\pi]}\|_{L^2} = 1$$

and, by (22.13) and support property of $\hat{f}$,

$$\langle \hat{f}, (2\pi)^{-1/2} \chi_{[-\pi,\pi]} \rangle = 1$$

We thus have equality in the Schwarz inequality, so

$$\hat{f} = (2\pi)^{-1/2} \chi_{[-\pi,\pi]}$$

which implies (22.16). □

This theorem has been applied in two ways:

(a) Lubinsky [**61**] noted that one can recover Theorem 20.3 from just Totik's result Theorem 17.3 without using the Lubinsky wiggle or Lubinsky's inequality.

(b) Avila–Last–Simon [**6**] have used this result to prove universality for ergodic Jacobi matrices with a.c. spectrum where $\mathfrak{e}$ can be a positive measure Cantor set.

23. Zeros: The Freud–Levin–Lubinsky Argument

In the final section of his book [**34**], Freud proved bulk universality under fairly strong hypotheses on measures on $[-1, 1]$ and noticed that it implied a strong result on local equal spacings of zeros. Without knowing of Freud's work, Simon, in a series of papers (one joint with Last) [**82, 83, 84, 57**], focused on this behavior, called it clock spacing, and proved it in a variety of situations (not using universality or the CD kernel). After Lubinsky's work on universality, Levin [**59**] rediscovered Freud's argument and Levin–Lubinsky [**59**] used this to obtain clock behavior in a very general context. Here is an abstract version of their result:

THEOREM 23.1. *Let μ be a measure of compact support on $\mathbb{R}$; let $x_0 \in \sigma(\mu)$ be such that for each A, for some c_n,*

$$\frac{K_{n-1}(x_0 + \frac{a}{nc_n}, x_0 + \frac{b}{nc_n})}{K_{n-1}(x_0, x_0)} \to \frac{\sin(\pi(b-a))}{\pi(b-a)} \tag{23.1}$$

uniformly for real a, b with $|a|, |b| \le A$. Let $x_j^{(n)}(x_0)$ denote the zeros of $p_n(x; d\mu)$ labelled so

$$\cdots < x_{-1}^{(n)}(x_0) < x_0 \le x_0^{(n)}(x_0) < x_1^{(n)}(x_0) < \cdots \tag{23.2}$$

Then

(1)

$$\limsup nc_n(x_0^{(n)} - x_0) \leq 1 \tag{23.3}$$

(ii) *For any J, for large n, there are zeros $x_j^{(n)}$ for all $j \in \{-J, -J+1, \dots, J-1, J\}$.*
(iii)

$$\lim_{n\to\infty} (x_{j+1}^{(n)} - x_j^{(n)})nc_n = 1 \qquad \textit{for each } j \tag{23.4}$$

REMARKS. 1. The meaning of $x_j^{(n)}$ has changed slightly from Section 6.

2. Only nc_n enters, so the "n" could be suppressed; we include it because one expects c_n as defined to be bounded above and below. Indeed, in all known cases, $c_n \to \rho(x_0)$, the derivative of the density of states. But see Remark 1 after Theorem 22.1 for cases where c_n might not have a limit.

3. See [**58**] for the OPUC case.

PROOF. Let $\tilde{x}_j^{(n)}(x_0)$ be the zeros of $p_n(x)p_{n-1}(x_0) - p_n(x_0)p_{n-1}(x)$ labelled as in (23.2) (with $\tilde{x}_0^{(n)}(x_0) = x_0$). By (23.1), we have $\tilde{x}_{\pm 1}^{(n)}(x_0)nc_n \to 1$ since $\sin(\pi a)/a$ is nonvanishing on $(-1,1)$ and vanishes at ± 1. The same argument shows $K_n(\tilde{x}_{\pm 1}^{(n)}, \tilde{x}_{\pm 1}^{(n)} + b/nc_n)$ is nonvanishing for $|b| < \frac{1}{2}$, and so there is at most one zero near $\tilde{x}_{\pm 1}^{(n)}$ on $1/nc_n$ scale. It follows by repeating this argument that

$$nc_n\tilde{x}_j^{(n)} \to j \tag{23.5}$$

for all j.

Since we have (see Section 6) that

$$x_0 \leq x_0^{(n)}(x_0) \leq \tilde{x}_1^{(n)}(x_0)$$

by interlacing, which implies (i) and similar interlacing gives (ii). Finally, (23.4) follows from the same argument that led to (23.5). □

24. Adding Point Masses

We end with a final result involving CD kernels—a formula of Geronimus [**37**, formula (3.30)]. While he states it only for OPUC, his proof works for any measure on $\mathbb{C}$ with finite moments. Let μ be such a measure, let $z_0 \in \mathbb{C}$, and let

$$\nu = \mu + \lambda\delta_{z_0} \tag{24.1}$$

for λ real and bigger than or equal to $-\mu(\{z_0\})$.

Since $X_n(z; d\nu)$ and $X_n(z; d\mu)$ are both monic, their difference is a polynomial of degree $n-1$, so

$$X_n(x; d\nu) = X_n(z; d\mu) + \sum_{j=0}^{n-1} c_j x_j(z; d\mu) \tag{24.2}$$

where

$$c_j = \int \overline{x_j(z; d\mu)}\, [X_n(z; d\nu) - X_n(z; d\mu)]\, d\mu \tag{24.3}$$

$$= \int \overline{x_j(z; d\mu)}\, X_n(z; d\nu)[d\nu - \lambda\delta_{z_0}] \tag{24.4}$$

$$= -\lambda\, \overline{x_j(z_0; d\mu)}\, X_n(z_0; d\nu) \tag{24.5}$$

where (24.4) follows from $x_j(\,\cdot\,, d\mu) \perp X_n(\,\cdot\,, d\mu)$ in $L^2(d\mu)$ and (24.5) from $x_j(\,\cdot\,, d\mu) \perp X_n(\,\cdot\,, d\nu)$ in $L^2(d\nu)$. Thus,

$$X_n(z; d\nu) = X_n(z; d\mu) - \lambda X_n(z_0; d\nu) K_{n-1}(z_0, z; d\mu) \tag{24.6}$$

Set $z = z_0$ and solve for $X_n(z_0; d\nu)$ to get:

THEOREM 24.1 (Geronimus [**37**]). *Let μ, ν be related by (24.1). Then*

$$X_n(z; d\nu) = X_n(z; d\mu) - \frac{\lambda X_n(z_0; d\mu) K_{n-1}(z_0, z; d\mu)}{1 + \lambda K_{n-1}(z_0, z_0; d\mu)} \tag{24.7}$$

This formula was rediscovered by Nevai [**67**] for OPRL, by Cachafeiro–Marcellán [**15, 16**], Simon [**81**] (in a weak form), and Wong [**105, 106**] for OPUC. For general measures on $\mathbb{C}$, the formula is from Cachafeiro–Marcellán [**17, 18**]. In particular, in the context of OPUC, Wong [**106**] noted that one can use the CD formula to obtain:

THEOREM 24.2 (Wong [**105, 106**]). *Let $d\mu$ be a probability measure on $\partial\mathbb{D}$ and let $d\tilde{\nu}$ be given by*

$$d\tilde{\nu} = \frac{d\mu + \lambda\delta_{z_0}}{1 + \lambda} \tag{24.8}$$

for $z_0 \in \partial\mathbb{D}$ and $\lambda \geq -\mu(\{z_0\})$. Then

$$\alpha_n(d\tilde{\nu}) = \alpha_n(d\mu) + \frac{(1 - |\alpha_n(d\mu)|^2)^{1/2}}{\lambda^{-1} + K_n(z_0, z_0; d\mu)} \overline{\varphi_{n+1}(z_0)}\, \varphi_n^*(z_0) \tag{24.9}$$

PROOF. Let

$$Q_n = \lambda^{-1} + K_n(z_0, z; d\mu) \tag{24.10}$$

Since

$$\Phi_{n+1}(z; d\tilde{\nu}) = \Phi_{n+1}(z; d\nu) \tag{24.11}$$

and

$$\alpha_n(d\tilde{\nu}) = \overline{-\Phi_{n+1}(0; d\tilde{\nu})} \tag{24.12}$$

(24.7) becomes

$$\alpha_n(d\tilde{\nu}) - \alpha_n(d\mu) = Q_n^{-1}\ \overline{\Phi_{n+1}(z_0)}\, \overline{K_n(z_0, 0; d\mu)} \tag{24.13}$$

By the CD formula in the form (3.23),

$$\overline{K_n(z_0, 0)} = \varphi_n^*(z_0)\, \overline{\varphi_n^*(0)} \tag{24.14}$$

$$= \frac{\varphi_n^*(z_0)}{\|\Phi_n\|} \tag{24.15}$$

since $\Phi_n^*(0) = 1$ and $\|\Phi_n^*\| = \|\Phi_n\|$. (24.9) then follows from $\|\Phi_{n+1}\|/\|\Phi_n\| = (1 - |\alpha_n|^2)^{1/2}$. □

To see a typical application:

COROLLARY 24.3. *Let z_0 be an isolated pure point of a measure $d\mu$ on $\partial\mathbb{D}$. Let $d\tilde{\nu}$ be given by (24.8) where $\lambda > -\mu(\{z_0\})$ (so z_0 is also a pure point of $d\tilde{\nu}$). Then for some $D, C > 0$,*

$$|\alpha_n(d\tilde{\nu}) - \alpha_n(d\mu)| \leq De^{-Cn} \tag{24.16}$$

PROOF. By Theorem 10.14.2 of [**81**],

$$|\varphi_n(z_0; d\mu)| \leq D_1 e^{-\frac{1}{2}Cn} \tag{24.17}$$

This plus (24.9) implies (24.16). □

This is not only true for OPUC but also for OPRL:

COROLLARY 24.4. *Let z_0 be an isolated pure point of a measure of compact support $d\mu$ on $\mathbb{R}$. Let $d\tilde{\nu}$ be given by (24.8) where $\lambda > -\mu(\{z_0\})$ so z is also a pure point of $d\tilde{\nu}$. Then for some $D, C > 0$,*

$$\text{(i)}\qquad \left|\frac{\kappa_n(d\nu)}{\kappa_n(d\mu)} - (1+\lambda)^{1/2}\right| \le De^{-Cn} \tag{24.18}$$

$$\text{(ii)}\qquad \|p_n(\,\cdot\,, d\nu) - (1+\lambda)^{1/2} p_n(\,\cdot\,, d\mu)\|_{L^2(d\nu)} \le De^{-Cn} \tag{24.19}$$

$$\text{(iii)}\qquad |a_n(d\nu) - a_n(d\mu)| \le De^{-Cn} \tag{24.20}$$

$$|b_n(d\nu) - b_n(d\mu)| \le De^{-Cn} \tag{24.21}$$

SKETCH. Isolated points in the spectrum of Jacobi matrices obey

$$|p_n(z_0)| \le D_1 e^{-C_1 n} \tag{24.22}$$

for suitable C_1, D_1 (see [**1, 25**]).

(24.7) can be rewritten for OPRL

$$\kappa_n(d\mu) P_n(x; d\nu) = p_n(x; d\mu) - \lambda p_n(x_0; d\mu) \frac{K_{n-1}(x_0, x; d\mu)}{1 + \lambda K_n(x_0, x_0; d\mu)} \tag{24.23}$$

Since

$$\|K_{n-1}(x_0, x; d\mu)\|^2_{L^2(d\mu)} = K_{n-1}(x_0, x_0; d\mu)$$

and

$$\int |K_{n-1}(x_0, x_0; d\mu)|^2 \, d\delta_{x_0} = K_{n-1}(x_0, x_0; d\mu)^2$$

and $K_{n-1}(x_0, x_0)$ is bounded (by (24.22)), we see that $\|K_{n-1}(x_0; \,\cdot\,; d\mu)\|_{L^2(d\nu)}$ is bounded. Thus, by (24.22) and (24.23),

$$\kappa_n(d\mu)\kappa_n(d\nu)^{-1} = (1+\lambda)^{-1/2} + O(e^{-C_1 n})$$

which leads to (24.18).

This in turn leads to (ii), and that to (iii) via (24.22), and, for example,

$$a_n(d\mu) = \int x p_n(x; d\mu) p_{n-1}(x; d\mu) \, d\mu \tag{24.24}$$

$$a_n(d\nu) = \int x p_n(x; d\nu) p_{n-1}(x; d\nu) \, d\nu \tag{24.25}$$

□

This shows what happens if the weight of an isolated eigenvalue changes. What happens if an isolated eigenvalue is totally removed is much more subtle—sometimes it is exponentially small, sometimes not. This is studied by Wong [**107**].

References

[1] S. Agmon, *Lectures on Exponential Decay of Solutions of Second-Order Elliptic Equations: Bounds on Eigenfunctions of N-body Schrödinger Operators*, Mathematical Notes, **29**, Princeton Univ. Press, Princeton, NJ; Univ. of Tokyo Press, Tokyo, 1982.

[2] L. V. Ahlfors, *Complex Analysis. An Introduction to the Theory of Analytic Functions of One Complex Variable*, McGraw–Hill, New York, 1978.

[3] A. Ambroladze, *On exceptional sets of asymptotic relations for general orthogonal polynomials*, J. Approx. Theory **82** (1995), 257–273.

[4] A. I. Aptekarev, *Asymptotic properties of polynomials orthogonal on a system of contours, and periodic motions of Toda chains*, Math. USSR Sb. **53** (1986), 233–260; Russian original in Mat. Sb. (N.S.) **125(167)** (1984), 231–258.

[5] F. V. Atkinson, *Discrete and Continuous Boundary Problems*, Academic Press, New York, 1964.

[6] A. Avila, Y. Last, and B. Simon, *Bulk universality and clock spacing of zeros for ergodic Jacobi matrices with a.c. spectrum*, in preparation.

[7] J. Avron and B. Simon, *Almost periodic Schrödinger operators, II. The integrated density of states*, Duke Math. J. **50** (1983), 369–391.

[8] D. Barrios Rolanía and G. López Lagomasino, *Ratio asymptotics for polynomials orthogonal on arcs of the unit circle*, Constr. Approx. **15** (1999), 1–31.

[9] M. Bello Hernández and G. López Lagomasino, *Ratio and relative asymptotics of polynomials orthogonal on an arc of the unit circle*, J. Approx. Theory **92** (1998), 216–244.

[10] A. Ben Artzi and T. Shalom, *On inversion of block Toeplitz matrices*, Integral Equations and Operator Theory **8** (1985), 751–779.

[11] C. Berg, *Fibonacci numbers and orthogonal polynomials*, to appear in J. Comput. Appl. Math.

[12] A. B. Bogatyrëv, *On the efficient computation of Chebyshev polynomials for several intervals*, Sb. Math. **190** (1999), 1571–1605; Russian original in Mat. Sb. **190** (1999), no. 11, 15–50.

[13] A. Borodin, *Biorthogonal ensembles*, Nuclear Phys. B **536** (1999), 704–732.

[14] J. Breuer, Y. Last, and B. Simon, in preparation.

[15] A. Cachafeiro and F. Marcellán, *Orthogonal polynomials and jump modifications*, in "Orthogonal Polynomials and Their Applications," (Segovia, 1986), pp. 236–240, Lecture Notes in Math., **1329**, Springer, Berlin 1988.

[16] A. Cachafeiro and F. Marcellán, *Asymptotics for the ratio of the leading coefficients of orthogonal polynomials associated with a jump modification*, in "Approximation and Optimization," (Havana, 1987), pp. 111–117, Lecture Notes in Math., **1354**, Springer, Berlin, 1988.

[17] A. Cachafeiro and F. Marcellán, *Perturbations in Toeplitz matrices*, in "Orthogonal Polynomials and Their Applications," (Laredo, 1987), pp. 139–146, Lecture Notes in Pure and Applied Math., **117**, Marcel Dekker, New York, 1989.

[18] A. Cachafeiro and F. Marcellán, *Perturbations in Toeplitz matrices: Asymptotic properties*, J. Math. Anal. Appl. **156** (1991) 44–51.

[19] M. J. Cantero, L. Moral, and L. Velázquez, *Measures and para-orthogonal polynomials on the unit circle*, East J. Approx. **8** (2002), 447–464.

[20] R. Carmona and J. Lacroix, *Spectral Theory of Random Schrödinger Operators. Probability and Its Applications*, Birkhäuser, Boston, 1990.

[21] M. P. Chebyshev, *Sur les valeurs limites des intégrales*, J. Math. Pures Appl. **19** (1874), 157–160.

[22] T. S. Chihara, *An Introduction to Orthogonal Polynomials*, Mathematics and Its Applications, **13**, Gordon and Breach, New York-London-Paris, 1978.

[23] E. B. Christoffel, *Über die Gaussische Quadratur und eine Verallgemeinerung derselben*, J. Reine Angew. Math. **55** (1858), 61–82.

[24] A. R. Collar, *On the reciprocation of certain matrices*, Proc. Roy. Soc. Edinburgh **59** (1939), 195–206.

[25] J. M. Combes and L. Thomas, *Asymptotic behaviour of eigenfunctions for multiparticle Schrödinger operators*, Comm. Math. Phys. **34** (1973), 251–270.

[26] D. Damanik, R. Killip, and B. Simon, *Perturbations of orthogonal polynomials with periodic recursion coefficients*, preprint.

[27] D. Damanik, A. Pushnitski, and B. Simon, *The analytic theory of matrix orthogonal polynomials*, Surveys in Approximation Theory **4** (2008), 1–85.

[28] G. Darboux, *Mémoire sur l'approximation des fonctions de très-grands nombres, et sur une classe étendue de développements en série*, Liouville J. (3) **4** (1878), 5–56; 377–416.

[29] S. Denisov and B. Simon, *Zeros of orthogonal polynomials on the real line*, J. Approx. Theory **121** (2003), 357–364.

[30] J. L. Doob, *Measure Theory*, Graduate Texts in Math., **143**, Springer-Verlag, New York, 1994.

[31] D. E. Edmunds and W. D. Evans, *Spectral Theory and Differential Operators*, Clarendon Press, Oxford, 1989.
[32] P. Erdős and P. Turán, *On interpolation. III. Interpolatory theory of polynomials*, Ann. of Math. (2) **41** (1940), 510–553.
[33] E. Findley, *Universality for locally Szegő measures*, to appear in J. Approx. Theory.
[34] G. Freud, *Orthogonal Polynomials*, Pergamon Press, Oxford-New York, 1971.
[35] P. A. Fuhrmann, *On a partial realization problem and the recursive inversion of Hankel and Toeplitz matrices*, Contemp. Math. **47** (1985), 149–161.
[36] J. S. Geronimo and W. Van Assche, *Orthogonal polynomials on several intervals via a polynomial mapping*, Trans. Amer. Math. Soc. **308** (1988), 559–581.
[37] Ya. L. Geronimus, *Orthogonal Polynomials: Estimates, Asymptotic Formulas, and Series of Polynomials Orthogonal on the Unit Circle and on an Interval*, Consultants Bureau, New York, 1961.
[38] D. J. Gilbert and D. B. Pearson, *On subordinacy and analysis of the spectrum of one-dimensional Schrödinger operators*, J. Math. Anal. Appl. **128** (1987), 30–56.
[39] I. C. Gohberg and N. Ya. Krupnik, *A formula for the inversion of finite Toeplitz matrices*, Mat. Issled **7(2)** (1972), 272–283. [Russian]
[40] I. C. Gohberg and A. A. Semencul, *On the inversion of finite Toeplitz matrices and their continuous analogs*, Mat. Issled **7(2)** (1972) 201–223. [Russian]
[41] L. Golinskii, *Quadrature formula and zeros of para-orthogonal polynomials on the unit circle*, Acta Math. Hungar. **96** (2002), 169–186.
[42] L. L. Helms, *Introduction to Potential Theory*, Pure and Applied Math., **22**, Wiley–Interscience, New York, 1969.
[43] H. Hochstadt, *On the theory of Hill's matrices and related inverse spectral problems*, Linear Algebra and Appl. **11** (1975), 41–52.
[44] S. Jitomirskaya and Y. Last, *Dimensional Hausdorff properties of singular continuous spectra*, Phys. Rev. Lett. **76** (1996), 1765–1769.
[45] S. Jitomirskaya and Y. Last, *Power-law subordinacy and singular spectra, I. Half-line operators*, Acta Math. **183** (1999), 171–189.
[46] S. Jitomirskaya and Y. Last, *Power law subordinacy and singular spectra, II. Line operators*, Comm. Math. Phys. **211** (2000), 643–658.
[47] K. Johansson, *Random growth and random matrices*, European Congress of Mathematics, Vol. I (Barcelona, 2000), pp. 445–456, Progr. Math., **201**, Birkhäuser, Basel, 2001.
[48] T. Kailath, A. Vieira, and M. Morf, *Inverses of Toeplitz, innovations and orthogonal polynomials*, SIAM Review **20** (1978), 106–119.
[49] S. Khan and D. B. Pearson, *Subordinacy and spectral theory for infinite matrices*, Helv. Phys. Acta **65** (1992), 505–527.
[50] R. Killip, A. Kiselev, and Y. Last, *Dynamical upper bounds on wavepacket spreading*, Am. J. Math. **125** (2003), 1165–1198.
[51] A. B. Kuijlaars and M. Vanlessen, *Universality for eigenvalue correlations from the modified Jacobi unitary ensemble*, Int. Math. Res. Not. **30** (2002), 1575–1600.
[52] N. S. Landkof, *Foundations of Modern Potential Theory*, Springer-Verlag, Berlin-New York, 1972.
[53] Y. Last, *On the measure of gaps and spectra for discrete 1D Schrödinger operators*, Comm. Math. Phys. **149** (1992), 347–360.
[54] Y. Last, *Spectral theory of Sturm–Liouville operators on infinite intervals: a review of recent developments*, in "Sturm–Liouville Theory," pp. 99–120, Birkhäuser, Basel, 2005.
[55] Y. Last and B. Simon, *Eigenfunctions, transfer matrices, and absolutely continuous spectrum of one-dimensional Schrödinger operators*, Invent. Math. **135** (1999), 329–367.
[56] Y. Last and B. Simon, *The essential spectrum of Schrödinger, Jacobi, and CMV operators*, J. Anal. Math. **98** (2006), 183–220.
[57] Y. Last and B. Simon, *Fine structure of the zeros of orthogonal polynomials, IV. A priori bounds and clock behavior*, Comm. Pure Appl. Math. **61** (2008) 486–538.
[58] E. Levin and D. S. Lubinsky, *Universality limits involving orthogonal polynomials on the unit circle*, Computational Methods and Function Theory **7** (2007), 543–561.
[59] E. Levin and D. S. Lubinsky, *Applications of universality limits to zeros and reproducing kernels of orthogonal polynomials*, J. Approx. Theory **150** (2008), 69–95.

[60] D. S. Lubinksy, *A new approach to universality limits involving orthogonal polynomials*, to appear in Ann. of Math.
[61] D. S. Lubinsky, *Universality limits in the bulk for arbitrary measures on compact sets*, to appear in J. Anal. Math.
[62] D. S. Lubinsky, *A new approach to universality limits at the edge of the spectrum*, to appear in Contemp. Math.
[63] A. A. Markov, *Démonstration de certaines inégalités de M. Tchébychef*, Math. Ann. **24** (1884), 172–180.
[64] A. Máté and P. Nevai, *Bernstein's inequality in L^p for $0 < p < 1$ and $(C, 1)$ bounds for orthogonal polynomials*, Ann. of Math. (2) **111** (1980), 145–154.
[65] A. Máté, P. Nevai, and V. Totik, *Extensions of Szegő's theory of orthogonal polynomials. III*, Constr. Approx. **3** (1987), 73–96.
[66] A. Máté, P. Nevai, and V. Totik, *Szegő's extremum problem on the unit circle*, Ann. of Math. **134** (1991), 433–453.
[67] P. Nevai, *Orthogonal polynomials*, Mem. Amer. Math. Soc. **18** (1979), no. 213, 185 pp.
[68] P. Nevai, *Géza Freud, orthogonal polynomials and Christoffel functions. A case study*, J. Approx. Theory **48** (1986), 167 pp.
[69] P. Nevai, V. Totik, and J. Zhang, *Orthogonal polynomials: their growth relative to their sums*, J. Approx. Theory **67** (1991), 215–234.
[70] F. Peherstorfer, *Orthogonal and extremal polynomials on several intervals*, in "Proc. Seventh Spanish Symposium on Orthogonal Polynomials and Applications (VII SPOA)" (Granada, 1991), J. Comput. Appl. Math. **48** (1993), 187–205.
[71] F. Peherstorfer, *Deformation of minimal polynomials and approximation of several intervals by an inverse polynomial mapping*, J. Approx. Theory **111** (2001), 180–195.
[72] F. Peherstorfer, *Inverse images of polynomial mappings and polynomials orthogonal on them*, in "Proc. Sixth International Symposium on Orthogonal Polynomials, Special Functions and their Applications" (Rome, 2001), J. Comput. Appl. Math. **153** (2003), 371–385.
[73] T. Ransford, *Potential Theory in the Complex Plane*, Press Syndicate of the University of Cambridge, New York, 1995.
[74] M. Reed and B. Simon, *Methods of Modern Mathematical Physics, II. Fourier Analysis, Self-Adjointness*, Academic Press, New York, 1975.
[75] C. Remling, *The absolutely continuous spectrum of Jacobi matrices*, preprint.
[76] W. Rudin, *Real and Complex Analysis*, 3rd edition, McGraw–Hill, New York, 1987.
[77] B. Simon, *The classical moment problem as a self-adjoint finite difference operator*, Adv. in Math. **137** (1998), 82–203.
[78] B. Simon, *Ratio asymptotics and weak asymptotic measures for orthogonal polynomials on the real line*, J. Approx. Theory **126** (2004), 198–217.
[79] B. Simon, *OPUC on one foot*, Bull. Amer. Math. Soc. **42** (2005), 431–460.
[80] B. Simon, *Orthogonal Polynomials on the Unit Circle, Part 1: Classical Theory*, AMS Colloquium Series, **54.1**, American Mathematical Society, Providence, RI, 2005.
[81] B. Simon, *Orthogonal Polynomials on the Unit Circle, Part 2: Spectral Theory*, AMS Colloquium Series, **54.2**, American Mathematical Society, Providence, RI, 2005.
[82] B. Simon, *Fine structure of the zeros of orthogonal polynomials, I. A tale of two pictures*, Electronic Transactions on Numerical Analysis **25** (2006), 328–268.
[83] B. Simon, *Fine structure of the zeros of orthogonal polynomials, II. OPUC with competing exponential decay*, J. Approx. Theory **135** (2005), 125–139.
[84] B. Simon, *Fine structure of the zeros of orthogonal polynomials, III. Periodic recursion coefficients*, Comm. Pure Appl. Math. **59** (2006) 1042–1062.
[85] B. Simon, *Equilibrium measures and capacities in spectral theory*, Inverse Problems and Imaging **1** (2007), 713–772.
[86] B. Simon, Rank one perturbations and the zeros of paraorthogonal polynomials on the unit circle, J. Math. Anal. Appl. **329** (2007), 376–382.
[87] B. Simon, *Two extensions of Lubinsky's universality theorem*, to appear in J. Anal. Math.
[88] B. Simon, *Weak convergence of CD kernels and applications*, to appear in Duke Math. J.
[89] B. Simon, *Szegő's Theorem and Its Descendants: Spectral Theory for L^2 Perturbations of Orthogonal Polynomials*, in preparation; to be published by Princeton University Press.

[90] M. Sodin and P. Yuditskii, *Almost periodic Jacobi matrices with homogeneous spectrum, infinite-dimensional Jacobi inversion, and Hardy spaces of character-automorphic functions*, J. Geom. Anal. **7** (1997), 387–435.
[91] H. Stahl and V. Totik, *General Orthogonal Polynomials*, in "Encyclopedia of Mathematics and its Applications," **43**, Cambridge University Press, Cambridge, 1992.
[92] T. Stieltjes, *Quelques recherches sur la théorie des quadratures dites mécaniques*, Ann. Sci. École Norm. Sup. (3) **1** (1884), 409–426.
[93] G. Szegő, *Orthogonal Polynomials*, Amer. Math. Soc. Colloq. Publ., **23**, American Mathematical Society, Providence, RI, 1939; 3rd edition, 1967.
[94] R. Szwarc, *A counterexample to subexponential growth of orthogonal polynomials*, Constr. Approx. **11** (1995), 381–389.
[95] M. Toda, *Theory of Nonlinear Lattices*, 2nd edition, Springer Series in Solid-State Sciences, **20**, Springer, Berlin, 1989.
[96] V. Totik, *Asymptotics for Christoffel functions for general measures on the real line*, J. Anal. Math. **81** (2000), 283–303.
[97] V. Totik, *Polynomial inverse images and polynomial inequalities*, Acta Math. **187** (2001), 139–160.
[98] V. Totik, *Christoffel functions on curves and domains*, in preparation.
[99] V. Totik, *Universality and fine zero spacing on general sets*, in preparation.
[100] W. F. Trench, *An algorithm for the inversion of finite Toeplitz matrices*, J. Soc. Industr. Appl. Math. **12** (1964), 515–522.
[101] W. F. Trench, *An algorithm for the inversion of finite Hankel matrices*, J. Soc. Industr. Appl. Math **13** (1965), 1102–1107.
[102] M. Tsuji, *Potential Theory in Modern Function Theory*, reprinting of the 1959 original, Chelsea, New York, 1975.
[103] J. L. Ullman, *On the regular behaviour of orthogonal polynomials*, Proc. London Math. Soc. (3) **24** (1972), 119–148.
[104] P. van Moerbeke, *The spectrum of Jacobi matrices*, Invent. Math. **37** (1976), 45–81.
[105] M.-W. L. Wong, *First and second kind paraorthogonal polynomials and their zeros*, J. Approx. Theory **146** (2007), 282–293.
[106] M.-W. L. Wong, *A formula for inserting point masses*, to appear in Proc. OPSFA (Marseille, 2007).
[107] M.-W. L. Wong, in preparation.
[108] J. Zhang, *Relative growth of linear iterations and orthogonal polynomials on several intervals*, Linear Algebra Appl. **186** (1993), 97–115.

Mathematics 253–37, California Institute of Technology, Pasadena, CA 91125, U.S.A.

E-mail address: `bsimon@caltech.edu`

URL: `http://www.math.caltech.edu/people/simon.html`

Proceedings of Symposia in Pure Mathematics
Volume **79**, 2008

A Saint-Venant Principle for Lipschitz Cylinders

Michael E. Taylor

Abstract. We show how a general program exposed by P. Lax applies to the study of the asymptotic behavior at infinity of a biharmonic function u on a half-infinite cylinder whose cross section $\mathcal{O}$ is a Lipschitz domain, given a mild bound on u and vanishing Dirichlet data on the lateral boundary. It is shown that such a solution u exists, given Dirichlet data on the base of the cylinder, $(f,g) \in H_0^1(\mathcal{O}) \oplus L^2(\mathcal{O})$, and that u has an asymptotic expansion in a series of progressively more rapidly exponentially decreasing terms. To carry out this analysis, we make use of results of B. Dahlberg, C. Kenig, G. Verchota, J. Pipher, and V. Adolfsson.

Dedicated to V. G. Maz'ya on the Occasion of his 70th Birthday

1. Introduction

Let $\mathcal{O} \subset \mathbb{R}^{n-1}$ be a bounded, strongly Lipschitz domain, and set $\Omega = \mathbb{R}^+ \times \mathcal{O} \subset \mathbb{R}^n$. Then Ω is a Lipschitz cylinder, with boundary $\partial\Omega = (\{0\} \times \mathcal{O}) \cup (\mathbb{R}^+ \times \partial\mathcal{O})$. We want to study the behavior of solutions to the following Dirichlet problem for the bi-Laplacian Δ^2, where $\Delta = \partial_y^2 + \Delta_x = \partial_y^2 + \sum_1^{n-1} \partial_{x_j}^2$. With $y \in \mathbb{R}^+$, $x \in \mathcal{O}$, we look for $u(y,x)$, solving

$$\begin{aligned} &\Delta^2 u = 0 \quad \text{on} \quad \Omega, \\ &u(0,x) = f(x), \quad \partial_y u(0,x) = g(x), \quad x \in \mathcal{O}, \\ &u(y,x) = 0, \quad \partial_\nu u(y,x) = 0, \quad x \in \partial\mathcal{O}, \ y \geq 0, \end{aligned} \tag{1.1}$$

where ∂_ν is the unit normal to $\partial\mathcal{O}$. We take

$$f \in H_0^1(\mathcal{O}), \quad g \in L^2(\mathcal{O}). \tag{1.2}$$

This Dirichlet problem for Δ^2 has a unique solution $u(y,x)$ in a class of functions we will specify later on, having some decay as $y \to \infty$. We will show that $u(y,x)$ has an asymptotic expansion as a series of progressively more rapidly exponentially decreasing terms; see (4.19)–(4.20) for a precise statement.

P. Lax [**3**] (cf. also [**4**], Chapter 26, and [**5**]) provided an abstract context in which to prove the existence of such an asymptotic expansion, for rather general families of y-independent elliptic PDE on infinite cylinders. An interesting aspect of the analysis in [**3**] is that no existence theorem was needed. On the other hand,

2000 *Mathematics Subject Classification.* 35J40, 35B40.
This work was supported by NSF grant DMS-0456861.

for specific problems like the one mentioned above, one is just as interested in the existence of non-exploding solutions as one is in their asymptotic behavior. We will build on several works on Lipschitz domains done in the 1980s and 1990s to obtain such a solution to (1.1)–(1.2). Then we derive the asymptotic behavior, using a variant of the methods of [**3**]. Actually, the existence result provides more structure, and this allows for a simpler endgame argument involving functional analysis and semigroups.

The plan of the rest of this paper is as follows. In §2 we replace Ω by $\Omega_R = [0,R] \times \mathcal{O}$, and study the solution to

$$\begin{aligned} \Delta^2 u = 0 \ \text{ on } \ \Omega_R, \\ u(0,x) = f(x), \quad \partial_y u(0,x) = g(x), \quad x \in \mathcal{O}, \\ u(y,x) = 0, \quad \partial_\nu u(y,x) = 0, \quad x \in \partial\mathcal{O},\ y \in [0,R], \\ u(R,x) = 0, \quad \partial_y u(R,x) = 0, \quad x \in \mathcal{O}, \end{aligned} \tag{1.3}$$

which is a Dirichlet problem for $\Delta^2 u = 0$ on the Lipschitz domain Ω_R. Results of [**2**] yield a unique solution to (1.3), with non-tangential maximal function bounds on u and ∇u in $L^2(\partial\Omega_R)$. We establish further properties of the solution u, making also use of results of [**6**], [**7**], and [**1**]. In §3 we use the results of §2 to find a solution to (1.1), also with non-tangential maximal function estimates, and satisfying

$$\int_0^\infty \int_{\mathcal{O}} |u(y,x)|^2 \, dx \, dy < \infty, \tag{1.4}$$

and other regularity properties and bounds, which we delineate there. In §4 we study the asymptotic behavior of $u(y,x)$ as $y \to \infty$. Following the Lax program, we study a semigroup of operators. In this case, we study the semigroup S^y on $H^1_0(\mathcal{O}) \oplus L^2(\mathcal{O})$, defined by

$$S^y \begin{pmatrix} f \\ g \end{pmatrix} = \begin{pmatrix} u(y) \\ \partial_y u(y) \end{pmatrix}, \tag{1.5}$$

where u is the solution to (1.1) constructed in §3.

In §5 we look at another semigroup, $\mathcal{T}^s : Y \to Y$, where

$$\begin{aligned} Y = \{u \in H^2(\Omega) : \Delta^2 u = 0,\ u \text{ and } \partial_N u = 0 \text{ on } \mathbb{R}^+ \times \partial\mathcal{O}\}, \\ \mathcal{T}^s u(y,x) = u(y+s,x). \end{aligned} \tag{1.6}$$

This set-up is more closely parallel to that of [**3**] than the consideration of S^y. We note parallel results for $\mathcal{T}^s$, also making use of the results of §§2–3.

Remark. It is interesting to compare (1.1) with the Dirichlet problem for Δ:

$$\begin{aligned} \Delta u = 0 \ \text{ on } \ \Omega, \\ u(0,x) = f(x), \quad x \in \mathcal{O}, \\ u(y,x) = 0, \quad x \in \partial\mathcal{O},\ y \geq 0. \end{aligned} \tag{1.7}$$

The method of separation of variables represents the non-exploding solution to (1.7) as

$$u(y,x) = \sum_{k=1}^\infty \hat{f}(k) e^{-\lambda_k y} \varphi_k(x), \tag{1.8}$$

where $\{\varphi_k : k \geq 1\}$ is an orthonormal basis of $L^2(\mathcal{O})$ consisting of Dirichlet eigenfunctions of $\Delta_x, \varphi_k \in H_0^1(\mathcal{O})$, $\Delta_x \varphi_k = -\lambda_k^2 \varphi_k$, $0 < \lambda_1 < \lambda_2 \leq \lambda_3 \nearrow \infty$, and $\hat{f}(k) = (f, \varphi_k)_{L^2(\mathcal{O})}$. In such a case, $\lambda_k \sim Ck^{1/(n-1)}$, $\sum_k \lambda_k^2 |\hat{f}(k)|^2 < \infty$, and (1.8) is both asymptotic and convergent, in $H_0^1(\mathcal{O})$. Such a separation of variables approach would work for solutions to $\Delta^2 u = 0$ if $\mathcal{O}$ were replaced by a compact manifold without boundary, or if the lateral boundary conditions in (1.1) were replaced by

$$u(y,x) = 0, \quad \Delta_x u(y,x) = 0, \quad x \in \partial\mathcal{O},\ y \geq 0, \tag{1.9}$$

but for (1.1) this method fails.

2. The Dirichlet problem on Ω_R

Here we fix $R \in (0, \infty)$ and discuss solutions to (1.3), which can be rephrased as

$$\Delta^2 u = 0 \ \text{ on } \ \Omega_R, \quad u\big|_{\partial\Omega_R} = f_R, \quad \partial_N u\big|_{\partial\Omega_R} = g_R, \tag{2.1}$$

where ∂_N is the unit normal to $\partial\Omega_R$. Here $f_R = f$ on $\{0\} \times \mathcal{O}$, $f_R = 0$ on the rest of $\partial\Omega_R$, while $g_R = g$ on $\{0\} \times \mathcal{O}$, $g_R = 0$ on the rest of $\partial\Omega_R$. The hypothesis (1.2) yields

$$f_R \in H^1(\partial\Omega_R), \quad g_R \in L^2(\partial\Omega_R). \tag{2.2}$$

Work of [**2**] yields a unique solution to (2.1), smooth in the interior of Ω_R, and satisfying

$$\|u^*\|_{L^2(\partial\Omega_R)} + \|(\nabla u)^*\|_{L^2(\partial\Omega_R)} \leq C\|f_R\|_{H^1(\partial\Omega_R)} + C\|g_R\|_{L^2(\partial\Omega_R)}. \tag{2.3}$$

Here, given a function v, continuous on the interior of Ω_R, we denote by v^* the nontangential maximal function of v,

$$\begin{aligned} v^*(x) &= \sup_{z \in \Gamma(x)} |v(z)|, \qquad x \in \partial\Omega_R, \\ \Gamma(x) &= \{z \in \Omega_R : \operatorname{dist}(z,x) \leq \kappa \operatorname{dist}(z, \partial\Omega_R)\}, \end{aligned} \tag{2.4}$$

for some fixed (large) positive κ.

The following additional information will be useful.

PROPOSITION 2.1. *The solution to (2.1) satisfying (2.2)–(2.3) also satisfies*

$$u \in H^{3/2}(\Omega_R). \tag{2.5}$$

PROOF. It is shown in Theorem 2.4 of [**7**] that, for such u,

$$\|\mathcal{A}(\nabla u)\|_{L^2(\partial\Omega_R)} \approx \|(\nabla u)^*\|_{L^2(\partial\Omega_R)}, \tag{2.6}$$

where

$$\mathcal{A}(\nabla u)(x) = \left(\int_{\Gamma(x)} \operatorname{dist}(x,z)^{2-n} |\nabla\nabla u(z)|^2 \, dz \right)^{1/2}. \tag{2.7}$$

One has

$$\int_{\partial\Omega_R} \chi_{\Gamma(x)}(z) \operatorname{dist}(x,z)^{2-n} \, dS(x) \approx \operatorname{dist}(z, \partial\Omega_R), \tag{2.8}$$

which, by Fubini's theorem, implies

$$\int_{\partial\Omega_R} \operatorname{dist}(z, \partial\Omega_R)\, |\nabla\nabla u(z)|^2\, dz \le C \int_{\partial\Omega_R} |\mathcal{A}(\nabla u)(x)|^2\, dS(x). \tag{2.9}$$

From here, Proposition S of [**1**] yields (2.5). □

We investigate higher regularity of u away from the top and bottom pieces of the boundary. We parametrize Ω_R by $z = (y, x),\ y \in [0, R],\ x \in \mathcal{O}$. Pick

$$\varphi = \varphi(y) \in C_0^\infty((0, R)). \tag{2.10}$$

PROPOSITION 2.2. *For u as in Proposition 2.1, φ as in (2.10),*

$$\varphi u \in H^{5/2-\varepsilon}(\Omega_R), \quad \forall\, \varepsilon > 0. \tag{2.11}$$

PROOF. First note that

$$\begin{aligned} \Delta^2(\varphi u) &= (\partial_y^2 + \Delta_x)^2(\varphi u) \\ &= (\partial_y^4 + 2\partial_y^2\Delta_x + \Delta_x^2)(\varphi u) \\ &= \varphi\Delta^2 u + Qu, \end{aligned} \tag{2.12}$$

where

$$\begin{aligned} Q &= [\partial_y^4, M_\varphi] + 2\Delta_x[\partial_y^2, M_\varphi] \\ &= Q_3(y, \partial_y) + \Delta_x Q_1(y, \partial_y), \end{aligned} \tag{2.13}$$

where $Q_j(y, \partial_y)$ have compact y-support in $y \in (0, R)$ and order j. We hence have

$$\Delta^2(\varphi u) = Qu \in H^{-3/2}(\Omega_R), \quad \varphi u\big|_{\partial\Omega_R} = \partial_N(\varphi u)\big|_{\partial\Omega_R} = 0, \tag{2.14}$$

the degree of regularity of Qu following from (2.5). Now Theorem 2.1 of [**1**] applies to (2.14), to give (2.11), with

$$\|\varphi u\|_{H^{5/2-\varepsilon}(\Omega_R)} \le C_\varepsilon \|Qu\|_{H^{-3/2}(\Omega_R)}. \tag{2.15}$$

□

In fact, Theorem 2.1 of [**1**] gives

$$\|\varphi u\|_{H^{s+4}(\Omega_R)} \le C_s \|\Delta^2(\varphi u)\|_{H^s(\Omega_R)}, \quad -\frac{5}{2} < s < -\frac{3}{2}. \tag{2.16}$$

We next establish the following result.

PROPOSITION 2.3. *Given u as in Proposition 2.1, φ as in (2.10), $m \in \mathbb{Z}^+$,*

$$\varphi\partial_y^m u \in H^{5/2-\varepsilon}(\Omega_R), \quad \forall\varepsilon > 0. \tag{2.17}$$

PROOF. Define $\partial_{y,h}$ by $\partial_{y,h} v(y, x) = h^{-1}\big[v(y + h, x) - v(y, x)\big]$. For small h have

$$\varphi\partial_{y,h} u \in H^{5/2-\varepsilon}(\Omega_R), \quad \partial_N(\varphi\partial_{y,h} u)\big|_{\partial\Omega_R} = 0, \tag{2.18}$$

and, since $\Delta^2(\partial_{y,h} u) = 0$,

$$\Delta^2(\varphi\partial_{y,h} u) = Q(\partial_{y,h} u), \ \text{ bounded in } \ H^{-3/2-\varepsilon}(\Omega_R). \tag{2.19}$$

Since $\varphi(y)\partial_{y,h} u$ has vanishing Dirichlet data on $\partial\Omega_R$, (2.16) yields

$$\|\varphi\partial_{y,h} u\|_{H^{5/2-\varepsilon}(\Omega_R)} \le C_\varepsilon, \tag{2.20}$$

with C_ε independent of h. Taking $h \to 0$, we have $\varphi\partial_{y,h}u \to \varphi\partial_y u$, and the bounds in (2.20) imply this convergence holds weak* in $H^{5/2-\varepsilon}(\Omega_R)$ and

$$\|\varphi\partial_y u\|_{H^{5/2-\varepsilon}(\Omega_R)} \leq C_\varepsilon.$$

This gives (2.17) for $m = 1$, and iterating this argument gives the result for larger m. □

The following corollary will be useful in §3.

COROLLARY 2.4. *Take u as in Proposition 2.1. Also choose $\psi = \psi(y) \in C^\infty([0, R])$ such that*

$$\psi(y) = 1 \;\; \textit{for} \;\; 0 \leq y \leq \frac{R}{3}, \quad \psi(y) = 0 \;\; \textit{for} \;\; \frac{2R}{3} \leq y \leq R. \tag{2.21}$$

Then

$$\Delta^2(\psi u) \in H^{1/2-\varepsilon}(\Omega_R), \quad \forall\, \varepsilon > 0. \tag{2.22}$$

PROOF. A calculation parallel to (2.12)–(2.13) gives

$$\Delta^2(\psi u) = \widetilde{Q}_3(y, \partial_y)u + \Delta_x \widetilde{Q}_1(y, \partial_y)u, \tag{2.23}$$

where $\widetilde{Q}_j(y, \partial_y)$ have compact y support in $(0, R)$ and order j. Then (2.22) follows from (2.17). □

3. The Dirichlet problem on Ω

To solve (1.1), with f, g satisfying (1.2), we do the following. Pick $R \in (0, \infty)$ and solve (1.3) on Ω_R. Call this solution v. Pick $\psi = \psi(y) \in C^\infty([0, R])$ such that (2.21) holds, so, by (2.22),

$$\Delta^2(\psi v) = F \in H^{1/2-\varepsilon}(\Omega_R), \tag{3.1}$$

with F vanishing for $y \geq 2R/3$. Extend F to Ω as 0 for $y \geq R$. We claim there is a unique w satisfying

$$w \in H_0^2(\Omega), \quad \Delta^2 w = F. \tag{3.2}$$

Then the solution to (1.1) is given by

$$u = \psi v - w. \tag{3.3}$$

To get (3.2), consider

$$L : H_0^2(\Omega) \longrightarrow H_0^2(\Omega)^*, \quad Lw = \Delta^2 w, \tag{3.4}$$

so

$$(Lw, w) = \|\Delta w\|^2_{L^2(\Omega)}, \quad \forall\, w \in H_0^2(\Omega). \tag{3.5}$$

Now

$$\begin{aligned} w \in H_0^2(\Omega) \Rightarrow (-\Delta w, w) &= (\nabla w, \nabla w) \\ &\geq \int_0^\infty \|\nabla_x w\|^2_{L^2(\mathcal{O})}\, dy \\ &\geq A\|w\|^2_{L^2(\Omega)}, \end{aligned} \tag{3.6}$$

where $A > 0$ is the smallest eigenvalue of $-\Delta_x$ on $L^2(\mathcal{O})$, with the Dirichlet boundary condition. Since $(-\Delta w, w) \leq \|\Delta w\|_{L^2}\|w\|_{L^2}$, we deduce that

$$\|\Delta w\|^2_{L^2(\Omega)} \geq A^2\|w\|^2_{L^2(\Omega)}, \quad \forall\, w \in H_0^2(\Omega). \tag{3.7}$$

Now the Friedrichs method produces a positive, self-adjoint operator $\mathcal{L}$, with

$$\mathcal{D}(\mathcal{L}^{1/2}) = H_0^2(\Omega), \tag{3.8}$$

and $\mathcal{L}$ is the self-adjoint extension of Δ^2 defined by the Dirichlet boundary condition. From (3.5) and (3.7) we have

$$\mathcal{L} - A^2 I \geq 0. \tag{3.9}$$

In particular, $\mathcal{L}$ is invertible on $L^2(\Omega)$, so we have (3.2), with

$$w = \mathcal{L}^{-1} F \in \mathcal{D}(\mathcal{L}) \subset \mathcal{D}(\mathcal{L}^{1/2}) = H_0^2(\Omega). \tag{3.10}$$

Once we have u in (3.3), the local regularity results of Propositions 2.2–2.3 readily extend. We get

$$\varphi \partial_y^m u \in H^{5/2-\varepsilon}(\Omega), \quad \forall\, \varepsilon > 0,\ m \in \mathbb{Z}^+, \tag{3.11}$$

whenever $\varphi = \varphi(y) \in C_0^\infty((0,\infty))$.

4. The semigroup S^y and asymptotic behavior

We define the semigroup

$$S^y : X \longrightarrow X, \quad X = H_0^1(\mathcal{O}) \oplus L^2(\mathcal{O}), \tag{4.1}$$

by

$$S^y \begin{pmatrix} f \\ g \end{pmatrix} = \begin{pmatrix} u(y) \\ \partial_y u(y) \end{pmatrix}, \tag{4.2}$$

where $u = u(y,x)$ is the solution to (1.1) constructed in §3. Given that

$$\partial_y^m u \in H^{5/2-\varepsilon}([a,b] \times \mathcal{O}), \quad \forall\, \varepsilon > 0,\ m \in \mathbb{Z}^+, \tag{4.3}$$

whenever $0 < a < b < \infty$, we see that S^y is strongly continuous in $y \in (0,\infty)$. Whether S^y is strongly continuous at $y = 0$ requires further investigation, but since our focus is on the behavior as $y \to \infty$, we leave this point.

From (4.3) we obtain

$$S^y : X \longrightarrow H^{5/2-\varepsilon}(\mathcal{O}) \oplus H^{5/2-\varepsilon}(\mathcal{O}), \tag{4.4}$$

for all $\varepsilon > 0$, whenever $y > 0$. In particular,

$$y > 0 \Longrightarrow S^y \ \text{ is compact on } \ X. \tag{4.5}$$

In particular, the spectrum of S^1 has only 0 as an accumulation point.

Proposition 4.1. *If $\zeta \in \mathrm{Spec}\, S^1$, then $|\zeta| < 1$.*

Proof. To begin, since S^1 is compact,

$$\begin{aligned} 0 \neq \zeta \in \operatorname{spec} S^1 &\Rightarrow \exists \text{ nonzero } \begin{pmatrix} f \\ g \end{pmatrix} \in X \text{ such that } S^1 \begin{pmatrix} f \\ g \end{pmatrix} = \zeta \begin{pmatrix} f \\ g \end{pmatrix} \\ &\Rightarrow S^k \begin{pmatrix} f \\ g \end{pmatrix} = \zeta^k \begin{pmatrix} f \\ g \end{pmatrix} \\ &\Rightarrow u(y+k,x) = \zeta^k u(y,x), \end{aligned} \tag{4.6}$$

where u is the solution to (1.1) constructed in §3. In such a case,

$$\int_k^\infty \int_{\mathcal{O}} |u(y,x)|^2\, dx\, dy = |\zeta|^{2k} \int_0^\infty \int_{\mathcal{O}} |u(y,x)|^2\, dx\, dy. \tag{4.7}$$

But by (3.2)–(3.3) we have $u \in L^2(\mathbb{R}^+ \times \mathcal{O})$, so the left side of (4.7) tends to 0 as $k \to \infty$. This forces $|\zeta| < 1$. □

To proceed, fix $\varepsilon = e^{-K} \in (0,1)$, and let P_ε be the projection of X onto the part of Spec S^1 lying inside $D_\varepsilon = \{\zeta \in \mathbb{C} : |\zeta| < \varepsilon\}$, i.e.,

$$P_\varepsilon = \frac{1}{2\pi i} \int_{\gamma_\varepsilon} (\zeta I - S^1)^{-1}\, d\zeta, \tag{4.8}$$

where $\gamma_\varepsilon = \partial D_\varepsilon$ or, if a point in Spec S^1 lies on ∂D_ε, take $\gamma_\varepsilon = \partial D_{\varepsilon-\eta}$ for appropriate $\eta << \varepsilon$. Then the range of $Q_\varepsilon = I - P_\varepsilon$ is a finite dimensional subspace X_ε of X, on which S^y acts as a semigroup. One has

$$S^y Q_\varepsilon = e^{-yA_\varepsilon}, \tag{4.9}$$

where A_ε is a linear operator on the finite dimensional space X_ε, with

$$\operatorname{Spec} A_\varepsilon = \{\lambda_1, \dots, \lambda_{N(\varepsilon)}\}, \quad \operatorname{Re}\lambda_j \geq B > 0, \tag{4.10}$$

where $B > 0$ is taken so that

$$\operatorname{Spec} S^1 \subset \{\zeta \in \mathbb{C} : |\zeta| \leq e^{-B}\}. \tag{4.11}$$

One can decrease ε and increase the list $\{\lambda_1, \dots, \lambda_{N(\varepsilon)}\}$; note that

$$\operatorname{Re}\lambda_j \to +\infty, \quad \text{as } j \to \infty. \tag{4.12}$$

Now for each λ_j there are associated eigenvectors of A_ε, and also possibly generalized eigenvectors, so we have (possibly with repetitions of some of the eigenvalues λ_j)

$$S^y Q_\varepsilon \begin{pmatrix} f \\ g \end{pmatrix} = \sum_{j=1}^{N(\varepsilon)} \sum_{\ell=1}^{M(j)} a_{j\ell}\, y^\ell e^{-\lambda_j y} \begin{pmatrix} \varphi_{j\ell}(x) \\ \psi_{j\ell}(x) \end{pmatrix}, \tag{4.13}$$

for certain functions $\varphi_{j\ell}(x)$ and $\psi_{j\ell}(x)$.

It remains to analyze, as $y \to \infty$,

$$Z_\varepsilon^y = S^y P_\varepsilon = P_\varepsilon S^y P_\varepsilon, \tag{4.14}$$

a semigroup on the range of P_ε. Note that

$$\operatorname{Spec} Z_\varepsilon^1 \subset \{\zeta \in \mathbb{C} : |\zeta| < \varepsilon\}. \tag{4.15}$$

The spectral radius formula implies

$$\limsup_{n\to\infty} \|Z_\varepsilon^n\|_{\mathcal{L}(X)}^{1/n} \leq \varepsilon, \tag{4.16}$$

hence

$$\|Z_\varepsilon^n\|_{\mathcal{L}(X)} \leq e^{-n(K-1)}, \quad \text{for } n \geq \nu(\varepsilon), \text{ sufficiently large.} \tag{4.17}$$

Now if $n+1 \leq y \leq n+2$, using the fact that $\{S^y : y \in [1,2]\}$ has uniformly bounded operator norm on X,

$$\|Z_\varepsilon^y\|_{\mathcal{L}(X)} \leq \|Z_\varepsilon^n\|_{\mathcal{L}(X)} \|Z^{y-n}\|_{\mathcal{L}(X)} \leq Ce^{-n(K-1)} \leq Ce^{-(y-2)(K-1)}. \tag{4.18}$$

We have the following conclusion (after some minor relabeling).

THEOREM 4.2. *Given* $\varepsilon = e^{-K} \in (0,1)$, *the solution* $u(y,x)$ *to (1.1) constructed in §3 has the behavior*

$$u(y,x) = \sum_{j=1}^{N(\varepsilon)} \sum_{\ell=1}^{M(j)} a_{j\ell}\, y^\ell e^{-\lambda_j y} \varphi_{j\ell}(x) + \Psi_\varepsilon(y,x), \tag{4.19}$$

where (4.12) holds and

$$\|\Psi_\varepsilon(y,\cdot)\|_{H_0^1(\mathcal{O})} = O(e^{-Ky}), \quad y \to \infty. \tag{4.20}$$

We have

$$\varphi_{j\ell} \in H^{5/2-\delta}(\mathcal{O}), \quad \forall\, \delta > 0. \tag{4.21}$$

Remark. Making one more use of (4.4), we can improve (4.20) to

$$\|\Psi_\varepsilon(y,\cdot)\|_{H^{5/2-\delta}(\mathcal{O})} = O(e^{-Ky}), \quad y \to \infty, \tag{4.22}$$

for each $\delta > 0$. Other bounds, involving L^p-Sobolev spaces and Besov spaces, can be deduced from regularity results of [**1**] and [**8**]. Of course, if $\partial\mathcal{O}$ has additional regularity, one has further estimates, in stronger norms.

5. Another semigroup

We define T^s for $s \ge 0$, acting on functions on Ω by

$$T^s u(y,x) = u(y+s,x). \tag{5.1}$$

We have T^s acting on various function spaces, such as $L^2(\Omega)$. Here we investigate the action on

$$Y = \{u \in H^2(\Omega) : \Delta^2 u = 0,\ u \text{ and } \partial_N u = 0 \text{ on } \mathbb{R}^+ \times \partial\mathcal{O}\}. \tag{5.2}$$

We denote the restriction of T^s to Y by $\mathcal{T}^s$.

The study of $\mathcal{T}^s : Y \to Y$ is more closely parallel to the general set-up of [**3**] than the study of $S^y : X \to X$. One advantage of $\mathcal{T}^s$ is that it is obviously a contraction semigroup on Y, strongly continuous in $s \in [0,\infty)$. On the other hand, an advantage of S^y is that it incorporates an existence result for the Dirichlet problem.

The next result, parallel to (4.5), makes use of results of §§2–3 as much as (4.5) does.

PROPOSITION 5.1. *For* $s > 0$, $\mathcal{T}^s : Y \to Y$ *is compact.*

PROOF. Consider

$$\rho u(x) = \begin{pmatrix} u(0,x) \\ \partial_y u(0,x) \end{pmatrix}, \quad \rho : Y \to H^{3/2}(\mathcal{O}) \cap H_0^1(\mathcal{O}) \oplus H^{1/2}(\mathcal{O}). \tag{5.3}$$

Since $X = H_0^1(\mathcal{O}) \oplus L^2(\mathcal{O})$, we see that

$$\rho : Y \to X \ \text{ is compact.} \tag{5.4}$$

Now, given $u \in Y,\ s > 0$, we have

$$\mathcal{T}^s u = T^s \Sigma \rho u, \tag{5.5}$$

where Σ is the solution operator to (1.1) constructed in §3. Results of §3 imply

$$T^s \Sigma : X \longrightarrow Y, \quad \forall\, s > 0, \tag{5.6}$$

so (5.5) represents $\mathcal{T}^s : Y \to Y$ as a composition of a compact operator and a continuous operator, for each $s > 0$. □

The next result parallels Proposition 4.1.

LEMMA 5.2. *If* $\zeta \in \operatorname{Spec} \mathcal{T}^1$, *then* $|\zeta| < 1$.

PROOF. Since $\mathcal{T}^1$ is compact on Y,

$$\begin{aligned} 0 \neq \zeta \in \operatorname{Spec} \mathcal{T}^1 &\Rightarrow \exists \text{ nonzero } u \in Y \text{ such that } \mathcal{T}^1 u = \zeta u \\ &\Rightarrow u(y+k,x) = \zeta^k u(y,x). \end{aligned} \tag{5.7}$$

Again (4.7) holds, and implies $|\zeta| < 1$ □

The path from here to the asymptotic expansion (4.19) is quite parallel to that taken in §4.

References

[1] V. Adolfsson and J. Pipher, The inhomogeneous Dirichlet problem for Δ^2 in Lipschitz domains, J. Funct. Anal. 199 (1998), 137–190.

[2] B. Dahlberg, C. Kenig and G. Verchota, The Dirichlet problem for the biharmonic equation in a Lipschitz domain, Ann. Inst. Fourier (Grenoble) 36 (1986), 109–135.

[3] P. Lax, A Phragmén-Lindelöf theorem in harmonic analysis and its applications to some questions in the theory of elliptic equations, Comm. Pure Appl. Math. 10 (1957), 361–389.

[4] P. Lax, Functional Analysis, J. Wiley, New York, 2002.

[5] P. Lax, Abstract Phragmén-Lindelöf theorem and Saint-Venant's principle, Abel Lecture, University of Oslo, May 2005.

[6] J. Pipher and G. Verchota, The Dirichlet problem in L^p for the biharmonic equation on Lipschitz domains, Amer. J. Math. 114 (1992), 923–972.

[7] J. Pipher and G. Verchota, Area integral estimates for the biharmonic operator in Lipschitz domains, Trans. AMS 327 (1991), 903–917.

[8] Z. Shen, The L^p Dirichlet problem for elliptic systems on Lipschitz domains, Math. Research Letters 13 (2006), 143–159.

MATHEMATICS DEPARTMENT, UNIVERSITY OF NORTH CAROLINA, CHAPEL HILL, NORTH CAROLINA 27599

E-mail address: `met@math.unc.edu`

Proceedings of Symposia in Pure Mathematics
Volume **79**, 2008

Wavelets in function spaces

Hans Triebel

Dedicated to Professor Vladimir Maz'ya on the occasion of his 70th birthday.

Abstract. The paper deals with wavelet bases in function spaces on Euclidean n-space, on the n-torus, and on diverse types of domains.

1. Introduction

This survey deals with the symbiotic relationship between (compactly supported) wavelets on $\mathbb{R}^n$ on the one hand and the recent theory of function spaces on $\mathbb{R}^n$, on (smooth and rough) domains and on manifolds in $\mathbb{R}^n$, governed by building blocks, on the other hand. We concentrate on detailed descriptions and outline preferably only those ideas which illuminate how closely some basic constructions of wavelet theory are interwoven with function spaces. The latter will be done in distinguished remarks called **Discussions**. We give references if the corresponding assertions are available in the literature. But most of the results are presented here for the first time. Then we must refer for (technical) details to forthcoming publications, especially to **[35]**.

Section 2 deals with the spaces A^s_{pq} where $A = B$ or $A = F$ on $\mathbb{R}^n$ and (briefly) on the n-torus $\mathbb{T}^n$ and their wavelet expansions under natural restrictions for the parameters involved. In Section 3 we describe wavelet systems and intrinsic wavelet bases for L_p-spaces on arbitrary domains. For a natural class of domains, called E-thick domains (covering in particular Lipschitz domains), we get in Section 4 common intrinsic wavelet bases for related scales of A^s_{pq}-spaces. This will be complemented in Section 5 by more specific assertions for corresponding A^s_{pq}-spaces on C^∞ manifolds and on C^∞ domains. In Section 6 we give some additional references and add a few comments.

2. Spaces on $\mathbb{R}^n$ and $\mathbb{T}^n$

2.1. Definitions. We use standard notation. Let $\mathbb{N}$ be the collection of all natural numbers and $\mathbb{N}_0 = \mathbb{N} \cup \{0\}$. Let $\mathbb{R}^n$ be Euclidean n-space where $n \in \mathbb{N}$. Put $\mathbb{R} = \mathbb{R}^1$, whereas $\mathbb{C}$ is the complex plane. Let $S(\mathbb{R}^n)$ be the usual Schwartz

1991 *Mathematics Subject Classification.* 46E35, 42C40.
Key words and phrases. function spaces, wavelet bases.

space and $S'(\mathbb{R}^n)$ be the space of all tempered distributions on $\mathbb{R}^n$. Furthermore, $L_p(\mathbb{R}^n)$ with $0 < p \le \infty$ is the standard quasi-Banach space with respect to the Lebesgue measure in $\mathbb{R}^n$, quasi-normed by

$$\|f\,|L_p(\mathbb{R}^n)\| = \left(\int_{\mathbb{R}^n} |f(x)|^p\,dx\right)^{1/p}$$

with the usual modification if $p = \infty$. As usual, $\mathbb{Z}$ is the collection of all integers and $\mathbb{Z}^n$ where $n \in \mathbb{N}$, denotes the lattice of all points $m = (m_1, \ldots, m_n) \in \mathbb{R}^n$ with $m_j \in \mathbb{Z}$. Let $\mathbb{N}_0^n$ where $n \in \mathbb{N}$, be the set of all multi-indices,

$$\alpha = (\alpha_1, \ldots, \alpha_n) \qquad \text{with} \quad \alpha_j \in \mathbb{N}_0 \quad \text{and} \quad |\alpha| = \sum_{j=1}^n \alpha_j.$$

If $x = (x_1, \ldots, x_n) \in \mathbb{R}^n$ and $\beta = (\beta_1, \ldots, \beta_n) \in \mathbb{N}_0^n$ then we put

$$x^\beta = x_1^{\beta_1} \cdots x_n^{\beta_n} \qquad \text{(monomials)}.$$

If $\varphi \in S(\mathbb{R}^n)$ then

$$\widehat{\varphi}(\xi) = (F\varphi)(\xi) = (2\pi)^{-n/2} \int_{\mathbb{R}^n} e^{-ix\xi}\,\varphi(x)\,dx, \qquad \xi \in \mathbb{R}^n, \tag{2.1}$$

denotes the Fourier transform of φ. As usual, $F^{-1}\varphi$ and $\varphi^\vee$ stand for the inverse Fourier transform, given by the right-hand side of (2.1) with i in place of $-i$. Here $x\xi$ denotes the scalar product in $\mathbb{R}^n$. Both F and F^{-1} are extended to $S'(\mathbb{R}^n)$ in the standard way. Let $\varphi_0 \in S(\mathbb{R}^n)$ with

$$\varphi_0(x) = 1 \text{ if } |x| \le 1 \quad \text{and} \quad \varphi_0(y) = 0 \text{ if } |y| \ge 3/2, \tag{2.2}$$

and let

$$\varphi_k(x) = \varphi_0(2^{-k}x) - \varphi_0(2^{-k+1}x), \qquad x \in \mathbb{R}^n, \quad k \in \mathbb{N}. \tag{2.3}$$

Then $\sum_{j=0}^\infty \varphi_j(x) = 1$ in $\mathbb{R}^n$ is a dyadic resolution of unity. The entire analytic functions $(\varphi_j \widehat{f})^\vee(x)$ make sense pointwise for any $f \in S'(\mathbb{R}^n)$.

Definition 2.1. *Let $\varphi = \{\varphi_j\}_{j=0}^\infty$ be the above dyadic resolution of unity.*
(i) *Let $0 < p \le \infty$, $0 < q \le \infty$, $s \in \mathbb{R}$. Then $B^s_{pq}(\mathbb{R}^n)$ is the collection of all $f \in S'(\mathbb{R}^n)$ such that*

$$\|f\,|B^s_{pq}(\mathbb{R}^n)\|_\varphi = \left(\sum_{j=0}^\infty 2^{jsq} \|(\varphi_j \widehat{f})^\vee\,|L_p(\mathbb{R}^n)\|^q\right)^{1/q} < \infty \tag{2.4}$$

(with the usual modification if $q = \infty$).
(ii) *Let $0 < p < \infty$, $0 < q \le \infty$, $s \in \mathbb{R}$. Then $F^s_{pq}(\mathbb{R}^n)$ is the collection of all $f \in S'(\mathbb{R}^n)$ such that*

$$\|f\,|F^s_{pq}(\mathbb{R}^n)\|_\varphi = \left\| \left(\sum_{j=0}^\infty 2^{jsq} \left|(\varphi_j \widehat{f})^\vee(\cdot)\right|^q\right)^{1/q} |L_p(\mathbb{R}^n)\right\| < \infty \tag{2.5}$$

(with the usual modification if $q = \infty$).

REMARK 2.2. The theory of these spaces may be found in [**27, 29, 32**]. In particular these spaces are independent of admitted resolutions of unity φ (equivalent quasi-norms). This justifies our omission of the subscript φ in (2.4), (2.5) in the sequel. As usual nowadays we write occasionally

$$A^s_{pq}(\mathbb{R}^n) \quad \text{with} \quad A \in \{B, F\} \tag{2.6}$$

if the assertion considered applies equally to $B^s_{pq}(\mathbb{R}^n)$ and $F^s_{pq}(\mathbb{R}^n)$ (always with $p < \infty$ for the F-spaces). We remind of a few special cases and properties referring for details to the above books, especially to [**32**, Section 1.2].

(i) Let $1 < p < \infty$ and $k \in \mathbb{N}_0$. Then

$$W^k_p(\mathbb{R}^n) = F^k_{p,2}(\mathbb{R}^n) \tag{2.7}$$

are the *classical Sobolev spaces*, usually equivalently normed by

$$\|f\,|W^k_p(\mathbb{R}^n)\| = \left(\sum_{|\alpha| \le k} \|D^\alpha f\,|L_p(\mathbb{R}^n)\|^p \right)^{1/p} \tag{2.8}$$

and based on the *Paley-Littlewood assertion* $L_p(\mathbb{R}^n) = F^0_{p,2}(\mathbb{R}^n)$ for the *Lebesgue spaces* $L_p(\mathbb{R}^n)$.

(ii) For all admitted s, p, q and $\sigma \in \mathbb{R}$,

$$I_\sigma A^s_{pq}(\mathbb{R}^n) = A^{s-\sigma}_{pq}(\mathbb{R}^n) \quad \text{with} \quad I_\sigma f = \left((1 + |\xi|^2)^{\sigma/2} \widehat{f} \right)^\vee$$

is an isomorphic map (equivalent quasi-norms). Nowadays one calls

$$H^s_p(\mathbb{R}^n) = I_{-s} L_p(\mathbb{R}^n) = F^s_{p,2}(\mathbb{R}^n), \qquad s \in \mathbb{R}, \quad 1 < p < \infty,$$

Sobolev spaces with the classical Sobolev spaces

$$W^k_p(\mathbb{R}^n) = H^k_p(\mathbb{R}^n), \qquad 1 < p < \infty, \quad k \in \mathbb{N}_0,$$

as special cases.

(iii) Let

$$(\Delta^1_h f)(x) = f(x+h) - f(x), \qquad \left(\Delta^{l+1}_h f\right)(x) = \Delta^1_h \left(\Delta^l_h f\right)(x),$$

where $x \in \mathbb{R}^n$, $h \in \mathbb{R}^n$, $l \in \mathbb{N}$, be the iterated differences in $\mathbb{R}^n$. Then the *Hölder-Zygmund spaces*

$$\mathcal{C}^s(\mathbb{R}^n) = B^s_{\infty\infty}(\mathbb{R}^n), \qquad s > 0,$$

can be equivalently normed by

$$\|f\,|\mathcal{C}^s(\mathbb{R}^n)\|_m = \sup_{x \in \mathbb{R}^n} |f(x)| + \sup |h|^{-s} \left|\Delta^m_h f(x)\right|, \quad 0 < s < m \in \mathbb{N},$$

where the second supremum is taken over all $x \in \mathbb{R}^n$ and $h \in \mathbb{R}^n$ with $0 < |h| \le 1$.

(iv) The spaces $B^s_{pq}(\mathbb{R}^n)$ according to Definition 2.1 are called *Besov spaces*. Let

$$0 < p, q \le \infty, \qquad n\left(\frac{1}{p} - 1\right)_+ < s < m \in \mathbb{N}, \tag{2.9}$$

where $a_+ = \max(a, 0)$ if $a \in \mathbb{R}$. Then $B^s_{pq}(\mathbb{R}^n)$ can be equivalently quasi-normed by

$$\|f\,|B^s_{pq}(\mathbb{R}^n)\|_m = \|f\,|L_p(\mathbb{R}^n)\| + \left(\int_{|h| \le 1} |h|^{-sq} \, \|\Delta^m_h f\,|L_p(\mathbb{R}^n)\|^q \, \frac{dh}{|h|^n} \right)^{1/q} \tag{2.10}$$

and

$$\|f\,|B^s_{pq}(\mathbb{R}^n)\|^*_m = \|f\,|L_p(\mathbb{R}^n)\| + \left(\int_0^1 t^{-sq} \sup_{|h|\le t} \|\Delta^m_h f\,|L_p(\mathbb{R}^n)\| \frac{dt}{t}\right)^{1/q} \tag{2.11}$$

(with the usual modification if $q = \infty$). If $1 < p < \infty$, $1 \le q \le \infty$, $s > 0$, then $B^s_{pq}(\mathbb{R}^n)$ are the *classical Besov spaces.*

Otherwise we refer to [**29**, Chapter 1] and [**32**, Chapter 1] where one finds the history of these spaces, further special cases and classical assertions.

2.2. Wavelets on $\mathbb{R}^n$. We suppose that the reader is familiar with wavelets on $\mathbb{R}^n$ of Daubechies type and the related multiresolution analysis. The standard references are [**11, 21, 23, 37**]. A short summary of some relevant aspects may also be found in [**32**, Section 1.7]. We give a brief description of some basic notation. As usual $C^u(\mathbb{R})$ with $u \in \mathbb{N}$ collects all (complex-valued) continuous functions on $\mathbb{R}$ having continuous bounded derivatives up to order u (inclusively). Let

$$\psi_F \in C^u(\mathbb{R}), \quad \psi_M \in C^u(\mathbb{R}), \qquad u \in \mathbb{N}, \tag{2.12}$$

be *real* compactly supported Daubechies wavelets with

$$\int_{\mathbb{R}} \psi_M(x)\, x^v\, dx = 0 \qquad \text{for all } v \in \mathbb{N}_0 \text{ with } v < u. \tag{2.13}$$

Recall that ψ_F is called the *scaling function* (father wavelet) and ψ_M the associated *wavelet* (mother wavelet). We extend these wavelets from $\mathbb{R}$ to $\mathbb{R}^n$ by the usual tensor procedure. Let $u \in \mathbb{N}$ and

$$G = (G_1, \ldots, G_n) \in G^0 = \{F, M\}^n$$

which means that G_r is either F or M. Let

$$G = (G_1, \ldots, G_n) \in G^j = \{F, M\}^{n*}, \qquad j \in \mathbb{N},$$

which means that G_r is either F or M where $*$ indicates that at least one of the components of G must be an M. Hence G^0 has 2^n elements, whereas G^j with $j \in \mathbb{N}$ has $2^n - 1$ elements. Let $L \in \mathbb{N}_0$ and

$$\Psi^j_{G,m}(x) = 2^{(j+L)n/2} \prod_{r=1}^n \psi_{G_r}\left(2^{j+L} x_r - m_r\right), \qquad G \in G^j, \quad m \in \mathbb{Z}^n, \tag{2.14}$$

where $j \in \mathbb{N}_0$. We always assume that ψ_F and ψ_M in (2.12) have L_2-norm 1.

PROPOSITION 2.3. *Let $L \in \mathbb{N}_0$ and $u \in \mathbb{N}$. Then*

$$\left\{\Psi^j_{G,m} : \ j \in \mathbb{N}_0, \ G \in G^j, \ m \in \mathbb{Z}^n\right\} \tag{2.15}$$

is an orthonormal basis in $L_2(\mathbb{R}^n)$.

REMARK 2.4. This is a cornerstone of (inhomogeneous) wavelet theory. We refer to the above-mentioned books. In $\mathbb{R}^n$ one may choose $L = 0$. But later on when it comes to domains then it will be important that one can choose $L \in \mathbb{N}_0$ large. The L_2-normalisation of $\Psi^j_{G,m}$ is natural. But for our later purposes it is convenient to switch to an *L_∞-normalisation.* To prepare what follows we remark that any $f \in L_2(\mathbb{R}^n)$ can be represented as

$$f = \sum_{j=0}^\infty \sum_{G \in G^j} \sum_{m \in \mathbb{Z}^n} \lambda^{j,G}_m\, 2^{-jn/2}\, \Psi^j_{G,m} \tag{2.16}$$

with

$$\lambda_m^{j,G} = \lambda_m^{j,G}(f) = 2^{jn/2} \int_{\mathbb{R}^n} f(x)\, \Psi^j_{G,m}(x)\, dx = 2^{jn/2} \left(f, \Psi^j_{G,m} \right) \tag{2.17}$$

and

$$\| f \,| L_2(\mathbb{R}^n) \| = \left(\sum_{j,G,m} 2^{-jn} \, |\lambda_m^{j,G}|^2 \right)^{1/2} . \tag{2.18}$$

(Recall that the $\Psi^j_{G,m}$'s are real).

2.3. Wavelet bases in $A^s_{pq}(\mathbb{R}^n)$. We extend the wavelet representation according to Proposition 2.3 and (2.16)-(2.18) from $L_2(\mathbb{R}^n)$ to $A^s_{pq}(\mathbb{R}^n)$ where $A = B$ or $A = F$ and $s \in \mathbb{R}$, $0 < p, q \leq \infty$ (with $p < \infty$ for the F-spaces). First we need a substitute of the sequence spaces ℓ_2 in (2.18). Let χ_{jm} be the characteristic function of a cube Q_{jm} in $\mathbb{R}^n$ with sides parallel to the axes of coordinates, centred at $2^{-j}m$ and with side-length 2^{-j+1} where $m \in \mathbb{Z}^n$ and $j \in \mathbb{N}_0$.

DEFINITION 2.5. *Let $s \in \mathbb{R}$, $0 < p \leq \infty$, $0 < q \leq \infty$. Then b^s_{pq} is the collection of all sequences*

$$\lambda = \left\{ \lambda_m^{j,G} \in \mathbb{C} : \; j \in \mathbb{N}_0, \; G \in G^j, \; m \in \mathbb{Z}^n \right\} \tag{2.19}$$

such that

$$\| \lambda \,| b^s_{pq} \| = \left(\sum_{j=0}^{\infty} 2^{j(s-\frac{n}{p})q} \sum_{G \in G^j} \left(\sum_{m \in \mathbb{Z}^n} |\lambda_m^{j,G}|^p \right)^{q/p} \right)^{1/q} < \infty$$

and f^s_{pq} is the collection of all sequences (2.19) such that

$$\| \lambda \,| f^s_{pq} \| = \left\| \left(\sum_{j,G,m} 2^{jsq} \, |\lambda_m^{j,G} \, \chi_{jm}(\cdot)|^q \right)^{1/q} | L_p(\mathbb{R}^n) \right\| < \infty$$

with the usual modifications if $p = \infty$ and/or $q = \infty$.

REMARK 2.6. One has $b^s_{pp} = f^s_{pp}$. With $s = 0$, $p = q = 2$, one gets the right-hand side of (2.18).

We use standard notation naturally extended from Banach spaces to quasi-Banach spaces. In particular $\{b_j\}_{j=1}^\infty \subset B$ in a complex quasi-Banach space is called a *basis* if any $b \in B$ can be uniquely represented as

$$b = \sum_{j=1}^{\infty} \lambda_j \, b_j, \qquad \lambda_j \in \mathbb{C} \quad \text{(convergence in } B\text{)}. \tag{2.20}$$

A basis $\{b_j\}_{j=1}^\infty$ is called an *unconditional basis* if for any rearrangement σ of $\mathbb{N}$ (one-to-one map of $\mathbb{N}$ onto itself) $\{b_{\sigma(j)}\}_{j=1}^\infty$ is again a basis and

$$b = \sum_{j=1}^{\infty} \lambda_{\sigma(j)} \, b_{\sigma(j)} \qquad \text{(convergence in } B\text{)} \tag{2.21}$$

for any $b \in B$ with (2.20). Standard bases of separable sequence spaces as considered in this paper are always unconditional. We refer to **[1]** for details about bases in Banach (sequence) spaces. Similarly as in (2.20), (2.21) we speak about *unconditional convergence* in $S'(\mathbb{R}^n)$ and in $A^s_{pq}(\mathbb{R}^n)$. *Local convergence* in $A^s_{pq}(\mathbb{R}^n)$ means convergence in $A^s_{pq}(K)$ for any ball K in $\mathbb{R}^n$. Let a^s_{pq} be either b^s_{pq} or f^s_{pq}. In any case $u \in \mathbb{N}$ in (2.12) and hence in (2.14) will be chosen in such a way that (independently of L)

$$\sum_{j=0}^{\infty} \sum_{G \in G^j} \sum_{m \in \mathbb{Z}^n} \lambda^{j,G}_m \, 2^{-jn/2} \, \Psi^j_{G,m}, \qquad \lambda \in a^s_{pq}, \tag{2.22}$$

converges unconditionally in $S'(\mathbb{R}^n)$ and locally in any $A^\sigma_{pq}(\mathbb{R}^n)$ with $\sigma < s$. This justifies to abbreviate (2.22) by

$$\sum_{j,G,m} \lambda^{j,G}_m \, 2^{-jn/2} \, \Psi^j_{G,m}, \qquad \lambda \in a^s_{pq}. \tag{2.23}$$

We use the nowadays standard abbreviations

$$\sigma_p = n \left(\frac{1}{p} - 1 \right)_+ \qquad \text{and} \qquad \sigma_{pq} = n \left(\frac{1}{\min(p,q)} - 1 \right)_+ \tag{2.24}$$

where $0 < p, q \le \infty$ and $b_+ = \max(b, 0)$ if $b \in \mathbb{R}$.

THEOREM 2.7. (i) *Let* $0 < p \le \infty$, $0 < q \le \infty$, $s \in \mathbb{R}$, *and let* $\Psi^j_{G,m}$ *be the wavelets in* (2.14) *with* $L \in \mathbb{N}_0$, *based on* (2.12), (2.13), *with*

$$u > \max(s, \sigma_p - s).$$

Let $f \in S'(\mathbb{R}^n)$. *Then* $f \in B^s_{pq}(\mathbb{R}^n)$ *if, and only if, it can be represented as*

$$f = \sum_{j,G,m} \lambda^{j,G}_m \, 2^{-jn/2} \, \Psi^j_{G,m}, \qquad \lambda \in b^s_{pq}, \tag{2.25}$$

unconditional convergence being in $S'(\mathbb{R}^n)$ *and locally in any space* $B^\sigma_{pq}(\mathbb{R}^n)$ *with* $\sigma < s$. *The representation* (2.25) *is unique,*

$$\lambda^{j,G}_m = 2^{jn/2} \left(f, \Psi^j_{G,m} \right) \tag{2.26}$$

and

$$I : \; f \mapsto \left\{ 2^{jn/2} \left(f, \Psi^j_{G,m} \right) \right\} \tag{2.27}$$

is an isomorphic map of $B^s_{pq}(\mathbb{R}^n)$ *onto* b^s_{pq}. *If, in addition,* $p < \infty$, $q < \infty$, *then* $\left\{ \Psi^j_{G,m} \right\}$ *is an unconditional basis in* $B^s_{pq}(\mathbb{R}^n)$.

(ii) *Let* $0 < p < \infty$, $0 < q \le \infty$, $s \in \mathbb{R}$, *and*

$$u > \max(s, \sigma_{pq} - s).$$

Let $f \in S'(\mathbb{R}^n)$. *Then* $f \in F^s_{pq}(\mathbb{R}^n)$ *if, and only if, it can be represented as*

$$f = \sum_{j,G,m} \lambda^{j,G}_m \, 2^{-jn/2} \, \Psi^j_{G,m}, \qquad \lambda \in f^s_{pq}, \tag{2.28}$$

unconditional convergence being in $S'(\mathbb{R}^n)$ *and locally in any space* $F^\sigma_{pq}(\mathbb{R}^n)$ *with* $\sigma < s$. *The representation* (2.28) *is unique with* (2.26). *Furthermore,* I *in* (2.27) *is an isomorphic map of* $F^s_{pq}(\mathbb{R}^n)$ *onto* f^s_{pq}. *If, in addition,* $q < \infty$, *then* $\left\{ \Psi^j_{G,m} \right\}$ *is an unconditional basis in* $F^s_{pq}(\mathbb{R}^n)$.

Discussion 2.8. As said in the Introduction there is a symbiotic relationship between some aspects of wavelets and the recent theory of function spaces based on building blocks. In particular the above wavelets $\Psi^j_{G,m}$ may serve simultaneously as atoms and as kernels of local means. Atomic representations in function spaces as used nowadays go essentially back to [**14, 15**]. But more details about the somewhat involved history of atoms may be found in [**29**, Section 1.9]. By the sharp version of atomic representations according to [**32**, Section 1.5.1] it follows that (2.25) is an atomic expansion based on the normalised *atoms*

$$a^G_{jm} = 2^{-j(s-\frac{n}{p})}\, 2^{-jn/2}\, \Psi^j_{G,m}, \qquad G \in G^j. \tag{2.29}$$

As far as the required cancellations for the atoms are concerned we remind of

$$\int_{\mathbb{R}^n} x^\beta\, \Psi^j_{G,m}(x)\, dx = 0 \qquad \text{if } j \in \mathbb{N} \text{ and } |\beta| < u, \tag{2.30}$$

as a consequence of (2.13) and (2.14). Then it follows from the atomic representation theorem that $f \in S'(\mathbb{R}^n)$, given by (2.25), belongs to $B^s_{pq}(\mathbb{R}^n)$ and

$$\|f\,|B^s_{pq}(\mathbb{R}^n)\| \le c\,\|\lambda\,|b^s_{pq}\|. \tag{2.31}$$

On the other hand, (2.26) may be considered as *local means* (appropriately interpreted)

$$\lambda^{j,G}_m = k^G_{jm}(f) = \int_{\mathbb{R}^n} k^G_{jm}(y)\, f(y)\, dy, \qquad k^G_{jm} = 2^{jn/2}\, \Psi^j_{G,m}. \tag{2.32}$$

Again cancellations for the kernels k^G_{jm} of type (2.30) are indispensable. Equivalent quasi-norms in function spaces $A^s_{pq}(\mathbb{R}^n)$ in terms of local means have some history going back to [**28**]. This has been presented in a more elaborated version in [**29**]. But a sharp assertion which can be used not only for function spaces of the above type on $\mathbb{R}^n$ but also also on smooth and rough domains has been obtained only recently in [**34**] (and [**35**, Section 1.1.3]) with [**31**] and [**32**, Section 3.1] as forerunners. Based on these observations one gets by (2.32) for $f \in B^s_{pq}(\mathbb{R}^n)$ that

$$\|\lambda\,|b^s_{pq}\| \le c\,\|f\,|B^s_{pq}(\mathbb{R}^n)\|.$$

By (2.31) both quasi-norms are equivalent. Similarly for the spaces $F^s_{pq}(\mathbb{R}^n)$. Of course these sketchy arguments are far from being a rigorous proof covering all technicalities. This may be found in [**32**, Section 3.1], [**34**] and will be the subject of [**35**]. We mainly wanted to illuminate

the double role played by the wavelets $\Psi^j_{G,m}$ as atoms and as kernels of local means.

This is the crucial observation, the rest are technicalities.

Remark 2.9. Proposition 2.3 is the starting point of the wavelet theory in function spaces. It is natural to ask whether (2.15) remains to be an (unconditional) basis in other spaces on $\mathbb{R}^n$. First candidates are $L_p(\mathbb{R}^n)$ with $1 < p < \infty$ but also related Sobolev spaces and classical Besov spaces. Something may be found in the above-mentioned books [**11, 21, 23, 37**]. One may also consult [**32**, Remarks 1.63, 1.65, 1.66, pp. 32-35] for more details and further references. An extension of this theory to all spaces $A^s_{pq}(\mathbb{R}^n)$ goes back to [**20, 31, 34**].

2.4. Wavelets on $\mathbb{T}^n$. By rule of thumb wavelets and function spaces on $\mathbb{R}^n$ have natural periodic counterparts on the n-torus $\mathbb{T}^n$,

$$\mathbb{T}^n = \{x = (x_1, \dots, x_n) \in \mathbb{R}^n : 0 \le x_r \le 1\} \tag{2.33}$$

(opposite points are identified in the usual way). We give a brief description for several reasons. There is a huge literature about periodic wavelets but only very little as far as wavelet bases in periodic function spaces is concerned, mostly restricted to $L_2(\mathbb{T})$, where $\mathbb{T} = \mathbb{T}^1$ is the 1-torus. We refer in this context to [**11**, pp. 304/305], [**21**, Section 7.5.1], [**37**, Section 2.5]. So it seems to be desirable to give precise definitions and assertions for the full scale of the periodic counterparts $A^s_{pq}(\mathbb{T}^n)$ of the above spaces $A^s_{pq}(\mathbb{R}^n)$. But there is also some use later on in Section 5 in connection with wavelet bases for some spaces $A^s_{pq}(\Omega)$ in (bounded C^∞) domains Ω in $\mathbb{R}^n$, where the boundary is diffeomorphic to an $(n-1)$-torus (with $n = 2$ as the case of preference).

Let $D'(\mathbb{T}^n)$ be the space of (periodic) distributions on $\mathbb{T}^n$. We assume that the reader is familiar with basic assertions about these distributions. Recall that $f \in D'(\mathbb{T}^n)$ can be represented as

$$f = \sum_{m \in \mathbb{Z}^n} a_m \, e^{i2\pi mx}, \qquad x \in \mathbb{T}^n, \tag{2.34}$$

where the Fourier coefficients $a_m \in \mathbb{C}$ are of at most polynomial growth,

$$|a_m| \le c(1 + |m|)^{\varkappa} \quad \text{for some } c > 0,\ \varkappa > 0, \text{ and all } m \in \mathbb{Z}^n.$$

The theory of periodic distributions and related periodic spaces $B^s_{pq}(\mathbb{T}^n)$ and $F^s_{pq}(\mathbb{T}^n)$ has some history which is not the subject of this survey. We rely on [**24**, Chapter 3] and [**27**, Chapter 9] where one finds also further references. Let $\{\varphi_j\}_{j=0}^\infty$ be the same dyadic resolution of unity in $\mathbb{R}^n$ as in (2.2), (2.3) and in Definition 2.1. Let $f \in D'(\mathbb{T}^n)$, given by (2.34), be extended periodically to $\mathbb{R}^n$. Then $f \in S'(\mathbb{R}^n)$ (using the same letter f) and

$$(\varphi_j \widehat{f})^\vee(x) = \sum_{m \in \mathbb{Z}^n} a_m \, \varphi_j(2\pi m) \, e^{i2\pi mx}$$

are trigonometrical polynomials. This justifies the following periodic counterpart of Definition 2.1.

DEFINITION 2.10. *Let $\varphi = \{\varphi_j\}_{j=0}^\infty$ be the above resolution of unity in $\mathbb{R}^n$.*
(i) *Let $0 < p \le \infty$, $0 < q \le \infty$, $s \in \mathbb{R}$. Then $B^s_{pq}(\mathbb{T}^n)$ is the collection of all $f \in D'(\mathbb{T}^n)$, given by* (2.34), *such that*

$$\|f \,|B^s_{pq}(\mathbb{T}^n)\|_\varphi = \left(\sum_{j=0}^\infty 2^{jsq} \left\| \sum_{m \in \mathbb{Z}^n} a_m \, \varphi_j(2\pi m) \, e^{i2\pi mx} \,|L_p(\mathbb{T}^n) \right\|^q \right)^{1/q} < \infty$$

(with the usual modification if $q = \infty$).
(ii) *Let $0 < p < \infty$, $0 < q \le \infty$, $s \in \mathbb{R}$. Then $F^s_{pq}(\mathbb{T}^n)$ is the collection of all $f \in D'(\mathbb{T}^n)$, given by* (2.34), *such that*

$$\|f \,|F^s_{pq}(\mathbb{T}^n)\|_\varphi = \left\| \left(\sum_{j=0}^\infty 2^{jsq} \left| \sum_{m \in \mathbb{Z}^n} a_m \, \varphi_j(2\pi m) \, e^{i2\pi mx} \right|^q \right)^{1/q} |L_p(\mathbb{T}^n) \right\| < \infty$$

(*with the usual modification if* $q = \infty$).

REMARK 2.11. One has a periodic version of Remark 2.2, including the special cases mentioned there.

There are natural periodic counterparts of the related sequence spaces and wavelet expansions. But the rigorous justification requires some care which may be found in [**34**] and to [**35**, Section 1]. We restrict ourselves to a description. Let $\Psi^j_{G,m}$ be the same wavelets as in (2.14) where we choose (and fix) $L \in \mathbb{N}_0$ such that

$$\operatorname{supp} \Psi^0_{G,0} \subset \{x \in \mathbb{R}^n : |x| < 1/2\}, \qquad G \in G^0 = \{F, M\}^n.$$

Let

$$\mathbb{P}^n_j = \{m \in \mathbb{Z}^n : 0 \le m_r < 2^{j+L}\}, \qquad j \in \mathbb{N}_0,$$

be the $2^{(j+L)n}$ lattice points in $2^{j+L}\mathbb{T}^n$. Let

$$\Psi^{j,\mathrm{per}}_{G,m}(x) = \sum_{l \in \mathbb{Z}^n} \Psi^j_{G,m}(x - l) = \sum_{l \in \mathbb{Z}^n} \Psi^j_{G,m+2^{j+L}l}(x), \qquad x \in \mathbb{T}^n, \tag{2.35}$$

with $j \in \mathbb{N}_0$ and $m \in \mathbb{P}^n_j$ be the periodic extension of the distinguished wavelets $\Psi^j_{G,m}$ with off-points $2^{-j-L}m \in \mathbb{T}^n$, restricted afterwards to $\mathbb{T}^n$. Then one has the following counterpart of Proposition 2.3.

PROPOSITION 2.12. *Let* $u \in \mathbb{N}$ *in* (2.12), (2.13) *and* (2.14), (2.35). *Then*

$$\left\{\Psi^{j,\mathrm{per}}_{G,m} : j \in \mathbb{N}_0,\ G \in G^j,\ m \in \mathbb{P}^n_j\right\}$$

is an orthonormal basis in $L_2(\mathbb{T}^n)$.

REMARK 2.13. Let

$$(f, g)_\pi = \int_{\mathbb{T}^n} f(x)\, g(x)\, dx$$

be the dual pairing in $(D(\mathbb{T}^n), D'(\mathbb{T}^n))$, appropriately interpreted. Similar as in (2.22), (2.23) (and with the same justification as there) we abbreviate

$$\sum_{j=0}^{\infty} \sum_{G \in G^j} \sum_{m \in \mathbb{P}^n_j} \lambda^{j,G}_m\, 2^{-jn/2}\, \Psi^{j,\mathrm{per}}_{G,m} = \sum_{j,G.m} \lambda^{j,G}_m\, 2^{-jn/2}\, \Psi^{j,\mathrm{per}}_{G,m} \tag{2.36}$$

in what follows. First we remark the obvious counterpart of (2.16)-(2.18): Any $f \in L_2(\mathbb{T}^n)$ can be represented by

$$f = \sum_{j,G,m} \lambda^{j,G}_m\, 2^{-jn/2}\, \Psi^{j,\mathrm{per}}_{G,m}$$

with

$$\lambda^{j,G}_m = \lambda^{j,G}_m(f) = 2^{jn/2} \left(f, \Psi^{j,\mathrm{per}}_{G,m}\right)_\pi$$

and

$$\|f\,|L_2(\mathbb{T}^n)\| = \left(\sum_{j,G,m} 2^{-jn}\, |\lambda^{j,G}_m|^2\right)^{1/2}.$$

(Recall that $\Psi^{j,\mathrm{per}}_{G,m}$ is real).

The periodic counterpart of the sequence spaces in Definition 2.5 can be described as follows.

DEFINITION 2.14. *Let $s \in \mathbb{R}$, $0 < p \le \infty$, $0 < q \le \infty$. Then $b^{s,\mathrm{per}}_{pq}$ is the collection of all sequences*

$$\lambda = \{\lambda^{j,G}_m \in \mathbb{C} : \; j \in \mathbb{N}_0, \; G \in G^j, \; m \in \mathbb{P}^n_j\} \tag{2.37}$$

such that

$$\|\lambda \,|b^{s,\mathrm{per}}_{pq}\| = \left(\sum_{j=0}^{\infty} 2^{j(s-\frac{n}{p})q} \sum_{G \in G^j} \left(\sum_{m \in \mathbb{P}^n_j} |\lambda^{j,G}_m|^p \right)^{q/p} \right)^{1/q} < \infty$$

and $f^{s,\mathrm{per}}_{pq}$ is the collection of all sequences (2.37) such that

$$\|\lambda \,|f^{s,\mathrm{per}}_{pq}\| = \left\| \left(\sum_{j,G,m} 2^{jsq} \, |\lambda^{j,G}_m \, \chi_{jm}(\cdot)|^q \right)^{1/q} |L_p(\mathbb{T}^n) \right\| < \infty$$

with the usual modifications if $p = \infty$ and/or $q = \infty$, where χ_{jm} is the characteristic function of a cube with the left corner $2^{-j-L}m$ and of side-length 2^{-j-L} (a subcube of $\mathbb{T}^n$).

After these preparations one gets now the following counterpart of Theorem 2.7. Recall the abbreviation (2.36). Furthermore, σ_p and σ_{pq} have the same meaning as in (2.24). Let $u \in \mathbb{N}$ be as in (2.12), (2.13) and (2.14), (2.35).

THEOREM 2.15. *Let $\left\{ \Psi^{j,\mathrm{per}}_{G,m} \right\}$ be the orthonormal basis in $L_2(\mathbb{T}^n)$ according to Proposition 2.12.*

(i) *Let $0 < p \le \infty$, $0 < q \le \infty$, $s \in \mathbb{R}$ and*

$$u > \max(s, \sigma_p - s).$$

Let $f \in D'(\mathbb{T}^n)$. Then $f \in B^s_{pq}(\mathbb{T}^n)$ if, and only if, it can be represented as

$$f = \sum_{j,G,m} \lambda^{j,G}_m \, 2^{-jn/2} \, \Psi^{j,\mathrm{per}}_{G,m}, \qquad \lambda \in b^{s,\mathrm{per}}_{pq}, \tag{2.38}$$

unconditional convergence being in $D'(\mathbb{T}^n)$ and in any space $B^\sigma_{pq}(\mathbb{T}^n)$ with $\sigma < s$. The representation (2.38) is unique,

$$\lambda^{j,G}_m = 2^{jn/2} \left(f, \Psi^{j,\mathrm{per}}_{G,m} \right)_\pi \tag{2.39}$$

and

$$I : \; f \mapsto \left\{ 2^{jn/2} \left(f, \Psi^{j,\mathrm{per}}_{G,m} \right)_\pi \right\} \tag{2.40}$$

is an isomorphic map of $B^s_{pq}(\mathbb{T}^n)$ onto $b^{s,\mathrm{per}}_{pq}$. If, in addition, $p < \infty$, $q < \infty$, then $\left\{ \Psi^{j,\mathrm{per}}_{G,m} \right\}$ is an unconditional basis in $B^s_{pq}(\mathbb{T}^n)$.

(ii) *Let $0 < p < \infty$, $0 < q \le \infty$, $s \in \mathbb{R}$ and*

$$u > \max(s, \sigma_{pq} - s).$$

Let $f \in D'(\mathbb{T}^n)$. Then $f \in F^s_{pq}(\mathbb{T}^n)$ if, and only if, it can be represented as

$$f = \sum_{j,G,m} \lambda^{j,G}_m \, 2^{-jn/2} \, \Psi^{j,\mathrm{per}}_{G,m}, \qquad \lambda \in f^{s,\mathrm{per}}_{pq}, \tag{2.41}$$

unconditional convergence being in $D'(\mathbb{T}^n)$ *and in any space* $F^\sigma_{pq}(\mathbb{T}^n)$ *with* $\sigma < s$. *The representation* (2.41) *is unique with* (2.39), *and* I *in* (2.40) *is an isomorphic map of* $F^s_{pq}(\mathbb{T}^n)$ *onto* $f^{s,\mathrm{per}}_{pq}$. *If, in addition,* $q < \infty$, *then* $\left\{\Psi^{j,\mathrm{per}}_{G,m}\right\}$ *is an unconditional basis in* $F^s_{pq}(\mathbb{T}^n)$.

Discussion 2.16. This is the direct and to some extent expected periodic counterpart of Theorem 2.7. Basically one extends functions and distributions $f \in D'(\mathbb{T}^n)$ periodically to $\mathbb{R}^n$. But these extended distributions do not belong to any space $A^s_{pq}(\mathbb{R}^n)$ with $p < \infty$ (with exception of $f = 0$). One has the same unpleasant effect if one periodises the $\mathbb{R}^n$-wavelets as in (2.35) with $x \in \mathbb{R}^n$ in place of $x \in \mathbb{T}^n$. This obstacle can be circumvented if one deals first with suitable weighted spaces on $\mathbb{R}^n$. Let

$$w_\alpha(x) = (1 + |x|^2)^{\alpha/2}, \qquad x \in \mathbb{R}^n, \quad \alpha \in \mathbb{R},$$

and

$$A^s_{pq}(\mathbb{R}^n, w_\alpha) = \left\{ f \in S'(\mathbb{R}^n) : \ w_\alpha f \in A^s_{pq}(\mathbb{R}^n) \right\},$$

naturally quasi-normed. There is a complete counterpart of Theorem 2.7 with the same wavelets $\Psi^j_{G,m}$ and the same restrictions for u and suitably modified sequence spaces. This goes back to **[16]** and may be found in **[32**, Section 6.2**]**. As for a refined version needed in the above context we refer to **[34]**. If $0 < p \le \infty$ and $\alpha < -n/p$ then

$$A^{s,\mathrm{per}}_{pq}(\mathbb{R}^n, w_\alpha) = \left\{ f \in A^s_{pq}(\mathbb{R}^n, w_\alpha) : \ f(\cdot) = f(\cdot - m), \ m \in \mathbb{Z}^n \right\}$$

is the closed subspace of $A^s_{pq}(\mathbb{R}^n, w_\alpha)$ consisting of the indicated periodic distributions on $\mathbb{R}^n$. It is isomorphic to $A^s_{pq}(\mathbb{T}^n)$. First one proves a representation of these distributions on $\mathbb{R}^n$ in terms of the wavelets $\Psi^{j,\mathrm{per}}_{G,m}$ according to (2.35) in $\mathbb{R}^n$. Reduction to $\mathbb{T}^n$ gives the above theorem. We refer to **[34]** and **[35]**.

3. Spaces on arbitrary domains

3.1. Definitions. The remaining sections of this survey deal with wavelet bases for function spaces on domains. First we fix some notation. Let Ω be an arbitrary domain in $\mathbb{R}^n$. *Domain means open set* without any further restrictions. Then $L_p(\Omega)$ with $0 < p \le \infty$ is the standard quasi-Banach space of all complex-valued Lebesgue measurable functions in Ω such that

$$\|f \,|L_p(\Omega)\| = \left(\int_\Omega |f(x)|^p \, dx \right)^{1/p}$$

(with the natural modification if $p = \infty$) is finite. As usual, $D(\Omega) = C^\infty_0(\Omega)$ stands for the collection of all complex-valued infinitely differentiable functions in $\mathbb{R}^n$ with compact support in Ω. Let $D'(\Omega)$ be the dual space of all distributions in Ω. Let $g \in S'(\mathbb{R}^n)$. Then we denote by $g|\Omega$ its restriction to Ω,

$$g|\Omega \in D'(\Omega) : \quad (g|\Omega)(\varphi) = g(\varphi) \text{ for } \varphi \in D(\Omega).$$

Recall that $A^s_{pq}(\mathbb{R}^n)$ stands by (2.6) both for $B^s_{pq}(\mathbb{R}^n)$ and $F^s_{pq}(\mathbb{R}^n)$ as introduced in Definition 2.1.

Definition 3.1. *Let* Ω *be an arbitrary domain in* $\mathbb{R}^n$ *with* $\Omega \neq \mathbb{R}^n$ *and let*

$$0 < p \le \infty, \quad 0 < q \le \infty, \quad s \in \mathbb{R},$$

with $p < \infty$ for the F-spaces.

(i) *Then*

$$A^s_{pq}(\Omega) = \left\{ f \in D'(\Omega) : \; f = g|\Omega \text{ for some } g \in A^s_{pq}(\mathbb{R}^n) \right\},$$

$$\|f\,|A^s_{pq}(\Omega)\| = \inf \|g\,|A^s_{pq}(\mathbb{R}^n)\|,$$

where the infimum is taken over all $g \in A^s_{pq}(\mathbb{R}^n)$ with $g|\Omega = f$.

(ii) *Let*

$$\widetilde{A}^s_{pq}(\overline{\Omega}) = \left\{ f \in A^s_{pq}(\mathbb{R}^n) : \; \operatorname{supp} f \subset \overline{\Omega} \right\}.$$

Then

$$\widetilde{A}^s_{pq}(\Omega) = \left\{ f \in D'(\Omega) : \; f = g|\Omega \text{ for some } g \in \widetilde{A}^s_{pq}(\overline{\Omega}) \right\},$$

$$\|f\,|\widetilde{A}^s_{pq}(\Omega)\| = \inf \|g\,|A^s_{pq}(\mathbb{R}^n)\|$$

where the infimum is taken over all $g \in \widetilde{A}^s_{pq}(\overline{\Omega})$ with $g|\Omega = f$.

REMARK 3.2. Part (i) is the usual definition of $A^s_{pq}(\Omega)$ by restriction. The spaces $\widetilde{A}^s_{pq}(\overline{\Omega})$ are closed subspaces of $A^s_{pq}(\mathbb{R}^n)$. They can be identified with $\widetilde{A}^s_{pq}(\Omega)$ if

$$\left\{ h \in A^s_{pq}(\mathbb{R}^n) : \; \operatorname{supp} h \subset \partial\Omega \right\} = \{0\}$$

what in general is not the case (especially not if $|\partial\Omega| > 0$). If Ω is a *bounded Lipschitz domain* according to Definition 4.1 below, then one has in a few cases intrinsic quasi-norms and characterisations. This applies in particular to the *classical Sobolev spaces*

$$F^k_{p,2}(\Omega) = W^k_p(\Omega), \qquad k \in \mathbb{N}_0, \quad 1 < p < \infty,$$

with

$$\|f\,|F^k_{p,2}(\Omega)\| \sim \|f\,|W^k_p(\Omega)\| = \left(\sum_{|\alpha| \le k} \|D^\alpha f\,|L_p(\Omega)\|^p \right)^{1/p}$$

in analogy to (2.7), (2.8), but also to the Besov spaces $B^s_{pq}(\Omega)$ with (2.9) and an Ω-version of (2.11). Details, references and also intrinsic characterisations of some spaces $F^s_{pq}(\Omega)$ may be found in [**32**, Theorems 1.118, 1.122, pp. 74/77].

3.2. Wavelet systems. It is the main aim of this survey to describe wavelet bases for function spaces on domains. For this purpose one asks for wavelet systems preserving as much as possible of the distinguished wavelet bases in $\mathbb{R}^n$ and $\mathbb{T}^n$ as used in the Propositions 2.3, 2.12 and in the function spaces $A^s_{pq}(\mathbb{R}^n)$ and $A^s_{pq}(\mathbb{T}^n)$.

Let Ω be an arbitrary domain in $\mathbb{R}^n$ with $\Omega \ne \mathbb{R}^n$. Balls in $\mathbb{R}^n$ centred at $x \in \mathbb{R}^n$ and of radius $\varrho > 0$ are denoted by $B(x, \varrho)$. If Γ^1 and Γ^2 are two sets in $\mathbb{R}^n$ then

$$\operatorname{dist}(\Gamma^1, \Gamma^2) = \inf \left\{ |x^1 - x^2| : \; x^1 \in \Gamma^1, \; x^2 \in \Gamma^2 \right\}. \tag{3.1}$$

Let for some positive numbers c_1, c_2, c_3,

$$\mathbb{Z}_\Omega = \left\{ x^j_r \in \Omega : \; j \in \mathbb{N}_0; \; r = 1, \ldots, N_j \right\} \tag{3.2}$$

where $N_j \in \overline{\mathbb{N}} = \mathbb{N} \cup \{\infty\}$ such that

$$|x^j_r - x^j_{r'}| \ge c_1\, 2^{-j}, \qquad j \in \mathbb{N}_0, \quad r \ne r', \tag{3.3}$$

and

$$\text{dist}\left(\bigcup_{r=1}^{N_j} B\left(x_r^j, c_2\, 2^{-j}\right), \Gamma\right) \geq c_3\, 2^{-j}, \qquad j \in \mathbb{N}_0, \tag{3.4}$$

where $\Gamma = \partial\Omega$ stands for the (non-empty) boundary. It is always assumed that the positive numbers c_1, c_2, c_3 are sufficiently small in dependence on Ω such that for any $j \in \mathbb{N}_0$ there are points with (3.2)-(3.4). Based on (2.12), (2.13) we extend now (2.14) to

$$\Psi_{G,m}^j(x) = 2^{(j+L)n/2} \prod_{r=1}^{n} \psi_{G_r}\left(2^{j+L} x_r - m_r\right), \qquad m \in \mathbb{Z}^n, \tag{3.5}$$

for all $j \in \mathbb{N}_0$, $G \in \{F, M\}^n$. Although not discussed in detail it is always assumed that $L \in \mathbb{N}_0$ in (3.5) is fixed such that for given (small) $\varepsilon > 0$,

$$\text{supp}\, \Psi_{G,m}^j \subset B\left(2^{-j-L}m, \varepsilon\, 2^{-j}\right), \qquad j \in \mathbb{N}_0, \quad m \in \mathbb{Z}^n. \tag{3.6}$$

Let $\overline{F} = \{F, \dots, F\} \in \{F, M\}^n$.

DEFINITION 3.3. *Let Ω be an arbitrary domain in $\mathbb{R}^n$ with $\Omega \neq \mathbb{R}^n$ and let $\mathbb{Z}_\Omega$ be as in* (3.2)-(3.4). *Let $L \in \mathbb{N}$ and $u \in \mathbb{N}$ be as in* (2.12), (2.13) *and* (3.5), (3.6). *Let $K \in \mathbb{N}$, $D > 0$ and $c_4 > 0$. Then*

$$\left\{\Phi_r^j :\ j \in \mathbb{N}_0;\ r = 1, \dots, N_j\right\} \qquad \textit{where} \quad N_j \in \overline{\mathbb{N}}, \tag{3.7}$$

$$\text{supp}\, \Phi_r^j \subset B\left(x_r^j, c_2\, 2^{-j}\right), \qquad j \in \mathbb{N}_0, \quad r = 1, \dots, N_j, \tag{3.8}$$

with $B\left(x_r^j, c_2 2^{-j}\right)$ as in (3.4), *is called a u-wavelet system* (*with respect to Ω*) *if it consists of the following three types of functions.*

(i) *Basic wavelets*

$$\Phi_r^0 = \Psi_{G,m}^0 \qquad \textit{for some} \quad G \in \{F, M\}^n, \quad m \in \mathbb{Z}^n. \tag{3.9}$$

(ii) *Interior wavelets*

$$\Phi_r^j = \Psi_{G,m}^j, \qquad j \in \mathbb{N}, \quad \text{dist}\,\left(x_r^j, \Gamma\right) \geq c_4\, 2^{-j}, \tag{3.10}$$

for some $G \in \{F, M\}^{n*}$, $m \in \mathbb{Z}^n$.

(iii) *Boundary wavelets*

$$\Phi_r^j = \sum_{|m-m'| \leq K} d_{m,m'}^j\, \Psi_{\overline{F},m'}^j, \qquad j \in \mathbb{N}, \quad \text{dist}\,\left(x_r^j, \Gamma\right) < c_4\, 2^{-j}, \tag{3.11}$$

for some $m = m(j,r)$ and $d_{m,m'}^j \in \mathbb{R}$ with

$$\sum_{|m-m'| \leq K} |d_{m,m'}^j| \leq D \quad \textit{and} \quad \text{supp}\, \Psi_{\overline{F},m'}^j \subset B\left(x_r^j, c_2\, 2^{-j}\right).$$

REMARK 3.4. By construction all wavelets Φ_r^j are real. Only $u \in \mathbb{N}$ is of interest. All other fixed numbers L, K, D and the constants $c_1, \dots, c_4$ are technical ingredients. By what follows they depend on u, but not on Ω. This may justify the above notation.

3.3. Wavelet bases in $L_p(\Omega)$. First we ask for counterparts of Propositions 2.3 and 2.12 in arbitrary domains Ω based on u-wavelet systems according to Definition 3.3.

DEFINITION 3.5. *Let Ω be an arbitrary domain in $\mathbb{R}^n$ with $\Omega \neq \mathbb{R}^n$ and let $u \in \mathbb{N}$. Then*

$$\left\{\Phi^j_r : \; j \in \mathbb{N}_0; \; r = 1, \ldots, N_j\right\} \qquad \textit{with} \quad N_j \in \overline{\mathbb{N}}$$

is called an orthonormal u-wavelet basis in $L_2(\Omega)$ if it is both a u-wavelet system according to Definition 3.3 and an orthonormal basis in $L_2(\Omega)$.

REMARK 3.6. In other words, for given $u \in \mathbb{N}$ one asks for constants $L, K, D, c_1, \ldots, c_4, d^j_{m,m'}$ such that one finds a related u-wavelet system as introduced in Definition 3.3 which is also an orthonormal basis in $L_2(\Omega)$.

PROPOSITION 3.7. *Let Ω be an arbitrary domain in $\mathbb{R}^n$ with $\Omega \neq \mathbb{R}^n$. For any $u \in \mathbb{N}$ there are orthonormal u-wavelet bases in $L_2(\Omega)$ according to Definition 3.5.*

REMARK 3.8. This proposition plays the same role for function spaces on domains as the Propositions 2.3, 2.12 for spaces on $\mathbb{R}^n$ and $\mathbb{T}^n$. The close connection between wavelets and function spaces resulting in the above assertions will be outlinend below in Discussion 3.12. There is a more or less obvious counterpart of (2.16)-(2.18). If $\{\Phi^j_r\}$ is an orthonormal u-wavelet basis in $L_2(\Omega)$ then any $f \in L_2(\Omega)$ can be represented as

$$f = \sum_{j=0}^{\infty} \sum_{r=1}^{N_j} \lambda^j_r \, 2^{-jn/2} \, \Phi^j_r \tag{3.12}$$

with

$$\lambda^j_r = \lambda^j_r(f) = 2^{jn/2} \left(f, \Phi^j_r\right) = 2^{jn/2} \int_\Omega f(x) \, \Phi^j_r(x) \, dx$$

and

$$\|f \,|L_2(\Omega)\| = \left(\sum_{j=0}^{\infty} \sum_{r=1}^{N_j} 2^{-jn} |\lambda^j_r|^2 \right)^{1/2}.$$

By (3.9)-(3.11) based on (3.5) the functions $2^{-jn/2}\Phi^j_r$ are L_∞-normalised what is convenient for our later considerations.

The extension of Proposition 3.7 and of the representation (3.12) to other function spaces on domains requires appropriate counterparts of the sequence spaces b^s_{pq} and f^s_{pq} in Definition 2.5. Let χ_{jr} be the characteristic functions of the balls $B(x^j_r, c_2\, 2^{-j})$ in (3.8).

DEFINITION 3.9. *Let Ω be an arbitrary domain in $\mathbb{R}^n$ with $\Omega \neq \mathbb{R}^n$ and let $\mathbb{Z}_\Omega$ be as in (3.2)-(3.4). Let $s \in \mathbb{R}$, $0 < p \leq \infty$, $0 < q \leq \infty$. Then $b^s_{pq}(\mathbb{Z}_\Omega)$ is the collection of all sequences*

$$\lambda = \left\{\lambda^j_r \in \mathbb{C} : \; j \in \mathbb{N}_0; \; r = 1, \ldots, N_j\right\}, \qquad N_j \in \overline{\mathbb{N}}, \tag{3.13}$$

such that

$$\|\lambda \,|b^s_{pq}(\mathbb{Z}_\Omega)\| = \left(\sum_{j=0}^{\infty} 2^{j(s-\frac{n}{p})q} \left(\sum_{r=1}^{N_j} |\lambda^j_r|^p \right)^{q/p} \right)^{1/q} < \infty$$

and $f^s_{pq}(\mathbb{Z}_\Omega)$ *is the collection of all sequences* (3.13) *such that*

$$\|\lambda \,|f^s_{pq}(\mathbb{Z}_\Omega)\| = \left\| \left(\sum_{j=0}^{\infty} \sum_{r=1}^{N_j} 2^{jsq} \, |\lambda^j_r \, \chi_{jr}(\cdot)|^q \right)^{1/q} |L_p(\Omega) \right\| < \infty$$

with the usual modification if $p = \infty$ *and/or* $q = \infty$.

REMARK 3.10. The structure of the sequence spaces $f^s_{pq}(\mathbb{Z}_\Omega)$ is somewhat complicated. A relevant discussion may be found in [**32**, Section 1.5.3]. One has

$$b^s_{pp}(\mathbb{Z}_\Omega) = f^s_{pp}(\mathbb{Z}_\Omega), \qquad s \in \mathbb{R}, \quad 0 < p \le \infty.$$

As usual $L^{\rm loc}_1(\Omega)$ collects all complex-valued Lebesgue-measurable functions in Ω such that

$$\int_\omega |f(x)| \, dx < \infty \quad \text{for any bounded domain } \omega \text{ with } \overline{\omega} \subset \Omega.$$

THEOREM 3.11. *Let* Ω *be an arbitrary domain in* $\mathbb{R}^n$ *with* $\Omega \neq \mathbb{R}^n$. *Let* $1 < p < \infty$ *and* $u \in \mathbb{N}$. *Let*

$$\{\Phi^j_r : \, j \in \mathbb{N}_0; \; r = 1, \ldots, N_j\} \qquad \textit{with} \quad N_j \in \overline{\mathbb{N}}, \tag{3.14}$$

be an orthonormal u-wavelet basis in $L_2(\Omega)$ *according to Proposition 3.7 and Definition 3.5. Then* $L_p(\Omega)$ *is the collection of all* $f \in L^{\rm loc}_1(\Omega)$ *which can be represented as*

$$f = \sum_{j=0}^{\infty} \sum_{r=1}^{N_j} \lambda^j_r \, 2^{-jn/2} \, \Phi^j_r, \qquad \lambda \in f^0_{p,2}(\mathbb{Z}_\Omega). \tag{3.15}$$

Furthermore, $\{\Phi^j_r\}$ *is an unconditional basis in* $L_p(\Omega)$. *If* $f \in L_p(\Omega)$ *then the representation* (3.15) *is unique with* $\lambda = \lambda(f)$,

$$\lambda^j_r(f) = 2^{jn/2} \left(f, \Phi^j_r\right) = 2^{jn/2} \int_\Omega f(x) \, \Phi^j_r(x) \, dx \tag{3.16}$$

and

$$I : \; f \mapsto \lambda(f) = \left\{ 2^{jn/2} \left(f, \Phi^j_r\right) \right\}$$

is an isomorphic map of $L_p(\Omega)$ *onto* $f^0_{p,2}(\mathbb{Z}_\Omega)$ *(equivalent norms).*

Discussion 3.12. Again there is a striking interplay between wavelets and building blocks in function spaces. We give an idea how to prove Proposition 3.7 and the above theorem. First one decomposes Ω in Whitney cubes Q_{lr}, centred at some points $2^{-l}m \in \Omega$ $(m \in \mathbb{Z}^n, l \in \mathbb{N}_0)$ and of side-length 2^{-l},

$$\Omega = \bigcup_{l=0}^{\infty} \bigcup_r Q_{lr}, \qquad \text{dist } (Q_{lr}, \partial\Omega) \sim 2^{-l} \text{ if } l \in \mathbb{N} \tag{3.17}$$

(modification for $l = 0$). Then

$$\|f \,|L_p(\Omega)\|^p = \sum_{l,r} \|f_{lr} \,|L_p(\mathbb{R}^n)\|^p, \qquad f_{lr} = \chi_{lr} f, \tag{3.18}$$

where χ_{lr} is the characteristic function of Q_{lr} and

$$\|g_{lr} \,|L_p(\mathbb{R}^n)\| = 2^{ln/p} \, \|f_{lr} \,|L_p(\mathbb{R}^n)\|, \qquad g_{lr}(x) = f_{lr}(2^{-l}x). \tag{3.19}$$

For the dilated functions g_{lr} one has a canonical situation which admits a uniform application of the $\mathbb{R}^n$-expansion according to Theorem 2.7. Pulling back one gets wavelet expansions for f_{lr} in a (small) neighbourhood of Q_{lr}. Clipping together these expansions (with some technical care) and using that the wavelets $\Psi^j_{G,m}$ in (2.14) are well-adapted to dyadic dilations and translations one gets expansions for f which are first steps towards (3.15). But there are some obstacles. In contrast to the dilated terms originating from $\Psi^j_{G,m}$ in(2.25) with $j \in \mathbb{N}$, the dilated starting terms originating from $\Psi^0_{G,m}$ do not fulfil the moment conditions needed for the interpretation as atoms in $L_p(\mathbb{R}^n)$. But this applies only to the starting terms and this difficulty can be removed by direct L_p-arguments. It is more serious that just these dilated starting terms spoil the desired orthogonality (in contrast to the other dilated wavelets). This is the point where the multiresolution structure of the wavelets is of great service. Let the origin in the one-dimensional case $n = 1$ be a breaking point with the dilation 2^{-l} at the left and the dilation 2^{-l-1} at the right. With ψ_G where $G \in \{F, M\}$ as in (2.12) one has near the origin the multiresolution property

$$\psi_G\left(2^{l+L}x - m\right) = \sum_{t\in\mathbb{Z}} c^G_{l,m,t}\, \psi_F\left(2^{l+1+L}x - t\right)$$

consisting of finitely many elements with

$$\operatorname{supp}\, \psi_F\left(2^{l+1+t}\cdot - t\right) \subset \operatorname{supp}\, \psi_G\left(2^{l+L}\cdot - m\right).$$

With some local orthogenalisation at the origin one can remove the disturbing terms $\psi_G\left(2^{l+L}\cdot - m\right)$ at the expense of $\psi_F\left(2^{l+1+L}\cdot - t\right)$. In case of $n = 1$ all breaking points are isolated endpoints of intervals and the above orthogonalisation can be done at each such point separately. If $n \geq 2$ then one can rely on the product structure of $\Psi^j_{G,m}$ in (2.14) which transfers the orthogonalisation from one direction to n directions (applying Fubini's theorem). Some care is necessary, especially in corner points and along edges. But all this can be done and results finally in the boundary wavelets Φ^j_r according to (3.11). In this way one proves first Proposition 3.7 and afterwards (with the help of the indicated atomic arguments) Theorem 3.11. The first step of this procedure (without the final orthogonalisation) was done in [**33**], where we denoted the outcome as a *para-basis*. The above versions of Proposition 3.7 and Theorem 3.11 are published here for the first time. Detailed proofs will be given in [**35**].

Discussion 3.13. The above arguments for the spaces $L_p(\Omega)$ in arbitrary domains Ω rely mainly on the *localisation* (3.18), the *homogeneity* (3.19) and the $\mathbb{R}^n$-wavelet theory. There is little hope that other spaces $A^s_{pq}(\Omega)$ and $\widetilde{A}^s_{pq}(\Omega)$ according to Definition 3.1 in arbitrary domains fit in this scheme. But there is a remarkable modification which even lays the foundation of all what follows. We sketch the basic ideas. Let σ_{pq} be as in (2.24) and let

$$0 < p < \infty, \quad 0 < q < \infty, \quad s > \sigma_{pq}. \tag{3.20}$$

Let Q_{lr} be the (roughly) indicated Whitney cubes in (3.17) and let $0 \leq \varrho_{lr} \in D(\Omega)$ be a related resolution of unity, say,

$$\operatorname{supp}\, \varrho_{lr} \subset 2Q_{lr} \subset \Omega, \qquad \sum_{lr} \varrho_{lr}(x) = 1 \quad \text{if} \quad x \in \Omega. \tag{3.21}$$

Then the *refined localisation space* $F^{s,\mathrm{rloc}}_{pq}(\Omega)$ consists of all $f \in L_1^{\mathrm{loc}}(\Omega)$ such that

$$\|f \,|F^{s,\mathrm{rloc}}_{pq}(\Omega)\|_{\varrho} = \Big(\sum_{l,r} \|\varrho_{lr} f \,|F^s_{pq}(\mathbb{R}^n)\|^p \Big)^{1/p} < \infty. \tag{3.22}$$

This quasi-Banach space is independent of admitted resolutions of unity $\varrho = \{\varrho_{lr}\}$ (equivalent quasi-norms). Furthermore for p, q, s in (3.20) one has the homogeneity

$$\|f(\lambda\cdot) \,|F^s_{pq}(\mathbb{R}^n)\| \sim \lambda^{s-\frac{n}{p}} \, \|f \,|F^s_{pq}(\mathbb{R}^n)\|, \qquad 0 < \lambda \le 1, \tag{3.23}$$

for all

$$f \in F^s_{pq}(\mathbb{R}^n) \quad \text{with} \quad \operatorname{supp} f \subset \{y \in \mathbb{R}^n : \ |y| < \lambda\}.$$

Then (3.22) (by definition) and (3.23) are the counterparts of (3.18), (3.19). One gets by the same outlined arguments as for $L_p(\Omega)$ that $\{\Phi^j_r\}$ where $u > \sigma_{pq}$ in (3.14) is a wavelet basis in $F^{s,\mathrm{rloc}}_{pq}(\Omega)$ with the expansions (3.15), (3.16) now with $\lambda \in f^s_{pq}(\mathbb{Z}_\Omega)$ in place of $f^0_{p,2}(\mathbb{Z}_\Omega)$. This theory started in **[33]** (where we denoted $F^{s,\mathrm{rloc}}_{pq}(\Omega)$ as $\overline{F}^s_{pq}(\Omega)$). There we got para-bases. As for final results in terms of the above orthogonal basis with the related sequence spaces $f^s_{pq}(\mathbb{Z}_\Omega)$ according to Definition 3.9 we refer to **[35]**.

4. Spaces on E-thick domains

4.1. Classes of domains. So far we described wavelet bases for function spaces A^s_{pq} on $\mathbb{R}^n$ and $\mathbb{T}^n$. In case of arbitrary domains Ω we have a satisfactory wavelet representation for $L_p(\Omega)$, $1 < p < \infty$, according to Theorem 3.11 and we indicated in Discussion 3.13 an extension of these assertions to refined localisation spaces $F^{s,\mathrm{rloc}}_{pq}(\Omega)$. But in general these spaces do not coincide with corresponding spaces $F^s_{pq}(\Omega)$ or $\widetilde{F}^s_{pq}(\Omega)$ according to Definition 3.1. All spaces $A^s_{pq}(\Omega)$, $\widetilde{A}^s_{pq}(\Omega)$ introduced there originate from the related spaces $A^s_{pq}(\mathbb{R}^n)$ which are governed by atoms and kernels of local means. But these elementary building blocks require not only some minimal smoothness but also some cancellations (moment conditions). This is well reflected by the wavelets $\Psi^j_{G,m}$ in $\mathbb{R}^n$ (and also in $\mathbb{T}^n$) playing a double role as atoms and kernels of local means as outlined in Discussion 2.8. In case of domains the basic wavelets Φ^0_r in (3.9) and the interior wavelets Φ^j_r in (3.10) fit in this scheme, but not the boundary wavelets in (3.11). To compensate this shortcoming we are going to introduce the natural class of E-thick domains covering as special cases the better known class of Lipschitz domains.

Tacitly we always assume that a domain (= open set) Ω in $\mathbb{R}^n$ is not empty, $\Omega \neq \emptyset$. Let $l(Q)$ be the side-length of a (finite) cube Q in $\mathbb{R}^n$ with sides parallel to the axes of coordinates. Let I be an (arbitrary) index set. Then

$$a_i \sim b_i \quad \text{for} \quad i \in I \qquad \text{(equivalence)}$$

for two sets of positive numbers $\{a_i : i \in I\}$ and $\{b_i : i \in I\}$ means that there are two positive numbers c_1 and c_2 such that

$$c_1 \, a_i \le b_i \le c_2 \, a_i \qquad \text{for all} \quad i \in I.$$

As usual, for $2 \le n \in \mathbb{N}$,

$$\mathbb{R}^{n-1} \ni x' \mapsto h(x') \in \mathbb{R} \tag{4.1}$$

is called a *Lipschitz function* (on $\mathbb{R}^{n-1}$) if there is a number $c > 0$ such that

$$|h(x') - h(y')| \le c\,|x' - y'| \quad \text{for all} \quad x' \in \mathbb{R}^{n-1}, \quad y' \in \mathbb{R}^{n-1}. \tag{4.2}$$

If h is infinitely differentiable and all derivatives are bounded then h is called a C^∞ function. The distance dist (Γ^1, Γ^2) between two sets in $\mathbb{R}^n$ has the same meaning as in (3.1).

DEFINITION 4.1. (i) *Let $n \in \mathbb{N}$. A domain (= open set) in $\mathbb{R}^n$ with $\Omega \neq \mathbb{R}^n$ and $\Gamma = \partial\Omega$ is said to be E-thick (exterior thick) if one finds for any interior cube $Q^i \subset \Omega$ with*

$$l(Q^i) \sim 2^{-j}, \quad \text{dist}\,(Q^i, \Gamma) \sim 2^{-j}, \qquad j \ge j_0 \in \mathbb{N}, \tag{4.3}$$

a complementing cube $Q^e \subset \Omega^c = \mathbb{R}^n \backslash \Omega$ with

$$l(Q^e) \sim 2^{-j}, \quad \text{dist}\,(Q^e, \Gamma) \sim \text{dist}\,(Q^i, Q^e) \sim 2^{-j}, \qquad j \ge j_0 \in \mathbb{N}. \tag{4.4}$$

(ii) *Let $2 \le n \in \mathbb{N}$. A Lipschitz graph domain (C^∞ graph domain) in $\mathbb{R}^n$ is the collection of all points*

$$(x', x_n) \quad \text{with} \quad x' \in \mathbb{R}^{n-1} \quad \text{and} \quad h(x') < x_n < \infty,$$

where $h(x')$ is a Lipschitz function according to (4.1), (4.2) (a C^∞ function).

(iii) *Let $2 \le n \in \mathbb{N}$. A bounded Lipschitz domain (bounded C^∞ domain) in $\mathbb{R}^n$ is a bounded domain Ω in $\mathbb{R}^n$ where the boundary Γ can be covered by finitely many open balls B_j in $\mathbb{R}^n$ with $j = 1, \ldots, J$, centred at Γ such that*

$$B_j \cap \Omega = B_j \cap \Omega_j \qquad \text{for} \quad j = 1, \ldots, J,$$

where Ω_j are rotations of suitable Lipschitz graph domains (C^∞ graph domains) in $\mathbb{R}^n$.

REMARK 4.2. The equivalence constants in (4.3), (4.4) are independent of j. In other words, a domain Ω is called E-thick if for any choice of positive numbers c_1, c_2, c_3, c_4 and $j_0 \in \mathbb{N}$ there are positive numbers c_5, c_6, c_7, c_8 such that one finds for each interior cube $Q^i \subset \Omega$ with

$$c_1\, 2^{-j} \le l(Q^i) \le c_2\, 2^{-j}, \qquad c_3\, 2^{-j} \le \text{dist}\,(Q^i, \Gamma) \le c_4\, 2^{-j},$$

$j \ge j_0$, an exterior cube $Q^e \subset \Omega^c$ with

$$c_5\, 2^{-j} \le l(Q^e) \le c_6\, 2^{-j}, \quad c_7\, 2^{-j} \le \text{dist}\,(Q^e, \Gamma) \le \text{dist}\,(Q^i, Q^e) \le c_8\, 2^{-j}, \tag{4.5}$$

$j \ge j_0$. One checks easily that Ω is E-thick if, and only if, one has (4.5) for some $c_5, \ldots, c_8$ for the standard Whitney cubes as in, say, [**25**, Theorem 3, p. 16, Theorem 1, p. 167]. By a *Lipschitz domain* we mean either a Lipschitz graph domain or a bounded Lipschitz domain. Quite obviously, any

Lipschitz domain in $\mathbb{R}^n$ is E-thick.

On the other hand, E-thick domains may be rather irregular. It may happen that $|\Gamma| > 0$. There are E-thick domains with fractal boundaries, for example, the snowflake curve. A discussion may be found in [**33, 35**]. In addition to E-thick domains we introduced in [**35**] also I-thick (interior thick) domains, where, roughly speaking, the above cubes Q^i and Q^e change their roles. Conditions of similar types (E-thick and I-thick) have been used several times in literature in connection with function spaces and PDE's. First we refer to the *corkscrew* property of *non-tangentially assessible domains* according to [**18**]. Details and generalisations (*domains of class S*) may be found in [**19**, pp. 4,8]. In connection with Sobolev and Poincaré inequalities in function spaces (preferably Sobolev spaces) conditions of the above type play a role resulting in *John domains* and *plump domains.* Details, references, examples and discussions may be found in [**12**, Section 4.3]. In connection with atomic representations of function spaces in rough domains we introduced in [**36**] (exterior and interior) *regular domains.* They are similar (but not identical) with the above (exterior and interior) domains and the other types of domains mentioned above. This may also be found in [**13**, Section 2.5]. In [**36, 13**] we referred also to other classes of domains in the literature.

4.2. Wavelet bases in $\widetilde{F}^s_{pq}(\Omega)$. As roughly indicated in Disussion 3.13 for the refined localisation spaces $F^{s,\mathrm{rloc}}_{pq}(\Omega)$ according to (3.20)-(3.22) in arbitrary domains Ω one has a counterpart of Theorem 3.11. If Ω is E-thick then these refined localisation spaces coincide with the spaces $\widetilde{F}^s_{pq}(\Omega)$ according to Definition 3.1. We formulate the outcome and discuss afterwards the key ideas. We incorporate now $\widetilde{F}^s_{\infty\infty}(\Omega) = \widetilde{B}^s_{\infty\infty}(\Omega)$. Let $f^s_{pq}(\mathbb{Z}_\Omega)$ be the sequence spaces as introduced in Definition 3.9 and let σ_{pq} be as in (2.24).

PROPOSITION 4.3. *Let Ω be an E-thick domain in $\mathbb{R}^n$ according to Definition 4.1(i) and let $\widetilde{F}^s_{pq}(\Omega)$ with*

$$0 < p \le \infty, \qquad 0 < q \le \infty, \qquad s > \sigma_{pq}, \tag{4.6}$$

($q = \infty$ if $p = \infty$) be the spaces as introduced in Definition 3.1. Let for $u \in \mathbb{N}$ with $u > s$,

$$\left\{\Phi^j_r : \ j \in \mathbb{N}_0;\ r = 1, \ldots, N_j\right\} \qquad \textit{with} \quad N_j \in \overline{\mathbb{N}}$$

be an orthonormal u-wavelet basis in $L_2(\Omega)$ according to Proposition 3.7 and Definition 3.5. Then

$$\widetilde{F}^s_{pq}(\Omega) \hookrightarrow L_v(\Omega), \qquad \max(1,p) < v \le \infty, \quad s - \frac{n}{p} > -\frac{n}{v}, \tag{4.7}$$

(what means $v = \infty$ if $p = \infty$). Furthermore, $f \in L_v(\Omega)$ is an element of $\widetilde{F}^s_{pq}(\Omega)$ if, and only if, it can be represented as

$$f = \sum_{j=0}^{\infty} \sum_{r=1}^{N_j} \lambda^j_r\, 2^{-jn/2}\, \Phi^j_r, \qquad \lambda \in f^s_{pq}(\mathbb{Z}_\Omega), \tag{4.8}$$

absolute (and hence unconditional) convergence being in $L_v(\Omega)$. If $f \in \widetilde{F}^s_{pq}(\Omega)$ then the representation (4.8) is unique with $\lambda = \lambda(f)$,

$$\lambda^j_r(f) = 2^{jn/2}\, \left(f, \Phi^j_r\right) = 2^{jn/2} \int_\Omega f(x)\, \Phi^j_r(x)\, dx \tag{4.9}$$

and

$$I: \ f \mapsto \lambda(f) = \left\{2^{jn/2}\, \left(f, \Phi^j_r\right)\right\}$$

is an isomorphic map of $\widetilde{F}^s_{pq}(\Omega)$ onto $f^s_{pq}(\mathbb{Z}_\Omega)$. If $p < \infty$, $q < \infty$ then $\{\Phi^j_r\}$ is an unconditional basis in $\widetilde{F}^s_{pq}(\Omega)$.

Discussion 4.4. First we remark that (4.7) is a continuous Sobolev embedding. It ensures that $f \in \widetilde{F}^s_{pq}(\Omega)$ admits an expansion of type (3.15) with $f^0_{v,2}(\mathbb{Z}_\Omega)$ in place of $f^0_{p,2}(\mathbb{Z}_\Omega)$ (locally if $p = \infty$). By Definition 3.1 one gets $f \in \widetilde{F}^s_{pq}(\Omega)$ from

$$f = g|\Omega \qquad \text{with} \quad g \in \widetilde{F}^s_{pq}(\overline{\Omega}) \subset F^s_{pq}(\mathbb{R}^n). \tag{4.10}$$

This reduces (4.8) to corresponding expansions in $\mathbb{R}^n$ as considered in Theorem 2.7 and Discussion 2.8. One does not need moment conditions for the atoms in (2.29) with Φ_r^j in place of $\Psi_{G,m}^j$. For the counterpart of the local means in (2.32), now given by (4.9),

$$\lambda_r^j(f) = \int_\Omega k_r^j(y)\, f(y)\, dy, \qquad k_r^j = 2^{jn/2}\, \Phi_r^j, \tag{4.11}$$

one needs moment conditions of type (2.30) with Φ_r^j in place of $\Psi_{G,m}^j$ and $u > s$. The kernels generated by the basic wavelets (3.9) and the interior wavelets (3.10) fit in this scheme, but not the kernles k_r^j with the boundary wavelets Φ_r^j as in (3.11). This is the point where the assumption that Ω is E-thick comes in naturally. Then supp $k_r^j \subset Q^i$ for some interior cube Q^i with (4.3). Let Q^e be a related complementing cube with (4.4). Then there is a function $\mathring{k}_r^j \in C^u(\mathbb{R}^n)$ with supp $\mathring{k}_r^j \subset Q^e$ such that

$$\widetilde{k}_r^j(x) = k_r^j(x) + \mathring{k}_r^j(x), \qquad x \in \mathbb{R}^n, \tag{4.12}$$

is an admitted $\mathbb{R}^n$-kernel having in particular the required moment conditions. Furthermore with g as in (4.10) (in particular supp $g \subset \overline{\Omega}$) one gets

$$\int_{\mathbb{R}^n} \widetilde{k}_r^j(x)\, g(x)\, dx = \int_\Omega k_r^j(y)\, f(y)\, dy = \lambda_r^j(f). \tag{4.13}$$

Now it might be at least plausible that the above proposition is closely connected with Theorem 2.7.

REMARK 4.5. There is a counterpart both of Proposition 4.3 and of Discussion 4.4 for the spaces $\widetilde{B}_{pq}^s(\Omega)$ with $0 < p, q \le \infty$ and $s > \sigma_p$. But all this will be covered by Theorem 4.8 below. The preference given to $\widetilde{F}_{pq}^s(\Omega)$ comes from the little history of the so-called *refined localisation property* which can be written as

$$F_{pq}^{s,\mathrm{rloc}}(\Omega) = \widetilde{F}_{pq}^s(\Omega) \quad \text{for } E\text{-thick domains } \Omega,$$

with s, p, q as in (4.6) where we outlined in Discussion 3.13 what is meant by $F_{pq}^{s,\mathrm{rloc}}(\Omega)$. We proved this property first for bounded C^∞ domains in [**30**, Theorem 5.14, pp. 60/61] and then for bounded Lipschitz domains in [**32**, Proposition 4.20, pp. 208/209].

4.3. Wavelet bases in $\overline{A}_{pq}^s(\Omega)$. We are now ready for the main result of Section 4. Let $A_{pq}^s(\Omega)$ and $\widetilde{A}_{pq}^s(\Omega)$ be the spaces introduced in Definition 3.1 for arbitrary domains Ω in $\mathbb{R}^n$ with $\Omega \neq \mathbb{R}^n$, considered as subspaces of $D'(\Omega)$. Let as before

$$\sigma_p = n\left(\frac{1}{p} - 1\right)_+ \quad \text{and} \quad \sigma_{pq} = n\left(\frac{1}{\min(p,q)} - 1\right)_+$$

where $b_+ = \max(b, 0)$ if $b \in \mathbb{R}$.

DEFINITION 4.6. *Let Ω be an E-thick domain in $\mathbb{R}^n$ according to Definition 4.1(i). Then*

$$\overline{F}_{pq}^s(\Omega) = \begin{cases} \widetilde{F}_{pq}^s(\Omega) & \text{if } 0 < p < \infty,\ 0 < q \le \infty,\ s > \sigma_{pq}, \\ F_{pq}^0(\Omega) & \text{if } 1 < p < \infty,\ 1 \le q < \infty,\ s = 0, \\ F_{pq}^s(\Omega) & \text{if } 0 < p < \infty,\ 0 < q \le \infty,\ s < 0, \end{cases} \tag{4.14}$$

and

$$\overline{B}^s_{pq}(\Omega) = \begin{cases} \widetilde{B}^s_{pq}(\Omega) & \text{if } 0 < p \le \infty,\ 0 < q \le \infty,\ s > \sigma_p, \\ B^0_{pq}(\Omega) & \text{if } 1 < p < \infty,\ 0 < q \le \infty,\ s = 0, \\ B^s_{pq}(\Omega) & \text{if } 0 < p \le \infty,\ 0 < q \le \infty,\ s < 0. \end{cases} \tag{4.15}$$

REMARK 4.7. We wish to extend the wavelet expansions according to Proposition 4.3 from $\widetilde{F}^s_{pq}(\Omega)$ to the above spaces $\overline{A}^s_{pq}(\Omega)$. At first glance this looks a little bit suspicious. All wavelets Φ^j_r according to Definition 3.3 have compact supports in Ω. Then one can hardly expect wavelet expansions of type (4.8) in spaces $A^s_{pq}(\Omega)$ having boundary values. This will be the subject of Section 5. However this possible counter-argument does not apply to the above spaces. This is quite obvious for the spaces $\widetilde{A}^s_{pq}(\Omega)$. Let $s < 0$ and, say, $1 < p < \infty$, $0 < q < \infty$. Then the restriction $D(\mathbb{R}^n)|\Omega$ of $D(\mathbb{R}^n)$ to Ω is dense in $A^s_{pq}(\Omega)$. But any function $\varphi \in D(\mathbb{R}^n)|\Omega$ can be approximated in $L_p(\Omega)$ by functions belonging to $D(\Omega)$. Since $s < 0$ this is also an approximation in $A^s_{pq}(\Omega)$. Hence $D(\Omega)$ is dense in this space. These arguments can be extended to other cases covered by the above definition.

In what follows the technicalities (unconditional and local convergence) are the same as in connection with Theorem 2.7. Let $f^s_{pq}(\mathbb{Z}_\Omega)$ and $b^s_{pq}(\mathbb{Z}_\Omega)$ be the sequence spaces according to Definition 3.9.

THEOREM 4.8. *Let Ω be an E-thick domain in $\mathbb{R}^n$ according to Definition 4.1(i). Let for $u \in \mathbb{N}$,*

$$\{\Phi^j_r :\ j \in \mathbb{N}_0;\ r = 1, \dots, N_j\} \qquad \text{with} \quad N_j \in \overline{\mathbb{N}},$$

be an orthonormal u-wavelet basis in $L_2(\Omega)$ according to Proposition 3.7 and Definition 3.5.

(i) *Let $\overline{F}^s_{pq}(\Omega)$ be the spaces in (4.14) and let*

$$u > \max(s, \sigma_{pq} - s), \qquad s \ne 0.$$

Then $f \in D'(\Omega)$ is an element of $\overline{F}^s_{pq}(\Omega)$ if, and only if, it can be represented as

$$f = \sum_{j=0}^{\infty} \sum_{r=1}^{N_j} \lambda^j_r\, 2^{-jn/2}\, \Phi^j_r, \qquad \lambda \in f^s_{pq}(\mathbb{Z}_\Omega), \tag{4.16}$$

unconditional convergence being in $D'(\Omega)$ and locally in any space $\overline{F}^\sigma_{pq}(\Omega)$ with $\sigma < s$. Furthermore, if $f \in \overline{F}^s_{pq}(\Omega)$ then the representation (4.16) is unique with $\lambda = \lambda(f)$,

$$\lambda^j_r = \lambda^j_r(f) = 2^{jn/2}\, \big(f, \Phi^j_r\big) \tag{4.17}$$

and

$$I :\ f \mapsto \lambda(f) = \Big\{2^{jn/2}\, \big(f, \Phi^j_r\big)\Big\} \tag{4.18}$$

is an isomorphic map of $\overline{F}^s_{pq}(\Omega)$ onto $f^s_{pq}(\mathbb{Z}_\Omega)$. If, in addition, $q < \infty$, then $\{\Phi^j_r\}$ is an unconditional basis in $\overline{F}^s_{pq}(\Omega)$.

(ii) *Let $\overline{B}^s_{pq}(\Omega)$ be the spaces in (4.15) and let*

$$u > \max(s, \sigma_p - s), \qquad s \ne 0.$$

Then $f \in D'(\Omega)$ is an element of $\overline{B}^s_{pq}(\Omega)$ if, and only if, it can be represented as

$$f = \sum_{j=0}^{\infty} \sum_{r=1}^{N_j} \lambda^j_r \, 2^{-jn/2} \, \Phi^j_r, \qquad \lambda \in b^s_{pq}(\mathbb{Z}_\Omega), \tag{4.19}$$

unconditional convergence being in $D'(\Omega)$ and locally in any space $\overline{B}^\sigma_{pq}(\Omega)$ with $\sigma < s$. Furthermore, if $f \in \overline{B}^s_{pq}(\Omega)$ then the representation (4.19) *is unique with $\lambda = \lambda(f)$ as in* (4.17)*, and I in* (4.18) *is an isomorphic map of $\overline{B}^s_{pq}(\Omega)$ onto $b^s_{pq}(\mathbb{Z}_\Omega)$. If, in addition, $p < \infty$, $q < \infty$, then $\{\Phi^j_r\}$ is an unconditional basis in $\overline{B}^s_{pq}(\Omega)$.*

Discussion 4.9. As indicated above, (4.19) with $s > \sigma_p$ and (4.16) with $s > \sigma_{pq}$ are atomic decompositions, no moment conditions are needed. But for the coefficients λ^j_r, interpreted as local means one needs moment conditions for the kernels k^j_r in (4.11). Since Ω is E-thick one can circumvent this difficulty by (4.11)-(4.13). If $s < 0$ then one does not need moment conditions for the kernels k^j_r, but for the atoms. However the necessary repair for the boundary wavelets Φ^j_r in (3.11) can be done in the same way as in (4.12). Otherwise the above theorem is published here for the first time. A detailed proof will be given in [**35**]. So far we excluded $s = 0$. Then the assumptions for Ω must be strengthened and some arguments from fractal analysis enter on the scene. We refer again to [**35**] for details. This covers in particular Lipschitz domains. Here is a corresponding formulation for this special case.

COROLLARY 4.10. *Let Ω be a Lipschitz domain in $\mathbb{R}^n$ with $n \geq 2$ according to Definition 4.1 or an interval if $n = 1$. Then Theorem 4.8 remains valid for all spaces $\overline{F}^s_{pq}(\Omega)$ in* (4.14) *and all spaces $\overline{B}^s_{pq}(\Omega)$ in* (4.15) (*including $s = 0$*).

REMARK 4.11. As mentioned in Remark 4.2 Lipschitz domains are E-thick.

5. Spaces on C^∞ domains

5.1. Preliminaries, spaces on manifolds. All constructions of wavelet bases so far in $\mathbb{R}^n$, $\mathbb{T}^n$, arbitrary domains and E-thick domains rely on dual pairings, say, $(D(\Omega), D'(\Omega))$, and the related interplay between atoms on the one hand and local means on the other hand. This is well reflected by the corresponding wavelets, say Φ^j_r in Definition 3.3, having compact supports in the underlying domain Ω. This type of arguments cannot be extended to spaces $A^s_{pq}(\Omega)$ having boundary values at $\Gamma = \partial\Omega$. We discussed this point briefly in Remark 4.7. In particular, there is no duality for these spaces within $(D(\Omega), D'(\Omega))$. One can try to circumvent these difficulties by decomposing $A^s_{pq}(\Omega)$ into an interior part, say, $\widetilde{A}^s_{pq}(\Omega)$, to which the above considerations can be applied, and trace spaces on Γ. In rough domains or for spaces $A^s_{pq}(\Omega)$ with $p < 1$ such an approach does not work. For this reason we restrict ourselves to bounded C^∞ domains Ω according to Definition 4.1 and to spaces $A^s_{pq}(\Omega)$ with $p \geq 1$, $q \geq 1$. Furthermore we give only an outline of a few basic ideas and assertions and shift a more elaborated presentation to a later occasion. We refer in particular to [**35**].

First we introduce function spaces on compact d-dimensional C^∞ manifolds Γ (without boundary), where $d \in \mathbb{N}$. Let $\{V_j, \psi_j\}_{j=1}^J$ with $J \in \mathbb{N}$ be an atlas of Γ

consisting of a covering $\Gamma = \bigcup_{j=1}^J V_j$ and homeomorphic maps

$$\psi_j : V_j \iff U_j = \psi_j(V_j) \subset \mathbb{R}^d, \qquad j = 1, \ldots, J, \tag{5.1}$$

onto open connected bounded sets in $\mathbb{R}^d$, say in the unit ball U in $\mathbb{R}^d$, with the usual compatibility conditions converting Γ into a compact C^∞ manifold. Then $D(\Gamma) = C^\infty(\Gamma)$ and $D'(\Gamma)$ have the usual meaning. Let $\{\chi_j\}_{j=1}^J \subset D(\Gamma)$ be a resolution of unity with

$$\operatorname{supp} \chi_j \subset V_j \qquad \text{and} \qquad \sum_{j=1}^J \chi_j(\gamma) = 1 \quad \text{if} \quad \gamma \in \Gamma.$$

If $f \in D'(\Gamma)$ then

$$(\chi_j f) \circ \psi_j^{-1} \in S'(\mathbb{R}^d), \qquad j = 1, \ldots, J.$$

DEFINITION 5.1. *Let $d \in \mathbb{N}$ and let Γ be a compact d-dimensional C^∞ manifold. Let $A \in \{B, F\}$ and $s \in \mathbb{R}$, $0 < p, q \le \infty$ ($p < \infty$ for the F-spaces). Then*

$$A^s_{pq}(\Gamma) = \left\{ f \in D'(\Gamma) : \ (\chi_j f) \circ \psi_j^{-1} \in A^s_{pq}(\mathbb{R}^d), \ j = 1, \ldots, J \right\}$$

and

$$\|f \,|A^s_{pq}(\Gamma)\| = \sum_{j=1}^J \left\| (\chi_j f) \circ \psi_j^{-1} \,|A^s_{pq}(\mathbb{R}^d) \right\|.$$

REMARK 5.2. It follows by standard arguments (diffeomorphic maps, pointwise multipliers) that these spaces are independent of compatible atlasses and related resolutions of unity. Basically one can transfer the wavelet expansions for the spaces $A^s_{pq}(\mathbb{R}^d)$ according to Theorem 2.7 via (5.1) to $A^s_{pq}(\Gamma)$ using that $\Psi^j_{G,m}$ in (2.25), (2.26) has compact support. The outcome is in general not a wavelet basis but a *wavelet frame* which is sufficient for many purposes. But in some cases one gets even bases. For details we refer again to **[35]**. We describe an example on which we rely afterwards.

REMARK 5.3. (**Wavelet bases on curves**). The boundary Γ of a bounded C^∞ domain Ω in the plane $\mathbb{R}^2$ (planar domain) consists of finitely many closed connected C^∞ curves, which are diffeomorphic to the 1-torus $\mathbb{T} = \mathbb{T}^1$ in (2.33). The wavelet bases from Proposition 2.12 and also the wavelet expansions according to Theorem 2.15 can be transferred to Γ, where the C^∞ distortion factors originating from the diffeomorphic maps can be incorporated in the wavelets. Then one gets orthonormal wavelet bases $\{\Phi^{j,\Gamma}_r\}$ on Γ similar as in (3.7), (3.8) based on the counterpart $\mathbb{Z}_\Gamma$ of (3.2), (3.3). Afterwards one obtains wavelet bases for all spaces $A^s_{pq}(\Gamma)$ of the same type as in Theorem 2.15 with $n = 1$.

5.2. Decompositions. Let Ω be a bounded C^∞ domain in $\mathbb{R}^n$ and let $\Gamma = \partial\Omega$ be its boundary. Let $A^s_{pq}(\Omega)$ and $\widetilde{A}^s_{pq}(\Omega)$ be as in Definition 3.1 and let $A^s_{pq}(\Gamma)$ be their counterparts on Γ as introduced in Definition 5.1. First we recall some more or less known trace assertions and decompositions for the spaces

$$A^s_{pq}(\Omega) \qquad \text{with} \quad 1 \le p < \infty, \quad 1 \le q < \infty, \quad s > 1/p.$$

The linear bounded trace operator tr_Γ,

$$\operatorname{tr}_\Gamma : \quad g \in A^s_{pq}(\Omega) \mapsto g|\Gamma \in L_p(\Gamma)$$

has the usual meaning. Let $K = [s - \frac{1}{p}]^-$ be the largest integer $K \in \mathbb{N}_0$ with $K < s - \frac{1}{p}$. Let ν be the (outer) normal on Γ. Then

$$\text{(5.2)} \qquad \mathrm{tr}_\Gamma^{s,p} : \quad g \mapsto \left\{ \mathrm{tr}_\Gamma \frac{\partial^k g}{\partial \nu^k} : 0 \le k \le K \right\}, \qquad K = [s - \frac{1}{p}]^-,$$

maps

$$B^s_{pq}(\Omega) \quad \text{onto} \quad \prod_{k=0}^{K} B_{pq}^{s-\frac{1}{p}-k}(\Gamma)$$

and

$$F^s_{pq}(\Omega) \quad \text{onto} \quad \prod_{k=0}^{K} B_{pp}^{s-\frac{1}{p}-k}(\Gamma).$$

Futhermore there are linear and bounded extension operators

$$\text{(5.3)} \qquad \mathrm{ext}_\Gamma^{s,p} : \quad \{g_0, \ldots, g_K\} \mapsto g$$

mapping

$$\text{(5.4)} \qquad \prod_{k=0}^{K} B_{pq}^{s-\frac{1}{p}-k}(\Gamma) \quad \text{into} \quad B^s_{pq}(\Omega)$$

and

$$\text{(5.5)} \qquad \prod_{k=0}^{K} B_{pp}^{s-\frac{1}{p}-k}(\Gamma) \quad \text{into} \quad F^s_{pq}(\Omega)$$

such that

$$\mathrm{tr}_\Gamma^{s,p} \circ \mathrm{ext}_\Gamma^{s,p} = \mathrm{id}, \quad \text{identity in} \quad \prod_{k=0}^{K} B_{pq}^{s-\frac{1}{p}-k}(\Gamma).$$

We refer to [**27**, p. 200] with corresponding assertions in [**26**] as a forerunner. But the extension operators constructed there are not good enough for our purpose. One needs *wavelet-friendly extension operators* extending functions g on Γ with supports in an ε-neighbourhood in Γ of some point $\gamma \in \Gamma$ into functions with supports in a corresponding ε-neighbourhood of γ in $\overline{\Omega}$. This can be done but will be shifted to [**35**]. Its paves the way to clip together wavelet bases of the spaces $\widetilde{A}^s_{pq}(\Omega)$ according to Theorem 4.8 with wavelet bases for $B^\sigma_{pq}(\Gamma)$ (if exist). This procedure relies on the following assertion.

Recall that $\mathring{A}^s_{pq}(\Omega)$ is the completion of $D(\Omega)$ in $A^s_{pq}(\Omega)$, whereas $\widetilde{A}^s_{pq}(\Omega)$ has the same meaning as in Definition 3.1.

PROPOSITION 5.4. *Let Ω be a bounded C^∞ domain in $\mathbb{R}^n$ according to Definition 4.1(iii) and let $\Gamma = \partial\Omega$. Let*

$$1 \le p < \infty, \qquad 1 \le q < \infty, \qquad 0 < s - \frac{1}{p} \notin \mathbb{N}_0.$$

Then

$$\text{(5.6)} \qquad \widetilde{A}^s_{pq}(\Omega) = \mathring{A}^s_{pq}(\Omega) = \left\{ f \in A^s_{pq}(\Omega) : \ \mathrm{tr}_\Gamma^{s,p} f = 0 \right\}$$

with $\mathrm{tr}_\Gamma^{s,p}$ *as in* (5.2). *Furthermore,*

$$\text{(5.7)} \qquad B^s_{pq}(\Omega) = \widetilde{B}^s_{pq}(\Omega) \times \prod_{k=0}^{K} B_{pq}^{s-\frac{1}{p}-k}(\Gamma)$$

and

$$F^s_{pq}(\Omega) = \widetilde{F}^s_{pq}(\Omega) \times \prod_{k=0}^{K} B_{pp}^{s-\frac{1}{p}-k}(\Gamma). \tag{5.8}$$

REMARK 5.5. If $1 \le p,q < \infty$ then (5.6) can be complemented by

$$\widetilde{A}^s_{pq}(\Omega) = \mathring{A}^s_{pq}(\Omega) = A^s_{pq}(\Omega), \qquad -1 < s - \frac{1}{p} < 0. \tag{5.9}$$

Both (5.6), (5.9) are covered by [**27**, p. 210] (and the related proof) with [**26**, pp. 317/318] as forerunners. The decomposition (5.7), (5.8) must be understood as follows. Let $\mathrm{ext}^{s,p}_\Gamma$ be as in (5.3)-(5.5). Then

$$P = \mathrm{ext}^{s,p}_\Gamma \circ \mathrm{tr}^{s,p}_\Gamma : \quad A^s_{pq}(\Omega) \hookrightarrow A^s_{pq}(\Omega)$$

is a projection and

$$\mathrm{ext}^{s,p}_\Gamma : \quad \prod_{k=0}^{K} B_{pq}^{s-\frac{1}{p}-k}(\Gamma) \iff PB^s_{pq}(\Omega)$$

$$\mathrm{ext}^{s,p}_\Gamma : \quad \prod_{k=0}^{K} B_{pp}^{s-\frac{1}{p}-k}(\Gamma) \iff PF^s_{pq}(\Omega)$$

are isomorphic maps. Furthermore, $\mathrm{id} - P$ is a projection of $A^s_{pq}(\Omega)$ onto $\widetilde{A}^s_{pq}(\Omega)$. It is just this decomposition of $A^s_{pq}(\Omega)$ into two complemented subspaces which paves the way to clip together wavelet expansions in $\widetilde{A}^s_{pq}(\Omega)$ and in $B^\sigma_{pq}(\Gamma)$.

5.3. Wavelet bases in $A^s_{pq}(\Omega)$. Let Ω be a bounded C^∞ domain in $\mathbb{R}^n$. We are looking for wavelet bases for the spaces covered by Proposition 5.4. In contrast to the wavelet bases considered so far, for example in Theorem 4.8, the boundary $\Gamma = \partial\Omega$ must come in now.

First we modify $\mathbb{Z}_\Omega$ in (3.2)-(3.4) and the sequence spaces according to Definition 3.9. We use the same notation as there. In particular, $B(x,\varrho)$ stands for a ball centred at $x \in \mathbb{R}^n$ and of radius $\varrho > 0$.

DEFINITION 5.6. *Let Ω be a bounded domain in $\mathbb{R}^n$ and let*

$$\mathbb{Z}^\Omega = \left\{ x^j_r \in \Omega : \ j \in \mathbb{N}_0; \ r = 1, \ldots, N_j \right\},$$

typically with $N_j \sim 2^{jn}$, such that for some $c_1 > 0$,

$$|x^j_r - x^j_{r'}| \ge c_1\, 2^{-j}, \qquad j \in \mathbb{N}_0, \quad r \ne r'.$$

Let χ_{jr} be the characteristic function of $B(x^j_r, c_2\, 2^{-j}) \cap \overline{\Omega}$ for some $c_2 > 0$. Let $s \in \mathbb{R}$, $0 < p,q \le \infty$. Then $b^s_{pq}(\mathbb{Z}^\Omega)$ is the collection of all sequences

$$\lambda = \left\{ \lambda^j_r \in \mathbb{C} : \ j \in \mathbb{N}_0; \ r = 1, \ldots, N_j \right\} \tag{5.10}$$

such that

$$\|\lambda\,|b^s_{pq}(\mathbb{Z}^\Omega)\| = \left(\sum_{j=0}^{\infty} 2^{j(s-\frac{n}{p})q} \left(\sum_{r=1}^{N_j} |\lambda^j_r|^p \right)^{q/p} \right)^{1/q} < \infty$$

and $f^s_{pq}(\mathbb{Z}^\Omega)$ *is the collection of all sequences* (5.10) *such that*

$$\|\lambda \,| f^s_{pq}(\mathbb{Z}^\Omega)\| = \left\| \left(\sum_{j=0}^{\infty} \sum_{r=1}^{N_j} 2^{jsq} \,|\lambda^j_r \,\chi_{jr}(\cdot)|^q \right)^{1/q} |L_p(\Omega) \right\| < \infty$$

(*modification if* $p = \infty$ *and/or* $q = \infty$).

REMARK 5.7. This is the counterpart of Definition 3.9. Recall that $C^u(\Omega)$ with $u \in \mathbb{N}_0$ is the collection of all complex-valued functions f in Ω such that all derivatives $D^\alpha f$ with $|\alpha| \le u$ can be extended continuouly to $\overline{\Omega}$. Then one gets the following modifications of the Definitions 3.3, 3.5.

DEFINITION 5.8. *Let* Ω *be a bounded* C^∞ *domain in* $\mathbb{R}^n$. *Let* $u \in \mathbb{N}$ *and let* $\mathbb{Z}^\Omega = \{x^j_r\}$ *be as in Definition 5.6.*

(i) *Then*

$$\{\Phi^j_r : \; j \in \mathbb{N}_0; \; r = 1, \ldots, N_j\} \subset C^u(\Omega)$$

is called a u-wavelet system (*with respect to* $\overline{\Omega}$) *if for some* $c_3 > 0$ *and* $c_4 > 0$,

$$\operatorname{supp} \Phi^j_r \subset B\left(x^j_r, c_3\, 2^{-j}\right) \cap \overline{\Omega}, \qquad j \in \mathbb{N}_0; \; r = 1, \ldots, N_j,$$

and

$$|D^\alpha \Phi^j_r(x)| \le c_4\, 2^{j|\alpha| + \frac{jn}{2}}, \qquad j \in \mathbb{N}_0; \; r = 1, \ldots, N_j,$$

for $x \in \overline{\Omega}$ *and* $0 \le |\alpha| \le u$.

(ii) *Let* $1 \le p, q < \infty$ *and* $s \in \mathbb{R}$. *A u-wavelet system* $\{\Phi^j_r\}$ *which is a unconditional basis in* $A^s_{pq}(\Omega)$ *is called a u-wavelet basis in* $A^s_{pq}(\Omega)$.

REMARK 5.9. For the spaces in Proposition 5.4 one cannot expect to find common u-wavelet bases originating from an orthonormal wavelet basis in $L_2(\Omega)$ as in Theorem 4.8. This may explain the difference of the above part (ii) and Definition 3.5.

After these preparations we can now formulate the main result of Section 5. As before we write $A^s_{pq}(\Omega)$ with $A \in \{B, F\}$ and similarly $a^s_{pq}(\mathbb{Z}^\Omega)$ with $a \in \{b, f\}$ if the assertion applies equally to the B-spaces and F-spaces.

THEOREM 5.10. *Let* $m \in \mathbb{N}_0$. *Let*

$$s = m + \sigma \quad \text{with} \quad 1 \le p < \infty \quad \text{and} \quad \frac{1}{p} - 1 < \sigma < \frac{1}{p}. \tag{5.11}$$

(i) *Let* $\Omega = I = (a, b)$ *with* $-\infty < a < b < \infty$ *be an open interval in* $\mathbb{R}$. *Then for any* $u \in \mathbb{N}$ *with* $u > m$ *there is a common u-wavelet basis according to Definition 5.8(ii) for all spaces* $A^s_{pq}(I)$ *with* $1 \le p, q < \infty$ *and* s *as in* (5.11). *Furthermore,*

$$f = \sum_{j=0}^{\infty} \sum_{r=1}^{N_j} \lambda^j_r(f)\, 2^{-j/2}\, \Phi^j_r$$

and $f \mapsto \lambda(f)$ *is an isomorphic map of* $A^s_{pq}(I)$ *onto* $a^s_{pq}(\mathbb{Z}^I)$.

(ii) *Let* Ω *be a bounded* C^∞ *domain in the plane* $\mathbb{R}^2$. *Then for any* $u \in \mathbb{N}$ *with*

$u > m$ there is a common u-wavelet basis according to Definition 5.8(ii) for all spaces $A^s_{pq}(\Omega)$ with $1 \le p, q < \infty$ and s as in (5.11). *Furthermore,*

$$f = \sum_{j=0}^{\infty} \sum_{r=1}^{N_j} \lambda^j_r(f)\, 2^{-j}\, \Phi^j_r$$

and $f \mapsto \lambda(f)$ is an isomorphic map of $A^s_{pq}(\Omega)$ onto $a^s_{pq}(\mathbb{Z}^\Omega)$.

Discussion 5.11. We give an idea how to prove part (ii). First one decomposes

$$A^s_{pq}(\Omega) = P A^s_{pq}(\Omega) \oplus \widetilde{A}^s_{pq}(\Omega), \qquad \widetilde{A}^s_{pq}(\Omega) = (\mathrm{id} - P)\, A^s_{pq}(\Omega),$$

according to Proposition 5.4 (modified by (5.9) if $m = 0$) and Remark 5.5. Then one clips together the wavelet bases from Theorem 4.8 and Remark 5.3 via wavelet friendly extension operators as discussed above. Details are shifted to a later occasion. We refer in particular to [**35**].

6. Comments

We add a few references and comments. As far as new results are concerned we refer for details and proofs to [**35**].

COMMENT 6.1. Some references for wavelet bases in function spaces on $\mathbb{R}^n$ and $\mathbb{T}^n$ have been given in Remark 2.9 and at the beginning of Section 2.4. Quite obviously wavelet bases for function spaces on intervals, cubes and Lipschitz domains attracted a lot of attention. We refer to [**21, 6, 7, 8, 9**] and in particular to [**5, 17**] and the literature given there. But the methods are different from what we outlined here.

COMMENT 6.2. In [**3, 4, 2**] Ciesielski and Figiel constructed common spline bases for Sobolev spaces and Besov spaces on compact d-dimensional C^∞ manifolds Γ (including the closure $\overline{\Omega}$ of bounded C^∞ domains Ω). The method is based on a rather sophisticated decomposition of Γ (or $\overline{\Omega}$) into finitely many domains which are diffeomorphic images of cubes in $\mathbb{R}^d$. Bases on cubes are shifted in this way to Γ and glued together. Similar procedures have been used to construct wavelet bases on (smooth) manifolds and on (smooth) domains admitting the required *domain decomposition*. Descriptions may be found in [**8**, Section 10], [**9**, Section 9], [**5**, p. 130] with a reference to the original paper [**10**]. It remains to be seen whether one can use these results in connection with the above considerations especially in Section 5.

COMMENT 6.3. One may ask whether one can extend Theorem 5.10 from one and two dimensions to higher dimensions based, for example, on the decompositions according to Proposition 5.4. But the topology of the connected components of bounded C^∞ domains in $\mathbb{R}^n$ with $n \ge 3$ is much more complicated than in case of bounded C^∞ domains in the plane $\mathbb{R}^2$. If one has wavelet bases in suitable spaces $B^\sigma_{pq}(\Gamma)$ with $\Gamma = \partial\Omega$ then one can argue as indicated in Discussion 5.11. If Γ (or one of its components) is diffeomorphic to an d-torus ($d = n-1$), then one can apply Theorem 2.15. If

$$\Gamma = \mathbb{S}^d = \big\{ x \in \mathbb{R}^{d+1} : |x| = 1 \big\}, \quad 2 \le d \in \mathbb{N},$$

is a d-sphere, $d = n-1$, or a diffeomorphic image of $\mathbb{S}^d$, then one can construct wavelet bases of the desired type in

$$B^\sigma_{pq}(\mathbb{S}^d) \quad \text{with} \quad 1 \le p,q < \infty, \quad \sigma > \frac{1}{p} - 1, \quad \sigma - \frac{l}{p} \notin \mathbb{N}_0,$$

for $l = 1, \ldots, d-1$. This can be done by induction with respect to dimensions starting from Theorem 5.10(ii) which causes the curious (and presumably unnatural) restrictions for σ.

COMMENT 6.4. By Definition 3.1 all spaces $A^s_{pq}(\Omega)$ on arbitrary domains Ω are defined by restriction of $A^s_{pq}(\mathbb{R}^n)$ to Ω. As mentioned in Remark 3.2 one recovers in case of bounded Lipschitz domains Ω the classical Sobolev spaces

$$W^k_p(\Omega) = \{f \in L_p(\Omega) : \ D^\alpha f \in L_p(\Omega), \ |\alpha| \le k\},$$

with

$$\|f \,|W^k_p(\Omega)\| = \left(\sum_{|\alpha| \le k} \|D^\alpha f \,|L_p(\Omega)\|^p \right)^{1/p}$$

where $1 < p < \infty$ and $k \in \mathbb{N}$. However for rough domains Ω having, for example, peaks and cusps these spaces do not fit in the scales $A^s_{pq}(\Omega)$. The standard reference for Sobolev spaces in rough domains is **[22]**. But one can say something. Let $\{\Phi^j_r\}$ be the same orthonormal u-wavelet basis for $L_p(\Omega)$ as in Theorem 3.11 where Ω is an arbitrary domain in $\mathbb{R}^n$. One can apply Theorem 3.11 to each $D^\alpha f \in L_p(\Omega)$ with $|\alpha| \le k$. Then it follows that $W^k_p(\Omega)$ is the collection of all $f \in L_p(\Omega)$ with its L_p-representation (3.15), (3.16) such that

$$\lambda(f)^k \in f^0_{p,2}(\mathbb{Z}_\Omega) \quad \text{where} \quad \lambda^j_r(f)^k = 2^{jn/2} \sum_{|\alpha| \le k} \left| (f, D^\alpha \Phi^j_r) \right|$$

(equivalent norms). Sometimes Φ^j_r are called *vaguelettes*. It remains to be seen if this observation is of any use.

COMMENT 6.5. We excluded in Proposition 5.4 and Theorem 5.10 the case $s - \frac{1}{p} \in \mathbb{N}_0$. But there are some negative results. Let again Ω be a bounded C^∞ domain in $\mathbb{R}^n$ and $1 < p,q < \infty$. Then $D(\Omega)$ is dense in $A^{1/p}_{pq}(\Omega)$. On the other hand there is no orthonormal u-wavelet basis according to Proposition 3.7 which is also a basis in $A^{1/p}_{pq}(\Omega)$ (in contrast to $\widetilde{A}^{1/p}_{pq}(\Omega)$ according to Theorem 4.8).

COMMENT 6.6. We mention a second negative result. By Remark 2.2(iv) the spaces $B^s_{pq}(\mathbb{R}^n)$ with p,q,s as in (2.9) can be equivalently quasi-normed by (2.10), (2.11). This is no longer valid if $s < n\left(\frac{1}{p} - 1\right)_+$. But one can define corresponding spaces as subspaces of $L_p(\mathbb{R}^n)$,

$$\mathbf{B}^s_{pq}(\mathbb{R}^n) = \left\{ f \in L_p(\mathbb{R}^n) : \ \|f \,|\mathbf{B}^s_{pq}(\mathbb{R}^n)\|_m < \infty \right\}$$

where

$$0 < p,q < \infty, \qquad 0 < s < m \in \mathbb{N}, \tag{6.1}$$

and

$$\|f \,|\mathbf{B}^s_{pq}(\mathbb{R}^n)\|_m = \|f \,|L_p(\mathbb{R}^n)\| + \left(\int_{|h| \le 1} |h|^{-sq} \, \|\Delta^m_h f \,|L_p(\mathbb{R}^n)\|^q \, \frac{dh}{|h|^n} \right)^{1/q}.$$

These are quasi-Banach spaces which are independent of m. References to the literature might be found in [**32**, Section 9]. One may ask whether these spaces have (wavelet) bases or frames for the full range of p, q, s in (6.1). But this is impossible since the coefficients in such expansions must be linear and continuous functionals in $\mathbf{B}^s_{pq}(\mathbb{R}^n)$. However one can prove that

$$\mathbf{B}^s_{pq}(\mathbb{R}^n)' = \{0\} \quad \text{if} \quad 0 < p < 1, \ 0 < q < \infty, \ 0 < s < n\left(\frac{1}{p} - 1\right).$$

References

[1] F. Albiac, N.J. Kalton. Topics in Banach space theory. Springer, New York, 2006

[2] Z. Ciesielski. Spline bases in classical function spaces on compact C^∞ manifolds, III. In: *Constructive Theory of Functions*, Proc. Intern. Conf. Varna. Publishing House Bulg. Acad. Sciences, Sofia, 1984, 214-223

[3] Z. Ciesielski, T. Figiel. Spline bases in classical function spaces on compact C^∞ manifolds, I. Studia Math. **76** (1983), 1-58

[4] Z. Ciesielski, T. Figiel. Spline bases in classical function spaces on compact C^∞ manifolds, II. Studia Math. **76** (1983), 95-136

[5] A. Cohen. Numerical analysis of wavelet methods. North-Holland, Elsevier, Amsterdam, 2003

[6] A. Cohen, W. Dahmen, R. DeVore. Multiscale decompositions on bounded domains. Trans. Amer. Math. Soc. **352** (2000), 3651-3685

[7] A. Cohen, W. Dahmen, R. DeVore. Adaptive wavelet techniques in numerical simulation. In: Encyclopedia Computational Mechanics. Wiley, Chichester, 2004, 1-64

[8] W. Dahmen. Wavelet and multiscale methods for operator equations. Acta Numerica **6** (1997), 55-228

[9] W. Dahmen. Wavelet methods for PDEs - some recent developments. Journ. Computational Appl. Math. **128** (2001), 133-185

[10] W. Dahmen, R. Schneider. Wavelets on manifolds I: construction and domain decomposition. SIAM Journ. Math. Anal. **31** (1999), 184-230

[11] I. Daubechies. Ten lectures on wavelets. CBMS-NSF Regional Conf. Series Appl. Math., SIAM, Philadelphia, 1992

[12] D.E. Edmunds, W.D. Evans. Hardy operators, function spaces and embeddings. Springer, Berlin, 2004

[13] D.E. Edmunds, H. Triebel. Function spaces, entropy numbers, differential operators. Cambridge Univ. Press, Cambridge, 1996 (sec. ed., 2008)

[14] M. Frazier, B. Jawerth. Decomposition of Besov spaces. Indiana Univ. Math. Journ. **34** (1985), 777-799

[15] M. Frazier, B. Jawerth. A discrete transform and decompositions of distribution spaces. Journ. Funct. Anal. **93** (1990), 34-170

[16] D.D. Haroske, H. Triebel. Wavelet bases and entropy numbers in weighted function spaces. Math. Nachr. **278** (2005), 108-132

[17] J.A. Hogan, J.D. Lakey. Time-frequency and time-scale methods. Birkhäuser, Boston, 2005

[18] D. Jerison, C. Kenig. Boundary behavior of harmonic functions in non-tangentially accessible domains. Advances Math. **146** (1982), 80-147

[19] C.E. Kenig. Harmonic analysis techniques for second order elliptic boundary value problems. CBMS, Regional Conf. Series Math. **83**. Amer. Math. Soc., Providence, 1994

[20] G. Kyriazis. Decomposition systems for function spaces. Studia Math. **157** (2003), 133-169

[21] S. Mallat. A wavelet tour of signal processing. Academic Press, San Diego, 1999 (sec. ed.)

[22] V.G. Maz'ya. Sobolev spaces. Springer, Berlin, 1985

[23] Y. Meyer. Wavelets and operators. Cambridge Univ. Press, Cambridge, 1992

[24] H.-J. Schmeisser, H. Triebel. Topics in Fourier analysis and function spaces. Wiley, Chichester, 1987

[25] E.M. Stein. Singular integrals and differentiability properties of functions. Princeton Univ. Press, Princeton, 1970

[26] H. Triebel. Interpolation theory, function spaces, differential operators. North-Holland, Amsterdam, 1978. (Sec. ed. Barth, Heidelberg, 1995)

[27] H. Triebel. Theory of function spaces. Birkhäuser, Basel, 1983

[28] H. Triebel. Characterizations of Besov-Hardy-Sobolev spaces: a unified approach. Journ. Approximation Theory **52** (1988), 162-203

[29] H. Triebel. Theory of function spaces II. Birkhäuser, Basel, 1992

[30] H. Triebel. The structure of functions. Birkhäuser, Basel, 2001

[31] H. Triebel. A note on wavelet bases in function spaces. In: Orlicz Centenary Vol., Banach Center Publ. **64**, Warszawa, Polish Acad. Sci., 2004, 193-206

[32] H. Triebel. Theory of function spaces III. Birkhäuser, Basel, 2006

[33] H. Triebel. Wavelet para-bases and sampling numbers in function spaces on domains. Journ. Complexity **23** (2007), 468-497

[34] H. Triebel. Local means and wavelets in function spaces. Proc. Conf. *Function Spaces 8*, Będlewo, 2006. Banach Center Publ. **79** (2008), 215-234

[35] H. Triebel. Function spaces and wavelets on domains. European Math. Soc. Publishing House, Zürich, 2008

[36] H. Triebel, H. Winkelvoß. Intrinsic atomic characterizations of function spaces on domains. Math. Zeitschr. **221** (1996), 647-673

[37] P. Wojtaszczyk. A mathematical introduction to wavelets. Cambridge Univ. Press, Cambridge, 1997

Mathematisches Institut, Friedrich-Schiller-Universität Jena, D-07737 Jena, Germany

E-mail address: `triebel@minet.uni-jena.de`

Proceedings of Symposia in Pure Mathematics
Volume **79**, 2008

Weighted norm inequalities with positive and indefinite weights

I. E. Verbitsky

This paper is dedicated to Vladimir Maz'ya on the occasion of his 70-th birthday.

ABSTRACT. Integral inequalities of the type

$$||Tu||_{L^q(d\omega)} \le C\,||u||_{L^p(d\sigma)}, \qquad u \in L^p(d\sigma),$$

and their variations are considered. Here ω, σ are positive measures on $\mathbf{R}^n$, and T is an integral operator with nonnegative kernel. A survey of some recent results is given with an emphasis on the so-called trace inequalities where $d\sigma = dx$ is Lebesgue measure. It reflects the impact of V. Maz'ya's pioneering work on this problem and its applications to spectral theory, linear and nonlinear PDE.

An important special case where $q = p = 2$ and $T = (1-\Delta)^{-1/2}$ is associated with the Schrödinger operator $H = -\Delta + \omega$ and the imbedding $W^{1,2}(\mathbf{R}^n) \subset L^2(\omega)$ if $\omega \ge 0$.

For "indefinite weights" ω, the following quadratic form inequality is equivalent to the fundamental notion of the relative form boundedness of the potential energy operator ω with respect to the kinetic energy operator $H_0 = -\Delta$:

$$|\langle \omega\, u,\, u\rangle| = \left|\int_{\mathbf{R}^n} |u|^2\,\omega\right| \le a\,||\nabla u||^2_{L^2(\mathbf{R}^n)} + b\,||u||^2_{L^2(\mathbf{R}^n)}, \quad u \in C_0^\infty(\mathbf{R}^n),$$

for some $a,\, b > 0$. Here ω is a locally integrable function which may change sign, or more generally, a real- or complex-valued distribution. This form boundedness problem, which is important to mathematical quantum mechanics, was recently solved by Maz'ya and Verbitsky [**MV2**]. It will be discussed below for general second order differential operators with distributional coefficients (not necessarily elliptic) in place of the Laplacian [**MV6**].

CONTENTS

1991 *Mathematics Subject Classification.* 31B15, 35J10, 35J60, 42B25.

Key words and phrases. Weighted norm inequalities, indefinite weights, nonlinear potentials, form boundedness, Schrödinger operators, general second-order differential operators.

The author was supported in part by NSF Grant DMS-0556309.

1. Introduction

We will consider weighted norm inequalities of the type

$$||Tf||_{L^q(d\omega)} \le C\,||f||_{L^p(d\sigma)}, \quad f \in L^p(d\sigma), \tag{1.1}$$

where ω and σ are positive Borel measures, and $0 < p, q \le \infty$, for a number of "positive" operators T. We are particularly interested in integral inequalities for such operators T as Riesz potentials $I_\alpha = (-\Delta)^{-\frac{\alpha}{2}}$ $(0 < \alpha < n)$, Bessel potentials $J_\alpha = (1-\Delta)^{-\frac{\alpha}{2}}$ $(\alpha > 0)$, and more general convolution operators on the Euclidean space $\mathbf{R}^n$ with radially decreasing kernels, as well as their dyadic models. Other important operators, e.g., various maximal operators, Green's potentials, Poisson integrals, etc. can be studied using similar approaches. (See, e.g., [**KS**], [**KV**], [**VW2**], [**V1**], [**V2**].) We will treat in detail the less studied "upper triangle" case where $q < p$ [**COV1**]–[**COV3**], [**MN**] which is intimately connected with modern nonlinear potential theory.

A special consideration will be given to the so-called trace inequalities of the type

$$||Tf||_{L^q(d\omega)} \le C\,||f||_{L^p(dx)}, \quad f \in L^p(\mathbf{R}^n), \tag{1.2}$$

with an arbitrary measure $d\omega$ on the left-hand side, and with Lebesgue measure dx on $\mathbf{R}^n$ in place of $d\sigma$ on the right. This important special case covers many classical inequalities, as well as a great number of applications.

A deep study of trace inequalities with an emphasis on necessary and sufficient conditions was pioneered by Vladimir Maz'ya starting in the early 1960s [**M1**], [**M2**].

Let $\Omega \subset \mathbf{R}^n$ be an arbitrary open set, and let e be a compact subset of Ω. We will need the Wiener capacity relative to the domain Ω defined by:

$$\operatorname{cap}(e,\Omega) = \inf\left\{\int_\Omega |\nabla u|^2\,dx : \quad u \in C_0^\infty(\Omega), \ u > 1 \text{ on } e\right\}. \tag{1.3}$$

For a nonnegative Borel measure ω on Ω, we set

$$c_1(\omega,\Omega) = \sup\left\{\int_{\mathbf{R}^n} |u(x)|^2\,d\omega \, : \, u \in C_0^\infty(\Omega), \ ||\nabla u||_{L^2(\Omega)} \le 1\right\} \tag{1.4}$$

and

$$c_2(\omega,\Omega) = \sup\left\{\frac{\omega(e)}{\operatorname{cap}(e,\Omega)} : \quad e \subset \Omega\right\}, \tag{1.5}$$

where the supremum above is over compact sets $e \subset \Omega$ of positive capacity. As was shown in [**M1**] (see also [**M4**], Sec. 2.5),

$$c_2(\omega,\Omega) \le c_1(\omega,\Omega) \le 4\,c_2(\omega,\Omega), \tag{1.6}$$

where the constants 1 and 4 are sharp. Measures ω obeying conditions of the type $c_2(\omega, \Omega) < \infty$ which characterizes the imbedding $L^{1,2}(\Omega) \subset L^2(\Omega, d\omega)$ will be called Maz'ya measures (or sometimes admissible measures).

The class of Maz'ya measures associated with (1.2) in the case of Riesz potentials $T = I_\alpha$ for $0 < \alpha < n$ on $\Omega = \mathbf{R}^n$ was characterized completely in terms of Riesz capacities by Adams, Dahlberg and Maz'ya (see [**AH**], [**M4**]) in the 1970s. Alternative characterizations were given later by Kerman and Sawyer [**KS**] who used more precise local energy conditions, and Maz'ya and Verbitsky [**MV1**] in terms of yet more localized pointwise inequalities. The latter turned out to be especially well adapted for applications to nonlinear PDE [**HMV**], [**KV**], [**PV1**]–[**PV3**], [**VW1**]. These results are discussed in Sec. 2.

In Sec. 3, we consider two weight inequalities and nonlinear potentials for the so-called dyadic model [**COV2**], [**COV3**], [**NTV**] where T is the integral operator

$$Tf(x) = \int_{R^n} K_{\mathcal{D}}(x, y)\, f(y)\, d\sigma(y) \tag{1.7}$$

with the kernel $K_{\mathcal{D}}(x,y) = \sum_{Q \in \mathcal{D}} K(Q) \chi_Q(x)\, \chi_Q(y)$. Here $\mathcal{D} = \{Q\}$ is the family of all dyadic cubes on $\mathbf{R}^n$, $K(Q)$ are arbitrary nonnegative constants, and χ_Q are characteristic functions of Q.

Weighted norm inequalities for general convolution operators

$$Tf(x) = \int_{R^n} k(|x-y|)\, f(y)\, d\sigma(y), \tag{1.8}$$

where $k = k(r)$ is an arbitrary positive nonincreasing function of $r > 0$, can be deduced from the dyadic model. This makes it possible to answer a question raised in [**AH**], Sec. 7.7, of how to define an analogue of Wolff's potential for general radially decreasing kernels. In Sec. 3 we give an appropriate definition of Wolff's potential, which is far from being obvious, and prove Wolff's inequality for radially decreasing kernels [**COV2**], [**COV3**]. We observe that Wolff's potentials and their modifications have become an important tool in modern theory of quasilinear and fully nonlinear PDE [**KiMa**], [**L**], [**MaZi**], [**PV1**]–[**PV3**], [**TW**].

In Sec. 4 we will present recent developments [**MV2**] concerning trace inequalities with "indefinite" weights of the type:

$$\left| \int_{\mathbf{R}^n} |u|^2\, \omega \right| \le c\, ||\nabla u||^2_{L^2(\mathbf{R}^n)}, \qquad u \in C_0^\infty(\mathbf{R}^n), \tag{1.9}$$

where $n \ge 3$, and their inhomogeneous analogues:

$$\left| \int_{\mathbf{R}^n} |u|^2\, \omega \right| \le a\, ||\nabla u||^2_{L^2(\mathbf{R}^n)} + b\, ||u||^2_{L^2(\mathbf{R}^n)}, \qquad u \in C_0^\infty(\mathbf{R}^n), \tag{1.10}$$

for some positive constants a, b, where $\omega \in D'(\mathbf{R}^n)$, $n \ge 1$. Here $D(\mathbf{R}^n) = C_0^\infty(\mathbf{R}^n)$, and the left-hand sides are understood in the sense of distributions.

Both inequalities (1.9) and (1.10) are important to the Schrödinger operator theory. The first inequality expresses the domination (in absolute value) of the potential energy associated with ω by the kinetic energy, while the second one is equivalent to the classical notion of the *relative form boundedness* of ω with respect to the kinetic energy operator $H_0 = -\Delta$. This concept is used in the so-called KLMN theorem which makes it possible to define a self-adjoint operator $H = H_0 + \omega$ so that the quadratic form domain $\mathcal{Q}(H)$ coincides with $\mathcal{Q}(H_0)$ provided ω is real-valued. See, e.g., [**EE**], [**RS**], [**Sch**]. It is worth mentioning that the quadratic form inequality (1.10) is equivalent to the boundedness of the operator $H: W^{1,2}(\mathbf{R}^n) \to W^{-1,2}(\mathbf{R}^n)$.

When the form bound $a > 0$ in (1.10) can be arbitrarily small (with b depending on a) ω is said to be *infinitesimally* form bounded relative to $\mathcal{L}$. This notion is used extensively in mathematical quantum mechanics (see [**RS**], [**Sch**]).

Another important version of these form boundedness properties occurs when one replaces the nonrelativistic kinetic energy, namely $||\nabla u||^2_{L^2(\mathbf{R}^n)}$, with its relativistic counterpart, $||(-\Delta+1)^{1/4}u||^2_{L^2(\mathbf{R}^n)}$ associated with the Sobolev space $W^{1/2,2}(\mathbf{R}^n)$ of order 1/2.

The form boundedness problem for $H = -\Delta + \omega$ where $\omega \in D'(\mathbf{R}^n)$, along with its infinitesimal version, was recently solved by Maz'ya and Verbitsky [**MV2**]–[**MV6**]. The main ideas discussed below can be expressed as follows: (1.9) holds for a distributional potential ω if and only if it holds for $|\nabla\Delta^{-1}\omega|^2$ in place of ω. This nonlinear transformation $\omega \to |\nabla\Delta^{-1}\omega|^2$ reduces the form boundedness problem for "indefinite weights" to the well-studied case of nonnegative weights. In Sec. 4 we will outline a proof of this form boundedness criterion for $H = -\Delta + \omega$ based on sharp estimates of equilibrium potentials and related weighted norm inequalities [**MV2**].

Similarly, the class of $\omega \in D'(\mathbf{R}^n)$ associated with the relativistic form boundedness of $\mathcal{H} = (-\Delta+1)^{1/2} + \omega$ is invariant under the transformation $\omega \to |(1-\Delta)^{-1/2}\omega|^2$. The relativistic problem can be reduced to a similar nonrelativistic one for the operator $H = -\Delta + \tilde{\omega}$ with a distributional potential $\tilde{\omega}$ on a higher dimensional Euclidean space as was shown in [**MV4**].

In [**MV5**] necessary and sufficient conditions were obtained for the *infinitesimal* form boundedness of the potential energy operator associated with ω with respect to the kinetic energy operator $H_0 = -\Delta$ on $L^2(\mathbf{R}^n)$. Here ω is an arbitrary real- or complex-valued potential (possibly a distribution). Furthermore, a related *form subordination property* of Trudinger type (see [**Tru**], and also [**Sim**] , [**RSS**], and the literature cited there) is characterized explicitly in [**MV5**].

More precisely, we will present in Sec. 5 a characterization of the class of potentials $\omega \in \mathcal{D}'(\mathbf{R}^n)$ which are $-\Delta$-form bounded with relative bound zero, i.e., for every $\epsilon > 0$, there exists $C(\epsilon) > 0$ such that

$$|\langle \omega u, u\rangle| \le \epsilon\,||\nabla u||^2_{L^2(\mathbf{R}^n)} + C(\epsilon)\,||u||^2_{L^2(\mathbf{R}^n)}, \quad \forall u \in C_0^\infty(\mathbf{R}^n). \tag{1.11}$$

For complex-valued ω, it follows that H is an m-sectorial operator on $L^2(\mathbf{R}^n)$ with $\mathcal{D}(H) \subset W^{1,2}(\mathbf{R}^n)$ ([**EE**], Sec. IV.4).

The characterization of (1.11) obtained in [**MV5**] uses only the functions $|\nabla(1-\Delta)^{-1}\,\omega|$ and $|(1-\Delta)^{-1}\,\omega|$, and is based on the representation:

$$\omega = \operatorname{div}\vec{\Gamma} + \gamma, \qquad \vec{\Gamma}(x) = -\nabla(1-\Delta)^{-1}\,\omega, \quad \gamma = (1-\Delta)^{-1}\,\omega. \tag{1.12}$$

In particular, $\vec{\Gamma} \in L^2_{\text{loc}}(\mathbf{R}^n)^n$, $\gamma \in L^1_{\text{loc}}(\mathbf{R}^n)$, and, when $n \ge 3$,

$$\lim_{\delta\to+0}\ \sup_{x_0\in\mathbf{R}^n} \delta^{2-n}\int_{B_\delta(x_0)} \left(|\vec{\Gamma}(x)|^2 + |\gamma(x)|\right)\,dx = 0, \tag{1.13}$$

once (1.11) holds. Here $B_\delta(x_0)$ is a Euclidean ball of radius δ centered at x_0.

In the opposite direction, it follows from the results of [**MV5**] that (1.11) holds whenever

$$\lim_{\delta\to+0}\sup_{x_0\in\mathbf{R}^n}\delta^{2r-n}\int_{B_\delta(x_0)}\left(|\vec{\Gamma}(x)|^2+|\gamma(x)|\right)^r dx=0, \tag{1.14}$$

for some $r>1$. Such potentials form a natural analogue of the Fefferman–Phong class [**F**] for the infinitesimal form boundedness problem, where cancellations between the positive and negative parts of ω come into play. It includes functions with highly oscillatory behavior as well as singular measures, and contains properly the class of potentials based on the original Fefferman–Phong condition where $|\omega|$ is used in (1.14) in place of $|\vec{\Gamma}|^2+|\gamma|$. Moreover, one can expand this class further using a sharp condition due to Chang, Wilson, and Wolff [**ChWW**] applied to $|\vec{\Gamma}|^2+|\gamma|$. In the proofs given in [**MV5**] considerable technical difficulties have been overcome using sharp estimates for powers of equilibrium potentials, factorization of functions in Sobolev spaces, and theory of A_p-weights, along with appropriate localization arguments, and good understanding of trace inequalities for nonnegative potentials ω.

In [**MV5**], we also study quadratic form inequalities of Trudinger type where $C(\epsilon)$ in (1.11) has power growth, i.e., there exists $\epsilon_0>0$ such that

$$|\langle\omega u,\,u\rangle|\le\epsilon\,||\nabla u||^2_{L^2(\mathbf{R}^n)}+c\,\epsilon^{-\beta}\,||u||^2_{L^2(\mathbf{R}^n)},\quad\forall u\in C_0^\infty(\mathbf{R}^n), \tag{1.15}$$

for every $\epsilon\in(0,\epsilon_0)$, where $\beta>0$. Such inequalities appear in studies of elliptic PDE with measurable coefficients, and have been used extensively in spectral theory of the Schrödinger operator [**AS**].

As it turns out, it is still possible to characterize (1.15) using only $|\vec{\Gamma}|$ and $|\gamma|$ defined by (1.12), provided $\beta>1$. In this case (1.15) holds if and only if both of the following conditions hold:

$$\sup_{\substack{x_0\in\mathbf{R}^n\\0<\delta<\delta_0}}\delta^{2\frac{\beta-1}{\beta+1}-n}\int_{B_\delta(x_0)}|\vec{\Gamma}(x)|^2\,dx<+\infty, \tag{1.16}$$

$$\sup_{\substack{x_0\in\mathbf{R}^n\\0<\delta<\delta_0}}\delta^{\frac{2\beta}{\beta+1}-n}\int_{B_\delta(x_0)}|\gamma(x)|\,dx<+\infty, \tag{1.17}$$

for some $\delta_0>0$. However, in the case $\beta\le1$ this is no longer true. For $\beta=1$, (1.16) has to be replaced with the condition that $\vec{\Gamma}$ is in the local BMO space, or respectively is Hölder-continuous of order $\frac{1-\beta}{1+\beta}$ if $0<\beta<1$.

In the homogeneous case $\epsilon_0=+\infty$, (1.15) is equivalent to the *multiplicative inequality*:

$$|\langle\omega\,u,\,u\rangle|\le C\,||\nabla u||^{2p}_{L^2(\mathbf{R}^n)}\,||u||^{2(1-p)}_{L^2(\mathbf{R}^n)},\quad\forall u\in C_0^\infty(\mathbf{R}^n), \tag{1.18}$$

where $p=\frac{\beta}{1+\beta}\in(0,\,1)$. In spectral theory, (1.18) is referred to as the form p-subordination property (see, e.g., [**RSS**], Sec. 20.4).

For *nonnegative* ω, where ω is a locally finite measure on $\mathbf{R}^n$, inequality (1.18) is known to hold if and only if

$$\omega\left(B_\delta(x_0)\right)\le c\,\delta^{n-2p}, \tag{1.19}$$

where the constant c does not depend on $\delta>0$ and $x_0\in\mathbf{R}^n$ ([**M3**], Sec. 1.4.7).

For general ω, we obtain the following result: If $p > \frac{1}{2}$, then (1.18) holds if and only if $\nabla \Delta^{-1}\omega$ lies in the Morrey space $\mathcal{L}^{2,\lambda}(\mathbf{R}^n)$, where $\lambda = n + 2 - 4p$. For $p = \frac{1}{2}$, it holds if and only if $\nabla \Delta^{-1}\omega \in \mathrm{BMO}(\mathbf{R}^n)$, and for $0 < p < \frac{1}{2}$, whenever $\nabla \Delta^{-1}\omega \in \mathrm{Lip}_{1-2p}(\mathbf{R}^n)$. These different characterizations are equivalent to (1.19) if ω is a nonnegative measure.

As a consequence, we are able to characterize those ω which obey an analogous inequality of Nash's type:

$$|\langle \omega u, u\rangle| \le C\,||\nabla u||^{2p}_{L^2(\mathbf{R}^n)}\,||u||^{2(1-p)}_{L^1(\mathbf{R}^n)}, \quad \forall u \in C_0^\infty(\mathbf{R}^n). \tag{1.20}$$

where $p \in (0, 1)$. In fact, the preceding inequality has two critical exponents, $p_* = \frac{n}{n+2}$ and $p^* = \frac{n+1}{n+2}$. For $0 < p < p_*$, (1.20) holds only if $\omega = 0$, whereas for $p = p_*$ it follows that $\omega \in L^\infty(\mathbf{R}^n)$, i.e., it is equivalent to Nash's inequality. For $p > p_*$, the validity of (1.20) is equivalent respectively to: $\nabla \Delta^{-1}\omega \in \mathrm{Lip}_{n+1-p(n+2)}(\mathbf{R}^n)$ if $p_* < p < p^*$; $\nabla \Delta^{-1}\omega \in \mathrm{BMO}(\mathbf{R}^n)$ if $p = p^*$; and $\nabla \Delta^{-1}\omega \in \mathcal{L}^{2,\lambda}(\mathbf{R}^n)$, where $\lambda = 3n + 2 - 2p(n + 2)$ if $p^* < p < 1$.

The form boundedness problem for the general second order differential operator

$$\mathcal{L} = \sum_{i,j=1}^{n} a_{ij}\,\partial_i\partial_j + \sum_{j=1}^{n} b_j\,\partial_j + c, \tag{1.21}$$

where a_{ij}, b_i, and c are real- or complex-valued distributions was solved in [**MV6**]. Here $\mathcal{L}$ is not even assumed to be elliptic.

We will discuss in Sec. 6 quite complicated necessary and sufficient conditions for the quadratic form inequality

$$|\langle \mathcal{L}\,u, u\rangle| \le a\,\langle -\Delta\,u, u\rangle + b\,||u||^2_{L^2(\mathbf{R}^n)}, \quad u \in C_0^\infty(\mathbf{R}^n), \tag{1.22}$$

to hold for some $a, b > 0$. It is easy to see that the symmetric part of the matrix (a_{ij}) must be uniformly bounded, and the skew-symmetric part reduces to the first-order terms. The main problem here is to investigate the interaction between the first-order and zero-order terms. The proofs make use of compensated compactness arguments (a vector-valued version of the *div-curl* lemma), along with the gauge transform involving powers of equilibrium potentials.

Applications to multidimensional Riccati's equations, quasilinear and fully nonlinear PDE, global estimates of Green's functions, etc., can be found in [**FV**], [**HMV**], [**KV**], [**MaZi**], [**PV1**]–[**PV3**].

2. Basic trace inequalities

We start with the following important theorem [**M3**].

THEOREM 2.1. (Maz'ya) *Let $1 < p < \infty$. Let Ω be an arbitrary open set in $\mathbf{R}^n$, and let ω be a nonnegative locally finite Borel measure on Ω. Then the inequality*

$$||u||_{L^p(\Omega, d\omega)} \le C\,||\nabla u||_{L^p(\Omega)}, \quad u \in C_0^\infty(\Omega), \tag{2.1}$$

holds if and only if, for any compact set $E \subset \Omega$,

$$\omega(E) \le C\,\mathrm{cap}_{1,p}(E, \Omega), \tag{2.2}$$

where C is a constant which is independent of E. Here $\mathrm{cap}_{1,p}(E,\Omega)$ *is the capacity defined by*

$$\mathrm{cap}_{1,p}(E,\Omega) = \inf\left\{\int_\Omega |\nabla u|^p\,dx:\ u(x)\geq 1 \text{ on } E,\ u\in C_0^\infty(\Omega)\right\}. \tag{2.3}$$

Theorem 2.1 has numerous applications in harmonic analysis, operator theory, function spaces, linear and nonlinear PDE's, etc. (see, e.g., [**AH**], [**FV**], [**M4**], [**MSh**], [**MaZi**], [**PV2**]).

For simplicity, we will only consider some analogues of Theorem 2.1 in the case $\Omega = \mathbf{R}^n$ for Riesz potentials defined by

$$I_\alpha f = (-\Delta)^{-\frac{\alpha}{2}} f = c(n,\alpha)\,(|\cdot|^{\alpha-n}\star f), \quad 0<\alpha<n,$$

where $c(n,\alpha)$ is a normalization constant. We also set

$$I_\alpha(f\,d\omega)(x) = c(n,\alpha)\int_{\mathbf{R}^n}\frac{f(y)}{|x-y|^{n-\alpha}}\,d\omega(y), \quad x\in\mathbf{R}^n, \tag{2.4}$$

for potentials with a Borel measure $d\omega$ in place of dx, and $I_\alpha\omega = I_\alpha(\mathbf{1}d\omega)$ if $f\equiv 1$ on $\mathbf{R}^n$. The Riesz capacity of a measurable set $E\subset\mathbf{R}^n$ is defined by

$$\mathrm{Cap}_{\alpha,p}(E) = \inf\left\{\int_\Omega |g|^p\,dx:\ I_\alpha g(x)\geq 1 \quad \text{on} \quad E, \quad g\in L^p_+(\mathbf{R}^n)\right\}. \tag{2.5}$$

In the case $\alpha=1$ it is known that $\mathrm{Cap}_{1,p}(E)\asymp \mathrm{cap}_{1,p}(E,\mathbf{R}^n)$ for compact sets E, where $\mathrm{cap}_{1,p}(\cdot,\mathbf{R}^n)$ is defined by (2.3), and constants of equivalence depend only on p (see [**AH**]).

The following theorem for $\alpha=1$ and $q=p$ is equivalent to Theorem 2.1 when $\Omega=\mathbf{R}^n$.

THEOREM 2.2. (D. Adams-Dahlberg-Maz'ya) *Let $1<p<\infty$ and let $0<\alpha<n$. Let ω be a nonnegative Borel measure on $\mathbf{R}^n$. Then the following statements are equivalent.*

(i) The inequality

$$||I_\alpha f||_{L^p(d\omega)} \leq C\,||f||_{L^p}, \quad f\in L^p(\mathbf{R}^n), \tag{2.6}$$

holds where C is a constant which is independent of f.

(ii) For every compact set $E\subset\mathbf{R}^n$,

$$\omega(E)\leq C\,\mathrm{Cap}_{\alpha,p}(E), \tag{2.7}$$

where C is a constant which is independent of E.

The statement (i)⇒(ii) in Theorem 2.3 is obvious; the converse follows from the so-called strong capacitary inequality

$$\int_0^\infty \mathrm{Cap}_{\alpha,p}(\{x:\,|I_\alpha f(x)|>t\})\,t^{p-1}\,dt \leq C\,||f||^p_{L^p}, \quad f\in L^p(\mathbf{R}^n),$$

discovered by V. Maz'ya in 1972 for $\alpha=1$. In a series of papers by D. Adams, B. Dahlberg and V. Maz'ya in the late 70-s, the preceding inequality was established for all $\alpha>0$ and $p>1$. Another proof valid for more general convolution operators is due to K. Hansson. (See [**AH**], [**M4**], [**H**].)

REMARK 2.3. Condition (2.7) combined with a standard estimate of the capacity from below, $\mathrm{Cap}_{\alpha,p}(E) \geq C\,|E|^{1-\frac{\alpha p}{n}}$, immediately yields the following well known sufficient condition: (2.6) holds if $\omega(E) \leq C\,|E|^{1-\frac{\alpha p}{n}}$, for every compact set E. This can be restated in terms of weak L^r spaces: $d\omega = \rho(x)\,dx$, where $\rho \in L^{r,\infty}(\mathbf{R}^n)$, for $r = \frac{n}{\alpha p}$.

A substantial improvement of the sufficient condition mentioned in Remark 2.3 was found by C. Fefferman and Phong [**F**].

THEOREM 2.4. *If* $d\omega = \rho(x)\,dx$ *where*

$$\int_B \rho^{1+\epsilon}\,dx \leq C\,|B|^{1-\frac{\alpha p(1+\epsilon)}{n}}, \qquad \epsilon > 0, \tag{2.8}$$

for every ball $B = B_r(x)$, *then (2.6) holds.*

A sharp version of the Fefferman–Phong condition is due to Chang, Wilson, and Wolff [**ChWW**].

REMARK 2.5. An improved version of both the Fefferman–Phong and Chang–Wilson–Wolff conditions where ω is not necessarily absolutely continuous with respect to Lebesgue measure is given in [**MV1**] (see Corollary 2.10 below).

It is easy to see that the capacitary condition (2.7) is equivalent via duality to the inequality

$$\int_{\mathbf{R}^n} [I_\alpha(\chi_E\,d\omega)]^{p'}\,dx \leq C\,\omega(E), \tag{2.9}$$

for every compact set $E \subset \mathbf{R}^n$, where $\frac{1}{p} + \frac{1}{p'} = 1$. It was noticed by Kerman and Sawyer [**KS**] that in this dual form it is enough to restrict oneself to $E = B$ where $B = B_r(x)$ is a ball (or cube) in $\mathbf{R}^n$.

THEOREM 2.6. (Kerman-Sawyer) *Let* $1 < p < \infty$ *and let* $0 < \alpha < n$. *Let* ω *be a nonnegative Borel measure on* $\mathbf{R}^n$. *Then the trace inequality (2.6) holds if and only if*

$$\int_B [I_\alpha(\chi_B\,d\omega)]^{p'}\,dx \leq C\,\omega(B), \tag{2.10}$$

for every ball $B = B_r(x)$.

The following theorem [**MV1**] shows that, in a sense, balls $B = B_r(x)$ in (2.10) may be replaced with single points x.

THEOREM 2.7. (Maz'ya-Verbitsky) *Let* $1 < p < \infty$ *and let* $0 < \alpha < n$. *Let* ω *be a nonnegative Borel measure on* $\mathbf{R}^n$. *Then the following statements are equivalent.*

(i) The trace inequality (2.6) holds.

(ii) For every compact set $E \subset \mathbf{R}^n$, *the capacitary inequality (2.7) holds.*

(iii) $I_\alpha\omega < \infty$ dx*-a.e. and*

$$I_\alpha[(I_\alpha\omega)^{p'}](x) \leq C\,I_\alpha\omega(x) \quad dx\text{-a.e.} \tag{2.11}$$

(iv) The trace inequality (2.6) holds with ω *replaced with the absolutely continuous measure* $d\nu = (I_\alpha\omega)^{p'}\,dx$, *or, equivalently,*

$$\nu(E) = \int_E (I_\alpha\omega)^{p'}\,dx \leq C\,\mathrm{Cap}_{\alpha,p}(E), \tag{2.12}$$

where C *is a constant which is independent of a compact set* E.

REMARK 2.8. A simple direct proof of Theorem 2.7 from which Theorem 2.6 is deduced as a corollary can be found in [**V4**] (see also [**VW2**]).

Another useful characterization of Maz'ya measures in terms of discrete Carleson measures was given in [**V4**].

COROLLARY 2.9. (Verbitsky) *Let* $1 < p < \infty$ *and let* $0 < \alpha < n$. *Let* ω *be a nonnegative Borel measure on* $\boldsymbol{R}^n$. *Then any one of the conditions (i)–(iv) of Theorem 2.7 is equivalent to the following property:*

(v) For every dyadic cube P,

$$\sum_{Q \subseteq P} \left(\frac{\omega(Q)}{|Q|^{1-\frac{\alpha}{n}}} \right)^{p'} |Q| \leq C\,\omega(P), \tag{2.13}$$

where the sum is taken over all dyadic cubes Q *contained in* P, *and* C *does not depend on* P.

The following sufficient conditions which are applicable to broader classes of ω than those considered in Remark 2.3 and Theorem 2.4 follow from Theorem 2.7 (iv).

COROLLARY 2.10. *(i) If* $I_\alpha \omega \in L^{s,\infty}(\boldsymbol{R}^n)$, *where* $s = \frac{n}{\alpha(p-1)}$, *then the trace inequality (2.6) holds.*

(ii) If $\epsilon > 0$ *and* $\int_B (I_\alpha \omega)^{(1+\epsilon)p'} dx \leq C\,|B|^{1-\frac{\alpha p(1+\epsilon)}{n}}$, *for every ball* B, *then (2.6) holds.*

We now consider the trace inequality

$$||I_\alpha f||_{L^q(d\omega)} \leq C\,||f||_{L^p}, \quad f \in L^p(\mathbf{R}^n), \tag{2.14}$$

for $q \neq p$. In the classical "lower triangle case" $q > p$ there is a definitive characterization of (2.14) due to D. Adams (see [**AH**], [**M4**]).

THEOREM 2.11. (D. Adams) *Let* $1 < p < q < \infty$. *Let* $0 < \alpha < \frac{n}{p}$. *Then the trace inequality* (2.14) *holds if and only if*

$$\omega(B_r(x)) \leq C\,r^{\frac{(n-\alpha p)q}{p}} \tag{2.15}$$

for all $r > 0$ *and* $x \in \boldsymbol{R}^n$.

REMARK 2.12. If $\alpha \geq \frac{n}{p}$ or $\frac{1}{q} < \frac{1}{p} - \frac{\alpha}{n}$ then (2.14) holds only for $\omega = 0$. When $\frac{1}{q} = \frac{1}{p} - \frac{\alpha}{n}$ and ω is Lebesgue measure Theorem 2.11 gives the celebrated Hardy–Littlewood–Sobolev inequality for fractional integrals.

We next define the following nonlinear potentials which play an important role in modern potential theory and applications to linear, quasilinear and fully nonlinear partial differential equations (see [**AH**], [**L**], [**KiMa**], [**MaZi**], [**TW**]). Let

$$V\omega(x) = V_{\alpha,p}\,\omega(x) = I_\alpha[(I_\alpha \omega)^{p'-1}](x), \tag{2.16}$$

and

$$W\omega(x) = W_{\alpha,p}\,\omega(x) = \int_0^\infty \left[\frac{\omega(B_r(x))}{r^{n-\alpha p}} \right]^{p'-1} \frac{dr}{r}. \tag{2.17}$$

We shall use the notion of the energy $\mathcal{E}(\omega) = \mathcal{E}_{\alpha,p}(\omega)$ of a measure ω on $\mathbf{R}^n$ defined by

$$\mathcal{E}(\omega) = ||I_\alpha \omega||_{L^{p'}}^{p'} = \int_{\mathbf{R}^n} V\omega \, d\omega. \tag{2.18}$$

Note that in the case $p = 2$ both $W\omega$ and $V\omega$ coincide with a classical linear potential:

$$V_{\alpha,2}\omega = W_{\alpha,2}\,\omega = I_{2\alpha}\omega. \tag{2.19}$$

The nonlinear potential $V_{\alpha,p}$ $(p \neq 2)$ was introduced and studied in detail by Havin–Maz'ya, Reshetnyak, and Meyers, and $W_{\alpha,p}$ by Adams–Meyers and Hedberg–Wolff (see [**AH**], [**HW**], [**M1**]). The latter is usually called Wolff's potential because of its prominent role in the following Wolff's inequality that appeared in [**HW**]:

$$\mathcal{E}_{\alpha,p}(\omega) = \int_{\mathbf{R}^n} V_{\alpha,p}\,\omega \, d\omega \le C \int_{\mathbf{R}^n} W_{\alpha,p}\,\omega \, d\omega, \tag{2.20}$$

where C is a constant which depends only on p, α, and n. The converse inequality is obvious since

$$W_{\alpha,p}\,\omega(x) \le C\, V_{\alpha,p}\,\omega(x), \qquad x \in \mathbf{R}^n, \tag{2.21}$$

where C is a constant which is independent of x and ω (see [**M1**], Sec. 7.2.2).

There are alternative proofs of Wolff's inequality (see [**AH**]. Sec. 4.5), but the original proof presented in [**HW**], which can be easily simplified using an elementary "integration by parts" lemma [**V4**], remains most direct and leads to important applications to nonlinear PDE [**PV2**], [**PV3**].

REMARK 2.13. It follows from Theorem 2.2 (in the dual form (2.9)) and Wolff's inequality that the trace inequality (2.14) for $q = p$ holds if $W_{\alpha,p}\,\omega$ is uniformly bounded. The converse is clearly not true.

We next consider the difficult case $q < p$. A capacitary characterization of (2.14) in this case was obtained by Maz'ya and Netrusov [**MN**].

THEOREM 2.14. (Maz'ya-Netrusov) *Let $1 < p < \infty$ and $0 < q < p < \infty$. Let ω be a positive Borel measure on $\mathbf{R}^n$ and let $\phi(t) = \inf\{\mathrm{Cap}_{\alpha,p}(F) : \omega(F) \ge t\}$. Then (2.14) holds if and only if*

$$\int_0^\infty \left[\frac{t^p}{\phi(t)^q}\right]^{\frac{1}{p-q}} \frac{dt}{t} < \infty.$$

An alternative noncapacitary characterization of (2.14) for $0 < q < p$ in terms of Wolff's potentials was given in [**COV1**], [**V4**].

THEOREM 2.15. (Cascante-Ortega-Verbitsky, $q > 1$; Verbitsky $q > 0$) *Let $1 < p < \infty$ and $0 < q < p < \infty$. Let ω be a positive Borel measure on $\mathbf{R}^n$. Then (2.14) holds if and only if*

$$W_{\alpha,p}\,\omega \in L^{\frac{q(p-1)}{p-q}}(d\omega). \tag{2.22}$$

All theorems stated above can be carried over to general convolution operators with nonnegative radial kernels [**COV2**], [**COV3**] (see Sec. 3), as well as to integral operators with "quasi-metric" kernels [**KV**].

3. Dyadic and radial trace inequalities and nonlinear potentials

In this section we concentrate on integral operators with general dyadic and radially nonincreasing kernels, as well as the corresponding nonlinear potentials, and two weight inequalities. Continuous versions for convolution operators with radial kernels are deduced from the dyadic models using averaging over shifts of the dyadic lattice (see [**COV3**]).

Let $\mathcal{D} = \{Q\}$ be the family of all dyadic cubes Q in $\mathbf{R}^n$, and $K : \mathcal{D} \to \mathbf{R}^+$. The kernel $K_{\mathcal{D}}(x, y)$ on $\mathbf{R}^n \times \mathbf{R}^n$ is defined by

$$K_{\mathcal{D}}(x,y) = \sum_{Q \in \mathcal{D}} K(Q)\, \chi_Q(x)\, \chi_Q(y), \tag{3.1}$$

where χ_Q is the characteristic function of $Q \in \mathcal{D}$.

Let σ be a locally finite positive Borel measure on $\mathbf{R}^n$, and let $f \in L^1_{\text{loc}}(d\sigma)$. We define the dyadic integral operator:

$$T_{K_{\mathcal{D}}}[f\, d\sigma](x) = \int_{\mathbf{R}^n} K_{\mathcal{D}}(x,y) f(y)\, d\sigma(y) = \sum_{Q \in \mathcal{D}} K(Q)\chi_Q(x) \int_Q f\, d\sigma. \tag{3.2}$$

In case $f \equiv 1$, we set

$$T_{K_{\mathcal{D}}}[\sigma](x) = \sum_{Q \in \mathcal{D}} K(Q)\, \sigma(Q)\, \chi_Q(x).$$

If $0 < q,\, p < +\infty$, and σ and ω are locally finite Borel measures on $\mathbf{R}^n$, the corresponding dyadic trace inequality is given by:

$$\int_{\mathbf{R}^n} |\, T_{K_{\mathcal{D}}}[f\, d\sigma]\,|^q\, d\omega \le C\, \|f\|^q_{L^p(d\sigma)}, \qquad f \in L^p(d\sigma). \tag{3.3}$$

Assume for a moment that $q,\, p > 1$. Duality then gives that (3.3) is equivalent to the inequality:

$$\int_{\mathbf{R}^n} |\, T_{K_{\mathcal{D}}}[g d\omega]|^{p'}\, d\sigma \le C\, \|g\|^{p'}_{L^{q'}(d\omega)}, \qquad g \in L^{q'}(d\omega). \tag{3.4}$$

The quantity on the left-hand side of (3.4) is a generalized version of the discrete energy of $d\nu = g d\omega$. For positive locally finite Borel measures ν and σ on $\mathbf{R}^n$, the discrete energy associated with ν and σ is defined by (cf. [**HW**]):

$$\mathcal{E}^{\mathcal{D}}_{K,\sigma}[\nu] = \int_{\mathbf{R}^n} (T_{K_{\mathcal{D}}}[\nu])^{p'}\, d\sigma = \int_{\mathbf{R}^n} \left(\sum_{Q \in \mathcal{D}} K(Q)\, \nu(Q)\, \chi_Q(x) \right)^{p'} d\sigma(x). \tag{3.5}$$

Fubini's theorem gives an alternative expression for $\mathcal{E}^{\mathcal{D}}_{K,\sigma}$:

$$\mathcal{E}^{\mathcal{D}}_{K,\sigma}[\nu] = \int_{\mathbf{R}^n} T_{K_{\mathcal{D}}}[(T_{K_{\mathcal{D}}}[\nu])^{p'-1} d\sigma]\, d\nu,$$

where $T_{K_{\mathcal{D}}}[(T_{K_{\mathcal{D}}}[\nu])^{p'-1} d\sigma]$ is a dyadic analogue of the nonlinear potential of Havin–Maz'ya originally defined for $d\sigma = dx$ and with Riesz kernels in place of $K_{\mathcal{D}}$ (see [**AH**], [**M4**]).

In the special case where $d\sigma$ is Lebesgue measure on $\mathbf{R}^n$, $K(Q) = \frac{1}{|Q|^{1-(\alpha/n)}}$, $0 < \alpha < n$ and $|Q|$ is the Lebesgue measure of Q, i.e., when $K_{\mathcal{D}}$ is a discrete Riesz kernel on $\mathbf{R}^n$, Hedberg and Wolff introduced a dyadic nonlinear potential defined by:

$$\mathcal{W}^{\mathcal{D}}_{\alpha,\,dx}[\nu](x) = \sum_{Q\in\mathcal{D}} \left(\frac{\nu(Q)}{\ell_Q^{n-\alpha p}} \right)^{p'-1} \chi_Q(x). \tag{3.6}$$

(Here ℓ_Q denotes the side length of Q.) A dyadic version of Wolff's inequality established in [**HW**] shows that, for $1 < p < +\infty$,

$$C_1\, \mathcal{E}^{\mathcal{D}}_{\alpha,\,dx}[\nu] \le \int_{\mathbf{R}^n} \mathcal{W}^{\mathcal{D}}_{\alpha,\,dx}[\nu](x)\, d\nu(x) \le C_2\, \mathcal{E}^{\mathcal{D}}_{\alpha,\,dx}[\nu]. \tag{3.7}$$

Consequently, the trace inequality (3.3) holds for $q = 1$, $1 < p < +\infty$, and $d\sigma = dx$ if and only if $\mathcal{W}^{\mathcal{D}}_{\alpha,\,dx}[\omega]$ is in $L^1(d\omega)$. For $1 < q < p < +\infty$, as was shown in [**COV1**], (3.3) holds if and only if $\mathcal{W}^{\mathcal{D}}_{\alpha,\,dx}[\omega] \in L^{\frac{q(p-1)}{p-q}}(d\omega)$.

We next define a suitable nonlinear potential associated with a pair of measures ν, σ and the kernel $K_{\mathcal{D}}$ so that it is applicable to characterization of the trace inequality (3.3).

Let ν and σ be positive locally finite Borel measures on $\mathbf{R}^n$. We denote by $\overline{K}(Q)$ the function

$$\overline{K}(Q)(x) = \frac{1}{\sigma(Q)} \sum_{Q'\subset Q} K(Q')\sigma(Q')\chi_{Q'}(x). \tag{3.8}$$

For $x \in \mathbf{R}^n$, we set

$$\mathcal{W}^{\mathcal{D}}_{K,\,\sigma}[\nu](x) = \sum_{Q\in\mathcal{D}} K(Q)\, \sigma(Q) \left(\int_Q \overline{K}(Q)(y) d\nu(y) \right)^{p'-1} \chi_Q(x). \tag{3.9}$$

It is worthwhile to observe that several other natural alternatives to $\mathcal{W}^{\mathcal{D}}_{K,\,\sigma}[\omega]$ discussed in [**COV2**] fail to satisfy the desired analogue of Wolff's inequality.

In [**COV2**], the following Wolff-type inequality was established for an arbitrary positive measure ν on $\mathbf{R}^n$, and dx in place of $d\sigma$:

$$C_1\, \mathcal{E}^{\mathcal{D}}_{K,\,dx}[\nu] \le \int_{\mathbf{R}^n} \mathcal{W}^{\mathcal{D}}_{K,\,dx}[\nu]\, d\nu \le C_2\, \mathcal{E}^{\mathcal{D}}_{K,\,dx}[\nu],$$

where C_1, C_2 are constants which do not depend on ν. Its generalization to a two-weight setting is also true [**COV3**].

THEOREM 3.1. *Let $K : \mathcal{D} \to \mathbf{R}^+$. Let $1 < p < +\infty$, and let ν and σ be locally finite positive Borel measures on $\mathbf{R}^n$. If $\mathcal{E}^{\mathcal{D}}_{K,\,\sigma}[\nu]$ and $\mathcal{W}^{\mathcal{D}}_{K,\,\sigma}[\nu]$ are defined respectively by* (3.5) *and* (3.9)*, then*

$$C_1\, \mathcal{E}^{\mathcal{D}}_{K,\,\sigma}[\nu] \le \int_{\mathbf{R}^n} \mathcal{W}^{\mathcal{D}}_{K,\,\sigma}[\nu](x)\, d\nu(x) \le C_2\, \mathcal{E}^{\mathcal{D}}_{K,\,\sigma}[\nu],$$

where C_1, C_2 are constants which do not depend on ν and σ.

We now state a two weight inequality proved in [**COV2**] for $1 < q < p$, and in [**COV3**] for all $0 < q < p$, $p > 1$.

THEOREM 3.2. *Let $K : \mathcal{D} \to \mathbf{R}^+$. Let $0 < q < p < +\infty$, $1 < p < +\infty$, and let ω and σ be locally finite positive Borel measures on $\mathbf{R}^n$.*

(i) *Suppose there exists a constant* $C > 0$ *such that the trace inequality*

$$\int_{\mathbf{R}^n} |T_{K_{\mathcal{D}}}[f d\sigma]|^q (x)\, d\omega(x) \le C \, \|f\|^q_{L^p(d\sigma)}, \qquad f \in L^p(d\sigma),$$

holds. Then $\mathcal{W}^{\mathcal{D}}_{K,\sigma}[\omega] \in L^{\frac{q(p-1)}{p-q}}(d\omega)$.

(ii) *Conversely, if* $\mathcal{W}^{\mathcal{D}}_{K,\sigma}[\omega] \in L^{\frac{q(p-1)}{p-q}}(d\omega)$ *then the preceding trace inequality holds provided the pair* (K, σ) *satisfies the dyadic logarithmic bounded oscillation condition* (DLBO)*:*

$$\sup_{x\in Q} \overline{K}(Q)(x) \le A \inf_{x \in Q} \overline{K}(Q)(x),$$

where A *does not depend on* $Q \in \mathcal{D}$.

If $q = 1$ then statement (ii) holds without the restriction $(K, \sigma) \in$ DLBO. (In this case Theorem 3.2 is, by duality, an immediate consequence of Theorem 3.1.)

The proofs of Theorems 3.1 and 3.2 are based on the following important lemma [**COV2**]. Let $1 < s < +\infty$, $\Lambda = (\lambda_Q)_{Q \in \mathcal{D}}$, $\lambda_Q \in \mathbf{R}^+$, and let σ be a positive locally finite Borel measure. We assume that $\lambda_Q = 0$ if $\sigma(Q) = 0$, and follow the convention that $0 \cdot \infty = 0$.

We define

$$A_1(\Lambda) = \int_{\mathbf{R}^n} \left(\sum_{Q \in \mathcal{D}} \frac{\lambda_Q}{\sigma(Q)} \chi_Q(x) \right)^s d\sigma(x),$$

$$A_2(\Lambda) = \sum_{Q\in\mathcal{D}} \lambda_Q \left(\frac{1}{\sigma(Q)} \sum_{Q' \subset Q} \lambda_{Q'} \right)^{s-1},$$

$$A_3(\Lambda) = \int_{\mathbf{R}^n} \sup_{x \in Q} \left(\frac{1}{\sigma(Q)} \sum_{Q' \subset Q} \lambda_{Q'} \right)^s d\sigma(x).$$

LEMMA 3.3. *Let* σ *be a positive locally finite Borel measure on* $\mathbf{R}^n$. *Let* $1 < s < \infty$. *Then there exist constants* $C_i > 0$, $i = 1, 2, 3$, *which depend only on* s, *such that, for any* $\Lambda = (\lambda_Q)_{Q\in\mathcal{D}}$, $\lambda_Q \in \mathbf{R}^+$,

$$A_1(\Lambda) \le C_1 \, A_2(\Lambda) \le C_2 \, A_3(\Lambda) \le C_3 \, A_1(\Lambda).$$

Theorem 3.1 is deduced from Lemma 3.3 in the case $\lambda_Q = K(Q)\omega(Q)\sigma(Q)$ and $s = p'$.

We next treat continuous versions of the above theorems for integral operators with radial kernels,

$$T_k[f \, d\sigma](x) = \int_{\mathbf{R}^n} k(|x - y|) \, f(y) \, d\sigma(y).$$

Here $k = k(r)$, $r > 0$, is an arbitrary lower semicontinuous nonincreasing positive function. The corresponding nonlinear potential is defined by

$$\mathcal{W}_{k,\sigma}[\omega](x) = \int_0^{+\infty} k(r) \, \sigma(B_r(x)) \left(\int_{B_r(x)} \overline{k}(r)(y) \, d\omega(y) \right)^{p'-1} \frac{dr}{r},$$

where

$$\overline{k}(r)(x) = \frac{1}{\sigma(B_r(x))} \int_0^r k(s) \, \sigma(B_s(x)) \, \frac{ds}{s},$$

for $x \in \mathbf{R}^n$, $r > 0$.

THEOREM 3.4. *Let $0 < q < p < +\infty$, $1 < p < \infty$, and let ω and σ be nonnegative Borel measures on $\boldsymbol{R}^n$. Assume that σ satisfies a doubling condition, and the pair (k, σ) has the following logarithmic bounded oscillation property* (LBO)*:*

$$\sup_{y \in B_r(x)} \overline{k}(r)(y) \le A \inf_{y \in B_r(x)} \overline{k}(r)(y), \tag{3.10}$$

where A does not depend on $x \in \boldsymbol{R}^n$, $r > 0$. Then the trace inequality

$$\int_{\boldsymbol{R}^n} | T_k[f d\sigma] |^q \, d\omega \le C \, \|f\|^q_{L^p(d\sigma)}, \qquad f \in L^p(d\sigma), \tag{3.11}$$

holds if and only if $\mathcal{W}_{k,\sigma}[\omega] \in L^{\frac{q(p-1)}{p-q}}(d\omega)$.

The (LBO) property is satisfied by all radially nonincreasing kernels in the case $d\sigma = dx$, and also by Riesz kernels $k(x) = |x|^{\alpha - n}$, $0 < \alpha < n$, if σ satisfies a reverse doubling condition (see [**COV2**]).

In particular, the following theorem holds for convolution operators $T_k[f] = k \star f$ and $d\sigma = dx$. We define the corresponding Wolff's potential by

$$\mathcal{W}_k[\omega](x) = \int_0^{+\infty} k(r) \, \overline{k}(r)^{\frac{1}{p-1}} \omega(B_r(x))^{\frac{1}{p-1}} \, r^{n-1} \, dr, \tag{3.12}$$

where

$$\overline{k}(r)(x) = \frac{1}{r^n} \int_0^r k(s) \, s^{n-1} \, ds, \tag{3.13}$$

for $x \in \mathbf{R}^n$, $r > 0$.

THEOREM 3.5. *Let $0 < q < p < +\infty$ and $1 < p < \infty$. Let ω be a nonnegative Borel measure on $\boldsymbol{R}^n$. Suppose $k = k(|x-y|)$, where $k(r)$ is a lower semicontinuous nonincreasing positive function on $\boldsymbol{R}_+$, and $\mathcal{W}_k[\omega]$ is defined by (3.12). Then the trace inequality*

$$||k \star f||_{L^q(d\omega)} \le C \, \|f\|_{L^p(\boldsymbol{R}^n)}, \qquad f \in L^p(\boldsymbol{R}^n), \tag{3.14}$$

holds if and only if

$$\mathcal{W}_k[\omega] \in L^{\frac{q(p-1)}{p-q}}(\omega). \tag{3.15}$$

REMARK 3.6. For Riesz kernels, a proof of Theorem 3.5 was given in [**V4**]. Some technical details related to passing from a discrete to continuous version using shifts of the dyadic lattice, as well as generalizations, can be found in [**COV3**].

REMARK 3.7. A characterization of (3.14) for Riesz or Bessel kernels in terms of capacities was given in [**MN**] (see Sec. 2).

The special case $q = 1$ of Theorem 3.5 leads by duality to Wolff's inequality for radially nonincreasing kernels [**COV2**], [**COV3**].

THEOREM 3.8. *Let $1 < p < \infty$. Let ω be a nonnegative Borel measure on $\boldsymbol{R}^n$. Suppose $k = k(|x - y|)$, where $k(r)$ is a lower semicontinuous nonincreasing positive function on $\boldsymbol{R}_+$, and $\mathcal{W}_k[\omega]$ is defined by (3.12). Then there exist positive constants C_1, C_2 which depend only on k, p and n such that*

$$C_1 \, ||k \star \omega||^{p'}_{L^{p'}(\boldsymbol{R}^n)} \le \int_{\boldsymbol{R}^n} \mathcal{W}_k[\omega] \, d\omega \le C_2 \, ||k \star \omega||^{p'}_{L^{p'}(\boldsymbol{R}^n)}. \tag{3.16}$$

Theorem 3.8 demonstrates that (3.12) is an appropriate definition of Wolff's potential for radially nonincreasing kernels. This solves a problem posed in [**AH**], p. 214.

4. Indefinite weights and Schrödinger operators

We start with some prerequisites for our main results. Let $D(\mathbf{R}^n) = C_0^\infty(\mathbf{R}^n)$ be the class of infinitely differentiable, compactly supported complex-valued functions, and let $D'(\mathbf{R}^n)$ denote the corresponding space of (complex-valued) distributions. By $L^{1,2}(\mathbf{R}^n)$ we denote the completion of $D(\mathbf{R}^n)$ in the Dirichlet norm $||\nabla u||_{L_2(\mathbf{R}^n)}$. It follows from Hardy's inequality that an equivalent norm on $L^{1,2}(\mathbf{R}^n)$ $(n \geq 3)$ is given by

$$||u||_{L^{1,2}(\mathbf{R}^n)} = \left[\int_{\mathbf{R}^n} (|x|^{-2}\,|u(x)|^2 + |\nabla u(x)|^2)\, dx\right]^{\frac{1}{2}}.$$

In this section, we assume that $n \geq 3$, since for the homogeneous space $L^{1,2}(\mathbf{R}^n)$ our results become vacuous if $n = 1$ and $n = 2$. Analogous results for inhomogeneous Sobolev spaces $W^{1,2}(\mathbf{R}^n)$ hold for all $n \geq 1$.

For $V \in D'(\mathbf{R}^n)$, consider the multiplier operator on $D(\mathbf{R}^n)$ defined by

$$< V\,u, v >:=< V, \bar{u}\,v >, \quad u,\, v \in D(\mathbf{R}^n), \tag{4.1}$$

where $< \cdot, \cdot >$ represents the usual pairing between $D(\mathbf{R}^n)$ and $D'(\mathbf{R}^n)$.

Let us denote by $L^{-1,2}(\mathbf{R}^n) = L^{1,2}(\mathbf{R}^n)^*$ the dual Sobolev space. If the sesquilinear form $< V\,\cdot, \cdot >$ is bounded on $L^{1,2}(\mathbf{R}^n) \times L^{1,2}(\mathbf{R}^n)$:

$$|< V\,u,\, v >| \leq c\,||\nabla u||_{L^2(\mathbf{R}^n)}\,||\nabla v||_{L^2(\mathbf{R}^n)}, \tag{4.2}$$

where the constant c is independent of $u,\, v \in D(\mathbf{R}^n)$, then $V\,u \in L^{-1,2}(\mathbf{R}^n)$, and the multiplier operator can be extended by continuity to all of the energy space $L^{1,2}(\mathbf{R}^n)$. (As usual, this extension is also denoted by V.)

We denote the class of multipliers $V :\; L^{1,2}(\mathbf{R}^n) \to L^{-1,2}(\mathbf{R}^n)$ by

$$M(L^{1,2}(\mathbf{R}^n) \to L^{-1,2}(\mathbf{R}^n)).$$

Note that the least constant c in (4.2) is equal to the norm of the multiplier operator:

$$||V||_{M(L^{1,2}(\mathbf{R}^n)\to L^{-1,2}(\mathbf{R}^n))} = \sup\,\{\,||Vu||_{L^{-1,2}(\mathbf{R}^n)} \;:\; ||u||_{L^{1,2}(\mathbf{R}^n)} \leq 1\}.$$

For $V \in M(L^{1,2}(\mathbf{R}^n) \to L^{-1,2}(\mathbf{R}^n))$, we will need to extend the form $< V, \bar{u}v >$ defined by the right-hand side of (4.1) to the case where both u and v are in $L^{1,2}(\mathbf{R}^n)$. This can be done by letting

$$< V\,u,\, v >= \lim_{N\to\infty} < V\,u_N,\, v_N >,$$

where $u = \lim_{N\to\infty} u_N$, and $v = \lim_{N\to\infty} v_N$, with $u_N, v_N \in D(\mathbf{R}^n)$. It is known that this extension is independent of the choice of u_N and v_N.

We now define the Schrödinger operator $H = H_0 + V$, where $H_0 = -\Delta$, on the energy space $L^{1,2}(\mathbf{R}^n)$. Since $H_0 :\; L^{1,2}(\mathbf{R}^n) \to L^{-1,2}(\mathbf{R}^n)$ is bounded, it follows that H is a bounded operator acting from $L^{1,2}(\mathbf{R}^n)$ to $L^{-1,2}(\mathbf{R}^n)$ if and only if $V \in M(L^{1,2}(\mathbf{R}^n) \to L^{-1,2}(\mathbf{R}^n))$. Clearly, (4.2) is equivalent to the boundedness of the corresponding quadratic form:

$$|< V\,u,\, u >| = |< V,\, |u|^2 >| \leq c\,||\nabla u||^2_{L^2(\mathbf{R}^n)},$$

where the constant c is independent of $u \in D(\mathbf{R}^n)$. If V is a (complex-valued) measure on $\mathbf{R}^n$, then this inequality can be recast in the form:

$$\left| \int_{\mathbf{R}^n} |u(x)|^2 \, V \right| \le c \, ||\nabla u||^2_{L^2(\mathbf{R}^n)}, \quad u \in D(\mathbf{R}^n). \tag{4.3}$$

For *positive* distributions (measures) V, this inequality is well studied (see Sections 1 and 2).

We now state the main result for arbitrary (complex-valued) distributions V. By $L^2_{\text{loc}}(\mathbf{R}^n)^n = L^2_{\text{loc}}(\mathbf{R}^n) \otimes \mathbf{C}^n$ we denote the space of vector-functions $\vec{\Gamma} = (\Gamma_1, \ldots, \Gamma_n)$ such that $\Gamma_i \in L^2_{\text{loc}}(\mathbf{R}^n)$, $i = 1, \ldots, n$.

THEOREM 4.1. *Let $V \in D'(\boldsymbol{R}^n)$. Then $V \in M(L^{1,2}(\boldsymbol{R}^n) \to L^{-1,2}(\boldsymbol{R}^n))$, i.e., the inequality*

$$| < V u, \, v > | \le c \, ||u||_{L^{1,2}(\boldsymbol{R}^n)} \, ||v||_{L^{1,2}(\boldsymbol{R}^n)} \tag{4.4}$$

holds for all $u, v \in L^{1,2}(\boldsymbol{R}^n)$, if and only if there is a vector-field $\vec{\Gamma} \in \mathbf{L}^2_{loc}(\boldsymbol{R}^n)$ such that $V = \operatorname{div} \vec{\Gamma}$, and

$$\int_{\boldsymbol{R}^n} |u(x)|^2 \, |\vec{\Gamma}(x)|^2 \, dx \le C \int_{\boldsymbol{R}^n} |\nabla u(x)|^2 \, dx, \tag{4.5}$$

for all $u \in D(\boldsymbol{R}^n)$. The vector-field $\vec{\Gamma}$ can be chosen in the form $\vec{\Gamma} = \nabla \, \Delta^{-1} V$.

REMARK 4.2. For $\vec{\Gamma} = \nabla \, \Delta^{-1} V$, the least constant C in the inequality (4.5) is equivalent to $||V||^2_{M(L^{1,2}(\mathbf{R}^n) \to L^{-1,2}(\mathbf{R}^n))}$.

The proof of the "if" part of Theorem 4.1 is easy as long as V is represented in the divergence form. This idea was discovered by mathematical physicists in the 1970s (see [**MV2**], p. 265). Indeed, suppose that $V = \operatorname{div} \vec{\Gamma}$, where $\vec{\Gamma}$ satisfies (4.5). Then using integration by parts and the Cauchy-Schwartz inequality we obtain:

$$\begin{aligned} | < V u, \, v > | &= | < V, \, \bar{u} \, v > | = | < \vec{\Gamma}, \, v \, \nabla \bar{u} > + < \vec{\Gamma}, \, \bar{u} \, \nabla v > | \\ &\le ||\vec{\Gamma} v||_{L^2(\mathbf{R}^n)^n} \, ||\nabla u||_{L^2(\mathbf{R}^n)} + ||\vec{\Gamma} u||_{L^2(\mathbf{R}^n)} \, ||\nabla v||_{L^2(\mathbf{R}^n)} \\ &\le 2\sqrt{C} \, ||\nabla u||_{L^2(\mathbf{R}^n)} \, ||\nabla v||_{L^2(\mathbf{R}^n)}, \end{aligned}$$

where C is the constant in (4.5).

The proof of the "only if" part of Theorem 4.1 given in [**MV2**] is much more complicated. It is based on a combination of methods of harmonic analysis and potential theory. We outline the main ideas in a series of lemmas and propositions stated below.

In the first lemma, which is only a preliminary estimate, it is shown that $\vec{\Gamma} = \nabla \, \Delta^{-1} V \in L^2_{\text{loc}}(\mathbf{R}^n)^n$, and a crude estimate of the average decay of $\vec{\Gamma}$ at infinity is given. We observe that expressions like $\nabla \, \Delta^{-1} V$ should be understood in a specific sense, e.g., in the sense of weak BMO convergence. The latter is discussed in detail in [**MV6**].

LEMMA 4.3. *Suppose that*

$$V \in M(L^{1,2}(\boldsymbol{R}^n) \to L^{-1,2}(\boldsymbol{R}^n)). \tag{4.6}$$

Then $\vec{\Gamma} = \nabla\,\Delta^{-1}V \in L^2_{loc}(\boldsymbol{R}^n)^n$, *and* $V = \operatorname{div}\vec{\Gamma}$ *in* D'. *Moreover, for any ball* $B_R(x_0)$ $(R>0)$ *and* $\epsilon > 0$,

$$\int_{B_R(x_0)} |\vec{\Gamma}(x)|^2\,dx \le C(n,\epsilon)\,R^{n-2+\epsilon}\,||V||^2_{M(L^{1,2}(\boldsymbol{R}^n)\to L^{-1,2}(\boldsymbol{R}^n))}, \tag{4.7}$$

where $R \ge \max\{1, |x_0|\}$.

The following statements are concerned with sharp estimates of equilibrium potentials associated with a set of positive capacity.

PROPOSITION 4.4. *Let* $\delta > \frac{1}{2}$ *and let* $P = P_e$ *be the equilibrium Newtonian potential of a compact set* $e \subset \boldsymbol{R}^n$ *of positive capacity. Then*

$$||\nabla P^\delta||_{L^2(\boldsymbol{R}^n)} = \frac{\delta}{\sqrt{2\delta - 1}}\,\sqrt{\operatorname{cap}(e)}. \tag{4.8}$$

REMARK 4.5. For $\delta \le \frac{1}{2}$, it is easy to see that $\nabla P^\delta \notin L^2(\mathbf{R}^n)$.

PROPOSITION 4.6. *Let* $\delta > 0$, *and let* v *be a real-valued function such that* $v \in L^{1,2}(\boldsymbol{R}^n)$. *Then*

$$||\nabla v||^2_{L^2(\boldsymbol{R}^n)} \le \int_{\boldsymbol{R}^n} |\nabla(v\,P^\delta)(x)|^2\,\frac{dx}{P(x)^{2\delta}} \le (\delta+1)(4\delta+1)\,||\nabla v||^2_{L^2(\boldsymbol{R}^n)}. \tag{4.9}$$

The proof of the preceding inequalities is based on multiple integration by parts, along with the following estimates for Newtonian potentials of positive (not necessarily equilibrium) measures ω.

PROPOSITION 4.7. *Let* ω *be a positive Borel measure on* $\boldsymbol{R}^n$ *such that* $P(x) = I_2\omega(x) \not\equiv \infty$. *Then the following inequalities hold:*

$$\int_{\boldsymbol{R}^n} v(x)^2\,\frac{|\nabla P(x)|^2}{P(x)^2}\,dx \le 4\,||\nabla v||^2_{L^2(\boldsymbol{R}^n)}, \qquad v \in D(\boldsymbol{R}^n), \tag{4.10}$$

and

$$\int_{\boldsymbol{R}^n} v(x)^2\,\frac{d\omega(x)}{P(x)} \le ||\nabla v||^2_{L^2(\boldsymbol{R}^n)}, \qquad v \in D(\boldsymbol{R}^n). \tag{4.11}$$

REMARK 4.8. The constants 4 and 1 respectively in (4.10) and (4.11) are sharp (see [**MV2**]).

REMARK 4.9. An inequality more general than (4.11), for Riesz potentials of order $\alpha \in (0,n)$ and L_p norms (with nonlinear Wolff's potential in place of $P(x)$), but with a different constant was proved in [**V4**].

We will also need the following proposition which is deduced from the facts that $P(x)^{2\delta}$ is an A_2 weight (this was proved earlier in [**MV1**]), and that the Riesz transforms are bounded in weighted L^2 spaces with such weights [**St**]. (See details in [**MV2**].)

PROPOSITION 4.10. *Let* $w = \Delta^{-1}\operatorname{div}\vec{\phi}$ *where* $\vec{\phi} \in D \otimes \boldsymbol{C}^n$. *Suppose that* $1 < 2\delta < \frac{n}{n-2}$. *Then*

$$\int_{\boldsymbol{R}^n} |\nabla w(x)|^2\,\frac{dx}{P(x)^{2\delta}} \le C(n,\delta)\int_{\boldsymbol{R}^n} |\vec{\phi}(x)|^2\,\frac{dx}{P(x)^{2\delta}}. \tag{4.12}$$

We now sketch the proof of the "only if" part of Theorem 4.1. Suppose that

$$|<V, u\,v>| \le ||V||_{M(L^{1,2}(\mathbf{R}^n)\to L_2^{-1}(\mathbf{R}^n))}\, ||\nabla u||_{L^2(\mathbf{R}^n)}\, ||\nabla v||_{L^2(\mathbf{R}^n)}.$$

Let $\vec{\phi} = (\phi_1, \ldots, \phi_n)$ be an arbitrary vector-field in $\mathcal{D}\otimes\mathbf{C}^n$, and let

$$w = \Delta^{-1}\operatorname{div}\vec{\phi} = -I_2 \operatorname{div}\vec{\phi}, \tag{4.13}$$

so that

$$\vec{\phi} = \nabla w + \vec{s}, \qquad \operatorname{div}\vec{s} = 0.$$

Note that $w \in C^\infty(\mathbf{R}^n)\cap L^{1,2}(\mathbf{R}^n)$, since

$$w(x) = O(|x|^{1-n}) \quad \text{and} \quad |\nabla w(x)| = O\left(|x|^{-n}\right) \quad \text{as} \quad |x|\to\infty.$$

Hence,

$$|<V,\, w>| = |<\vec{\Gamma},\, \phi>| = \left|\int_{\mathbf{R}^n} \vec{\Gamma}(x)\cdot\vec{\phi}(x)\,dx\right|$$

where by Lemma 4.3, $\vec{\Gamma} \in L^2_{\mathrm{loc}}(\mathbf{R}^n)^n$. We pick δ so that $1 < 2\delta < \frac{n}{n-2}$, and factorize $w(x) = u(x)\,v(x)$, where

$$u(x) = P(x)^\delta, \qquad v = \frac{w(x)}{P(x)^\delta}.$$

Consequently,

$$|<\vec{\Gamma}, \vec{\phi}>| \le ||V||_{M(L^{1,2}(\mathbf{R}^n)\to L_2^{-1}(\mathbf{R}^n))}\, ||\nabla P^\delta||_{L^2(\mathbf{R}^n)}\, ||\nabla v||_{L^2(\mathbf{R}^n)}.$$

By Proposition 4.6

$$||\nabla v||^2_{L_2(\mathbf{R}^n)} \le \int_{\mathbf{R}^n} |\nabla(vP^\delta)(x)|^2\, \frac{dx}{P(x)^{2\delta}} = \int_{\mathbf{R}^n} |\nabla w(x)|^2\, \frac{dx}{P(x)^{2\delta}} < \infty.$$

From this, applying Proposition 4.4, we obtain:

$$\left|\int_{\mathbf{R}^n} \vec{\Gamma}(x)\cdot\vec{\phi}(x)\,dx\right| \le \delta\,(1-2\delta)^{-\frac{1}{2}}\, ||V||_{M(L^{1,2}(\mathbf{R}^n)\to L_2^{-1}(\mathbf{R}^n))}$$
$$\times\, \operatorname{cap}(e)^{\frac{1}{2}} \left(\int_{\mathbf{R}^n} |\nabla w(x)|^2\, \frac{dx}{P(x)^{2\delta}}\right)^{\frac{1}{2}}.$$

Notice that by Proposition 4.10,

$$\int_{\mathbf{R}^n} |\nabla w(x)|^2\, \frac{dx}{P(x)^{2\delta}} \le C(n,\delta) \int_{\mathbf{R}^n} |\vec{\phi}(x)|^2\, \frac{dx}{P(x)^{2\delta}}.$$

Hence,

$$\left|\int_{\mathbf{R}^n} \vec{\Gamma}(x)\cdot\vec{\phi}(x)\,dx\right| \le C(n,\delta)\, ||V||_{M(L^{1,2}(\mathbf{R}^n)\to L_2^{-1}(\mathbf{R}^n))}$$
$$\times\, \operatorname{cap}(e)^{\frac{1}{2}} \left(\int_{\mathbf{R}^n} |\vec{\phi}(x)|^2\, \frac{dx}{P(x)^{2\delta}}\right)^{\frac{1}{2}}.$$

From the preceding inequality we deduce

$$\left(\int_{\mathbf{R}^n} |\vec{\Gamma}(x)|^2\, P(x)^{2\delta}(x)\,dx\right)^{\frac{1}{2}} \le C(n,\delta)\, ||V||_{(M(L^{1,2}(\mathbf{R}^n)\to L^{-1,2}(\mathbf{R}^n))}\, \operatorname{cap}(e)^{\frac{1}{2}}.$$

Note that P is the equilibrium potential of e, and hence $P(x) \ge 1$ dx-a.e. on e. Thus,

$$\int_e |\vec{\Gamma}(x)|^2\,dx \le C(n,\delta)\operatorname{cap}(e)$$

for every compact set $e \subset \mathbf{R}^n$, and by Theorem 2.2 this gives (4.5), which completes the proof of Theorem 4.1.

There is an analogue of Theorem 4.1 in terms of $(-\Delta)^{-1/2}V$.

THEOREM 4.11. *Under the assumptions of Theorem 4.1, it follows that*

$$V \in M(L^{1,2}(\boldsymbol{R}^n) \to L^{-1,2}(\boldsymbol{R}^n))$$

if and only if $(-\Delta)^{-1/2}V \in M(L^{1,2}(\boldsymbol{R}^n) \to L^2(\boldsymbol{R}^n))$.

5. Infinitesimal form boundedness, Trudinger's and Nash's inequalities

Throughout this section we will be using the following notation and conventions. We denote by $W^{1,2}(\mathbf{R}^n)$ the Sobolev space of weakly differentiable functions on $\mathbf{R}^n$ $(n \geq 1)$ such that

$$||u||_{W^{1,2}(\mathbf{R}^n)} = ||u||_{L^2(\mathbf{R}^n)} + ||\nabla u||_{L^2(\mathbf{R}^n)} < +\infty,$$

and by $W^{-1,2}(\mathbf{R}^n) = W^{1,2}(\mathbf{R}^n)^*$ the dual Sobolev space. For a compact set $e \subset \mathbf{R}^n$, the capacity associated with $W^{1,2}(\mathbf{R}^n)$ is defined by

$$\operatorname{cap}(e) = \inf \left\{ ||u||^2_{W^{1,2}(\mathbf{R}^n)} : \quad u \in C_0^\infty(\mathbf{R}^n), \quad u(x) > 1 \text{ on } e \right\}.$$

For $0 < r < \infty$, we denote by $L^r_{\text{unif}}(\mathbf{R}^n)$ all $f \in L^r_{\text{loc}}(\mathbf{R}^n)$ such that

$$||f||_{L^r_{\text{unif}}} = \sup_{x_0 \in \mathbf{R}^n} ||\chi_{B_1(x_0)} f||_{L^r(\mathbf{R}^n)} < \infty.$$

By $L^r(\mathbf{R}^n)^n = L^r(\mathbf{R}^n) \otimes \mathbf{C}^n$ we denote the class of vector fields $\vec{\Gamma} = \{\Gamma_j\}_{j=1}^n :$ $\mathbf{R}^n \to \mathbf{C}^n$, such that $\Gamma_j \in L^r(\mathbf{R}^n)$, $j = 1, 2, \ldots, n$, and use similar notation for other vector-valued function spaces.

By $M^+(\mathbf{R}^n)$ we denote the class of nonnegative locally finite Borel measures on $\mathbf{R}^n$. If $V \in \mathcal{D}'(\mathbf{R}^n)$ is nonnegative, i.e., coincides with $\omega \in M^+(\mathbf{R}^n)$, we write $\int_{\mathbf{R}^n} |u(x)|^2 \, d\omega$ in place of $\langle V, |u|^2 \rangle = \langle Vu, u \rangle$ for the quadratic form associated with the distribution V, if $u \in C_0^\infty(\mathbf{R}^n)$. Sometimes we will use $\int_{\mathbf{R}^n} |u(x)|^2 \, V(x) \, dx$ in place of $\langle Vu, u \rangle$ even if V is not in $L^1_{\text{loc}}(\mathbf{R}^n)$.

We set

$$m_B(f) = \frac{1}{|B|} \int_B f(x) \, dx$$

for a ball $B \subset \mathbf{R}^n$, and denote by $\text{BMO}(\mathbf{R}^n)$ the class of $f \in L^r_{\text{loc}}(\mathbf{R}^n)$ for which

$$\sup_{x_0 \in \mathbf{R}^n, \, \delta > 0} \frac{1}{|B_\delta(x_0)|} \int_{B_\delta(x_0)} |f(x) - m_{B_\delta(x_0)}(f)|^r \, dx < +\infty,$$

for any $1 \leq r < +\infty$. It follows from the John–Nirenberg inequality that this definition does not depend on the choice of $r \in [1, +\infty)$.

We will also need an inhomogeneous version of $\text{BMO}(\mathbf{R}^n)$ (the so-called local BMO) which we denote by $\text{bmo}(\mathbf{R}^n)$. It can be defined in a similar way as the set of $f \in L^r_{\text{unif}}(\mathbf{R}^n)$ such that the preceding condition holds for $0 < \delta \leq 1$ (see [**St**], p. 264).

The Morrey space $\mathcal{L}^{r,\lambda}(\mathbf{R}^n)$ $(r > 0, \lambda > 0)$ consists of $f \in L^r_{\text{loc}}(\mathbf{R}^n)$ such that

$$\sup_{x_0 \in \mathbf{R}^n, \, \delta > 0} \frac{1}{|B_\delta(x_0)|^{\frac{\lambda}{n}}} \int_{B_\delta(x_0)} |f|^r \, dx < +\infty.$$

In the corresponding inhomogeneous analogue, we set $0 < \delta \le 1$ in the preceding inequality. It will be clear from the context which version of the Morrey space is used.

We now state the main results of [**MV5**].

THEOREM 5.1. *Let $V \in \mathcal{D}'(\boldsymbol{R}^n)$, $n \ge 2$. Then the following statements hold.*

(i) Suppose that V is represented in the form:

$$V = \operatorname{div} \vec{\Gamma} + \gamma, \tag{5.1}$$

where $\vec{\Gamma} \in L^2_{\text{loc}}(\boldsymbol{R}^n)^n$ and $\gamma \in L^1_{\text{loc}}(\boldsymbol{R}^n)$ satisfy respectively the conditions:

$$\lim_{\delta \to +0} \sup_{x_0 \in \boldsymbol{R}^n} \sup_{u} \frac{\int_{B_\delta(x_0)} |\vec{\Gamma}(x)|^2 \, |u(x)|^2 \, dx}{||\nabla u||^2_{L^2(B_\delta(x_0))}} = 0, \tag{5.2}$$

$$\lim_{\delta \to +0} \sup_{x_0 \in \boldsymbol{R}^n} \sup_{u} \frac{\int_{B_\delta(x_0)} |\gamma(x)| \, |u(x)|^2 \, dx}{||\nabla u||^2_{L^2(B_\delta(x_0))}} = 0, \tag{5.3}$$

where $u \in C_0^\infty(B_\delta(x_0))$, $u \not\equiv 0$. Then V is infinitesimally form bounded with respect to $-\Delta$, i.e., for every $\epsilon > 0$ there exists $C(\epsilon) > 0$ such that (1.11) holds.

(ii) Conversely, suppose V is infinitesimally form bounded with respect to $-\Delta$. Then V can be represented in the form (5.1) so that both (5.2) and (5.3) hold. Moreover, one can set $\vec{\Gamma} = -\nabla(1-\Delta)^{-1}V$ and $\gamma = (1-\Delta)^{-1}V$ in the representation (5.1).

REMARK 5.2. In the statement of Theorem 5.1 one can replace conditions (5.2) and (5.3) with the equivalent condition where $|(1-\Delta)^{-\frac{1}{2}}V|^2$ is used in place of $|\vec{\Gamma}|^2$ in (5.2).

The importance of Theorem 5.1 is in the means it provides for deducing explicit criteria of the infinitesimal form boundedness in terms of the *nonnegative* locally integrable functions $|\vec{\Gamma}|^2$ and $|\gamma|$.

THEOREM 5.3. *Let $V \in \mathcal{D}'(\boldsymbol{R}^n)$, $n \ge 2$. The following statements are equivalent:*

(i) V is infinitesimally form bounded with respect to $-\Delta$.

(ii) V has the form (5.1) where $\vec{\Gamma} = -\nabla(1-\Delta)^{-1}V$, $\gamma = (1-\Delta)^{-1}V$, and the measure $\omega \in M^+(\boldsymbol{R}^n)$ defined by

$$d\omega = \left(|\vec{\Gamma}(x)|^2 + |\gamma(x)|\right) dx \tag{5.4}$$

has the property that, for every $\epsilon > 0$, there exists $C(\epsilon) > 0$ such that

$$\int_{\boldsymbol{R}^n} |u(x)|^2 \, d\omega \le \epsilon \, ||\nabla u||^2_{L^2(\boldsymbol{R}^n)} + C(\epsilon) \, ||\nabla u||^2_{L^2(\boldsymbol{R}^n)}, \quad \forall u \in C_0^\infty(\boldsymbol{R}^n). \tag{5.5}$$

(iii) For ω defined by (5.4),

$$\lim_{\delta \to +0} \sup_{P_0 :\, \operatorname{diam} P_0 \le \delta} \frac{1}{\omega(P_0)} \sum_{P \subseteq P_0} \frac{\omega(P)^2}{|P|^{1-\frac{2}{n}}} = 0, \tag{5.6}$$

where P, P_0 are dyadic cubes in $\boldsymbol{R}^n$, i.e., sets of the form $2^i(k + [0, 1)^n)$, where $i \in \boldsymbol{Z}$, $k \in \boldsymbol{Z}^n$.

(iv) For ω *defined by* (5.4),

$$\lim_{\delta\to+0}\ \sup_{e:\,\mathrm{diam}\,e\le\delta} \frac{\omega(e)}{\mathrm{cap}\,(e)} = 0, \tag{5.7}$$

where e *denotes a compact set of positive capacity in* $\boldsymbol{R}^n$.

(v) For ω *defined by* (5.4),

$$\lim_{\delta\to+0}\ \sup_{x_0\in\boldsymbol{R}^n} \frac{\left\|\omega_{B_\delta(x_0)}\right\|^2_{W^{-1,\,2}(\boldsymbol{R}^n)}}{\omega(B_\delta(x_0))} = 0, \tag{5.8}$$

where $\omega_{B_\delta(x_0)}$ *is the restriction of* ω *to the ball* $B_\delta(x_0)$.

(vi) For ω *defined by* (5.4),

$$\lim_{\delta\to+0}\ \sup_{x,\,x_0\in\boldsymbol{R}^n} \frac{G_1\star\left(G_1\star\omega_{B_\delta(x_0)}\right)^2(x)}{G_1\star\omega_{B_\delta(x_0)}(x)} = 0, \tag{5.9}$$

where $G_1\star\omega = (1-\Delta)^{-\frac{1}{2}}\omega$ *is the Bessel potential of order* 1.

It is worth noting that although Theorem 5.3 holds in the two-dimensional case, its proof requires certain modifications in comparison to $n \ge 3$. In the one-dimensional case, the infinitesimal form boundedness of the Sturm-Liouville operator $H = -\frac{d^2}{dx^2} + V$ on $L^2(\mathbf{R}^1)$ is actually a consequence of the form boundedness.

THEOREM 5.4. *Let* $V \in \mathcal{D}'(\boldsymbol{R}^1)$. *Then the following statements are equivalent.*

(i) V *is infinitesimally form bounded with respect to* $-\frac{d^2}{dx^2}$.

(ii) V *is form bounded with respect to* $-\frac{d^2}{dx^2}$, *i.e.,*

$$|\langle V\,u,\,u\rangle| \le \mathrm{const}\,||u||^2_{W^{1,\,2}(\boldsymbol{R}^1)}, \quad \forall u\in C_0^\infty(\boldsymbol{R}^1).$$

(iii) V *can be represented in the form* $V = \frac{d\Gamma}{dx} + \gamma$, *where*

$$\sup_{x\in\boldsymbol{R}^1}\int_x^{x+1}\left(|\Gamma(x)|^2 + |\gamma(x)|\right)dx < +\infty. \tag{5.10}$$

(iv) Condition (5.10) *holds where*

$$\Gamma(x) = \int_{\boldsymbol{R}^1}\mathrm{sign}\,(x-t)\,e^{-|x-t|}\,V(t)\,dt, \qquad \gamma(x) = \int_{\boldsymbol{R}^1} e^{-|x-t|}\,V(t)\,dt$$

are understood in the distributional sense.

The statement (iii)⇒(i) in Theorem 5.4 is known ([**Sch**], Theorem 11.2.1), whereas (ii)⇒(iv) follows from [**MV2**].

We now state a characterization of the form subordination property (1.15). It was formulated originally in [**Tru**], in the form of the inequality:

$$|\langle Vu,\,u\rangle| \le \epsilon\,||\nabla u||^2_{L^2(\mathbf{R}^n)} + c\,\epsilon^{-\nu}\,||u||^2_{L^1(\mathbf{R}^n)}, \quad \forall u\in C_0^\infty(\mathbf{R}^n), \tag{5.11}$$

for $V \ge 0$. Such V are called ϵ^ν*-compactly bounded* in [**Tru**]. It follows from Nash's inequality that (1.15) yields (5.11) with $\nu = \frac{n+2}{2}\beta + \frac{n}{2}$; the converse is also true, provided $\nu > \frac{n}{2}$, and is deduced using a localization argument. In the critical case $\nu = \frac{n}{2}$, (5.11) holds if and only if $V \in L^\infty(\mathbf{R}^n)$, while for $0 < \nu < \frac{n}{2}$, it holds only if $V = 0$.

Necessary and sufficient conditions for (1.15), or equivalently (5.11) with $\nu = \frac{n+2}{2}\beta + \frac{n}{2}$ (see [**MV5**]), can be formulated in terms of Morrey-Campanato spaces using mean oscillations of the functions $\vec{\Gamma}$ and γ which have appeared in Theorems 5.1–5.4.

THEOREM 5.5. *Let $V \in \mathcal{D}'(\boldsymbol{R}^n)$, $n \geq 2$, and let $0 < \beta < +\infty$.*

(i) Suppose there exists $\epsilon_0 > 0$ such that (1.15) *holds for every $\epsilon \in (0, \epsilon_0)$. Then V can be represented in the form*

$$V = \operatorname{div} \vec{\Gamma} + \gamma, \tag{5.12}$$

where $\vec{\Gamma} = -\nabla(1-\Delta)^{-1}V \in L^2_{\text{loc}}(\boldsymbol{R}^n)^n$ and $\gamma = (1-\Delta)^{-1}V \in L^1_{\text{loc}}(\boldsymbol{R}^n)$.

Moreover, there exists $\delta_0 > 0$ such that

$$\int_{B_\delta(x_0)} |\vec{\Gamma}(x) - m_{B_\delta(x_0)}(\vec{\Gamma})|^2 \, dx \leq c \, \delta^{n-2\frac{\beta-1}{\beta+1}}, \quad 0 < \delta < \delta_0, \tag{5.13}$$

$$\int_{B_\delta(x_0)} |\gamma(x)| \, dx \leq c \, \delta^{n-\frac{2\beta}{\beta+1}}, \quad 0 < \delta < \delta_0, \tag{5.14}$$

where c does not depend on $x_0 \in \boldsymbol{R}^n$ and δ_0. Furthermore, $\vec{\Gamma} \in L^2_{\text{unif}}(\boldsymbol{R}^n)^n$ if $\beta \geq 1$, and $\vec{\Gamma} \in L^\infty(\boldsymbol{R}^n)^n$ if $0 < \beta < 1$.

(ii) Conversely, if V is given by (5.12) *where $\vec{\Gamma} \in L^2_{\text{loc}}(\boldsymbol{R}^n)^n$, $\gamma \in L^1_{\text{loc}}(\boldsymbol{R}^n)$ satisfy* (5.13), (5.14) *for all $0 < \delta < \delta_0$ then there exists $\epsilon_0 > 0$ such that* (1.15) *holds for all $0 < \epsilon < \epsilon_0$.*

REMARK 5.6. (a) In the case $\beta = 1$, it follows that (5.13) holds if and only if $\vec{\Gamma} \in \text{bmo}(\mathbf{R}^n)^n$. In other words, $V \in \text{bmo}_{-1}(\mathbf{R}^n)$, where $\text{bmo}_{-1}(\mathbf{R}^n)$ can be defined as the space of distributions f that can be represented in the form $f = \operatorname{div} \vec{g}$ where $\vec{g} \in \text{bmo}(\mathbf{R}^n)^n$. We observe that $\text{bmo}_{-1}(\mathbf{R}^n) = F^{2,\infty}_{-1}(\mathbf{R}^n)$, where $F^{p,q}_\alpha$ stands for the scale of inhomogeneous Triebel–Lizorkin spaces (see, e.g., [**KT**], [**T**]).

(b) In the case $0 < \beta < 1$, (5.13) holds if and only if $\vec{\Gamma}$ is Hölder-continuous:

$$|\vec{\Gamma}(x) - \vec{\Gamma}(x')| \leq c \, |x - x'|^{\frac{1-\beta}{\beta+1}}, \qquad |x - x'| < \delta_0.$$

For $\beta > 1$, (5.13) holds if and only if $\vec{\Gamma}$ lies in the inhomogeneous Morrey space $\mathcal{L}^{2, \, n-2\frac{\beta-1}{\beta+1}}(\mathbf{R}^n)^n$, i.e.,

$$\int_{B_\delta(x_0)} |\vec{\Gamma}(x)|^2 \, dx \leq c \, \delta^{n-2\frac{\beta-1}{\beta+1}}, \qquad 0 < \delta < \delta_0.$$

These statements follow from the known characterizations of Morrey spaces. Note that, according to (5.14), $\gamma \in \mathcal{L}^{1, \, n-\frac{2\beta}{\beta+1}}(\mathbf{R}^n)$.

REMARK 5.7. (a) An immediate consequence of Theorem 5.5 is that, for all $\beta > 0$, (1.15) is equivalent to the following localized energy condition:

$$||(1-\Delta)^{-\frac{1}{2}}(\eta_{\delta, \, x_0} V)||^2_{L^2(B_\delta(x_0))} \leq c \, \delta^{n-2\frac{\beta-1}{\beta+1}}, \quad 0 < \delta < \delta_0, \; x_0 \in \mathbf{R}^n,$$

where $\eta_{\delta, \, x_0}(x) = \eta\left(\delta^{-1}(x - x_0)\right)$; here η is a smooth cut-off function such that $\eta \in C^\infty(B_1(0))$, $0 \leq \eta \leq 1$, and $\eta = 1$ on $B_{\frac{1}{2}}(0)$.

(b) A similar energy condition,

$$||(1-\Delta)^{-\frac{1}{2}}(\eta_{\delta, \, x_0} V)||^2_{L^2(\mathbf{R}^n)} \leq c \, \delta^{n-2\frac{\beta-1}{\beta+1}}, \quad 0 < \delta < \delta_0, \; x_0 \in \mathbf{R}^n,$$

is sufficient, but generally *not necessary* in the case $n = 2$.

We next state a criterion for the multiplicative inequality (1.18) to hold, which is equivalent to a homogeneous version of (1.15) with $\epsilon_0 = +\infty$ and $p = \frac{\beta}{\beta+1}$.

THEOREM 5.8. *Let $V \in \mathcal{D}'(\boldsymbol{R}^n)$, $n \geq 2$, and let $0 < p < 1$.*

(i) Suppose that (1.18) *holds. Then V can be represented in the form*

$$V = \operatorname{div} \vec{\Gamma}, \tag{5.15}$$

where $\vec{\Gamma} = \nabla \Delta^{-1} V$, and one of the following conditions hold:

$$\vec{\Gamma} \in \mathrm{BMO}(\boldsymbol{R}^n)^n \quad \textit{if} \quad p = \tfrac{1}{2}; \quad \vec{\Gamma} \in \mathrm{Lip}_{1-2p}(\boldsymbol{R}^n)^n \quad \textit{if} \quad 0 < p < \tfrac{1}{2}; \tag{5.16}$$

$$\int_{B_\delta(x_0)} |\vec{\Gamma}(x)|^2 \, dx \leq c \, \delta^{n+2-4p}, \qquad \textit{if} \ \tfrac{1}{2} < p < 1; \tag{5.17}$$

where c does not depend on $x_0 \in \boldsymbol{R}^n$, $\delta > 0$.

(ii) Conversely, if V is given by (5.15) *where $\vec{\Gamma} \in L^2_{\mathrm{loc}}(\boldsymbol{R}^n)^n$ and satisfies* (5.16), (5.17), *then* (1.18) *holds.*

REMARK 5.9. In Theorem 5.8, the "antiderivative" $\vec{\Gamma} = \nabla \Delta^{-1} V$ can be replaced with $(-\Delta)^{-\frac{1}{2}} V$. Furthermore, as a corollary we deduce that (1.18) holds if and only if $V \in \mathrm{BMO}_{-1}(\mathbf{R}^n) = \dot{F}^{2,\infty}_{-1}(\mathbf{R}^n)$ for $p = \frac{1}{2}$, where $\mathrm{BMO}_{-1}(\mathbf{R}^n)$ is defined as above in the case of its inhomogeneous counterpart $\mathrm{bmo}_{-1}(\mathbf{R}^n)$. (See, e.g., [**KT**] where this space is thoroughly studied in the context of Navier–Stokes equations.) For $0 < p < \frac{1}{2}$, (1.18) holds if and only if $V \in \dot{B}^{\infty,\infty}_{-2p}(\mathbf{R}^n)$ for $0 < p < \frac{1}{2}$. Here $\dot{F}^{r,q}_{\alpha}$ and $\dot{B}^{r,q}_{\alpha}$ are homogeneous Triebel–Lizorkin and Besov spaces respectively (see [**T**]).

In the case $p = \frac{1}{2}$, statement (ii) of Theorem 5.8 (sufficiency of the condition $\vec{\Gamma} \in \mathrm{BMO}$) is equivalent via the $\mathcal{H}^1$ – BMO duality to the inequality

$$||u \, \nabla u||_{\mathcal{H}^1(\mathbf{R}^n)} \leq c \, ||u||_{L^2(\mathbf{R}^n)} \, ||\nabla u||_{L^2(\mathbf{R}^n)}, \qquad \forall u \in C_0^\infty(\mathbf{R}^n). \tag{5.18}$$

Here $\mathcal{H}^1(\mathbf{R}^n)$ is the real Hardy space on $\mathbf{R}^n$ ([**St**]).

The preceding estimate yields the following vector-valued inequality which is used in studies of the Navier–Stokes equation, and is related to the *compensated compactness* phenomenon [**CLMS**]:

$$||(\vec{u} \cdot \nabla) \, \vec{u}||_{\mathcal{H}^1(\mathbf{R}^n)} \leq c \, ||\vec{u}||_{L^2(\mathbf{R}^n)} \, ||\nabla \vec{u}||_{L^2(\mathbf{R}^n)}, \qquad \operatorname{div} \vec{u} = \vec{0}, \tag{5.19}$$

for all $\vec{u} \in C_0^\infty(\mathbf{R}^n)^n$.

6. Form boundedness of general second order differential operators

In this section we discuss analytic characterizations of form boundedness for the general second order differential operator

$$\mathcal{L} = \sum_{i,\,j=1}^{n} a_{ij} \, \partial_i \partial_j + \sum_{j=1}^{n} b_j \, \partial_j + c, \tag{6.1}$$

where a_{ij}, b_i, and c are real- or complex-valued distributions, on the Sobolev space $W^{1,\,2}(\mathbf{R}^n)$, and its homogeneous counterpart $L^{1,\,2}(\mathbf{R}^n)$. These results were obtained in [**MV6**].

One of the motivations is to give a criterion for the relative form boundedness of the operator $\vec{b} \cdot \nabla + V$ with distributional coefficients $\vec{b}$ and V with respect to the Laplacian Δ on $L^2(\mathbf{R}^n)$. This ensures, in view of the so-called KLMN Theorem (see [**EE**], Theorem IV.4.2; [**RS**], Theorem X.17), that $\mathcal{L} = \Delta + \vec{b} \cdot \nabla + V$ can be defined, under appropriate smallness assumptions on $\vec{b}$ and V, as an m-sectorial operator on $L^2(\mathbf{R}^n)$ so that its quadratic form domain coincides with $W^{1,2}(\mathbf{R}^n)$.

In particular, we can deduce a characterization of the relative form boundedness for the magnetic Schrödinger operator

$$\mathcal{M} = (i\,\nabla + \vec{a})^2 + V, \tag{6.2}$$

with arbitrary vector potential $\vec{a} \in L^2_{\mathrm{loc}}(\mathbf{R}^n)^n$, and $V \in D'(\mathbf{R}^n)$ on $L^2(\mathbf{R}^n)$ with respect to Δ.

This approach is based on factorization of functions in Sobolev spaces and integral estimates of potentials of equilibrium measures discussed above, combined with compensated compactness arguments, commutator estimates, and the idea of gauge invariance. We are able to treat general second order differential operators, and establish an explicit Hodge decomposition for form bounded vector fields. It is worth mentioning that in this decomposition, the irrotational part of the vector field is subject to a more stringent condition than its divergence-free counterpart.

We observe that no additional assumptions (like ellipticity) are imposed on the coefficients of $\mathcal{L}$. In particular, without loss of generality we can and will assume that the principal part of $\mathcal{L}$ is in the divergence form, i.e.,

$$\mathcal{L}\,u = \operatorname{div}(A\,\nabla u) + \vec{b} \cdot \nabla u + V\,u, \qquad u \in C_0^\infty(\mathbf{R}^n), \tag{6.3}$$

where $A = (a_{ij})_{i,j=1}^n \in D'(\mathbf{R}^n)^{n\times n}$, $\vec{b} = (b_j)_{j=1}^n \in D'(\mathbf{R}^n)^n$, and $V \in D'(\mathbf{R}^n)$.

We will present necessary and sufficient conditions on A, $\vec{b}$, and V which guarantee the boundedness of the sesquilinear form associated with $\mathcal{L}$:

$$|\langle \mathcal{L}\,u, \, v\rangle| \le C\,||u||_{L^{1,2}(\mathbf{R}^n)}\,||v||_{L^{1,2}(\mathbf{R}^n)}, \tag{6.4}$$

where the constant C does not depend on $u, v \in C_0^\infty(\mathbf{R}^n)$. Here $L^{1,2}(\mathbf{R}^n)$ is the completion of (complex-valued) $C_0^\infty(\mathbf{R}^n)$ functions with respect to the norm $||u||_{L^{1,2}(\mathbf{R}^n)} = ||\nabla u||_{L^2(\mathbf{R}^n)}$.

Equivalently, we will characterize all A, $\vec{b}$, and V such that

$$\mathcal{L}: \ L^{1,2}(\mathbf{R}^n) \to L^{-1,2}(\mathbf{R}^n) \tag{6.5}$$

is a bounded operator, where $L^{-1,2}(\mathbf{R}^n) = L^{1,2}(\mathbf{R}^n)^*$ is a dual Sobolev space.

In the special case where A, $\vec{b}$ and V are locally integrable, the form boundedness of $\mathcal{L}$ may be expressed in the form of the integral inequality

$$\left| \int_{\mathbf{R}^n} (-(A\,\nabla u)\cdot \nabla \bar{v} + \vec{b}\cdot \nabla u\,\bar{v} + V\,u\,\bar{v})\,dx \right| \le C\,||u||_{L^{1,2}(\mathbf{R}^n)}||v||_{L^{1,2}(\mathbf{R}^n)}, \tag{6.6}$$

where the constant C does not depend on $u, v \in C_0^\infty(\mathbf{R}^n)$. Sometimes it will be convenient to write (6.4) in this form even for distributional coefficients a_{ij}, b_j, and V.

To state our main results, it is convenient to introduce the class of Maz'ya measures $\mathfrak{M}_+^{1,2}(\mathbf{R}^n)$, i.e., nonnegative Borel measures ω on $\mathbf{R}^n$ which obey the trace inequality

$$\int_{\mathbf{R}^n} |u|^2\,d\omega \le C\,||u||^2_{L^{1,2}(\mathbf{R}^n)}, \qquad u \in C_0^\infty(\mathbf{R}^n), \tag{6.7}$$

where the constant C does not depend on u. We will call measures $\omega \in \mathfrak{M}^{1,2}_+(\mathbf{R}^n)$ admissible. For admissible $V(x)\,dx$ with nonnegative density $V \in L^1_{\text{loc}}(\mathbf{R}^n)$, we will write $V \in \mathfrak{M}^{1,2}_+(\mathbf{R}^n)$.

Inequalities of this type (with ω possibly singular with respect to Lebesgue measure) have been thoroughly studied. A straightforward consequence of (6.7) is that if $\omega \in \mathfrak{M}^{1,2}_+(\mathbf{R}^n)$ then

$$\int_{|x-y|<r} d\omega(y) \le \text{const}\; r^{n-2}, \tag{6.8}$$

for all $r > 0$, $x \in \mathbf{R}^n$, if $n \ge 3$, and $\omega = 0$ if $n = 1, 2$ (see e.g. [**M4**], Sec. 2.4).

A close sufficient condition on $V \in L^1_{\text{loc}}(\mathbf{R}^n)$, $V \ge 0$, which ensures that $V \in \mathfrak{M}^{1,2}_+(\mathbf{R}^n)$, is provided by the Fefferman–Phong class

$$\int_{|x-y|<r} V^{1+\epsilon}\,dy \le \text{const}\, r^{n-2(1+\epsilon)}, \tag{6.9}$$

where $\epsilon > 0$, and the constant does not depend on $r > 0$, $x \in \mathbf{R}^n$.

A complete characterization of the class of admissible measures $\mathfrak{M}^{1,2}_+(\mathbf{R}^n)$ can be expressed in several equivalent forms discussed above. These criteria employ various degrees of localization of ω, and each of them has its own advantages depending on the area of application.

We now state the main form boundedness criterion [**MV6**]. For $A = (a_{ij})$, let $A^t = (a_{ji})$ denote the transposed matrix, and let $\text{Div}\colon D'(\mathbf{R}^n)^{n\times n} \to D'(\mathbf{R}^n)^n$ be the row divergence operator defined by

$$\text{Div}(a_{ij}) = \Big(\sum_{j=1}^{n} \partial_j\, a_{ij}\Big)_{i=1}^{n}. \tag{6.10}$$

THEOREM 6.1. *Let $\mathcal{L} = \text{div}\,(A\,\nabla\cdot) + \vec{b}\cdot\nabla + V$, where $A \in D'(\boldsymbol{R}^n)^{n\times n}$, $\vec{b} \in D'(\boldsymbol{R}^n)^n$ and $V \in D'(\boldsymbol{R}^n)$, $n \ge 2$. Then the following statements hold.*

(i) The sesquilinear form of $\mathcal{L}$ is bounded, i.e., (6.4) holds if and only if $\frac{1}{2}\,(A + A^t) \in L^\infty(\boldsymbol{R}^n)^{n\times n}$, and $\vec{b}$ and V can be represented respectively in the form

$$\vec{b} = \vec{c} + \text{Div}\,F, \qquad V = \text{div}\,\vec{h}, \tag{6.11}$$

where F is a skew-symmetric matrix field such that

$$F - \tfrac{1}{2}\,(A - A^t) \in \text{BMO}(\boldsymbol{R}^n)^{n\times n}, \tag{6.12}$$

whereas $\vec{c}$ and $\vec{h}$ belong to $L^2_{\text{loc}}(\boldsymbol{R}^n)^n$, and obey the condition

$$|\vec{c}|^2 + |\vec{h}|^2 \in \mathfrak{M}^{1,2}_+(\boldsymbol{R}^n). \tag{6.13}$$

(ii) If the sesquilinear form of $\mathcal{L}$ is bounded, then $\vec{c}$, F, and $\vec{h}$ in decomposition (6.11) can be determined explicitly by

$$\vec{c} = \nabla(\Delta^{-1}\text{div}\,\vec{b}), \qquad \vec{h} = \nabla(\Delta^{-1}\,V), \tag{6.14}$$

$$F = \Delta^{-1}\text{curl}\,[\vec{b} - \tfrac{1}{2}\,\text{Div}\,(A - A^t)] + \tfrac{1}{2}(A - A^t). \tag{6.15}$$

where

$$\Delta^{-1}\text{curl}\,[\vec{b} - \tfrac{1}{2}\,\text{Div}\,(A - A^t)] \in \text{BMO}(\boldsymbol{R}^n)^{n\times n}, \tag{6.16}$$

and

$$|\nabla(\Delta^{-1}\text{div}\,\vec{b})|^2 + |\nabla(\Delta^{-1}\,V)|^2 \in \mathfrak{M}^{1,2}_+(\boldsymbol{R}^n). \tag{6.17}$$

REMARK 6.2. Condition (6.16) in statement (ii) of Theorem 6.1 may be replaced with

$$\vec{b} - \operatorname{Div} \tfrac{1}{2}(A - A^t) \in \mathrm{BMO}_{-1}(\mathbf{R}^n)^n, \tag{6.18}$$

which ensures that decomposition (6.11) holds.

REMARK 6.3. In the case $n = 2$, we will show that (6.4) holds if and only if $\frac{1}{2}\,(A + A^t) \in L^\infty(\mathbf{R}^2)^{2\times 2}$, $\vec{b} - \frac{1}{2}\operatorname{Div}(A - A^t) \in \mathrm{BMO}_{-1}(\mathbf{R}^2)^2$, and $V = \operatorname{div}\vec{b} = 0$.

REMARK 6.4. Expressions like $\nabla(\Delta^{-1}\operatorname{div}\vec{b})$, $\operatorname{Div}(\Delta^{-1}\operatorname{curl}\vec{b})$, and $\nabla(\Delta^{-1}V)$ used above which involve nonlocal operators are defined in the sense of distributions. This is possible, as is shown in [**MV6**], since $\Delta^{-1}\operatorname{div}\vec{b}$, $\Delta^{-1}\operatorname{curl}\vec{b}$, and $\Delta^{-1}V$ can be understood as the limits in the sense of the weak BMO-convergence (see [**St**], p. 166) of, respectively, $\Delta^{-1}\operatorname{div}(\psi_N\vec{b})$, $\Delta^{-1}\operatorname{curl}(\psi_N\vec{b})$, and $\Delta^{-1}(\psi_N V)$ as $N \to +\infty$. Here ψ_N is a smooth cut-off function supported on $\{x : |x| < N\}$, and the limits above do not depend on the choice of ψ_N.

It follows from Theorem 6.1 that $\mathcal{L}$ is form bounded on $L^{1,2}(\mathbf{R}^n) \times L^{1,2}(\mathbf{R}^n)$ if and only if the symmetric part of A is essentially bounded, i.e., $\frac{1}{2}(A + A^t) \in L^\infty(\mathbf{R}^n)^{n\times n}$, and $\vec{b}_1 \cdot \nabla + V$ is form bounded, where

$$\vec{b_1} = \vec{b} - \tfrac{1}{2}\operatorname{Div}(A - A^t). \tag{6.19}$$

In particular, the principal part $\mathcal{P}u = \operatorname{div}(A\,\nabla u)$ is form bounded if and only if

$$\tfrac{1}{2}(A + A^t) \in L^\infty(\mathbf{R}^n)^{n\times n}, \tag{6.20}$$

$$\operatorname{Div}\tfrac{1}{2}(A - A^t) \in \mathrm{BMO}_{-1}(\mathbf{R}^n)^n. \tag{6.21}$$

A simpler condition with $\frac{1}{2}(A - A^t) \in \mathrm{BMO}(\mathbf{R}^n)^{n\times n}$ in place of (6.21) is sufficient, but generally not necessary, unless $n \le 2$.

Thus, the form boundedness problem for the general second order differential operator in the divergence form (6.3) is reduced to the special case

$$\mathcal{L} = \vec{b}\cdot\nabla + V, \qquad \vec{b} \in D'(\mathbf{R}^n)^n, \quad V \in D'(\mathbf{R}^n). \tag{6.22}$$

As a corollary of Theorem 6.1, we deduce that, if $\vec{b}\cdot\nabla + V$ is form bounded, i.e., for all $u,\ v \in C_0^\infty(\mathbf{R}^n)$,

$$\left|\int_{\mathbf{R}^n} (\vec{b}\cdot\nabla u\;\overline{v} + V\,u\,\overline{v})\,dx\right| \le C\,||u||_{L^{1,2}(\mathbf{R}^n)}||v||_{L^{1,2}(\mathbf{R}^n)}, \tag{6.23}$$

then the Hodge decomposition

$$\vec{b} = \nabla(\Delta^{-1}\operatorname{div}\vec{b}) + \operatorname{Div}(\Delta^{-1}\operatorname{curl}\vec{b}) \tag{6.24}$$

holds, where $\Delta^{-1}\operatorname{curl}\vec{b} \in \mathrm{BMO}(\mathbf{R}^n)^{n\times n}$, and

$$\int_{|x-y|<r} [\,|\nabla(\Delta^{-1}\operatorname{div}\vec{b})|^2 + |\nabla(\Delta^{-1}V)|^2\,]\,dy \le \text{const}\ r^{n-2}, \tag{6.25}$$

for all $r > 0$, $x \in \mathbf{R}^n$, in the case $n \ge 3$; in two dimensions, it follows that $\operatorname{div}\vec{b} = V = 0$.

We observe that condition (6.25) is generally stronger than $\Delta^{-1}\operatorname{div}\vec{b} \in \mathrm{BMO}$ combined with $\Delta^{-1}V \in \mathrm{BMO}$, while the divergence-free part of $\vec{b}$ is characterized by $\Delta^{-1}\operatorname{curl}\vec{b} \in \mathrm{BMO}$, for all $n \ge 2$.

A close sufficient condition of the Fefferman–Phong type can be stated in the following form:

$$(6.26) \qquad \int_{|x-y|<r} [\,|\nabla(\Delta^{-1}\mathrm{div}\,\vec{b})|^2 + |\nabla(\Delta^{-1}\,V)|^2\,]^{1+\epsilon}\,dy \le \mathrm{const}\,r^{n-2(1+\epsilon)},$$

for some $\epsilon > 0$ and all $r > 0$, $x \in \mathbf{R}^n$. This is a consequence of Theorem 6.1 coupled with (6.9), where $|(\Delta^{-1}\mathrm{div}\,\vec{b})|^2+|\nabla(\Delta^{-1}\,V)|^2$ is used in place of V. Sharper conditions of the Chang–Wilson–Wolff type are readily deduced from Theorem 6.1 by combining it with the results of [**ChWW**].

It is worth mentioning that the class of potentials obeying (6.26) is substantially broader than its subclass

$$(6.27) \qquad \int_{|x-y|<r} (|\vec{b}|^2 + |V|)^{1+\epsilon}\,dy \le \mathrm{const}\,r^{n-2(1+\epsilon)}.$$

The sufficiency of the preceding condition for (6.23) is deduced by a direct application of the original Fefferman–Phong condition and Schwarz's inequality.

More generally, (6.23) clearly follows from a cruder estimate,

$$(6.28) \qquad \int_{\mathbf{R}^n} |u|^2\,(|\vec{b}|^2 + |V|)\,dx \le \mathrm{const}\,||u||^2_{L^{1,2}(\mathbf{R}^n)}, \quad u \in C_0^\infty(\mathbf{R}^n),$$

which is equivalent to $|\vec{b}|^2 + |V| \in \mathfrak{M}_+^{1,2}(\mathbf{R}^n)$.

However, by replacing (6.23) with (6.28), one strongly reduces the class of admissible vector fields $\vec{b}$ and potentials V. Various examples of this phenomenon are given in [**MV1**], [**MV6**].

The main difficulty in the proof of Theorem 6.1 is the interaction between the quadratic forms associated with $V - \frac{1}{2}\,\mathrm{div}\,\vec{b}$ and the divergence free part of $\vec{b}$. To overcome this difficulty, one needs to distinguish the class of vector fields $\vec{b}$ such that the commutator inequality

$$(6.29) \qquad \left|\int_{\mathbf{R}^n} \vec{b}\cdot(u\,\nabla\bar{v} - \bar{v}\,\nabla u)\,dx\right| \le \mathrm{const}\,||u||_{L^{1,2}(\mathbf{R}^n)}\,||v||_{L^{1,2}(\mathbf{R}^n)}$$

holds for all $u, v \in C_0^\infty(\mathbf{R}^n)$. In the important special case of irrotational fields where $\vec{b} = \nabla f$, the preceding inequality is equivalent to the boundedness of the commutator $[f,\,\Delta]$ acting from $L^{1,2}(\mathbf{R}^n)$ to $L^{-1,2}(\mathbf{R}^n)$.

A complete characterization of those $\vec{b}$ which obey (6.29) is obtained using the idea of the gauge transformation ([**RS**], Sec. X.4):

$$\nabla \to e^{-i\lambda}\,\nabla\,e^{+i\lambda},$$

where the gauge λ is a real-valued function which lies in $L^{1,2}_{\mathrm{loc}}(\mathbf{R}^n)$.

The problem of choosing an appropriate gauge is known to be highly nontrivial. In the present paper, λ is picked in a very specific form:

$$\lambda = \tau\,\log\,(P\omega), \qquad 1 < 2\tau < \tfrac{n}{n-2}, \quad n \ge 3,$$

where τ is a constant, and $P\omega = (-\Delta)^{-1}\omega$ is the Newtonian potential of the equilibrium measure ω associated with an arbitrary compact set e of positive capacity.

We will verify that, with this choice of λ, the energy space $L^{1,2}(\mathbf{R}^n)$ is gauge invariant, and the irrotational part $\vec{c} = \nabla(\Delta^{-1}\mathrm{div}\,\vec{b})$ of $\vec{b}$ obeys

$$\int_e |\vec{c}|^2\,dx \le \mathrm{const}\,\mathrm{cap}\,(e),$$

where the constant does not depend on e. This is known to be equivalent to $|\vec{c}|^2 \in \mathfrak{M}^{1,2}_+(\mathbf{R}^n)$. In addition, a careful analysis shows that $F = \Delta^{-1}\mathrm{curl}\,\vec{b}$ belongs to BMO, and $\vec{b} = \vec{c} + \mathrm{Div}\,F$. These conditions combined turn out to be necessary and sufficient for (6.29).

In [**MV6**] applications are given to the magnetic Schrödinger operator $\mathcal{M}$ defined by (6.2). It is shown that $\mathcal{M}$ is form bounded if and only if both $V + |\vec{a}|^2$ and $\vec{a} \cdot \nabla$ are form bounded. Thus, the form boundedness criterion for $\mathcal{M}$ can be deduced from Theorem 6.1.

These results are extended to the Sobolev space $W^{1,2}(\mathbf{R}^n)$. In particular, necessary and sufficient conditions are given ([**MV6**], Theorem 5.1) for the boundedness of the general second order operator

$$\mathcal{L} : W^{1,2}(\mathbf{R}^n) \to W^{-1,2}(\mathbf{R}^n).$$

This solves the *relative form boundedness* problem for $\mathcal{L}$, and consequently for the magnetic Schrödinger operator $\mathcal{M}$, with respect to the Laplacian on $L^2(\mathbf{R}^n)$. The proofs involve an inhomogeneous version of the div-curl lemma (see [**MV6**], Lemma 5.2).

Other fundamental properties of quadratic forms associated with general differential operators, e.g., relative compactness, infinitesimal form boundedness, inequalities of Trudinger's and Nash's type can be characterized using a similar approach (see [**MV6**]).

References

[AH] D.R. Adams and L.I. Hedberg, *Function Spaces and Potential Theory*, Springer–Verlag, Berlin–Heidelberg–New York, 1996

[AS] M. Aizenman and B. Simon, *Brownian motion and Harnack inequality for Schrödinger operators*, Comm. Pure Appl. Math. **35** (1982) 209–273.

[COV1] C. Cascante, J.M. Ortega, and I.E. Verbitsky, *On imbedding potential spaces into $L^q(d\omega)$*, Proc. London Math. Soc. **80** (2000), 391–414.

[COV2] C. Cascante, J.M. Ortega, and I.E. Verbitsky, *Nonlinear potentials and two weight trace inequalities for general dyadic and radial kernels*, Indiana Univ. Math. J. **53** (2004), 845–882.

[COV3] C. Cascante, J.M. Ortega, and I.E. Verbitsky, *On $L^p - L^q$ trace inequalities*, J. London Math. Soc **74** (2006), 497–511.

[ChWW] S.-Y.A. Chang, J.M. Wilson, and T.H. Wolff, *Some weighted norm inequalities concerning the Schrödinger operators*, Comment. Math. Helv. **60** (1985), 217–246.

[CLMS] R. Coifman, P.L. Lions, Y. Meyer and S. Semmes, *Compensated compactness and Hardy spaces*, J. Math. Pures Appl., **72** (1993), 247–286.

[EE] D.E. Edmunds and W.D. Evans, *Spectral Theory and Differential Operators*, Clarendon Press, Oxford, 1987.

[F] C. Fefferman, *The uncertainty principle*, Bull. Amer. Math. Soc., **9** (1983), 129–206

[FV] M. Frazier and I.E. Verbitsky, *Global exponential bounds for Green's function and the conditional gauge*, preprint (2008).

[H] K. Hansson, *Imbedding theorems of Sobolev type in potential theory*, Math. Scand., **45** (1979), 77–102.

[HMV] K. Hansson, V.G. Maz'ya and I.E. Verbitsky, *Criteria of solvability for multidimensional Riccati's equations,* Arkiv för Matem. **37** (1999), 87–120.

[HW] L.I. Hedberg and Th.H. Wolff, *Thin sets in nonlinear potential theory*, Ann. Inst. Fourier (Grenoble), **33** (1983), 161–187.

[KV] N.J. Kalton and I.E. Verbitsky, *Nonlinear equations and weighted norm inequalities*, Trans. Amer. Math. Soc. **351** (1999), 3441–3497.

[KS] R. Kerman and E.T. Sawyer, *The trace inequality and eigenvalue estimates for Schrödinger operators*, Ann. Inst. Fourier (Grenoble) **36** (1986), 207–228.

[KiMa] T. Kilpeläinen and J. Malý, *The Wiener test and potential estimates for quasilinear elliptic equations*, Acta Math., **172** (1994), 137–161.

[KT] H. Koch and D. Tataru, *Well-posedness for the Navier-Stokes equations*, Adv. Math., **157** (2001), 22–35.

[L] D.A. Labutin, *Potential estimates for a class of fully nonlinear elliptic equations*, Duke Math. J. **111** (2002), 1–49.

[MaZi] J. Malý and W.P. Ziemer, *Fine Regularity of Solutions of Elliptic Partial Differential Equations*, Amer. Math. Soc., Providence, R.I., 1997

[M1] V.G. Maz'ya, *Classes of domains and embedding theorems for functional spaces*, Dokl. Akad. Nauk SSSR, **133** (1960), 527–530.

[M2] V.G. Maz'ya, *On the theory of the n-dimensional Schrödinger operator*, Izv. Akad. Nauk SSSR, ser. Matem., **28** (1964), 1145–1172.

[M3] V.G. Maz'ya, *On certain integral inequalities for functions of many variables*, Probl. Math. Anal., **3**, Leningrad Univ. (1972), 33–68. English transl.: J. Soviet Math., **1** (1973), 205–234.

[M4] V.G. Maz'ya, *Sobolev Spaces*, Springer–Verlag, Berlin–Heidelberg–New York, 1985.

[MN] V.G. Maz'ya and Y. Netrusov, *Some counterexamples for the theory of Sobolev spaces on bad domains*, Potential Analysis, **4** (1995), 47–65.

[MSh] V.G. Maz'ya and T.O. Shaposhnikova, *The Theory of Multipliers in Spaces of Differentiable Functions*, Monographs and Studies in Mathematics, **23**, Pitman, Boston–London, 1985.

[MV1] V.G. Maz'ya and I.E. Verbitsky, *Capacitary inequalities for fractional integrals, with applications to partial differential equations and Sobolev multipliers*, Arkiv för Matem., **33** (1995), 81–115

[MV2] V.G. Maz'ya and I.E. Verbitsky, *The Schrödinger operator on the energy space: boundedness and compactness criteria*, Acta Math., **188** (2002), 263–302

[MV3] V.G. Maz'ya and I.E. Verbitsky, *Boundedness and compactness criteria for the one-dimensional Schrödinger operator*, In: Function Spaces, Interpolation Theory and Related Topics. Proc. Jaak Peetre Conf., Lund, Sweden, August 17-22, 2000. Eds. M. Cwikel, A. Kufner, G. Sparr. De Gruyter, Berlin, 2002, 369–382

[MV4] V.G. Maz'ya and I.E. Verbitsky, *The form boundedness criterion for the relativistic Schrödinger operator*, Ann. Inst. Fourier **54** (2004), 317–339.

[MV5] V.G. Maz'ya and I.E. Verbitsky, *Infinitesimal form boundedness and Trudinger's subordination for the Schrödinger operator*, Invent. Math. **162** (2005), 81–136.

[MV6] V.G. Maz'ya and I.E. Verbitsky, *Form boundedness of the general second order differential operator*, Comm. Pure Appl. Math. **59** (2006), 1286–1329.

[NTV] F. Nazarov, S. Treil, and A. Volberg, *The Bellman functions and two-weight inequalities for Haar multipliers*, J. Amer. Math. Soc. **12** (1999), 909–928.

[PV1] N.C. Phuc and I.E. Verbitsky, *Local integral estimates and removable singularities for quasilinear and Hessian equations with nonlinear source terms*, Comm. PDE **31** (2006), 1779–1791.

[PV2] N.C. Phuc and I.E. Verbitsky, *Quasilinear and Hessian equations of Lane-Emden type*, to appear in Ann. Math.

[PV3] N.C. Phuc and I.E. Verbitsky, *Singular quasilinear and Hessian equations and inequalities*, preprint (2008).

[RS] M. Reed and B. Simon, *Methods of Modern Mathematical Physics. II: Fourier Analysis, Self-Adjointness*, Academic Press, New York–London, 1975.

[RSS] G.V. Rozenblum, M.A. Shubin, and M.Z. Solomyak, *Spectral Theory of Differential Operators*, Encyclopaedia of Math. Sci., **64**. Partial Differential Equations VII. Ed. M. A. Shubin. Springer-Verlag, Berlin–Heidelberg, 1994.

[S] E.T. Sawyer, *Two weight norm inequalities for certain maximal and integral operators*, Trans. Amer. Math. Soc., **308** (1988), 533–545

[Sch] M. Schechter, *Operator Methods in Quantum Mechanics*, North-Holland, Amsterdam – New York – Oxford, 1981.

[Sim] B. Simon, *Schrödinger semigroups*, Bull. Amer. Math. Soc. **7** (1982), 447–526.

[StW] E.M. Stein and G. Weiss, *Introduction to Fourier Analysis on Euclidean Spaces*, Princeton University Press, Princeton, New Jersey, 1971.

[St] E.M. Stein, *Harmonic Analysis: Real-Variable Methods, Orthogonality, and Oscillatory Integrals*, Princeton University Press, Princeton, NJ, 1993

[T] H. Triebel, *Theory of Function Spaces. II*, Monographs in Mathematics, **84**, Birkhäuser Verlag, Basel, 1992.

[Tru] N.S. Trudinger, *Linear elliptic operators with measurable coefficients*, Ann. Scuola Norm. Sup. Pisa **27** (1973), 265–308.

[TW] N.S. Trudinger and X.J. Wang, *On the weak continuity of elliptic operators and applications to potential theory*, Amer. J. Math. **124** (2002), 369–410.

[V1] I.E. Verbitsky, *Weighted norm inequalities for maximal operators and Pisier's theorem on factorization through* $L^{p,\infty}$, Integr. Equat. Operator Theory, **15** (1992), 121–153.

[V2] I.E. Verbitsky, *Imbedding and multiplier theorems for discrete Littlewood–Paley spaces*, Pacific J. Math., **176** (1996), 529–556.

[V3] I.E. Verbitsky, *Superlinear equations, potential theory, and weighted norm inequalities*, Nonlinear Analysis, Function Spaces and Applications, Proceedings of the Spring School, **5**, Prague, May 31 - June 6, 1998, 1–47.

[V4] I.E. Verbitsky, *Nonlinear potentials and trace inequalities*, in: The Maz'ya anniversary collection, Vol. 2. Proceedings of the International Conference on Functional Analysis, Partial Differential Equations and Applications. Rostock, Germany, August 31 - September 4, 1998. Operator Theory: Advances and Applications, **110**, Birkhäuser Verlag, 1999, 323–343

[VW1] I.E. Verbitsky and R.L. Wheeden, *Weighted inequalities for fractional integrals and applications to semilinear equations*, J. Funct. Anal., **129** (1995), 221–241.

[VW2] I.E. Verbitsky and R.L. Wheeden, *Weighted norm inequalities for integral operators*, Trans. Amer. Math.Soc. **350** (1998), 3371–3391.

School of Mathematics, University of Birmingham, Birmingham, B15 2TT, United Kingdom

E-mail address: I.E.Verbitsky@bham.ac.uk

Department of Mathematics, University of Missouri, Columbia, MO 65211, USA

E-mail address: igor@math.missouri.edu

Proceedings of Symposia in Pure Mathematics
Volume **79**, 2008

THE MIXED PROBLEM FOR HARMONIC FUNCTIONS IN POLYHEDRA OF $\mathbb{R}^3$

MOISES VENOUZIOU AND GREGORY C. VERCHOTA

ABSTRACT. R. M. Brown's theorem on mixed Dirichlet and Neumann boundary conditions is extended in two ways for the special case of polyhedral domains. A (1) more general partition of the boundary into Dirichlet and Neumann sets is used on (2) manifold boundaries that are not locally given as the graphs of functions. Examples are constructed to illustrate necessity and other implications of the geometric hypotheses.

1. INTRODUCTION

In [Bro94] R. M. Brown initiated a study of the *mixed boundary value problem* for harmonic functions in *creased Lipschitz domains* Ω with data in the Lebesgue and Sobolev spaces $L^2(\partial\Omega)$ and $W^{1,2}(\partial\Omega)$ (with respect to surface measure ds) taken in the strong pointwise sense of nontangential convergence.

At the end of his article Brown poses a question concerning a certain topologic-geometric difficulty not included in his solution: Can the mixed problem be solved in the (infinite) pyramid of $\mathbb{R}^3$, $|X_1| + |X_2| < X_3$, when Neumann and Dirichlet data are chosen to alternate on the faces? In this article we avoid the geometric difficulties of what can be called Lipschitz faces or facets and provide answers in the case of compact polyhedral domains of $\mathbb{R}^3$. Some other recent approaches to the mixed problem for second order operators and systems in polyhedra can be found in [MR07] [MR06] [MR05] [MR04] [MR03] [MR02] and [Dau92].

Consider a compact polyhedron of $\mathbb{R}^3$ with the property that its interior Ω is connected. Ω will be termed a *compact polyhedral domain.* Suppose its boundary $\partial\Omega$ is a connected 2-manifold. Such a domain Ω need not be a Lipschitz domain. Partition the boundary of Ω into two disjoint sets N and D, for Neumann data and Dirichlet data respectively, so that the following is satisfied.

(1.1) (i) N is the union of a number (possibly *zero*) of closed faces of $\partial\Omega$.

(ii) $D = \partial\Omega \setminus N$ is nonempty.

(iii) Whenever a face of N and a face of D share a 1-dimensional edge as boundary, the dihedral angle measured in Ω between the two faces is *less* than π.

Date: June 9, 2008.

2000 *Mathematics Subject Classification.* 35J30,35J40.

The second author gratefully acknowledges partial support provided by the National Science Foundation through award DMS-0401159

E-mail address: gverchot@syr.edu

The L^2-*polyhedral mixed problem* for harmonic functions is

(1.2)

Given $f \in W^{1,2}(\partial\Omega)$ and $g \in L^2(N)$ show there exists a solution to $\triangle u = 0$ in Ω such that

(i) $u \rightarrow^{n.t.} f$ *a.e.* on D.

(ii) $\partial_\nu u \rightarrow^{n.t.} g$ *a.e.* on N.

(iii) $\nabla u^* \in L^2(\partial\Omega)$.

Here ∇u^* is the *nontangential maximal function* of the *gradient* of u. Generally for a function w defined in a domain G

$$w^*(P) = \sup_{X \in \Gamma(P)} |w(X)|, P \in \partial G.$$

For a choice of $\alpha > 0$ *nontangential approach regions* for each $P \in \partial G$ are defined by

$$\Gamma(P) = \{X \in G : |X - P| < (1+\alpha)dist(X, \partial G)\} \tag{1.3}$$

Varying the choice of α yields nontangential maximal functions with comparable $L^p(\partial G)$ norms $1 < p \leq \infty$ by an application of the Hardy-Littlewood maximal function. Therefore α is suppressed. In general when $w^* \in L^p(\partial G)$ is written it is understood that the nontangential maximal function is with respect to cones determined by the domain G. The *outer unit normal* vector to Ω (or a domain G) is denoted $\nu = \nu_P$ for *a.e.*$P \in \partial\Omega$ and the limit of (ii) is understood as

$$\lim_{\Gamma(P) \ni X \rightarrow P} \nu_P \cdot \nabla u(X) = g(P)$$

and similarly for (i).

A consequence of solving (1.2) is that the gradient of the solution has well defined nontangential limits at the boundary *a.e.*

In addition, as Brown points out, solving the mixed problem yields *extension operators* $W^{1,2}(D) \rightarrow W^{1,2}(\partial\Omega)$ by $f \mapsto u|_{\partial\Omega}$ where u is a solution to the mixed problem with $u|_D = f$. Consequently *problem* (1.2) *cannot be solved for all* $f \in W^{1,2}(D)$ *when* D *and* N *are defined as on the boundary of the pyramid.* For example, since the pyramid is Lipschitz at the origin so that Sobolev functions on its boundary project to Sobolev functions on the plane, solving (1.2) implies that a local $W^{1,2}$ function exists in $\mathbb{R}^2$ that is identically 1 in the first quadrant and identically *zero* in the third. Such a function necessarily restricts to a local $W^{\frac{1}{2},2}$ function on any straight line through the origin. But a step function is not locally in $W^{\frac{1}{2},2}(\mathbb{R})$. The boundary domain D (and its projection) do not satisfy the *segment property* commonly invoked to show the two Sobolev spaces $H_1(D)$ and $W^{1,2}(D)$ equal [Agm65] [GT83].

The admissible Sobolev functions on D must then be those that have extensions to $W^{1,2}(\partial\Omega)$. Or equivalently, the admissible Sobolev functions on D are the restrictions of $W^{1,2}(\partial\Omega)$ functions. We introduce the following *norm* on the *space of restrictions of* $W^{1,2}(\partial\Omega)$ *functions* f *to* D

$$\|f\|_D^2 = \inf_{\tilde{f}|_D = f} \int_{\partial\Omega} \tilde{f}^2 + |\nabla_t \tilde{f}|^2 ds$$

Here $\widetilde{f}$ denotes all $W^{1,2}(\partial\Omega)$ functions that restrict to f on D, and ∇_t denotes the tangential gradient. That this is a norm follows by arguments such as: Given $f \in W^{1,2}(\partial\Omega)$ and a real number a, the functions $a\widetilde{f}$ form a subset of all extensions $\widetilde{af}$ of $(af)|_D$ so that $\|af\|_D \leq |a|\|f\|_D$, and thus likewise $\|f\|_D \leq |a|^{-1}\|af\|_D$ when $a \neq 0$.

This normed space is complete by using the standard completeness proof for Lebesgue spaces: Given a Cauchy sequence $\{f_j\}$ let j_k be such that $\|f_i - f_j\|_D < 2^{-k}$ for all $i, j \geq j_k$ and define $g_k = f_{j_{k+1}} - f_{j_k}$. Then there exists an extension $\widetilde{g_k}$ such that $\|\widetilde{g_k}\|_{W^{1,2}(\partial\Omega)} < 2^{-k}$. Extensions of $f_{j_{n+1}}$ may then be defined by $\widetilde{f_{j_1}} + \sum_{k=1}^{n} \widetilde{g_k}$ Cauchy in $W^{1,2}(\partial\Omega)$. Completeness will follow. The Banach space of restrictions to D is undoubtedly the generally smaller Sobolev space $H_1(D)$ (e.g. [Fol95] p. 220), but this will not be pursued further.

A homogeneous Sobolev semi-norm on D is defined by

$$\|f\|_{D^o}^2 = \inf_{\widetilde{f}|_D = f} \int_{\partial\Omega} |\nabla_t \widetilde{f}|^2 ds \tag{1.4}$$

When $\partial\Omega$ is connected the following *scale invariant* theorem is established in the Section 2.

Theorem 1.1. *Let $\Omega \subset \mathbb{R}^3$ be a compact polyhedral domain with connected 2-manifold boundary $\partial\Omega = D \cup N$ satisfying the conditions* (1.1). *Then given $f \in W^{1,2}(\partial\Omega)$ and $g \in L^2(N)$ there exists a unique solution u to the mixed problem* (1.2). *In addition there is a constant C independent of u such that*

$$\int_{\partial\Omega} (\nabla u^*)^2 ds \leq C \left(\|f\|_{D^o}^2 + \int_N g^2 ds \right)$$

In the following section it is proved that a change from Dirichlet to Neumann data on a single face is necessarily prohibited when the change takes place across the graph of a Lipschitz function. The strict convexity condition of (1.1) is also shown to be necessary. In the final section compact polyhedra are discussed for which the set N is necessarily *empty*.

2. Proof of Theorem 1.1

The estimates that follow are scale invariant. Therefore to lighten the exposition a bit it will be assumed, when working near any vertex of the boundary of the compact polyhedron $\overline{\Omega}$, that the vertex is at least a distance of 4 units from any other vertex. Because $\partial\Omega$ is assumed to be a 2-manifold it will also be assumed that each edge that does not contain a given vertex v as an endpoint is at least 4 units from v and similarly each face. Consequently, by another application of the manifold condition, the picture that emerges is that the *truncated cones*

$$\mathcal{C}(v, r) = \{X \in \overline{\Omega} : |v - X| \leq r\}$$

for any vertex v and $0 \leq r \leq 4$ are homeomorphic to the closed ball $\mathbb{B}^3$ while the *cone bases*

$$\mathcal{B}(v, r) = \{X \in \overline{\Omega} : |v - X| = r\}$$

are homeomorphic to the closed disc $\mathbb{B}^2$.

Define

$$\Omega_r = \Omega \setminus \bigcup_v \mathcal{C}(v,r),\ 0 < r < 2$$

where the finite union is over all boundary vertices. Then each Ω_r is a Lipschitz domain (see, for example, §12.1 of [VV06] and Theorem 6.1 of [VV03] for a proof and generalizations in dimensions $n \geq 3$). Likewise the interiors of the *arches* defined by

$$\mathcal{A}(v,r,R) = \{X \in \overline{\Omega} : r \leq |v - X| \leq R\}\ ,\ 0 < r < R < 4 \tag{2.1}$$

are Lipschitz domains.

In general neither of these kinds of domains have a uniform Lipschitz nature as $r \to 0$. Therefore the following *polyhedral Rellich identity* of [VV06] will be of use. It is proved as in [JK81] by an application of the Gauss divergence theorem, but with respect to the vector field

$$W := \frac{X}{|X|}, X \in \mathbb{R}^3 \setminus \{0\}$$

when the origin is on the boundary of the domain.

Lemma 2.1. *Let* $\mathcal{A}$ *be any* arch (2.1) *of the polyhedral domain* $\Omega \subset \mathbb{R}^3$ *and suppose* u *is harmonic in* $\mathcal{A}$ *with* $\nabla u^* \in L^2(\partial\mathcal{A})$. *Then, taking the vertex* v *to be at the origin*

$$2\int_{\mathcal{A}} (W \cdot \nabla u)^2 \frac{dX}{|X|} = \int_{\partial\mathcal{A}} \nu \cdot W|\nabla u|^2 - 2\partial_\nu u W \cdot \nabla u ds \tag{2.2}$$

Lemma 2.2. *With* $\mathcal{A} = \mathcal{A}(v,r,R)$ *and* u *as in Lemma 2.1*

(2.3)

$$2\int_{\mathcal{A}} (W \cdot \nabla u)^2 \frac{dX}{|X|} \leq \int_{\mathcal{B}(v,R)} |\nabla u|^2 ds + 2\int_{\mathcal{B}(v,r)} (W \cdot \nabla u)^2 ds + 2\int_{\partial\Omega\cap\mathcal{A}} |\partial_\nu u||\nabla_t u| ds$$

Proof. The term $\nu \cdot W$ on the right of (2.2) is negative on $\mathcal{B}(v,r)$ and vanishes on $\partial\Omega$. Likewise the second integrand on the right of (2.2) is a perfect square on $\mathcal{B}(v,r)$, the negative of a square on $\mathcal{B}(v,R)$, and $W \cdot \nabla u$ is a tangential derivative on $\partial\Omega$. □

The partition $D \cup N = \partial\Omega$ induces a decomposition of the Lipschitz boundaries $\partial\Omega_r$ into a Dirichlet part, a Neumann part, and bases $\mathcal{B}(v,r)$ of the cones removed from Ω. Define

$$N_r = (N \cap \partial\Omega_r) \bigcup_v \mathcal{B}(v,r)$$

and

$$D_r = \partial\Omega_r \setminus N_r.$$

This partition of $\partial\Omega_r$ *satisfies the requirements of a creased domain in* [Bro94]. See [VV06] pp. 586-587. (Including the bases in the Dirichlet part would also satisfy the requirements.) It will therefore be possible to invoke Brown's *existence* results in the domains Ω_r.

Similarly, *arches* $\mathcal{A} = \mathcal{A}(v,r,R)$ *are creased Lipschitz domains* with

$$N_r^R(v) = (N \cap \partial\mathcal{A}(v,r,R)) \cup \mathcal{B}(v,r) \cup \mathcal{B}(v,R)$$

and

$$D_r^R = \partial\mathcal{A} \setminus N_r^R$$

for each vertex v.

Brown's estimate from [Bro94] Theorem 2.1 is not scale invariant. However, the following special case is.

Theorem 2.3. (R. M. Brown) *Let $G \subset \mathbb{R}^n$ be a* creased *Lipschitz domain with $\partial G = D \cup N$. Then there exists a unique solution u to the mixed problem* (1.2) *for data f identically* zero *and $g \in L^2(N)$. Furthermore there is a constant C determined only by the scale invariant geometry of G, D and N and independent of g such that*

$$\int_{\partial G} (\nabla u^*)^2 ds \leq C \int_N g^2 ds$$

As is

Theorem 2.4. (R. M. Brown) *Let $G \subset \mathbb{R}^n$ be a* creased *Lipschitz domain with $\partial G = D \cup N$. Suppose that D is* connected. *Then there is a constant C such that for all harmonic functions u with $\nabla u^* \in L^2(\partial G)$*

$$\int_{\partial G} (\nabla u^*)^2 ds \leq C \left(\int_D |\nabla_t u|^2 ds + \int_N (\partial_\nu u)^2 ds \right)$$

Proof. Subtracting from u its mean value over D allows the Poincaré inequality over the connected set D. The conclusion still applies to u. □

Lemma 2.5. *Let $\Omega \subset \mathbb{R}^3$ be a compact polyhedral domain with 2-manifold boundary partitioned as $\partial\Omega = D \cup N$. Let v be a vertex and let j be a natural number. Suppose u is harmonic in the arch $\mathcal{A}(v, 2^{-j}, 2)$ with $\nabla u^* \in L^2(\partial\mathcal{A})$ and u* vanishing *on $D^2_{2^{-j}}$. Then there is a constant C independent of j so that*

$$\int_{\partial\Omega\cap\mathcal{A}(v,2^{-j},2)} |\nabla u|^2 ds \leq$$
$$C \left(\int_{\partial\Omega\cap N_{2^{-j}}} (\partial_\nu u)^2 ds + \int_{\mathcal{B}(v,2^{-j})} (W \cdot \nabla u)^2 ds + \int_{\mathcal{A}(v,1,2)} |\nabla u|^2 dX \right)$$

Proof. For natural numbers $k \leq j$ and real numbers $1 \leq t \leq 2$ the arches $\mathcal{A}_{k,t} := \mathcal{A}(v, t2^{-k}, t2^{1-k})$ are geometrically *similar* Lipschitz domains. Therefore by the scale invariance of Brown's Theorem 2.3 above

$$\int_{\partial\Omega\cap\mathcal{A}_{k,t}} |\nabla u|^2 ds \leq C \int_{N_{t2^{-k}}^{t2^{1-k}}} (\partial_\nu u)^2 ds$$

with C independent of k. Take v to be the origin. For each k, integrating in $1 \leq t \leq 2$ and observing that $\nu = W$ or $-W$ on any cone base $\mathcal{B}$

$$\frac{1}{2} \int_{\partial\Omega\cap(\mathcal{A}_{k,1}\cup\mathcal{A}_{k,2})} |\nabla u|^2 ds \leq$$
$$2C \left(\int_{N\cap(\mathcal{A}_{k,1}\cup\mathcal{A}_{k,2})} (\partial_\nu u)^2 ds + \int_{\mathcal{A}_{k,1}\cup\mathcal{A}_{k,2}} (W \cdot \nabla u)^2 \frac{dX}{|X|} \right)$$

Summing on $k = 1, 2, \ldots, j$ and using Lemma 2.2 on the arch $\mathcal{A}(v, 2^{-j}, R)$ for each $1 \leq R \leq 2$ together with the vanishing of u on $D^2_{2^{-j}}$ again

$$\frac{1}{2}\int_{\partial\Omega\cap\mathcal{A}(v,2^{-j},2)} |\nabla u|^2 ds \leq 4C(\int_{\partial\Omega\cap N_{2^{-j}}} (\partial_\nu u)^2 ds + \int_{\mathcal{B}(v,R)} |\nabla u|^2 ds$$
$$+ 2\int_{\mathcal{B}(v,2^{-j})} (W\cdot\nabla u)^2 ds + 2\int_{\partial\Omega\cap N^R_{2^{-j}}} |\partial_\nu u||\nabla_t u| ds + \int_{\mathcal{A}(v,1,2)} |\nabla u|^2 dX)$$

An application of Young's inequality ($2ab \leq \frac{1}{\epsilon}a^2 + \epsilon b^2$) allows the square of the tangential derivatives in the second to last term to be hidden on the left side and the normal derivatives to be incorporated in the first right side integral. Integrating in $1 \leq R \leq 2$ yields the final inequality. □

By the same arguments, but using Theorem 2.4 and then Young's inequality in suitable ways for the D portion and the N portion of the last integral of Lemma 2.2, the next lemma is proved. For a given vertex, $D \cap \mathcal{C}(v, R_1)$ is connected if and only if any $D \cap \mathcal{A}(v, r, R_2)$ is connected.

Lemma 2.6. *Let $\Omega \subset \mathbb{R}^3$ be a compact polyhedral domain with 2-manifold boundary partitioned as $\partial\Omega = D \cup N$. Let v be a vertex and let j be a natural number. Suppose $D \cap \mathcal{C}(v, 2)$ is* connected *and u is harmonic in the arch $\mathcal{A}(v, 2^{-j}, 2)$ with $\nabla u^* \in L^2(\partial\mathcal{A})$. Then there is a constant C independent of j so that*

$$\int_{\partial\Omega\cap\mathcal{A}(v,2^{-j},2)} |\nabla u|^2 ds \leq C\left(\int_{D_{2^{-j}}} |\nabla_t u|^2 ds + \int_{\partial\Omega\cap N_{2^{-j}}} |\partial_\nu u|^2 ds\right) +$$
$$C\left(\int_{\mathcal{B}(v,2^{-j})} (W\cdot\nabla u)^2 ds + \int_{\mathcal{A}(v,1,2)} |\nabla u|^2 dX\right)$$

Let v be a vertex of the compact polyhedral domain Ω and consider the collection of nontangential approach regions $\Gamma(P)$ for $G = \Omega$ and parameter α (1.3) with $P \in \partial\Omega \cap \mathcal{C}(v, 4)$. By scale invariance each approach region can be *truncated* to a region

$$\Gamma^T(P) = \{X \in \Gamma(P) : |X - P| < (1+\alpha)dist(X, \partial\mathcal{A}(v, r/2, 2r))\},\ \ |v - P| = r$$

so that the collections $\{\Gamma^T(P) : r \leq |v - P| \leq 2r\}$ can be extended in a uniform way to systems of nontangential approach regions *regular* in the sense of Dahlberg [Dah79] for the arches $\mathcal{A}(v, r/2, 4r)$.

Denote by w^T the nontangential maximal function of w with respect to the truncated cones Γ^T.

Denote the Hardy-Littlewood maximal operator on $\partial\Omega$ by $\mathcal{M}$. See, for example, [Ste70] pp.10-11 or [VV03] pp.501-502 for polyhedra.

For α large enough a geometric argument shows that there is a constant independent of P and w such that

$$w^*(P) \leq C\mathcal{M}(w^T)(P) + \max_K |w|,\ \ P \in \partial\Omega \cap \mathcal{C}(v, 4) \tag{2.4}$$

where K is a compactly contained set in the Lipschitz domain Ω_2.

Using Theorems 2.3 and 2.4 to estimate the truncated maximal functions introduces into the proofs of Lemmas 2.5 and 2.6 a doubling of the dyadic arches and therefore one dyadic term that is not immediately hidden by Young's inequality. Thus by the same proofs

Lemma 2.7. *With the same hypotheses as Lemma 2.5 there is a constant C independent of j so that*

$$\int_{\partial\Omega\cap\mathcal{A}(v,2^{1-j},2)} \left(\nabla u^T\right)^2 ds - \frac{1}{2}\int_{\partial\Omega\cap\mathcal{A}(v,2^{-j},2^{1-j})} |\nabla u|^2 ds \le$$
$$C\left(\int_{\partial\Omega\cap N_{2-j}} (\partial_\nu u)^2 ds + \int_{\mathcal{B}(v,2^{-j})} (W\cdot\nabla u)^2 ds + \int_{\Omega_1} |\nabla u|^2 dX\right)$$

Lemma 2.8. *With the same hypotheses as Lemma 2.6 there is a constant C independent of j so that*

$$\int_{\partial\Omega\cap\mathcal{A}(v,2^{1-j},2)} \left(\nabla u^T\right)^2 ds - \frac{1}{2}\int_{\partial\Omega\cap\mathcal{A}(v,2^{-j},2^{1-j})} |\nabla u|^2 ds \le$$
$$C\left(\int_{D_{2-j}} |\nabla_t u|^2 ds + \int_{\partial\Omega\cap N_{2-j}} |\partial_\nu u|^2 ds + \int_{\mathcal{B}(v,2^{-j})} (W\cdot\nabla u)^2 ds + \int_{\Omega_1} |\nabla u|^2 dX\right)$$

Remark 2.9. Lemmas 2.5 and 2.6 apply to the negative terms of Lemmas 2.7 and 2.8. Consequently those terms may be removed from the inequalities.

2.1. **The regularity problem.** The *regularity problem* is the mixed problem for $\partial\Omega = D$.

Theorem 2.10. *Let $\Omega \subset \mathbb{R}^3$ be a compact polyhedral domain with 2-manifold connected boundary. Then for any $f \in W^{1,2}(\partial\Omega)$ the regularity problem is uniquely solvable and the estimate for the solution u*

$$\int_{\partial\Omega} |\nabla u^*|^2 ds \le C\int_{\partial\Omega} |\nabla_t f|^2 ds$$

holds with C independent of f.

Proof. For each Ω_{2-j} there is a unique solution u_j to the mixed problem with $u_j = f$ on D_{2-j} and $\partial_\nu u_j = 0$ on N_{2-j} by Brown's existence result [Bro94]. By definition of the truncated approach regions in each vertex cone $\mathcal{C}(v,4)$ the regions may be extended to a regular system of truncated approach regions for the $\partial\Omega\cap\partial\Omega_1$ part of the boundary. Thus the truncated nontangential maximal function can be defined there. By Lemma 2.8 and Remark 2.9, summing over all vertices, using analogous estimates on the local Lipschitz boundary of $\partial\Omega$ outside of the vertex cones and using $W\cdot\nabla u_j = 0$ on the bases $\mathcal{B}(v,2^{-j})$,

$$\int_{D_{2^{1-j}}} \left(\nabla u_j^T\right)^2 ds \le C\left(\int_{D_{2-j}} |\nabla_t f|^2 ds + \int_{\Omega_1} |\nabla u_j|^2 dX\right) \tag{2.5}$$

with C independent of j.

Subtracting from u_j the mean value m_f of f over $\partial\Omega$ does not change (2.5). Thus Poincaré (see [VV06]p.639 for polyhedral boundaries) can be applied over $\partial\Omega$ with constant independent of j in

$$\int_{\Omega_1} |\nabla u_j|^2 dX \le \int_{D_{2-j}} (u_j - m_f)\partial_\nu u_j ds = \int_{D_{2-j}} (f - m_f)\partial_\nu u_j ds \le \tag{2.6}$$
$$C_\epsilon \int_{\partial\Omega} |\nabla_t f|^2 ds + \epsilon\int_{D_{2-j}} |\nabla u_j|^2 ds$$

Applying Lemma 2.6 to the part of the integral over the regions $D_{2^{-j}}^{2^{1-j}}$ and using $W \cdot \nabla u_j = 0$ again

$$\epsilon \int_{D_{2^{-j}}} |\nabla u_j|^2 ds \leq \epsilon C \left(\int_{D_{2^{-j}}} |\nabla_t f|^2 ds + \int_{\Omega_1} |\nabla u_j|^2 dX \right) + \epsilon \int_{D_{2^{1-j}}} |\nabla u_j|^2 ds$$

so that (2.6) yields

$$\frac{1}{2} \int_{\Omega_1} |\nabla u_j|^2 dX \leq (C_\epsilon + \epsilon C) \int_{\partial\Omega} |\nabla_t f|^2 ds + \epsilon \int_{D_{2^{1-j}}} |\nabla u_j|^2 ds$$

for all ϵ chosen small enough depending on C but not on j. Using this in (2.5) for ϵ chosen small enough gives

$$\int_{D_{2^{1-j}}} \left(\nabla u_j^T\right)^2 ds \leq C' \int_{\partial\Omega} |\nabla_t f|^2 ds \tag{2.7}$$

with the constant independent of j.

Given any compact subset of Ω, (2.7) together with $u_j = f$ on $D_{2^{-j}}$ for all j implies there exists a subsequence so that both u_{j_k} and ∇u_{j_k} converge uniformly on the compact set to a harmonic function u and its gradient respectively. A diagonalization argument gives pointwise convergence on all of Ω. Intersecting a compact subset K with the truncated approach regions yields compactly contained regions and corresponding maximal functions $\nabla u_{j_k}^{T,K} \to \nabla u^{T,K}$ uniformly. Thus by (2.7) and then monotone convergence, as Ω is exhausted by compact subsets K,

$$\int_{\partial\Omega} \left(\nabla u^T\right)^2 ds \leq C \int_{\partial\Omega} |\nabla_t f|^2 ds \tag{2.8}$$

See [JK82] for these arguments.

A difficulty with the setup here is that the $\int_{D_{2^{1-j}}} \left(\nabla(u_j - u_k)^T\right)^2 ds$ for $k > j$ do not *a priori* have better bounds than the right side of (2.7). However, (2.7) together with weak convergence in $L^2(\partial\Omega_{2^{-j}})$ and pointwise convergence on the bases $\mathcal{B}(v, 2^{-j})$ shows that for each j and every $X \in \Omega_{2^{-j}}$ a subsequence of

$$u_k(X) = \int_{\partial\Omega_{2^{-j}}} \mathcal{P}_j^X u_k ds = \int_{D_{2^{-j}}} \mathcal{P}_j^X f ds + \sum_v \int_{\mathcal{B}(v,2^{-j})} \mathcal{P}_j^X u_k ds$$

converges to $u(X)$, perforce with Poisson representation that must be an extension from $D_{2^{-j}}$ of f. Here $\mathcal{P}_j^X$ is the Poisson kernel for the Lipschitz polyhedral domain $\Omega_{2^{-j}}$ and may be seen to be in $L^2(\partial\Omega_{2^{-j}})$ by Dahlberg [Dah77]. Consequently u has nontangential limits f on $\partial\Omega$, and by (2.4) and (2.8) the theorem is proved. □

2.2. **The mixed problem with vanishing Dirichlet data.**

Theorem 2.11. *Let $\Omega \subset \mathbb{R}^3$ be a compact polyhedral domain with 2-manifold connected boundary. Then for any $g \in L^2(N)$ there is a unique solution u to the mixed problem* (1.2) *that vanishes on D and has Neumann data g on N. Further*

$$\int_{\partial\Omega} (\nabla u^*)^2 ds \leq C \int_N g^2 ds$$

Proof. Again by [Bro94] there exists a unique solution u_j in $\Omega_{2^{-j}}$ to the mixed problem so that $\partial_\nu u_j = g$ on $\partial\Omega \cap N_{2^{-j}}$, $W \cdot \nabla u_j = 0$ on the $\mathcal{B}(v, 2^{-j})$ and $u_j = 0$ on $D_{2^{-j}}$. Lemma 2.7 and Remark 2.9 imply

$$\int_{\partial\Omega\cap\partial\Omega_{2^{1-j}}} \left(\nabla u_j^T\right)^2 ds \le C\left(\int_{\partial\Omega\cap N_{2^{-j}}} g^2 ds + \int_{\Omega_1} |\nabla u_j|^2 dX\right)$$

A Poincaré inequality independent of j is also needed here and is supplied by the following lemma. Polyhedral domains are naturally described as simplicial complexes. See for definitions and notations [Gla70] [RS72] [VV03] [VV06] or others.

Lemma 2.12. *Suppose u is harmonic in $\Omega_{2^{-j}}$ with $\partial_\nu u = g$ on $\partial\Omega \cap N_{2^{-j}}$, $\partial_\nu u = 0$ on the $\mathcal{B}(v, 2^{-j})$ and $u = 0$ on $D_{2^{-j}}$. Then*

$$\int_{\Omega_{2^{-j}}} |\nabla u|^2 \, dX \le C \int_{\partial\Omega\cap N_{2^{-j}}} g^2 ds$$

with C independent of j.

Proof. By Green's first identity and Young's inequality

$$\int_{\Omega_{2^{-j}}} |\nabla u|^2 dX = \int_{\partial\Omega\cap N_{2^{-j}}} u \, \partial_\nu u ds \le C_\epsilon \int_{\partial\Omega\cap N_{2^{-j}}} g^2 ds + \epsilon \int_{\partial\Omega\cap N_{2^{-j}}} u^2 ds \tag{2.9}$$

The polyhedron $\overline{\Omega}$ can be realized as a finite homogeneous simplicial 3-complex. A cone $\mathcal{C}(v, 1)$ is then the intersection of the ball $|X| \le 1$ with the *star* $St(v, \overline{\Omega})$ in the 3-complex $\overline{\Omega}$ of the vertex v. Each 2-simplex σ^2 of $St(v, \overline{\Omega})$ that is also contained in N is contained in a unique 3-simplex $\sigma^3 \in St(v, \overline{\Omega})$. Let B denote the *unit* vector in the direction from the *barycenter* of σ^3 to v. Then $\sigma^2 \cap \{|X| \le 1\}$ may be projected into the sphere $|X| = 1$ along lines parallel to B by $Q \mapsto Q + t_Q B$ onto a set contained in $\sigma^3 \cap \mathcal{B}(v, 1)$. The sets $\{Q + tB : Q \in \sigma^2 \cap N^1_{2^{-j}}(v)$ and $0 \le t \le t_Q\}$ are contained in $\sigma^3 \cap \mathcal{A}(v, 2^{-j}, 1)$. Thus by the fundamental theorem of calculus for each $Q \in \partial\Omega \cap N^1_{2^{-j}}(v)$ and integrating $ds(Q)$

$$\int_{\partial\Omega\cap N^1_{2^{-j}}(v)} u^2 ds \le C\left(\int_{\mathcal{A}(v,2^{-j},1)} |\nabla u|^2 dX + \int_{\mathcal{B}(v,1)} u^2 ds\right) \tag{2.10}$$

where the constant depends only on the projections, i.e. only on the finite geometric properties of the complex that realizes $\overline{\Omega}$ and not on j.

By the fundamental theorem, the connectedness of Ω_1 and the vanishing of u on the fixed nonempty set D

$$\int_{\partial\Omega_1} u^2 ds \le C \int_{\Omega_1} |\nabla u|^2 dX$$

This together with (2.10) implies

$$\epsilon \int_{\partial\Omega\cap N_{2^{-j}}} u^2 ds \le \epsilon C \int_{\Omega_{2^{-j}}} |\nabla u|^2 \, dX$$

and ϵ can be chosen independently of j so that (2.9) yields the lemma. □

The lemma yields the analogue of (2.7)

$$\int_{\partial\Omega\cap\partial\Omega_{2^{1-j}}} \left(\nabla u_j^T\right)^2 ds \le C \int_{\partial\Omega\cap N_{2^{-j}}} g^2 ds \tag{2.11}$$

Continuing to argue as in the proof of Theorem 2.10 , this and the vanishing of the u_j on D_{2-j} produces a harmonic function u defined in Ω that is the pointwise limit of a subsequence of the u_j. In addition u satisfies

$$\int_{\partial\Omega} \left(\nabla u^T\right)^2 ds \leq C \int_N g^2 ds$$

which in turn yields the maximal estimate of the theorem.

To show that u assumes the correct data, (2.11) along with weak L^2-convergence, pointwise convergence and the Poisson representation in each Ω_{2-j} proves as before that u vanishes nontangentially on D. By constructing a Neumann function (possible by [JK81]) in analogy to the Green function, or by using the invertibility of the classical layer potentials [Ver84], a Neumann representation of u in each Ω_{2-j} can be obtained so that $\frac{\partial u}{\partial \nu} = g$ nontangentially on N can be deduced by the same arguments.

Uniqueness follows from Green's first identity valid in polyhedra when $\nabla u^* \in L^2$. □

2.3. **Proof of Theorem 1.1.** Recall the definition of the homogeneous Sobolev semi-norm (1.4).

Lemma 2.13. *When $\partial\Omega$ is connected and $\|f\|^2_{D^o} = 0$, f is identically constant on D.*

The lemma says that f equals the same constant value on each component of D.

Proof. Because the semi-norm equals *zero* there is a sequence of extensions $\widetilde{f}_j$ of f and a sequence of numbers m_j so that by Poincaré in the second inequality

$$\int_D (f - m_j)^2 ds \leq \int_{\partial\Omega} (\widetilde{f}_j - m_j)^2 ds \leq C \int_{\partial\Omega} |\nabla_t \widetilde{f}_j|^2 ds \to 0$$

□

Proof of Theorem 1.1. Choose an extension $\widetilde{f}$ of f so that $\int_{\partial\Omega} |\nabla_t \widetilde{f}|^2 ds \leq 2\|f\|^2_{D^o}$. This is always possible by the lemma. Then from Theorem 2.10 there is a unique solution u_D with regularity data $\widetilde{f}$ and $\int_{\partial\Omega} (\nabla u^*_D)^2 ds \leq C\|f\|^2_{D^o}$. From Theorem 2.11 there is a unique solution u_N vanishing on D, with Neumann data $g - \partial_\nu u_D$ on N, and

$$\int_{\partial\Omega} (\nabla u^*_N)^2 ds \leq C\left(\int_N (\partial_\nu u_D)^2 ds + \int_N g^2 ds\right) \leq C\left(\|f\|^2_{D^o} + \int_N g^2 ds\right)$$

The solution is $u = u_D + u_N$. Theorem 2.11 established uniqueness. □

3. On violations of the postulates for the partition $\partial\Omega = D \cup N$

When D is empty the mixed problem is the Neumann problem and solvable for any data that has mean value *zero* on the boundary [Ver01]. We consider the two remaining postulates.

3.1. N **is the union of a number (possibly** ***zero*****) of closed faces of** $\partial\Omega$**.** Solving the mixed problem means that every $W^{1,2}(D)$ function has a $W^{1,2}(\partial\Omega)$ *extension.* This observation raises the possibility that the mixed problem might be solvable when a given (open) face $\mathcal{F}$ has nonempty intersection both with D and with N in such a way that $D \cap \mathcal{F}$ is an *extension domain.* Here we will only consider the possibility that this extension domain has a Lipschitz boundary [Ste70] and show that *the mixed problem is never solvable when this condition on the partition occurs.*

Let $\phi : \mathbb{R} \to \mathbb{R}$ be a Lipschitz continuous function $y = \phi(x)$ with $\|\phi'\|_\infty \leq M$. Choose a point p_0 on the graph $(x, \phi(x))$ in the plane and consider the rectangle with width 2 parallel to the x-axis, length $8M$ and *center* p_0. Locate the origin directly below p_0 and M units from the bottom of the rectangle. Here it will be convenient to name the region N that is *strictly below* the graph and contained in the rectangle. Call its complement in the rectangle D. Let (x, y, t) be the rectangular coordinates of $\mathbb{R}^3$ with *origin* coinciding with the origin of the plane. Let Z be the open right circular cylinder of $\mathbb{R}^3$ with center p_0 that intersects the plane in precisely the (open) rectangle.

The domain $\Omega = Z \setminus D \subset \mathbb{R}^3$ is *regular* for the Dirichlet problem. This follows by the Wiener test applied to each of the points of $\partial\Omega = \partial Z \cup D$ and the observation that the Newtonian capacity in $\mathbb{R}^3$ of a disc from the plane is proportional to its radius. See, for example, [Lan72] p. 165. Here the Lipschitz (or NTA) condition is also used. Consequently the Green function, $g = g^0$ for Ω with pole at the origin, is *continuous* in $\overline{\Omega} \setminus \{0\}$.

Approximating Lipschitz domains to Ω are constructed as follows. For each $\tau > 0$ define Lipschitz surfaces with boundary (the graph of ϕ) by

$$D_\tau = \{p + s(p - \tau e_3) : p \text{ is on the graph of } \phi \text{ and } 0 \leq s\} \cap Z$$

Here e_3 is the standard basis vector perpendicular to the xy-plane. Denote by H_τ the part of Z between D and D_τ. Then $\Omega_\tau = \Omega \setminus \overline{H}_\tau = Z \setminus \overline{H}_\tau$ are Lipschitz domains. Denote by g_τ the Green function for Ω_τ with pole at τe_3.

Lemma 3.1. *(i)* $-\partial_t g(x, y, t)$ *for* $t > 0$ *has continuous boundary values* $\partial_\nu g := -\lim_{t \downarrow 0} g(x, y, t)/t$ *at every point of* D *for which* $y > \phi(x)$*.*

(ii) $\int_D (\partial_\nu g)^2 ds = +\infty$*.*

(iii) $\partial_t g(x, y, 0) = 0$ *at every point of* $N \setminus \{0\}$*.*

(iv) $\partial_\nu g \in L^2(\partial Z)$*.*

Proof. (i) follows by Schwarz reflection while (iii) follows by the symmetry in t of Ω and g. The maximum principle shows that the Green function for Z dominates from below the Green function for Ω, $g_Z \leq g \leq 0$. On ∂Z both Green functions vanish so that $\partial_\nu g_Z \geq \partial_\nu g \geq 0$ while $\partial_\nu g_Z$ is square integrable there, establishing (iv).

D. S. Jerison and C. E. Kenig's Rellich identity for harmonic measure ([JK82] Lemma 3.3) is valid on any *Lipschitz* domain G that contains the *origin.* It is

$$(n-2)w_G(0) = \int_{\partial G} (\partial_\nu g_G(Q))^2 \nu \cdot Q ds(Q)$$

with respect to the vector field X. Here $g_G(X) = F(X) + w_G(X)$ is the Green function for G, and F is the fundamental solution for Laplace's equation. Denote by w_τ, w and w_Z the corresponding harmonic functions for the Ω_τ, Ω and Z Green functions respectively.

By $Z \supset Z \setminus D = \Omega \supset \Omega_\tau$ and the maximum principle

$$\partial_\nu g_\tau \leq \partial_\nu g_Z \text{ on } \partial\Omega_\tau \setminus D \setminus D_\tau$$

$$0 < \partial_\nu g_\tau \leq \partial_\nu g \text{ on } D$$

and

$$w_Z < w < w_\tau \text{ in } \Omega_\tau \tag{3.1}$$

For $Q \in D$ and $\nu = \nu_Q$ the outer unit normal to Ω_τ, $\nu \cdot (Q - \tau e_3) = \tau$, while for $Q \in D_\tau$, $\nu \cdot (Q - \tau e_3) = 0$. Formulating the Rellich identity with respect to the vector field $X - \tau e_3$ and using these facts ($n = 3$)

$$w_\tau(\tau e_3) = \int_{\partial\Omega_\tau \setminus D \setminus D_\tau} (\partial_\nu g_\tau)^2 \nu \cdot (Q - \tau e_3) ds + \tau \int_D (\partial_\nu g_\tau)^2 ds \leq$$

$$\int_{\partial Z} (\partial_\nu g_Z)^2 \nu \cdot (Q - \tau e_3) ds + \tau \int_D (\partial_\nu g)^2 ds = w_Z(\tau e_3) + \tau \int_D (\partial_\nu g)^2 ds$$

so that

$$\frac{w(\tau e_3) - w_Z(\tau e_3)}{\tau} < \frac{w_\tau(\tau e_3) - w_Z(\tau e_3)}{\tau} \leq \int_D (\partial_\nu g)^2 ds$$

and (ii) follows from (3.1) and $\tau \downarrow 0$. □

For $\delta > 0$ define smooth subdomains of Ω

$$G_\delta = \{g < -\delta\}.$$

$\partial G_\delta \to \partial\Omega$ uniformly. The $\partial_\nu g \mid_{\partial G_\delta} ds$ are a collection of probability measures on $\mathbb{R}^3$ that have *harmonic measure for Ω at the origin* as weak-$*$ limit.

By $G_\delta \uparrow \Omega$, Green's first identity, and monotone convergence

$$\int_{\Omega \setminus B_r} |\nabla g|^2 dX < \infty \tag{3.2}$$

for all balls centered at the origin.

With ϕ, N, D and Z as above define the half-cylinder domain $Z_+ = \{(x, y, t) \in Z : t > 0\}$. Then $D \cup N \subset \partial Z_+ \cap \{t = 0\}$.

Lemma 3.2. *Suppose $\Delta u = 0$ in Z_+, $\nabla u^* \in L^2(\partial Z_+)$, $\partial_\nu u \to^{n.t.} 0$ a.e. on N, and $u \to^{n.t.} 0$ a.e. on D. Let $Y \subset Z$ be a scaled cylinder centered at p_0 with $dist(\partial Y, \partial Z) > 0$. Let Y_+ be the corresponding half-cylinder. Then $u \in C(\overline{Y}_+)$.*

Proof. The hypothesis on ∇u^* implies $u^* \in L^2(\partial Z_+)$ so that u and ∇u have non-tangential limits *a.e.* on ∂Z_+ [Car62] [HW68]. Extend u to the bottom component of $Z \setminus D \setminus N$ by $u(x, y, t) = u(x, y, -t)$. By the vanishing of the Neumann data on N, $\Delta u = 0$ in the sense of distributions in the domain $\Omega = Z \setminus D$ and then classically.

Fix $d > 0$ and suppose $X \in \overline{Y}_+$ is of the form $X = (x, y, d)$ for $y \geq \phi(x) - Md$. Denote 3-balls of radius and distance to D *comparable to* d by B_d. Denote 2-discs in ∂Z_+ with radius *comparable to* d by Δ_d and let $⨍$ denote integral average. Then by the mean value theorem, the fundamental theorem of calculus, the *a.e.* vanishing of u on D and the geometry of the nontangential approach regions

$$|u(X)| \le \fint_{B_d(X)} |u| \le Cd(\fint_{\Delta_d(x,y+2Md,0)} \nabla u^* ds)$$

where C depends only on M. By absolute continuity of the surface integrals and $\nabla u^* \in L^2$ there is a function $\eta(d) \to 0$ as $d \to 0$ so that $\int_{\Delta_d} (\nabla u^*)^2 ds \le \eta(d)$ for all $\Delta_d \subset \partial Z_+$. Consequently the Schwarz inequality now yields $|u(X)| \le C\eta(d)$.

Suppose now X is of the form $X = (x, \phi(x) - Md, t)$ for $0 \le t \le d$. Because u has been extended

$$|u(X)| \le \fint_{B_d(X)} |u| \le (\fint_{B_d(x,\phi(x)-Md,d)} |u|) + d(\fint_{\Delta_d(x,\phi(x)-Md,0)} \nabla u^* ds)$$

and $|u(X)| \le 2C\eta(d)$. The lemma follows. □

Partition ∂Z_+ by $N_+ = \overline{N}$, $D_+ = \partial Z_+ \backslash \overline{N}$ and $\partial Z_+ = N_+ \cup D_+$. For $3/4 > r > 0$ let Z^r be the scaled cylinder centered at p_0 of width $2r$ and length $8Mr$. Define the corresponding half-cylinders Z^r_+ with

$$N^r_+ = N_+ \cap \partial Z^r_+$$

(not a scaling of N_+) and

$$D^r_+ = \partial Z^r_+ \setminus N^r_+$$

With this partition Z^r_+ is called a *split cylinder with Lipschitz crease.*

By (3.2) *and the Fubini theorem,* $g \in W^{1,2}(\partial Z^r_+ \setminus \{t = 0\})$ *for a.e.* r.

Proposition 3.3. *Let* g *be the Green function for* $\Omega = Z \setminus D$ *with pole at the origin. For almost every* $\frac{3}{4} > r > 0$ *there exists no solution* u *to the* L^2*-mixed problem* (1.2) *in the split cylinder with Lipschitz crease* Z^r_+ *with boundary values* $u \to^{n.t.} g \in W^{1,2}(D^r_+)$ *and* $\partial_\nu u \to \partial_\nu g = 0$ *on* N^r_+.

Proof. Suppose instead that there is such a solution u with $\nabla u^* \in L^2(\partial Z^r_+)$. Then the first paragraph of the proof of Lemma 3.2 applies and, in particular, u extends to $Z^r \setminus D$ evenly and harmonically across N^r_+. The Dirichlet data that u takes *a.e.* on D^r_+ is a continuous function, as is the *Dirichlet* data that u takes (continuously) on N^r_+. The Dirichlet data u takes *a.e.* on ∂Z^r_+ will be shown to be a continuous function if it can be shown to be continuous across the boundary ∂N^r_+ of the surface N^r_+. Lemma 3.2, scaled to apply to the split cylinders here, shows that the Dirichlet data is continuous across the Lipschitz crease part of ∂N^r_+. The same argument used there works on the other parts: Suppose $dist(X, \partial Z^r) = d$ for $X \in N^r_+$. Let $\Delta_d \subset \partial Z^r \cap D^r_+$ be a disc approximately a distance d from $X + de_3$. Then

$$|u(X) - \fint_{\Delta_d} gds| \le |\fint_{B_d(X)} u(Y) - u(Y + de_3)dY| + |\fint_{B_d(X+de_3)} u(Y) - \fint_{\Delta_d} gds| \le C\eta(d)$$

and the continuity across ∂N^r_+ follows from the continuity of g and $\eta(d) \to 0$.

Thus the data u takes *a.e.* on ∂Z^r_+ is a continuous function. Since also $u^* \in L^2(\partial Z^r_+)$ it follows that $u \in C(\overline{Z^r_+})$. The evenly extended u is then continuous on $\overline{Z^r}$, harmonic in $Z^r \setminus D$ with the same Dirichlet data as g on $\partial(Z^r \setminus D)$. The maximum principle implies $u = g$.

Let g_r denote the Green function for $Z^r \setminus D$ with pole at a point $\{P\}$ of N^r_+. Again g_r is continuous in $\overline{Z^r} \setminus \{P\}$. Let $B \subset \overline{B} \subset Z^r \setminus D$ be a ball centered at P. Then by the maximum principle $cg \geq g_r$ on $\overline{Z^r \setminus B}$ for some constant c. By this domination, the vanishing of both g and g_r on $D^r_+ \cap \{t = 0\}$ and (ii) of Lemma 3.1 applied to g_r, it follows that $\partial_\nu g$ which is not in $L^2(D)$ can neither be square integrable over the smaller set $D^r_+ \cap \{t = 0\}$. Since $u = g$ this contradicts the assumption on the nontangential maximal function of the gradient. □

The nonsolvability of the L^2-mixed problem in the split cylinders can be extended to nonsolvability in any polyhedron that has a Lipschitz graph crease on any face by a *globalization argument.* Let g and r be as in the Proposition. By using the approximating domains $Z^r \cap G_\delta$ as $\delta \to 0$, the Green's representation

$$g(X) = \int_{\partial Z^r} \partial_\nu F^X g - F^X \partial_\nu g ds - \int_{D \cap \overline{Z^r}} F^X d\mu^0, X \in Z^r \setminus D$$

can be justified where μ^0 is harmonic measure for $\Omega = Z \setminus D$ at the origin and F is the fundamental solution for Laplace's equation. Let $\chi \in C_0^\infty(\mathbb{R}^3)$ be a cut-off function that is supported in a ball contained in Z^r centered at p_0, and is identically 1 in a concentric ball B^r with smaller radius. Then define

$$u(X) = -\int_{D \cap \overline{Z^r}} F^X \chi d\mu^0$$

harmonic in $\mathbb{R}^3$ outside *supp*$(\chi) \cap D$. Similarly $g(X) - u(X)$ is harmonic inside B^r. Consequently $\nabla u^* \notin L^2(B^r \cap D)$ by applying a scaled (ii) of Lemma 3.1 to g again. Also

$$\text{(3.3)} \quad u(X) = -\int_{D \cap \overline{Z^r}} F^X(Q)\left(\chi(Q) - \chi(X)\right) d\mu^0(Q) - \chi(X) \int_{\partial Z^r} \partial_\nu F^X g - F^X \partial_\nu g ds + \chi(X) g(X)$$

The last term has bounded Neumann data on N and vanishing Dirichlet data on D. The Cauchy data of the middle term is smooth and compactly supported on $D \cup N$. For any $X \notin D$ the *gradient* of the first term is bounded by a constant, depending on χ, times

$$\left| \int_{D \cap \overline{Z^r}} F^X(Q) d\mu^0(Q) \right| \leq -F^X(0) + g^X(0) \leq \frac{1}{4\pi |X|}$$

Here g^X is the (negative) Green function for $\Omega = Z \setminus D$ with pole at X. Thus the first term is Lipschitz continuous on $D^r_+ \cup N^r_+$. Altogether u has bounded Neumann data on N and Lipschitz continuous data on D while $\nabla u^* \notin L^2(D)$. Finally $\nabla u \in L^2_{loc}(\mathbb{R}^3)$ by (3.3) since this is true for χg.

Thus whenever a split cylinder Z^r_+ can be contained in a polyhedral domain so that $\partial Z^r_+ \cap \{t = 0\}$ is contained in a face and so that the Lipschitz crease is part of the boundary between the Dirichlet and Neumann parts of the polyhedral boundary, then the harmonic function u just constructed is defined in the entire polyhedra domain. Its properties suffice to compare it with any solution w in the class $\nabla w^* \in L^2$ by Green's first identity $\int |\nabla u - \nabla w|^2 dX = \int (u - w)\partial_\nu (u - w) ds$. Regardless of the nature of the partition away from Z^r_+, when w has the same data as does u it must be concluded, as in Proposition 3.3, that it is identical to u. This establishes

Theorem 3.4. *Let $\Omega \subset \mathbb{R}^3$ be a compact polyhedral domain with partition $\partial\Omega = D \cup N$. Let $\mathcal{F}$ be an open face of $\partial\Omega$ such that $\mathcal{F} \cap D$ is a Lipschitz domain of $\mathcal{F}$ with nonempty complement $\mathcal{F} \cap N$. Then there exist mixed data (1.2) for which there are no solutions u in the class $\nabla u^* \in L^2(\partial\Omega)$.*

3.2. **Whenever a face of N and a face of D share a 1-dimensional edge as boundary, the dihedral angle measured in Ω between the two faces is *less* than π.** Continue to denote points of $\mathbb{R}^3$ by $X = (x, y, t)$. Define D to be the upper half-plane of the xy-plane. Introduce polar coordinates $y = r\cos\theta$ and $t = r\sin\theta$, let $\pi \le \alpha < 2\pi$ and define N to be the half-plane $\theta = \alpha$. The crease is now the x-axis.

Define

$$b(X) = r^{\frac{\pi}{2\alpha}} \sin(\frac{\pi}{2\alpha}\theta)$$

for X above $D \cup N$. These are Brown's counterexample solutions for nonconvex plane sectors [Bro94]. The Dirichlet data vanishes on D while the Neumann vanishes on N, and $\nabla b^* \notin L^2$.

These solutions are globalized to a compact polyhedral domain with interior dihedral angle α:

Denote by Θ the intersection of a (large) ball centered at the origin and the domain above $D \cup N$. Then $b(X)$ is represented in Θ by

$$b(X) = \int_{\partial\Theta\setminus D} \partial_\nu F^X b ds - \int_{\partial\Theta\setminus N} F^X \partial_\nu b ds$$

Let $\chi \in C_0^\infty(\mathbb{R}^3)$ be a cut-off function as before , but centered at the *origin* on the crease. Define

$$u(X) = \int_N \partial_\nu F^X \chi b ds - \int_D F^X \chi \partial_\nu b ds$$

As before, u is harmonic everywhere outside $supp(\chi) \cap (D \cup N)$ and $\nabla u^* \notin L^2\,(supp(\chi) \cap (D \cup N))$. Also

$$\begin{aligned}(3.4) \quad u(X) = &\int_{N\cap\overline{\Theta}} \partial_\nu F^X(Q)\,(\chi(Q) - \chi(X))\,b(Q)ds(Q) \\ &- \int_{D\cap\overline{\Theta}} F^X(Q)\,(\chi(Q) - \chi(X))\,\partial_\nu b(Q)ds(Q) \\ &- \chi(X) \int_{\partial\Theta\setminus N\setminus D} \partial_\nu F^X b - F^X \partial_\nu b ds + \chi(X)b(X)\end{aligned}$$

Again the boundary values around the support of χ are the issue. The last two terms are described just as the middle and last after (3.3). The *gradient* of the second term is *bounded* because the integral over D can be no worse than, for example, $\int_0^1 dx \int_0^1 \frac{1}{\sqrt{x^2+r^2}} \frac{dr}{r^\beta} < \infty$ for any $\beta < 1$ (e.g. $\beta = 1 - \frac{\pi}{2\alpha}$).

For a $\frac{\partial}{\partial X_j}$ derivative define tangential derivatives (in Q) to any surface with unit normal ν by $\partial_i^t = \nu_i \partial_j - \nu_j \partial_i$. Then by the harmonicity of F away from X and the divergence theorem in Θ, the $\frac{\partial}{\partial X_j}$ derivative of the first integral equals the sum in i of

$$\int_{N\cap\overline{\Theta}} \partial_i F^X \partial_i^t\,((\chi - \chi(X))b)\,ds$$

plus integrals over $\partial\Theta \setminus N \setminus D$ (b vanishes on D) that will all be *bounded* since X is near the support of χ. When the tangential derivative falls on b the integral is bounded like the second term of (3.4). The remaining integral has boundary values in every L^p for $p < \infty$ by singular integral theory. (In fact, it too is bounded by a closer analysis, thus making it consistent with the example from Section 3.1.)

Finally $\nabla u \in L^2_{loc}(\mathbb{R}^3)$ by its now established properties and the corresponding property for b. The argument using Green's first identity as at the end of Section 3.1 is justified and

The solutions u can now be placed in polyhedral domains that have interior dihedral angles greater than or equal to π and provide mixed data for which no L^2-solution can exist.

4. Polyhedral domains that admit only the trivial mixed problem

Consider the L^2-mixed problem for the unbounded domain exterior to a compact polyhedron. When the polyhedron is convex the requirement of postulate (iii) of (1.1) eliminates all but the trivial partition from the class of well posed mixed problems. In this case we will say that the *exterior* problem is *monochromatic.*

The mixed problem for a compact polyhedral domain can also be monochromatic for the *interior* problem. An example is provided by the regular compound polyhedron that is the union of 5 equal regular tetrahedra with a common center, a picture of which may be found as Number 6 on Plate III between pp.48-49 of H. S. M. Coxeter's book [Cox63]. An elementary arrangement of plane surfaces that elucidates the *local* element of this phenomenon is found upon considering the domain of $\mathbb{R}^3$ that is the union of the upper half-space together with all points (x, y, t) with (x, y) in the first quadrant of the plane, i.e. the union of a half-space and an infinite wedge. The boundary consists of 3 faces: the 4th quadrants of both the xt and yt-planes and the piece of the xy-plane outside of the 1st quadrant of the xy-plane. The requirement of postulate (iii) is met only by the *negative* t-axis. But no color change is possible there because any color on either of the 4th quadrants must be continued across the *positive* x or y-axis to the 3rd face of the boundary. On the other hand, a color change is possible for the complementary domain and is possible for the exterior domain to the compound of 5 tetrahedra.

Is there a polyhedral surface with a finite number of faces for which both interior and exterior mixed problems are monochromatic?

References

[Agm65] Shmuel Agmon, *Lectures on elliptic boundary value problems*, Prepared for publication by B. Frank Jones, Jr. with the assistance of George W. Batten, Jr. Van Nostrand Mathematical Studies, No. 2, D. Van Nostrand Co., Inc., Princeton, N.J.-Toronto-London, 1965. MR MR0178246 (31 #2504)

[Bro94] R.M. Brown, *The mixed problem for Laplace's equation in a class of Lipschitz domains*, Comm. Part. Diff. Eq. **19** (1994), no. 7-8, 1217–1233.

[Car62] Lennart Carleson, *On the existence of boundary values for harmonic functions in several variables*, Ark. Mat. **4** (1962), 393–399 (1962). MR MR0159013 (28 #2232)

[Cox63] H. S. M. Coxeter, *Regular polytopes*, Second edition, The Macmillan Co., New York, 1963. MR MR0151873 (27 #1856)

[Dah77] Björn E. J. Dahlberg, *Estimates of harmonic measure*, Arch. Rational Mech. Anal. **65** (1977), no. 3, 275–288. MR 57 #6470

[Dah79] ———, *On the Poisson integral for Lipschitz and C^1-domains*, Studia Math. **66** (1979), no. 1, 13–24. MR 81g:31007

[Dau92] Monique Dauge, *Neumann and mixed problems on curvilinear polyhedra*, Integral Equations Operator Theory **15** (1992), no. 2, 227–261. MR MR1147281 (93e:35025)

[Fol95] Gerald B. Folland, *Introduction to partial differential equations*, second ed., Princeton University Press, Princeton, NJ, 1995. MR MR1357411 (96h:35001)

[Gla70] L. C. Glaser, *Geometrical combinatorial topology*, vol. 1, Van Nostrand Reinhold, 1970.

[GT83] David Gilbarg and Neil S. Trudinger, *Elliptic partial differential equations of second order*, second ed., Grundlehren der Mathematischen Wissenschaften [Fundamental Principles of Mathematical Sciences], vol. 224, Springer-Verlag, Berlin, 1983. MR MR737190 (86c:35035)

[HW68] R. A. Hunt and R. L. Wheeden, *On the boundary values of harmonic functions*, Trans. Amer. Math. Soc. **132** (1968), 307–322.

[JK81] David S. Jerison and Carlos E. Kenig, *The Neumann problem on Lipschitz domains*, Bull. Amer. Math. Soc. (N.S.) **4** (1981), no. 2, 203–207. MR 84a:35064

[JK82] D. S. Jerison and C. E. Kenig, *Boundary value problems on Lipschitz domains*, MAA Studies in Mathematics, Studies in Partial Differential Equations **23** (1982), 1–68.

[Lan72] N. S. Landkof, *Foundations of modern potential theory*, Springer-Verlag, New York, 1972, Translated from the Russian by A. P. Doohovskoy, Die Grundlehren der mathematischen Wissenschaften, Band 180. MR MR0350027 (50 #2520)

[MR02] V. G. Maz'ya and J. Rossmann, *Point estimates for Green's matrix to boundary value problems for second order elliptic systems in a polyhedral cone*, ZAMM Z. Angew. Math. Mech. **82** (2002), no. 5, 291–316. MR MR1902258 (2003d:35002)

[MR03] Vladimir G. Maz'ya and Jürgen Roßmann, *Weighted L_p estimates of solutions to boundary value problems for second order elliptic systems in polyhedral domains*, ZAMM Z. Angew. Math. Mech. **83** (2003), no. 7, 435–467. MR MR1987897 (2004g:35061)

[MR04] V. Mazya and J. Rossmann, *Schauder estimates for solutions to boundary value problems for second order elliptic systems in polyhedral domains*, Appl. Anal. **83** (2004), no. 3, 271–308. MR MR2033239 (2004k:35098)

[MR05] V. Maz′ya and J. Rossmann, *Pointwise estimates for Green's kernel of a mixed boundary value problem to the Stokes system in a polyhedral cone*, Math. Nachr. **278** (2005), no. 15, 1766–1810. MR MR2182091 (2007b:35269)

[MR06] V. G. Maz′ya and J. Rossmann, *Schauder estimates for solutions to a mixed boundary value problem for the Stokes system in polyhedral domains*, Math. Methods Appl. Sci. **29** (2006), no. 9, 965–1017. MR MR2228352 (2007f:35220)

[MR07] V. Maz'ya and J. Rossmann, *L_p estimates of solutions to mixed boundary value problems for the Stokes system in polyhedral domains*, Math. Nachr. **280** (2007), no. 7, 751–793. MR MR2321139

[RS72] C. P. Rourke and B. J. Sanderson, *Introduction to piecewise-linear topology*, Springer-Verlag, 1972.

[Ste70] Elias M. Stein, *Singular integrals and differentiability properties of functions*, Princeton Mathematical Series, No. 30, Princeton University Press, Princeton, N.J., 1970. MR 44 #7280

[Ver84] G. C. Verchota, *Layer potentials and regularity for the Dirichlet problem for Laplace's equation in Lipschitz domains*, J. Funct. Anal. **59** (1984), no. 3, 572–611.

[Ver01] Gregory C. Verchota, *The use of Rellich identities on certain nongraph boundaries*, Harmonic analysis and boundary value problems (Fayetteville, AR, 2000), Amer. Math. Soc., Providence, RI, 2001, pp. 127–138. MR 1 840 431

[VV03] Gregory C. Verchota and Andrew L. Vogel, *A multidirectional Dirichlet problem*, J. Geom. Anal. **13** (2003), no. 3, 495–520. MR 2004b:35060

[VV06] ———, *The multidirectional Neumann problem in $\mathbb{R}^4$*, Math. Ann. **335** (2006), no. 3, 571–644. MR MR2221125 (2007f:35053)

215 Carnegie, Syracuse University, Syracuse NY 13244
E-mail address: gverchot@syr.edu

Titles in This Series

For a complete list of titles in this series, visit the
AMS Bookstore at **www.ams.org/bookstore/**.